WITHDRAWN FROM STOCK
0 566 735 6
D02967762

APPROXIMATION THEORY III

Academic Press Rapid Manuscript Reproduction

Proceedings of a Conference Honoring
Professor George G. Lorentz
Held at The University of Texas at Austin
January 8–12, 1980

APPROXIMATION THEORY III

edited by

E. W. Cheney

The University of Texas at Austin

1980

ACADEMIC PRESS

A Subsidiary of Harcourt Brace Jovanovich, Publishers

New York London Toronto Sydney San Francisco

COPYRIGHT © 1980, BY ACADEMIC PRESS, INC.
ALL RIGHTS RESERVED.
NO PART OF THIS PUBLICATION MAY BE REPRODUCED OR TRANSMITTED IN ANY FORM OR BY ANY MEANS, ELECTRONIC OR MECHANICAL, INCLUDING PHOTOCOPY, RECORDING, OR ANY INFORMATION STORAGE AND RETRIEVAL SYSTEM, WITHOUT PERMISSION IN WRITING FROM THE PUBLISHER.

ACADEMIC PRESS, INC.
111 Fifth Avenue, New York, New York 10003

United Kingdom Edition published by
ACADEMIC PRESS, INC. (LONDON) LTD.
24/28 Oval Road, London NW1 7DX

Library of Congress Cataloging in Publication Data

Main entry under title:

Approximation theory III.

1. Approximation theory—Congresses. 2. Lorentz, G.G. I. Lorentz, G. G. II. Cheney, Elliott Ward, Date
QA297.5.A67 511'.4 80-19723
ISBN 0-12-171050-5

PRINTED IN THE UNITED STATES OF AMERICA

80 81 82 83 9 8 7 6 5 4 3 2 1

This volume is dedicated by his many friends, admirers, colleagues, and students to

Professor George G. Lorentz

on the occasion of his seventieth birthday (February 12, 1980)

CONTENTS

CONTRIBUTORS

Numbers in parentheses indicate pages on which authors' contributions appear.

Abatzoglou, Theagines (151), 6721 Cory Drive, Huntington Beach, California 92647

Amir, D. (157), Department of Mathematics, Tel-Aviv University, Tel-Aviv, Israel

Anselone, Phillip (163), Department of Mathematics, Oregon State University, Corvalis, Oregon 97331

Ansorge, R. (169), Institut für Angewandte Mathematik, der Universität Hamburg, Bundestrasse, 55, 2000 Hamburg, Federal Republic of Germany

Askey, Richard A. (175), Department of Mathematics, University of Wisconsin, Madison, Wisconsin 53706

Barrar, R.B. (183), Department of Mathematics, University of Oregon, Eugene, Oregon 97403

Bartelt, Martin W. (187), Department of Mathematics, Christopher Newport College, Newport News, Virginia 23606

Baszinski, G. (193), Rechenzentrum, University of Bochum, Bochum, Federal Republic of Germany

Beatson, Rick (199), Department of Mathematics, The University of Texas, Austin, Texas 78712

Becker, Michael (207), FB Mathematik, Universität Kaiserlautern, Pfaffenbergstrasse 95, 6700 Kaiserslautern, Federal Republic of Germany

Berens, Herbert (1, 213), Mathematisches Institut, Universität Erlangen-Mürnberg, Erlangen, Federal Republic of Germany

Blatt, Hans-Peter (221, 223), Fakultät für Mathematik und Informatik, Universität Mannheim, 6800 Mannheim, Federal Republic of Germany

Blatter, Jörg (229), Instituto de Matemática e Estatistica, Universidade de São Paulo, Caixa Postal 20570, São Paulo, Brazil

Bogar, Gary (233), Department of Mathematics, Montana State University, Bozeman, Montana 59715

Böhmer, Klaus (237), Institut für Praktische Mathematik, Englerstrasse 2, D-75 Karlsruhe 1, Federal Republic of Germany

Bojanic, R. (243), Department of Mathematics, Ohio State University, Columbus, Ohio 43210

Borosh, I. (249), Department of Mathematics, Texas A & M University, College Station, Texas 77843

Brosowski, Bruno (255), Johann Wolfgang Goethe-Universität, Fachbereich Mathematik, Frankfurt (M), Federal Republic of Germany

de Bruin, M.G. (261), Department of Mathematics, University of Amsterdam, Amsterdam, Netherlands

Bultheel, A. (267), Applied Mathematics Division, University of Leuven, Heverlee, Belgium

Burchard, H.G. (273), Department of Mathematics, Oklahoma State University, Stillwater, OK 74074 and Institut f. Angew. Mathematik der Universität Bonn/SFB 72, WegelerstraBe 6, 5300 Bonn, Federal Republic of Germany

Byrnes, J.S. (279), Department of Mathematics, University of Massachusetts, Boston, Massachusetts 02125

Chalmers, B.L. (285), Department of Mathematics, University of California, Riverside, California 92521

Cheney, E.W. (729), Department of Mathematics, The University of Texas, Austin, Texas 78712

Chui, Charles K. (299, 305), Department of Mathematics, Texas A & M University, College Station, Texas 77843

Collatz, Lothar (311), Institut für Angewandte Mathematik, Universität Hamburg, BundesstraBe 55, 2000 Hamburg 13, Federal Republic of Germany

Delvos, F.J. (193, 321), Fachbereich Mathematik, University of Siegen, Siegen, Federal Republic of Germany

Deutsch, Frank (327), Department of Mathematics, Pennsylvania State University, University Park, Pennsylvania 16802

DeVore, R. (213), Department of Mathematics, University of South Carolina, Columbia, South Carolina 29208

Dierieck, Claude (335), Phillips Research Labs, Ave. Van Becelaer 2, Box 8, B-1170 Brussels, Belgium

Ditzian, Z. (341), Department of Mathematics, University of Alberta, Edmonton, Alberta, Canada

Doedel, E.J. (349), Computer Science Department, Condordia University, Montreal, Quebec, Canada

Donner, Klaus (355), Mathematisches Institut, Universität Erlanger-Nürnberg, Erlangen, Federal Republic of Germany

Drols, Wolfgang (361), Fachbereich 11, Mathematik, Lotharstrasse 65, 4100 Duisburg 1, Federal Republic of Germany

Dunham, Charles B. (367), Department of Computer Science, The University of Western Ontario, London, Ontario, Canada

Dyn, Nira (371), Department of Mathematical Sciences, Tel-Aviv University, Tel-Aviv, Israel

Eggert, Norman (377), Department of Mathematics, Montana State University, Bozeman, Montana 59715

Ellacott, S.W. (383), Department of Mathematics, Brighton Polytechnic, Brighton, Sussex, England

Ferguson, Le Baron O. (389), Department of Mathematics, University of California, Riverside, California 92521

de Figueiredo, Rui J.P. (937), Department of Mathematical Science, Rice University, Houston, Texas 77001

Fisher, Stephen D. (405), Department of Mathematics, Northwestern University, Evanston, Illinois 60201

Flösser, Hans O. (409), Fachbereich Mathematik-Informatik, Universität-Gesamthochschule Paderborn, Federal Republic of Germany

Foley, Thomas A. (419), Department of Computer Science and Statistics, California Polytechnic State University, San Luis Obispo, California 95407

Franchetti, Carlo (425), Department of Mathematics, University of Genoa, Genoa, Italy

V. Golitschek, Manfred (429), Institut für Angewandte Mathematik, der Universität Wurzburg, Am Hubland, 8700 Wurzburg, Federal Republic of Germany

Golomb, Michael (435), Division of Mathematical Science, Purdue University West Lafayette, Indiana 47907

Gonska, Heinz H. (443), Department of Mathematics, University of Duisburg, Duisburg, Federal Republic of Germany

Gorenflo, Rudolf (449), Institut für Mathematik III, Freie Universität Berlin, Berlin, Federal Republic of Germany

Gragg, William B. (455), Department of Mathematics, University of Kentucky, Lexington, Kentucky 40506

Greville, T.N.E. (461), Mathematics Research Center, University of Wisconsin, Madison, Wisconsin 53706

Gutknecht, Martin H. (467), Seminar für Angewandte Mathematik, Eidgenössische Technische Hochschule, Zürich, Switzerland

Haber, Seymour (473), Mathematical Analysis Division, Center for Applied Mathematics, National Bureau of Standards, Washington, District of Columbia 20234

Handscomb, D.C. (481), Computing Laboratory, Oxford University, Oxford, England

Hartley, Peter (485), Department of Mathematics, Coventry (Lanchester) Polytechnic, Coventry, England

Hasson, Maurice (491), Department of Mathematics, Texas A & M University, College Station, Texas 77843

Haussmann, Werner (495), Department of Mathematics, University of Duisburg, Duisburg, Federal Republic of Germany

Henry, Myron S. (507), Department of Mathematics, Montana State University, Bozeman, Montana 59717

Hilpert, Martin (449), Institut für Mathematik III, Freie Universität Berlin, Berlin, Federal Republic of Germany

Holland, A.S.B. (513), Department of Mathematics, University of Calgary, Calgary, Alberta, Canada

Hollig, K. (517), Institut für Angewandte Mathematik der Universität Bonn, Bonn, Federal Republic of Germany

Holmes, Richard B. (523), M.I.T. Lincoln Laboratory, Lexington, Massachusetts 02173

Höllig, K. (517), Institut f. Angew. Mathematik der Universität Bonn/SFB 72, WegelerstraBe 6, 5300 Bonn, Federal Republic of Germany

Ismail, Mourad E.H. (175), Department of Mathematics, Arizona State University, Tempe, Arizona 85281

Jakimovski, Amnon (531, 537), Department of Mathematical Sciences, Tel-Aviv University, Tel-Aviv, Israel

Jeppson, Ron (233), Department of Mathematics, Montana State University, Bozeman, Montana 59715

Jerome, Joseph (435), Department of Mathematics, Northwestern University, Evanston, Illinois 60201

Jory, Virginia V. (543), School of Mathematics, Georgia Institute of Technology, Atlanta, Georgia 30332

Kammler, David W. (549), Department of Mathematics, Southern Illinois University, Carbondale, Illinois 62901

Kaufman, Jr., Edwin H. (555), Department of Mathematics, Central Michigan University, Mount Pleasant, Michigan 48859

Kaufman, Robert (561), Department of Mathematics, University of Illinois, Urbana, Illinois 61801

Kenderov, Petar (327), Institute of Mathematics, P.O. Box 373, 1090 Sofia, Bulgaria

Kimchi, Esther (565), Department of Mathematics, Fordham University, Bronx, New York 10458

Lambert, Joseph M. (571), Department of Mathematics, Pennsylvania State University, University Park, Pennsylvania 16802

Larkin, F.M. (577), Department of Computing and Information Science, Queen's University, Kingston, Ontario, Canada

Leviatan, D. (583), Department of Mathematics, Tel-Aviv University, Tel-Aviv, Israel

Light, William A. (589), Department of Mathematics, University of Lancaster, Lancaster, England

Loeb, Henry (21, 183) Department of Mathematics, University of Oregon, Eugene, Oregon 97403

Lorentz, G.G. (595), Department of Mathematics, The University of Texas, Austin, Texas 78712

Lorentz, Rudolph A. (595), Gesellschaft für Mathematik und Datenverarbeitung, 5205 St. Augustin 1, Federal Republic of Germany

Luke, Yudell L. (601), Department of Mathematics, University of Missouri, Kansas City, Missouri 64110

Lund, John (377), Department of Mathematics, Montana State University, Bozeman, Montana 59715

Lutterodt, Clement H. (603), Department of Mathematics, University of South Florida, Tampa, Florida 33620

Lyche, Tom (611), Institutt for Informatikk, Universitetet i Oslo, Oslo, Norway

Madych, W.R. (615), Department of Mathematics, Iowa State University, Ames, Iowa 50011

Mansfield, Lois (623), Department of Applied Mathematics and Computer Science, University of Virginia, Charlottesville, Virginia 22901

Mason, J.C. (629), Department of Mathematics and Ballistics, Royal Military College of Science, Shrivenham, Swindon, England

Mazhar, S.M. (243), Department of Mathematics, University of Kuwait, Kuwait, Kuwait

McAllister, D.F. (757), Department of Computer Science, North Carolina State University, Raleigh, North Carolina 27650

McCabe, John (637), The Mathematical Institute, University of St. Andrews, St. Andrews, Fift, Scotland

McCoy, Peter A. (643), Department of Mathematics, United States Naval Academy, Annapolis, Maryland 21402

McLaughlin, H.W. (647), Department of Mathematical Sciences, Rensselaer Polytechnic Institute, Troy, New York 12181

Metcalf, F.T. (285), Department of Mathematics, University of California, Riverside, California 92521

Micchelli, Charles A. (405), IBM Thomas J. Watson Research Center, Yorktown Heights, New York 10598

Mühlbach, G. (655), Lehrstuhl F. Praktische Mathematik, Universität Hannover, Welfengarten 1, 3000 Hannover 1, Federal Republic of Germany

Müller, Manfred W. (661), Lehrstuhl Mathematik III, Universität Dortmund, Dortmund, Federal Republic of Germany

Nashed, M.Z. (667), Department of Mathematical Science, University of Delaware, Newark, Delaware 19711

Nessel, Rolf J. (207), Lehrstuhl A. für Mathematik, Technische Hochschule Aachen, Templegraben 64, 5100 Aachen, Federal Republic of Germany

Neuman, Edward (675), Institute of Computer Science, University of Wroclaw, Place Grunwaldski 2/4, 50-384, Wroclaw, Poland

Nevai, Paul G. (679), Department of Mathematics, Ohio State University, Columbus, Ohio 43210

Nielson, Gregory M. (419), Department of Mathematics, Arizona State University, Tempe, Arizona 85281

Nürnberger, Günther (687), Institut für Angewandte Mathematik, Universität Erlangen-Nürnberg, Martenstrasse 3, 8520 Erlangen, Federal Republic of Germany

Oden, J.T. (693), Department of Engineering Mechanics, The University of Texas at Austin, Austin, Texas 78712

Opfer, Gerhard (699), Institut für Angewandte Mathematik, Universität Hamberg, Bundesstrabe 55, 2000 Hamburg 13, Federal Republic of Germany

Osborne, M.R. (705), Department of Statistics, Institute of Advanced Studies, Australian National University, Canberra, Australia

Papini, P.L. (711), Istituto Matematico, Università di Bologna, Bologna, Italy

Paur, S.O. (715), Department of Mathematics, North Carolina State University, Raleigh, North Carolina 27650

Pence, Dennis D. (721), Department of Mathematics, University of Vermont, Burlington, Vermont 05401

Phillips, G.M. (513), Mathematical Institute, University of St. Andrews, St. Andrews, Scotland

Posdorf, H. (193, 321), Rechenzentrum, University of Bochum, Bochum, Federal Republic of Germany

Reddien, G.W. (349), Department of Mathematics, Southern Methodist University, Dallas, Texas 75275

Reimer, Manfred (723), Abteilung Mathematik, Universität Dortmund, Dortmund, Federal Republic of Germany

Respess, John (729), Department of Mathematics, The University of Texas, Austin, Texas 78712

Rhoades, B.E. (735), Department of Mathematics, Indiana University, Bloomington, Maryland 47401

Riemenschneider, S.D. (741), Department of Mathematics, University of Alberta, Edmonton, Alberta, Canada

Rivlin, T.J. (75), Department of Mathematical Sciences, IBM Thomas J. Watson Research Center, Yorktown Heights, New York 10598

van Rossum, H. (747), Department of Mathematics, Universiteit van Amsterdam, Amsterdam, Netherlands

Roulier, J.A. (757), Department of Mathematics, North Carolina State University, Raleigh, North Carolina 27650

Runck, P.O. (763), Institut für Mathematik, Johannes Kepler Universität, Linz, Austria

Russell, D.C. (531), Department of Mathematics, York University, Downsview, Ontario, Canada

Saff, E.B. (769), Department of Mathematics, University of South Florida, Tampa, Florida 33620

Sahney, B.N. (783), Department of Mathematics, University of Calgary, Calgary, Alberta, Canada

Schaback, Robert (795), Lihrstühle für Numerische und Angewandte Mathematik der Universität Göttingen, D-3400, Göttingen, Federal Republic of Germany

Schempp, Walter (803), Lehrstühl für Mathematik I, Universität Siegen, Siegen, Federal Republic of Germany

Scherer K. (517), Institut für Angewandte Mathematik der Universität Bonn, Bonn, Federal Republic of Germany

Schmidt, Darrell (805), Department of Mathematical Sciences, Oakland University, Rochester, Michigan 48063

Schoenberg, I.J., (811), Mathematics Research Center, University of Wisconsin-Madison, Madison, Wisconsin 53706

Schurer, F. (823), Department of Mathematics, Eindhoven University of Technology, Eindhoven, Netherlands

Shapiro, Harold S. (89), Department of Mathematics, Royal Institute of Technology, Stockholm, Sweden

Sharma, A. (741), Department of Mathematics, University of Alberta, Edmonton, Alberta, Canada

Shekhtman, Boris (829), Department of Mathematics, Kent State University, Kent, Ohio 44242

Shisha, O. (279), Department of Mathematics, University of Rhode Island, Kingston, Rhode Island 02881

Sikkema, P.C. (837), Department of Mathematics, University of Technology, Delft, Netherlands

Sinclair, Annette (841), Department of Mathematics, Purdue University, Lafayette, Indiana 47907

Singer, I (687), Institutut de Mathematica, Bucuresti, Romania

Singh, S.P. (783), Department of Mathematics and Statistics, Memorial University of Newfoundland, St. John's, Newfoundland, Canada

Sinwel, H.F. (763), Institut für Mathematik, Johannes Kepler Universität, Linz, Austria

Smith, P.W. (847), Department of Mathematics, University of Alberta, Edmonton, Alberta, Canada

Smith, Philip W. (305), Department of Mathematics, Texas A&M University, College Station, Texas 77843

Sommer, Manfred (223), Institut für Angewandte Mathematik, Universität Erlangen-Nürnberg, Martenstrasse 3, 8520 Erlangen, Federal Republic of Germany

Sreedharan, V.P. (857), Department of Mathematics, Michigan State University, East Lansing, Michigan 48824

Steutel, F.W. (827), Department of Mathematics, Eindhoven University of Technology, Eindhoven, Netherlands

Stieglitz, Michael (537), Mathematisches Institut I der Universität Karlsruhe, Karlsruhe, Federal Republic of Germany

Strauss, Hans (865), Institüt für Angewandte Mathematik, Unversität Erlangen-Nürnberg, Erlangen, Federal Republic of Germany

Summers, W.H. (871), Department of Mathematics, University of Arkansas, Fayetteville, Arkansas 72701

Swetits, John J. (877), Department of Mathematical Sciences, Old Dominion University, Norfolk, Virginia 23508

Szabados, Josef (881), Hungarian Academy of Sciences, Budapest, Hungary

Taylor, Gerald D. (555), Department of Mathematics, Colorado State University, Fort Collins, Colorado 80523

Taylor, P.J. (513), Department of Mathematics, University of Stirling, Stirling, Scotland

Ullman, Joseph L. (769, 889), Department of Mathematics, University of Michigan, Ann Arbor, Michigan 48109

Varga, R.S. (769), Department of Mathematics, Kent State University, Kent, Ohio 44242

Varma, A.K. (881), Department of Mathematics, University of Florida, Gainesville, Florida 32601

Vaughan, Hubert (461),* Sidney, Australia

Wachspress, E.L. (897), Knolls Atomic Power Laboratory, Schenectady, New York 12301

Wahba, Grace (905), Department of Statistics, University of Wisconsin, Madison, Wisconsin 53705

Wallin, Hans (913'), Department of Mathematics, University of Umeå, Umeå, Sweden

Ward, Joseph D. (305), Department of Mathematics, Texas A&M University, College Station, Texas 77843

Weinstein, Stanley E. (507), Department of Mathematical Sciences, Old Dominion University, Norfolk, Virginia 23508

Werner, Helmut, (125), Institut für Numerische und instrumentelle, Mathematik und Rechenzentrum, Westfälische Wilhelms Universität, Münster, Federal Republic of Germany

Wood, B. (919), Department of Mathematics, University of Arizona, Tucson, Arizona 85721

Wuytack, L. (267), Department of Mathematics, University of Antwerp, B-2610 Wilrijk, Belgium

Zacharski, J.J. (647), Department of Mathematical Sciences, Rensselaer Polytechnic Institute, Troy, New York 12181

Zalik, R. A. (927), Department of Mathematics, Auburn University, Auburn, Alabama 36830

Zeller, Karl (495), Department of Mathematics, University of Tübingen, Tübingen, Federal Republic of Germany

Zugler, Z. (157), Department of Mathematics, Technion-Israel Institute of Technology, Haifa, Israel

Zielke, Roland (933), Universitaet Osnabrueck, Osnabrueck, Federal Republic of Germany

*Deceased

PREFACE

On January 8, 1980, approximately 130 mathematicians from all over the world gathered in Austin to take part in the seventieth birthday celebration of Professor George G. Lorentz. The scientific part of this celebration consisted in a four-day symposium on approximation theory, underwritten by the National Science Foundation. This volume represents the official proceedings of that conference. And contains contributions from almost all of the participants. Because of the large number of contributors, the papers of the proceedings had to be severely limited in length. But the organizing committee (Larry Schumaker and I) invited a few longer papers to provide a backbone for the conference and for these proceedings; these contributions are placed first in the book.

The success of the conference was due to the enthusiasm and cooperation of our many participants, and to the efforts of many individuals working behind the scenes. It is a pleasure to acknowledge their help and to thank them here.

First, Dr. Larry Schumaker shared in all the early planning of the conference, although his political campaign for the U.S. Senate curtailed his later involvement. Dr. William Pell of the National Science Foundation was instrumental in obtaining our main financial support. Dr. Peter Flawn, President of The University, kindly made additional funds available for the incidental expenses of the conference. Dr. William S. Livingston, Dean of the Graduate School, provided a grant from the university's research fund to help meet the travel expenses of some participants. Dr. James W. Daniel, Chairman of the Mathematics Department, generously put departmental resources at our disposal. Although the entire administrative staff of the department was involved one way or another with the conference, I would like to thank especially Ms. Nita Goldrick and Mrs. Marsha Beierschmitt for their help during the conference and afterward in preparing the proceedings for publication. The staff of Academic Press merits our warm thanks for undertaking to publish the proceedings and for overseeing the entire project.

BEST APPROXIMATION IN HILBERT SPACE

Hubert Berens

Mathematisches Institut
Universität Erlangen-Nürnberg
Erlangen, BRD

For any nonempty subset of an inner product space its metric projection is a monotone set-valued mapping. Indeed, this property characterizes such spaces. It is the aim of the paper to make the reader acquainted with this aspect of the theory of best approximation in inner product spaces and to point out some of its principal consequences.

0. Introduction

0.1. Let H be a Hilbert space over $\mathbb{R}$ with inner product $\langle \cdot,\cdot \rangle$. Its norm is denoted by $|\cdot|$. By H_w we denote the Hilbert space endowed with the weak topology.

If K is an arbitrary nonempty subset of H, the metric projection $P_K : H \to 2^K$ is defined by

$$H \ni x \mapsto P_K(x) = \{k \in K : |x - k| = d_K(x)\},$$

where $d_K : H \to \mathbb{R}$ is the distance function.

$$\mathcal{D}(P_K) = \{x \in H : P_K(x) \neq \emptyset\}$$

is said to be the domain of P_K.

It is convenient to identify P_K with its graph in $H \times H$. Writing $(x;k) \in P_K$ thus means $k \in P_K(x)$. P_K^{-1} denotes the inverse of P_K.

We say that the set $K \subset H$ is proximinal if $\mathcal{D}(P_K) = H$ and K is said to be chebychev if for all $x \in H$ $\#\{P_K(x)\} = 1$.

Copyright © 1980 by Academic Press, Inc.
All rights of reproduction in any form reserved.
ISBN: 0-12-171050-5

0.2. The paper is divided into two parts. In the first part we study the uniqueness problem, while in the second one we discuss rational L^2-approximation as an example of nonconvex approximation in a Hilbert space. In this connection we take up the notion of a reach of a set, as defined by H. Federer, and show its usefulness by considering the problem from the geometrical viewpoint.

In [7] D. Braess gave an excellent account on nonlinear approximation. This article may be considered as a reflection on his section on nonlinear mean-square approximation. The paper does not claim completeness, and the author would be grateful for making him aware of any omissions and incorrectnesses.

1. Approximation by Closed Convex Sets

1.1. It is well known that a nonempty closed convex set K in H is a chebychev set and that for each $x \in H \setminus K$ $(x,k) \in P_K$ if, and only if, $k + (x-k)^{\perp}$ is a supporting hyperplane of K at k, i.e.,

$$\forall\ k' \in K \qquad < k' - k, x - k > \ \leq 0 .$$

It follows that

$$\forall\ (x,k),(x',k') \in P_K \qquad |k - k'|^2 \leq \ < k-k', x - x' > ,$$

which in turn implies that P_K is a contraction.

A simple exercise shows that the converse is also true. More generally, we have

THEOREM. _For a set_ K _in_ H _the following statements are equivalent_:

(i) K _is a closed convex set_.

(ii) K is a chebychev set and P_K is a contraction.

(iii) K is a chebychev set and P_K is a continuous, even a demi-continuous mapping. The latter means $P_K : H \to H_w$ is continuous.

(iv) K is a chebychev set and P_K is a weakly outer radial continuous mapping, i.e.,

$$\forall\ (x,k) \in P_K \qquad \underset{t \to 1+}{\text{w-lim}}\ P_K(k + t(x - k)) = k.$$

(v) P_K is maximally monotone.

The chain of implications (i) ⇒ (ii) ⇒ (iii) ⇒ (iv) is obviously true. Extending results of F. A. Ficken and V. Klee, E. Asplund [4] proved in 1969 the implication (iii) ⇒ (i), while already in 1967 L. P. Vlasov [22] had established the equivalence between (i) and (iv), see also [23; Chap. IV]. The implications (iv) ⇒ (v) ⇒ (i) were given by H. Berens and U. Westphal [6] in 1977.

For any set K in H its metric projection P_K is a monotone mapping, as will be seen in the following section. It is the aim of the chapter to study this aspect of best approximation in a Hilbert space in greater detail. Among others, we shall present a proof of the theorem by using purely Hilbert space arguments.

1.2. As we just stated, for any nonempty set K in H P_K is monotone, i.e.,

$$\forall\ (x,k),(x',k') \in P_K \qquad 0 \le \langle k - k', x - x' \rangle.$$

Indeed, we have $|x - k|^2 \le |x - k'|^2$ and $|x' - k'|^2 \le |x' - k|^2$. Adding up the inequalities and arranging all the terms on the right-hand side of the inequality sign, gives the desired statement. More generally, P_K is <u>cyclically monotone</u>, i.e.,

$$\forall\ (x_o,k_o),(x_1,k_1),\dots,(x_n,k_n) \in P_K, \qquad n \in \mathbb{N},$$

$$0 \leq \sum_{j=0}^{n} < x_{j+1} - x_j, k_{j+1} >, \text{ where } (x_{n+1},k_{n+1}) = (x_o,k_o).$$

It follows from a result of R. T. Rockafellar on cyclically monotone operators, see e.g. H. Brezis [9; p. 38], that there exists a proper, lower semi-continuous (l. s. cont.) convex function $\varphi_K : H \to \mathbb{R} \cup \{\infty\}$ such that $P_K \subset \partial\varphi_K$, $\partial\varphi_K$ denotes the subdifferential of φ_K. In general, φ_K is not uniquely determined, naturally, up to an additive constant. If, however, K is a closed set in H then it is. Following Asplund, we set

$$H \ni x \mapsto \varphi_K(x) = \sup\{< x,k > - \frac{|k|^2}{2} : k \in K\}$$
$$= \frac{|x|^2}{2} - \frac{d_K^2(x)}{2}.$$

In particular, $\varphi_K : H \to \mathbb{R}$ is continuous and convex and

$$\varphi_K(x) = < x,k > - \frac{|k|^2}{2} \text{ for a pair } (x,k) \in H \times K$$
$$\Leftrightarrow (x,k) \in P_K.$$

The subdifferential $\partial\varphi$ of a proper, l. s. cont. convex function φ on H is <u>maximally monotone</u>, i.e., there is no proper monotone extension of $\partial\varphi$ in $H \times H$. For details see [9; Chap. II]. At this point we just want to mention that for each $x \in \mathcal{D}(\partial\varphi)$ $\partial\varphi(x)$ is closed and convex. In our situation, $\partial\varphi_K$ defines a maximal extension of P_K, being unique for closed sets K in H. Clearly, for each $x \in H$ $\partial\varphi_K(x) \neq \emptyset$.

Although the monotony of P_K for an arbitrary subset K in H follows from Asplund's considerations, it was P. Kenderov [17] who formulated and proved this property explicitly in 1975. For convex subsets, however, it was well known, see e.g.

§ 32 in Part V of R. B. Holmes' lecture notes [16]. The author and U. Westphal were also motivated by Asplund's paper to carry out their investigations.

In the following section we want to give a characterization of $\partial\varphi_K$ which shows that in an infinite dimensional Hilbert space it is to some extent the more appropriate object to deal with than the metric projection P_K.

1.3. Let K be a nonempty set in H. For an $x \in H$ consider the set

$$\Phi_K(x) = \bigcap_{r>d_K(x)} \overline{co}\{b_r(x) \cap K\}.$$

Clearly, it is a nonempty closed convex set and

$$\forall\ x \in H \quad \overline{co}P_K(x) \subset \Phi_K(x) \subset \bar{b}_{d_K(x)}(x).$$

Here $b_r(x)$ and $\bar{b}_r(x)$ denote the open and closed balls centered at x with radius $r \in \mathbb{R}^+$, respectively, while $\overline{co}\{\cdot\}$ denotes the closed convex hull of the set under consideration. Moreover, it is not difficult to prove

PROPOSITION. <u>The set-valued operator Φ_K on H into itself is cyclically monotone. It is maximally monotone and contained in $\partial\varphi_K$. Hence,</u>

$$\forall\ x \in H \quad \partial\varphi_K(x) = \bigcap_{r>d_K(x)} \overline{co}\{b_r(x) \cap K\}.$$

Using further properties of maximally monotone operators we obtain:

$\Phi_K : H \to H_w$ <u>is upper semi-continuous</u> (u. s. cont.) <u>and continuous exactly on</u>

$$U_\Phi = \{x \in H : \#\{\Phi_K(x)\} = 1\}$$

<u>which is in a dense G_δ-set in</u> H. <u>Moreover,</u>

$$\forall\ x \in H \quad \Phi_K(x) = \bigcap_{\delta>0} \{\Phi_K(x') : |x - x'| < \delta \ \underline{\text{and}}\ x' \in \mathcal{U}_\Phi\}.$$

Making use of a result of S. B. Stečkin [20] in 1963 which states that for a closed set K in H the complement of the uniqueness set

$$\mathcal{U}_P = \{x \in H : \#\{P_K(x)\} = 1\}$$

is a set of first category in H, we obtain

$$\forall\ x \in H \quad \Phi_K(x) = \bigcap_{\delta>0} \{P_K(x') : |x - x'| < \delta \text{ and } x' \in \mathcal{U}_P\}.$$

It follows from this statement that for closed sets K in H φ_K is uniquely determined up to an additive constant.

In § 19 of Part II of Holmes' notes it is proved that for a closed and convex set K in H $d_K^2/2 : H \to \mathbb{R}$ is a continuously differentiable, convex function and that

$$\nabla \frac{d_K^2}{2} = I - P_K.$$

This statement is obviously wrong for arbitrary subsets K in H. However, the generalized subdifferential

$$\partial \frac{d_K^2}{2} = I - \Phi_K$$

nicely extends the results, as will be seen in Section 1.5. below.

Independently from the author, in [15] C. Franchetti and P. L. Papini introduced the mapping $\Phi_K \subset H \times H$, even in a normed vector space, and studied its properties especially for sets K in H with bounded complement.

1.4. Proof of the implication (iv) → (v): The statement (v) means: $P_K = \Phi_K$. Let $(x,k) \in P_K$, $x_t = k + t(x - k)$, $t > 0$, and $k_t = P_K(x_t)$. Vlasov's assumption then reads $\text{w-}\lim_{t\to 1+} k_t = k$.

By monotony, for each $\eta \in \Phi_K(x)$

$$0 \leq < k_t - \eta, x_t - x > = (t-1) < k_t - \eta, x - k >.$$

Dividing the inequality by $t-1$ and letting $t \to 1+$, we obtain

$$0 \leq < k - \eta, x - k > \quad \text{or} \quad |k - x|^2 \leq < \eta - x, k - x >$$

which is true if, and only if, $\eta = k$.

1.5. Proof of the implication (v) $\to$ (i): Following Brezis' notes we can conclude from his treatment of the evolution equation in Chap. III

PROPOSITION. *Let* k *be a closed set in* H. *For each* $x \in H \setminus K$ *there is a uniquely determined strong solution* $\mathbb{R}^+ \cup \{0\} \ni t \mapsto$ $\mapsto u(t)$ *of*

$$u' + \Phi_K(u) - u \ni 0 \quad \text{and} \quad u(0) = x. \tag{*}$$

Furthermore, for each $t \in \mathbb{R}^+ \cup \{0\}$ $u'_+(t)$ *exists an*

$$u'_+(t) = (I - \Phi_K)^{\circ}(u(t)) := u(t) - P_{\Phi_K(u(t))}(u(t)),$$

the function $t \mapsto u'_+(t)$ *is continuous from the right and*

$$|u'_+(t)| \leq e^t |(I - \Phi_K)^{\circ}(x)|.$$

Finally, for each $T \in \mathbb{R}^+$ $[0,T] \ni t \mapsto d_K^2(u(t))/2$ *is absolutely continuous and*

$$\frac{d}{dt} \frac{d_K^2(u(t))}{2} = |u'_+(t)|^2 \qquad \text{a.e.}$$

The notion of a strong solution $t \mapsto u(t)$ means that for each $T \in \mathbb{R}^+$ $[0,T] \ni t \mapsto u(t)$ is absolutely continuous and the differential inclusion is satisfied a.e.

We would like to remark that (*) can be rewritten as

$$u' \in \partial \frac{d_K^2(u)}{2} \quad \text{and} \quad u(0) = x,$$

showing that it is a gradient inclusion.

Let us assume that for a given element $x \in H \setminus K$

$$\exists\ T \in \mathbb{R}^+ \ni \ |u'(t)| = d_K(u(t)) \text{ a.e. on } [0,T],$$

i.e., for almost all $t \in [0,T]$ $P_K(u(t)) = \Phi_K(u(t))$, then

$$\frac{d_K^2(u(t))}{2} - \frac{d_K^2(x)}{2} = \int_0^t d_K^2(u(\tau))d\tau, \qquad 0 \le t \le T,$$

or

$$d_K(u(t)) = d_K(x)e^t.$$

Moreover,

$$d_K(u(t)) - d_K(x) = |u(t) - x| \le \int_0^t |u'(\tau)|d\tau = \\ = d_K(u(t)) - d_K(x)$$

giving

$$d_K(u(t)) = d_K(x) + |u(t) - x|, \qquad 0 \le t \le T.$$

Consequently, there is a unique element $k \in K$ such that for each $0 \le t \le T$ $P_K(u(t)) = \Phi_K(u(t)) = \{k\}$ and

$$u(t)) = k + (x - k)e^t.$$

Clearly, the solution can be extended to $-\infty < t < T$.

To prove the implication (v) → (i), let us first remark that the statement (v) immediately implies that K is a chebychev set. If $(x,k) \in P_K$, $x \notin K$, then the solution $t \mapsto u(t)$ of (*) satisfies for almost all $t \in \mathbb{R}^+ \cup \{0\}$ the condition $|u'(t)| = d_K(u(t))$, giving

$$u(t) = k + (x - k)e^t, \qquad t \in \mathbb{R}.$$

Thus K is a <u>sun</u>. A chebychev sun however has to be closed and convex.

The method of proof goes back to H. Federer in 1959. In Section 4 of his paper [14] on curvature measures he introduced and studied sets of positive reach in $\mathbb{R}^n$. A set $K \subset \mathbb{R}^n$ is said to have <u>positive reach at its point</u> k if there exists a positive radius r such that $b_r(k) \subset \mathcal{U}_{P_K}$. The largest possible radius is called the <u>reach</u> <u>of</u> K <u>at</u> k and is denoted by reach(k;,K). If the set K has positive reach at each element $k \in K$ and if reach(k;K) $\geq$ const. > 0 on K, then K is said to have <u>positive</u> <u>reach</u> and inf{reach(k;K) : $k \in K$} is said to be the <u>reach</u> of K, in notation reach(K). Clearly, such a set is closed.

Let K be a set of positive reach in $\mathbb{R}^n$. Setting

$$K_{reach} = \{x \in \mathbb{R}^n : d_K(x) < \text{reach}(K)\},$$

it is proved in Theorem 4.8 of [14] that the metric projection P_K is locally lipschitz-continuous on $K_{reach} \setminus K$ and that the solution $t \to u(t)$

$$u' = u - P_K(u) \quad \text{and} \quad u(0) = x \in K_{reach} \setminus K$$

is given by $u(t) = k + (x-k)e^t$, $-\infty < t < T_x$, where

$$P_K(x) = \{k\} \text{ and } T_x = \log d_K(u(T_x))/d_K(x) =$$
$$= \log \text{reach}(K)/d_K(x).$$

The proof of the latter part is just the one given above.

Federer further remarked that the sets K in $\mathbb{R}^n$ having reach(K) = ∞ are exactly the closed and convex sets, reproving Motzkin's theorem, see Section 1.6. below.

There are other proofs of the implication (v) $\Rightarrow$ (i) available, see [6]. The one presented here, however, seems to be the most pleasing, at least to the author. In addition, it shall give a motivation to study the dynamical system associated with the approximation problem.

1.6. In 1935 Th. Motzkin stated and proved that a chebychev set in $\mathbb{R}^n$ is closed and convex. The priority of the statement goes actually to L. Bunt who has established the result in his doctoral thesis in 1934 *). For an infinite dimensional Hilbert space it seems however that one needs some additional assumption for a chebychev set to be convex. The weakest known condition seems to be Vlasov's weak outer radial continuity assumption on the metric projection.

Already in 1961 V. Klee [18] made the conjecture that there are chebychev sets in an infinite dimensional Hilbert space, possibly nonseparable, which are nonconvex. To support his conjecture he constructed a semi-chebychev set the complement of which is open, bounded, and convex. Clearly, such a set does not exist in $\mathbb{R}^n$. Asplund in his paper even proved: If Klee's conjecture is true then there exists a chebychev set having an open, bounded, and convex complement.

If Klee's conjecture is true, then there exists a proximinal, even a chebychev set K in H such that

$$\text{for some element } x \in H \quad \overline{co}P_K(x) \subsetneq \Phi_K(x).$$

To support Klee's conjecture the author tried unsuccessfully to construct such a set.

A solution to the problem:

$$\text{prove or disprove, } \forall \text{ proximinal set } K \subset H \quad \overline{co}P_K = \Phi_K,$$

seems to be of independent interest.

1.7. To illustrate the problem, consider the space $\ell_2(\mathbb{N})$. If $x \in \ell_2(\mathbb{N})$, we shall denote its series expansion with respect to the natural orthonormal basis $\{e_j\}_{j\in\mathbb{N}}$ by $\sum \alpha_j e_j$.

We take $K_o = \{e_j\}_{j\in\mathbb{N}}$. It follows from

$$\ell_2(\mathbb{N}) \ni x \mapsto \varphi_{K_o}(x) = \sup_j \alpha_j - \frac{1}{2}$$

*) The author would like to thank Professor F. Deutsch for making him aware of Bunt's thesis.

that

$$\mathcal{D}(P_{K_o}) = \{x \in \ell_2(\mathbb{N}) : \exists\ j \in \mathbb{N} \ni \alpha_j \geq 0\}.$$

and that

$$\ell_2(\mathbb{N}) \ni x \mapsto \Phi_{K_o}(x) = \begin{cases} \overline{co}P_{K_o}(x), \text{ if } \exists\ j \in \mathbb{N} \ni \alpha_j > 0, \\ \overline{co}\{0, e_j : \forall\ j \in \mathbb{N} \ni \alpha_j = 0\}, \\ \qquad \text{if } \forall\ j \in \mathbb{N}\ \ \alpha_j \leq 0. \end{cases}$$

To obtain a proximinal extension K of K_o by just adding one point $y = \sum \beta_j e_j$ to K_o, the coefficients of y have to satisfy the condition: for each $j \in \mathbb{N}$ $\beta_j \leq 0$ and $\sum \beta_j^2 \leq 1$. However, for each such extension it turns out that $\overline{co}P_K = \Phi_K$.

The author admits that the example is not very sophisticated, but going through the details shows nevertheless the trouble one has to cope with. Due to Stečkin's result any perturbation of a closed set K in H has a severe impact on the approximation behaviour of the set which is difficult to handle. In addition, our knowledge on existence sets, even in a Hilbert space is quite meager.

1.8. We want to conclude the chapter by giving several sufficient conditions on a subset K in H to guarantee that $\overline{co}P_K = \Phi_K$. In $\mathbb{R}^n$ these are the closed sets. For an infinite dimensional Hilbert space H we have

PROPOSITION. <u>Each of the following conditions is sufficient to conclude for a subset</u> K <u>in</u> H <u>that</u> $\overline{co}P_K = \Phi_K$ <u>is true</u>:

(i) K <u>is an approximatively compact set</u>, <u>i.e.</u>, <u>for each</u> $x \in H \setminus K$ <u>each sequence</u> $\{k_j\}_{j \in \mathbb{N}}$ <u>in</u> K <u>satisfying</u> $d_K(x) = \lim_j |x - k_j|$ <u>has a cluster point in</u> K.

(ii) K <u>is proximinal and</u> P_K <u>is u. s. continuous</u>.

(iii) K <u>is proximinal and</u> $\overline{co}P_K : H \to H_w$ <u>is u. s. const</u>.

(iv) K <u>is closed and</u>

$$\forall\ x \in H \quad \overline{co}P_K(x) = \bigcap_{\delta>0} \overline{co}\{P_K(x') : |x - x'| < \delta \text{ and } x' \in \mathcal{U}_P\}.$$

(v) <u>for each closed halfspace</u> M <u>in</u> H K $\cap$ M <u>is proximinal</u>.

Asplund proved in [4] that each chebychev set in H satisfying condition (v) is necessarily closed and convex.

Finally let us remark that the set K in the example in Section 1.7 is neither approximately compact nor does it satisfy condition (v) in $\ell_2(\mathbb{N})$ for the appropriate choice of y.

2. Approximation by Sets of Positive Reach

2.1. Approximation by rational functions in $L^2[-1,1]$ seems to be one of the best studied examples of approximation by a nonconvex set in an infinite dimensional Hilbert space.

By $P_n[-1,1]$, $n \in \mathbb{N}_o$, we denote the space of real polynomials of degree $\leq n$ restricted to the interval $[-1,1]$ and by $\mathring{P}_n^+[-1,1]$ the interior of the cone of the positive polynomials in $P_n[-1,1]$. For $n \in \mathbb{N}_o$ and $m \in \mathbb{N}$ $R_m^n[-1,1]$ denotes the space of all rational functions $r = p/q$ on $[-1,1]$ where $p \in P_n[-1,1]$ and $q \in \mathring{P}_m^+[-1,1]$. Setting for a point $a \in \mathbb{R}^{n+m+1}$

$$p(a;x) = \sum_o^n a_j x^j \quad \text{and} \quad q(a;x) = 1 + \sum_1^m a_{n+j} x^j,$$

$R_m^n[-1,1]$ is the smooth image of the open set

$$A_m^n = \{a \in \mathbb{R}^{n+m+1} : q(a) \in \mathring{P}_m^+[-1,1]\}$$

in $L^2[-1,1]$ under the mapping

$$A_m^n \ni a \to r(a) = \frac{p(a)}{q(a)}.$$

The restriction of r to the open subset

$$_{\text{norm}}A_m^n = \{a \in \mathbb{R}^{n+m+1} : r'(a) \in L(\mathbb{R}^{n+m+1}, L^2[-1,1]) \text{ is injective}\}$$

defines a homeomorphism of ${}_{\text{norm}}A^n_m$ onto the <u>normal</u> functions of $\mathcal{R}^n_m[-1,1]$, in notation ${}_{\text{norm}}\mathcal{R}^n_m[-1,1]$. Thus ${}_{\text{norm}}\mathcal{R}^n_m[-1,1]$ is a smooth hypersurface in $L^2[-1,1]$; in addition it is an open and dense subset of $\mathcal{R}^n_m[-1,1]$.

In 1961 N. V. Efimov and S. B. Stečkin [13] proved that $\mathcal{R}^n_m[-1,1]$ is an approximatively compact subset of $L^2[-1,1]$. It follows that it is proximinal and that $\overline{co}P_{\mathcal{R}^n_m} = \Phi_{\mathcal{R}^n_m}$, in particular that $\mathcal{R}^n_m[-1,1]$ is nonchebychev. Actually, in [13] the two named authors formulated the notion of approximative compactness in normed vector spaces and they established the existence and nonuniqueness of best rational approximation in $L^p[-1,1]$, $1 \leq p < \infty$.

In 1967 E. W. Cheney and A. A. Goldstein [10] picked up the problem of rational approximation in $L^2[-1,1]$. They proved among others that each normal rational function is the element of best approximation (el. of b. appr.) of some $f \in L^2[-1,1] \setminus \setminus \mathcal{R}^n_m[-1,1]$, while on the other hand

$$\forall\ f \in L^2 \setminus \mathcal{R}^n_m \qquad P_{\mathcal{R}^n_m}(f) \subset {}_{\text{norm}}\mathcal{R}^n_m .$$

In particular,

$$d_{\mathcal{R}^{n-1}_{m-1}}(f) > d_{\mathcal{R}^n_m}(f).$$

A third important paper on the subject was published in 1973 by J. M. Wolfe [24]. Wolfe proved that

$$\mathcal{U}_{\mathcal{R}^n_m} \cap L^2 \setminus \mathcal{R}^n_m \ \ \underline{\text{is open and dense in}}\ L^2$$

and that

$$P_{\mathcal{R}^n_m} \ \ \underline{\text{is locally lipschitz-continuous on this set}}.$$

The latter part was actually proved by him in [25], which appeared three years later; a second proof is due to Th. Abatzoglou [2] in 1978. These statements are especially important from the numerical point of view.

Wolfe further observed that

$\mathcal{R}_m^n$ <u>has a positive reach for each</u> $r \in {}_{norm}\mathcal{R}_m^n$.

For nonnormal elements in $\mathcal{R}_m^n[-1,1]$ its reach need not be positive. Thus there is an open neighborhood of ${}_{norm}\mathcal{R}_m^n[-1,1]$, namely,

$$\cup \{b_{reach(r;\mathcal{R}_m^n)}(r) : r \in {}_{norm}\mathcal{R}_m^n\}$$

which belongs to $\mathcal{U}_{P_{\mathcal{R}_m^n}}$. Moreover, if $f \in b_{reach(r;\mathcal{R}_m^n)}(r)$ for some $r \in {}_{norm}\mathcal{R}_m^n[-1,1]$ and if r^* is an el. of <u>local</u> b. appr. of f in $b_{reach(r;\mathcal{R}_m^n)}(r) \cap \mathcal{R}_m^n$, then r^* is the el. of <u>global</u> b. appr. of f in $\mathcal{R}_m^n[-1,1]$.

The last statement, as weak as it may seem, has to be contrasted with the following result of Wolfe which has been extended by D. Braess [8] in 1975:

$\forall\ N \in \mathbb{N}\ \exists\ f \in L^2 \setminus \mathcal{R}_m^n$ <u>having more than</u> N <u>el. of local b. appr. in</u> $\mathcal{R}_m^n$,

a very disturbing statement for the numerical analyst.

As an upper estimate for the reach of $\mathcal{R}_m^n[-1,1]$ at its normal function r, we have

$$reach(r;\mathcal{R}_m^n) \leq \min\{\rho(r;\mathcal{R}_m^n), fd(r)/2\},$$

where $\rho(r;\mathcal{R}_m^n)$ is the metric radius of curvature of $\mathcal{R}_m^n[-1,1]$ at r and fd(r) is its folding.

Let $T(r;\mathcal{R}_m^n)$, resp. $N(r;\mathcal{R}_m^n)$, denote the <u>tangential</u>, resp. <u>normal</u>, <u>cone of</u> $\mathcal{R}_m^n[-1,1]$ <u>at</u> r. Since ${}_{norm}A_m^n \ni a \mapsto r(a) = p(a)/q(a)$ defines a smooth embedding of ${}_{norm}\mathcal{R}_m^n[-1,1]$ in $L^2[-1,1]$, for an $a \in {}_{norm}A_m^n$ $T(r(a);\mathcal{R}_m^n)$ is given by the subspace $r'(a)(\mathbb{R}^{n+m+1}) = \mathcal{P}_{n+m+1}/\{q(a)\}^2$, while $N(r(a);\mathcal{P}_m^n)$ is its orthogonal complement. Following J. Rice [19; Chap. 11] we denote by $\rho(r,y;R_m^n)$ the <u>metric radius of curvature of</u> $R_m^n[-1,1]$ <u>at</u> r <u>in the direction</u> y, y being a normalized element of $N(r;R_m^n)$. Here

$$\frac{1}{\rho(r;y;R_m^n)} = \max_{\substack{h\in\mathbb{R}^{n+m+1}\\ |h|=1}} \frac{< y,r''(a)(h,h) >}{|r'(a)(h)|^2}$$

which has been shown by Th. Abatzoglou [24] in 1978. The <u>metric radius of curvature</u> of $R_m^n[-1,1]$ <u>at</u> r is then defined by

$$\frac{1}{\rho(r;R_m^n)} = \sup\{\frac{1}{\rho(r;y;R_m^n)}\ ;\ y\in N(r;R_m^n) \text{ and } |y| = 1\}.$$

The inverse of $\rho(r;R_m^n)$ is said to be the <u>metric curvature</u> of R_m^n at r and is denoted by $\sigma(r;R_m^n)$.

$R_m^n[-1,1]$ is a connected family in $L^2[-1,1]$. The <u>folding</u> of $R_m^n[-1,1]$ at r is defined by

$$fd(r) = \sup\{\delta\in\mathbb{R}^+ : b_t(r)\cap R_m^n \text{ is connected } \forall\ 0 < t \leq \delta\}.$$

The value fd(r) measures how much $R_m^n[-1,1]$ turns back towards r.

Abatzoglou showed that for the family $R_1^o[-1,1]$ its folding is equal to infinity for each of its elements, nor is it difficult to calculate the metric radius of curvature for an $r\in R_1^o[-1,1]\setminus\{0\}$. The zero element has zero reach. These quantities are not known for the functions r in $R_m^n[-1,1]$ for general n and m in $\mathbb{N}$, nor are there estimates of their reach available, estimates from below as well as from above.

2.2. Since ${}_{norm}R_m^n[-1,1]$ is a smooth $(n+m+1)$-dimensional hypersurface in $L^2[-1,1]$ and since $R_m^n[-1,1]\setminus{}_{norm}R_m^n[-1,1]$ interacts only in the trivial way:

$$\forall\ r\in R_m^n\setminus{}_{norm}R_m^n \qquad P_{R_m^n}^{-1}(r) = \{r\},$$

the study of the best approximation of rational functions in $L^2[-1,1]$ has served as a pilot for the study of related families in $L^2[-1,1]$ such as families of exponential sums or spline functions with free knots or "even" for the study of smooth manifolds with or without boundary in an abstract Hilbert space. A substantial part of the research on this subject was done in the middle of the seventies by a group of mathematicians at the Texas A & M University associated with C. K. Chui and P. W. Smith. They studied in particular the concepts of metric curvature and folding of a manifold in a Hilbert space and, more generally, in a normed vector space. One of their central results published in 1976 by C. K. Chui, E. R. Rozema, P. W. Smith and J. D. Ward [12] reads

THEOREM. *Let* K *be a smooth manifold (with boundary) in* H. *If the metric curvature is bounded on compact subsets of* K *and if for each* $k \in K$ $fd(k) > 0$, *then* K *has a positive reach for each* $k \in K$.

Recently, Th. Abatzoglou picked up the subject again. In [1] and [3] he studied the global approximation behaviour of a C^2-manifold in $\mathbb{R}^n$ and in an abstract Hilbert space, respect.; see also the theorem stated below. In his discussion he sticks closer to the concepts of differential geometry than the gentlemen named above.

Different results on the uniqueness of best approximation in $L^2[0,1]$ by piecewise linear functions have been established by D. L. Barrow, C. K. Chui, P. W. Smith and J. D. Ward [5] in 1977. The author would not miss the opportunity to make the reader aware of their interesting paper.

2.3. Let us recall a result of Wolfe which states that a function $f \in L^2[-1,1] \setminus R^n_m[-1,1]$ may have many elements of local b. appr. in $R^n_m[-1,1]$. If. however, f is placed within the reach of one of its el. of local b. appr., say r, then r is its unique el. of global b. appr.

At a conference in 1966 E. W. Cheney and A. A. Goldstein already discussed the problem under which conditions upon an element $x \in H \setminus K$, K a sufficiently smooth manifold, is an el. of local b. appr. of x its global one. See [11] and also J. Spieß [20]. Here we would like to point out a result of H. Federer which in some way complements the results of Cheney, Goldstein, and Spieß.

Let K be a closed nonempty set in H and let $k \in K$. The _tangential cone of_ K _at_ k is defined by

$$T(k;K) = \{u \in H : \underline{\lim}_{t \to 0+} \frac{d_K(k+tu)}{t} = 0\}$$

(The definition is equivalent to the one usually attributed to A. Dubovitskii and A. Milyutin, see [16; Part II, § 16].) T(k;K) is a closed cone with vertex at the origin. Its dual cone N(k;K), defined by $\{v \in H : \langle v,u \rangle \leq 0, u \in T(k;K)\}$, is the _normal cone of_ K _at_ k. N(k;K) is closed and convex.

We would like to denote by $N_A(k;K)$ the _cone with vertex at the origin generated by_ $P_K^{-1}(k) - k$. Since $P_K^{-1}(k)$ is convex, so is $N_A(k;K)$, and it is well known that $N_A(k;K) \subset N(k;K)$. In general the inclusion is proper. If, however, k is a point of positive reach then we have equality. More precisely, we have

PROPOSITION. _Let_ K _be a proximinal set of_ H _and let_ $\overline{co}P_K = \Phi_K$. _If_ K _has positive reach at_ k, _then_ $N_A(k;K) = N(k;K)$. _Moreover,_ T(k;K) _is convex and dual to_ N(k;K).

The proposition shows the strength of the notion of posi-ive reach of a set K at the element k by pointing out the correlation between the approximation behaviour of K at k on the one hand and the geometrical structure of K at k on the other hand. More precisely, we have the following

COROLLARY. *Under the assumptions of the proposition we can conclude:*
If $k \in K$ *is an el. of local b. appr. of some element* $x \in H \setminus K$ *and if* K *has positive reach at* k, *then* k *is the unique el. of global b. appr. of* $x_\lambda = (1-\lambda)k + \lambda x, \lambda \in \mathbb{R}^+$, *for all values of* λ *for which* x_λ *belongs to the reach of* K *at* k.

A more precise result was stated and proved by Abatzoglou, loc. cit., in the following

THEOREM. *Let* K *be a connected, complete* n-*dimensional* C^3-*manifold in* H. *Suppose for some* $x \in H$ *and* $k \in K$ $x - k \in N(k;K)$ *and* $\sup\{\sigma(k') : |k-k'| \leq 2R\} \leq 1/R$. *If* $|x-k| < \min\{R, fd(k)/2\}$, *then* k *is the unique el. of b. appr. of* x *in* K.

2.4. It was not the intention of the author to give a full account of what is known in nonconvex best approximation in a Hilbert space but rather to make the reader acquainted the subject by studying the family $R^n_m[-1,1]$ of rational functions in $L^2[-1,1]$ as an example. Secondly, the author wanted to draw the reader's attention to Federer's notion of a reach of a set and to show its relevance within the theory. For the latter Abatzoglou really deserves the credit.

References

[1] Abatzoglou, Th., The minimum norm projection on C^2-manifolds. Trans. Amer. Math. Soc. 243 (1978), 115-122.

[2] Abatzoglou, Th., The Lipschitz-continuity of the metric projection. Preprint 1978, 11 pages.

[3] Abatzoglou, Th., Unique best approximation from a C^2-manifold in a Hilbert space. Preprint 1978, 16 pages.

[4] Asplund, E., Chebychev sets in Hilbert space. Trans. Amer. Math. Soc. 144 (1969), 235-240.

[5] Barrow, D. L., C. K. Chui, P. W. Smith and J. D. Ward, Unicity of best L_2 approximation by second-order splines with variable knots. Bull. Amer. Math. Soc. 83 (1977), 1049-1050.

[6] Berens, H. und U. Westphal, Kodissipative metrische Projektionen in normierten linearen Räumen. In "Linear Spaces and Approximation", ed. by P. L. Butzer and B. Sz.-Nagy, ISNM 40, Basel 1978, 120-130.

[7] Braess, D., Nonlinear Approximation. In "Approximation Theory II", ed. by G. G. Lorentz, C. K. Chui and L. L. Schumaker, New York 1976, 49-77.

[8] Braess, D., On rational L_2-approximation. J. Approximation Theory 18 (1976), 136-151.

[9] Brezis, H., Opérateurs Maximaux Monotones et semi-groupes de contractions dans les espaces de Hilbert. North-Holland Mathematics Studies 5, Amsterdam 1973.

[10] Cheney, E. W. and A. A. Goldstein, Mean-square approximation by generalized rational functions. Math. Z. 95 (1967), 232-241.

[11] Cheney, E. W. and A. A. Goldstein, A note on nonlinear approximation theory. In "Numerische Mathematik, Differentialgleichungen, Approximationstheorie", ed. by L. Collatz, G. Meinardus and H. Unger, ISNM 9, Basel 1968, 251-255.

[12] Chui, C. K., E. R. Rozema, P. W. Smith and J. D. Ward, Metric curvature, folding, and unique best approximation. Siam J. Math. Anal. 7 (1976), 436-449.

[13] Efimov, N. V. and S. B. Stečkin, Approximative compactness and Chebychev sets. Dokl. Akad. Nauk SSSR 140 (1961), 522-524 (Russian) = Soviet Math. Dokl. 2 (1961), 1226-1228.

[14] Federer, H., Curvature measures. Trans. Amer. Math. Soc. 93 (1959), 418-491.

[15] Franchetti, C. and P. L. Papini, Approximation properties of sets with bounded complements. Preprint 1979, 35 pages.

[16] Holmes, R. B., A course on Optimization and Best Approximation. Springer Lecture Notes in Mathematics 257, Berlin 1972.

[17] Kenderov, P. S., A note on multivalued monotone mappings. C. R. Acad. Bulgare Sci. 28 (1975), no. 5, 583-584.

[18] Klee, V., Remarks on nearest points in normed linear spaces. In "Proceedings of the Colloquium on Convexity, Copenhagen 1965." Copenhagen 1967, 168-176.

[19] Rice, J. R., The Approximation of Functions, Vol. I and II. Reading 1964 and 1969.

[20] Spieß, J., Uniqueness theorems for nonlinear L_2-approximation problems. Computing 11 (1973), 327-335.

[21] Stečkin, S. B., Approximative properties of sets in linear normed spaces. Rev. Roumaine Math. Pur. Appl. 8 (1963), 5-18 (Russian).

[22] Vlasov, L. P., On Chebychev sets. Dokl. Akad. Nauk SSSR 173 (1967), 491-494 (Russian) = Soviet Math. Dokl. 8 (1968), 401-409.

[23] Vlasov, L. P., Approximative properties of sets in normed linear spaces. Uspehi Mat. Nauk 28 (1973), 3-66 (Russian) = Russian Math. Surveys 28 (1973), 1-66.

[24] Wolfe, J. M., On the uniqueness of nonlinear approximation in smooth spaces. J. Approximation Theory 12 (1974), 165-181.

[25] Wolfe, J. M., Differentiability of nonlinear best approximation operators in a real inner product space. J. Approximation Theory 16 (1976), 341-346.

THE MONOSPLINE OF LEAST NORM AND RELATED PROBLEMS

Henry Loeb *

Department of Mathematics
University of Oregon
Eugene, Oregon

I. INTRODUCTION

In this paper we will describe the present status of some of the problems associated with characterizing the monosplines of least norm and some related problems. We hope, perhaps more importantly, to impart "some of the flavor" of the ideas which were used to solve several of these problems.

A list of the people who have made recent contributions to this area would have to include Barrar, Braess, Bojanov, Jetter Karlin, Lange, Micchelli, Pinkus, Strauss, Werner and Schumaker.

It is well known that the monosplines of least norm are associated with the solution of certain optimization problems such as the one of securing optimal integration formulas for

*Written during a stay at the University of Münster.

Copyright © 1980 by Academic Press, Inc.
All rights of reproduction in any form reserved.
ISBN: 0-12-171050-5

various L_p norms [7]. This matter will not be pursued in this paper, rather, one is refered to the excellent article of Schoenberg [26].

Let K(t,y) be a sufficiently smooth real-valued function of two variables where t ranges over a real interval which contains [0,1] and y ranges also over [0,1]. The following notation is employed,

$$K^{(j)}(t,y) = \frac{\partial^j}{\partial y^j} K(t,y) \quad .$$

Then for a set of given odd integers, $\{m_i\}_{i=1}^r$, and an integer $n \geq 1$, we consider the class of all monosplines of the form

$$(1) \quad M(t) = \int_0^1 K(t,y)dy + \sum_{j=0}^{n-1} a_j K^{(j)}(t,0) + \sum_{i=1}^{r} \sum_{j=0}^{m_i-1} a_{ij} K^{(j)}(t,y_i) \quad ,$$

where $0 < y_1 < y_2 < \cdots < y_r < 1$ and $\{a_j, a_{ij}\} \subset R$. For each $1 \leq p \leq \infty$ one seeks to characterize the monospline M(t) of the form (1) of least $L_p[0,1]$ norm. Note that the number of parameters associated with each monospline is N, where

$$N = n + \sum_{i=1}^{r} (m_i + 1) \quad .$$

Thus each monospline M(t) is defined by a N - dimensional real parameter vector,

$$A = (a_0,\ldots,a_{n-1},\ a_{1,0},\ldots,a_{1,m_1-1},\ a_{2,0},\ldots,a_{r,m_r-1},\ \xi_1,\xi_2,\ldots,\xi_r),$$

and we write

$$M(A,t) \equiv M(t) \quad .$$

Throughout the remainder of this paper we shall endow $K(x,y)$ with various total positivity properties. Ideally we would like to use assume only that $t_1 < \ldots < t_s$ and $y_1 < \ldots < y_s$ implies that

$$K\begin{pmatrix} t_1,\ldots,t_s \\ y_1,\ldots,y_s \end{pmatrix} \doteq \det\left\{K(t_i,y_j);\ i,j=1,\ldots,s\right\} \geq 0$$

in order to obtain all the desired principal results, that is the existence, characterization, and uniqueness of the monospline of least norm. The possibility of this being correct is probably remote; although in the parallel problem for perfect splines and $p=\infty$, one is able to obtain the desired results [1]. In this regard, there seems to be a "Folk Theorem" which states that for non-linear L_p problems, the L_∞ problem is the most tractable.Basically we shall be dealing with two types of kernels: first, the polynomial spline kernel,

$$K(t,y) = \frac{(t-y)_+^{n-1}}{(n-1)!}$$

where $n \geq 2$ and

$$x_+^k = \begin{cases} x^k & x \geq 0 \\ 0 & x < 0 \end{cases} \quad ;$$

secondly, an <u>extended</u> <u>totally</u> <u>positive</u> <u>kernel</u> $K(t,y)$. This means that $t_1 \leq \cdots \leq t_s$ and $y_1 \leq \cdots \leq y_s$ implies that

$$K^*\begin{pmatrix} t_1,\ldots,t_s \\ y_1,\ldots,y_s \end{pmatrix} > 0$$

For the extended definition $K^*\binom{t_1,\ldots,t_s}{y_1,\ldots,y_s}$ of the determinent $K\binom{t_1,\ldots,t_s}{y_1,\ldots,y_s}$, one is refered to Karlin [16, Pg.13].

II. POLYNOMIAL MONOSPLINES

In this section we consider the spline kernel $K(t,y) = \dfrac{(t-y)_+^{n-1}}{(n-1)!}$ where $n \geq 2$ and the prescribed multiplicities $\{m_i\}_{i=1}^r$ of the knots (see (1)) are restricted by the inequalities, $m_i \leq n$ $(i=1,\ldots,r)$.

The existence and characterization problems were first solved by Karlin [18], Powell [25] for $L_2[0,1]$ where the monosplines have simple knots $(m_i \equiv 1)$. Their results have been extended to $1 \leq p < \infty$ and to all odd multiplicities [5, 6, 19].

Theorem 1 There exists a polynomial monospline of the form (1) of minimal $L_p[0,1]$ norm $(1 \leq p < \infty)$. Further any minimal solution must have the property; $a_{i,m_i-1} < 0$ $(i=1,\ldots,r)$. (Indeed Micchelli [23] has shown that $a_{i,j} < 0$ for even j's)

The main difficulty one faces in the existence proof is how to demonstrate that some of the knots cannot coalesce and produces higher multiplicities when a minimizing sequence is considered. Indeed a technique for "seperating the knots" in this situation and producing a monospline of smaller norm is needed. An alternative to this procedure for $p = \infty$ and multiple knots is to consider the <u>extended totally positive kernel</u> obtained by applying the Gaussian transform to this spline kernel for $\varepsilon > 0$. Then one can produce a optimal monospline with respect to this kernel, $M^{(\varepsilon)}(t)$. Letting $\varepsilon \downarrow 0$ yields a minimal polynomial monospline. We shall discuss this technique in more detail in the next section.

The main tool in many of the characterization, investigations is the fundamental determinent relationship for splines, Karlin, Ziegler [15, pg. 503]:

<u>Theorem 2</u> Let $\{t_i, y_i\}$ $i=1,\ldots,s$ satisfy:

(a) $t_1 \leq \cdots \leq t_s$ and $y_1 \leq \cdots \leq y_s$,

(b) whenever $\ell \geq 1$ of the t_i's and $m \geq 1$ of the y_i's have the same value, then $\ell + m \leq n + 1$

(c) No more than n consecutive y_i's or t_i's, can coincide.

Then,

$$K^*\begin{pmatrix} t_1,\ldots,t_s \\ y_1,\ldots,y_s \end{pmatrix} \geq 0$$

with strict inequality iff

$$t_{i-n} < y_i < t_i \qquad i = 1,\ldots,s \quad ,$$

Indeed if a monospline has a full set of zeros, N to be exact, the Jacobian of the monospline with respect to the corresponding parameter set A evaluated at the zeros is nonsingular. This point will be stressed in the next section.

The uniqueness question as is usually the case in nonlinear settings, generates a much more formidable problem. Recently in [12] Jetter and Lange proved that if all $m_i = 1$ then there was a unique polynomial monospline of minimal L_2 norm. As is the usual situation when a crack appears in the wall, Jetter [13] then gave an uniqueness proof for $p = 1$ and again for simple knots. Recently we received a preprint from Strauss [29] in which he demonstrated the result for $L_1[0,1]$ and all odd multiplicities. Finally Bojanov [7] "completed the circle", by establishing the uniqueness result for $1 < p < \infty$ and all odd multiplicities. An essential ingredient in all these proofs is very nice technique which allows one to bound the second derivative of the optimal monosplines. This technique is due to Lange and appears in his thesis [21].

This technique is strictly tailored for splines; thus it appears that it cannot be used for extended totally positive kernels. Indeed in this direction for $1 \leq p < \infty$ the only known result is negative. Braess [9] has shown that in general the optimal monospline, for the reproducing kernel space H_2 of the unit circle, is not unique. However existence and characterization proofs have been given by Karlin [20] and Barrar, Loeb [4, 5].

For $p = \infty$, the problem has been solved completely. In 1960 Johnson [14] established the existence, characterization and uniqueness of the polynomial monospline of least norm where only simple knots ($m_i = 1$) are permitted. (See also Karlin [19]) His characterization and uniqueness proofs, depended heavily on Theorem 2. Indeed the demonstration of uniqueness follows the pattern. Armed with the characterization theorem (see Theorem 3), one shows that if one has two distinct optimal monosplines, the difference of these two monosplines has too many zeros. For multiplicities greater than one, another device is needed.

Barrar and Loeb in 1976 established the three desired properties for polynomial monosplines of odd multiplicity.

Theorem 3 [2] There exist one and only one polynomial monospline of minimal $L_\infty[0,1]$ norm. The optimal monospline is uniquely characterized by a set of $N+1$ points, $0 = t_0 < t_1 < \dots < t_{N-1} < t_N = 1$ such that

$$M(t_i) = (-1)^{N+i} \|M\|_\infty \qquad i = 0,1,\dots,N$$

where $\|M\|_\infty = \max_{x \in [0,1]} |M(x)|$.

In contrast to the other norms, Theorem 3, in its entirety, has been derived for extended totally positive kernels, [3]. In the next section we will show how this result and a smoothing technique can be used to establish Theorem 3.

III. EXTENDED TOTALLY POSITIVE MONOSPLINES

In this section in order to illustrate the flavor of our work on the L_∞ problem we outline the crucial portions of the development. For the proofs one is referred to [3].

Our work on the extended totally positive version of Theorem 3 is in the spirit of the differential equation approach of Fitzgerald, Schumaker [11].

We first pose the following interpolation problem.

Given: $\{e_i\}_{i=0}^{N}$ where $\operatorname{sgn}(e_k - e_{k-1}) = (-1)^{N+k}$ $k = 1,\dots,N$

Find: An E.T.P. Monospline, $M(t)$, of the form (1), $E > 0$, and a set of $N+1$ points, $0 = t_0 < t_1 < \dots < t_N = 1$ such that

$$(2)\qquad \begin{aligned} M(t_i) &= E\,e_i \qquad & i = 0,1,\dots,N \\ M'(t_i) &= 0 \qquad & i = 1,\dots,N-1 \ . \end{aligned}$$

Indeed it will shown that there is exactly one solution to (2).

We impose the additional restriction on $K(t,y)$ that

$$K(x,0) \equiv C \neq 0 \ .$$

Let

$$\Lambda = \Big\{A \in R^N:\ 0 < \xi_1 < \dots < \xi_r < 1:\ M'(A,t) \text{ has } N-1 \text{ distinct zeros, } 0 < t_1(A) < \dots < t_{N-1}(A) < 1\Big\}$$

For each $A \in \Lambda$, consider the system of N+1 differential equations,

$$\frac{d}{d\tau} M(A(\tau),t_k(A(\tau))) = \frac{d}{d\tau} E(\tau)e_k - d_k \qquad k = 0,1,\ldots,N$$

in the $N+1$ unknowns $(A(\tau),E(\tau))$ with initial conditions $(A(0),E(0)) = (A,0)$ where $t_o(A(\tau)) \equiv 0$, $t_N(A(\tau)) \equiv 0$, $M(A,t_k(A)) = d_k$ $(k = 0,1,\ldots,N)$, and of course

$$M'(A(\tau),t_k(A(\tau))) = 0 \qquad k = 1,\ldots,N-1 .$$

It is easy to see that the solutions satisfy

$$M(A(\tau),t_k(A(\tau))) = E(\tau)e_k + (1-\tau)d_k \qquad k = 0,1,\ldots,N .$$

Infact if the solution can be extended to the interval $[0,1]$, the desired solution is attained at $\tau = 1$; that is,

$$(3) \quad M(A(1),t_k(A(1))) = E(1)e_k \qquad (k = 0,1,\ldots,N).$$

The proof that one can extend the solution to $[0,1]$ is developed in [3]. Using a similar type of differential equation technique [10] , it be can shown that Λ is connected.

For each $A \in \Lambda$, let $F(A) = (A(1),E(1))$. Hence from the theory of differential equations, it follows that $F(\Lambda)$ is connected. Now consider

$$V = \left\{(A,E): E > 0 : M(A,t) \text{ and } E \text{ satisfy } (2)\right\} .$$

One can verify that $(A,E) \in V \Rightarrow F(A) = (A,E)$. Thus F maps Λ onto V and V is connected. Consider the nonlinear systems of equation (2) again in the unknowns (A,E),

$$M(A,t_k(A)) = E\, e_k \qquad k = 0,1,\ldots,N$$

where

$$M'(A,t_k(A)) = 0 \qquad k = 1,\ldots,N-1 .$$

In [3] it has been established that at a solution of the system, the Jacobian matrix is non-singular. Thus by the Implicit Function Theorem, it follows that each point of V is an isolated point of V. Since V is connected, this can only mean that V consists of a singleton. Thus there is one and only one solution to (2).

If we let $e_k = (-1)^{k+N} \qquad (k = 0,1,\ldots,N)$,

this result translates into:

Theorem 4 There is a unique extended totally positive monospline $M(A,t)$ such that for some unique set,
$0 = t_0 < t_1 < \ldots < t_{N-1} < t_N = 1$,

$$(4) \quad M(A,t_k) = (-1)^{k+N} \|M(A,\cdot)\|_\infty \qquad (k = 0,\ldots,N)$$

It is well known, for example see [22, pg. 145] that (4) is a necessary condition that $M(A,t)$ be of minimal uniform norm. Further in [3] it has been established that there is a monospline of the form (1) of minimal uniform norm. Thus the

E.T.P. version of Theorem 3 follows.

In this section we will first illustrate the utility of a smoothing technique, by showing how one can obtain the uniqueness portion of Theorem 3 from Theorem 4.

Let $K(t,y;0) = \frac{(t-y)_+^{n-1}}{(n-1)!}$ and for a fixed $\delta > 0$ define for each $|\varepsilon| > 0$

$$\hat{K}(t,y;\varepsilon) = \frac{1}{\sqrt{2\pi}\,|\varepsilon|} \int_{-\delta}^{1+\delta} e^{-\frac{(t-z)^2}{2\varepsilon^2}} K(z,y;0)dz .$$

It is well known that for $|\varepsilon| > 0$, the kernel $\hat{K}(t,y;\varepsilon)$ is ETP for all real t and y [16, pg.15]. Further the kernel,

$$K(t;y;\varepsilon) = \frac{\hat{K}(t,y;\varepsilon)}{\hat{K}^{(n-1)}(t,0;\varepsilon)}$$

is also ETP for $|\varepsilon| > 0$ [15, pg. 511].

For exposition purposes, we assume that each $m_i \leq n-2$. We claim that (2) has at most one solution for the spline kernel $K(t,y;0)$. We argue by contradiction. Let's assume that (2) has two distinct solutions $(A^{(1)},E^{(1)})$ and $(A^{(2)},E^{(2)})$.

For any ε, let $M(A,t;\varepsilon)$ be a arbitrary monospline of the form (1) for the kernel $K(t,x;\varepsilon)$. Consider the system of (N+1) non-linear equation

$$(5) \quad M(A,t_i;\varepsilon) = E\,e_i = 0 \qquad i = 0,1,\ldots,N$$

where

$$M'(A,t_i,\varepsilon) = 0 \qquad i = 1,\ldots,N-1$$

near the solution $(A,E) = (A^{(1)},E^{(1)})$ for $\varepsilon = 0$. We think of ε as a parameter for the non-linear system (5). Employing Theorem 2 and the fact that $M'(A^{(1)},t)$ has a full set of zeros, it can be shown that the Jacobian matrix of the system with respect to (A,E) is non-singular at $(A,E) = (A^{(1)},E^{(1)})$ for $\varepsilon = 0$. Using the convergence properties of the Gaussian transform [16, pg. 15] and again the fact that $M'(A^{(1)},t;0)$ has a full set of zeros it follows that for (A,ε) close to $(A^{(1)},0)$, $M'(A,t;\varepsilon)$ has a full complement of zeros,

$$0 < t_1(A;\varepsilon) < \ldots < t_{N-1}(A;\varepsilon) < 1 \quad .$$

Thus we can invoke the implicit function theorem to conclude that there exists an $\varepsilon_1 > 0$ such that $|\varepsilon| < \varepsilon_1$ implies that there is a solution of (5), $(A^{(1)}(\varepsilon),E^{(1)}(\varepsilon))$ close to $(A^{(1)},E^{(1)})$. A similar situation can be created for $(A^{(2)},E^{(2)})$. Thus for small $|\varepsilon| > 0$ there are two distinct solutions $(A^{(1)}(\varepsilon),E^{(1)}(\varepsilon))$ and $(A^{(2)}(\varepsilon),E^{(2)}(\varepsilon))$ to (5). This contradicts the uniqueness of the solution of (2) for a extended totally positive kernel. Thus the uniqueness of the interpolation problem for polynomial monosplines has been demonstrated and we clearly have also verified the uniqueness portion of Theorem 4 for this setting.

For polynomial Monosplines, Braess [8] has shown that if $M(A,t)$ is optimal with respect to the uniform norm, then it must alternate N times; that is, there are points,

$$0 = t_0 < t_1 < \ldots < t_{N-1} < t_N = 1$$

such that

$$(5)\quad M(A,t_i) = (-1)^{i+N}\, ||M(A,\cdot)||_\infty \qquad (i=0,1,\ldots,N)$$

Thus we have sketched the proofs of the uniquness and characterization portion of Theorem 3.

Next we outline the existence proof for the polynomial Monospline problem. For $[a,b] \subset (0,1)$, we first seek the function of the form (1) of minimal $L_\infty[a,b]$ norm. Towards this goal for $\varepsilon > 0$, consider the transform $K(t,y;\varepsilon) = \dfrac{\hat{K}(t,y;\varepsilon)}{\hat{K}^{(n-1)}(t,0;\varepsilon)}$ where

$$\hat{K}(t,y;\varepsilon) = \frac{1}{\sqrt{2\pi}\,\varepsilon}\int_0^1 e^{-\frac{(t-z)^2}{2\varepsilon^2}} K(z,y;0)\,dz \quad ,$$

and $K(z,y;0) = \dfrac{(z-y)_+^{n-1}}{(n-1)!}$, which is certainly extended totally positive. By the results of Meinardus [22, pg. 145] for $\varepsilon > 0$ an optimal monospline $M^{(\varepsilon)}(t)$ for the kernel $K(t,y;\varepsilon)$, must satisfy (5) with $a \leq t_0 < t_1 < \ldots < t_{N-1} < t_N \leq b$. Thus by the variation diminshing properties of the Gaussian transform [15, pg. 233], $M_0^{(\varepsilon)}(t)$ has at least N sign changes in $(0,1)$. $M_0^{(\varepsilon)}(t)$ is the polynomial monospline obtained from $M^{(\varepsilon)}(x)$ by replacing $K(t,y;\varepsilon)$ by $K(t,y;0)$ in the expression for $M^{(\varepsilon)}(x)$. We now invoke the compactness result of Micchelli [23]

Theorem 5 There is an $K > 0$ such that for any polynomial monospline of the form (1) which has N distinct zeros in $(0,1)$, its coefficients $\left\{a_i\right\}_{i=0}^{n-1}$, $\left\{a_{ij}\right\}_{j=0,i=1}^{m_i-1\;r}$ are each bounded in magnitude by K.

Hence by going to a subsequence it follows that we can assume that the respective coefficient of $M^{(\varepsilon)}(t)$ converge to a set $\{a_i\}_{i=0}^{n-1}$ $\left\{\{a_{ij}\}_{j=0}^{m_i-1}\right\}_{i=1}^{r}$ and the knots converge to $\{\xi_i\}_{i=1}^{r}$. Because of the convergence properties of the Gaussian transform, for the subsequence,

$$M^{(\varepsilon)}(t) \to M(t) = \int_0^1 K(t,y;0)dy + \sum_{i=0}^{n-1} a_i K^{(i)}(t,0;0) + \\ + \sum_{i=1}^{r} \sum_{j=0}^{m_i-1} a_{ij} K^{(j)}(t,\xi_i;0)$$

uniformly on [a,b]. It is easy to see then that M(t) is an optimal polynomial monospline over [a,b] and that it alternates N times. Thus for each $[a,b] \subset (0,1)$ we have an optimal monospline $M_{[a,b]}(t)$. Letting $a \downarrow 0$ and $b \uparrow 1$, and applying the same techniques, yields a limit monospline which is optimal over [0,1]. The reasons for our circuitous route are two-fold: first, only for $t \in [0,1]$ is $\int_0^1 K(t,y)dy = \frac{t^n}{n!}$; secondly, the convergence properties of the Gaussian transform.

IV. RELATED PROBLEMS

In this section we briefly discuss some related problems. In the differential equation approach to the polynomial monospline problem it is clear that one needs the following result.

<u>Theorem 6</u> <u>Fundamental theorem of Algebra for Monospline with Multiple Knots in the Simple Zero Case</u>. Micchelli [23], see also Karlin, Schumaker [17]. For each set, $0 < t_1 < \ldots < t_N < 1$

there is a unique polynomial monospline, $M(t)$ of the form (1) such that

$$M(t_i) = 0 \qquad i = 1,\ldots,N .$$

In some of the related optimization problems one is interested in the extension of Theorem 6 to the multiple zero case. Recently Barrar and I have made some progress on this problem. During this conference Professor Barrar will report on this work.

In some recent papers Strauss [27, 28] has shown the relationship between the so-called perfect splines and the $L_1[0,1]$ polynomial monospline problem. Further Micchelli and Pinkus [24] have extended some of the results for polynomial perfect splines to the totally positive setting and have developed some relationships between n-widths in $L_\infty[0,1]$ and these splines. Specifically the extended setting is the following.

Consider the family of functions $\{k_1,\ldots,k_r\}$ and the kernel $K(x,y)$ where

$$k_i : [0,1] \to R \qquad i = 1,\ldots,r$$

$$K(x,y) : [0,1] \times [0,1] \to R$$

For each set $1 \le i_1 < \ldots < i_s \le r$, $0 < y_1 < \ldots < y_\ell < 1$, $0 < t_1 < t_2 < \ldots < t_{s+\ell} < 1$, define

$$K\begin{pmatrix} i_1,\ldots,i_s,y_1,\ldots,y_\ell \\ t_1,\ldots,t_s,t_{s+1},\ldots,t_{s+\ell} \end{pmatrix} = \det\begin{pmatrix} k_{i_1}(t_1) & \ldots & k_{i_1}(t_{s+\ell}) \\ \vdots & & \\ k_{i_s}(t_1) & \ldots & k_{i_s}(t_{s+\ell}) \\ K(t_1,y_1) & \ldots & K(t_{s+\ell},y_1) \\ \vdots & \ldots & \vdots \\ K(t_1,y_\ell) & \ldots & K(t_{s+\ell},y_\ell) \end{pmatrix}$$

We assume that the functions are sufficiently smooth and satisfy

1. For each set of points, $0 < y_1 < \ldots < y_m < 1$, the functions $\{k_1(\cdot),\ldots,k_r(\cdot),\ K(\cdot,y_1),\ldots,K(\cdot,y_m)\}$ are independent

2. $k_1,\ldots,k_r$ form a Tchebycheff system over $(0,1)$; that is, $0 < x_1 < \ldots < x_r < 1 \Rightarrow K\begin{pmatrix} 1,\ldots,r \\ x_1,\ldots,x_r \end{pmatrix} > 0$.

3. Each determinent

$$K\begin{pmatrix} i_1,\ldots,i_s,y_1,\ldots,y_\ell \\ t_1,\ldots,t_s,t_{s+1},\ldots,t_{s+\ell} \end{pmatrix}$$

of the above form is non-negative.

Then a totally positive perfect spline with $n-r$ interior knots is any function of the form

$$(6)\quad p(t) = \sum_{i=1}^{r} a_i\, k_i(t) + \sum_{j=0}^{n-r} (-1)^j \int_{\xi_j}^{\xi_{j+1}} K(t,y)\,dy$$

where $0 = \xi_0 < \xi_1 < \ldots < \xi_{n-r} < \xi_{n-r+1} = 1$.

In a recent paper [1], Barrar and Loeb have developed some of the interpolating properties of perfect splines of the form (6). Further in the $L_\infty[0,1]$ case they have shown

Theorem 7 Assume that $r \geq 2$, $k_1 \equiv 1$, and $\frac{d}{dt} k_2(t) > 0$. Then there is a unique perfect spline of the form (6) which is of minimal $L_\infty[0,1]$ norm.

It is uniquely characterized by a set of $n+1$ points $0 = x_1 < \ldots < x_n < x_{n+1} = 1$ such that

$$p(x_i) = (-1)^{i-r+1} \|p\|_\infty \qquad i = 1,\ldots,n+1 .$$

The smoothing technique which were developed in the last section plays an important role in the proof of this theorem.

In closing I would like to thank Professor H. Werner for the courtesies extended to me during my stay at the University of Münster.

REFERENCES

1. Barrar, R.B., Loeb, H.L., The Fundamental Theorem of Algebra and the Interpolating Envelope for Totally Positive Perfect Splines, to appear J. Approx. Theory.

2. Barrar, R.B., Loeb, H.L., On Monosplines with Odd Multiplicity of Least Norm, J. Analyse Math. 33 (1978), 12-38.

3. Barrar, R.B., Loeb, H.L., Werner, H., On the Uniqueness of the Best Uniform Extended Totally Positive Monospline, to appear J. Approx. Theory.

4. Barrar, R.B. and Loeb, H.L., Multiple zeros and applications to optimal linear functionals, Numer. Math. 25 (1976), 257-262.

5. Barrar, R.B., Loeb, H.L., On a Non-Linear Characterization Problem for Monosplines, J. Approx. Theory 18 (1976), 220-240.

6. Bojanov, B.D., Existence and characterization of monosplines of least L_p deviation, to appear Proceedings of the International Conference on Constructive Function Theory, Blagoevgrad 1977.

7. Bojanov, B.D., Uniqueness of the Monosplines of the Least Deviation, Numerical Integration ISNM 15, Edited by G. Hämmerlin, Birkhäuser Verlag, Basel 1979.

8. Braess, D., Chebyshev approximation by spline functions with free knots, Numer. Math. 17 (1971), 357-366.

9. Braess, D., On the Nonuniqueness of Monosplines with Least L_2-Norm, J. Approx. Theory 12 (1974), 91-93.

10. Cavaretta, A.S. Jr., Oscillatory and zero properties for perfect splines and monosplines, J. Analyse Math. 28 (1975), 41-59.

11. Fitzgerald, C.H. and Schumaker, L.L., A differential equation approach to interpolation at extremal points, J. Analyse Math. 22 (1969), 117-134.

12. Jetter, K.,Lange, G., Die Eindeutigkeit L_2-Optimaler Polynomialer Monosplines, Math. Z. 158 (1978), 23-34.

13. Jetter, K., L_1-Approximation verallgemeinerter konvexer Funktionen durch Splines mit freien Knoten, Math. Z. 164 (1978', 53-66.

14. Johnson, R.S., On monosplines of least deviation, Trans. Amer. Math. Soc. 96 (1960), 458-477.

15. Karlin, S.J., Total Positivity, Vol. 1, Stanford Univ. Press 1968.

16. Karlin, S.J., Studden, W.J., Tchebycheff systems with applications in analysis and statistics, Interscience Publishers New York, 1966.

17. Karlin, S.J., Schumaker, L.L., The fundamental theorem of algebra for Tchebysheffian monosplines, J. Analyse Math. 20 (1967), 233-270.

18. Karlin, S.J., The Fundamental Theorem of Algebra for Monosplines Satisfying Certain Boundary Conditions and Applications to Optimal Quadrature Formals, Approximation with Special Emphasis on Spline Functions, I.J. Schoenberg, Editor, Academic Press, New York 1969, 467-484.

19 Karlin, S.J., A Global Improvement Theorem for Polynomial Monosplines, Studies in Spline Functions and Approximation Theory, Karlin,S., Micchelli, C.A., Pinkus, A., Schoenberg, I.J., Editors, Academic Press, New York 1976, 67-82.

20. Karlin, S.J., On a Class of Best Nonlinear Approximation Problems and Extended Monosplines, Studies in Spline Functions and Approximation Theory, Karlin, S., Micchelli, C.A., Pinkus, A., Schoenberg, I.J., Editors, Academic Press, New York 1976, 19-66.

21. Lange, G., Beste und optimale definite Quadraturformeln, Dissertation, Clausthal-Zellerfeld 1977.

22. Meinardus, G., Approximation of Functions Theory and Numerical Methods, Springer-Verlag, Berlin (1967).

23. Micchelli, C.A., The fundamental theorem of algebra for monosplines with multiplicities, Linear Operations and Approximation, P.L. Butzer et al. (eds.), Birkhäuser Verlag, Basel 1972, 419-430.

24. Micchelli, C.A. and Pinkus, A., On n-Widths in L^∞, Trans. of Amer. Math. Soc. 234 (1977), 139-174.

25. Powell, M.J., On best L_2 spline approximations, Differentialgleichungen, Approximationstheorie, ISNM 9, Birkhäuser Verlag, Basel 1968, 317-337.

26. Schoenberg, I.J., Monosplines and quadrature formula, Theory and Applications of Spline Functions, 157-207, Editor, Greville, T.N.E., Academic Press, New York 1969.

27. Strauss, H., Approximation mit Splinefunktionen und Quadraturformeln, Spline Functions, Karlsruhe 1975, Lecture Notes in Mathematics 501, Springer-Verlag, Berlin 1976.

28. Strauss, H., Untersuchungen über Quadraturformeln, Numerische Methoden der Approximationstheorie, Oberwolfach, ISNM 42, Edited by Collatz, L., Meinardus, G., Werner, H., Birkhäuser-Verlag, Basel 1978.

29. Strauss, H., Optimale Quadraturformeln und Perfektsplines, Pre-Print.

PROBLEMS FOR INCOMPLETE POLYNOMIALS

G. G. Lorentz

Department of Mathematics
The University of Texas
Austin, Texas

1. Problems proposed at the Tampa meeting

In December 1976, a meeting devoted to Padé and Rational Approximation had been organized at Tampa, Florida, by E.B. Saff and R.S. Varga. In my talk at this meeting [13] I outlined several problems concerning "incomplete polynomials", -- that is, polynomials in x with several coefficients equal to zero. Simplest types are

$$P_n(x) = x^k Q_m(x) , \qquad k+m=n , \tag{1.1}$$

where Q_m is a polynomial of degree m, and

$$P_n(x) = \sum_{i=0}^{p} a_i x^{k_i} , \qquad 0 \le k_0 < \ldots < k_p = n . \tag{1.2}$$

One can require k to be large in (1.1), or p be much smaller than n in (1.2). Basic problems of approximation theory breathe a new life if approximating polynomials are restricted to be of type (1.1) or (1.2). I had formulated several problems of this kind in my talk at Tampa. There exists now a considerable literature concerning the problems. Some of them I solved myself, more have been solved by others, in particular by Saff and Varga themselves. One conjecture remains open (see §5). In this paper I want to review what has been achieved so far.

Copyright © 1980 by Academic Press, Inc.
All rights of reproduction in any form reserved.
ISBN: 0-12-171050-5

A natural question to start with is whether the Weierstrass approximation theorem is valid for an arbitrary function $f \in C[0,1]$ if approximating polynomials are of type (1.1), with large k. Of course, one must assume that $f(0) = 0$. The question has a simple answer: approximation is possible if one assumes that $k = o(n)$; it is not possible if for infinitely many n, $k \geq \theta n$, where $0 < \theta < 1$ is a constant. The polynomials, which converge to $f(x)$ uniformly, are of course, bounded, and the reason why the theorem of Weierstrass is not applicable here is simply that $|P_n(x)| \leq 1$ on $[0,1]$ and $k \geq \theta n$ for a sequence of polynomials (1.1) together imply

$$\lim_{n \to \infty} P_n(x) = 0, \qquad 0 \leq x \leq \Delta \tag{1.3}$$

for some $\Delta > 0$. This is enforced convergence to zero. It is easy to prove (1.3), but the best value of Δ is less obvious. As a first interesting theorem (Lorentz [13]) we have:

Theorem 1.1. *If for a sequence of polynomials* $P_n(x)$ *of type* (1.1) *one has*

$$|P_n(x)| \leq 1, \qquad 0 \leq x \leq 1 \tag{1.4}$$

and $k \geq n\theta$ *for some* θ *with* $0 < \theta < 1$, *then*

$$\lim_{n \to \infty} P_n(x) = 0, \qquad 0 \leq x < \theta^2, \tag{1.5}$$

uniformly on compact subsets of $[0, \theta^2)$.

We prove this, with extensions, in §2. Later Saff and Varga [19] showed that $\Delta = \theta^2$ is the best possible in this theorem. Their first approach to this statement was by means of certain sequences of Jacobi polynomials. Very interesting is their second approach [20,22]: They define "constrained Chebyshev polynomials" of degree $n = k + m$, as monic polynomials of type (1.1) which deviate least from zero. These polynomials provide a proof of the same result, but their properties are of great independent interest (see §3). The negative theorem means simply that for each θ, $0 < \theta < 1$ there is a sequence of polynomials $P_n(x)$ satisfying the conditions of Theorem 1.1 which, for each $\delta > 0$, does not converge to zero for $x \in (\theta^2, \theta^2 + \delta]$. (It is not known whether one can achieve this even for $x = \theta^2$.)

If the sequence does not converge necessarily to zero, to what can it converge? My conjectures (see also [15]) were as follows:

Conjecture 1. *For each* θ, $0 < \theta < 1$ *and each* $f \in C(\theta^2, 1]$ *there is a sequence of polynomials* (1.1) *with* $k \geq n\theta$ *with the property*

$$\lim_{n \to \infty} P_n(x) = f(x) \tag{1.6}$$

uniformly on compact subsets of $(\theta^2, 1]$.

Conjecture 2. *For each* θ, $0 < \theta < 1$ *and each* $f \in C[\theta^2, 1]$ *there is a sequence of* $P_n(x)$ *for which* (1.6) *holds uniformly on* $[\theta^2, 1]$.

Conjecture 1 has been proved by v. Golitschek [5] and by Saff and Varga [21] (see §4). The truth of Conjecture 2 is doubtful, but the possible behavior of polynomials $P_n(x)$ in neighborhoods of θ has not yet been clarified.

The next question discussed in [13] concerned polynomials of type (1.1) which are polynomials of best approximation. Let $f \in C[-1, +1]$ be fixed (probably with $f(0) = 0$) and let P_n stand for its polynomial of best approximation of degree $\leq n$. Can all or many of them be of type (1.1)? One easily proves that for two polynomials P_{n-1}, P_n of best approximation to f, the difference $P_n - P_{n-1}$ has n distinct zeros in

(-1,+1), or vanishes identically. Hence, if $P_{n-1} \neq P_n$, these polynomials cannot be both of type (1.1) with $k \geq 2$.

If $k \geq 2$ is not possible for all n, maybe it is possible for infinitely many n? This has been answered positively in my paper [14]. In fact, one can achieve

$$k = k(n) \geq c \log n , \quad \text{with} \quad 0 < c < \frac{1}{\log 3} \tag{1.7}$$

for infinitely many n. It is not known whether this is best possible (see §5).

In this connection, it is natural to ask: is it possible that $k \geq 1$ for all n? This is certainly the case when f (and consequently all P_n) is odd. Is this the only possibility? This my conjecture [13] proved to be the hardest to crack. Saff and Varga [23] proved it for entire functions f of a certain class, Borosh and Chui [1] added many interesting remarks (see §5).

About the degree of approximation by incomplete polynomials not much is known. M. Hasson [6,7] has investigated the degree of approximation of a function $f \in C[0,1]$ or $f \in C[-1,1]$, when just one power x^k of x is not allowed to appear in $P_n(x)$, that is, when $p = n - 1$ in (1.2). Surprisingly, the degree of approximation is sensitive for this omission, if f is smooth enough. There are precise relations between k and the degree of smoothness of f required (see §6).

The degree of approximation of $f \in C[0,1]$ by polynomials which contain only powers x^{λ_i} for a prescribed sequence $0 \leq \lambda_0 < \lambda_1 < \dots$ was a subject of deep investigations of D.J. Newman and many other mathematicians. In [4] simple formulas are given which allow in this situation to estimate the degree of approximation of f in terms of its modulus of continuity. There must be interesting applications of these formulas. Some rather trivial applications can be found in [13]. Somewhat isolated is an interesting note of Newman and Rivlin [16], which discusses the possibility (or impossibility) of approximation of x^n on $[0,1]$ by polynomials $P_t(x)$ of degree $t = t(n)$ which is much smaller than n.

We return to polynomials (1.2) with fixed p and with $k_p < n$, and take $f(x) = x^n$ on $[0,1]$. Which selection of the exponents k_i gives the best approximation to x^n? In the L_2-norm this can be easily answered for there exist explicit formulas for the distance from x^n to the subspace spanned by the x^{k_i}. In this case the best result is achieved by k_i which are as close as possible to n, that is, for $k_i = n - p + i + 1$, $i = 1,\ldots,p$. I conjectured that this is true also for spaces C and L_p, $1 \leq p < +\infty$. For the first space this has been proved by Borosh, Chui and Ph. Smith [2]. Later simple proofs, which belong to the theory of Chebyshev (or of Descartes) systems, and do not depend on the metric of the space, have been found by Pinkus and Ph. Smith (see §7).

2. Enforced convergence to zero

I have proved Theorem 1.1 of §1 in [13] by means of a saturation theorem for the Poisson integral. The proof was rather artificial. A better use of the Poisson integral for this purpose is possible. It is even likely that in some situations this method is superior to other methods and also to one described below. We shall develop here an approach, which works in the complex region and has interesting applications (Kemperman and Lorentz [9]).

In the sequel, $Q_m(z)$ will denote a polynomial of degree m with complex coefficients. We shall assume that on $[-1,+1]$ the polynomial satisfies

$$|Q_m(x)| \leq C\Omega(x) \quad \text{for} \quad x \in [-1,+1] , \tag{2.1}$$

where

$$Q(z) = \prod_{k=1}^{N} |z-c_k|^{-s_k} \quad . \tag{2.2}$$

Here C is a positive constant, the s_k are real constants and the c_k are complex constants.

Basic for us is the transformation

$$z = \frac{1}{2}(w+w^{-1}) , \qquad 0 < |w| \le 1 , \tag{2.3}$$

which maps the disc $|w| \le 1$ $(w \ne 0)$ onto the complex plane z. Its inverse is $w = z \pm (z^2-1)^{\frac{1}{2}}$. In fact, w is unique and satisfies $|w| < 1$ if $z \notin [-1,+1]$. Otherwise, we have $z = \cos t$ with real t, and we can select $w = e^{it}$ or e^{-it}.

In a similar way, for $k = 1,\dots,N$ we define the complex constants w_k by

$$c_k = \frac{1}{2}(w_k + w_k^{-1}) , \qquad 0 < |w_k| \le 1 , \tag{2.4}$$

with an arbitrary but fixed of the two choices for w_k in case when $c_k \in [-1,+1]$.

We see from (2.3) and (2.4) that

$$z - c_k = -(2w_k)^{-1}(1-w_k w)(1-w_k w^{-1}) \quad . \tag{2.5}$$

If $|w| = 1$ then $w^{-1} = \bar{w}$ so that

$$|z-c_k| = |(2w_k)^{-1}(1-w_k w)(1-\bar{w}_k w)| , \quad \text{provided} \quad |w| = 1 \quad . \tag{2.6}$$

Theorem 2.1. *Suppose that* (2.1) *holds for a polynomial* $Q_m(z)$ *of degree* m. *Then*

$$|Q_m(z) < C|w|^{-m} \prod_{k=1}^{N} |(2w_k)^{-1}(1-w_k w)(1-\bar{w}_k w)|^{-s_k} \tag{2.7}$$

holds for all complex z. *An equivalent inequality is*

$$|Q_m(z)| \le C\Omega(z)|w|^{-m} \prod_{k=1}^{N} \left| \frac{w^{-1}(w-w_k)}{1-\bar{w}_k w} \right|^{s_k} . \tag{2.8}$$

Proof. That (2.7) and (2.8) are equivalent follows immediately from (2.2) and (2.5). In proving (2.7), one may assume that $C = 1$. Consider the pair of analytic functions

$$f(w) = w^m Q_m(\tfrac{1}{2}(w + w^{-1}))$$

and

$$g(w) = \prod_{k=1}^{N} \{(2|w_k|)^{-1}(1-w_k w)(1-\bar{w}_k w)\}^{s_k} .$$

Here, $f(w)$ is an entire function, in fact, a polynomial of degree $2m$. Since $|w_k| \le 1$, the function $g(w)$ is analytic for $|w| \le 1$, except that $|w_k| = 1$ leads to the singularities w_k and w_k^{-1} on $|w| = 1$. One can make g unique by requiring that $g(w)$ be real and positive for small real values w.

Observe that (2.7) (with $C = 1$) is equivalent to $|f(w)g(w)| \le 1$ for $|w| \le 1$. Hence, it suffices to show that $|f(w)\, g(w)| \le 1$ for $|w| = 1$. And the latter follows immediately from (2.1), (2.2) and (2.6). □

We give an application with $N = 1$. Let $0 < a < 1$, s and $C > 0$ denote real constants.

Theorem 2.2. *If a polynomial* Q_m *of degree* m *satisfies*

$$|Q_m(x)| \le Cx^{-s} \quad \text{for} \quad 0 < a^2 \le x \le 1 \tag{2.9}$$

then we have for all complex values z *that*

$$|Q_m(z)| \le C|w|^{-m} \left| \frac{1+a}{2} + \frac{1-a}{2} w \right|^{-2s} , \tag{2.10}$$

where w *is defined by*

$$z = \frac{1+a^2}{2} + \frac{1-a^2}{2} \cdot \frac{w+w^{-1}}{2} \; ; \quad |w| \leq 1 \; . \tag{2.11}$$

An equivalent upper bound is

$$|Q_m(z)| \leq C|z|^{-s}|w|^{-m} \cdot \left|\frac{(1+a)+(1-a)w^{-1}}{(1+a)+(1-a)w}\right|^s . \tag{2.12}$$

Proof. Let us introduce

$$x = \frac{1-a^2}{2} x^* + \frac{1+a^2}{2} \; ; \quad Q_m(x) = Q_m^*(x^*) \; .$$

It follows from (2.9) that

$$|Q_m^*(x^*)| \leq C^*|x^*-c^*|^{-s} \quad \text{when} \quad -1 \leq x^* \leq +1 \; ,$$

where

$$C^* = C\left(\frac{1-a^2}{2}\right)^{-s} \; ; \quad c^* = -(1+a^2)/(1-a^2) \; .$$

Applying (2.7) with $N=1$, we find that

$$|Q_m^*(\tfrac{1}{2}(w+w^{-1}))| \leq C^*|w|^{-m}|(2w_1)^{-1}(1-w_1w)^2|^{-s} \quad \text{for} \quad |w| \leq 1 \; . \tag{2.13}$$

Here, w_1 is defined by (2.4) with $c_1 = c^*$. We see from (2.4) that w_1 is a real number, in fact $w_1 = -(1-a)/(1+a)$. Consequently, (2.13) reduces to (2.10) with w as defined by (2.11). Next observe that (2.11) implies

$$z = \left[\frac{1+a}{2} + \frac{1-a}{2} w\right] \cdot \left[\frac{1+a}{2} + \frac{1-a}{2} w^{-1}\right] ,$$

showing that (2.10) and (2.12) are equivalent.

Remark. If $z = x$ is real with $a^2 \leq x \leq 1$ then $|w| = 1$ so that the bound (2.12) coincides with the assumed inequality $|Q_m(x)| \leq Cx^{-s}$.

We can now extend Theorem 1.1 to the entire complex plane. We can even replace condition (1.4) by

$$|P_n(x)| \le 1 \quad \text{for} \quad \theta^2 \le x \le 1 , \tag{2.14}$$

where θ, $0 < \theta < 1$ is a fixed constant. Suppose further that P_n has $x = 0$ as a zero of order $s(n)$ and that

$$s(n)/n \to \theta \quad \text{as} \quad n \to \infty . \tag{2.15}$$

(A polynomial P_n, satisfying (2.15), will be called a polynomial of type θ.)

Theorem 2.3. *Let $\Gamma = \Gamma(\theta)$ be the bounded and open set of the z-plane defined by*

$$z = \frac{1+\theta^2}{2} + \frac{1-\theta^2}{2}\,\frac{w+w^{-1}}{2} , \tag{2.16}$$

where w satisfies both $|w| \le 1$ and

$$|w|^{-1}\cdot\left|\frac{(1-\theta)+(1+\theta)w}{(1+0)+(1-\theta)w}\right|^{\theta} < 1 . \tag{2.17}$$

If the polynomials $P_n(x)$ of degree n satisfy (2.14) and (2.15) then

$$\lim_{n\to\infty} P_n(z) = 0 \quad \textit{for each} \quad z \in \Gamma(\theta) . \tag{2.18}$$

The latter convergence is uniform and exponentially fast for each compact subset of $\Gamma(\theta)$. Similarly for each sequence of derivatives $\{P_n^{(k)}(z)\}_{n=1}^{\infty}$, ($k$ fixed).

Proof. Let $m = m(n) = n - s(n)$, (thus, $m(n)/n \to 1-\theta$ as $n \to \infty$), and let $\theta(n) = s(n)/n$. One can write

$$P_n(x) = x^{s(n)} Q_{m(n)}(x) ,$$

where Q_m is a polynomial of degree m. By (2.14), Q_m satisfies (2.9) with $C=1$ and $a=\theta$. Applying (2.12), one finds that

$$|P_n(z)| \leq |w|^{-m(n)} \cdot \left|\frac{(1+\theta)+(1-\theta)w^{-1}}{(1+\theta)+(1-\theta)w}\right|^{s(n)} .$$

Here, w is defined as in (2.16). An equivalent form is

$$|P_n(z)|^{1/n} \leq |w|^{-1} \cdot \left|\frac{(1-\theta)+(1+\theta)w}{(1+\theta)+(1-\theta)w}\right|^{\theta(n)} . \tag{2.19}$$

The stated assertions are now an immediate consequence of (2.15) and (2.19). The derivative $P_n^{(k)}(z)$ is easily handled by using its Cauchy integral representation.

This theorem appears also in Saff and Varga [19].

Remark 1. The open set $\Gamma(\theta)$ in the complex z-plane intersects the real axis in an interval $(-\phi(\theta),\theta^2)$ about the origin. Here,

$$\phi(\theta) = \frac{1-\theta^2}{2} \cdot \frac{u+u^{-1}}{2} - \frac{1+\theta^2}{2} , \tag{2.20}$$

where the number $0 < u < (1-\theta)/(1+\theta)$ satisfies

$$\{(1-\theta)-(1+\theta)u\}/\{(1+\theta)-(1-\theta)u\} = u^{1/\theta} .$$

For instance, $\phi(2/3) = 1/2$; $\phi(1/2) = 1/8$; $\phi(1/3) = -\frac{1}{3} + \frac{2}{9}\sqrt{3} = .051567$.

Remark 2. Exponential convergence in Theorem 2.3 allows to weaken considerably the condition (2.14). For example, let $M_n \geq 1$ and $M_n^{1/n} \to 1$. Then already

$$|P_n(x)| \leq M_n \quad \text{for} \quad \theta^2 \leq x \leq 1 \tag{2.21}$$

together with (2.15) implies (2.18). This is revealed by the transformation $t = M_n^{-s} x$, which maps polynomials $P_n(x)$ into $P_n^*(t)$, satisfying $|P_n^*(t)| \leq 1$ on $[\theta^2, 1-\delta]$ for each $\delta > 0$.

Saff and Varga formulate the following theorem, which is a version of Theorem 1.1 or 2.3 and has the same proof (see [20]):

Theorem 1.1*. *If a polynomial* $P_n \not\equiv 0$ *of degree* $\leq n$ *has a zero of order* $s(n) \geq n\theta$ *at* $x = 0$, *and if for some* ξ, $0 \leq \xi \leq 1$, $|P_n(\xi)| = \|P_n\|$, *then* $\xi > \theta^2$.

One could call $[0, \theta^2]$ the "region of influence" of the zero $x = 0$.

Lachance, Saff and Varga ([11],[18, Theorem 2.1]) study polynomials P_n which have zeros of high order at both endpoints of the interval:

$$
\begin{aligned}
&P_n(x) = (1-x)^{s_1}(1+x)^{s_2} Q_m(x), \quad n = s_1 + s_2 + m \\
&s_1 \geq n\theta_1, \quad s_2 \geq n\theta_2 .
\end{aligned}
\tag{2.22}
$$

Theorem 2.4. *For polynomials* (2.22), *let* $|P_n(x)| \leq 1$ *on* $[\alpha, \beta]$. *Then*

$$\lim P_n(x) = 0 \quad \text{for} \quad x \in [-1, \alpha) \cup (\beta, 1],$$

where, with $\sigma = \theta_2 + \theta_1$, $\delta = \theta_2 - \theta_1$,

$$\alpha = \sigma\delta - \sqrt{(1-\sigma^2)(1-\delta^2)}, \quad \beta = \sigma\delta + \sqrt{(1-\sigma^2)(1-\delta^2)} . \tag{2.23}$$

This can be probably obtained from our main Theorem 2.1.

Another application of this theorem is a simple proof of a theorem of Teljakovskii [9] which asserts that under certain conditions, for which the approximation (for some $\alpha > 0$, $M > 0$) of $f \in C[-1,+1]$

$$
\begin{cases}
|f(x) - P_n(x)| \leq M\Delta_n(x)^\alpha, \quad -1 \leq x \leq 1, \quad n = 1, 2, \\[2mm]
\Delta_n(x) = \max\left(\dfrac{1}{n^2}, \dfrac{\sqrt{1-x^2}}{n}\right)
\end{cases}
$$

is known, one even has (with another $M^* > 0$)

$$|f(x)-P_n^*(x)| \leq M^* \left(\frac{\sqrt{1-x^2}}{n}\right)^\alpha .$$

The key lemma here is that the inequality

$$|P_n(x)| \leq M\Delta_n(x)^\alpha , \quad -1 \leq x \leq 1 \tag{2.24}$$

implies

$$|P_n(x)| \leq M^* \left(\frac{\sqrt{1-x^2}}{n}\right)^\alpha , \quad -1 \leq x \leq 1 . \tag{2.25}$$

Indeed, (2.24) implies (2.25) for $|x| \leq 1 - n^{-2}$, because in this case $n^{-2} \leq (1-x^2)^{-\frac{1}{2}} n^{-1}$. Thus one has to extend the inequality (2.25) from the interval $[-1+n^{-2}, 1-n^{-2}]$ to $[-1,+1]$ by means of Theorem 2.1. This is possible because the product in (2.8) proves to be bounded on the difference of the two intervals.

3. Constrained Chebyshev polynomials of Saff and Varga

In several papers, Saff and Varga (sometimes with Lachance, Ullman), contributed importantly to our subject of incomplete polynomials. Their work centers on the inverses to the main theorems of §2, but has developed into theory (or theories) of its own. It is hardly possible to do complete justice to these investigations and reproduce the proofs, which are often technical, depending upon properties of correctly chosen Jacobi polynomials and on complex methods. We give a selection of their results.

In Saff and Varga [19], they prove that θ^2 is best possible in Theorem 1.1, they find the number $\phi(\theta)$ of (2.20), and they give an independent proof of Theorem 2.3. Their method is to majorize incomplete polynomials P_n of type θ by certain expressions depending on the Jacobi polynomials $P_k^{-\frac{1}{2}, 2s-\frac{1}{2}}$, $k = 0,1,\ldots,n-s$, $s \geq \theta n$. These expressions are estimated in the complex plane by the method of steepest descent.

Important to the inverse problem and interesting in themselves are the "constrained Chebyshev polynomials" $T_{s,m}$ for the interval $[0,1]$, introduced in Saff and Varga [20] and studied also in [11],[18],[22]. They share with the ordinary Chebyshev polynomials many of their properties. (It is not known whether the $T_{s,m}$ are orthogonal with some weight.)

We use the following notations. For $n = s+m$, let $I_{s,m}$ denote the class of incomplete polynomials

$$P_{s,m}(x) = \sum_{k=s}^{n} a_k x^k , \qquad \theta(s,m) = \frac{s}{n} . \tag{3.1}$$

We define $Q_{m,s}$ as the solution of the extremal problem

$$E_{s,m} = \min\{\|P_{ms}\|_{[0,1]} : P_{ms} \in I_{m,s}, a_n = 1\} = \|Q_{ms}\| . \tag{3.2}$$

One shows that the solution is unique for each pair s,m, and that the value $|Q_{m,s}(x)| = E_{s,m}$ is achieved at exactly $m+1$ distinct points

$$\theta(s,m)^2 \leq \xi_0^{s,m} < \ldots < \xi_m^{s,m} = 1 ,$$

with alternation of signs of $Q(\xi_k^{s,m})$. The constrained Chebyshev polynomials are

$$T_{s,m} = Q_{s,m}/E_{s,m} . \tag{3.3}$$

Theorem 3.1. *For* $P \in I_{s,m}$, *inequality* $\|P\|_{[0,1]} \leq M$, *or every* $|P(\xi_k^{s,m})| \leq M$, $k = 0,\ldots,n$ *implies*

$$|P(x)| \leq M|T_{s,m}(x)| \quad , \quad x \notin (\xi_0^{s,m},1) , \tag{3.4}$$

$$|P^{(\nu)}(x)| \leq M|T_{s,m}^{(\nu)}(x)| , \quad \nu = 1,2,\ldots, \quad x \notin (0,1) . \tag{3.5}$$

Let $\mu_{s,m}$ be the unique negative x with $|T_{s,m}(x)| = 1$. Then, if

$$\theta_i = \theta(s_i, m_i) \to \theta , \quad 0 < \theta < 1 , \quad n_i \to \infty , \tag{3.6}$$

we have, with $\phi(\theta)$ from (2.20)

$$\lim \xi_0^{s,m} = \theta^2 , \qquad \lim \mu_{s,m} = -\phi(\theta) . \tag{3.7}$$

Theorem 3.2. *If* (3.6) *holds, then*

$$\lim_{i\to\infty} E_{s_i,m_i}^{1/n_i} = \frac{1}{4}(1+\theta)^{1+\theta}(1-\theta)^{1-\theta} . \tag{3.8}$$

If w is the function of z given by (2.16) we have

Theorem 3.3. *Under the assumption* (3.6),

$$\lim_{i\to\infty} \left| T_{s_i,n_i}(z) \right|^{1/n_i} = |w| \left| \frac{(1-\theta)w + (1+\theta)}{(1+\theta)w + (1-\theta)} \right|^{\theta} = G(w,\theta) . \tag{3.9}$$

These results are obtained by comparing T_{s_i,n_i} to certain Jacobi polynomials, and by studying $\log T_{s_i n_i}(z)$ in the complex plane, using normal families of subharmonic functions. They establish again Theorems 2.2 and 2.3 of §2, and show that the statements are the best possible. This is a much simpler approach than in [19], but for direct theorems our approach, explained in §2, seems to be more basic.

Constrained Chebyshev polynomials with zeros of high order at both endpoints of the interval have also been studied. They lead to Theorem 2.4 (see Lachance, Saff and Varga [11]).

Degree of approximation $E_{s,m}$ can be estimated also for values of s,m which do not satisfy $s/n \to \theta$. See inequalities in [22, Th. 3.1].

In [2], Borosh, Chui and Smith investigate the degree of approximation of x^n, in the uniform norm on $[0,1]$, by polynomials containing powers $x^{n-k},\dots,x^{n-1}$. In the notation of (3.2), this is $E_{n-k,k}$. They proved, for each fixed k:

$$E_{n-k,k} \approx n^k \quad \text{for} \quad n \to \infty . \tag{3.10}$$

We write $a_n \approx b_n$ for the weak equivalence, which means the existence of two constants $0 < C' \le C < +\infty$ for which $C' \le a_n/b_n \le C$. Saff and Varga [22] improve (3.10) to

Theorem 3.4. *One has*

$$\lim_{n\to\infty} E_{n-k,k}\, n^k = \varepsilon_k , \qquad k = 0,1,2,\dots \tag{3.11}$$

where ε_k *is defined as the infimum, for all polynomials* P_{k-1} *of degree* $k-1$,

$$\varepsilon_k = \inf\{\|e^{-t}(t^k - P_{k-1}(t))\|_{[0,+\infty]}\} . \tag{3.12}$$

The reason for this fact is that the extremal polynomials $Q_{n-k,k}$ in (3.2), which are the L_∞-analogues of Jacobi polynomials, are related by a limit relation to extremal polynomials, $t^k - P_{k-1}$ in (3.12), which are the L_∞-version of the Laguerre polynomials.

For more general polynomials of type (1.2) one has a similar relation (see [22])

Theorem 3.5. *Let the integers* $0 \le \mu_1 < \dots < \mu_k < m$ *be fixed, and for each non negative integer* n, *set*

$$E_n = \inf\{\|x^n(x^m - \sum_{j=1}^{k} c_j x^{\mu_j})\|_{[0,1]} , \quad c_j \in \mathbb{R} , \quad j = 1,\dots,k\} . \tag{3.13}$$

Then

$$\lim_{n\to\infty} n^k E_n = \frac{\varepsilon_k}{k!} \prod_{j=1}^{k} (m - \mu_j) . \tag{3.14}$$

In the paper [18] of Saff, Ullman and Varga, which is to appear in this volume, the authors discuss the following electrostatic problem:

Problem Let fixed charges of amounts θ_1, θ_2 be placed at $x=1$, $x=-1$, with $0 \leq \theta_1$, $0 \leq \theta_2$, $0 < \theta_1 + \theta_2 < 1$. How must the remaining charge of the amount $1 - \theta_1 - \theta_2$ be continuously distributed in $[-1,+1]$ in order to have equilibrium?

The anwere is that the continuous charge lies entirely in the interval $[\alpha,\beta]$, α,β being given by (2.23). Its density is

$$\delta(t) = \sqrt{(t-a)(b-t)} / \pi(1-t^2) , \quad a \leq t \leq b .$$

As a tool a lemma is proved, which describes the asymptotic density of zeros of polynomials, satisfying certain conditions. Polynomials $T_{s,m}$ meet these conditions, and we obtain

Theorem 3.6. *Let (3.6) be satisfied for a sequence of constrained Chebyshev polynomials* $T_{s_i,m_i}(x)$. *Then their zeros, except that at* $x=0$, *have no limit points outside of* $[\theta^2,1]$. *If* $N_i(c,d)$ *denotes the total number of zeros of* T_{s_i,m_i} *in an interval* $(c,d) \subset [\theta^2,1]$, *then*

$$\lim_{i\to\infty} \frac{N_i(c,d)}{n_i} = \frac{1}{\pi} \int_c^d \frac{1}{x} \sqrt{\frac{x-\theta^2}{1-x}}\, dx .$$

Finally, we shall mention the paper of Lachance, Saff and Varga [12] which treats similar extremal problems for polynomials with zeros on the unit circle in the complex plane.

4. Theorems of Weierstrass type

That Conjecture 1 of §1 is true, has been proved independently by Saff and Varga [21] and v. Golitschek [5]. It is interesting that proofs of these authors, sent to me just after their completion, arrived within ten days of each other!

Theorem 4.1. *For each function* $f \in C[0,1]$ *which vanishes on* $[0,\theta^2]$, $0 < \theta < 1$, *there is a sequence of incomplete polynomials of type* θ,

$$P_n(x) = \sum_{k=s}^{n} a_k x^k , \qquad s = s(n) \geq n\theta , \tag{4.1}$$

which converges to f *uniformly on* $[0,1]$.

Lemma 4.2. *For each* f *of the theorem and each* $\varepsilon > 0$, $\delta > 0$, *there exists, for all large* n *an incomplete polynomial of type* $\theta(1-\delta)$ *with*

$$\|f-P_n\| < \varepsilon . \tag{4.2}$$

If $F \in C[0,1]$ vanishes on $[0,\Delta]$, $0 < \Delta < 1$, then its Bernstein polynomials

$$B_m(F,y) = \sum_{m\Delta < j \leq m} F\left(\frac{j}{m}\right)\binom{m}{j} y^j (1-y)^{m-j} \tag{4.3}$$

are incomplete polynomials of type Δ which converge to F. We must exclude some of the lower powers of y from (4.3). This is achieved by the known lemma [4, p. 125]

Lemma 4.3. *Let* j *be given, let* $0 < j < s < n$. *There exist polynomials* $Q_{jsn}(x) = \sum_{k=s}^{n} c_k x^k$ *such that*

$$\|x^j - Q_{jsn}(x)\|_{[0,1]} \leq \prod_{k=s}^{n} \frac{k-j}{k+j} . \tag{4.4}$$

The last product does not exceed

$$\frac{(s-j)(s-j+1)\dots(n-j)}{(s+j)(s+j+1)\dots(n+j)} = \frac{(s-j)\dots(s-j+1)}{(n-j+1)\dots(n+j)}$$
$$\leq \frac{s}{n}\,\frac{s^2-(j-1)^2}{(n+1)^2-(j-1)^2}\cdots\frac{s^2}{(n+1)^2} \leq \left(\frac{s}{n}\right)^{2j} . \tag{4.5}$$

For the polynomial B_m we have

$$B_m(F,y) = \sum_{m\Delta<k\leq m} b_{km}\, y^k , \tag{4.6}$$

where

$$|b_{km}| \leq \|F\| \sum_{j=0}^{k} \binom{m}{j}\binom{m-j}{k-j} = \|F\| \sum_{j=0}^{k} \frac{m(m-1)\dots(m-k+1)}{j!\,(k-j)!}$$
$$= \|F\| \binom{m}{k} 2^k \leq \|F\| \frac{m^k 2^k}{k^k} \leq \|F\| \left(\frac{2em}{k}\right)^k , \tag{4.7}$$

by Stirling's inequality.

If we take $F = f$, $m = n$ and $\Delta = \theta^2$, we can eliminate the undesirable powers of y from $B_m(F)$, by means of Lemma 4.3, but we do not obtain sufficiently good approximation (because in the sum (4.11) below we would have $(2e)^k$ instead of $(2e)^{k/p}$).

Instead, we apply this argument to the function

$$F(y) = f(x) , \quad x = y^{1/p} ,$$

where p is a large integer with the property

$$(2e)^{1/p} < (1-2\delta)^{-2} . \tag{4.8}$$

For the function F, $\Delta = \theta^{2p}$; we take $m = [n/p]$. After the substitution $y = x^p$, the polynomial $B_m(F,y)$ becomes

$$B_n^*(x) = \sum_{m\Delta<k\leq m} b_{km}\, x^{kp} , \tag{4.9}$$

and we have

$$\|f-B_n^*\| \to 0 \quad \text{for} \quad n \to \infty . \tag{4.10}$$

In (4.9), we replace x^{kp} for $kp < n\theta(1-\delta)$, by the polynomial $Q_{kp,s,n}$ according to Lemma 4.3, taking for s the smallest integer $s \geq n\theta(1-\delta)$. We obtain then a polynomial P_n of degree $\leq n$, which will be a polynomial of type $\theta(1-\delta)$. For large n, $s/n < \theta(1-2\delta)$, and for $k > m\Delta$ we have $m/k < \theta^{-2p}$. The estimates (4.4), (4.5), and (4.7) yield

$$\begin{aligned} \|B_n^*-P_n\| &\leq \sum_{m\Delta<k\leq m} |b_{mk}| \left(\frac{s}{n}\right)^{2kp} \\ &\leq \|f\| \sum_{k>m\Delta} \left(\frac{2em}{k}\right)^k \left(\frac{s}{n}\right)^{2kp} \\ &\leq \|f\| \sum_{k>m\Delta} \left((2e)^{\frac{1}{p}}(1-2\delta)^2\right)^{2kp} \to 0 \quad \text{for} \quad n \to \infty . \end{aligned} \tag{4.11}$$

With (4.10) we obtain Lemma 4.2.

It is easy to derive from Lemma 4.2 the theorem by means of the substitution $x = t(1-\delta)^2$.

For functions with a jump at θ^2 we derive a corollary (see also [21, Theorem 2.21]):

Corollary 4.4. *For each function f on $[0,1]$ which is continuous on $[\theta^2,1]$ and vanishes on $[0,\theta^2)$ there exist: (i) a sequence of incomplete polynomials P_n of types $\theta_n \to \theta$, $\theta_n \leq \theta$, which converges uniformly to f on $[\theta^2,1]$, (ii) a sequence of incomplete polynomials P_n of type θ which converges to f uniformly on compact subsets of $[0,1] \setminus (\theta^2)$.*

The above proof follows the lines of v. Golitschek's [5]. He derives also similar results for "Λ-polynomials" of the form $\sum a_k x^{\lambda_k}$ for real and complex λ_k, and estimates the degree of approximation in Theorem 4.1. Saff and Varga [21] have a different approach, they start with the degree of approximation of 1 on $[0,1]$ in the L_2 metric by polynomials in $x^s,\dots,x^n$, $s \geq \theta n$, derived from the well known proof of Muntz' theorem.

5. Zeros of polynomials of best approximation

We would like to construct a function $f \in C[-1,+1]$, some of whose polynomials of best approximation have a zero of high order at 0. Let T_n stand for the Chebyshev polynomial of degree n. Relevant here is

Lemma 5.1. *Let $f(x)$ be the odd function on $[-1,+1]$, defined by*

$$f(x) = \sum_{j=1}^{\infty} b_j T_{n_j}(x) , \quad \sum |b_j| < +\infty , \tag{5.1}$$

and let the n_j be odd integers so that n_{j-1} divides n_j. Then the partial sums of the series (5.1) are polynomials of best approximation of f.

If all $b_j \geq 0$, this is a remark of S. Bernstein, known since 1912. Its proof is trivial. The proof of my lemma is different, but also simple.

By means of this lemma one proves (Lorentz [14])

Theorem 5.2. *There exists a function $f \in C[-1,+1]$ and a constant $c > 0$ with the property that for infinitely many n, the polynomial P_n of best approximation to f is of the form (1.1) with $k = k(n) \geq c \log n$.*

For the proof we take the function f in the form

$$f(x) = \sum_{k=1}^{\infty} b_k T_{3^k}(x) , \quad |b_k| \leq k^{-2} . \tag{5.2}$$

The coefficients of the odd powers $x^{2\ell+1}$ of x in $T_{3^k}(x)$ are known, and for fixed ℓ and large k they are asymptotically simple. Therefore the condition that $x, x^3, \ldots, x^{2\ell+1}$ have coefficients zero in $\sum_{k=1}^{p} b_k T_{3^k}$ is a system of linear non-homogeneous equations for the b_k. We not only have to show that these equations have solutions, but also have to ensure that the solutions b_k are small enough to satisfy $|b_k| \leq k^{-2}$. This is done by some estimates of minors of Vandermonde determinants

$$V = V(x_1, \ldots, x_n) = \det |x_k^{i-1}| . \tag{5.3}$$

One of the results needed is

Lemma 5.3. *Let V_t be a minor of V obtained by removing from V t rows $i_1 < \ldots < i_t$ and t columns $k_1 < \ldots < k_t$. Let $|x_k - x_\ell| \geq 1$ for $k \neq \ell$. Then one has*

$$|V_t/V| \leq \prod_{j \neq k_1, \ldots, k_t} (1 + |x_j|)^t / \prod_{\substack{j \neq k_1, \ldots, k_t \\ s=1, \ldots, t}} |x_j - x_{k_s}| . \tag{5.4}$$

K. Zeller remarked that Lemma 5.1 and its proof remain true for functions

$$f(x) = x + \sum_{j=2}^{\infty} b_j T_{n_j}(x)^{k_j} , \tag{5.5}$$

if n_j increases repidly and k_j are not too large (see [14]). He also noticed that this implies

Theorem 5.4. *There exists an odd function $f(x)$ on $[-1,+1]$, which has polynomials of best approximation with zero of arbitrary high order at $x = 0$.*

Proof. In (5.2), we take $k_2 = 1$, $k_3 = 3$, $k_5 = 5, \dots$. For odd n, $T_n(x)$ has as its lowest term $\pm nx$, hence $T_n(x)^k$ has $\pm n^k x^k$. We then define the b_j in (5.5) inductively: b_2 is so selected that $x + b_2 T_{n_2}(x)$ has no term with x; then b_3 so that $x + b_2 T_{n_2} + b_3 T_{n_3}^3$ has no term with x^3; and so on. Because of the above remark, one can enforce (by taking large n_j) the necessary condition $\sum |b_j| < +\infty$. □

Since this proof is much simpler than that of Theorem 5.2, D.J. Newman has asked me, what is the best result that can be achieved by this approach. At that time I thought that one can get as far as $k \geq c \log n / \log\log n$ (this is formulated as Theorem 2 in [13]). But I am not sure of this now.

It is strange that nobody before [13] has made the

Conjecture 3. *If all polynomials* P_n *of best approximation to* $f \in C[-1,+1]$ *satisfy*

$$P_n(0) = 0 \; , \quad n = 0, 1, \dots, \tag{5.6}$$

then f *is odd.*

This is still open. However, Saff and Varga [23] proved that the conjecture is correct if f is an entire function of exponential type τ with $0 \leq \tau < \pi/2$. One can even weaken the assumption (5.6) to $P_{2n+1}(0) = 0$, $n = 0, 1, 2, \dots$. Originally, I had also the following conjecture:

Conjecture 4. *If for a function* $f \in C[-1,+1]$ *and some* α, $-1 < \alpha < 1$ *one has*

$$P_n(\alpha) = 0 \; , \quad \alpha \neq 0 \; , \quad n = 0, 1, \dots \tag{5.7}$$

then f *is identically zero.*

Saff and Varga [23] pointed out that the conjecture is not true. Indeed, if $T_m(x)$ is any Chebyshev polynomial with a zero $\alpha \neq 0$ $(-1 < \alpha < 1)$, then (5.7) holds for all n: for $n < m$ because $P_n \equiv 0$, and for $n \geq m$ because $P_n = T_m$. One can also find functions which are not Chebyshev polynomials, which refute the conjecture [23]. For those who want to save the conjecture, one can suggest:

Conjecture 4*. *Relations (5.7) imply that α is a zero of a Chebyshev polynomial.*

In a paper [1] of this volume, Borosh and Chui make interesting remarks about Conjectures 3 and 4 in the L_2 metric.

Similar to Conjecture 3 are the following two conjectures by Borosh (see [13],[23]):

Conjecture 5. *If for some* $f \in C[-1,+1]$

$$P_{2k}(x) = P_{2k+1}(x), \quad k = 0,1,\ldots, \tag{5.8}$$

then f *is even.*

Conjecture 6. *If for some* $f \in C[-1,+1]$ *with* $f(0) = 0$,

$$P_{2k+1}(x) = P_{2k+2}(x), \quad k = 0,1,\ldots \tag{5.9}$$

then f *is odd.*

6. Incomplete polynomials of M. Hasson

In [7] Hasson considers polynomials P_n which are restricted not to contain just one fixed power of x, say x^k, $k \geq 1$. He would like to compare the degree of approximation $E_n^k(f)$ of a function $f \in C[a,b]$ by polynomials of this kind with its unrestricted degree of approximation $E_n(f)$. The answer is different if 0 is an interior point of $[a,b]$ or one of the endpoints. The following theorems, formulated here in a way different from [7], answer this question completely.

Theorem 6.1. *For each integer* $k \geq 1$ *there is a constant* $C > 0$ *with the property that for each function* $f \in C^k[-1,+1]$ *with* $a_k = f^{(k)}(0)/k! \neq 0$ *one has*

$$E_n^k(f) \geq Cn^{-k}(|a_k| + o(1)), \quad n \to \infty . \tag{6.1}$$

Under the same assumptions,

$$E_n(f) \leq \text{Const } n^{-k} E_n(f^{(k)}) = o(n^{-k}), \tag{6.2}$$

hence

$$E_n^k(f)/E_n(f) \to +\infty, \quad n \to \infty . \tag{6.3}$$

Next theorem shows that the assumption $f \in C^k$ is best possible:

Theorem 6.2. *For every integer* $N \geq 0$ *there exists a function* $g \in C^N[-1,+1]$ *for which*

$$\overline{\lim_{n\to\infty}} \{E_n^k(g)/E_n(g)\} < +\infty \tag{6.4}$$

for all $k > N$.

For the interval $[0,1]$ there are similar theorems:

Theorem 6.1*. *For each* $k = 1,2,\ldots$ *there is a constant* C^* *with the property that* $f \in C^{2k}[0,1]$ *and* $a_k = f^{(k)}(0)/k! \neq 0$ *imply*

$$E_n^k(f) \geq C^* n^{-2k}(|a_k| + o(1)) . \tag{6.5}$$

Theorem 6.2*. *For every integer* $N \geq 0$ *there exists a function* $g \in C^N[0,1]$ *for which*

$$\overline{\lim_{n\to\infty}} \{E_n^{(k)}(g)\} < +\infty \quad \text{for all} \quad k > [N/2] .$$

As an example, we prove Theorem 6.1. This depends on the following fact

Theorem 6.3. *For the approximation on* $[-1,+1]$ *one has, for* $k = 1, 2, \ldots$

$$E_n^k(x^k) \approx n^{-k} . \tag{6.6}$$

Proof. We first show that

$$E_n^k(x^k) \le Cn^{-k} , \quad C = C(k) > 0 . \tag{6.7}$$

It is sufficient to establish this when $n \equiv k \pmod 2$. In this case the coefficient c_n^k of x^k in the Chebyshev polynomial $T_n(x)$ is not zero, and satisfies

$$|c_n^k| \ge Kn^k , \quad n = 1, 2, \ldots$$

with some $K = K(k) > 0$ (see for example [17, p. 32]). Then (6.7) follows from

$$E_n^k(x^k) \le \|x^k - (x^k - T_n(x)/c_n^k)\| \le |c_n^k|^{-1}\|T_n\| = |c_n^k|^{-1} .$$

The estimation from the other side follows from

Lemma 6.4. *For each interval* $[-a, a]$ *there is a constant* M *with the following property. If for some polynomial* P_n, *and an integer* k

$$\|x^k - P_n(x)\|_{[-a,a]} \le Cn^{-k} , \tag{6.8}$$

then also

$$\|x^k - P_n(x)\| \ge CM\, n^{-k} . \tag{6.9}$$

Proof. (a) The statement of the lemma is true for $k = 1$. Indeed, let $\|P_n(x)-x\| \le Cn^{-1}$, $P_n'(0) = 0$, then by applying twice the Bernstein-Markov inequality, we get, with some b, $0 < b < a$,

$$\|P_n''(x)\|_{[-b,b]} \le C_1 n .$$

Since $P_n'(0) = 0$, by Taylor's formula we obtain from this

$$|P_n(x)-P_n(0)| \le \frac{1}{2}|x| , \qquad |x| \le \alpha n^{-1} , \tag{6.10}$$

where $\alpha = (2C_1)^{-1}$. We claim that

$$\|P_n(x)-x\|_{[-\alpha n^{-1}, \alpha n^{-1}]} \ge \frac{1}{4}\alpha n^{-1} .$$

If $|P_n(0)| \ge \frac{1}{4}\alpha n^{-1}$, we are through. In the opposite case, from (6.10),

$$|P_n(\alpha n^{-1})| \le \frac{1}{4}\alpha n^{-1} + \frac{1}{2}\alpha n^{-1} = \frac{3}{4}\alpha n^{-1} , \ |P_n(\alpha n^{-1})-\alpha n^{-1}| \ge \frac{1}{4}\alpha n^{-1} .$$

(b) To prove the general case, let $P_n^{(k)}(0) = 0$ and $\|P_n-x^k\|_{[-a,a]} \le Cn^{-k}$. By $k-1$ applications of the Bernstein-Markov inequality,

$$\|P_n^{(k-1)}-k!x\|_{[-b,b]} \le D_k n^{k-1}\|P_n-x^k\|_{[-a,a]} \le D_k Cn^{-1} . \tag{6.11}$$

Hence by case (a),

$$\|P_n^{(k-1)}-k!x\|_{[-b,b]} \ge D_k MC n^{-1} ,$$

and from (6.11),

$$\|P_n-x^k\|_{[-a,a]} \ge MC n^{-k} . \ \square$$

We can now prove Theorem 6.1. Let a_n^k be the coefficient of x^k in the polynomial P_n of best approximation of f on $[-1,+1]$. Then we have

$$E_n^k(a_n^k x^k) = E_n^k(f - f + a_n^k x^k) \le E_n^k(f) + E_n^k(f - a_n^k x^k)$$
$$= E_n^k(f) + E_n(f) \le 2E_n^k(f) \ .$$

By Theorem 6.3,

$$E_n^k(f) \ge Cn^{-k}|a_n^k| \ge Cn^{-k}(|a_k| - |a_k - a_n^k|) \ .$$

It remains to observe that

$$k!\,|a_k - a_n^k| = |f^{(k)}(0) - P_n^{(k)}(0)| \le E_{n-k}(f^{(k)}) \le E_{n-k}(f^{(k)}) \to 0 \quad \text{for} \quad n \to \infty \ . \tag{6.12}$$

Theorem 6.1* has a similar proof. Here one has, for the approximation on $[0,1]$,

$$E_n^k(x^k) \approx n^{-2k} \ , \qquad n \to \infty \ . \tag{6.13}$$

This is easier to prove than (6.6), because it depends only on some properties of the Chebyshev polynomials. However, if we want to follow the above proof of Theorem 6.1, at the place (6.12) we need deeper properties of polynomials of best approximation. Hasson develops these in [6]. The result used can be formulated as follows [6, Theorem 2.4].

Theorem 6.5. *Let* $N \ge 0$ *be an integer and* $[a,b]$ *an interval. There exists a constant* $C_0 > 0$ *with the following property. If* $f \in C^N[a,b]$, *and if* P_n *is the polynomial of best approximation of* f *on* $[a,b]$, *then for all* $0 \le k \le N/2$ *and* $n \ge N$,

$$\|P^{(k)} - f^{(k)}\| \le C_0 E_{n-2k}(f^{(2k)}) \ . \tag{6.14}$$

7. Selection of best powers

Let k, N with $k < N$ be given positive integers. We fix k integers Λ: $0 \leq \lambda_1 < \ldots < \lambda_k < N$ and consider the degree of approximation $E_\Lambda(x^N)$ of x^N on $[a,b]$, $0 \leq a < b$ by polynomials $P(x) = \sum_{i=1}^{k} a_i x^{\lambda_i}$. For which selection of the set Λ is this degree of approximation minimal? This question can be asked in the C or the L_p norm. My conjecture in [13] has been that this happens when the λ_i are as close to N as possible, that is, when $\lambda_i = N - k + i - 1$, $i = 1, \ldots, k$. For the space C this has been proved by Borosh, Chui and Smith [2]. They have used properties of positive kernels of S. Karlin and an argument similar to the Revez algorithm.

Later, Pinkus proved this theorem for arbitrary Descartes systems (see papers of Shisha [24] and of Ph. Smith [25]). The proof is based on an interesting comparison lemma (Theorem 7.1 below), due apparently independently to Pinkus and Smith. It allows to establish the minimality with respect to the norm in any reasonable Banach function space.

The principle involved in the conjecture ("one should approximate likes by likes") has been confirmed in different other situations. For example, Shisha [24] proves results similar to Theorem 7.2 for trigonometric functions.

A set of continuous functions $\{u_i\}_{i=1}^N \subset C[a,b]$ is called a Descartes system on $[a,b]$ (see [8, p. 25]) if for each p, $1 \leq p \leq N$ there is a sign $\varepsilon_p = \pm 1$ with the property that for all integers $1 \leq \lambda_1 < \ldots < \lambda_p \leq N$ and all selections of knots $a < x_1 < \ldots < x_p < b$, the determinants

$$D = \begin{vmatrix} u_{\lambda_1}(x_1), & u_{\lambda_2}(x_1), & \ldots, & u_{\lambda_p}(x_1) \\ \ldots & \ldots & \ldots & \ldots \\ u_{\lambda_1}(x_p), & u_{\lambda_2}(x_p), & \ldots, & u_{\lambda_p}(x_p) \end{vmatrix} \tag{7.1}$$

are all different from zero and have the same sign ε_p. Thus, the notions of a Chebyshev system and of a Descartes system intersect; a Descartes system is a Chebyshev system if all $\varepsilon_p = 1$, $p = 1,\ldots,N$. The system of powers $1, x, \ldots, x^N$ is a Descartes system on $(0, +\infty)$.

Let $\{u_i\}_i^N$ be a Descartes system on (a,b), let $k < N-1$. We would like to compare the two polynomials

$$\begin{cases} P(x) = u_N + \sum_{i=1}^{k} a_i u_{\lambda_i} , \\ \\ Q(x) = u_N + \sum_{i=1}^{k} b_i u_{\mu_i} , \end{cases} \tag{7.2}$$

assuming that the integers λ_i, μ_i satisfy $1 \le \mu_1 < \ldots < \mu_k < N$, $1 \le \lambda_1 < \ldots < \lambda_k < N$.

<u>Theorem 7.1</u>. (Comparison lemma). *If one has* $\mu_i \le \lambda_i$, $i = 1,\ldots,k$, *if the polynomials* P, Q *are not identical and if they interpolate zero at the same points* $a < x_1 < \ldots < x_k < b$:

$$P(x_i) = Q(x_i) = 0 , \quad i = 1,\ldots,k , \tag{7.3}$$

then $|P(x)| < |Q(x)|$ *for* $a < x < b$, $x \ne x_i$, $i = 1,\ldots,k$. *In addition,* $Q - P$ *changes sign at the points* x_i.

For an elegant proof, see [25].

We use this lemma to approximate u_N best by sums $S_\Lambda(x) = \sum_{i=1}^{k} a_i u_{\lambda_i}(x)$, where $k < N-1$, Λ: $1 \le \lambda_1 < \ldots < \lambda_k$, in the norm $\|\cdot\|$ of the space $L_p[a,b]$, $1 \le p < +\infty$, or $C[a,b]$.

For fixed Λ, sum S_Λ which approximates u_N best is defined by

$$E_\Lambda(u_N) = \|u_N - S_\Lambda\| = \min \|u_N - \sum_1^k a_i u_{\lambda_i}\| . \tag{7.4}$$

Theorem 7.2. (Pinkus, see [24]). *For* $\Lambda^*: N-k,\dots,N-1$ *and* $\Lambda \neq \Lambda^*$ *we have*

$$E_{\Lambda^*}(u_N) < E_\Lambda(u_N) \ . \tag{7.4}$$

Proof. Let S_Λ be the best approximation to u_N in the L_p - or C-norm on $[a,b]$. It is known that S_Λ interpolates u_N at k distinct points of (a,b), $a < x_1 < \dots < x_k < b$. Then $Q = u_N - S_\Lambda$ satisfies $Q(x_i) = 0$, $i = 1,\dots,k$. If $\Lambda \neq \Lambda^*$, we define the sum S_{Λ^*} by solving the equations $P(x_i) = 0$, $i = 1,\dots,k$, $P = u_N - S_{\Lambda^*}$. By Theorem 7.1,

$$|u_N(x) - S_{\Lambda^*}(x)| < |u_N(x) - S_\Lambda(s)| \ , \quad x \neq x_i \ , \quad i = 1,\dots,k \ .$$

This implies that

$$\|u_N - S_{\Lambda^*}\| < \|u_N - S_\Lambda\| = E_\Lambda(u_N) \ ,$$

hence we have (7.5).

N. Dyn and E. Kimchi note that this argument is good for all Banach function spaces with monotone norm. One uses their theorems from [10].

Examining the proof of Theorem 7.1 one sees that, for a given k in (7.3), it involves only determinants (7.1) of order $p = k$ and $p = k+1$, with $\lambda_p = N$. This is equivalent to assume that for some signs ε_p, $\varepsilon_{p+1} = \pm 1$, both $u_{\lambda_1},\dots,u_{\lambda_{p-1}}, \varepsilon_p u_N$ and $u_{\lambda_1},\dots,u_{\lambda_p}, \varepsilon_{p+1} u_N$ are Chebyshev systems. To have applications of Theorem 7.2 to the functions $1, x, \dots, x^n, f(x)$ we need a good sufficient condition for the functions

$$x^{\lambda_1}, \dots, x^{\lambda_p}, f, \quad 0 \leq \lambda_1 < \dots < \lambda_p \tag{7.6}$$

to form a Chebyshev system on $(0, a)$. We offer

Example 7.3. *Let* $\Delta_1 = \lambda_1$, $\Delta_k = \lambda_k - \lambda_{k-1} - 1$, $k = 2,\dots,p$, *let* D_k *be the differential operators*

$$D_k g = \frac{d}{dx}[x^{-\Delta_k} g(x)] , \quad k = 0,\dots,p . \tag{7.7}$$

Then (7.6) *is a Chebyshev system on* $(0,a)$ *if*

$$D_p \cdots D_1 f(x) > 0 \quad \text{on} \quad (0,a) . \tag{7.8}$$

Proof. In the well-known M.G. Krein's representation of a Chebyshev system (see [8, p. 381, (2.1)]) by means of positive functions ω_k,

$$u_k(x) = \omega_1(x) \int_0^x \omega_2(t_2) \cdots \int_0^{t_{k-1}} \omega_k(t_k)dt_k \cdots dt_2 , \quad k = 1,\dots,p+1,$$

we take $\omega_k(x) = x^{\Delta_k}$, $k = 1,\dots,p$, $\omega_{p+1}(x) = g(x)$. Then (7.8) is simply the condition that $g(x) > 0$ on $(0,a)$. □

Shisha [24, Theorem 3] offers sufficient conditions of a similar type:

Example 7.4. *System* (7.6) *is a Chebyshev system if there exists an integer* $N > \lambda_p$ *for which*

$$(x^{k-N} f(x))^{(k)} \geq 0 , \quad 0 < x < a , \quad k = 0,\dots,p , \tag{7.9}$$

with strict inequality for $k = p$.

One has here infinitely many inequalities, if one does not know how to select N. We shall show that (7.9) implies (7.8), at least for $p = 1,2$. For $p = 1$, (7.8) is $-\lambda_1 f + xf' > 0$ on $(0,a)$, and (7.9) for $N = \lambda_1 + 1$ is the same inequality plus $f(x) \geq 0$. For $p = 2$, condition (7.8) is

$$\lambda_1\lambda_2 f - (\lambda_1 + \lambda_2 - 1)xf' + x^2 f'' > 0 , \tag{7.10}$$

while conditions (7.9) with $N = \lambda_2 + 1$ are

$$\begin{cases} f \geq 0 , \quad -\lambda_2 f + xf' \geq 0 \\ \lambda_2(\lambda_2 - 1)f - 2(\lambda_2 - 1)xf' + x^2 f'' > 0 . \end{cases} \tag{7.11}$$

Multiplying the second inequality with $(\lambda_2 - \lambda_1 - 1)$ and adding to the third one gets (7.10).

REFERENCES

[1] I. Borosh and C.K. Chui, A problem of Lorentz on approximation by incomplete polynomials, Approximation Theory III, (E.W. Cheney, ed.), (1980), 249-254.

[2] I. Borosh, C.K. Chui and P.W. Smith, Best uniform approximation from a collection of subspaces, Math. Z., 156 (1977), 13-18.

[3] I. Borosh, C.K. Chui and P.W. Smith, On approximation of x^N by incomplete polynomials, J. Approx. Theory, 24 (1978), 227-235.

[4] R.P. Feinerman and D.J. Newman, Polynomial Approximation, Williams and Wilkins Co., Baltimore 1974.

[5] M.v. Golitschek, Approximation by incomplete polynomials, to appear in J. Approx. Theory.

[6] M. Hasson, Derivatives of the algebraic polynomials of best approximation, in print in J. Approx. Theory.

[7] M. Hasson, Comparison between the degrees of approximation by lacunary and ordinary algebraic polynomials, in print in J. Approx. Theory.

[8] S.J. Karlin and W.J. Studden, Tchebycheff systems with applications in analysis and statistics, Interscience Publishers, New York 1966.

[9] J.H.B. Kemperman and G.G. Lorentz, Bounds for polynomials with applications, Indagationes Math., 88 (1979), 13-26.

[10] E.Kimchi and N. Richter-Dyn, Restricted range approximation of k-convex functions in monotone norms, SIAM J. Numer. Anal., 15 (1978), 1030-1038.

[11] M. Lachance, E.B. Saff and R.S. Varga, Bounds for incomplete polynomials vanishing at both endpoints of an interval, pp. 421-437, in: Constructive Approaches to Mathematical Models, (C.V. Coffman and G.J. Fix, editors), Academic Press, New York, 1979.

[12] M. Lachance, E.B. Saff and R.S. Varga, Inequalities for polynomials with a prescribed zero, Math. Z., 168 (1979), 105-116.

[13] G.G. Lorentz, Approximation by incomplete polynomials (Problems and results), pp. 289-302 in: Padé and Rational Approximation (Saff and Varga, editors), Academic Press, New York, 1977.

[14] G.G. Lorentz, Incomplete polynomials of best approximation, Israel J. Math., 29 (1978), 132-140.

[15] G.G. Lorentz, Problem 10: Approximation by algebraic polynomials, pp. 664-665, in: Linear Spaces and Approximation, P.L. Butzer and B. Sz.-Nagy, editors, ISNM 40, Birkhäuser Verlag, Basel 1978.

[16] D.J. Newman and T.J. Rivlin, Approximation of monomials by lower degree polynomials, Aequationes Math., 14 (1976), 451-455.

[17] T.J. Rivlin, The Chebyshev polynomials, John Wiley and Sons, New York, 1974.

[18] E.B. Saff, J.L. Ullman and R.S. Varga, Incomplete polynomials: an electrostatic approach. Approximation Theory III, (1980), (E.W. Cheney, editor), 769-782.

[19] E.B. Saff and R.S. Varga, The sharpness of Lorentz' theorem on incomplete polynomials, Trans. Amer. Math. Soc., 249 (1979), 163-186.

[20] E.B. Saff and R.S. Varga, On incomplete polynomials, pp. 281-298, in: Numerische Methoden der Approximationstheorie, (L. Collatz, G. Meinardus, H. Werner, editors), ISNM 42, Birkhäuser Verlag, Basel, 1978.

[21] E.B. Saff and R.S. Varga, Uniform approximation by incomplete polynomials, Internat. J. Math. and Math. Sci., 1 (1978), 407-420.

[22] E.B. Saff and R.S. Varga, On incomplete polynomials II, to appear in Pacific J. Math.

[23] E.B. Saff and R.S. Varga, Remarks on some conjectures of G.G. Lorentz, to appear in J. of Approx. Theory.

[24] O. Shisha, Tchebycheff systems and best partial bases, to appear in Pacific J. Math.

[25] P.W. Smith, An improvement theorem for Descartes systems, Proc. Amer. Math. Soc., 70 (1978), 26-30.

BEST UNIFORM APPROXIMATION BY POLYNOMIALS IN THE COMPLEX PLANE

T. J. Rivlin
Mathematical Sciences Dept.
T. J. Watson Research Center
Yorktown Heights, New York

What follows is a selective survey of the problem of best uniform approximation by polynomials to continuous functions in the complex domain. The emphasis will be on characterization and uniqueness questions, with examples. Many details omitted here may be found in the book of Smirnov and Lebedev (10).

I. CHARACTERIZATIONS OF BEST APPROXIMATIONS

Suppose we are given a compact set, B, in the complex plane containing at least n+2 distinct points and a function $f \epsilon C(B)$. Let P_n denote the space of all polynomials of degree at most n having complex numbers as coefficients. We say that $p^* \epsilon P_n$ is a best uniform approximation to f on B out of P_n if

$$\| f - p^* \| := \max_{z \epsilon B} | f(z) - p^*(z) | \leq \| f - p \|$$

for all $p \epsilon P_n$, and we put $E_n(f) = \| f - p^* \|$. The existence of a best approximation can be shown as follows: since the zero function is in P_n it suffices to consider $p \epsilon P_n$ satisfying $\| f - p \| \leq \| f \|$, and hence $\| p \| \leq 2 \| f \|$. Thus the continuous function of $(a_0, \ldots, a_n)$,

Copyright © 1980 by Academic Press, Inc.
All rights of reproduction in any form reserved.
ISBN: 0-12-171050-5

$\| f(z)-(a_0 + a_1 z + \dots + a_n z^n) \|$ attains its minimum on the compact set $\| a_0 + a_1 z + \dots + a_n z^n \| \leq 2 \| f \|$.

In order to characterize best approximations we rely on the notion of extremal signature as developed for the more general setting of approximation out of finite dimensional linear subspaces by Rivlin and Shapiro (6). (See also Rivlin (8)). We present a sketch of this approach to the problem at hand.

Def. 1. A <u>signature</u> is a continuous function, S, whose domain is a closed subset of B and whose range is a subset of the unit circle in the complex plane. The domain of S is denoted by d(S). S' is a <u>subsignature</u> of S if S' is the restriction of S to a closed subset of d(S).

Def. 2. A signature S is said to be <u>associated</u> with $g \in C(B)$ if $d(S) \subseteq E(g;B) := \{z \in B: |g(z)| = \|g\|\}$ and $S(z) = \operatorname{sgn} g(z)$. (Recall that $\operatorname{sgn} z = z|z|^{-1}$ for $z \neq 0$ and $\operatorname{sgn} 0 = 0$.) S[g] means that the signature S is associated with g.

Def. 3. A signature S is said to be extremal for P_n if there exist complex numbers $s_1,\dots,s_k$ and distinct points $z_1,\dots,z_k$ of d(S) such that

$$\operatorname{sgn} s_j = \overline{S(z_j)}, \; j = 1,\dots,k \tag{1}$$

and

$$\sum_{j=1}^{k} s_j p(z_j) = 0, \text{ all } p \in P_n. \tag{2}$$

$s_1,\dots,s_k$ are called the <u>weights</u> for S. With no loss of generality we frequently normalize the weights so that $\sum_{j=1}^{k} |s_j| = 1$.

Def. 4. An extremal signature, S, is called <u>primitive</u> if it has no extremal subsignature, S', with $d(S') \subset d(S)$.

Note that in Def. 3 we must have $k \geq n+2$, since if $k \leq n+1$ there exists $p \in P_n$ such that $p(z_j) = \bar{s}_j$, $j = 1,\ldots,k$ contradicting (2).

We can now state our basic characterization theorem.

Theorem A. p^* is a best approximation out of P_n to $f \in C(B)$ $(f \notin P_n)$ if, and only if, there exists an extremal signature for P_n, $S[f - p^*]$, with $d(S)$ consisting of k points where k satisfies $n+2 \leq k \leq 2n+3$.

For the proof see Rivlin (8).

An easy consequence of Theorem A is

Theorem B. (Kolmogorov Criterion) p^* is a best approximation to $f \in C(B)$ if, and only if,

$$\min_{z \in E(f-p^*;B)} \mathrm{Re}[(\mathrm{sgn}\, \overline{f(z)-p^*(z)})\ p(z)] \leq 0 \tag{3}$$

for all $p \in P_n$.

Remark. Consider the condition

$$\min_{z \in d(S[f-p^*])} \mathrm{Re}[(\mathrm{sgn}\overline{f(z) - p^*(z)})\ p(z)] \leq 0, \tag{4}$$

for all $p \in P_n$. (4) $\Rightarrow$ (3) (since $d(S[f - p^*]) \subseteq E(f - p^*;B)$). But if (3) holds then p^* is a best approximation to f. Hence according to Theorem A

$$\mathrm{Re} \sum_{j=1}^{k} |s_j| (\mathrm{sgn}\overline{(f(z_j) - p^*(z_j)})p(z) = 0,$$

where $\{z_1,\ldots,z_k\} = d(S[f - p^*])$, for all $p \in P_n$, and (4) holds. Thus (4) and (3) are equivalent.

We also have

Theorem C. (Skeleton Theorem). If p^* is a best approximation out of P_n to f on B it is also a best approximation on d(S) consisting of k points of B, where S is an extremal signature for P_n and $k \leq 2n + 3$.

Proof. Apply Theorem A, with the same S and B replaced by d(S).

How can we find extremal signatures for P_n? In particular, given distinct points $z_1,...,z_k$ how can we determine $s_1,...,s_k$ so that (2) holds? A useful tool in this endeavor is due to Videnski. (See Smirnov and Lebedev (10)).

Theorem D. $s_1,...,s_k$, $n + 2 \leq k \leq 2n + 3$ satisfy (2) if

$$s_j = \frac{u(z_j)}{v'(z_j)}, \quad j = 1,...,k \tag{5}$$

where $v(z) = (z-z_1)...(z-z_k)$ and u is any (suitably normalized) polynomial in P_{k-n-2} such that $u(z_j) \neq 0$, $j = 1,...,k$. Moreover, if $s_1,...,s_k$, $n + 2 \leq k \leq 2n + 3$ satisfy (2) then (5) holds for some $u \epsilon P_{k-n-2}$.

For example, suppose that $k = n+2$, $v(z) = (z-z_1)...(z - z_{n+2})$ and $u(z) \equiv c \equiv |c| e^{it}$ where $|c| = (\Sigma \frac{1}{|v'(z_j)|})^{-1}$. Then

$$s_j = \frac{c}{v'(z_j)}, \quad j = 1,...,n + 2 \tag{6}$$

satisfies (2) and an extremal signature with $\{z_1,...,z_{n+2}\}$ as domain is given by

$$S(z_j) = e^{-it} \operatorname{sgn} \frac{1}{v'(z_j)} = :e^{-it} e^{i\phi_j}, \quad j = 1,...,n + 2.$$

Conversely, any extremal signature, S, with $d(S) = \{z_1,...,z_{n+2}\}$ has weights

$$s_j = \frac{b}{v'(z_j)}, \; j = 1,\ldots,n+2$$

for some $b = |c|e^{i\tau}$ and hence

$$S(z_j) = e^{-i\tau}e^{i\phi_j}, \; j = 1,\ldots,n+2.$$

Thus extremal signatures with domain consisting of n+2 points (which are necessarily primitive extremal signatures) are "uniquely" determined by the domain.

In particular, when $z_1,\ldots,z_{n+2}$ are real, the weights may be chosen to be real and alternate in sign. When we consider approximation of real-valued functions on $B \subset R$ we thus recover the familiar alternation characterization of best approximations.

If $k > n+2$ there are many extremal signatures generated by (5) and each is associated with the best approximation of some f (see Smirnov and Lebedev (10)).

Example 1. If

$$z_j = e^{i\left(\frac{\alpha+2\pi j}{n+2}\right)}, \; j = 1,\ldots,n+2$$

then

$$s_j = \frac{1}{n+2} z_j e^{-i\alpha}, \; j = 1,\ldots,n+2$$

satisfies (2) and $S(z_j) = e^{i\alpha}\bar{z}_j$, $j = 1,\ldots,n+2$ is an extremal signature for P_n.

Example 2. $B = D : = \{z: |z| \leq 1\}$ and $f = z^{n+1}$. $p^* \equiv 0$ is a best approximation, since, in the notation of Example 1 (for $\alpha = 0$), $z_j \epsilon E(z^{n+1};B)$, $j=1,\ldots,n+2$ and $S(z_j) = \bar{z}_j = z_j^{n+1}$, $j = 1,\ldots,n + 2$. A best approximation on D to $z^{n+1} + tz^n$ out of P_{n-1} is described in Ryzakov (9).

Example 3. Let $B = G$, a simple closed curve in the plane, and suppose $f \epsilon C(G)$. Suppose $q \epsilon P_n$ satisfies $|f(z) - q(z)| = \| f - q \| = M > 0$ for all $z \epsilon G$, and $w(z) = f(z) - q(z)$ traverses the circle $|w| = M$ at least $n+1$ times as z traverses G in the positive direction once. Then q is a best approximation to f on G.

To see this note that $E(f-q;G) = G$. As z traverses G once in the positive direction arg $(f(z)\text{-}q(z))$ increases (or decreases) by $2\pi m$, $m \geq n + 1$. But arg $p(z)$ increases by at most $2\pi n$ for any $p \epsilon P_n$, $(p \neq 0)$. Thus the image of G under

$$\mathrm{sgn}\overline{(f(z) - q(z))}\, p(z)$$

for $p \neq 0$ is a closed curve which encircles the origin at least once, hence cuts the negative real axis. (3) now implies that q is a best approximation to f on G.

This same observation in the case that f is analytic in $|z| < 1$ and continuous in D is in the literature (cf. Trefethen (11) and references given there). Note that it implies that 0 is a best approximation to z^m on D for any $m > n$.

Suppose $n > 0$. Then

$$f(z) = \frac{z-a}{1-\bar{a}z} = \frac{z-a}{1-\bar{a}z}\,(1-(\bar{a}z)^n) + \bar{a}^n z^n \frac{z-a}{1-\bar{a}z} .$$

Example 3 shows that

$$p^*(z) = \frac{z-a}{1-\bar{a}z}(1-(\bar{a}z)^n)$$

is a best approximation to f out of P_n on $C : |z| = 1$ (if $|a| \neq 1$) and hence (by the maximum principle) on D (if $0 < |a| < 1$). When n=0, $p^* = 0$.

Example 4. Let C(r) be the ellipse in the z-plane defined by

$$z = \frac{1}{2}(w + \frac{1}{w}), |w| = r, \ 0 < r \leq 1.$$

Let

$$w_j = r e^{\frac{j\pi i}{n+1}}, j = 1,...,2n + 2$$

and $z_j = (w_j + w_j^{-1})/2$ be the corresponding points of C(r). Theorem D implies that if $v(z) = (z-z_1)...(z-z_{2n+2})$

$$s_j = \frac{U_n(z_j)}{v'(z_j)} = \frac{2^{2n}}{n+1}(r^{n+1} + \frac{1}{r^{n+1}})^{-1}(-1)^j, j = 1,...,2n + 2,$$

(where U_n is the Chebyshev polynomial of the second kind) are the (unnormalized) weights of an extremal signature for P_n, S, with $d(S) = \{z_1,...,z_{2n+2}\}$, defined by

$$S(z_j) = (-1)^j, j = 1,...,2n + 2.$$

Now

$$T_{n+1}(z_j) = \frac{1}{2}(r^{n+1} + \frac{1}{r^{n+1}})(-1)^j, \; j = 1,...,2n+2,$$

where T_{n+1} is the Chebyshev polynomial, which has leading coefficient 2^n, and satisfies

$$\max_{z \in C(r)} |T_{n+1}(z)| = \frac{1}{2}(r^{n+1} + \frac{1}{r^{n+1}}).$$

Hence a best approximation to $f(z) = z^{n+1}$ on $B = C(r)$ (and therefore on the closure of the inside of $C(r)$) out of P_n is $p^*(z) = z^{n+1} - 2^{-n}T_{n+1}(z)$, for all r, $0 < r \leq 1$.

If we adopt the notation $T_j(z;B)$ for a polynomial in P_j with leading coefficient 1 which has least norm on B among all monic polynomials in P_j (and call it a Chebyshev polynomial for B) then we have shown that $T_j(z;C(r)) = 2^{-(j-1)}T_j(z)$ for $0 < r \leq 1$ (a familiar fact for $r=1$) and $T_j(z;D) = z^j$.

If $z_1,...,z_{n+2}$ are distinct complex numbers then

$$T_{n+1}(z;\{z_1,...,z_{n+2}\}) = \Sigma_{i=1}^{n+2} |s_i| \frac{v(z)}{z - z_1},$$

where the s_i are given in (6). Note that when $z_1,...,z_{n+2}$ are the vertices of a regular polygon we have $T_{n+1}(z;\{z_1,...,z_{n+2}\}) = z^{n+1}$.

If $z_1,...,z_n$ are points of the plane then the lemniscate

$$L_m = |z - z_1| ... |z - z_n| = m$$

is a simple closed curve, for $m > m_0$, with $z_1,...,z_n$ inside it. The conditions of Example 3 are in force and we conclude that $(z-z_1)...(z-z_n) = T_n(z;L_m)$. Indeed, note that <u>every</u> monic polynomial is a Chebyshev polynomial for some domain!

Finally, we remark that the Chebyshev polynomials for circular sectors are studied in Geiger and Opfer (4).

2. UNIQUENESS QUESTIONS

The uniqueness of a best approximation to f out of P_n is an immediate consequence of Theorem A. For if p and q are best approximations to f out of P_n, then if $s_1,\ldots,s_k$ are the weights for an extremal $S[f - p]$ with $d(S) = \{z_1,\ldots,z_k\}$ whose existence is affirmed by Theorem A

$$\|f - p\| = \sum_{i=1}^{k} |s_i| \, |f(z_i) - p(z_i)| = \Sigma s_i(f(z_i) - p(z_i)) = \Sigma s_i(f(z_i) - q(z_i))$$

$$\leq \Sigma |s_i| \, |f(z_i) - q(z_i)| \leq \|f - q\| = \|f - p\|.$$

Therefore, both inequalities are equalities and we have $|f(z_i) - q(z_i)| = \|f - q\| = \|f - p\|$ and $\text{sgn}[f(z_i) - q(z_i)] = \text{sgn}\bar{s}_i = \text{sgn}\,[f(z_i) - p(z_i)]$, $i=1,\ldots,k$. Hence $f(z_i) - q(z_i) = f(z_i) - p(z_i)$, $i = 1,\ldots,k$ and $p-q$ has $k \geq n + 2$ zeros in B so that $p \equiv q$.

There is a quantitative notion which is somewhat stronger than uniqueness. Namely the best approximation p^* to f out of P_n is called <u>strongly unique</u> if

$$\|f - p\| \geq \|f - p^*\| + \gamma \, \|p - p^*\|$$

for some positive constant $\gamma(f,B)$ and all $p \epsilon P_n$. Newman and Shapiro (5) (cf. Cheney (2)) proved that real continuous functions have a strongly unique best approximation out of the subspace of real polynomials of degree at most n on any compact subset of the real line. They also showed that in the case of complex-valued functions, which we are considering,

$$\|f - p\| \geq \|f - p^*\| + \gamma_1(\|f - p\| - \|f - p^*\|)^{1/2} + \gamma_2\|p - p^*\|$$

for positive constants γ_1,γ_2. However, as Cline (3) has observed: if $B = \{-1,1\}$, $f(z) \equiv z$ and $n = 0$ then the (unique) best approximation to f out of P_0 on B is $p^* = 0$. But it is not strongly unique. For consider

$p = ic$, $c \geq 0$, then strong uniqueness would require, for some $\gamma > 0$, and all $c > 0$,

$$\| f - p \| = (1 + c^2)^{\frac{1}{2}} \geq 1 + \gamma c,$$

which is false for $c = \gamma$. (This example may be recast as follows: Let $w = \phi(z)$ map the open unit disc conformally onto the interior of the ellipse $4u^2 + v^2 = 4$, with $\phi(1) = 1$ and $\phi(-1) = -1$. Then the best approximation to ϕ on $|z| \leq 1$ by a constant is 0, uniquely, but not strongly so.)

As we shall soon see strong uniqueness <u>does</u> hold in many complex best approximation problems and so further study of this phenomenon seems warranted.

Bartelt and McLaughlin (1) show that a best approximation p^*, to f is strongly unique, if, and only if,

$$\min_{z \in E(f-p^*;B)} \operatorname{Re}[\operatorname{sgn}\overline{(f(z) - p^*(z))}\; p(z)] < 0 \tag{7}$$

for all $p \in P_n$ satisfying $\| p \| = 1$. (Compare (3).)

We observe, therefore, that in the situation of Example 3 above strong uniqueness holds. Example 4 also exhibits strong uniqueness. However, as Cline's example shows not all Chebyshev polynomials are strongly unique.

Indeed, Cline's example is easily generalized. Suppose that B consists of the n+2 roots of unity

$$z_j = e^{\frac{2\pi i j}{n+2}}, \; j = 1,\ldots,n + 2.$$

Then we claim that the best approximation to <u>any</u> f out of P_n is not strongly unique. For we know that for any $p \in P_n$ we have

$$\sum_{j=1}^{n+2} z_j p(z_j) = 0,$$

hence if we choose $q \epsilon P_n$ to satisfy

$$q(z_j) = \frac{i}{z_j}, \; j = 1,\dots,n+1$$

then $z_{n+2}\, q(z_{n+2}) = -\,i(n+1)$, and we have

$$\text{Re}\; z_j q(z_j) = 0, \; j = 1,\dots,n+2.$$

But $E(f - p^*;B) = B$ and, as we saw in Example 1

$$\text{sgn}\; \overline{f(z_j) - p^*(z_j)} = z_j, \; j = 1,\dots,n+2$$

Thus for $p = q$ (7) does not hold.

Finally we remark that we provided (Rivlin and Weiss (7)) conditions which imply that the partial sums of the power series expansion of a function analytic in $|z| < 1$ and continuous in D are its best approximations of appropriate degree. Examination of the proof given shows that these best approximations are strongly unique.

REFERENCES

1. Bartelt, M. W., and McLaughlin, H. W., Characterizations of strong unicity in approximation theory, J. of Approx. Theory, 9 255-266 (1973).

2. Cheney, E. W., Introduction to Approximation Theory, McGraw-Hill, N. Y., 1966.

3. Cline, A. K., Lipschitz conditions on uniform approximation operators, J. of Approx. Theory, 8, 160-172 (1973).
4. Geiger, C., and Opfer, G., Complex Chebyshev polynomials on circular sectors, J. of Approx. Theory, 24, 93-118 (1978).
5. Newman, D. J., and Shapiro, H. S., Some theorems on Cebysev approximation, Duke Math. J. 30, 673-682 (1963).
6. Rivlin, T. J., and Shapiro, H. S., A unified approach to certain problems of approximation and minimization, J. Soc. Indust. Appl. Math., 9, 670-699 (1961).
7. Rivlin T. J., and Weiss, B., Some best polynomial approximations in the plane, Duke Math. J., 35, 475-482 (1968).
8. Rivlin, T. J., The Chebyshev Polynomials, John Wiley & Sons, N. Y. (1974).
9. Ryzakov, I. Ju., An analog for the disk of Zolotarev's problem, Soviet Math. Dokl. 15, 882-885 (1974).
10. Smirnov, V. I., and Lebedev, N. A., Functions of a Complex Variable: Constructive Theory, M.I.T. Press, Cambridge, Mass. (1968).
11. Trefethen, L. N., Near Circularity of the Error Curve in Complex Chebyshev Approximation, Research Report No. 79-03, Seminar fuer Angewandte Mathematik ETH, CH-8092, Zuerich (August 1979).

ON SOME FOURIER AND DISTRIBUTION-THEORETIC METHODS IN APPROXIMATION THEORY

Dedicated to G.G. Lorentz
on his seventieth birthday

Harold S. Shapiro
Department of Mathematics
Royal Institute of Technology
Stockholm, Sweden

0. INTRODUCTION

The purpose of this talk is to illustrate how certain ideas and methods, developed since the 'thirties mainly to prove results about partial differential equations, apply to approximation theory. I shall discuss in detail five sample topics, which should convey the flavor of the development and the scope of these methods (I shall strive for clarity rather than generality, and readily limit myself to special cases when generalizations are more or less obvious):

(1) Hölder (also called Lipschitz) classes and other related function spaces, from the standpoint of harmonic analysis.

Copyright © 1980 by Academic Press, Inc.
All rights of reproduction in any form reserved.
ISBN: 0-12-171050-5

(2) Approximating a harmonic function by a harmonic polynomial on a domain in $\mathbb{R}^d$, in L^2 norm.

(3) Approximating a smooth function by a polynomial on a domain in $\mathbb{R}^d$, in L^2 norm (estimation of the remainder in terms of norms of derivatives).

(4) Proof of a closedness theorem arising in the mathematical theory of computerized tomography.

(5) An inequality of K. Friedrichs concerning conjugate harmonic functions on a plane domain.

Let us be a bit more specific. Topic (1), the only one we discuss in some detail, deals with a sophisticated way of defining Hölder classes of order α (α an arbitrary real number) of tempered distributions on $\mathbb{R}^d$ or on the torus $\mathbb{T}^d$, the "Fourier definition". For $\alpha > 0$ the classes Λ_α thus defined coincide locally with the Hölder classes as defined traditionally by means of finite differences, or by means of Poisson integrals or other integral kernels. The point of the Fourier definition is that once a few key lemmas are established nearly all the main properties of Hölder classes follow very easily (these include e.g. the action of singular integrals, interpolation theorems, and direct and inverse theorems concerning approximation by trigonometric polynomials). The same ideas can easily be extended to Besov spaces and Sobolev spaces, however we shall not carry out this program here.

Topic (2) is a beautiful introduction to the interplay of ideas from Fourier analysis, functional analysis and complex function theory as applied to p.d.e. and approximation theory. Runge-type theorems concerning polynomial approximation of solutions of the Laplace and other partial differential equations were established by Malgrange [32, 33] in the 'fifties (see also Lax [30]). Obtaining stronger approximations in $L^2(\Omega)$ norm involves one in new difficulties involving subtle aspects of Sobolev spaces, especially approximation by test functions with compact support in Ω.

The topics (3), (4), (5), despite their superficial diversity, are closely linked in that they may all be treated by the same method. This method (but only as applied to topic (4)) has appeared in print in the work of Pedersen, Smith and Solmon [38]. Similar ideas have been elaborated and extended by W. Abramczuk, a doctoral student at Stockholm University, in connection with problem (3). What is involved is a judicious combination of two deep theorems: a theorem due to Ehrenpreis and Malgrange concerning the existence of a distribution solution to a linear system of PDE with constant coefficients [17, 32, 33, 29, 37] and a "coercivity up to the boundary" theorem of K.T. Smith [47] for very strongly elliptic systems of PDE. Precise statements of these theorems are given below; for the moment let us describe topics (3), (4) and (5) a little more closely.

Topic (3) deals with approximation of a smooth function u on a suitably regular bounded domain $\Omega \subset \mathbb{R}^d$, by polynomials of degree at most n - 1, in L^2 norm. The result is that this can be done with an error not exceeding C times the Sobolev seminorm

$$\|u\|_{n,2} = \sum_{|\alpha|=n} \|D^{\alpha}u\|_{L^2(\Omega)}$$

where C is a constant depending only on Ω. (We have used throughout customary multi-index notations as e.g. in Adams [1], and also the notations of [1] for Sobolev spaces.)

The case $n = 1$ and Ω a cube is classical and goes back to works of Poincaré, Friedrichs and others. The cases $n > 1$, Ω "arbitrary", and L^p norms occur implicitly in Sobolev [48] and explicitly in Morrey [35]. Estimates of this kind, apart from their intrinsic interest, are useful in connection with finite element methods and multivariate quadrature formulas (cf. the "Bramble-Hilbert lemma" [10]). By duality they lead to the existence of multivariate Peano kernel theorems as shown in [45]. Recently Dupont and Scott [16] further developed Sobolev's elementary method and proved constructive variants (as well as generalizations) of the above approximation theorem, with estimates of C. Their identities immediately yield also constructive multivariate Peano kernel theorems. Despite this we have included an "abstract" treatment of topic (3) here, mainly to illustrate the general method, which also yields some generalizations that go beyond [16].

Topic (4) has its origins in a paper of Logan and Shepp [31] dealing with a certain reconstruction method for computerized tomography. The mathematical context is as follows. Let Ω denote the open unit disc in $\mathbb{R}^2$ and for each θ, $0 \leq \theta < \pi$, denote by R_θ the set of functions in $L^2(\Omega)$ which are "constant on each line segment in Ω which makes an angle θ with the x-axis" (the quotation marks are to indicate that a couple of "almost all" qualifiers have to be inserted for a precise statement). It is easy to prove: $\mathbb{R}_\theta$ is a closed subspace of $L^2(\Omega)$. A problem encountered in [31] was: Given $0 \leq \theta_1 < \theta_2 < \ldots < \theta_n < \pi$, is $R_{\theta_1} \oplus R_{\theta_2} \oplus \ldots \oplus R_{\theta_n}$ closed in $L^2(\Omega)$? An affirmative answer was given by Hamaker and Solmon [20], and shortly afterwards considerably generalized versions were proved by Pedersen, Smith and Solmon [38], Svensson [52], and Boman [8] among others. We shall present here the solution given in [38].

Topic (5) concerns the following inequality of Friedrichs [18]. *Let* f = u + iv *be analytic and square integrable in a bounded plane domain* Ω, *with mean value zero* (u *and* v *are real*). *Then*

$$\iint_{\Omega} u^2 \, dx \, dy \leq C \iint_{\Omega} v^2 \, dx \, dy \qquad (*)$$

where C *depends only on* Ω, *provided* Ω *satisfies suitable regularity conditions*.

As for the regularity conditions of Friedrichs we shall not state them here. They *imply*, however, that Ω satisfies an interior cone condition. We show below, as an almost immediate consequence of results of Smith *et al*, *that* (*) *holds if* Ω *satisfies an interior cone condition*. Thus, a *stronger* version of Friedrichs' inequality than the original (whose proof required many pages) is a corollary of *general* theorems about p.d.e.!

I acknowledge with pleasure my indebtedness to several colleagues and students: especially Jan Boman, who in discussions over a period of many years has helped me to get oriented in this field; Wojciech Abramczuk for discussions of topic (3) and permission to quote unpublished results; Lars-Inge Hedberg for numerous references; Lars Svensson concerning topics (4) and (5); and Ron Devore for informing me of the work of Dupont and Scott. Also, much of what is written below under topic (1) is from a manuscript I prepared some years ago for a projected joint paper with Nestor Rivière; the presentation there benefitted from conversations with him.

1. TOPIC (1) — HÖLDER SPACES

1.1 Motivation. In various problems arising in differential equations, approximation theory, calculus of variations, etc., it is necessary to be able to describe degrees of regularity (smoothness) of a function stronger than continuity. One natural way to do this is by means of the modulus of continuity: for a (say, complex-valued) function f defined and continuous on a compact metric space with distance function d(x, y), this is the function

$$\omega_f(t) = \sup |f(x) - f(y)| , \; d(x, y) \leq t \quad (t > 0) . \tag{1.1}$$

Because of uniform continuity, $\omega_f(t)$ tends to zero as $t \to 0+$, and the rapidity with which it does so is a measure of the "regularity" of f. In particular the Lipschitz classes Lip α are defined by:

$$\text{Lip } \alpha = \{f : \omega_f(t) = O(t^\alpha) , \; t \to 0+\} \quad (\alpha > 0) . \tag{1.2}$$

Suppose henceforth that the underlying metric space is some subset of $\mathbb{R}^d$. Although formally meaningful for all $\alpha > 0$, Lip α contains only constant functions when $\alpha > 1$. Also, when $\alpha = 1$, it is not always the "right" class (see below). Therefore one must seek some other way to classify functions possessing a "high" degree of regularity. For functions on $\mathbb{R}^d$, one can introduce the classes C^m ($m = 0, 1, 2, \ldots$) of functions with continuous partial derivatives up to order m, or for greater refinement the classes $C^{m,\beta}$ ($0 < \beta \leq 1$) of functions in C^m whose partial derivatives of order m belong to Lip β. This very natural approach, common in the older literature, turned out not to be altogether adequate for the needs of modern analysis, especially when $\beta = 1$.

We may illustrate this with the following assertion concerning approximation by trigonometric polynomials, which combines well-known results of Jackson, Bernstein and de la Vallée Poussin (see for example [12, 42, 43, 53]). Let T_N denote the set of trigonometric polynomials of order not exceeding N, in one variable, and f a continuous complex-valued 2π-periodic function on $\mathbb{R}$. A necessary and sufficient condition that

$$\inf_{g \in T_N} \|f - g\|_\infty = O(N^{-\alpha}) \quad , \quad N \to \infty \tag{1.3}$$

where α is positive and not an integer ($\|\cdot\|_\infty$ denoting the supremum norm) is

$$f \in C^{m,\beta} \; ; \; m = [\alpha] \; , \; \beta = \alpha - m \; . \tag{1.4}$$

(Here $[\alpha]$ denotes the greatest integer not exceeding α.)

Now, when α is a positive integer, (1.3) and (1.4) are not equivalent, and in fact, denoting by Λ_α the set of f satisfying (1.3) (this is equivalent to the more usual definitions of this symbol), we have for every $\varepsilon > 0$

$$C^{\alpha-1,1-\varepsilon} \underset{\neq}{\supset} \Lambda_\alpha \underset{\neq}{\supset} C^{\alpha-1,1} \; .$$

Thus, the function classes Λ_1, Λ_2, ..., arising naturally in approximation theory, appear anomalous from the standpoint of the classes $C^{m,\beta}$. It was a relatively late discovery of Zygmund that Λ_1 coincides with the set of all continuous 2π-periodic functions f satisfying

$$\sup_{x \in R} |f(x-t) - 2f(x) + f(x+t)| = O(t) \quad , \quad t \to 0+ \; . \tag{1.5}$$

The classes Λ_2, Λ_3, ... may be characterized in an analogous fashion, involving difference operators of suitably high order.

From a modern standpoint the classes Λ_α, $\alpha = 1, 2, \ldots$ are by no means "anomalous". The scale of spaces $\{\Lambda_\alpha\}_{\alpha>0}$ is in many ways more natural than the $C^{m,\beta}$, although their definition by means of (1.3) is not the handiest (for instance, it is hard to deduce from (1.3) that Λ_α is stable under Hilbert transformation, or that fractional differentiation of order β maps Λ_α onto $\Lambda_{\alpha+\beta}$). The alternative definition using difference operators is no better in these respects. The present section is devoted to another definition of Λ_α, which with small modifications is applicable also to most other regularity classes currently used in analysis. It is based on the behaviour, for small positive a, of the convolution of a given function with the normalized dilations

$$\varphi_{(a)}(x) = a^{-n}\,\varphi(x/a)$$

of "test function" φ of a rather general type. This definition can also (via Fourier transformation) be formulated in terms of the Fourier transform of the given function, hence we refer to it as "the Fourier method".

Before embarking on the details, some explanatory words may be in order for the reader inexperienced in this field, as to why a bewildering array of function spaces has invaded analysis. L^p analogs of the $C^{m,\beta}$ classes arose in problems on mean approximation, singular integrals, and a priori estimates in p.d.e. Moreover, various new spaces came into focus on the basis of known ones as intermediate spaces, duals, spaces of multipliers, restrictions to lower-dimensional manifolds (trace spaces), or through Fourier transformation, fractional integration and differentiation, etc. Hence there naturally arose a new branch of analysis to study systematically the inclusion relations and transformation properties of all these spaces. Of the many techniques known to us for this study the Fourier method illustrated below seems to us far and away superior to all others, so long as we deal with functions defined on the whole space $\mathbb{R}^d$ (or the whole torus $\mathbb{T}^d$).

1.2 Hölder classes – the Fourier method.

Let $\hat{G}$ denote the class of functions $\hat{\varphi}$ on $C^\infty(\mathbb{R}^d)$ such that

$\hat{\varphi}$ vanishes on neighborhoods of 0 and ∞ (1.6)

$\hat{\varphi}$ satisfies the "ray condition": for every x, $|x| = 1$ (1.7)

there exists $\rho > 0$ such that $\hat{\varphi}(\rho x) \neq 0$. Then, the class G of inverse Fourier transforms of functions in $\hat{G}$ is a class of rapidly decreasing C^∞ functions on $\mathbb{R}^d$. For any distribution u on $\mathbb{R}^d$, and $a > 0$, we define its normalized dilation $u_{(a)}$ by

$$\langle u_{(a)}, \varphi\rangle = \langle u, \varphi_a\rangle \ , \ \forall\varphi \in C_0^\infty(\mathbb{R}^d)$$

where $\varphi_a(t) = \varphi(at)$.

Definition. For each $\varphi \in G$, tempered distribution u and real number α,

$$\|u\|_{\alpha;\varphi} = \sup_{a>0} a^{-\alpha} \|u * \varphi_{(a)}\|_\infty \ . \tag{1.8}$$

A key point in connection with this definition is:

Lemma 1. For any φ_1, φ_2 in G the seminorms $\|\cdot\|_{\alpha;\varphi_1}$ and $\|\cdot\|_{\alpha;\varphi_2}$ are equivalent.

This follows at once from results in [9] or [43], since φ_1 is expressible as a finite sum of dilations of φ_2 convolved with integrable functions, and vice versa. Thus, henceforth we select some fixed $\varphi_0 \in G$ and compute seminorms using it exclusively. For notational convenience we then write $\|\cdot\|_\alpha$ in place of $\|\cdot\|_{\alpha;\varphi_o}$ or when confusion is possible, we write in place of $\|u\|_\alpha$ either $\|u\|_{\Lambda_\alpha}$ or $\|u;\Lambda_\alpha\|$.

Definition. Λ_α _is the set of tempered distributions_ u _on_ $\mathbb{R}^d$ _such that_ $\|u;\Lambda_\alpha\|$ _is finite_.

Observe that $\|u;\Lambda_\alpha\| = 0$ if u is a polynomial. The converse is also true, and more generally also the following proposition, whose proof again is a simple consequence of ideas in [9, 43]:

Lemma 2. _Let_ u _be a tempered distribution in_ Λ_α, _where_ $\alpha > 0$. _Then_, u _is a polynomial plus a function uniformly continuous on_ $\mathbb{R}^d$ _which belongs to_ Λ_α _in the classical sense_.

What we mean by "Λ_α in the classical sense" is that for some (and hence each) integer $m > \alpha$ the differences of order m and step length a in an arbitrary direction are uniformly $O(a^\alpha)$. Note that "classical Λ_α" in terms of the class just defined, is $L^\infty \cap \Lambda_\alpha$. If u is restricted _a priori_ to be a continuous function bounded on $\mathbb{R}^d$, the norm $\|u;\Lambda_\alpha\|$ is equivalent to the Hölder α-norm as defined by finite differences. In this case it does not matter whether in (1.8) we take $\sup_{a>o}$ or $\sup_{o<a\leq 1}$; the Λ_α classes defined by these are identical.

Let's illustrate the method by proving

Theorem (_Calderón and Zygmund_ [14]). _Let_ u _be a bounded continuous function in_ $\Lambda_\alpha(\mathbb{R}^d)$ _where_ $\alpha > 0$. _Let_ k _be a tempered distribution on_ $\mathbb{R}^d$ _such that_ $\hat{k} \in C^\infty(\mathbb{R}^d \setminus \{0\})$ _and_ $\hat{k}$ _is homogeneous of degree_ 0 _and nonvanishing on the unit sphere_. _Then_ $u * k \in \Lambda_\alpha(\mathbb{R}^d)$.

<u>Proof</u>. Pick a test function $\varphi \in G$. Now, $(u * k) * \varphi_{(a)}$ $= u * (k * \varphi_{(a)}) = u * (k_{(a)} * \varphi_{(a)})$ (since k is homogeneous of degree -d, hence $k = k_{(a)}) = u * (k * \varphi)_{(a)}$.

Now $\psi = k * \varphi$ also is in G (just check that $\hat{\psi}$ satisfies (1.6) and (1.7)) so $\|u * \psi_{(a)}\|_\infty = O(a^\alpha)$ and hence also $\|(u * k) * \varphi_{(a)}\|_\infty = O(a^\alpha)$, i.e. $u * k \in \Lambda_\alpha$.

This completes the proof.

The switch from one test function in G to another is the key to the convenience of this method. We now give a few more examples.

<u>Theorem 1.1</u> (<u>Essentially, S. Bernstein</u>, $0 < \alpha < 1$; <u>A. Zygmund</u> $\alpha \geq 1$). <u>A necessary and sufficient condition that</u> $f \in L^\infty \cap \Lambda_\alpha(\mathbb{R}^d)$ <u>is, for every</u> $\lambda \geq 1$ <u>there exists</u> F_λ <u>whose Fourier transform is in</u> C^∞ <u>with support in</u> $B_\lambda = \{x : |x| \leq \lambda\}$ such that

$$\|f - F_\lambda\|_\infty \leq A\lambda^{-\alpha} \tag{1.9}$$

<u>where</u> A <u>is independent of</u> λ. <u>The infimum of admissible</u> A <u>in</u> (1.9) <u>defines an equivalent norm on</u> $L^\infty \cap \Lambda_\alpha(\mathbb{R}^d)$.

<u>Outline of proof</u>. Assume $f \in L^\infty \cap \Lambda_\alpha(\mathbb{R}^d)$ and let k satisfy: $\hat{k} \in C^\infty(\mathbb{R}^d)$, $\hat{k}(x) = 0$ <u>for</u> $|x| \geq 1$, $\hat{k}(x) = 1$ <u>for</u> $|x| \leq 1/2$. Then $F_\lambda = f * k_{(1/\lambda)}$ has Fourier transform supported in B_λ and satisfies (1.9). Conversely, suppose $f \in L^\infty(\mathbb{R}^d)$ admits the approximants F_λ in (1.9). Choose $\hat{\varphi} \in \hat{G}$ supported in $\{x : 1 \leq |x| \leq 2\}$. Then

$$\|f * \varphi_{(a)}\|_\infty = \|(f - F_\lambda) * \varphi_{(a)}\|_\infty \leq C\,\|f - F_\lambda\|_\infty$$

where λ has been chosen equal to $1/a$, and the last expression is $O(a^\alpha)$ by (1.9), which gives $f \in \Lambda_\alpha$.

Corollary 1. _Let_ $f, g \in L^\infty \cap \Lambda_\alpha(\mathbb{R}^d)$, $\alpha > 0$. _Then_ fg _belongs to_ $\Lambda_\alpha(\mathbb{R}^d)$ _and_

$$\|fg;\Lambda_\alpha\| \leq C\,(\|f\|_\infty\ \|g;\Lambda_\alpha\| + \|g\|_\infty\ \|f;\Lambda_\alpha\|) \tag{1.10}$$

where C _is a constant depending only on_ m.

Proof. Let k, F_λ, G_λ be as in the last proof. Then $\|F_\lambda\|_\infty \leq \|k\|_1\ \|f\|_\infty$, $\|G_\lambda\|_\infty \leq \|k\|_1\ \|g\|_\infty$. Observe that $(F_\lambda G_\lambda)\hat{} = \hat{F}_\lambda * \hat{G}_\lambda$ is supported in $B_{2\lambda}$, and (all unspecified norms being sup norms)

$$\|fg - F_\lambda G_\lambda\| \leq \|f\|\cdot\|g - G_\lambda\| + \|G_\lambda\|\cdot\|f - F_\lambda\|$$

$$\leq \|f\|\cdot C_1\ \|g;\Lambda_\alpha\|\lambda^{-\alpha} + \|k\|_1\ \|g\|\cdot C_1\ \|f;\Lambda_\alpha\|\lambda^{-\alpha}$$

$$= C_2(\ \|f\|_\infty\ \|g;\Lambda_\alpha\| + \|g\|_\infty\ \|f;\Lambda_\alpha\|)\ (2\lambda)^{-\alpha}\ .$$

Now, replace λ by $\lambda/2$ and apply the theorem. We conclude that the Λ_α norm of fg satisfies (1.10) with $C = C_2$.

Corollary 2. $L^\infty \cap \Lambda_\alpha(\mathbb{R}^d)$ _and_ $\Lambda_\alpha(\mathbb{T}^d)$ _are_ (_after eventual renorming by equivalent norms_) _Banach algebras_.

Outline of proof. We have remarked that for $f \in L^\infty(\mathbb{R}^d)$ and $\alpha > 0$, $\|f;\Lambda_\alpha\|$ dominates $\|f\|_\infty$, hence (1.10) implies

$$\|fg;\Lambda_\alpha\| \leq C_3\ \|f;\Lambda_\alpha\|\cdot\|g;\Lambda_\alpha\|\ . \tag{1.11}$$

The case of $\Lambda_\alpha(\mathbb{T}^d)$ is similar (observe that distributions in $\Lambda_\alpha(\mathbb{T}^d)$ with $\alpha > 0$ are necessarily L^∞).

The next theorem has a very simple proof (by _this_ method!) which is left to the reader.

Theorem 1.2. _If_ $u \in \Lambda_\alpha$ _then_ $D_i u \in \Lambda_{\alpha-1}$ (_here_ $D_i = \frac{\partial}{\partial x_i}$, _understood distributionally_). _Conversely, if_ $D_i u \in \Lambda_\alpha$ _for_ $i = 1, 2, \ldots, d$ _then_ $u \in \Lambda_{\alpha+1}$.

Corollary. $\Lambda_\alpha \subset C^n$ _if_ $\alpha > n$.

To prepare for the next theorem we observe that our definition of Λ_α can be carried through with $\|\cdot\|_\infty$ norms replaced by the corresponding $\|\cdot\|_p$ norms on $\mathbb{R}^d$, $1 \le p < \infty$. In this way we can define classes $\Lambda_\alpha^p(\mathbb{R}^d)$. To avoid a slight technical complication, we state the next theorem in the periodic case:

Theorem 1.3 (Zygmund [57], Stein and Zygmund [51], Herz [27]). *The necessary and sufficient condition that a distribution* u *on* $\mathbb{T}^d$ *satisfy* $u * f \in \Lambda_\alpha$ *whenever* $f \in \Lambda_\alpha$ *is* $u \in \Lambda_o^1$.

Note that the condition on u is independent of α. The formulations in [57, 51] are different in appearance than the above, since there finite-difference definitions of Hölder classes are used.

We prove the sufficiency, and leave the other half to the reader. Thus, suppose $f \in \Lambda_\alpha$ and $u \in \Lambda_o^1$. Let $\varphi \in G$. Then also $\psi = \varphi * \varphi$ is in G and we have

$$(u * f) * \psi_{(a)} = (u * f) * (\varphi * \varphi)_{(a)}$$

$$= (u * \varphi_{(a)}) * (f * \varphi_{(a)}) \text{ , hence}$$

$$\|(u * f) * \psi_{(a)}\|_\infty \le \|u * \varphi_{(a)}\|_1 \cdot \|f * \varphi_{(a)}\|_\infty \le Ca^\alpha$$

and hence $u * f \in \Lambda_\alpha$. This proves the sufficiency. Similarly, generalizations to convoluters from Λ_α^r to Λ_β^s are easily proven.

As a final example we choose a theorem of Peetre [40] on intermediate spaces, formulated however without explicit reference to that concept.

Theorem 1.4 (Peetre). *Let* $\beta > \alpha > 0$. *A necessary and sufficient condition for* $f \in L^\infty \cap \Lambda_\alpha(\mathbb{R}^d)$ *is*: *for every* $\varepsilon > 0$ *there exists* $f_\varepsilon \in L^\infty \cap \Lambda_\beta(\mathbb{R}^d)$ *such that* $\|f - f_\varepsilon\|_\infty \le \varepsilon$ *and* $\|f_\varepsilon\|_{\Lambda_\beta} \le C\varepsilon^{\frac{\alpha-\beta}{\alpha}}$, *where* C *does not depend on* ε.

The theorem expresses that, in Peetre's sense, $L^\infty \cap \Lambda_\alpha$ is an interpolation space between L^∞ and $L^\infty \cap \Lambda_\beta$. The sufficiency is nearly trivial: suppose such f_ε exist, and choose $\varphi \in G$. Then

$$\|f * \varphi_{(a)}\|_\infty \leq \|\varphi\|_1 \varepsilon + \|f_\varepsilon * \varphi_{(a)}\| \leq C_1\varepsilon + C_2 a^\beta \varepsilon^{\frac{\alpha-\beta}{\alpha}} .$$

This holds for every a and ε. Choosing $\varepsilon = a^\alpha$ we get $\|f * \varphi_{(a)}\|_\infty \leq C_3 a^\alpha$ showing that $f \in \Lambda_\alpha$. The proof in the other direction is along similar lines, but less trivial and we omit it.

Using Peetre's theorem I deduced several years ago (unpublished) the following result on composition of Hölder functions:

Theorem 1.5. Let $1 \leq \alpha < \beta$. Then, for every $f \in \Lambda_\beta(\mathbb{R})$ and $g \in \Lambda_\alpha(\mathbb{R}^d)$ the composed function $f \circ g$ lies in $\Lambda_\alpha(\mathbb{R}^d)$.

This result can be improved. For $\alpha = 1$, I. Wik showed by an elementary argument (unpublished) that

$$\omega_2(f\,;\,t) = O\left(t(\log(1/t))^{-1}\right)$$

suffices, and this is the best possible condition on the modulus of smoothness of f. Presumably his elementary method works best for all results of the type of Theorem 1.5.

1.3 Some historical comments concerning the above method

The first step towards defining Hölder classes by means of convolution integrals is in work of Hardy and Littlewood (see [50] for references). Let f denote a bounded uniformly continuous function on $\mathbb{R}$, and consider its Poisson integral

$$u(x,\ y) = \frac{1}{\pi}\int_{-\infty}^{\infty} f(\xi) \cdot \frac{y}{(x-\xi)^2 + y^2}\, d\xi\ . \tag{1.12}$$

Hardy and Littlewood proved: f <u>is in</u> Lip α $(0 < \alpha \leq 1)$ <u>if and only if</u>

$$\frac{\partial u}{\partial x} = O(y^{\alpha-1}) \quad , \ y \to 0+ \tag{1.13}$$

<u>uniformly for</u> $x \in \mathbb{R}$. (Rather, they proved the periodic version of this, but $\mathbb{R}$ is a more convenient setting for our present purpose.)

It is convenient to reformulate this slightly. Differentiating (1.12) gives (writing u_x for $\frac{\partial u}{\partial x}$)

$$u_x(x, y) = -\frac{2}{\pi}\int_{-\infty}^{\infty} f(\xi) \cdot \frac{(x - \xi)y}{((x - \xi)^2 + y^2)^2}\, d\xi \ .$$

The last integral is the convolution of f with the function (of the variable x)

$$xy(x^2 + y^2)^{-2} = y^{-2}k(x/y) \ ,$$

where $k(x) = x(1 + x^2)^{-2}$. Thus, in terms of the notation

$$h_{(a)}(x) = a^{-1}h(x/a) \ , \ a > 0$$

the condition (1.13) takes the form

$$\|f * k_{(y)}\|_\infty = O(y^\alpha) \quad , \ y \to 0+ \tag{1.14}$$

where $*$ denotes convolution on $\mathbb{R}$. With the theorem thus reformulated, one is naturally inclined to ask, what is the class of functions f described by (1.14) when for $k(x)$, instead of the above choice, we take some other integrable function? In fact, another theorem of Hardy and Littlewood asserts that for $0 < \alpha < 1$, Lip α is just the class of f for which

$$\frac{\partial u}{\partial y} = O(y^{\alpha-1}) \quad , \ y \to 0+ \tag{1.15}$$

which can in turn be recast in the form (1.14), but now with the "kernel" $k(x) = (1 - x^2)(1 + x^2)^{-2}$. Thus one can make these observations:

Let

$$k_1(x) = x(1 + x^2)^{-2} \tag{1.16}$$

$$k_2(x) = (1 - x^2)(1 + x^2)^{-2} , \quad x \in \mathbb{R} . \tag{1.17}$$

Then, with either of the choices $k = k_1$, $k = k_2$, *(1.14) characterizes the class* Lip α, *provided* $0 < \alpha < 1$. *Moreover, (1.3) with* $\alpha = 1$ *characterizes the class* Lip 1 *with the choice* $k = k_1$, *but not with the choice* $k = k_2$.

A series of further results involving conditions of the type (1.13) but with higher-order mixed partial derivatives in place of $\frac{\partial u}{\partial x}$ leads to characterizations of higher regularity classes by conditions of the type (1.14) with appropriate kernels k. For a systematic approach to regularity classes based on the Poisson integral formalism, see [50], and references there to Taibleson's work.

A complete analysis of the condition (1.14) for *general* kernels $k \in L^1$ (fine enough to explain, for instance, the above noted discrepancy at $\alpha = 1$ between the kernels (1.16) and (1.17)) was first given in [42, Chapter 5], although earlier studies by P.L. Butzer and his collaborators, devoted in part to the closely related problem of "saturation", should also be mentioned here (see [12] for references). Roughly, the conclusion reached is that the class described by (1.14) can be determined effectively from the asymptotic behaviour of the Fourier transform $\hat{k}(x)$ near $x = 0$ (for the most complete development of this theory, see Boman [7]). An analogous situation obtains for functions defined on $\mathbb{R}^n$, and with L^p norms in place of sup norms. Thus, the way is opened to a general calculus of "regularity classes" based on conditions of type (1.14).

Important steps towards such a calculus were made independently by Calderón [13] in 1964, who considered functions $f \in L^p(\mathbb{R}^n)$ such that $g(a) = \|f * \varphi_{(a)}\|_p$ decays in some specified way as $a \to 0+$, whereby φ is an element of $C^\infty(\mathbb{R}^n)$. Calderón's axiomatics allow great flexibility, requiring that g should belong to a suitable Banach lattice of functions on $\mathbb{R}^+$. A weakness of Calderón's formalism is that his test functions φ were required to have radial symmetry, which causes technical inconveniences and also rules out e.g. the possibility to define Sobolev spaces by means of them. His method has been further developed by, among others Herz [27], Heidemann [26] and Torchinsky [54, 55].

J. Peetre in papers and lecture notes from 1964 onwards (see [39] for references) proposed definitions of Lipschitz and Besov spaces similar to those of Calderón, using test functions of class C^∞ with compact support in $\mathbb{R}^n \setminus \{0\}$ and satisfying the condition

$$\sum_{m=-\infty}^{\infty} \hat{\varphi}(2^m x) = 1 \;, \quad x \neq 0 \;.$$

This approach seems to trace its ancestry to earlier work of Hörmander in partial differential equations, and of Cotlar on singular integrals. As a basis for a general calculus of regularity classes, Peetre's formalism (like Calderón's) still needs to be liberated from inessential restrictions on the test function, but once the need for this was recognized the step was fairly easy to carry out. This came as a by-product of our theory [42, 44] of "generalized moduli of continuity", which extended the above-sketched Hardy-Littlewood results for general kernels. A complete account of Hölder (also Besov, Sobolev, ...) spaces in their own right from the standpoint of the present exposition seems not to exist as yet.

We want to stress that "the Fourier method" is not the creation of any one person or school, it has evolved at the confluence of a number of independent streams. In conclusion, let us further bolster this view by pointing out two interesting landmarks:

(a) For 2π-periodic functions $f \sim \Sigma\, \hat{f}(n)e^{in\theta}$ the Fourier characterization of Λ_α becomes

$$\max_\theta \left|\sum_n \hat{\varphi}(na)\hat{f}(n)e^{in\theta}\right| \leq Ca^\alpha$$

where $\hat{\varphi} \in \hat{G}$. It is enough to take here $a = 2^{-m}$, $m = 1, 2, \ldots$. But this is just a "smooth cut-off" version of a classical characterization of Λ_α based on dyadic blocks of the Fourier series, used by Hardy and Littlewood.

(b) The "ray condition" (1.7) occurs in Hörmander's book [28], see equation (2.4.6) on p. 46, in connection with generalizations of a lemma of Friedrichs concerning equivalent norms in certain Sobolev-like spaces (Theorem 2.4.1 of [28]); for references to earlier work of Friedrichs and Hörmander see [28, p. 34]. Here the treatment is entirely geared to L^2 estimates, adequate for the applications in view; these works constitute one of the "streams" that prepared the way for later developments of the Fourier method, especially with regard to the researches of Peetre.

The modification of the above theory to deal with e.g. Sobolev spaces is quite clear: take e.g. the definition of $W^{m,p}(\mathbb{R}^d)$ when $m = 1$. To this end, let $\varphi^j \in C_o^\infty(\mathbb{R}^d)$ have Fourier transform $(\varphi^j)\hat{}(\xi) = \xi_j$ on some neighborhood of 0. Then $f \in L^p(\mathbb{R}^d)$ is in $W^{1,p}(\mathbb{R}^d)$ if and only if

$$\|f * \varphi^j_{(a)}\|_p = O(a) \ , \ j = 1,2, \ldots, d \ .$$

2. TOPIC (2) – APPROXIMATING HARMONIC FUNCTIONS BY HARMONIC POLYNOMIALS

Theorem 2.1. Let Ω be a bounded open simply-connected domain in $\mathbb{R}^d$, which has the segment property. Let u be a function harmonic in Ω belonging to $L^2(\Omega)$. Then, there exists a sequence of harmonic polynomials $\{p_n\}$ such that

$$\lim_{n\to\infty} \|u - p_n\|_{L^2(\Omega)} = 0 .$$

Remarks. We use the term "segment property" as in [2, p. 11]. Various weaker conditions would suffice in Theorem 2.1, see below. The interesting question to what extent harmonic functions can be replaced by solutions of other p.d.e. or systems of p.d.e. shall not be gone into here [22, 41].

The earliest theorem of the type of Theorem 2.1 seems to be that of Carleman [15] concerning L^2 approximation of an analytic function in a Jordan domain by polynomials. The earliest reference we can find to Theorem 2.1 is Babuška [3], where it is treated en passant (Theorem 6.3). Actually, Theorem 2.1 combines two separate features: (i) approximating u by a function $\tilde{u}$ harmonic on a neighborhood of $\bar{\Omega}$, and (ii) approximating $\tilde{u}$ by harmonic polynomials. To deal with step (i) alone can be done much more simply than in the following proof, cf. [41].

Outline of proof. Let $v \in L^2(\Omega)$ satisfy

$$\int_\Omega vp\, dx = 0 \text{ , all harmonic polynomials } p \tag{2.1}$$

Let now

$$W = \{\xi \in \mathbb{C}^d : \xi_1^2 + \ldots + \xi_d^2 = 0\} .$$

For every $x \in \mathbb{R}^d$ and $\xi \in W$, the harmonic function $x \mapsto \exp(-i\xi \cdot x)$ is uniformly approximable on Ω by harmonic polynomials, hence from (2.1),

$$\int_\Omega v(x) \exp(-i\xi \cdot x)\, dx = 0 \ , \quad \underline{\text{all}}\ \xi \in W \ ,$$

where $\xi \cdot x$ denotes the inner product in $\mathbb{C}^d$.

Define

$$\tilde{v}(x) = \begin{cases} v(x) & , \ x \in \Omega \\ 0 & , \ x \in \mathbb{R}^d \setminus \Omega \ . \end{cases}$$

Then, (2.2) says that the Fourier transform V of $\tilde{v}$, conceived as a function on $\mathbb{C}^d$ (where it is holomorphic) vanishes on the variety W. Since W is irreducible if $d \geq 3$ (and the union of two irreducible varieties if $d = 2$) it follows that

$$V(\xi) = (\xi_1^2 + \ldots + \xi_d^2)Y(\xi) \tag{2.3}$$

for a certain function Y holomorphic on $\mathbb{C}^d$ (see [19]; a simple elementary proof can also be given in the special case at hand).

Now, by the Paley-Wiener-Schwartz theorem [56, p. 161], V satisfies certain estimates which characterize the (complex!) Fourier transforms of tempered distributions with compact support (we could use here the classical Plancherel theorem, but prefer a more general argument which applies to the L^p case). It is now fairly easy to show from (2.3) that Y also satisfies these estimates, and so by the P-W-S theorem in the opposite direction Y <u>is the (complex) Fourier transform of a tempered distribution</u> y, <u>with compact support in</u> $\mathbb{R}^d$. From (2.3) we see that $\Delta y = -\tilde{v}$. Thus, y is a harmonic function outside $\bar{\Omega}$ and, vanishing on a neighborhood of ∞, it vanishes outside $\bar{\Omega}$. Thus, $\operatorname{supp} y \subset \bar{\Omega}$. Also, since $\Delta y \in L^2(\mathbb{R}^d)$, we have using standard theory of elliptic operators [5, 28]

$$y \in W^{2,2}_{loc}(\mathbb{R}^d) \ . \tag{2.4}$$

Now comes a crucial point in the proof: (2.4) together with supp $y \subset \bar{\Omega}$ imply $y \in W^{2,2}_0(\Omega)$. This deduction uses regularity assumptions about Ω. It can be proved by a simple modification of the argument in [2, Theorem 2.1]. (But much greater generality is possible, for a full discussion see Hedberg [23].)

The rest is easy. We can find a sequence of functions $\{\varphi_n\} \subset C^\infty_0(\Omega)$ which tend to $-y$ in the norm of the space $W^{2,2}(\Omega)$. Hence $\Delta\varphi_n \to -\Delta y = v$ in $L^2(\Omega)$, and so, for any harmonic function u in $L^2(\Omega)$,

$$\int_\Omega uv \, dx = \lim_{n\to\infty} \int u\Delta\varphi_n \, dx = 0 \ . \tag{2.5}$$

Since any $v \in L^2(\Omega)$ satisfying (2.1) satisfies (2.5), the conclusion of the theorem follows by duality.

What I find most remarkable in the above proof is this: at a crucial stage, we have to show that a function in $W^{2,2}(\mathbb{R}^d)$ with support in $\bar{\Omega}$, where Ω is a "reasonable" domain, is approximable by smooth functions with compact support in Ω. Thus, the original approximation problem involving harmonic polynomials was transformed into this novel one, involving only C^∞ regularity and therefore allowing local methods e.g. cut-off functions, partitions of unity, to be employed. Proving analyticity theorems with the aid of such methods seems to me a very remarkable and unexpected development. Technique has here moved worlds beyond such classical methods as those in [15]. For more on these developments see [21, 22, 41]. Reccommended survey articles are [24, 25]. For additional background concerning the harmonic analysis and holomorphy aspects ot the above proof see [4, 17, 28, 29, 32, 33, 37, 56].

3. TOPIC (3) – POLYNOMIAL APPROXIMATION OF SMOOTH FUNCTIONS IN $L^2(\Omega)$

Throughout this and the next two sections we shall always assume, unless specified otherwise, that Ω is bounded and satisfies an _(interior) cone property_ [1, p. 66]. Under these assumptions we now prove

Theorem 3.1. _Let_ $u \in C^\infty(\Omega)$. _Then, there exists_ $p \in P_{n-1}$ _such that_

$$\|u - p\|_{L^2(\Omega)} \le C \sum_{|\alpha|=n} \|D^\alpha u\|_{L^2(\Omega)} . \tag{3.1}$$

(Here $n \ge 1$ is an integer, P_{n-1} denotes the polynomials of degree at most $n - 1$, and C is a constant depending on Ω and n.)

From the abstract point of view, P_{n-1} enters the play as the intersection of the null-spaces of the operators $\{D^\alpha\}_{|\alpha|=n}$. Let $\{\alpha_1, \dots \alpha_m\}$ be some ordering of all the multi-indices of rank n. Then, the above theorem is equivalent to

Theorem 3.2. _Consider the map_

$$T : u \mapsto (D^{\alpha_1}u, \dots D^{\alpha_m}u) \tag{3.2}$$

from $W^{n,2}(\Omega)$ _to_ $L^2(\Omega)^m$. _The range of_ T _is closed_.

Indeed, that the first version implies the second is all but obvious, once we observe that (3.1) holds (by approximation) also for $u \in W^{n,2}(\Omega)$ and with the norm on the left side taken in the space $W^{n,2}(\Omega)$. In the other direction, ker T is P_{n-1} and T induces a bijective linear map from $W^{n,2}(\Omega)/\ker T$ onto the Banach space range T which is isometrically embedded in $L^2(\Omega)^m$. By Banach's theorem the inverse map is continuous, and this implies (3.1).

To prove the theorem (in its second version) we require two general theorems, which we now recall. First, however, some notation. We consider a general linear system of partial differential equations with constant coefficients:

$$\sum_{k=1}^{s} P_{jk}(D)u_k = f_j \ , \ j = 1, 2, \ldots r \tag{3.3}$$

where $D = (D_1, \ldots D_d)$ and the P_{jk} are polynomials in d letters. In general the f_j will be distributions in Ω.

There is an obvious necessary condition for the solvability of (3.3) (i.e. for the existence of distributions $u_1, \ldots u_s$ solving (3.3)). Let $Q_1, \ldots, Q_r$ be any polynomials such that

$$\sum_{j=1}^{r} Q_j P_{jk} = 0 \ , \ k = 1, 2, \ldots s \ . \tag{3.4}$$

Then clearly

$$\sum_{j=1}^{r} Q_j(D) f_j = 0 \ . \tag{3.5}$$

Thus, any r-tuple of polynomials satisfying the s identities (3.4) must satisfy (3.5). These are called the compatibility conditions for the system (3.3).

Theorem A (Ehrenpreis, Malgrange) *If Ω is convex and the compatibility conditions hold then there exist distributions u_k such that (3.3) hold. If the f_j are of finite order, the u_k may be taken of finite order.*

We refer to [17, 32, 33, 37] for proofs. A perhaps more accessible treatment of a less general version is [29, p. 198], Theorem 7.6.13.

Corollary. *Let $P_1, \ldots P_m$ denote homogeneous polynomials in $\xi = (\xi_1, \ldots, \xi_d)$ such that the system of equations*

$P_j(\xi) = 0$, $(j = 1, \ldots, m)$ has for $\xi \in \mathbb{C}^d$ only the root $\xi = 0$. Let Ω be any simply-connected open set in $\mathbb{R}^d$, and let $f_1, \ldots, f_m$ denote distributions in Ω such that the system

$$P_j(D)u = f_j \quad (j = 1, 2, \ldots, m) \tag{3.6}$$

satisfies the compatibility conditions (i.e. in this case, $P_j(D)f_k = P_k(D)f_j$ for all j, k). Then, there is a distribution u in Ω such that (3.6) holds.

Outline of proof. The condition on the roots of the system of equations $P_j(\xi) = 0$ $(j = 1, 2, \ldots, m)$ implies, by the Hilbert Nullstellensatz, that every polynomial in d variables of degree greater than or equal to some integer N belongs to the ideal generated by the P_j. Consequently all solutions of the homogeneous system $P_j(D)u = 0$ $(j = 1, 2, \ldots, m)$ are polynomials of degree less than N, in particular the solutions of the homogeneous system comprise a finite-dimensional vector subspace of $C^\infty(\Omega)$. Using this it is easy to check that if we have solutions to (3.6) in two open subsets U_1, U_2 of Ω we can piece together a solution in $U_1 \cup U_2$. Since the E-M theorem can be applied to any ball in Ω, we deduce that (3.6) is solvable in any open $U \subset \Omega$ which is contained in a finite union of balls, hence in any open $U \subset \Omega$ with compact closure. Considering now a nested sequence $U_n \subset \Omega$ each of which has compact closure, such that $\bigcup_{n=1}^{\infty} U_n = \Omega$, we can construct for each n a solution u_n in Ω_n; these u_n can now be adjusted so that $u_m = u_n$ on Ω_n when $m > n$, which implies the desired conclusion.

We require next a general theorem of Smith. Whereas the E-M theorem gives a sufficient condition for a system of PDE to have a distribution solution, Smith's theorem is a so-called coercivity theorem: it tells us that (under suitable restrictions, a very strong kind of ellipticity) any existing solution must have a certain degree of regularity, depending on the amount of regularity possessed by the f_j.

Theorem B (Smith [47]) Let $\Omega \subset \mathbb{R}^d$ be bounded and satisfy an interior cone condition. Suppose the polynomials P_{jk} in (3.3) are homogeneous and satisfy

(i) $\deg P_{jk} = m_j - n_k$ for certain integers m_j, n_k

(ii) For every $\xi \in \mathbb{C}^d$, $\xi \neq 0$ the matrix $\|P_{jk}(\xi)\|$ has rank s (the number of columns in the matrix, or of unknown functions). If then $f_j \in W^{q-m_j,p}(\Omega)$ $(j = 1, 2, \ldots r)$ and $\{u_k\}$ are distributions in Ω satisfying (3.3) then $u_k \in W^{q-n_k,p}(\Omega)$. This conclusion holds for every integer q and $1 < p < \infty$.

Corollary. Under the above hypotheses, the map

$$T: (u_1, u_2, \ldots, u_s) \mapsto \left(\sum_{k=1}^{s} P_{1k}(D)u_k, \ldots, \sum_{k=1}^{s} P_{rk}(D)u_k\right)$$

from

$$X \triangleq W^{q-n_1,p}(\Omega) \times \ldots \times W^{q-n_s,p}(\Omega)$$

to

$$Y \triangleq W^{q-m_1,p}(\Omega) \times \ldots \times W^{q-m_r,p}(\Omega)$$

has a closed range.

Deduction of the Corollary. Let $(v_1, \ldots v_r) = v$ be a point of Y which is in the closure of the range of T. We have to show that the system $Tu = v$ has a solution $u = (u_1, \ldots, u_s)$ in X. Now, since v lies in the closure of range T with respect to the Y topology it does so with respect to the weaker distribution topology, and therefore the system

$$\sum_{k=1}^{s} P_{jk}(D)u_k = v_j \quad (j = 1, 2, \ldots r) \tag{3.7}$$

satisfies the compatibility conditions. Therefore, by Theorem A, for each ball in Ω there is a distribution solution $u = (u_1, \ldots, u_s)$. Now, by a reasoning like that in the Corollary to Theorem A (adapted however to _systems_, cf. [38, 47]) we deduce that (3.7) has a distribution solution u in Ω. By Theorem B, u lies in X, hence v is in range T, which therefore is closed.

As a special case,

Corollary 2. _The hypotheses are the same as in the corollary to Theorem A, plus the assumptions that_ Ω _is bounded and satisfies an interior cone condition_. _Then, the map_

$$T: u \mapsto (P_1(D)u, \ldots P_m(D)u)$$

from $W^{n,p}(\Omega)$ _to_ $W^{n-d_1,p}(\Omega) \times \ldots \times W^{n-d_m,p}(\Omega)$ _has closed range, for every integer_ n _and_ $1 < p < \infty$. _Here_ d_j = _degree of_ P_j.

The theorem of this section, Theorem 3.2, evidently follows at once. The following generalization of Corollary 2 is also of interest:

Conjecture 3.3 (_W. Abramczuk, unpublished_.) _The conclusion of Corollary 2 remains valid even if the hypotheses on the_ P_j _are weakened to read_: _the system of equations_ $P_j(\xi) = 0$ $(j = 1, 2, \ldots m)$ _has no real solution except_ $\xi = 0$.

As an example, consider the case of _one_ polynomial, $P(\xi) = \xi_1^2 + \ldots + \xi_d^2$. Then $P(D) = \Delta$, the Laplace operator. The conjucture asserts that Δ, as an operator from $W^{2,p}(\Omega)$, to $L^p(\Omega)$, $1 < p < \infty$, has a closed range (in fact, one can easily show directly, in this simple case, that Δ is surjective). The analogue of Theorem 3.1 corresponding to this is

Corollary. *If* Ω *is bounded and satisfies an interior cone condition, and* $u \in W^{2,2}(\Omega)$, *there is a harmonic function* $h \in L^2(\Omega)$ *such that*

$$\|u - h\|_{L^2(\Omega)} \leq C \|\Delta u\|_{L^2(\Omega)}$$

where C *depends only on* Ω.

One weakness with the methods we have used in this section is, it is unclear what happens when $p = 1$ or $p = \infty$. The constructive approach in [16] allows these cases to be handled. Thus, in [16] it is proved that, provided Ω obeys some mild regularity condition, each $f \in C^\infty(\Omega)$ admits a representation

$$f(x) = p_{n-1}(x) + \sum_{|\alpha|=n} \int_\Omega k_\alpha(x, y)(D^\alpha f)(y)dy . \tag{3.8}$$

Here p_{n-1} is a polynomial of degree at most $n - 1$, given constructively by a linear averaging process applied to the $(n - 1)^{th}$ order Taylor partial sums of f with respect to a variable initial point in Ω, and the k_α are explicitly known kernels. Applying Hölder's inequality to (3.8) yields Theorem (3.1). Again, if R denotes a linear functional annihilating polynomials of degree $n - 1$, we get from (3.8) a Peano-type representation formula

$$Rf = \sum_{|\alpha|=n} \int_\Omega h_\alpha(x, y)(D^\alpha f)(y)\, dy \tag{3.9}$$

where $h_\alpha(x, y) = R_x k_\alpha(x, y)$. These kernels h_α appear to have essentially optimal regularity properties among kernels which represent R as in (3.9). Results of this type were proved nonconstructively in [45]. We remark finally that integral identities related to those in [16] were employed in [6, 11, 49]; see also [34].

4. TOPIC (4) – A CLOSURE PROBLEM FROM THEORETICAL TOMOGRAPHY

In this section Ω denotes a bounded open convex domain in $\mathbb{R}^2$. Given $0 \leq \theta < \pi$, R_θ denotes the set of $u \in L^2(\Omega)$ whose restriction to almost every line making an angle θ with the positive x-axis is a.e. constant. In other terms, R_θ is the set of weak L^2 solutions of the equation

$$(\cos\theta)\frac{\partial u}{\partial x} + (\sin\theta)\frac{\partial u}{\partial y} = 0 .$$

Clearly, R_θ is a closed subspace of $L^2(\Omega)$.

Theorem 4.1 If $0 \leq \theta_1 < \theta_2 < \dots < \theta_n < \pi$, $R_{\theta_1} \oplus \dots \oplus R_{\theta_n}$ is closed in $L^2(\Omega)$.

Proof [38]. Let

$$P_j(\xi) = (\cos\theta)\xi_1 + (\sin\theta)\xi_2 , \quad j = 1, 2, \dots n .$$

Consider the map

$$T: (u_1, u_2, \dots, u_n) \mapsto \left(P_1(D)u_1, \dots P_n(D)u_n, u_1 + \dots + u_n\right)$$

from $X \triangleq L^2(\Omega)^n$ to $Y \triangleq W^{-1,2}(\Omega)^n \times L^2(\Omega)$. The matrix describing T (cf. the preceding section) is

$$\begin{bmatrix} P_1(\xi) & 0 & \dots & 0 \\ 0 & P_2(\xi) & \dots & 0 \\ & & \ddots & \vdots \\ & & & P_n(\xi) \\ 1 & 1 & \dots & 1 \end{bmatrix}$$

Now, the polynomials P_j are pairwise linearly independent, so if $\xi \in \mathbb{C}^2$ is not $(0, 0)$, at most one of the $P_j(\xi)$ can vanish. Consequently the rank of this matrix is n, and we can apply the Corollary to Theorem B to deduce that range T is closed. Therefore, by Banach's theorem, T is an open map, hence it maps the closed set $R_1 \times R_2 \times \ldots \times R_m \subset X$ onto a closed set. Hence the set of all points $(0, 0, \ldots 0, v)$ where $v \in R_1 \oplus \ldots \oplus R_n$ is closed in Y, which implies $R_1 \oplus \ldots \oplus R_n$ is closed in $L^2(\Omega)$.

5. TOPIC (5) – AN INEQUALITY DUE TO K. FRIEDRICHS

The following generalizes the main theorem in [18].

Theorem 5.1. *Let Ω be a bounded simply connected open set in $\mathbb{R}^2$ satisfying an interior cone condition. There is a constant C depending only on Ω such that*

$$\iint_\Omega u^2 \, dx \, dy \leq C \iint_\Omega v^2 \, dx \, dy \tag{5.1}$$

holds for every analytic function $f = u + iv$ on Ω which is square-integrable and has mean value zero.

Remarks. The following argument yields also the analogous L^p inequality, $1 < p < \infty$ with $C = C(\Omega, p)$. The restriction to simply-connected Ω is not essential: by a technique developed in [46] one can extend the above theorem to domains which are finite unions of such Ω as figure in Theorem 5.1.

Proof of theorem. Let H denote the Hilbert space $L^2(\Omega)$, A the closed subspace consisting of analytic functions, and $A^\square = \{f \in H : \bar{f} \in A\}$. Then $A \cap A^\square$ is the subspace of constant functions. It suffices to prove

$$A \oplus A^\square \text{ is closed in } H \tag{5.2}$$

since that is equivalent to A and $A^\square$ making a non-zero angle, i.e. to the existence of a constant $\rho < 1$ such that for all f, g in A such that $\iint_\Omega f \, dx \, dy = \iint_\Omega g \, dx \, dy = 0$ one has

$$|\iint_\Omega fg\,dx\,dy|^2 \le \rho \iint_\Omega |f|^2 dx\,dy \iint_\Omega |g|^2 dx\,dy \,.$$

But if this holds then, taking $f = g$ we have

$$\mathrm{Re} \iint_\Omega f^2 dx\,dy \le \rho \iint |f|^2 dx\,dy$$

for all $f \in A$ of mean value zero, which implies (5.1). Hence, it suffices to prove (5.2). But this is a consequence of the Corollary to Theorem B, the proof being thus nearly identical to that of Theorem 4.1. We have namely to prove that the map

$$(F, G) \mapsto \left(\frac{\partial F}{\partial \bar{z}}, \frac{\partial G}{\partial z}, F + G\right)$$

from $L^2(\Omega) \times L^2(\Omega)$ to $W^{-1,2}(\Omega) \times W^{-1,2}(\Omega) \times L^2(\Omega)$ has a closed range. If we write $F = u_1 + iv_1$, $G = u_2 + iv_2$ where u_j, v_j are real this comes down to showing that the map

$$(u_1, v_1, u_2, v_2) \mapsto \left[\frac{1}{2}\left(\frac{\partial u_1}{\partial x} - \frac{\partial v_1}{\partial y}\right), \frac{1}{2}\left(\frac{\partial u_1}{\partial y} + \frac{\partial v_1}{\partial x}\right), \frac{1}{2}\left(\frac{\partial u_2}{\partial x} + \frac{\partial v_2}{\partial y}\right), \frac{1}{2}\left(\frac{\partial u_2}{\partial y} - \frac{\partial v_2}{\partial x}\right), u_1 + u_2, v_1 + v_2\right]$$

has closed range. The matrix describing this map is, after multiplying through by 2,

$$\begin{bmatrix} \xi_1 & -\xi_2 & 0 & 0 \\ \xi_2 & \xi_1 & 0 & 0 \\ 0 & 0 & \xi_1 & \xi_2 \\ 0 & 0 & \xi_2 & -\xi_1 \\ 2 & 0 & 2 & 0 \\ 0 & 2 & 0 & 2 \end{bmatrix}$$

and we have only to check that its rank is 4 for complex ξ_1, ξ_2 not both zero. Now, if $\xi_1^2 + \xi_2^2 \ne 0$, the first four rows are linearly independent, whereas if $\xi_1^2 + \xi_2^2 = 0$ but $\xi_1 \ne 0$ the first, third, fifth and sixth rows are.

REFERENCES

1. Adams, R.A., Sobolev Spaces, Academic Press, 1975.

2. Agmon, S., Lectures On Elliptic Boundary Value Problems, Van Nostrand Math. Studies 2, 1965.

3. Babuška, I., Stability of the domain of definition with respect to the fundamental problems in the theory of p.d.e. ..., part II, Czechoslovak Math. J. 11 (86) (1961) 165-203 (Russian).

4. Berenstein, C.A. and M. Dostal, Analytically Uniform Spaces And Their Applications To Convolution Equations, Lecture Notes In Math., vol. 256, Springer-Verlag, 1972.

5. Bers, L., F. John and M. Schechter, Partial Differential Equations (Part II, Elliptic Equations), Wiley-Interscience, 1964.

6. Besov, O.V., V.P. Ilyin, and S.M. Nikolski, Integral Representation of Functions and Embedding Theorems, Nauka, Moscow, 1975 (Russian).

7. Boman, J., Equivalence of generalized moduli of continuity, Arkiv för mat. 1980 (in press).

8. ——— , On the range of the X-ray transform, seminar held at Stockholm University, February 1980 (written version in preparation).

9. ——— and H.S. Shapiro, Comparison theorems for a generalized modulus of continuity, Arkiv för mat. 9 (1971) 91-116.

10. Bramble, J., and S. Hilbert, Estimation of linear functionals on Sobolev spaces with applications to Fourier transforms and spline interpolation, SIAM J. Numer. Anal. 7 (1970) 112-124.

11. Burenkov, V.I., Sobolev's integral representation and Taylor's formula, Trudy Mat. Inst. Steklova 131 (1974) 33-38.

12. Butzer, P., and R. Nessel, Fourier Analysis And Approximation, Birkhäuser Verlag, 1971.

13. Calderón, A., Intermediate spaces and interpolation, the complex method, Studia Math. 24 (1964) 113-190.

14. ——— and A. Zygmund, On the existence of certain singular integrals, Acta Math. 88 (1952) 85-139.

15. Carleman, T., Über die Approximation analytischer Funktionen durch lineare Aggregate von vorgegebenen Potenzen, Ark. Mat. Astr. Fys. 17 (1923) 1-30.

16. Dupont, T., and R. Scott, Polynomial approximation of functions in Sobolev spaces, preprint, 1979.

17. Ehrenpreis, L., Fourier Analysis In Several Complex Variables, Wiley-Interscience, 1970.

18. Friedrichs, K., On certain inequalities and characteristic value problems for analytic functions and for functions of two variables, Trans. A.M.S. 41 (1937) 321-364.

19. Gunning, R., and H. Rossi, Analytic Functions Of Several Complex Variables, Prentice-Hall, 1965.

20. Hamaker, C., and D. Solmon, The angles between the null spaces of X-rays, J. Math. Anal. Appl. 62 (1978) 1-23.

21. Havin, V.P., Approximation in the mean by analytic functions, Dokl. Akad. Nauk SSSR 178 (1968) 1025-1028 (Russian).

22. Hedberg, L.-I., Approximation in the mean by solutions of elliptic equations, Duke Math. J. 40 (1973) 9-16.

23. ———, Two approximation problems in function spaces, Ark. för mat. 16 (1978) 51-81.

24. ——— , Approximation in L^p by analytic and harmonic functions, Proc. Symp. Pure Math. 35, Part 1 (1979) 377-382.

25. ——— , Spectral synthesis and stability in Sobolev spaces, in proceedings of the "Special Year In Harmonic Analysis" held at Univ. of Maryland 1979, Lecture Notes in Math., Springer-Verlag, 1980 (to appear).

26. Heidemann, N.J.H., Duality and fractional integration in Lipschitz spaces, Studia Math. 50 (1974) 65-85.

27. Herz, C., Lipschitz spaces and Bernstein's theorem on absolutely convergent Fourier transforms, J. Math. Mech. 18 (1968) 283-324.

28. Hörmander, L., Linear Partial Differential Operators, Third Revised Printing, Springer-Verlag, 1969.

29. ——— , An Introduction To Complex Analysis In Several Variables, Van Nostrand, 1966.

30. Lax, P., A stability theorem for solutions of abstract differential equations ..., Comm. Pure Appl. Math. 9 (1956) 747-766.

31. Logan, B.F., and L. Shepp, Optimal reconstruction of a function from its projections, Duke Math. J. 42 (1975) 645-659.

32. Malgrange, B., Existence et approximation des solutions des équations aux dérivées partielles et les équations de convolution, Ann. Inst. Fourier 6 (1955) 271-355.

33. ——— , Sur les systèmes differentiels à coefficients constants, Sém. sur les équations aux derivées partielles, Collège de France, 1961-2.

34. Meinguet, J., Structure et estimation de coefficients d'erreur, RAIRO Analyse Numérique, v. 11, 1977, pp.355-368.

35. Morrey, C., Jr., Multiple Integrals In The Calculus Of Variations, Springer-Verlag, 1966.

36. Nikolski, S.M., Approximation of Functions Of Several Variables And Embedding Theorems, Nauka, Moscow, 1969 (Russian).

37. Palamodov, V.P., Linear Differential Operators With Constant Coefficients, translated from Russian, Springer-Verlag, 1970.

38. Pedersen, B., K. Smith and D. Solmon, Sums of plane waves, and the range of the Radon transform, Math. Annalen 243 (1979) 153-161.

39. Peetre, J., New Thoughts On Besov Spaces, Duke Univ. Press, 1976.

40. ——— , Espaces d'interpolation et théorème de Soboleff, Ann. Inst. Fourier 16 (1966) 279-317.

41. Polking, J.C., Approximation in L^p by solutions of elliptic partial differential equations, Amer. J. Math. 94 (1972) 1231-1244.

42. Shapiro, H.S., Smoothing And Approximation Of Functions, Van Nostrand Reinhold, 1969.

43. ——— , Topics In Approximation Theory, Lecture Notes In Math., Vol. 187, Springer-Verlag, 1971.

44. ——— , A Tauberian theorem related to approximation theory, Acta Math. 120 (1968) 279-292.

45. ——— , Integral representation of remainder functionals in one and several variables, in "Multivariate Approximation", D.S. Handscomb ed., Academic Press, 1978 pp. 69-82.

46. ——— , Some inequalities for analytic functions integrable over a plane domain, research report, Royal Inst. of Technology, 1979 (to appear in the proceedings of the conference on function spaces and approximation held in Gdansk, Poland, 1979).

47. Smith, K., Formulas to represent functions by their derivatives, Math. Annalen 188 (1970) 53-77.

48. Sobolev, S.L., Applications Of Functional Analysis In Mathematical Physics, Amer. Math. Soc. translation, 1963.

49. ——— , Introduction To The Theory Of Cubature Formulae, Nauka, Moscow, 1974 (Russian).

50. Stein, E., Singular Integrals And Differentiability Properties Of Functions, Princeton 1970.

51. ——— and A. Zygmund, Boundedness of translation invariant operators on Hölder and L^p spaces, Annals of Math. 85 (1967) 337-349.

52. Svensson, L., When is the vector sum of closed subspaces closed? An example from computerized tomography, research report, Royal Inst. of Technology, 1980.

53. Timan, A.F., Theory Of Approximation Of Functions Of A Real Variable, translated from Russian, Pergamon, 1963.

54. Torchinsky, A., Singular integrals in the spaces $\Lambda(B, X)$, Bull. Amer. Math. Soc. 78 (1972) 1015-1019.

55. ——— , On a mean value inequality, Bull. Amer. Math. Soc. 81 (1975) 950-952.

56. Yoshida, K., Functional Analysis, Second printing corrected, Springer-Verlag, 1966.

57. Zygmund, A., On the preservation of classes of functions, J. Math. Mech. 8 (1959) 889-896.

Comments and Errata

Page 4, line -3. Here and later on, in Theorem 4.1 on page 29 and in equation (5.2) on page 30, $\oplus$ denotes vector sum of subspaces which may have a larger intersection than $\{0\}$ (it has been better to write $+$).

Page 19. Theorem 2.1. Here, and in the Corollary on page 24, by "simply connected" open set in $\mathbf{R}^d$ is meant one with connected complement (this deviates from the standard usage when $d \neq 2$).

Page 29, line -1 and top of page 30. R_j denotes here R_{θ_j}. Also, there is a gap in the argument: T being an open map, it does not automatically map closed sets to closed sets. We have, however, the following:

Proposition 1. Let T *be a linear map from the Banach space* X *to the Banach space* Y *with closed range. Let* S *denote a closed subspace of* X. *Then* TS *is closed if and only if the image of* S *in* $X/\ker T$ *under the natural map is closed. Necessary and sufficient for this is that* $S + \ker T$ *is closed in* X.

Proof. Without loss of generality we can assume Range $T = Y$. We have $T = \tilde{T} \circ \pi$ in accordance with the diagram

$$T: X \xrightarrow{\pi} X/\ker T \xrightarrow{\tilde{T}} Y$$

where π denotes the natural map of X to the quotient space $X/\ker T$ and $\tilde{T}$ is defined by $\tilde{T}(x + \ker T) = Tx$. Since $\tilde{T}$ is a homeomorphism, we see that TS is closed if and only if πN is closed, where N denotes ker T. Therefore, Proposition 1 follows from

Proposition 2. *Let X be a Banach space and N, S any closed subspaces thereof. Let π denote the natural map of X to X/N. Then, πS is closed in X/N if and only if N + S is closed in X.*

Proof. It is easily verified that $\pi^{-1}(\pi S) = N + S$. Since π is continuous, N + S is closed if πS is. Conversely, if N + S is closed then $(N + S)^c$ is open. But $(N + S)^c = \pi^{-1}((\ S)^c)$. Since π is an open map, $\pi\pi^{-1}((\pi S)^c) = (\pi S)^c$ is open, hence πS is closed.

A consequence of Proposition 1 is

Corollary. *Under the hypotheses of Proposition 1, if ker T has finite dimension, or finite codimension, then TS is closed in Y.*

Since, in the proof of Theorem 4.1, T has finite-dimensional kernel (this is a corollary of Smith's results in the argument given is thus justifiable.

Afterward. On the basis of criticism of the present paper which has been circulated in manuscript, I see I have perhaps not sufficiently emphasized two points:

(i) The aim of this paper is purely expository. I claim no originality for the ideas herein. Conversely, those named on page 5 bear no responsibility for possible bias or errors in the presentation.

(ii) In topics (2) and (3) a model problem has in each case been chosen (namely, Theorem 2.1 and Theorem 3.1) as a vehicle to illustrate certain powerful methods. These methods would not be called for if only the model problem were at stake; there are quite elementary proofs available for each of these theorems. However, generalizations of these theorems (e.g. to systems of p.d.e. in the case of Theorem 2.1, or to general differential operators in the case of Theorem 3.1 (cf. Corollary 2 on page 26) really seem to require the heavy machinery.

THE DEVELOPMENT OF NON LINEAR SPLINES AND THEIR APPLICATIONS

Helmut Werner

Institut für Numerische und instrumentelle Mathematik
und Rechenzentrum
Westfälische Wilhelms Universität
Münster / Westfalen, Germany

1. Introduction

Splines may be considered an offspring of approximation theory. Although the antecedents of splines appeared in the literature more than four decades ago, it is generally agreed that the formal birth of splines occurred in the 1940's and it is closely associated with the name of Professor I. J. Schoenberg. They have grown into a mighty subject with an almost uncountable number of papers appearing each year. To verify this one is referred to the bibliography of Schurer and van Rooij on this field.

Let us first define what we mean by a spline in the context of this paper.

Copyright © 1980 by Academic Press, Inc.
All rights of reproduction in any form reserved.
ISBN: 0-12-171050-5

Definition: Let $I := [x_-, x_+]$ denote a real interval and $\{x_o, \dots, x_N\}$ with $x_j < x_{j+1}$ denote a set of points of I, sometimes called knots. They may be fixed or variable. Let k and m denote positive numbers with $m > k$. Then a spline u is a function from $C^k(I)$ which satisfies

$$u\Big|_{[x_j, x_{j+1}]} = t(x; x_j, c_o, \dots, c_m) \tag{1.1}$$

for every $j = 0, \dots, N-1$.

Here t is a function depending on $x \in I$ and the specified parameters which vary over an admissible domain*) It is convenient in particular for theoretical considerations to assume that the parameters coincide with the values and derivatives of the spline at one of the end-points. E.g. we set

$$c_i = \frac{d^i}{dx^i} t(x; x_j, c_o, \dots, c_m)\Big|_{x = x_j} \quad \text{for } i=0, \dots, m . \tag{1.2}$$

Frequently the functions t are smoother than the function u which is only of class C^{m-d} globally , $d =$ defect.

In order to satisfy the global smoothness condition imposed on u over I, the parameters of course must be coupled. Further conditions will be needed to complement these relations so that we finally arrive at a set of equations and conditions, the number of which matches the number of parameters and will lead, hopefully, to a well determined problem.

The family of splines thus depends upon the chosen function t. If t is a linear function of the parameters

*) One could even take different generating functions in different subintervals $[x_j, x_{j+1}]$.

$c_o,\ldots,c_m$ the resulting family is called linear, in spite of the fact that working with these splines leads to many phenomena that are typical for nonlinear problems. Obviously we will speak of nonlinear splines if t is nonlinear in the designated parameters.

Examples of t are rational functions in x and in the parameters, exponential functions, and algebraic functions as we will see below. It is clear that this freedom of selection allows one to very closely pattern the spline family to the function to be approximated.

Historically the use of nonlinear splines started out from interpolation and Tschebyscheff-approximation problems. In a number of papers by Arndt, Braess, Loeb, Meder, Schaback, Schomberg, and Werner, as early as 1973, the parameters were usually chosen so that the defect of the order of differentiability was only 1, in other words, $k = m-1$. An attempt by Baumeister to characterize nonlinear splines by the use of variational methods has not produced as general a family as the one described above and doesn't appear to have opened up new vistas. A more heuristic approach to nonlinear splines is given by Späth. He uses them in connection with smoothing of data by means of graphical devises, e.g. optical displays of computers.

An attempt to use nonlinear splines in connection with the numerical integration of functions which are singular or have a singularity close to the interval of integration has been reported on in the papers by Werner, Wuytack, and Zwick.

In 1967 Loscalzo and Talbot used polynomial splines in solving ordinary differential equations numerically. Later

Schoenberg and Varga returned to this topic. In Münster this idea was considered again by Arndt, Runge, Werner and applied in the context of nonlinear splines to obtain those results and generalizations which we will focus on in this paper. In talks at conferences in Copenhagen and Calgary a summary on the state of the art with respect to interpolation, approximation, and numerical integration was given. Therefore we will concentrate here on the treatment of differential equations by means of nonlinear splines. Particular attention will be given to the asymptotic behavior of the errors and to extrapolation techniques.

2. Numerical Treatment of Ordinary Differential Equations by Means of Splines

Consider the following problem

$$\begin{aligned} y' &= f(x,y) \\ y(x_o) &= y_o \quad , \end{aligned} \tag{2.1}$$

here we assume that $f : G \to \mathbb{R}$ is given in an appropriate domain and has suitable differentiability properties. Furthermore the point (x_o, y_o) should be contained in the domain of f. We look for a function $y : I \to \mathbb{R}$ which is continuously differentiable, solves the differential equation and possesses the prescribed initial value. For simplicity we will usually assume $I := [0,1]$.

Numerical methods to solve differential equations can be split into discrete ones and continuous ones. Here we are concerned with the continuous methods, because we do not only calculate the approximation at some fixed points of the interval but we describe functions which are approximations of y. Using splines has the following advantages as compared to discrete methods:

1. The family of splines used can be adjusted to a priori known properties of the solutions of the differential equation at hand, for instance to the singularities of those solutions.
2. We gain high accuracy with comparatively small numerical work, get a stable numerical procedure, comparatively small round off errors and thrifty use of calculation time. Asymptotic expansions make it feasible to apply extrapolation techniques.
3. We cannot pursue here but remark that use of splines is particularly advisable in cases in which one has to use the solution at intermediate points, not only at the grid points. Such applications arise in connection with difference differential equations when the delay is not fixed but may vary in a way which is not a priori known.

We briefly outline the proposed method. We assume that in each interval the splines depend on four parameters. The splines should be of class C^2. Furthermore it is no restriction to assume that

$$x_o = 0 \quad \text{and} \quad x_j = j \cdot h \tag{2.2}$$

for a given h, $h = {}^1/_N > 0$. Let $I_j := [x_{j-1}, x_j]$. The algorithm proceeds in the following way.

Start: Define

$$u(x_o,h) = y_o \tag{2.3}$$
$$D^i u(x_o,h) = y^i(x_o) = \frac{d^{i-1}}{dx^{i-1}} f(x,y(x))\Big|_{x=x_o} \qquad i=1,2.$$

These initial data are given in terms of f and the initial condition only.

Iterative definition of u in the interval I_{j+1}. Assume that u_j, u_j', u_j'' are already given by initial data or the most recent set of parameters. To determine u(x,h) in I_{j+1} find the last parameter u_j''' from the equation

$$u'(x_{j+1},h) = f(x_{j+1},u(x_{j+1},h)). \tag{2.4}$$

The resulting spline segment u(x,h) in I_{j+1} then defines u_{j+1}, u_{j+1}', u_{j+1}'' so that the iteration can be continued unless the stopping condition is satisfied:

Either I is covered or the graph (x,u(x,h)) is not in the domain of f for $x \in I_{j+1}$.

3. Linear Families

In case that t is of the class of cubic polynomials the theory was worked out by Loscalzo and Talbot [3]. Generalizations were given by Schoenberg and Varga 1969 [7]. We summarize the results on linear splines as they are given in a definitive study by Mülthei, 1979 [4].

Consider polynomial splines that are made up piecewise of polynomials of degree m and that are globally of the smoothness class C^q where $q = m-d$, and let d denote the defect with respect to differentiability of the spline. In classical cases $d = 1$.

The previously stated four parameter method generalizes in the following way.

On the left-hand side of the interval I_{j+1} $q+1$ values of the function and its derivatives are prescribed by the initial conditions or the values brought forward from the adjacent left-hand interval I_j

$$u_j^{(k)} = u^{(k)}(x_j - 0,h) \quad \text{for } k = 0,\ldots,q \qquad (3.1).$$

Then d equations for derivatives at the right-hand end-point of the interval are used to determine the remaining parameters

$$u^{(k)}(x_{j+1} - 0,h) = D^{k-1} f(x,u(x,h))\Big|_{x \to x_{j+1}} \qquad (3.2)$$

for $k = 1,\ldots,d$.

Under appropriate regularity conditions concerning f and for sufficiently small h the $m+1$ relations uniquely determine a polynomial in I_{j+1}. With above notations we have the following result, comp. Mülthei [4].

i) The above sketched method is divergent if

$$m \geq 2d + 2 .$$

For d=1 this was already observed by Loscalzo and Talbot.

ii) Case $m \leq 2d+1$.

1) If $m \leq 2d$, therefore $q \leq d$, then one has some kind of convergent implicit one step method.

$$u_{j+1} - u_j$$

is a linear combination of the $q+1$ data at the left and the d values $f(x,y(x)), \frac{d}{dx} f(\ldots), \ldots$ at the right end-point of the interval I_{j+1}.

The order of consistency is given at the knots by

$$y^{(i)}(x_j) - u^{(i)}(x_j,h) = O(h^m)$$

for $i = 0,\ldots,d$ and $\forall j$,

and for the whole interval by

$$y^{(i)}(x) - u^{(i)}(x,h) = O(h^{\min(m,m+1-i)})$$

for $i = 0,\ldots,m$ and $\forall x \in I$.

2) If $m = 2d+1$, d odd, then one has an implicit two step method which expresses $u_{j+1} - u_{j-1}$ as a combination as described before with evaluations at x_{j-1}, x_j, x_{j+1}. This time the order of consistency is at the knots

$$y^{(i)}(x_j) - u^{(i)}(x_j,h) = O(h^{m+1})$$

for $i = 0,\ldots,d$ and $\forall j$,

for the whole interval

$$y^{(i)}(x) - u^{(i)}(x,h) = O(h^{m+1-i})$$

for $i = 0,\ldots,m$ and $x \in I$.

These methods may be looked upon as generalizations of the

Milne-Simpson method.

If d is even one has to use special arguments that are also developed by Mülthei.

4. Non Linear Families

Similar results were obtained for the nonlinear splines in the four parameter case by H. Werner 1979 [10]. A generalization to higher order differential equations was given by H. Arndt 1979 [1]. One has to solve the initial value problem

$$y^{(n)} = f(x,y,y',\dots,y^{(n-1)}) \tag{4.1}$$

with the initial values

$$y^{(i)}(x_o) = y_o^i \qquad \text{for} \quad i = 0,\dots,n-1 \tag{4.2}$$

under appropriate technical conditions. Then the previously stated algorithm can be generalized in an obvious way.

We will use a non linear spline that depends upon $k+2$ parameters, $k \geq n$, and that is globally of class $C^k[I]$. The procedure is the same as before. To start, $k+1$ parameters are obtained from (4.2) and (4.1) and its derivatives (if necessary). The last parameter is used to satisfy

$$u^{(n)}(x_{j+1};x_j,c_o,\dots,c_{k+1}) = f(x_{j+1},u(x_{j+1},\dots),\dots,u^{(n-1)}(x_j,\dots)) \tag{4.3}$$

at first for $j=0$, and then iteratively for $j=1,\dots,N-1$.

In these last cases the parameters $c_o, \ldots, c_k$ for I_{j+1} are obtained by the continuity of the derivatives of u up to the order k, evaluated at the knot x_j. Stopping takes place if the interval I is covered or the solution leaves the domain of f.

Arndt proves the following facts [1].

i) If $k > n+1$ then the method is instable.

ii) If $k \leq n+1$ we obtain (under appropriate technical assumptions) the following orders of convergence:

$$
\begin{aligned}
u^{(i)}(x,h)-y^{(i)}(x) &= O(h^{\min(2,n+2-i)}) \\
&\qquad \text{for } i=0,\ldots,n+2 \quad \text{if } k=n\ , \\
u^{(i)}(x,h)-y^{(i)}(x) &= O(h^{\min(4,n+3-i)}) \\
&\qquad \text{for } i=0,\ldots,n+3 \quad \text{if } k=n+1\ .
\end{aligned}
\tag{4.4}
$$

These results are the generalization of those facts presented in Werner [10] for k = 2. They correspond to the case that in (4.1) the function f only depends upon x and $y^{(n-1)}$ and one secures the lower order derivatives and u itself from $u^{(n-1)}$ by integrations.

In (4.1) these functions act like completely continuous pertubations.

We now return to the question which type of splines one should choose.

5. Properties of the Solution of Algebraic Differential Equations

To be specific we will consider here only nonlinear differential equations of the form

$$y' = p(x,y) \; , \; p(x,y) = y^m p_m(x) + y^{m-1} p_{m-1}(x) + \ldots + y \cdot p_1(x) + p_o(x) . \quad (5.1)$$

By the classical theory of Painlevé, compare e.g. Hille [2], one gets the following expansion for a solution that is singular at a point x*.

$$y(t) = c \cdot t^{\mu} (1 + c_1 t^{\gamma} + c_2 t^{\gamma+|\mu|} + c_3 t^{\gamma+2|\mu|} + \ldots) \quad (5.2)$$

with $t = x - x^*$.

Substituting the series expansion of the solution and its derivative into (5.1) one finds the following values for the exponents and the coefficients

$$\mu = \frac{1}{1-m}$$

$$c^{m-1} = \frac{\mu}{p_m(x^*)} \; , \quad \text{if} \quad p_m(x^*) \neq 0 \quad (5.3)$$

Furthermore, γ will be a multiple of $|\mu|$.

We see that for $m > 1$ the exponent μ will be negative, indeed

$$-1 \leq \mu < 0.$$

6. The Selection of the Spline Family

The above form of a solution of the differential equation in the neighborhood of a singularity x^* suggests the following choice of t for the generation of the splines:

$$u(x_j+z,h) = t(z;x_j,u,\dots,u''') \quad \text{for} \quad 0 \le z \le h \quad \text{with}$$

$$t(z;x,u,u',u'',u''') = u + u' \cdot \frac{b}{\alpha}\left[(1+\frac{z}{b})^{\alpha} - 1\right], \quad \alpha \neq 0,1 \tag{6.1}$$

where $z := x - x_j$, with $z = 0$ at x_j. α and β are used instead of u'' and u''', but it is easy to establish the connection. From

$$t'(z;\dots) = u' \cdot (1+\frac{z}{b})^{\alpha-1}$$

$$t^{(i)}(z;\dots) = \frac{\alpha-i+1}{b} \cdot t^{(i-1)}(0;\dots) \cdot \left[1+\frac{z}{b}\right]^{\alpha-i} \quad \text{for} \quad i > 1 \tag{6.2}$$

we obtain

$$\left.\begin{aligned} u''/u' &= \frac{\alpha-1}{b} \\ u'''/u'' &= \frac{\alpha-2}{b} \end{aligned}\right\} \text{hence} \quad \begin{aligned} \frac{1}{b} &= u''/u' - u'''/u'' \\ &= \frac{(u'')^2 - u'.u'''}{(u')^2} \cdot \frac{u'}{u''}, \end{aligned} \tag{6.3}$$

which transforms into

$$\frac{1}{b} = -\left[\ell n\left(\frac{u''(z;\dots)}{u'(z;\dots)}\right)\right]'_{|z=0}, \tag{6.4}$$

$$\frac{1}{\alpha-1} = \frac{u'(0;\dots)}{u''(0;\dots)} \cdot \frac{1}{b}. \tag{6.5}$$

If $\alpha < 0$ the above function becomes singular if $1 + \frac{z}{b}$ tends to zero, i.e. at the point

$$\underline{x}(x) = x - b(x) \quad . \tag{6.6}$$

This fact will later be used to estimate the location of the singularity of the approximated function y. The approximation in this case may be an <u>osculating</u> spline, i.e. y, y', y'', y''' are used to replace u, u', u'', u'''.

If y satisfies the differential equation $y' = f(x,y)$ then

$$\frac{y''}{y'} = f_y(x,y) + f_x(x,y)/f(x,y) \tag{6.7}$$

and similarly the logarithmic derivative depends on f . Remarkably, if we take the trouble of evaluating the derivatives of f analytically then we get the quantities b, α with the same accuracy as y, i.e. if numerical integration is performed,

$$y(x) \sim u(x) + O(h^4) \quad . \tag{6.8}$$

This means very high accuracy for the derivatives and associated terms. A warning - the coefficient of the O-Relation may become large if x approaches the singularity x^*.

In this case we use the exponent α as a free parameter. Since the exponent μ for the expansion of y at the singularity is available immediately from the differential equation one may fix $\alpha = \mu$ and instead of α introduce another free parameter, say, by adding a linear term. We have considered such examples in our programs but will report on these results at some other place.

We will see by numerical evidence that the error of numerical integration influences the error of the singularity estimate less than the error inherent to our procedure. (This may be called the *systematic error*.) Therefore we replace the numerical solution $u(x,h)$ by the "theoretical" solution $y(x)$ of the differential equation in our analysis.

In a recent paper [11] it was shown, how the osculating spline could even be used to get asymptotically an estimate of the interval where the singularity x^* of the solution to an initial value problem lies.

Suppose that $\underline{x}(x)$ of (6.6) is a monotonically increasing concave function of x in some neighborhood $x_o \leq x < x^*$. Furthermore let x and $x-h$, $h > 0$ belong to this interval. Letting $\Delta_t(x,x-h)\underline{x}(t)$ be the first difference quotient of $\underline{x}$ we set

$$\overline{x}(x) = x - \frac{b(x)}{1-\Delta_t(x,x-h)\underline{x}(t)} \quad . \tag{6.9}$$

(Assume the denominator to be position.)

Then

$$\underline{x}(x) < x^* < \underline{x}(x) \quad . \tag{6.10}$$

Remark: What is important here is that $\underline{x}(x)$ has some analytical property which allows us to deduce that such an interval exists, for example the existence of a series expansion with leading coefficients of known sign.

In section 9 we will return to the calculation of x^* by an extrapolation technique. We will point out why it is more difficult for initial value problems than for boundary value problems.

7. An Algorithm for a Second Order Boundary value Problem with Singular Boundary Conditions

Consider the differential equation

$$y'' = f(x,y,y') \quad \text{in} \quad x \in (0,1) \tag{7.1}$$

with the prescribed boundary conditions:

$$y(0) = y_o \tag{7.2}$$

and

$$\lim_{x \to 1} |y(x)| = \infty \tag{7.3}$$

To characterize the solution of this problem (assuming its existence) we set out to find $y'(0)$. The solution is then obtained by integrating an initial value problem, using e.g. the spline technique described before which is also good in the neighborhood of $x = 1$.

To determine $y'(0)$ we propose the following <u>shooting</u> <u>algorithm.</u>

Let $h = 1/N$, where $N \in \mathbb{N}$.

Select an "appropriate" C and use the initial value spline technique to calculate

$$\begin{aligned} &u(x;h,C) \quad \text{with} \\ &u(0;h,C) = y_0 \\ &u'(0;h,C) = C \end{aligned} \tag{7.4}$$

By applying the Regula falsi, one can obtain a C such that

$$u(x;h,C) \quad \text{in} \quad [1-2h,1-h]$$

is singular at $x = 1$, i.e.

$$b = -2h\ . \tag{7.5}$$

Then the derivative of u at O gives the estimate

$$y_0' \approx u'(0;h,C)\ . \tag{7.6}$$

To solve the initial value problem for the second order differential equation by the spline technique, we use the integral of (6.1) to find

$$t(z;x,h) = u+u'z + u''\frac{b^2}{\alpha(\alpha-1)} \cdot \left[(1+\frac{z}{b})^{\alpha} - \frac{\alpha}{b} z - 1 \right], \text{ for } \alpha \neq 0,1,2. \tag{7.7}$$

For such elements in every subinterval $[jh,(j+1)h]$, the spline shall be a function of class $C^3(I)$.

Obviously $t^{(i)}(0;\ldots) = u^{(i)}$ for $i = 0,1,2$.

Again the relation between α,b and u''', u^{IV} is easily established.

$$\begin{aligned} u''(z;x,h) &= u''(0;\ldots)(1+\frac{z}{b})^{\alpha-2} \\ u'''(z;x,h) &= u''(0;\ldots)(1+\frac{z}{b})^{\alpha-3} \cdot \frac{\alpha-2}{b} \end{aligned} \tag{7.8}$$

The numbers α and b are found in a like manner as before. We observe in particular

$$\frac{\alpha-2}{b} = \frac{u'''(0;x,h)}{u''(0;x,h)} \quad . \tag{7.9}$$

As was stated before, the parameter u^{IV} in addition to $u,\ldots,u'''$ (or equivalently the parameters α,b to u,u',u") should be chosen to satisfy

$$u''(h;jh,h) = f(jh,u(jh,\ldots),u'(jh,\ldots)) \quad . \tag{7.10}$$

Consider the interval [1-2h,1-h], i.e. $j = N-2$. If the condition (7.10) is met we have

$$u''(h;jh,h) = u''(0;jh,h) \cdot (1 + \frac{h}{b})^{\alpha-2} \tag{7.11}$$

We now realize that u"(h;jh,h) is nothing else than the value of the (global) spline derivative u"((j+1)h).

Eliminating $\alpha-2$ by (7.9) and introducing the above stated condition (7.5) (which has to be fulfilled by the choice of C) we end up with the condition

$$\left(\frac{u''(1-h)}{u''(1-2h)}\right)^{\frac{u''(1-2h)}{h\cdot u'''(1-2h)}} = (1 + \frac{h}{b})^{\frac{b}{h}} = 4 \quad . \tag{7.12}$$

As we will see, this equation is the key in the analysis of the convergence of the shooting technique.

To simplify the error analysis we consider a "first order boundary value problem" for which we have the same set of relations.

8. Modelproblem for First Order Differential Equations

Consider $y' = f(x,y)$ in $(o,1)$. Determine (if possible) y_o such that

$$\begin{aligned} & y(x) \quad \text{with} \quad y(0) = y_o \quad \text{satisfies} \\ & \lim_{x\to 1} |y(x)| = \infty . \end{aligned} \tag{8.1}$$

Technically we proceed as before, compare [12]. For fixed $N \in \mathbb{N}$ let $h = 1/N$.

A shooting technique is applied to select $C = C(h)$ such that the spline solution $u(x,h)$ with

$$\begin{aligned} & u(0,h) = C(h) \quad \text{satisfies} \\ & \lim_{x\to 1} |u(x,h)| = \infty . \end{aligned} \tag{8.2}$$

We take $C(h)$ as approximation to y_o and ask for the asymptotic behaviour of $C(h) - y_o$.

Examples have shown a sometimes unexpected high order of convergence with respect to h, see [12].

Refering to the previous considerations, we concentrate on the systematic error of the spline method.

1) Let $y_h(x)$ denote that solution of the differential equation which satisfies

$$y_h(0) = C(h) . \tag{8.3}$$

The singularity will probably not occur exactly at $x = 1$ but in a neighborhood, say at

$$x^* = 1 + \sigma(h) . \tag{8.4}$$

2) We assume that there is a differentiable relation between $y_h(0)$ and σ as independent quantity such that

$$\frac{\partial y_h(0)}{\partial \sigma} \neq 0 . \qquad (8.5)$$

3) Finally we <u>explicitly</u> assume that the shift $\sigma(h)$ of the singularity primarily depends upon the systematic error, not the round off and the propagated error.

Practically $y_h(x)$ is used instead of $u(x,h)$ in the following considerations.

A consequence of these assumptions is the fact that we may study the dependence between h and σ instead of h and $y_h(0)$.

Following section 4 the parameter u''' or its equivalent is determined from

$$u'(jh+h;h) = f(jh+h,u(jh+h,h)). \qquad (8.6)$$

The equation reduces to

$$\frac{u'_{j+1}}{u'_j} = \left(1+\frac{h}{b}\right)^{\alpha-1} \qquad \text{with } u'_j := u'(jh,h), \text{ etc.} \qquad (8.7)$$

Again elimination of α by means of b furnishes

$$\left(\frac{u'(jh+h,h)}{u'(jh,h)}\right)^{\frac{u'_j}{h\cdot u''_j}} = \left(1+\frac{h}{b}\right)^{\frac{b}{h}} . \qquad (8.8)$$

Specialization to $j = N-2$ and the condition $b = -2h$ yield the relation for the dependence of σ on h. Now we replace u by y_h, taking into account that

$$u_j' = f(x_j,u_j) \doteq f(x_j,y_h(x_j)) = y_h'(x_j) . \tag{8.9}$$

We thus find the defining equation

$$F(\sigma,h) = \left(\frac{y_h'(x_{N-1})}{y_h'(x_{N-2})}\right)^{\frac{y_h'(x_{N-2})}{hy_h''(x_{N-2})}} = 4 . \tag{8.10}$$

This is equivalent to the relation (7.12) of the second order differential equation if the second derivatives are replaced by the first order ones.

We now introduce

$$t = x - x_h^* = x - (1+\sigma(h)) \qquad \text{and}$$

$$y_h(t) = c\cdot t^{\mu}(1 + c_1 t^{|\mu|} + c_2 t^{2|\mu|} + c_3 t^{3|\mu|} + \ldots), \tag{8.11}$$

$$y_h'(t) = c\cdot\mu t^{\mu-1} + |\mu|\cdot c_2\cdot ct^{|\mu|-1} + 2|\mu|\cdot c_3 ct^{2|\mu|-1} + \ldots . \tag{8.12}$$

It is possible that some of the c_j might vanish.

Observation: 1. Since the defining equation contains only first and second order derivatives the second term of the expansion of $y_h(t)$, i.e. $c\cdot c_1$ does not influence $F(\sigma,h)$. Therefore in all expansions of F the difference between the first and second exponent of t is (at least) $2|\mu|$.

2. The coefficients c, c_j may be dependent upon σ. Since x_{N-2}, x_{N-1} correspond to $t_{N-2} = -2h-\sigma$, $t_{N-1} = -h-\sigma$ one may expand F to find that

$$\sigma = h^{1+\gamma}\cdot\text{const} + o(h^{1+\gamma}) . \tag{8.13}$$

Here γ denotes the difference of the exponents of the first and second term of y_h' for which the coefficients are different from zero at $x^* = 1$.

The details are given in [12]. It is also pointed out there that it might seem that the numerical convergences of σ and hence $y_h(0)$ is of higher order than $1+\gamma$. This may be due to the fact that the coefficient of $h^{1+\gamma}$ in the expansion of σ is relatively small as compared to that of $h^{1+\gamma+|\mu|}$ (or any higher one). In any case Richardson's extrapolation is a feasable way to improve the convergence of

$$y_h(0) = y_o + d_1 \cdot h^{1+\gamma} + d_2 h^{1+\gamma\ |\mu|} + \ldots \quad , \tag{8.14}$$

with $h = h_o,\ h_o/2,\ h_o/4, \ldots$.[For numerical examples see [12].]

9. Singularity Estimations of the Solutions of Initial Value Problems

We may ask how the solution y of an initial value problem can be continued up to a singular point. Assume that the differential equation has the form (5.1). Therefore we use splines from the family defined by (6.1). The approximating spline may be iteratively continued until we reach an interval $I_{j+1} := [x_j, x_{j+1}]$ such that the analytic continuation of $u|_{I_{j+1}}$ will become singular within I_{j+2}, i.e. b found in I_{j+1} satisfies $-2h \leq b_j < -h$. Then

$$x = x_j - b_j \tag{9.1}$$

provides an estimate for the singularity of y.

If we repeat the calculations with a different step size, say $h/2$, we may get a slightly different b at the same point x_j. Therefore we should write

$$b = b(x_j,h) \quad .$$

It can be shown that

$$b(x_j,h) \to b(x_j) = \left[\ln\left(\frac{y''}{y'}\right) \right]'_{\big| x=x_j} \quad \text{for} \quad h \to 0 \tag{9.2}$$

and that this convergence is linear in h for fixed x_j.[Compare with [11].] Hence we may apply the extrapolation procedure:

$$b(x) \doteq b(x,0) = 2b(x,h/2) - b(x,h) \quad . \tag{9.3}$$

We search for an x such that $b(x,0) = 0$. As before we use the expansion (5.2) of $y(t)$ to see by (6.4) and (9.2) that there is an appropriate expansion

$$b(t) = t \cdot (1 + a_1\, t^{\gamma} + a_2\, t^{\gamma+|\mu|} + \ldots) \quad .$$

Unfortunately the value t_j that corresponds to x_j is unknown, in fact,it is exactly what we want to calculate. Inverting this expansion we find

$$t = b + d_1 \cdot b^{1+\gamma} + d_2 \cdot b^{1+\gamma+|\mu|} + \ldots \tag{9.4}$$

with unknown coefficients d_j.

For $b := b(x_j,0),\ b(x_{j+1},0),\ldots$ the corresponding values t_j, t_{j+1} satisfy $t_j - t_{j+1} = x_j - x_{j+1}$ which is equal to the step size h. If we use (9.4) for several values of j we may eliminate the quantities $t_j,\ldots,$ calculate the coefficients

$d_1, d_2, \ldots$ and then use (9.4) to calculate t_j.

This procedure replaces the Richardson's extrapolation process that we used in the case of the boundary value problem.

Practically, the elimination of the coefficients d_j can be achieved in the following way. If x_j is added to both sides of (9.4) then

$$x^* = x_j + t_j = \tilde{x}(x_j) + d_1 \cdot b(x_j, 0)^{1+\gamma} + \ldots \quad , \qquad (9.5)$$
$$\text{with} \quad \tilde{x}(x_j) := x_j + b(x_j, 0) \quad .$$

Subtracting x^* on both sides, we get the homogeneous equations

$$A_n \cdot \begin{pmatrix} 1 \\ d_1 \\ d_2 \\ \vdots \end{pmatrix} = \begin{pmatrix} 0 \\ \vdots \\ 0 \end{pmatrix}$$ for the quantities $1, d_1, d_2, \ldots, d_n$

where A_n denotes an $(n+1) \times (n+1)$ matrix

$$\begin{pmatrix} \tilde{x}(x_j) - x^* & b(x_j, 0)^{1+\gamma} & \ldots & b(x_j, 0)^{1+\gamma+|\mu|\cdot(n-1)} \\ \vdots & \vdots & & \vdots \\ \tilde{x}(x_{j+n}) - x^* & b(x_{j+n}, 0)^{1+\gamma} & \ldots & b(x_{j+n}, 0)^{1+\gamma+|\mu|(n-1)} \end{pmatrix} .$$

Since the first component of the above solution of the homogeneous system is equal to 1, we have $\det A_n = 0$.

Expanding this determinant one finds,

$$\tilde{x}^* \cdot \sum_{i=0}^{n} A_{j1} = \sum_{i=0}^{n} \tilde{x}(x_{j+i}) \cdot A_{j1} \quad , \qquad (9.6)$$

the quantities A_{j1} being the cofactors of the determinantal expansion of the first column.

Before giving an example we remark once more that we have made the assumption that the error of the x* estimation is primarily due to systems error and that the calculation of the values u(x,h) by the spline technique is sufficiently accurate to interchange y and u.

As an example we choose the initial value problem

$$y' = 1 + y^2 + y^4$$

$$y(0) = 1 \quad .$$

Its solution can be found by elementary integration. The solution has a singularity at $x^* = \frac{\pi}{12} \cdot \sqrt{3} - \frac{\ln 3}{4} \doteq .178\,796\,769$. In [12] inclusion techniques were applied to this example. We will show here how well the extrapolation performs in this regard.

For h = .015 625 we find

j	x_j	$-b(x_j,h)$	$-b(x_j,h/2)$	$-b(x_j,0)$	$\tilde{x}(x_j)$
8	.125	.051 534	.051 146	.050 758	.175 758
9	.140 625	.037 296	.036 944	.036 592	.177 217
10	.156 25	.022 364	.022 190	.022 016	.178 266

Remarks: It is convenient to work with -b since b is always negative.

With the formula (9.6) we obtained for

$$n = 1 : (j=9,10) \qquad \tilde{x}^* \doteq .179\,053 \qquad \tilde{x}^* - x^* \doteq .257 \cdot 10^{-3} ,$$

$$n = 2 : (j=8,9,10) \qquad \tilde{x}^* \doteq .178\,951 \qquad \tilde{x}^* - x \doteq .155 \cdot 10^{-3} .$$

If we take

$h = 0.003\ 906\ 25$ then the values are

j	x_j	$-b(x_j,h)10^3$	$-b(x_j,h/2)10^3$	$-b(x_j,0)10^3$	$\tilde{x}(x_j)$
42	.164 0625	14.564 08	14.537 38	14.510 68	.178 573 18
43	.167 968 75	10.746 29	10.725 05	10.703 81	.178 672 56
44	.171 875	6.895 70	6.882 51	6.869 32	.178 744 32 .

For $n = 1$: $(j=43,44)$ $\quad \tilde{x}^* \doteq .178\ 809\ 896 \quad \tilde{x}^* - x^* \doteq 1.3127 \cdot 10^{-5}$

For $n = 2$: $(j=42,43,44)$ $\quad \tilde{x}^* \doteq .178\ 800\ 637 \quad \tilde{x}^* - x^* \doteq .386 \cdot 10^{-5}$.

We numerically confirm the observation of [12] that the first coefficient in the expansion (9.4) of $t(b)$ is relatively small,

$$t(b) = b(1 + d_1 b^{2/3} + d_2 b^{4/3} + \ldots)$$

$$d_1 \doteq 0.170\ 238 \quad , \quad d_2 \doteq 1.567\ 376 \quad .$$

REFERENCES

[1] Arndt, H., Lösung von gewöhnlichen Differentialgleichungen mit nichtlinearen Splines, Num. Math. 33, 323-333 (1979).

[2] Hille, E., Ordinary differential equations in the complex domain, J. Wiley & Sons, New York-London-Sydney-Toronto, (1976).

[3] Loscalzo, F.R., and Talbot, T.D., Spline function approximations for solutions of ordinary differential equations, SIAM J. Numer. Anal. 4, 433-445 (1967).

[4] Mülthei, H.N., Splineapproximationen von beliebigem Defekt zur numerischen Lösung gewöhnlicher Differentialgleichungen, Teil I und II, Numer. Math. 32, 146-157, 343-358 (1979).

[5] Rooij van, P.L.J., and Schurer, F., A bibliography on spline functions II, TH Report 73-WSK-01, Technological University Eindhoven, Netherlands.

[6] Runge, R., Lösung von Anfangswertproblemen mit Hilfe nichtlinearer Klassen von Spline-Funktionen, Dissertation Münster, 1972.

[7] Schoenberg, I.J. (ed.), Approximations with special emphasis on spline functions, Academic Press, New York-London, (1969).

[8] Varga, R., Error bounds for Spline Interpolation, p. 367-388 in [7], (1969).

[9] Werner, H., Neuere Entwicklungen auf dem Gebiete der nichtlinearen Splinefunktionen, ZAMM 58, 86-95 (1978).

[10] Werner, H., An introduction to non-linear splines, in Polynomial and Spline Interpolation, B.N. Sahney (ed.), Reidel, Dordrecht - Boston-London, (1979).

[11] Werner, H., Extrapolationsmethoden zur Bestimmung der beweglichen Singularitäten von Lösungen gewöhnlicher Differentialgleichungen, in Numer. Math., R. Ansorge, K. Glashoff, B. Werner (ed.), ISNM 49, p. 159-176 (1979).

[12] Werner, H., Spline Functions and the Numerical Solution of Differential Equations, North-Holland Publishing-Company, Amsterdam-New York-Oxford, 1980, to appear.

[13] Werner, H., and Wuytack, L., Nonlinear Quadrature Rules in the Presence of a Singularity, Comp. & Maths. with Appls., Vol. 4, p. 237-245, Pergamon Press Ltd. (1978).

[14] Werner, H., and Zwick, D., Algorithms for numerical integration with regular splines, Schriftenreihe des Rechenzentrums der Universität Münster, Nr. 27 (1977).

UNIQUENESS AND CONTINUITY OF BEST APPROXIMATIONS

Theagenis Abatzoglou

6721 Cory Drive
Huntington Beach, California

1. INTRODUCTION

We consider the metric projection P defined in a Banach space X by $P(x) = \{m \mid \|x-m\| = \inf_{m' \in M} \|x-m'\|\}$ where $M \subset X$.

We impose sufficient conditions on X and M so that $P(x)$ is a singleton. We also obtain explicit estimates on the modulus of continuity of P around points x where P is again a singleton. These results depend on the notion of directional curvature of M and they also generalize and improve similar results obtained in (8), (10) and (11). An improvement of the classical estimate $\|P(y) - P(x)\| \leq \|y - x\|$ for the case when M is convex is also obtained.

2. Basic Finite Dimensional Results

E. Asplund was able to obtain the following deep result, see (7).

<u>Theorem 2.1</u>. Let M be a closed subset of R^n with Euclidean norm; then P is Frechet differentiable a.e. .

Proof. Let $K(x) = \frac{1}{2}\|x\|^2 - \frac{1}{2}\|x - P(x)\|^2$, then K is convex and $K'(x) = P(x)$ a.e. By a theorem of A. D. Alexandrov, see (7), $K''(x)$ exists a.e. in the Frechet sense and therefore P is Frechet differentiable a.e.

T. Abatzoglou in (1) obtained a similar result for a finite dimensional Banach space whose norm satisfies
$m\|y\|^2 \leq \nabla^2\|x\|^2(y)^{(2)} \leq M\|y\|^2$ for every $x \neq 0$ and every y,

Copyright © 1980 by Academic Press, Inc.
All rights of reproduction in any form reserved.
ISBN: 0-12-171050-5

where $0 < m \leq M$. It is not known whether in $\ell^P_{(n)}$ the map P onto any closed set is Frechet differentiable a.e.

3. Continuity and Differentiability of P in Terms of the Curvature of M

If M is a closed convex set in a Hilbert space then we have the classical estimate $\|P(y) - P(x)\| \leq \|y - x\|$. We will show that by taking into account the curvature of M we will be able to improve this result and at the same time obtain estimates for $\|P(y) - P(x)\|$ for non-convex M.

Example 1. Let $S_R = \{m \mid \|m\| = R\} = M$, then it can be easily shown that P is differentiable for $x \neq 0$, $\|x\| \neq R$ and that

$\|P'(x)\| = R/(R + \|x - P(x)\|)$ for $\|x\| > R$,

$\|P'(x)\| = R/(R - \|x - P(x)\|)$ for $\|x\| < R$.

We will now define the directional radius of curvature of M at m along the direction v, $\rho(m,v)$. Suppose that $m = P(x) \neq x$; set $v = \frac{x-m}{\|x-m\|}$. Next consider the line $m+tv$, $t \in R$, and points $\mu \in M$ close to m such that $|t| = \|m+tv - \mu\|$ for some finite t. If the last equation holds for no finite t, set $t = \infty$. We now define $\rho(m,v)$ by

$$\rho(m,v) = \lim_{\varepsilon \downarrow 0} t_\varepsilon \quad \text{where} \quad t_\varepsilon = \left[\sup_{0<\|\mu-m\|<\varepsilon} \left\{\frac{1}{t} \,\Big|\, |t| = \|m-\mu+tv\|\right\}\right]^{-1}.$$

The directional radius of curvature was introduced in a less general form of J. R. Rice in (10). He and E. Rozema, P. W. Smith, J. Ward and C. Chui were able to obtain a number of interesting results which motivated the work of present author, see (10) and (11). In (2) we find an analytic formula for $\rho(m,v)$ when M is a C^2 manifold in a Hilbert space. As it turns out then $\frac{1}{\rho(m,v)} = \max_{\|w\|=1} \frac{\langle Aw,w\rangle}{\langle Bw,w\rangle}$ where $m = f(a)$, $B = f'(a)^T f'(a)$, $A = \left(\langle v, \frac{\partial^2 f(a)}{\partial t_i \partial t_j}\rangle\right)$ and f represents M around m. In differential geometric terms $\frac{1}{\rho(m,v)}$ is the maximal principal curvature of M at m on the direction v.

The following theorem can be deduced from (3) and (12).

Theorem 3.1. Let M be an approximatively compact, n-dimensional, C^2 manifold in a Hilbert space H. Then

$P'(x) = f'(a)(B-rA)^{-1}f'(a)^T$ $\forall x$ in an open dense set in H and $\|P'(x)\| = \frac{\rho(m,v)}{\rho(m,v)-r}$. Here we assume that $m = P(x)$ and $r = x-m$.

It is also possible to estimate the modulus of continuity of P even if M is not convex or smooth. The method is purely geometric, see (2).

Theorem 3.2. Let M be a closed set in H. Assume that $m = P(x)$ is a singleton and that $\lim_{y\to x} P(y) = P(x)$. Then

$$\overline{\lim_{y\to x}} \frac{\|P(y) - P(x)\|}{\|y - x\|} \leq 2 \frac{\rho(m,v)}{\rho(m,v)-r}.$$

In the same vein it was shown in (2).

Theorem 3.3. Let M be a closed convex set in H. Then

$$\overline{\lim_{y\to x}} \frac{\|P(y) - P(x)\|}{\|y - x\|} \leq \frac{2\rho(m,v)}{2\rho(m,v)-r}.$$

We remark that since $\rho(m,v) \leq 0$ for convex sets Theorem 3.3 improves the usual estimate $\|P(y) - P(x)\| \leq \|y - x\|$. Also by examples in (2), Theorems 3.2 and 3.3 can be seen to be sharp.

Results similar to Theorems 3.1, 3.2 and 3.3 can be obtained in a Banach space X whose norm is C^2 and which satisfies $\nabla^2\|x - P(x)\|(y)^2 \geq m\|y\|^2$ for every y and $m > 0$. For this see (3).

It is also possible to obtain analogous results if the Banach space X is uniformly convex and uniformly smooth. Again a geometric proof is used based on the radius of curvature of M at $P(x)$, see (4).

Theorem 3.4. Let M be an approximatively compact set in X, assume $P(x)$ is a singleton; then

$$\|P(y) - P(x)\| \leq C_1\delta^{-1}(C_2\rho(C_3\|y-x\|)) + C_4\|y-x\|$$

where C_1, C_2, C_3, C_4 are constants depending on m, x and $\rho(m,v)$. Here δ and ρ are the moduli of uniform convexity and uniform smoothness respectively. By specializing to $L^p(d\mu)$ we can get:

Corollary 3.1. In $L^p(d\mu)$, with the additional assumptions that $\frac{\rho(m,v)}{\|x-m\|} \notin [1,2)$ we have

$$\overline{\lim_{y\to x}} \frac{\|P(y)-P(x)\|}{\|y-x\|^{p/2}} \leq \frac{4r^{2-p/2}}{\sqrt{p(p-1)}} \left(\frac{2\rho(m,v)}{\rho(m,v)-r}\right)^{p/2}, \quad 1 < p < 2$$

$$\overline{\lim_{y\to x}} \frac{\|P(y)-P(x)\|}{\|y-x\|^{2/p}} \leq 2r^{\frac{p-2}{p}} \sqrt{p(p-1)} \left(\frac{2\rho(m,v)}{\rho(m,v)-r}\right)^{2/p}, \quad 2 < p < \infty$$

These results are best possible up to a constant; see (8) for an example that shows this.

4. Uniqueness of Best Approximations from Smooth Manifolds

Given a C^2 manifold M in a Hilbert space H and x in H we consider $f(m) = \|x-m\|$ where m is in M. We would like to know when is a critical point of f a global minimum? A condition guaranteeing that m is a local minimum is $\rho(m,v) \notin [o,r]$. First we treat the case when M is a curve. The following theorem appearing in (5) turns out to be quite useful.

<u>Theorem 4.1</u>. Let γ be a C^2 curve in H such that $\|\gamma'(t)\| = 1$, $\|\gamma''(t)\| \leq 1/R$ for all t, $\|\gamma(0)\| = R$ and $\gamma(0) \perp \gamma'(0)$. Then 1) $\|\gamma(t)\| \geq R$, for $|t| \leq \pi R$

2) $\frac{d}{dt}\|\gamma(t)-\gamma(0)\| \geq 0$, for $0 < t \leq \pi R$ and $\|\gamma(\pm\pi R)-\gamma(0)\| \geq 2R$.

If $x = \alpha\gamma(0)$, $0 < \alpha < 1$, Theorem 4.1 tells us that $\gamma(0)$ is a "temporary best approximation" to x until time πR. Further knowledge of γ for $|t| > \pi R$ should inform us whether $\gamma(0) = P(x)$. For this purpose the following 2 definitions are quite appropriate, these also appear in (6), (10) and (11).

1. <u>Definition</u>. The metric curvature $\frac{1}{\rho(m)}$ of M at m is defined by $\frac{1}{\rho(m)} = \sup_v \frac{1}{\rho(m,v)}$ where $v \perp M$ at m.

<u>Example 1</u>. Let M be a C^2 curve γ with $\|\gamma'(t)\| = 1$. Then $\frac{1}{\rho(m)} = \|\gamma''(t)\|$ where $m = \gamma(t)$.

<u>Example 2</u>. Let γ be a geodesic on M with $\|\gamma'(t)\| = 1$. Then $\|\gamma''(t)\| \leq \frac{1}{\rho(m)}$ = metric curvature of M at m.

2. <u>Definition</u>. The folding $\phi(m)$ of M at m is defined by $\phi(m) = \sup\{s | B(m,r) \cap M$ is connected $\forall r \leq s\}$.

<u>Example 3</u>. Let $M \equiv f(a,b,c) = \frac{ax+b}{cx+1} = m \in L^2[-1,1]$. Then $\phi(m) = \infty$.

The next theorem which is in (5) gives us a geometric idea on how small the folding can be.

Theorem 4.2. Let M be a C^1, n-dimensional manifold in a Banach space X with smooth norm. Suppose $B(m,\phi(m)+\varepsilon) \cap M$ is compact for some $\varepsilon > 0$. Let F be the embedding of M into X. Then $\phi(m) \geq \inf\|m-F(\bar{x})\|$ where $\bar{x}$ is a critical point of $\|m-F(\bar{x})\|$ and $F(\bar{x}) \neq m$.

Now we can present a precise estimate of how close must a point be from a manifold so that it may have a unique best approximation. This estimate appears in (5) and for related results see (10) and (11).

Theorem 4.3. Let M be a C^2, complete, connected, n-dimensional manifold embedded in a Hilbert space H. Suppose x is in H, m is in M and $x-m \perp M$ at m. Assume $\frac{1}{R} \geq \sup_{\mu \in M}\{\frac{1}{\rho(\mu)} \Big| \|\mu-m\| \leq 2R\}$. Then if $\|x-m\| < \min\{R,\frac{\phi(m)}{2}\}$, m is the unique best approximation from M to x.

Example 4. Let M be the curve $\gamma(t) = \frac{1}{1-tx} \in L^2[-1,1]$. It can be shown that $\sqrt{\frac{18}{5}} = \rho(\gamma(0)) \leq \rho(\gamma(t))$ and $\phi(\gamma(t)) = \infty$ for each t. Then it follows from Theorem 4.3 that if for some $f \in L^2[-1,1]$, $\min_t\|f-\gamma(t)\| < \frac{1}{3}\sqrt{\frac{5}{2}}$ then f has a unique best approximation from γ.

We would like an extension of Theorem 4.3 into Banach spaces. This can be accomplished by obtaining Theorem 4.3 first in the case when M is a curve and then for manifolds that are spanned by geodesics with sufficient regularity conditions. The proof uses a generalization of Theorem 4.1 in Banach spaces which satisfy Property C.

Property C. Let S be the unit sphere of a Banach space X and γ be a C^1 curve such that $\|\gamma'(t)\| = 1$, $\rho(\gamma(t)) \leq 1$ for all t and γ is tangent to S at $p = \gamma(0)$. Then there is an $\varepsilon > 0$ independent of p such that $\|\gamma(t)\| \geq 1$ for $|t| \leq \varepsilon$.

A class of spaces that satisfy Property C is characterized by positive definite Hessian of the norm.

Another property that we require the Banach space satisfies is that of "finite escape time" from balls. The "finite escape time property" is contained in the following theorem obtained by S. Nelson and T. Abatzoglou, see (9).

Theorem 4.4. Let γ be a curve in X such that $\gamma(0) = 0$ and $\|\gamma'(t)\| = 1$ for all t. For each t find a z, $\|z\| = 1$, z contained in $\mathrm{span}\{\gamma(t), \gamma'(t)\}$ such that $\nabla\|z\|(\gamma'(t)) = 0$. Suppose that $\|\gamma(t) \pm z\| \geq 1$ for each t, then we get

1. $\|\gamma(\pi)\| \geq 2$ if X = Hilbert space
2. $\|\gamma(T)\| \geq 2$ under the following assumptions on X.

a. X is smooth and uniformly convex with modulus of uniform convexity δ.

b. $\int_0^1 \frac{\delta^{-1}(s)}{s}\, ds < \infty$

c. There is a $M > 0$ such that for any z, w with $\|z\| = 1$ and $\nabla\|z\|(w) = 0$ we have $\|z+3w\| - 1 \leq M(\|z+w\| - 1)$.

If the above conditions are satisfied we get

$$T \leq \int_0^2 \frac{2M+2}{2-r}\, \delta^{-1}\left(\frac{2-r}{2M+2}\right) dr$$

Example 5. $X = L^p(d\mu)$, $1 < p < \infty$, satisfies the requirements of Theorem 4.4.

The problem of establishing precise conditions for unique best approximations from smooth manifolds in Banach spaces remains open and its solution should be very fruitful.

REFERENCES

1. Abatzoglou, T., Finite Dimensional Banach Spaces with a.e. Differential Metric Projection, Proc. Amer. Math. Soc., to appear.
2. Abatzoglou, T., The Lipschitz Continuity of the Metric Projection, J. Approximation Theory, 26(1979), 212-218.
3. Abatzoglou, T., The Metric Projection on C^2 Manifolds in Banach Spaces, J. Approximation Theory, 26(1979), 204-211.
4. Abatzoglou, T., Continuity of Metric Projections in Uniformly Convex and Uniformly Smooth Banach Spaces, submitted.
5. Abatzoglou, T., Unique Best Approximations from a C^2 Manifold in Hilbert Space, Pacific J. Math., to appear.
6. Abatzoglou, T. and Nelson, S., Unique Best Approximations from C^2 Manifolds in Banach Spaces, submitted.
7. Asplund, E., Differentiability of the Metric Projection in Finite Dimensional Euclidean Space, Proc. Amer. Math. Soc., 38(1973), 218-219.
8. Bjornestal, B. O., Local Lipschitz Continuity of the Metric Projection Operator, preprint.
9. Nelson, S. and Abatzoglou, T., Curvature and Escape Times from a Ball for Curves in a Banach Space, submitted.
10. Rice, J., Nonlinear Approximation. II. Curvature in Minkowski Geometry and Local Uniqueness, Trans. Amer. Math. So.c, 128(1967), 437-459.
11. Rozema, E. and Smith, P., Nonlinear Approximation in Uniformly Smooth Banach Spaces, Trans. Amer. Math. Soc., 188(1974), 199-211.
12. Wolfe, J., Differentiability of Nonlinear Best Approximation Operators in a Real Inner Product Space, J. Approximation Theory, 16(1976), 341-346.

CHARACTERIZATIONS OF RELATIVE CHEBYSHEV CENTERS

D. Amir

Tel-Aviv University
Tel-Aviv, Israel

Z. Ziegler
Technion
Haifa, Israel

Let E be a normed linear space, let A, $A \subset E$ be a bounded set, and let $G \subset E$ be an arbitrary set.

Define, for all $x \in E$

$$S(x,r) = \{y \; ; \; \|x-y\| \leq r\}$$

$$r(x,A) = \inf \{r \; ; \; A \subset S(x,r)\}$$

$$r(G,A) = \inf \{r(x,A) \; ; \; x \in G\}$$

With this notation, the set Z(G; A) defined by

$$Z(G;\, A) = \{x \in G \; ; \; r(x,A) = r(G,A)\}$$

is called the <u>relative Chebyshev center of A in G.</u> Note that Z(G; A) is the set of centers in G of spheres of minimal radius covering A. When G is the whole space E, we have the Chebyshev center of A, a concept introduced by Garkavi [5] and studied by several authors in recent years. If $A = \{y\}$, a singleton, then the relative Chebyshev center is exactly the set of best approximants to y from G.

The extensively studied concept of simultaneous best approximation (e.g.[4],[6],[7],[8],[10]) fits as a special case in the study of relative centers.

Our investigations focus on three basic goals of theoretical and practical value:

1. Describe the situation where the relative center reduces to a singleton.

2. Characterize the relative center, with particular attention given to the case where the relative center reduces to a singleton.

3. Construct an algorithm yielding an element of the relative center.

Copyright © 1980 by Academic Press, Inc.
All rights of reproduction in any form reserved.
ISBN: 0-12-171050-5

We start by developing the connections between structural properties of the relative centers and convexity properties of the spaces. We first introduce a concept akin to strict convexity.

Definition 1: Let F be a subspace of E. E is strictly convex with respect to F if its unit sphere contains no segment parallel to F., i.e.,

$$\{\|x\| = \|y\| = \|\tfrac{x+y}{2}\| = 1 \ , \ x-y \in F\} \Rightarrow x = y.$$

We then have

Assertion 2: The following statements are equivalent:

(i) E is strictly convex with respect to F.

(ii) For each compact K, $K \subset E$, the center Z(F; K) is at most a singleton.

(iii) For each pair $x,y \in E$, the center Z(F; x,y) is at most a singleton.

Unfortunately, some interesting and practical situations do not fit into this framework. For example, we can prove the following assertions:

Assertion 3: Let μ be any measure. Then $L_1(\mu)$ is not strictly convex with respect to F if dim F ≥ 2. If μ is atomless, then $L_1(\mu)$ is not strictly convex with respect to any subspace.

Assertion 4: The space $C_0(T)$ is not strictly convex with respect to any subspace F with dim F ≥ 2.

Assertion 5: The space $(C[a,b], \| \ \|_1)$ of continuous functions with the L_1-norm is not strictly convex with respect to F if dim F ≥ 2.

A related characterization is obtained via the concept of uniform convexity of E with respect to every direction in F (Uced-F). Generalizing the work of Day, James and Swaminathan [3] we prove that the relative center of every bounded $A \subset E$ with respect to F is at most a singleton iff E is Uced-F. This is also a very restrictive concept.

We next confine our attention to the special case where A = {x,y}. This has been extensively studied by many authors, from a different perspective, under the heading of best simultaneous approximation. The following results can be proved:

Assertion 6: Let $u \in Z\{F; x,y\}$. Then exactly one of the following alternatives holds:

a. $\|u-x\| = \|u-y\|$.

b. $u \in Z(F; x)$ and $\|u-x\| > \|u-y\|$,

c. $u \in Z(F; y)$ and $\|u-y\| > \|u-x\|$.

On the basis of this result we can devise a procedure for finding an element of Z(F; x,y) if an element of Z(F; x) and an element of Z(F; y) can be found, provided that

$$Z(F; x,y) \cap [u,v] \neq \phi \qquad (*)$$

for all $u \in Z(F; x)$, $v \in Z(F; y)$.

When studying the structure of pairs $F,E,F \subset E$ where property (*) holds, a useful concept is that of "homogeneously embedded" subspaces. We recall that F is a proximinal subspace of E if $P(F; x) \neq \emptyset$ for all $x \in E$.

Definition 7: Let F be proximinal in E. We say that F is homogeneously embedded in E if

$$\{0 \in Z(F; x) \ ; \ y,z \in F \ ; \ \|y\| = \|z\|\} \Rightarrow \|x-y\| = \|x-z\|$$

This is an appropriate concept, since we have

Assertion 8: Let F be homogeneously embedded in E. Then (*) holds for all $x,y \in E$. If F is separable then (*) is equivalent to homogeneous embedding.

The property of homogeneous embedding is essentially a pre-Hilbert property and therefore does not help in answering our questions for $C[a,b]$ or L_1. Some results demonstrating the inapplicability of the concept to these spaces are given by the following assertions:

Assertion 9: Let T be a compact metric space. The only nontrivial homogeneously embedded finite dimensional subspaces of $C(T)$ are the one dimensional subspaces spanned by functions v, with $|v(t)| \equiv 1$.

Assertion 10: Let $E = (C(T) \ , \ \| \ \|_{L_1(\mu)})$, where T is a connected locally Hausdorff space T and μ is a Radon measure. Then E has no nontrivial homogeneously embedded subspace.

Details of the foregoing analysis, including proofs and additional results, can be found in [1].

We next proceed to deal with the $C[a,b]$ case which lies outside the framework of the foregoing analysis.

We are searching for u yielding the minimum for

$$\min_{u \in F} \{\mathrm{Max}(\|f-u\| \ ; \ f \in A)\}$$

We recall (see e.g. [4]) that if A is compact, then $r(g,A) = r(g;A_U,A_L)$, where $A_U(t) = \sup\{f(t); f \in A\}$, $A_L(t) = \inf\{f(t); f \in A\}$. Hence, we may restrict attention to $Z(F; f,g)$, $f \geq g$. Since we are interested in questions of uniqueness, the natural framework would be n-unisolvent approximating families (see,e.g.[12] for definitions and properties). We recall the following definitions:

Definition 11: Let u,f,g be given, $f \geq g$. Then $\{t_i\}_1^k$ is a $(u; f,g)$-alternance if

$$f(t_{2i}) - u(t_{2i}) = u(t_{2j+1}) - g(t_{2j+1}) = r(u; f,g), \text{ for all } i,j$$

or

$$u(t_{2i}) - g(t_{2i}) = f(t_{2j+1}) - u(t_{2j+1}) = r(u; f,g), \text{ for all } i,j \ .$$

Each point where $f(t) - u(t) = r(u; f,g)$ is called a (+)-point, and a point where $u(t) - g(t) = r(u; f,g)$ is called a (-)-point. Both kinds are called (e)-points. A point that is simultaneously a (+)-point and a (-)-point is called a straddle point.

The following theorem is due to Dunham [4].

Assertion 12: Let $f \geq g$, and let F be n-unisolvent. Then $u \in Z(F; f,g)$ if, and only if either (i) $(u; f,g)$ has a straddle point, or (ii) there exists an (n+1)-point alternance.

This theorem characterizes the center, but does not provide an insight to the uniqueness question. We can easily prove that if there exist n straddle points then $Z(F; f,g)$ is a singleton. However, this is far from necessary. For example, if $F = [1,x]$, and t_0 is a straddle point such that $f'(t_0) = g'(t_0)$, then $Z(F; f,g)$ is a singleton. This phenomenon motivates the following definitions.

Definition 13: The family $F = (F(\bar{a}; t) ; \bar{a} \in S \subset R^n)$ is called extended n-unisolvent if for each set of n Hermite data

$$t_i, i=1,\dots, k \; ; \; y_i^j, i=1,\dots k, \; j=0,\dots, n_i-1 \; ; \; \Sigma_1^k n_i = n,$$

there exists a unique $\bar{a}$ such that

$$F^{(j)}(\bar{a}; t_i) = y_i^j \quad , \; i=1,\dots k \; ; \; j=0,1,\dots n_i-1.$$

Definition 14: A straddle point t_0 is of deficiency k if k Hermite type data are imposed on the extremal functions at t_0.

Definition 15: Let $\tilde{u} \in Z(F; f,g)$ and let $\tilde{t}$ be a straddle point of deficiency k. If $\tilde{u}^{(k)}(\tilde{t}) = g^{(k)}(\tilde{t})$ and $\tilde{t}$ is not a cluster point of (+)-points we call $\tilde{t}$ a (-)-boundary point. Similarly for (+)-boundary points.

Definition 16:: Let $\tilde{u} \in Z(F; f,g)$ and let $\tilde{t}$ be a straddle point, of deficiency k, which is a cluster point of (+)-points. Let m be the largest integer, $0 \leq m \leq n-k$, such that

$$\tilde{u}^{(j)}(\tilde{t}) = g^{(j)}(\tilde{t}) \quad , \; j = 1,\dots, k+m-1,$$

then $h = k + m$ is called the total deficiency of $\tilde{t}$.

Definition 17: Let $\bar{X} = (x_1,\dots,x_k)$ be a vector. Then $S^+(\bar{X})$ = the maximal number of sign changes of the components, where zeros are assigned (+) or (-) values. Let $\alpha(t)$ be a function. Then

$$S^+(\alpha) = \sup[(S^+(\alpha(t_1),\dots, \alpha(t_k))]$$

over all finite ordered k-tuples $(t_1,\dots, t_k)$.

Using these notions, we define for each $u \in Z(F; f,g)$ a function $\alpha(t) = \alpha[x(t)]$ associated with u, as a composition of two mappings, $\alpha(x)$ and $x(t)$, as follows:

$$x(t) = \begin{cases} t + H_t + \ell_t, & \text{if t is not a straddle point} \\ \bigcup_{j=0}^{h_t-1} \{t + H_t + \ell_t + j\}, & \text{if t is a non-boundary straddle point} \\ \bigcup_{j=0}^{h_t} \{t + H_t + \ell_t + j\}, & \text{if t is a boundary straddle point} \end{cases}$$

Here h_t is the total deficiency of t, $H_t = \Sigma_{s<t} h_s$, and ℓ_t is the number of boundary straddle points that are smaller than t.

$$\underline{\alpha}(s) = \begin{cases} +1 & \text{if } x^{-1}\{s\} \text{ is a (+)-point} \\ -1 & \text{if } x^{-1}\{s\} \text{ is a (-)-point} \\ +1 & \text{if } s=t+H_t+\ell_t+h_t, \text{ where } t\neq 1 \text{ is a (+)-boundary straddle point,} \\ & \text{or if } s=1+H_1+\ell_1 \text{ when } t=1 \text{ is a (+)-boundary straddle point.} \\ -1 & \text{if } s=t+H_t+\ell_t+h_t, \text{ where } t\neq 1 \text{ is a (-)-boundary straddle point,} \\ & \text{or if } s=1+H_1+\ell_1, \text{ when } t=1 \text{ is a (-)-boundary straddle point.} \\ 0 & \text{if } x^{-1}\{s\} \text{ is a straddle point which is not a boundary straddle point.} \end{cases}$$

The complete characterization of the case of uniqueness is then given by

Assertion 18: $Z(F; f,g)$ is a singleton if and only if

either i) $\Sigma_{i=1}^{k} h_i \geq n$, where h_i are the total deficiencies

or ii) There exists $u^* \in Z(F; f,g)$ such that $S^+(\alpha) \geq n$.

The details and proofs of the results described in the foregoing analysis of the C[a,b] case can be found in [2].

We consider next the case of C(X), where $X=\{t_i\}_1^k$ consists of a finite number of points and F is a Chebyshev system. The procedure for constructing an element of the center consists of 3 stages: 1) Identify candidates for straddle points. These are the points where $\|f-g\|$ is achieved. 2) Use the technique (see [9]) of weighted approximation with interpolation. Here the interpolation is required at the points chosen in step 1.

The function G(t) to be approximated passes through $\frac{f(t_i)+g(t_i)}{2}$, $i=1,\ldots k$. The weight at t_i is $1/[g(t_i)-f(t_i)+r]$, where r is $\|f-g\|$. If the error is ≤ 1, we have found an element of the center. If the error is > 1, points are not straddle points, then there exists a unique element in the center and we pass to: Step 3) consisting of using a modified Remez exchange algorithm.

We finally proceed to devise an algorithm for finding an element of $Z(F; f,g)$ where F is a Chebyshev system and the underlying space is C[a,b]. We choose a system of partitions of [a,b], with the norm of the partition tending to 0. At each stage we identify candidates for straddle points and proceed through steps 1)-2) as above. If the error is ≤ 1, we then proceed to construct the best restricted range approximation [14] to the function G(t), where the range at t_i is $[f(t_i)-r, g(t_i)+r]$. If for all partitions, starting from some n, there are no straddle points, then the sequence of unique elements of the center for the n-th partition will converge to the center. If, on the other hand, there exists a sequence of partitions with straddle points, then the sequence of best restricted range approximations can be shown (by employing results from [11]) to converge to an element of the center.

Details of the results of this portion will be published separately.

REFERENCES

1. Amir, D., and Ziegler, Z., Relative Chebyshev centers in normed linear spaces; Part 1, to appear in *J. Approx. Th.*
2. Amir, D., and Ziegler, Z., Relative Chebyshev centers in normed linear spaces; Part 2, submitted to *J. Approx. Th.*
3. Day, M. M., James, R. C, and Swaminathan, S., Normed linear spaces that are uniformly convex in every direction, *Can. J. Math. 23* (1971), 1051-1059.
4. Dunham, C. B., Simultaneous Chebyshev approximation of functions on an interval, *Proc. Amer. Math. Soc. 18* (1967), 472-477.
5. Garkavi, A. L., The Chebyshev center and the convex hull of a set, *Usp. Mat. Nauk. 18* (1964), 139-145.
6. Holland, A. S. B., McCabe, J. H., Phillips, G. M., and Sahney, B. N., Best Simultaneous L_1 approximation, *J. Approx. Th. 24* (1978),361-365.
7. Laurent, P. J., and Tuan, P. D., Global approximation of a compact set by elements of a convex set in a normed space,*Num.Math.15*(1970),137-150.
8. Ling, W. H., McLaughlin, H. W., and Smith, M. L., Approximation of random functions, *J. Approx. Th. 20* (1977), 10-22.
9. Loeb, H. L., Moursund, D. G., Schumaker, L. L., and Taylor, G. D., Uniform Generalized Weight Function Polynomial Approximation with Interpolation, *SIAM J. Num. Anal. 6* (1969), 284-293.
10. Phillips, G. M., McCabe, J. H., and Cheney, E. W., On Simultaneous Chebyshev Approximation, *JAT 27* (1979), 93-98.
11. Rivlin, T. J., and Cheney, E. W., A comparison of Uniform Approximations on an interval and a finite subset thereof, *SIAM J. Numer. Anal. 3* (1966), 311-320.
12. Rice, J. R.,"The approximation of functions", vol. 1, Addison-Wesley, Palo Alto, Ca., (1964).
13. Rozema, E. R., and Smith, P. W., Global approximation with bounded coefficients, *J. Approx. Th. 16* (1976), 162-174.
14. Schumaker, L. L., and Taylor, G. D., On Approximation by polynomials having restricted ranges, II, *SIAM J. Numer. Anal. 6* (1969), 31-36.

PERTURBATIONS OF OUTER INVERSES

P. M. Anselone
Department of Mathematics
Oregon State University
Corvallis, Oregon

M. Z. Nashed[1]
Department of Mathematical Sciences
University of Delaware
Newark, Delaware

1. Introduction

An outer inverse B^{ϕ} of a linear operator B satisfies $B^{\phi}BB^{\phi} = B^{\phi}$ or, equivalently, $B^{\phi}B = I$ on the range of B^{ϕ}. Outer inverses are useful in the approximate solution of linear operator equations $Bx = y$ when B is not one-to-one. This study compares bounded outer inverses of linear operators A and B under reasonable conditions on A and B. Special constructions of outer inverses are devised for this purpose. Applications include convergence results for approximate solutions of compact operator equations. Outer inverses have useful properties that circumvent the instability that usually arises in the approximation of inner inverses $B^{\psi}(BB^{\psi}B = B)$ and of generalized inverses.

2. Continuous Dependence of Outer and Generalized Inverses in Banach Spaces

Let X be a Banach space, and $[X]$ the space of all

[1]Partially supported by NSF Grant MCS-79-04480.

Copyright © 1980 by Academic Press, Inc.
All rights of reproduction in any form reserved.
ISBN: 0-12-171050-5

bounded linear operators on X. We let $A, B, C, \ldots \in [X]$, M, S be closed subspaces of X, and denote projectors by $E, F, P, Q \in [X]$. The restriction of B to M is denoted by B_M. Let $B \in [X]$, and assume that the null space of B and the closure of the range of B admit topological complements, i.e.,

$$X = N(B) \oplus M, \quad QX = M,$$

$$X = \overline{BX} \oplus S, \quad FX = \overline{BX}.$$

The generalized inverse $B^{\dagger}_{Q,F} := B_M^{-1}F$ has (maximal) domain $D(B^{\dagger}) = BX \dotplus S$. Clearly $B^{\dagger}_{Q,F}$ is a function of B, Q and F. It follows easily that $B^{\dagger}$ is characterized by the equations $B^{\dagger}BB^{\dagger} = B^{\dagger}$, $B^{\dagger}B = Q$, and $BB^{\dagger} = F$, which together imply $BB^{\dagger}B = B$. Note that $Bx = y$ implies $B^{\dagger}y = Qx$. Hence, $y \in BX$ and $x := B^{\dagger}y$ imply $y = Bx$. $B^{\dagger} \in [X]$ if and only if BX is closed. Nashed [6] has shown that $B^{\dagger}$ is continuous in Q, F. However, $B^{\dagger}$ is <u>not</u> continuous in B, even for matrices. This is one of the main points of departure of the theory of generalized inverses from that of inverse operators. The set of all invertible operators in $[X]$ is an open set in $[X]$, so is the set of all left (right) invertible operators. Inner inverses in $[X]$ are quite unstable: for a large class of operators A which have a bounded inner inverse, there are compact operators of infinite rank, so that $A + K$ has no bounded inner inverse no matter how small $||K||$ is chosen. This cannot happen for outer inverses (see [6]): the set of all operators in $[X]$ with bounded outer inverses is open in $[X]$. This result motivates our study of perturbations and approximations for

<u>outer</u> inverses rather than for inner inverses (see also [5] for some applications of outer inverses to nonlinear analysis). This study is closely related to the approach which was initiated by Moore and Nashed [2] and further investigated in [3]. The essence of their approach is to seek a <u>variant</u> $B^{\#}$ of $B^{\dagger}$ which has some properties of $B^{\dagger}$, $B^{\#} \in [X]$, and $B^{\#}$ is continuous in B. This variant turns out to be always an outer inverse of B, but not necessarily an inner inverse on all of X. Using the setting and results of [3], it follows in particular that if $\{C_n \in [X]\}$ is a sequence such that $C_n B C_n - C_n \to 0$, $BC_n B - B \to 0$, $C_n B \to Q$ and $BC_n \to F$ in the uniform operator topology, then $\{C_n\}$ converges to $B^{\dagger}$. In the case of matrices, if $A_n \to A$ and $\operatorname{rank} A_n = \operatorname{rank} A$ for $n \geq N$, then $C_n := A_n^{\dagger}$ provides a sequence with the above properties. We shall not require these properties, since we are only interested in outer inverses. We do not require BX to be closed.

3. Outer Inverses and Regularization

Henceforth all operators are bounded. Let B^{ϕ} be an outer inverse of $B: B^{\phi}BB^{\phi} = B^{\phi}$. Clearly

$$B^{\phi}BB^{\phi} = B^{\phi},\ B^{\phi}X = M \Longleftrightarrow B^{\phi}B_M = I_M.$$

$Q := B^{\phi}B$, $F := BB^{\phi}$, and $B^{\phi}X =: M$ imply

$Q^2 = Q$, $QX = M$, $N(Q) \supset N(B)$, M is closed and $X = M \oplus N(Q)$, where $N(B)$ denotes the null space of B. Similarly, $F^2 = F$, $FX = BM$, $N(F) = N(B^{\phi})$, BM is closed and $X = BM \oplus N(B^{\phi})$. Note that FX is a <u>proper</u> subspace of BX; unless B^{ϕ} is also an inner inverse. This can be used to our advantage. In particular we can choose outer inverses with finite dimensional ranges. An appropriate choice of outer inverses is tantamount to some regularization. For example, a truncated singular function expansion of the generalized inverse of a compact operator K in a Hilbert space is an outer inverse of K but not an inner inverse.

4. Comparison of A^ϕ and B^ϕ

Let $A^\phi A = P$, $AA^\phi = E$, $B^\phi B = Q$ and $BB^\phi = F$. Then

$$B^\phi - A^\phi = B^\phi(A - B)A^\phi + B^\phi(I - E) - (I - Q)A^\phi.$$

This identity with outer inverses replaced by generalized inverses has been used by Wedin [7] and Nashed [4] for developing some aspects of perturbation theory of generalized inverses of matrices and operators, respectively. It follows easily that

$$B^\phi(I - E) = 0 \Longleftrightarrow N(A^\phi) \subset N(B^\phi)$$
$$(I - Q)A^\phi = 0 \Longleftrightarrow A^\phi X \subset B^\phi X.$$

Thus $B^\phi - A^\phi = B^\phi(A - B)A^\phi$ if $N(A^\phi) \subset N(B^\phi)$ and $A^\phi X \subset B^\phi X$. This perturbation result yields the following theorem on the approximation of outer inverses.

Theorem. Suppose that $N(A^\phi) \subset N(A_n^\phi)$, $A^\phi X \subset A_n^\phi X$, and $\{A_n^\phi\}$ is uniformly bounded. Then $A_n \to A$ implies $A_n^\phi \to A^\phi$ (pointwise), and $||A_n - A|| \to 0$ implies $||A_n^\phi - A^\phi|| \to 0$.

5. Construction of Bounded Outer Inverses with Prescribed Range

Given B, M and $P^2 = P$ with $PX = M$, we seek B^ϕ such that $B^\phi X = M$. Let C be such that $CB = P - H$, where $HM \subset M$ and $||H_M|| = \delta < 1$. Then $CB_M = I_M - H_M$ and $B^\phi := (I - H)_M^{-1}C$ exists. Clearly $B^\phi X = M$ and $||B^\phi - C|| \leq \frac{\delta||C||}{1-\delta}$. Also the projector $Q := B^\phi B$ is close to P: $||P - Q|| \leq \frac{\delta||I - P||}{1 - \delta}$.

This permits a construction of an outer inverse of an operator from an outer inverse of another operator, with both outer inverses having the same range. Given B, and given A^ϕ with $A^\phi A = P$, $A^\phi X = M = PX$, we seek B^ϕ such that $B^\phi X = M$. Let $A^\phi B = P - H$, $HM \subset M$. Suppose $||H_M|| = \delta < 1$. Then there exists $B^\phi := (I_M - H_M)^{-1}A^\phi$ and $B^\phi X = M$. Furthermore

$$||A^\phi - B^\phi|| \le \frac{\delta||A^\phi||}{1 - \delta}$$

and
$$||P - Q|| \le \frac{\delta||I - P||}{1 - \delta},$$

where $Q := B^\phi B$.

6. Setting for Integral Operators

The preceding perturbation and approximation results, combined with the theory of collectively compact operators (see [1]) provide a basis for an approximation theory for outer inverses for integral operators. In the setting of Section 4, given $A = I - K$, $B = I - L$ and $H = P - CB$, it follows that

$$(I - K)^\phi = P + (I - K)^\phi K.$$

Let $C = P + (I - K)^\phi L$. Then $CB = P - H$, where $H = (I - K)^\phi (L - K)L$. Let $\delta = ||H_M|| < 1$.

Then there exists

$$(I - L)^\phi = (I_M - H_M)^{-1}A^\phi(I + L - K),$$

$$(I - L)^\phi - (I - K)^\phi = (I_M - H_M)^{-1}[H(I - K)^\phi + (I - K)^\phi(L - K)],$$

and

$$||(I - L)^\phi y - (I - K)^\phi y||$$
$$\le \frac{\delta||(I - K)^\phi y|| + ||(I - K)^\phi(L - K)y||}{1 - \delta}.$$

An important application is when $K_n \to K$ (pointwise), K is compact and $\{K_n\}$ collectively compact. We seek approximations to $(I - K)x = y$ using $(I - K_n)x_n = y$ when $I - K$ is not invertible. This setting is especially relevant to eigenvalue problems.

References

[1] Anselone, P.M., Collectively Compact Operator Approximation Theory, Prentice-Hall, Englewood Cliffs, New Jersey, 1971.

[2] Moore, R.H., and Nashed, M.Z., Approximations of generalized inverses of linear operators in Banach spaces, in "Approximation Theory" (G.G. Lorentz, ed.), pp. 425-428, Academic Press, New York, 1973.

[3] __________, Approximations to generalized inverses of linear operators, SIAM J. Appl. Math. 27 (1974), 1-16.

[4] Nashed, M.Z., ed., Generalized Inverses and Applications, Academic Press, New York, 1976; especially pp. 193-243, 325-396.

[5] Nashed, M.Z., Generalized inverse mapping theorems, in "Nonlinear Analysis in Abstract Spaces" (V. Lakshmikantham, ed.), pp. 210-245, Academic Press, New York, 1978.

[6] __________, On the perturbation theory for generalized inverse operators in Banach spaces, in "Functional Analysis Methods in Numerical Analysis" (M.Z. Nashed, ed.), pp. 180-195, Springer-Verlag, New York, 1979.

[7] Wedin, P.Å., Perturbation theory for pseudo-inverses, BIT, 13 (1973), pp. 217-232.

APPROXIMATION OF SET-VALUED OPERATORS BY SEQUENCES OF SET-VALUED COLLECTIVELY COMPACT OPERATORS

Rainer Ansorge [1]

Institut für Angewandte Mathematik
der Universität Hamburg

Dedicated to Professor G.G. Lorentz on his 70th birthday

I. INTRODUCTION

From the point of view of applications, the investigation of ordinary differential equations with discontinuous terms turns out to become more and more important. The theoretical as well as the numerical treatment of such problems requires the use of suitable non-classical definitions of what will be called a solution.

For first order initial value problems, a definition was given by Filippow (3), where - in a first step - the originally given differential equation was transformed into a set-valued differential equation. A corresponding concept can be given for certain boundary value problems, but let us consider here initial value problems only.

[1] Present address: Institut für Angewandte Mathematik,
Universität Hamburg,
Bundesstr. 55, D- 2000 Hamburg, W-Germany

Copyright © 1980 by Academic Press, Inc.
All rights of reproduction in any form reserved.
ISBN: 0-12-171050-5

A very clear access to the theoretical and numerical treatment of classical one-valued differential equations can be described by means of operator approximation by sequences of collectively compact operators (see p.e.(1)). The attempt to use this access also for set-valued differential equations, leads very naturally to the necessity of the introduction and investigation of set-valued collectively compact operators.

A brief introduction to this scope and applications to initial value problems are presented here. A more complete draft will appear soon in Computing (2).

II. SET-VALUED COLLECTIVELY COMPACT OPERATORS

Let (X,ϱ) be a metric space, $B \subset X$ closed, $B \neq \emptyset$.

Def.: A sequence $(T_n)_{n\in\mathbb{N}}$ of set-valued operators $(B \to X)$ is said to be pointwise convergent to the set-valued operator $T(B \to X)$ if

$$\forall x \in B: \quad y_n \in T_n x, \quad y_n \to y \Rightarrow y \in Tx.$$

Def.: $(T_n)_{n\in N}$ is said to be continously convergent to T on B if

$$\forall x \in B: \quad (x_n)_{n\in\mathbb{N}} \subset B, \ x_n \to x, \ y_n \in T_n x_n, \ y_n \to y \Rightarrow y \in Tx.$$

Th.1: If the operators T_n $(n=1,2,\ldots)$ are equi-upper-semicontinuous and if $(T_n)_{n\in\mathbb{N}}$ is pointwise convergent to T on B, then $(T_n)_{n\in\mathbb{N}}$ is even continuously convergent to T on B.

Here, an operator is called upper-semicontinuous on B if

$$\forall x \in B, \ \forall \varepsilon > 0, \ \exists \delta = \delta(\varepsilon,x) > 0: \ Tx' \subset [Tx]_\varepsilon, \ \forall x' : \varrho(x',x) < \delta$$

$$([Tx]_\lambda = \{ y \in X \mid \inf_{\eta\in Tx} \varrho(y,\eta) < \lambda \}).$$

We omit the proof of Theorem 1 as well as the proofs of Theorems 2,3 because these proofs are very similar to the proofs of the corresponding theorems for one-valued operators which were presented in (1).

<u>Def.:</u> The operators T_n $(n = 1,2,...)$ are called collectively compact if $\overline{\bigcup_{n\in\mathbb{N}} T_n B}$ is compact.

<u>Th.2:</u> Assume

a) $T_n (n = 1,2,...)$ collectively compact

b) $\forall\, n \in \mathbb{N} \quad \exists\, x_n\; B: x_n \in T_n\, x_n$

c) $T_n \rightarrow T$ continuously on B

Then: $\exists\, x \in B : x \in Tx$.

Moreover: The set of fixed points x_n of $T_n (n = 1,2,...)$ mentioned in assumption b) of Theorem 1 is uniformly set-convergent to a non-empty subset of the set of fixed points of T, according to

<u>Th.3:</u> For N sufficiently large, all fixed points $x_n \in B$ of T_n, $n \geq N$, are located in given neighborhoods of the fixed points of T.

Theorem 3 yields the numerical potentialities, particularly if the fixed point $x \in Tx$ is unique; in this case Theorem 3 says $x_n \rightarrow x$.

III. SET-VALUED INITIAL VALUE PROBLEMS

Consider the problem

$$\begin{gathered} \dot{y} \in F(t,y) \quad t_o \leq t \leq t_o + a \\ y(t_o) = y_o. \end{gathered} \tag{1}$$

We want to construct a solution $y \in X = C[t_o, t_o+a]$. F is assumed to be a mapping from $[t_o, t_o+a] \times \mathbb{R}$ into the set of closed subsets of $\mathbb{R}$ which has the properties

1) $$F(t,z) = \bigcap_{\delta>0} \overline{\text{Kon} \bigcup_{\substack{|t'-t|<\delta \\ |z'-z|<\delta}} F(t',z')} \neq \emptyset \tag{2}$$

(Kon A means the smallest convex subset of $\mathbb{R}$ which contains A)

2) $F(t,z)$ uniformly bounded for all $(t,z) \in [t_o, t_o+a] \times \mathbb{R}$. (3)

Def.: $y \in X$ is called a solution of (1) if there is a Lebesgue-integrable function ξ such that

$$y(t) = y_o + \int_{t_o}^{t} \xi(s)ds, \quad \xi(s) \in F(s,y(s)) \text{ a.e. .}$$

The fixed point problem corresponding to (1) is $x \in Tx$ with

$$Tx = \left\{ y \mid y(t) = y_o + \int_{t_o}^{t} \xi(s)ds, \ \xi(s) \in F(s,x(s)) \text{ a.e.} \right\}.$$

The discrete problems is assumed to be

$$x_n \in T_n x_n$$

with

$$T_n x = \left\{ y_n \mid y_n(t) = y_o + \int_{t_o}^{t} \xi_n(s)ds, \quad \xi_n \in H_n x \quad \text{a.e.} \right\}$$

where

$$H_n x = \left\{ \xi_n \mid \xi_n(t) = \begin{cases} \xi_n^o & t_n^o \leq t < t_n^1 \\ \xi_n^1 \quad \text{for} & t_n^1 \leq t < t_n^2, \ \xi_n^i \in G_n^i x \\ \vdots & \\ \xi_n^{n-1} & t^{n-1} \leq t < t_n^n \end{cases} \right\},$$

$$G_n^i x = \overline{\text{Kon} \bigcup_{t_n^i \leq t < t_n^{i+1}} F(t,x(t))} \quad (i = 0,1,\ldots,n-1), \ t_n^i = \frac{ia}{n}.$$

Hence, ξ_n are step functions. For numerical purposes, the numbers $\xi_n^i \in G_n^i x$ can be chosen arbitrarily, p.e. by random processes.

<u>Th.4:</u> The operators T, T_n (n = 1,2,...) of this section fulfill the assumptions of Theorem 2 and also Theorem 3 applies.

Proof (withaut going into details):

1) $T_n x \neq \emptyset, \ \forall x \in X, \quad \forall n \in \mathbb{N}$ (from $F(t,x) \neq \emptyset$).

2) T_n (n = 1,2,...) collectively compact (from the uniform boundedness of F via the Arzelà-Ascoli-theorem)

3) $F(t',z') \subset [F(t,z)]_\varepsilon, \quad \forall (t',z') : |t'-t|<\delta, \ |z'-z|<\delta,$
$\delta = \delta(\varepsilon,t,z) > 0$ (from (2),(3)).

Hence, $\forall x \ \exists \xi$ measurable: $\xi(s) \in F(s,x(s))$ a.e. (Filippow), i.e. $Tx \neq \emptyset, \ \forall x \in X$.

4) Let $x_n \rightarrow x$, $y_n \in T_n x_n$, $y_n \rightarrow y$.

$y_n \in T_n x_n \Rightarrow \exists \xi_n \in H_n x_n : y_n(t) = y_o + \int_{t_o}^{t} \xi_n(s)ds.$

y is absolutely continuous, hence

$\exists \xi : y(t) = y_o + \int_{t_o}^{t} \xi(s)ds.$

$$\xi_n \in H_n x_n \Rightarrow \xi(s) \in \overline{\text{Kon} \bigcup_{t_{n_o}^i \leq t \leq t_{n_o}^{i+1}} \bigcup_{n \geq n_o} F(t,x_n(t))} \text{ a.e.}$$

for $t_{n_o}^i \leq s \leq t_{n_o}^{i+1}$ (i=0,...,n-1)

for all n_o sufficiently large.

Hence $\xi(s) \in F(s,x(s))$ a.e., i.e.

$$y \in Tx$$

($\Rightarrow T_n \rightarrow T$ continuously, from Theorem 2).

5) The existence of at least one fixed point y_n of every T_n can constructively be proven, p.e. by (numerical) construction via a set-valued explicit Euler-method

$$y_n(t_n^o) = y_o \quad (t_n^o = t_o, \ \forall n \in \mathbb{N})$$

$$y_n(t_n^{i+1}) = y_n(t_n^i) + \frac{a}{n}\xi_n^i, \quad \xi_n^i \in F(t_n^i, y_n(t_n^i)) \text{ (p.e. random)}$$

$$(i = 0,1,\ldots,n-1).$$

Remark: This proof is nothing else than a generalization and more structured version of one of the well-known proofs of Peano's existence theorem for one-valued initial value problems. The proof includes the uniform set-convergence of the set-valued Euler-method and can be extended to other numerical methods (p.e. set-valued Runge-Kutta-methods) by construction of other operators T_n.

Moreover, these investigations also hold - in a very similar way - for the theoretical and numerical treatment of certain set-valued boundary value problems, p.e. such which arise from certain physical systems with constraints.

REFERENCES

1. Anselone, P.M. and Ansorge, R., Numer. Functional Anal. Optim. (to appear).
2. Ansorge, R. and Taubert, K., Computing (to appear)
3. Filippov, A.F. (1961) Amer. Math. Soc. Transl. 42, 199

THE ROGERS q-ULTRASPHERICAL POLYNOMIALS[1]

Richard A. Askey[2]
Mourad E. H. Ismail[3]

INTRODUCTION

It is well known that the Chebychev polynomials play an important role in approximation theory. They are denoted by $T_n(x)$ and $U_n(x)$ with $T_n(\cos\theta) = \cos n\theta$, $U_n(\cos\theta) = \frac{\sin(n+1)\theta}{\sin\theta}$. They are orthogonal polynomials with respect to the weight functions $(1-x^2)^{-1/2}$ and $(1-x^2)^{1/2}$ respectively. One set of polynomials that connects these polynomials are the ultraspherical polynomials $C_n^\lambda(x)$, which are orthogonal with respect to $(1-x^2)^{\lambda-1/2}$ on $[-1,1]$ when $\lambda > -1$. There is a second set of polynomials which can be used to connect these polynomials. These are polynomials $p_n^\beta(x)$ which are orthogonal on $[-1,1]$ with respect to the weight function $(1-x^2)^{1/2}\{(1+\beta)^2-4\beta x^2\}^{-1}$. When $\beta = 0$ this is $(1-x^2)^{1/2}$ and when $\beta = 1$ it is $\frac{1}{4}(1-x^2)^{-1/2}$. The polynomials in this case can be found from the Bernstein-Szegö theory (9) and they were given explicitly by Geronimus (4).

Long before the Bernstein-Szegö theory was developed, L. J. Rogers introduced an extension of the ultraspherical polynomials, and contained in his polynomials are the polynomials of Geronimus. Rogers used his polynomials to establish the

[1]This research was partially supported by the National Science Foundation.
[2]University of Wisconsin, Madison, Wisconsin 53706
[3]Arizona State University, Tempe, Arizona 85281

Copyright © 1980 by Academic Press, Inc.
All rights of reproduction in any form reserved.
ISBN: 0-12-171050-5

Rogers-Ramanujan identities, see (1) and (8) for details and references. The Rogers polynomials $C_n(x;\beta|q)$ are generated by

$$(1-q^{n+1})\, C_{n+1}(x;\beta|q) = 2x(1-\beta q^n)\, C_n(x;\beta|q) - (1-\beta^2 q^{n-1}) C_{n-1}(x;\beta|q), \qquad (1.1)$$

when $n > 0$, $-1 < q < 1$ and

$$C_0(x;\beta|q) = 1,\ C_1(x;\beta|q) = 2x(1-\beta)/(1-q). \qquad (1.2)$$

As $q \to 1^-$, $C_n(x;q^\lambda|q) \to C_n^\lambda(x)$. The polynomials $C_n(x;\beta|q)$ are called continuous q-ultraspherical polynomials, (1). We shall establish the orthogonality relation

$$\int_{-1}^{1} C_n(x;\beta|q)\, C_m(x;\beta|q)\, w_\beta(x)\, (1-x^2)^{-1/2} dx \qquad (1.3)$$

$$= \frac{2\pi(\beta^2;q)_n(\beta;q)_\infty(\beta;q)_\infty}{(1-\beta q^n)(q;q)_n(\beta^2;q)_\infty(q;q)_\infty}\, \delta_{m,n}\ ,\ 0 < q < 1$$

where

$$w_\beta(\cos\theta) = \frac{(e^{2i\theta};q)_\infty(e^{-2i\theta};q)_\infty}{(\beta e^{2i\theta};q)_\infty(\beta e^{-2i\theta};q)_\infty}\ ,\ -1 < \beta < 1 \qquad (1.4)$$

and

$$(a;q)_\infty = \prod_0^\infty (1-q^n a) \text{ and } (a;q)_n = \prod_1^n (1-aq^{j-1})\ , \qquad (1.5)$$

as a consequence of the fairly general methods of (2). These methods essentially go back to Markoff but the deepest use of them was given by Pollaczek (7). We also use a recent result of Nevai (5). This procedure is briefly outlined in section 2. In section 3 we use the aforementioned method to establish (1.3). There are also other cases of orthogonality, treated in section 3, when the measure contains point masses.

2. Orthogonal Polynomials

If a sequence of polynomials $\{p_n(x)\}$ satisfies a three term recurrence relation

$$p_{n+1}(x) = (A_n x + B_n) p_n(x) - C_n p_{n-1}(x), \; n > 0, \tag{2.1}$$

and the initial conditions

$$p_o(x) = 1, \; p_1(x) = A_o x + B_o , \tag{2.2}$$

and if $A_n A_{n-1} C_n > 0$ for $n > 0$ then there exists a finite positive measure μ such that

$$\int_{-\infty}^{\infty} p_n(x) p_m(x) \, d\mu(x) = \{(A_o/A_n) \prod_1^n C_k\} \delta_{m,n}. \tag{2.3}$$

Furthermore the support of $d\mu$ is infinite, Chihara (3). The above conditions are necessary as well as sufficient. The polynomials of the second kind $p_n^*(x)$ satisfy the recursion (2.1) and the following initial conditions

$$p_n^*(x) = 0, \; p_1^*(x) = A_o , \tag{2.4}$$

hence $p_n^*(x)$ is a polynomial of degree n-1. When the support of $d\mu$ is bounded, a theorem of Markoff asserts that

$$\int_{-\infty}^{\infty} \frac{d\mu(t)}{z-t} = \lim_{n\to\infty} p_n^*(z)/p_n(z) \; , \; z \notin \text{support of } d\mu, \tag{2.5}$$

Szegö (9) or Chihara (3). The measure μ can be recovered from the Perron-Stieltjes inversion formula

$$F(z) = \int_{-\infty}^{\infty} \frac{d\mu(t)}{z-t} \text{ implies}$$

$$\mu(t_2) - \mu(t_1) = \lim_{\varepsilon\to 0} + \int_{t_1}^{t_2} \frac{F(t-i\varepsilon) - F(t+i\varepsilon)}{2\pi i} dt. \tag{2.6}$$

Sometimes it is easier to recover the absolutely continuous component of the measure from the following theorem of Nevai (5).

Theorem 1. If $\sum_{1}^{\infty} \{ \left|\frac{B_n^2}{A_o C_1 \cdots C_n}\right|^{\frac{1}{2}} + \left|\left(\frac{C_{n+1}}{A_n A_{n+1}}\right)^{\frac{1}{2}} - \frac{1}{2}\right| \}$ converges then for almost every x in the support of $d\mu$ the limiting relation

$$\limsup_n \{\mu'(X)\sqrt{1-x^2}\, \tilde{p}_n^2(x)\} = 2/\pi, \tag{2.7}$$

holds, where $\{\tilde{p}_n(x)\}$ is the orthonormal set associated with $\{p_n(x)\}$ and $\mu'(x)$ is the absolutely continuous component of $d\mu$.

The following method of Darboux is very useful in determining the asymptotic behavior of $p_n(x)$ and $p_n^*(x)$ which provide the clue to computing the measure μ explicitly.

Theorem 2. (Darboux Method). Let $f(z) = \sum_0^\infty f_n z^n$ be analytic in $|z| < r$ and assume that there is a comparison function $g(z) = \sum_0^\infty g_n z^n$ such that $g(z)$ is analytic in $|z| < r$ and $f(z) - g(z)$ is continuous on $|r| = r$. Then $f_n = g_n + o(r^{-n})$. Darboux' method in the above form follows from the Riemann Lebesgue lemma, Olver (6).

3. Orthogonality Relations

We first derive a generating function for $C_n(x;\beta|q)$. Let

$$\sum_0^\infty C_n(x;\beta|q)t^n = C(x,t). \tag{3.1}$$

Multiply (1.1) by t^{n+1}, add the resulting equations for $n > 0$ and use (1.2) to get

$$C(x,t) = \left(\frac{1-2x\beta t+\beta^2 t^2}{1-2xt+t^2}\right) C(x,qt). \tag{3.2}$$

Set

$$1 - 2xt + t^2 = (1-ta)(1-tb), \quad |a| \le |b|. \tag{3.3}$$

It is easy to see that $|a| = |b|$ if and only if $x \in [-1,1]$. Iterating (3.2) yields

$$C(x,t) = (\beta ta;q)_\infty(\beta tb;q)_\infty/\{(ta;q)_\infty(tb;q)_\infty\}, \tag{3.4}$$

since $C(x,tq^n) \to C(x,0) = 1$, as $n \to \infty$. The above formal manipulations can be justified because the right side of (3.4) is an analytic function of t in $|t| < a$. We now apply Darboux method to (3.4). A comparison function, when $x \notin [-1,1]$, is

$$(a^2\beta;q)_\infty(\beta;q)_\infty(1-tb)^{-1}/\{(a^2;q)_\infty(q;q)_\infty\},$$

since $ab = 1$. This and theorem 2 lead to

$$C_n(x;\beta|q) \sim \frac{(a^2\beta;q)_\infty(\beta;q)_\infty}{(a^2;q)_\infty(q;q)_\infty} b^n, \quad n \to \infty, \; x \notin [-1,1]. \tag{3.5}$$

The polynomials of the second kind can be similarly treated. The generating function

$$C^*(x,t) = \sum_0^\infty C_n^*(x;\beta|q)t^n, \tag{3.6}$$

satisfies the q difference equation

$$C^*(x,t) = \frac{(1-q)t\, C_1^*(x;\beta|q)}{(1-at)(1-bt)} + \frac{(1-\beta at)(1-\beta bt)}{(1-at)(1-bt)} C^*(x,qt),$$

whose solution, subject to $C^*(x,0) = 0$, is

$$C^*(x,t) = 2t(1-\beta) \sum_0^\infty q^k \frac{(\beta at;q)_k(\beta bt;q)_k}{(at;q)_{k+1}(bt;q)_{k+1}}. \tag{3.7}$$

Therefore we have

$$C_n^*(x;\beta|q) \sim \frac{2b(1-\beta)}{(1-a^2)} b^n \sum_0^\infty \frac{(\beta a^2;q)_k(\beta;q)_k}{(a^2q;q)_k(q;q)_k} q^k, \tag{3.8}$$

$$n \to \infty, \; x \notin [-1,1].$$

The boundedness of the support of $d\mu$ follows from theorem 2.2, page 111 in (3). Applying Markoff's theorem we obtain

$$\int_{-\infty}^{\infty} \frac{d\mu(t)}{x-t} = 2b(1-\beta)\frac{(qa^2;q)_\infty(q;q)_\infty}{(a^2\beta;q)_\infty(\beta;q)_\infty}\sum_0^\infty \frac{(\beta a^2;q)_k(\beta;q)_k}{(a^2q;q)_k(q;q)_k} q^k. \quad (3.9)$$

The necessary condition $A_nA_{n-1}C_n > 0$, $n > 0$ is equivalent to $-1 < \beta < q^{-1/2}$ when $0 < q < 1$; $-1 < q < 0$, and $\beta > -1$ when $q = 0$, [1]. We may assume $\beta \neq 1$ because

$$\lim_{\beta\to 1}\frac{(1-q^n)}{2(1-\beta)} C_n(\cos\theta;\beta|q) = \cos(n\theta),\ n > 0,$$

$$C_o(\cos\theta;\beta|q) = 1.$$

When $0 < q < 1$, $-1 < \beta < 1$, the right side of (3.9) has no poles, hence the measure has no point masses. When $0 < q < 1$, $1 < \beta < q^{-1/2}$ the right side of (3.9) has only one pole because $|a^2| \le 1$. This pole is at $a = \pm\beta^{-1/2}$, so

$$x = \pm(\beta^{1/2} + \beta^{-1/2})/2.$$

The point masses J at these poles is the residue (in x),

$$J = \frac{1}{2}\beta(q/\beta;q)_\infty(\beta-1)/(q\beta;q)_\infty. \quad (3.10)$$

When $x = \cos\theta \in (-1,1)$, a,b are $\exp(\pm i\theta)$ and the comparison function

$$\frac{(\beta;q)_\infty}{(q;q)_\infty}\left\{\frac{(\beta e^{2i\theta};q)_\infty}{(1-te^{-i\theta})(e^{2i\theta};q)_\infty} + \frac{(\beta e^{-2i\theta};q)_\infty}{(1-te^{i\theta})(e^{-2i\theta};q)_\infty}\right\}$$

implies the asymptotic formula

$$C_n(\cos\theta,\beta|q) \sim 2\left|\frac{(\beta;q)_\infty(\beta e^{2i\theta};q)_\infty}{(q;q)\ \ (e^{2i\theta};q)}\right|\cos(n\theta-\phi), \quad (3.11)$$

$$0 < \theta < \pi,$$

where

$$\phi = \arg\{(\beta;q)_\infty(\beta e^{2i\theta};q)_\infty/(e^{2i\theta};q)_\infty\}.$$

The asymptotic behavior of the orthonormal polynomials may be obtained from (2.3) and (3.11). The assumptions in Nevai's theorem are satisfied and (2.3), (2.7), and (3.11) establish

$$\mu'(x)\sqrt{1-x^2} = \frac{1}{2\pi}\,\frac{(q;q)_\infty(\beta^2;q)_\infty(e^{2i\theta};q)_\infty(e^{-2i\theta};q)_\infty}{(\beta;q)_\infty(\beta q;q)_\infty(\beta e^{2i\theta};q)_\infty(\beta e^{-2i\theta};q)_\infty},$$

$$x = \cos\theta$$

which gives the orthogonality relation (1.3). Recall that the point masses (3.10) must be added to the orthogonality relation when $1 < \beta < q^{-1/2}$. The orthogonality relation when $1 < \beta < q^{-1/2}$ is the following

$$\frac{(q;q)_\infty(\beta^2;q)_\infty}{(\beta;q)_\infty(\beta q;q)_\infty}\int_{-1}^{1} C_n(x;\beta|q)C_m(x;\beta|q)\,\frac{dx}{\sqrt{1-x^2}}$$

$$+\ \pi\beta\,\frac{(\beta-1)(q/\beta;q)_\infty}{(q\beta;q)_\infty}\{C_n(\xi;\beta|q)C_m(\xi;\beta|q)$$

$$+\ C_n(-\xi;\beta|q)C_m(-\xi;\beta|q)\}$$

$$= \frac{(1-\beta)(\beta^2;q)_n}{(1-\beta q^n)(q;q)_\infty}\,\delta_{m,n},$$

where

$$\xi = \tfrac{1}{2}(\beta^{1/2} + \beta^{-1/2}).$$

The two remaining cases, $-1 < \beta < -q^{-1}$ when $-1 < q < 0$, and $\beta > +1$ with $q = 0$ can be similarly treated and the corresponding measures do have point masses.

REFERENCES

1.) R. Askey and M. E. H. Ismail, A generalization of the ultraspherical polynomials, to appear in a book dedicated to the memory of P. Turan.

2.) R. Askey and M. E. H. Ismail, Recurrence relations, continued fractions and orthogonal polynomials, to appear.

3.) T. Chihara, An Introduction to Orthogonal Polynomials, Gordon and Breach, New York, 1978.

4.) Ja. Geronimus, On a set of polynomials, Ann. Math. (2) 31 (1930), 681-686.

5.) P. Nevai, Orthogonal polynomials, Memoirs of the Amer. Math. Soc., volume 213, 1979.

6.) F. W. J. Olver, Asymptotics and Special Functions, Academic Press, New York, 1974.

7.) F. Pollaczek, Sur une généralisation des polynômes de Jacobi, Memorial des Sciences Mathematique, volume 131, 1956.

8.) L. J. Rogers, Third Memoir on the expansion of certain infinite products, Proc. Lond. Math. Soc., 26 (1894-1895), 15-32.

9.) G. Szegö, Orthogonal Polynomials, American Math. Soc. Colloquium Publications 23, Fourth Edition, Providence, 1975.

MULTIPLE NODE GAUSSIAN THEORY FOR WEAK TCHEBYCHEFF SYSTEMS

R. B. Barrar
H. L. Loeb

Department of Mathematics
University of Oregon
Eugene, Oregon

Quadrature formulas of the form

$$Q(u) = \Sigma_{i=0}^{M_0-1} a_i u^{(i)}(0) + \Sigma_{i=1}^{s}\Sigma_{j=0}^{m_i-1} a_{ij} u^{(j)}(\xi_i) + \Sigma_{i=0}^{M_1-1} b_i u^{(i)} \tag{1}$$

where

$$0 \leq \xi_1 \leq \cdots \leq \xi_s \leq 1 \tag{2}$$

are valid for $Q(u_i) = \int_0^1 u_i(x)dx \quad i = 1,\cdots,N$; when the $u_i(x) \quad i = 1,\cdots,N$ form a Tchebycheff subspace U. This result can easily be obtained by extending known results of Barrow (1) and Karlin and Pinkus (3).

It is the purpose of the present paper to study quadrature formulas of the form (1) and (2) when the $u_i(x)$, $i = 1,\cdots,N$ are only assumed to be a weak Tchebycheff system, and in particular to apply our results to polynomial monosplines.

To treat the general case of a weak Tchebycheff system we follow Burchard (2) by considering the convexity cone K(U) defined as

Copyright © 1980 by Academic Press, Inc.
All rights of reproduction in any form reserved.
ISBN: 0-12-171050-5

$$K(U) = \{f : 0 < t_1 < \cdots < t_{N+1} < 1 \Rightarrow U\begin{pmatrix} 1, \cdots, N, f \\ t_1, \cdots, t_N, t_{N+1} \end{pmatrix} \geq 0\}$$

where $U\begin{pmatrix} 1, \cdots, \nu \\ t_1, \cdots, t_\nu \end{pmatrix} = \det\{u_i(t_j); i,j = 1, \cdots, \nu\}$.

We make the following basic assumption on the cone:

I. For each set, $0 < t_1 < t_2 < \cdots < t_s < 1$,

$$U[t_1, \cdots, t_s] \equiv \{(f(0), \cdots, f^{(M_o - 1)}(0), f(t_1), \cdots, f^{(m_1)}(t_1), f(t_2), \cdots, f^{(m_s)}(t_s), f(1), \cdots, f^{(M_1 - 1)}(1)) : f \in K(U)\}$$

contains a basis for R^N.

Using this basic assumption, we are able to show that if one applies the Gaussian transform

$$u_i(x; \epsilon) = \frac{1}{|\epsilon|\sqrt{2\pi}} \int_{-\delta}^{1+\delta} e^{-\frac{1}{2}\left(\frac{\eta - x}{\epsilon}\right)^2} u_i(\eta) d\eta$$

to the weak Tchebycheff system, then applies the Barrow results to the $u_i(x;\epsilon)$, and finally takes the limit as $\epsilon \downarrow 0$, one obtains:

Theorem 1. Let $\{u_i\}_{i=1}^N$ be a basis for the N-dimensional weak Tchebycheff subspace U of $C^k[-\delta, 1 + \delta]$ for some $\delta > 0$. Let $K(U)$ satisfy basic assumption I. Finally we are given a set of positive odd integers $\{m_i\}_{i=1}^s$ and two non-negative integers M_o and M_1 which satisfy the relationships:

a) $N = M_o + M_1 + \Sigma_{i=1}^s (m_i + 1)$.

b) $k \geq \max\{\max_{1 \leq i \leq s} m_i, \max_{0 \leq i \leq 1} (M_i - 1)\}$.

Then if $Q(u)$ is a positive functional, there exists a unique quadrature formula of the form (1) with

$$0 < \xi_1 < \cdots < \xi_s < 1.$$

As an application of these results we are able to prove the following generalization of the fundamental theorem of algebra for monosplines.

Theorem 2. Let $\Phi_p(x,y) = \frac{(x-y)_+^{p-1}}{(p-1)!}$ where $p \geq 3$. Further we are given: A non-negative integer $M_o \leq p$, odd integers $\{m_i\}_{i=1}^s$, and positive integers $\{n_i\}_{i=1}^r$ which satisfy the restrictions:

a) $N = s + \Sigma_{i=1}^s m_i + M_o = \Sigma_{i=1}^r n_i$

b) If $m = \max\limits_{1\leq i\leq s} \{m_i\}$ and $n = \max\limits_{1\leq i\leq r} n_i$, then

$m + n \leq p - 1$.

Then for each set of r distinct numbers, $0 < x_1 < x_2 < \cdots < x_r \leq 1$, there is a unique monospline of the form,

$$M(x) = \int_0^1 \Phi_p(x,y)dy - \sum_{j=0}^{M_o-1} a_j \Phi_p^{(j)}(x,0)$$

$$- \Sigma_{i=1}^s \sum_{j=0}^{m_i-1} a_{ij} \Phi_p^{(j)}(x,\xi_i),$$

such that

$$M^{(\ell)}(x_\nu) = 0 \quad \ell = 0,1,\cdots,n_\nu - 1, \; i = 1,\cdots,r.$$

Here $\Phi_p^{(j)}(x,\xi) = \frac{\partial^j}{\partial \xi^j} \Phi_p(x,\xi)$ and $0 \leq \xi_1 \leq \cdots \leq \xi_s \leq 1$.

Full details will appear elsewhere.

REFERENCES

1. Barrow, D., On multiple node Gaussian quadrature formulae, Math. Comp. 32, 1978, pp. 431-439.

2. Burchard, H., Interpolation and approximation by generalized convex functions, Ph.D. dissertation, Purdue University, West Lafayette, IN, 1968.

3. Karlin, S. and Pinkus, A., Gaussian quadrature formulae with multiple nodes, in Studies in Spline Functions and Approximation Theory, S. Karlin et al (eds.), Academic Press, New York, 1976, pp. 113-141.

ON STRONG UNICITY AND A CONJECTURE OF HENRY AND ROULIER

Martin W. Bartelt

Department of Mathematics
Christopher Newport College
Newport News, Virginia

Darrell Schmidt

Department of Mathematics
Oakland University
Rochester, Michigan

This note surveys some of the history of strong unicity and discusses some old and new results on the conjecture of Henry and Roulier about strong unicity constants.

Let $T_n(f)$ denote the best Chebyshev approximation to a function $f \in C[a,b]$ from Π_n the algebraic polynomials of degree n or less. Let Π denote the set of all algebraic polynomials.

Newman and Shapiro (1) introduced the concept of strong unicity (strong uniqueness) by demonstrating for a given $f \in C[a,b]$ the existence of a constant γ, $0 < \gamma \leq 1$, such that for all $p \in \Pi_n$,

$$||f-p|| \geq ||f-T_n(f)|| + \gamma \; ||T_n(f)-p|| \tag{1}$$

Actually Newman and Shapiro showed that (1) holds when approximating from any Haar set. Strong unicity, equation (1), does not hold when approximating in an L^pnorm - $1 < p < \infty$. The largest constant γ in (1) is called the strong unicity constant and denoted here by $\gamma_n(f)$. Bartelt and McLoughlin (2) showed that for $f \notin \Pi_n$,

Copyright © 1980 by Academic Press, Inc.
All rights of reproduction in any form reserved.
ISBN: 0-12-171050-5

$$\gamma = \inf_{\substack{p \in \Pi_n \\ p \neq 0}} \max_{x \in E_n(f)} \frac{f(x)-T_n(f)(x)}{||f-T_n(f)||} \frac{p(x)}{||p||} \tag{2}$$

where $E_n(f)$ denotes the set of extreme points

$$E_n(f) = \{x: |f(x) - T_n(f)(x)| = ||f - T_n(f)||\}.$$

Cline (3) found a suitable constant $1/K_n$ where (1) holds with $1/K_n$ replacing γ and thus $1/K_n(f) \leq \gamma_n(f)$. He let $x_0 < x_1 < \ldots < x_{n+1}$ be an alternate for $f - T_n(f)$ and for $j = 0, \ldots, n+1$ he let q_j be the unique function in Π_n satisfying $q_j(x_i) = \text{sgn}(f(x_i) - T_n(f)(x_i))$, $i = 0, \ldots, n + 1, i \neq j$. Then

$$K_n = \max_{0 \leq j \leq n+1} ||q_j|| \tag{3}$$

Bartelt and Henry (4) showed that sometimes, $\gamma_n(f) > \max 1/K_n(f)$ where the max is taken over all possible alternates. In the special case where E_n contains exactly $n + 2$ points, Henry and Roulier (5) observed that $\gamma_n = 1/K_n$.

Cline (3) was studying Lipschitz constants for the best approximation operator. Freud (6) using interpolating polynomials similar to those Cline used, had already shown that the best approximation operator T_n satisfies

$$||T_n(f) - T_n(g)|| \leq \lambda_f ||f-g|| \tag{4}$$

for some constant λ_f depending on f and for all $g \in C[a,b]$. Cheney (7) related strong unicity and Lipschitz constants by showing that $2/\gamma(f)$ is a suitable Lipschitz constant in (4) i.e., $\lambda_f \leq 2/\gamma(f)$.

Now γ (or the Lipschitz constant) has been studied as a function of f, as a function of the interval $[a,b]$ and as a function of the approximating subspace. For example, Henry and Roulier (8) sought a uniform Lipschitz constant as the interval changed and n and f were fixed.

It is known (5) that there is no uniform strong unicity constant for fixed $n \geq 1$ and $[a,b]$ for all the functions $f \in C[a,b]$. (Also see (9-12) for related results.) This note is concerned with a uniform strong unicity constant as the approximating subspace changes and $[a,b]$ and f remain fixed.

Poreda (13) wondered which functions f satisfied $\gamma_n(f)$ is bounded away from 0 for all n. Let B denote the set of all $f \in C[a,b]$ such that there exists a $\delta = \delta(f)$ such that $\gamma_n(f) \geq \delta > 0$ for all n.

Conjecture of Henry and Roulier $B = \Pi$

Clearly $\Pi \subseteq B$ and Poreda (13) showed that $B \neq C(I)$. We have shown (14) that B, just like Π, is of first category in $C(I)$. Since $\gamma(\alpha f) = \gamma(f)$ for any real number α and for $n \geq N$, $\gamma_n(f+p) = \gamma_n(f)$ for any $p \in \Pi_N$ (2), the set B satisfies $\alpha B \subseteq B$ and $B + \Pi \subseteq B$. But it is not known if B is closed under addition or multiplication. It is not true that $\gamma(f+g) = \gamma(f) + \gamma(g)$. In fact there is an example where $\gamma(f+g) < \min(\gamma(f), \gamma(g))$.

If f is a polynomial, then $\gamma_n(f) = 1$ for all sufficiently large n. We have shown the converse. If $\gamma_n(f) = 1$ for any $n \geq 1$ then, f is a polynomial. Thus the conjecture can be restated as: There exists a constant $\delta > 0$ such that $\gamma_n(f) \geq \delta > 0$ for all n if and only if $\gamma_n(f) = 1$ for all sufficiently large n.

Results related to the conjecture often depend on the distribution of $E_n(f)$ in $[a,b]$ and on $|E_n(f)|$ the cardinality of $E_n(f)$. For example Schmidt (15) showed that:

Theorem 1 If $|E_n(f)| = n + 2$ for infinitely many n, then $f \notin B$.

Theorem 2 If $E_n(f) \cap [c,d] = \emptyset$ for some non-degenerate interval $[c,d] \subseteq [a,b]$ for infinitely many n, then $f \notin B$.

Theorem 1 does not solve the conjecture since it has been proven (14) that there do exist functions $f \notin \Pi$ for which $|E_n(f)| > n + 2$ for all n.

The results of Henry and Roulier (5) use the interpolatory characterization (3) in the case $|E_n(f)| = n + 2$. For $|E_n(f)| > n + 2$, we obtain a partial characterization using other interpolating polynomials. Without loss we can assume $T_n(f) \neq T_{n+1}(f)$. Then $e_n(f)(x) = f(x) - T_n f(x)$ can have no more than $n + 2$ points of alternation in $E_n(f)$ even though $|E_n(f)|$ can be greater than $n + 2$. Decompose $E_n(f)$ into $n + 2$ non-empty subsets

$$E^0, E^1, \ldots, E^{n+1}$$

where (i) $E^{(i)}$ is compact, $i = 0, \ldots, n + 1$, (ii) $\max E^{(i)} < \min E^{i+1}$, $i = 0, \ldots, n$, (iii) $\sigma(x) = \operatorname{sgn} e_n(f)(x)$ is constant on $E^{(i)}$, $i = 0, \ldots, n + 1$, and (iv) $\operatorname{sgn}_{E^i} e_n(f)(x) = - \operatorname{sgn}_{E^{i+1}} e_n(f)(x)$, $i = 0, \ldots, n$. It is shown that there exist $q_{nj} \in \Pi_n$ and for $j = 0, \ldots, n + 1$, $i \neq j$ there are points $y^i_{nj} \in E^i$ such that $\sigma(x) \cdot q_{nj}(x) \leq 1$ for $x \in E_n(f)$ and $\sigma(y^i_{nj})\, q_{nj}(y^i_{nj}) = 1$.

<u>Theorem 3</u> (14) Let $f \in C[a,b]\backslash\Pi$ and $T_n(f) \neq T_{n+1}(f)$, then

$$1/\gamma_n(f) \geq \max_{0 \leq j \leq n+1} ||q_{nj}||$$

This is similar to characterization (3) and leads to

<u>Theorem 4</u> (14) Let $f \in C[a,b]\backslash\Pi$. If $|E_n(f)| \leq n + 4$ for infinitely many n, then $f \notin B$.

Let n_k denote an integer n for which $T_n(f) \neq T_{n+1}(f)$ and let Δ_{n_k} be the largest diameter of the sets E^i in (5) in the decomposition of E_{n_k}.

Theorem 5 (14) If $\liminf_{k\to\infty} n_k \Delta_{n_k} = 0$, then $f \notin B$.

This generalizes the result in Theorem 1 since there

$|E_n(f)| = n + 2$ and thus $\Delta_{n_k} = 0$.

REFERENCES

1. D. J. Newman and H. S. Shapiro, Some theorems on Chebyshev approximation, Duke Math. J. 30 (1963), 673-681.

2. M. W. Bartelt and H. W. McLaughlin, Characterizations of strong unicity in approximation theory, J. Approximation Theory 9 (1973), 255-266.

3. A. K. Cline, Lipschitz conditions on uniform approximation operators, J. Approximation Theory 8 (1973), 160-172.

4. M. W. Bartelt and M. S. Henry, Continuity of the strong unicity constant on C(X) for changing X, J. Approximation Theory (to appear) 1980.

5. M. S. Henry and J. A. Roulier, Lipschitz and strong unicity constants for changing dimention, J. Approximation Theory 22 (1978), 85-94.

6. G. Freud, Eine Ungleichung Fur Tschebyscheffsche Approximations Polynome, Acta Scientarium Mathematicarum 19 (1958), 162-164.

7. E. W. Cheney, "Introduction to Approximation Theory," McGraw-Hill, New York, 1966.

8. M. S. Henry and J. A. Roulier, Uniform Lipschitz Constants on Small Intervals. J. Approximation Theory 21 (1977), 224-235.

9. M. S. Henry and L. R. Huff, On the behavior of the strong unicity constant for changing dimension, J. Approximation Theory (to appear).

10. M. S. Henry and D. Schmidt, Continuity theorems for the product approximation operator, in "Theory of Approximation with Applications (A. G. Law and B. N. Sahney, Eds.), Academic Press, New York, 1976.

11. A. Kroo, The continuity of best approximations, Acta Mathematica Scientiarum Hungaricae, 30 (1977), 175-188.

12. S. O. Paur and J. A. Roulier, Lipschitz and Strong unicity constants on subintervals, Abstract in International Conference on Approximation theory, University of Texas at Austin, January 8-12, 1980.

13. S. J. Poreda, Counterexamples in best approximation, Proceedings Amer. Math. Soc. 56 (1976), 167-171.

14. M. W. Bartelt and D. Schmidt, On Poreda's problem for strong unicity constants, submitted, 1980.

15. D. Schmidt, On an unboundedness conjecture for strong unicity constants, J. Approximation Theory 24 (1978), 216-223.

REPRESENTATION FORMULAS FOR CONFORMING BIVARIATE INTERPOLATION

G. Baszenski
H. Posdorf

Rechenzentrum, University of Bochum
Bochum, W. Germany

F.J. Delvos

University of Siegen
Siegen, W. Germany

I. INTRODUCTION

MELKES [10] proved the following theorem.

THEOREM 1: Let $S=[x_o,x_1]\times[y_o,y_1]$, $N=2n$ or $N=2n+1$, $n\in N_o$, $F\in C^N(S)$. Then there exists exactly one polynomial

$$p_{2N+1}\in M_{2N+1}:=\{\alpha(x,y):\ \alpha(x,y)=\sum_{i,j=o}^{2N+1}\alpha_{ij}x^iy^j,\ \alpha_{ij}=o \text{ for } 2n+2\leq i,j\leq 2N+1,$$

$$D_x^k\alpha(x_\kappa,\cdot),\ D_y^k\alpha(\cdot,y_\kappa)\in\Pi_{2N-2k+1},\ k\leq n,\ \kappa=o,1\}$$

satisfying

$$D_x^iD_y^jp_{2N+1}(x_\kappa,y_\lambda)=D_x^iD_y^jF(x_\kappa,y_\lambda),\ \kappa,\lambda\in\{o,1\},\ o\leq i+j\leq N.$$

Ex.: For N=o,1,2 we will consider the nodal configurations where the usual notation for nodal point function and derivative evaluations is adopted; a dot denotes a function value, a single circle the values of all derivatives of order one, a double circle the values of all derivatives of order less than or equal to two.

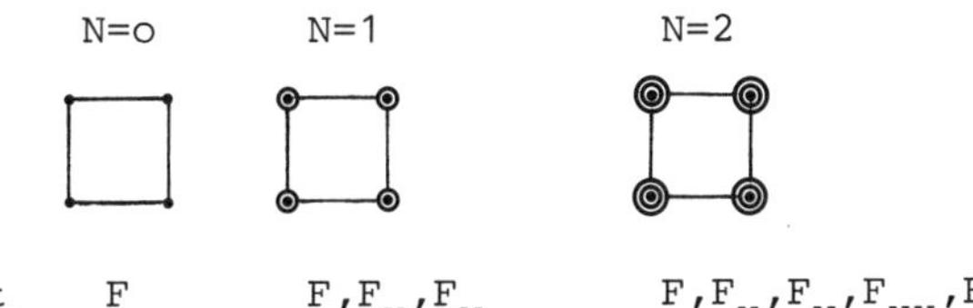

Nodal point evaluations: F $\quad F,F_x,F_y \quad$ 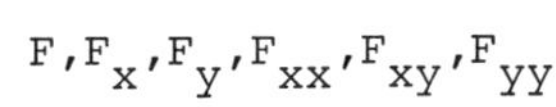

Copyright © 1980 by Academic Press, Inc.
All rights of reproduction in any form reserved.
ISBN: 0-12-171050-5

For N=o $p_1 \in M_1 = \Pi_{1,1}$ is the bilinear tensor product Lagrange interpolation polynomial, for N=1 $p_3 \in M_3$ represents the ADINI-element of Finite Elements; both elements are C^o-conforming. For N=2 $p_5 \in M_5$ is C^1-conforming.
In general the further elements have $C^{[N/2]}$-conformity (where rectangles with sides parallel to the coordinate axes are composed to polygons, MELKES [1o]).
MELKES called this interpolation problem reduced Hermite interpolation since the number of nodes is reduced when compared with the well known tensor product Hermite interpolation.
It is the aim of this paper to derive explicit representation formulas for the interpolation polynomial p_{2N+1} for all $N \in \mathbb{N}_o$.
For simplicity let in the following $S=[o,1] \times [o,1]$.

II. A BOOLEAN CHARACTERIZATION OF p_{2N+1}

We first recall some properties of univariate Hermite interpolation. It is well known that, for any k-times differentiable function f=f(x) defined on [o,1], there exists a unique polynomial $H_k(f)$ of degree less than or equal 2k+1 which interpolates to $f, \ldots, D^k f$ at the points u=o,1; i.e.

$$H_k: C^k[o,1] \to \Pi_{2k+1}, \quad D^i H_k(f)(u) = D^i f(u), \quad o \le i \le k, \ u=o,1.$$

PHILLIPS [11] derived for the two point Hermite interpolation polynomial $H_k(f)$ the representation formula

$$H_k(f)(x) = \sum_{u=o}^{1} \sum_{i=o}^{k} D^i f(u) q_{i,k}^{(u)}(x)$$

where

$$q_{i,k}^{(o)}(x) = \frac{x^i}{i!}(1-x)^{k+1} \sum_{s=o}^{k-i} \binom{k+s}{s} x^s, \quad q_{i,k}^{(1)}(x) = (-1)^i q_{i,k}^{(o)}(1-x).$$

Furthermore, we need the following definition: For $i,n \in \mathbb{N}_o$, $i \le n$ let

$$P_{i,n}: C^i[o,1] \to \operatorname{Im} P_{i,n} \subset \Pi_{2n+1}, \quad P_{i,n}(f)(x) = \sum_{u=o}^{1} D^i f(u) q_{i,n}^{(u)}(x).$$

Thus, we obtain

LEMMA 1: H_k and $P_{i,n}$ are projectors holding the relations

(i) $P_{i,n}P_{j,m}=\delta_{ij}P_{i,n}$, $i\leq n,\ i,j\leq m$,

(ii) $P_{i,n}H_k=H_kP_{i,n}=P_{i,n}$, $i\leq n\leq k$,

(iii) $H_kH_l=H_lH_k=H_k$, $k\leq l$.

We use the projectors H_k and $P_{i,n}$ to define the projectors H_k^x, H_k^y, $P_{i,n}^x$, $P_{i,n}^y$ by the method of parametric extensions, GORDON [6], so that

$$H_k^x(F)(x,y)=\sum_{u=o}^{1}\sum_{i=o}^{k}D_x^iF(u,y)q_{i,k}^{(u)}(x),$$

$$H_k^y(F)(x,y)=\sum_{u=o}^{1}\sum_{i=o}^{k}D_y^iF(x,u)q_{i,k}^{(u)}(y),$$

$$P_{i,n}^x(F)(x,y)=\sum_{u=o}^{1}D_x^iF(u,y)q_{i,n}^{(u)}(x),$$

$$P_{i,n}^y(F)(x,y)=\sum_{u=o}^{1}D_y^iF(x,u)q_{i,n}^{(u)}(y),$$

where $F=F(x,y)$ is a sufficiently smooth function on S.
In the sequel let $n = [N/2]$.

We now apply the concept of commutative Boolean sum interpolation developed by GORDON [6,7] to construct the projector of conforming reduced Hermite interpolation. Since

$$\{P_{o,n}^x,P_{o,n}^y,\ldots,P_{n,n}^x,P_{n,n}^y,H_n^x,H_n^y,\ldots,H_N^x,H_N^y\}$$

is a set of commuting projectors (cf. Lemma 1 and GORDON-CHENEY [8]) we are able to define the Boolean sum projector

$$B_{2N+1}=\bigoplus_{\mu=o}^{n}P_{\mu,n}^xH_{N-\mu}^y\oplus\bigoplus_{\mu=o}^{n}H_{N-\mu}^xP_{\mu,n}^y\ .$$

The interpolation properties of B_{2N+1} are given in

THEOREM 2: Suppose that $F\in C^N(S)$, $u,v\in\{o,1\}$, $o\leq i+j\leq N$. Then

$$D_x^iD_y^jB_{2N+1}(F)(u,v)=D_x^iD_y^jF(u,v).$$

The proof of Theorem 2 makes use of two facts: The first is the absorptive condition of Boolean sum projectors and the second are the interpolation properties of the product projectors $P_{i,n}^xH_{N-i}^y$; $H_{N-i}^xP_{i,n}^y$.

The result of Theorem 2 suggest that B_{2N+1} be called the projector of conforming reduced Hermite interpolation. Our next purpose is to determine the function precision of B_{2N+1}.

THEOREM 3: The function precision of B_{2N+1} is given by

$$\text{Im } B_{2N+1} = M_{2N+1} .$$

Because of Theorem 1 and Theorem 2 we merely have to prove $\text{Im } B_{2N+1} \subset M_{2N+1}$. Using the isomorphism theorem of GORDON [6] ($\text{Im } P \oplus Q = \text{Im } P + \text{Im } Q$; $\text{Im } PQ = \text{Im } P \cap \text{Im } Q$) the proof of this statement is a rather technical verification of the properties of polynomials belonging to M_{2N+1}.
Thus we can formulate

THEOREM 4: Let $F \in C^N(S)$ and $p_{2N+1} \in M_{2N+1}$ the uniquely determined conforming reduced Hermite interpolation polynomial, then $p_{2N+1} = B_{2N+1}(F)$.

III. REPRESENTATION FORMULAS FOR p_{2N+1}

Using the Boolean construction of p_{2N+1} we are able to derive explicit representation formulas for p_{2N+1}.

THEOREM 5: Let $i,j,N \in \mathbb{N}_o$, $o \leq i+j \leq N$, $F \in C^N(S)$, and

$$\Phi_{i,j}^{(2N+1)}(x,y) := \begin{cases} q_{i,n}^{(o)}(x) q_{j,N-i}^{(o)}(y) + q_{i,N-j}^{(o)}(x) q_{j,n}^{(o)}(y) \\ -q_{i,n}^{(o)}(x) q_{j,n}^{(o)}(y) \quad , \quad o \leq i,j \leq n, \\ q_{i,n}^{(o)}(x) q_{j,N-i}^{(o)}(y) \quad , \quad o \leq i \leq n < j \leq N-i, \\ q_{i,N-j}^{(o)}(x) q_{j,n}^{(o)}(y) \quad , \quad o \leq j \leq n < i \leq N-j. \end{cases}$$

Then the following representation formula for p_{2N+1} is valid:

$$\begin{aligned} p_{2N+1}(x,y) = \sum_{i=o}^{n} \sum_{j=o}^{N-i} (& D_x^i D_y^j F(o,o) \Phi_{i,j}^{(2N+1)}(x,y) \\ & + D_x^i D_y^j F(1,o) \Phi_{i,j}^{(2N+1)}(1-x,y)(-1)^i \\ & + D_x^i D_y^j F(o,1) \Phi_{i,j}^{(2N+1)}(x,1-y)(-1)^j \\ & + D_x^i D_y^j F(1,1) \Phi_{i,j}^{(2N+1)}(1-x,1-y)(-1)^{i+j}). \end{aligned}$$

Proof: The proof of this representation theorem uses the property that the Boolean sum of commuting projectors having the property (i) of Lemma 1 coincides with the ordinary sum of these projectors as well as the property (ii) of Lemma 1.

$$B_{2N+1} = \bigoplus_{i=o}^{n} P^x_{i,n} H^y_{N-i} \oplus \bigoplus_{j=o}^{n} H^x_{N-j} P^y_{j,n} = \sum_{i=o}^{n} P^x_{i,n} H^y_{N-i} \oplus \sum_{j=o}^{n} H^x_{N-j} P^y_{j,n}$$

$$= \sum_{i=o}^{n} P^x_{i,n} H^y_{N-i} + \sum_{j=o}^{n} H^x_{N-j} P^y_{j,n} - \sum_{i=o}^{n} \sum_{j=o}^{n} P^x_{i,n} P^y_{j,n}$$

$$= \sum_{i=o}^{n} \sum_{j=o}^{N-i} P^x_{i,n} P^y_{j,N-i} + \sum_{j=o}^{n} \sum_{i=o}^{N-j} P^x_{i,N-j} P^y_{j,n}$$

$$- \sum_{i=o}^{n} \sum_{j=o}^{n} P^x_{i,n} P^y_{j,n}$$

Regrouping the terms in the above sums yields:

$$B_{2N+1} = \sum_{i=o}^{n} \sum_{j=o}^{n} (P^x_{i,n} P^y_{j,N-i} + P^x_{i,N-j} P^y_{j,n} - P^x_{i,n} P^y_{j,n})$$

$$+ \sum_{i=o}^{n} \sum_{j=n+1}^{N-i} P^x_{i,n} P^y_{j,N-i} + \sum_{j=o}^{n} \sum_{i=n+1}^{N-j} P^x_{i,N-j} P^y_{j,n}$$

thus completing the proof.¤

REFERENCES

1. Adini, A., Clough, R.W., "Analysis of Plate Bending by the Finite Element Method". Nat.Sci.Found. Rept. G 7337 (1961), Univ. of California.
2. Baszenski, G., Delvos, F.J., Posdorf, H., "Boolean Methods in Bivariate Reduced Hermite Interpolation". ISNM 51, 11-29 (1979).
3. Delvos, F.J., Posdorf, H., "N-th Order Blending". Lecture Notes in Mathematics 571, 53-64 (1977).
4. Delvos, F.J., Posdorf, H., "A Representation Formula for Reduced Hermite Interpolation". ISNM 42, 124-137 (1978).
5. Delvos, F.J., Posdorf, H., "Boolesche zweidimensionale Lagrange Interpolation". Computing 22, 311-323 (1979).
6. Gordon, W.J., "Distributive Lattices and the Approximation of Multivariate Functions". Proc.Symp. Madison, Wisc. 1969, 223-277.
7. Gordon, W.J., "Blending Function Methods of Bivariate and Multivariate Interpolation and Approximation". SIAM, J.Num.Anal. 8, 158-177 (1971).

8. Gordon, W.J., Cheney, E.W., "Bivariate and Multivariate Interpolation with Noncommutative Projectors". ISNM 4o, 381-387 (1977).
9. Lancaster, P., Watkins, D.S., "Interpolation in the Plane and Rectangular Finite Elements". ISNM 31, 125-145 (1976).
1o. Melkes, F., "Reduced Piecewise Bivariate Hermite Interpolations". Num.Math. 19, 326-34o (1972).
11. Phillips, G.M., "Explicit Forms for Certain Hermite Approximations". BIT 13, 177-18o (1973).
12. Watkins, D.S., "A Conforming Rectangular Plate Element". The Mathematics of Finite Elements and Applications II, 77-83 (1976).
13. Watkins, D.S., "On the Construction of Conforming Rectangular Plate Elements". Int.Journ.Num.Meth.Eng. 1o, 925-933 (1976).
14. Watkins, D.S., Lancaster, P., "Some Families of Finite Elements". J.Inst. Maths Applics 19, 385-397 (1977).

JOINT APPROXIMATION OF A FUNCTION AND ITS DERIVATIVES

Rick Beatson

Department of Mathematics
The University of Texas
Austin, Texas

The classical method of proving Jackson's theorem for $f \in C[-1,1]$ is to use a convolution operator

$$K(f,x) = (f * k)(x) = (1/\pi)\int_{-\pi}^{\pi} f(\cos t)k(\cos(\theta - t))dt\,, \quad x = \cos\theta\,. \qquad (1)$$

In this paper it is shown that the cone of j-convex functions is invariant under K iff k is j-convex. Several applications are given; e.g. Jackson-Timan theorems for monotone approximation, and an $O(n^{-2\alpha})$ bound on the uniform error in monotone approximation of j-convex $(j \geq 2)$ functions in Lip α, $0 < \alpha \leq 1/2$.

Throughout $\|.\|_\infty$ will denote the uniform norm on $[-1,1]$. A function f will be called j-convex if $f \in C[-1,1]$ and all its j-th forward differences are non-negative. Clearly the cone of j-convex functions is closed under uniform limits. The cases $j = 0,1$ of the Theorem below are known (Fekete [3], Roulier [8] and Senderovizh [9] prove the $j = 1$ result).

THEOREM 1. Let $k \in C[-1,1]$ and j be a non-negative integer. The cone of j-convex functions is invariant under the operator $K(f) = f * k$ iff k is j-convex.

Copyright © 1980 by Academic Press, Inc.
All rights of reproduction in any form reserved.
ISBN: 0-12-171050-5

Proof. Necessity: Suppose the cone of j-convex functions is invariant under K. Consider the de La Vallée Poussin kernel, $\{f_n\}$, where $f_n = c_n(1+x)^n$ and $\int_{-\pi}^{\pi} f_n(\cos\theta)d\theta = \pi$. Each f_n is j-convex and so, by hypothesis is $f_n * k$. But $k * f_n$ converges uniformly to k as $n \to \infty$. Hence k is j-convex.

Sufficiency: The key to the proof is a relationship between the convolution (1) and convolution structures for the ultraspherical polynomials discovered by Hirschman [4]. Generalizations to the Jacobi polynomials have been given by Askey and Wainger, and by Gasper. References and the definition and properties of the convolutions, $**$, may be found in Bavinck [1]. Note that $**$ is a function of the order, $(j-1/2,\ j-1/2)$, of the Jacobi polynomials. $R_n^{(j-1/2,j-1/2)}$ will denote the Jacobi polynomial of degree n and order $(j-1/2,\ j-1/2)$.

For $j = 0$, $f*k = (2/\pi)f ** k$. For $f,k \in C^i[-1,1]$, $i \geq j > 0$

$$(f*k)^{(j)} = [2^{2j+1}/(\pi\cdot 1\cdot 3\cdot\ \cdot(2j-1))]f^{(j)} ** k^{(j)} \quad . \tag{2}$$

To show (2) for f and k polynomial compute the series in $\{R_n^{(j-1/2,j-1/2)}\}$ of both sides Now write general k and f as limits in the norm, $|||h|||_i = \max(\|h^{(i)}\|_\infty,\ \|h^{(i-1)}\|_\infty, \ldots, \|h\|_\infty)$ of sequences of polynomials, $\{f_n\}$, $\{k_n\}$. The continuity properties of the convolutions $**$ imply $f_n^{(j)} ** k_n^{(j)} \to f^{(j)} ** k^{(j)}$ uniformly, $0 \leq j \leq i$. Then theorems about uniformly converging sequences show $f*k \in C^{(i)}[-1,1]$ and satisfies (2).

The sufficiency for $f,k \in C^j[-1,1]$ is immediate from (2) since the convolution $**$ of two non-negative functions is non-negative. Write other j-convex f,k as uniform limits of sequences of j-convex polynomials $\{f_n\}$, $\{k_n\}$. Then $f_n * k_n$ is j-convex and $f_n * k_n \to f*k$ uniformly, implying $f*k$ is j-convex.

Theorem 1 may be applied to yield a new Jackson type theorem for monotone approximation. Much work has been done in this area by Lorentz, Lorentz and Zeller, Roulier, DeVore and others; see the survey of Chalmers and Taylor [2]. All linear j-convexity preserving operators are limited to $O(n^{-2})$ approximation on $\{x^j, x^{j+1}, x^{j+2}\}$; for if $L_{n+j}: C[-1,1] \to \pi_{n+j}$ is j-convexity preserving then

$$H_n(f,x) = (d/dx)^j [L_{n+j}[\int_{-1}^{x} \int_{-1}^{t_j} \cdots \int_{-1}^{t_2} f(t_1)dt_1 \cdots dt_j]] ,$$

is positive, and Korovkin's theorem limits $\{H_n\}$ to $O(n^{-2})$ approximation on $\{1, x, x^2\}$.

THEOREM 2. Let j be a positive integer. There exists an M_j such that for each $f \in C[-1,1]$ and $n = 0,1,2,\ldots$, there exists a $p_n \in \pi_n$ with p_n i-convex for any $i \in \{0,1,\ldots,j\}$ for which f is i-convex, and $|f(x) - p_n(x)| \leq M_j \omega(f, \Delta_n(x))$, $-1 \leq x \leq 1$. [$\Delta_0(x) = 1$ and for $n \geq 1$ $\Delta_n(x) = \max(\sqrt{1-x^2}/n,\ 1/n^2)$.]

Proof. Theorem 1 and well-known results (see e.g. Lorentz [6, pp. 65-68]) show it is sufficient to construct a kernel $\{k_n\}$ with $k_n \in \pi_n$, k_n i-convex for $0 \leq i \leq j$, $\int_{-\pi}^{\pi} k_n(\cos t)dt = \pi$ and $\int_0^{\pi} k_n(\cos t)t^2 dt \approx n^{-2}$. [$a_n \approx b_n$ iff $a_n = O(b_n)$ and $b_n = O(a_n)$.] To find such a kernel start with a sequence of non-negative polynomials $\{h_n \in \pi_n\}_{n=0}^{\infty}$ satisfying $\int_0^{\pi} h_n(\cos t)t^i dt \approx n^{-i}$, $i = 0,1,\ldots,2j+2$. Such sequences exist, see e.g. [6, pp. 55-57]. Then apply the following lemma j-times.

LEMMA 3. Let $\{h_n\}$ be a sequence of algebraic polynomials with each h_n i-convex for $i = 0,1,\ldots,m$ and $\int_0^{\pi} h_n(\cos t)t^i dt \approx n^{-i}$, $i = 0,1,\ldots,2k$. Then $\{g_n\}$ defined by $g_{n+1}(x) = \int_{-1}^{x} h_n(s)ds$ is a sequence of algebraic polynomials with each g_n i-convex for $i = 0,1,\ldots,m+1$ and $\int_0^{\pi} g_n(\cos t)t^i dt \approx n^{-i-2}$, $i = 0,1,\ldots,2k-2$.

Proof. Integration by parts and the inequality $(2/\pi)x \leq \sin x \leq x$, $0 \leq x \leq \pi/2$.

For convex (i.e. 2-convex) functions in Lip α, $0 < \alpha \leq 1/2$, the order of approximation given by Jackson's theorem is not best possible:

THEOREM 4. Let j be a positive integer. There exists a constant M_j and for each convex $f \in \mathrm{Lip}_M\alpha$, $0 < \alpha \leq 1/2$, a sequence of polynomials $\{p_n\}$, $p_n \in \pi_n$, i-convex for any $i \in \{0,1,\ldots,j\}$ for which f is i-convex with $\|f-p_n\|_\infty \leq M_j M n^{-2\alpha}$, $n = 1,2,\ldots$.

Proof. It is sufficient to show that for $g(\theta) = f(\cos\theta)$, $\omega(g,n^{-1}) \leq CMn^{-2\alpha}$. This since the i-convexity preserving operators of Theorem 2 give $O(\omega(g,n^{-1}))$ approximation to f (see e.g. [6, p. 56]).

Fix $n \geq 1$. Let $\varepsilon > 0$. Approximate f by an $h \in C^1[-1,1]$ which has the same i-convexity properties as f for $i \in \{0,1,\ldots,j\}$ and satisfies $\|f-h\|_\infty \leq \varepsilon M\delta^\alpha$ where $\delta = 1 - \cos(1/n)$. Then $|h'(x)| \leq (1+2\varepsilon)M(1-|x|)^{\alpha-1}$, $x \in [-1+\delta, 1-\delta]$. This since there exists an $\alpha \in [-1+\delta,1-\delta]$ so that $|h'(x)|$ decreases on $(-1+\delta,\alpha)$ and increases on $(\alpha,1-\delta)$. Applying the mean value theorem for $x \in (\alpha,1-\delta)$, $|h'(x)| \leq (1+2\varepsilon)M(1-x)^{\alpha-1} \leq (1+2\varepsilon)M(1-|x|)^{\alpha-1}$. Similarly for $x \in (-1+\delta,\alpha)$, $|h'(x)| \leq (1+2\varepsilon)M(1+x)^{\alpha-1} \leq (1+2\varepsilon)M(1-|x|)^{\alpha-1}$. Hence, by continuity, the inequality holds on $[-1+\delta,1-\delta]$.

Now consider $t(\theta) = h(\cos\theta)$. Using the result above $|\dot{t}(\theta)| \leq (1+2\varepsilon)M(1-|x|)^{\alpha-1}\sqrt{1-x^2} \leq (1+2\varepsilon)\sqrt{2}\,M\delta^{\alpha-\frac{1}{2}}$, for $x = \cos\theta \in [-1+\delta, 1-\delta]$. Since also $\cos([-n^{-1}, n^{-1}]) = [1-\delta, 1]$ and $\cos([\pi-n^{-1}, \pi+n^{-1}]) = [-1, -1+\delta]$ it follows that

$$\omega(t, n^{-1}) \leq (1+2\varepsilon)\sqrt{2}\,Mn^{-1}\delta^{\alpha-\frac{1}{2}} + M(1+2\varepsilon)\delta^{\alpha} .$$

Using $(2/\pi^2)n^{-2} \leq 1 - \cos(1/n) = \delta \leq n^{-2}/2$ and taking a limit as $\varepsilon \downarrow 0$, it follows that $\omega(g, n^{-1}) \leq CMn^{-2\alpha}$, where C does not depend on n, f, or α.

Using an analogous argument, the Jackson order, $O(n^{-\alpha})$, can also be improved to $O(n^{-2\alpha})$ for f m-convex $(m > 2)$ in Lip α, $0 < \alpha \leq \frac{1}{2}$. In these analogous theorems the constants M_j depend on f. The details are omitted.

Theorem 1 identifies those convolution operators which are j-convexity preserving. It is natural then to seek "good" or even "best" non-negative j-convex polynomial kernels. One "good" non-negative increasing kernel is

$$k_n(\cos t) = \frac{1}{2\pi}\int_{t-\pi/(n+2)}^{t+\pi/(n+2)} \left[\frac{\sin(\frac{\pi}{n+2})\cos(\frac{n+2}{2})s}{\cos s - \cos\pi/(n+2)}\right]^2 ds$$

$$= \frac{1}{2} + \left[\frac{n+2}{2\pi}\sin(\frac{2\pi}{n+2})\right]\cos t + \rho_{2,n}\cos 2t + \ldots + \rho_{n,n}\cos nt ,$$

derived from the Fejér-Korovkin kernel. The adjective "good" is justified by largeness of the coefficient, $\rho_{1,n}$, of $\cos t$ in the expansion of $k_n(\cos t)$. Indeed $1 - \rho_{1,n} \leq (2/3)\pi^2(n+2)^{-2}$, $n = 0,1,\ldots$. This kernel and the quantitative Korovkin theorem [5, pp. 72-73] yield a Jackson type theorem for 1-convex approximation with small explicit constants:

THEOREM 5. If $f \in C[-1,1]$ is increasing then for each $n = 0,1,2,\ldots$, there is an increasing $p_n \in \pi_n$ with $\|f-p_n\| \leq (1+\pi^2/\sqrt{3})\omega(f,1/(n+2))$.

This improves the quantitative but not the qualitative aspects of some of the known theorems. The first Jackson theorem for 1-convex-approximation is the discrete operator proof of Lorentz and Zeller [7]. Their proof contains a construction which motivated the construction of k_n above.

It will now be shown that k_n is indeed increasing as claimed. The π periodicity of $\cos^2$ shows that $(d/dt)k_n(\cos t)$ equals

$$A(n)\cos^2\left(\left(\frac{n+2}{2}\right)t+\frac{\pi}{2}\right)\left[\left(\cos\left(t+\frac{\pi}{n+2}\right)-\cos\left(\frac{\pi}{n+2}\right)\right)^{-2} - \left(\cos\left(t-\frac{\pi}{n+2}\right)-\cos\left(\frac{\pi}{n+2}\right)\right)^{-2}\right]$$

where $A(n)$ is positive. Also

$$\left[\cos\left(t-\frac{\pi}{n+2}\right)-\cos\left(\frac{\pi}{n+2}\right)\right]^2 - \left[\cos\left(t+\frac{\pi}{n+2}\right)-\cos\left(\frac{\pi}{n+2}\right)\right]^2 = \sin\left(\frac{2\pi}{n+2}\right)[\sin 2t - 2\sin t]$$

is non-positive on $[0,\pi]$. Hence $k_n(\cos t)$ decreases on $[0,\pi]$ and $k_n(x)$ increases on $[-1,1]$.

There are, of course, many possible definitions of "best" kernels. The following is one natural definition: Let $U_n \subset \pi_n$ be any closed set with the property that

$k \in U_n$ implies $k(x) \geq 0$ for $x \in [-1,1]$ and $\int_{-\pi}^{\pi} k(\cos t) = \pi$.

Then $k^* \in U_n$ will be called a best element of U_n if $\rho_1(k^*) \geq \rho_1(k)$ for all $k \in U_n$. Here $\rho_1(f)$ is the coefficient of $\cos t$ in the Fourier cosine expansion of $f(\cos t)$.

For

$$S_n = \{k \in \pi_n : k \text{ is } j\text{-convex}, \ j = 0,1,2,\dots, \text{ and } \int_{-\pi}^{\pi} k(\cos t)dt = \pi\},$$

we find the best element is the de La Vallée Poussin kernel $f_n(x) = c_n(1+x)^n$. For if $k_n \in S_n$ the convexity conditions imply $k_n(x) = a_0 + a_1(1+x) + \dots + a_n(1+x)^n$ with $a_0, a_1, \dots, a_n \geq 0$. Using the normalizing condition k_n is a convex linear combination of $f_0, f_1, \dots, f_n$. Since $\rho_1(f_i) = i/(i+1)$, the result follows. This can be restated as a Korovkin type theorem. Namely: If $\{K_n\}$, $k_n \in \pi_n$, is a sequence of convolution operators, i-convexity preserving for all i, then it cannot be that both $\|1-K_n(1)\|_\infty = o(n^{-1})$ and $\|x-K_n(x)\|_\infty = o(n^{-1})$.

For

$$M_n = \{k \in \pi_n : k(x) \geq 0 \text{ and } k'(x) \geq 0, \ x \in [-1,1] \text{ and } \int_{-\pi}^{\pi} k(\cos t)dt = \pi\},$$

the problem of finding a best kernel is equivalent, via a representation theorem of F. Riesz, to a certain generalized eigenvalue problem. The best kernels have been computed for values of n up to 25 but no nice closed form is apparent.

Finally consider the approximation of the derivatives of f by the derivatives of $K(f)$. Let $k \in C^j[-1,1]$, $j \geq 1$, and define pseudo norms

$$_j\|K\| = \sup\{\|K(f)^{(j)}\|_\infty : f \in C^j[-1,1] \text{ and } \|f^{(j)}\|_\infty \leq 1\} .$$

For j-convexity preserving K it is easy to see that the supremum on the right above is achieved for $x^j/j!$. Hence $_j\|K\|$ is the coefficient of T_j in the expansion of k in a series of Chebyshev polynomials of the first kind. More generally, (2) and known facts about ** (see [1]) imply

$$_j\|K\| = [2/(\pi\cdot 1\cdot 3\cdot \ \cdot(2j-1))]\int_{-1}^{1} |k^{(j)}(x)|(1-x^2)^{j-1/2}dx . \tag{4}$$

For $j = 0$ there is the well known formula

$$_o\|K\| = (2/\pi)\int_{-1}^{1} |k(x)|(1-x^2)^{-1/2}dx .$$

(4) has many applications to convergence and, via the Principle of Uniform Boundedness, divergence theorems.

REFERENCES

1. Bavinck, H., Jacobi Series and Approximation, Mathematical Centre Tracts, #39, Amsterdam 1972.
2. Chalmers, B.L. and Taylor, G.D., "Uniform Approximation with Constraints", Jber. d. Dt. Math.-Verein, 81(1979), 49-86.
3. Fekete, M., "On Certain Classes of Periodic Functions and Their Fourier Series", Bull. Res Council of Israel, 7F(1958), 103-112.
4. Hirschman, I.I. Jr., "Harmonic Analysis and Ultraspherical Polynomials", in Symposium of the Conference on Harmonic Analysis, Cornell, 1956.
5. Korovkin, P.P., Linear Operators and Approximation Theory, Hindustan, Delhi, 1960.
6. Lorentz, G.G., Approximation of Functions, Holt, New York, 1966.
7. Lorentz, G.G. and Zeller, K.L., "Degree of Approximation by Monotone Polynomials I", J. Approx. Theory, 1(1968), 501-504.
8. Roulier, J., "Linear Operators Invariant on the Nonnegative Monotone Functions", SIAM J. Numer. Anal., 8(1971), 30-35.
9. Senderovizh, R., "On Convexity Preserving Operators", SIAM J. Math. Anal., 9(1978), 157-159.

ON THE GLOBAL APPROXIMATION BY KANTOROVITCH POLYNOMIALS

Michael Becker
Rolf J. Nessel

Lehrstuhl A für Mathematik
RWTH Aachen, Germany

DEDICATED TO PROFESSOR G.G. LORENTZ ON THE OCCASION OF HIS SEVENTIETH BIRTHDAY, IN HIGH ESTEEM

This note surveys some recent results concerning the global approximation by Kantorovitch polynomials in $L^p(0,1)$.

In 1912 Bernstein [8] introduced the polynomials

$$B_n f(x) := \sum_{k=0}^{n} f(\tfrac{k}{n})\, p_{k,n}(x), \qquad p_{k,n}(x) := \binom{n}{k} x^k (1-x)^{n-k},$$

and proved that one has uniform convergence for any $f \in C[0,1]$, the space of functions continuous on $[0,1]$ with norm $\|f\|_{C[0,1]} := \max_{u\in[0,1]} |f(u)|$. In order to treat integrable functions, Kantorovitch [17] in 1930 suggested the following modification

$$K_n f(x) := \sum_{k=0}^{n} \Big[(n+1) \int_{k/(n+1)}^{(k+1)/(n+1)} f(u)\, du \Big] p_{k,n}(x) .$$

The first result concerning strong convergence and thus the starting point for our considerations was given in 1937 by our birthday celebrant, Prof. Lorentz, in his "Kandidatdissertation" [19]: He proved that for any $f \in L^1 := L^1(0,1)$ with $\|f\|_1 := \int_0^1 |f(u)|\, du$

$$\lim_{n\to\infty} \|K_n f - f\|_1 = 0. \tag{1}$$

For a treatment in L^p, $1<p<\infty$, see [10; 20,p.31f], in fact, then even

Copyright © 1980 by Academic Press, Inc.
All rights of reproduction in any form reserved.
ISBN: 0-12-171050-5

dominated convergence holds as was shown in 1951 by Butzer [10; 11] in his thesis under supervision of Prof. Lorentz in Toronto.

Since then the problem of the rate of convergence in (1) is posed. There is an extensive literature concerning local results connected with work of K.DeLeeuw, Z.Ditzian, C.P.May, M.W. Müller, and others. In the following, however, we would like to discuss only those aspects which treat the problem globally on the whole interval, including the endpoints.

In order to have an idea of the type of results we would like to have for Kantorovitch polynomials in L^p-metric, let us briefly recall relevant assertions for Bernstein polynomials in C-metric. With $\Delta_h^2 f(x) := f(x+h) - 2f(x) + f(x-h)$ and $\varphi(x) := x(1-x)$

<u>Result I.</u> For $0<\alpha\leqslant 1$ and $f\in C[0,1]$ there are equivalent:

(i) $\quad \|\varphi^{-\alpha}[B_n f - f]\|_{C[0,1]} = O(n^{-\alpha})$,

(ii) $\quad \|\Delta_h^2 f\|_{C[h,1-h]} = O(h^{2\alpha})$.

For the saturation case $\alpha=1$ see Lorentz [21]. The nonsaturated cases $0<\alpha<1$ were treated by Berens-Lorentz [7] who indeed offered two methods of proof for the inverse part (i) $\Rightarrow$ (ii): A first method using intermediate spaces, covering the whole range $0<\alpha<1$, and a second more elementary one for $0<\alpha<1/2$, further developed to all $0<\alpha<1$ in [1].

A different kind of result may be obtained by moving the weight φ from the approximation to the smoothness assertion.

<u>Result II.</u> For $0<\alpha\leqslant 1$ and $f\in C[0,1]$ there are equivalent:

(i) $\quad \|B_n f - f\|_{C[0,1]} = O(n^{-\alpha})$,

(ii) $\quad \|\varphi^{\alpha}\Delta_h^2 f\|_{C[h,1-h]} = O(h^{2\alpha})$.

For $\alpha=1$ see Lorentz-Schumaker [22] (via generalized convexity) or [2] (via abstract Trotter-type theorems). Concerning $0<\alpha<1$ we refer to [7] for precursory material in connection with the intermediate space method which was then completed by Ditzian [12]. For a parallel treatment via the elementary method see [3]. For assertions intermediate between Results I and II see Ditzian [13], again completing relevant portions of [7].

The hope that one may carry out such a program for Kantorovitch polynomials was first confirmed by the following saturation theorem of Maier [23; 24].

Theorem M. For $f\in L^1$ there are equivalent:

(i) $\|K_n f - f\|_1 = O(n^{-1})$,

(ii) $f \in S := \{f\in L^1;\ f\in AC_{loc}(0,1) \text{ and } \varphi f'\in BV[0,1]\}$.

Here $AC_{loc}(0,1)$ is the set of functions absolutely continuous on any compact subinterval of $(0,1)$, and $BV[0,1]$ the space of functions which are of bounded variation globally on $[0,1]$. Let us mention that it is the direct part (ii) $\Rightarrow$ (i) which is the most interesting one: Its proof is based on the fact that one may derive (i) for the test-function $\log x$. The corresponding result for $1<p<\infty$ was given by Riemenschneider [27] (cf. [25]). The characterization (ii) of Thm. M is obviously only suitable for the saturation case, in other words, setting

$$D(\alpha) := \{f\in L^1;\ \|K_n f - f\|_1 = O(n^{-\alpha})\}, \tag{2}$$

it only makes sense for the integral case $\alpha=1$: It does not suggest any candidate for fractional α. Thus one is interested in further characterizations of S which might lead to conjectures even for $0<\alpha<1$. Here we mention ($\|g\|_{BV+C} := \|g\|_{BV} + \|g\|_C$)

Theorem 1. With $F(x) := \int_0^x f(u)\,du$ one has $f\in S$ iff

$$\|\varphi\,\Delta_h^2 F\|_{BV+C[h,1-h]} = O(h^2).$$

In view of $\|K_n f - f\|_1 = \|B_{n+1}F - F\|_{BV[0,1]}$ this corresponds to the case $\alpha=1$ of Result II. For a proof of Thm. 1 see [4], also for an analog in terms of a different type of second difference.

Let us return to the problem of characterizing the approximation class $D(\alpha)$ for the nonsaturated values $0<\alpha<1$. There is an abstract characterization given by Grundmann [15] in terms of the associated K-functional

$$K(L^1,S;f,t) := \inf_{g\in S}\{\|f-g\|_1 + t\,\|\varphi g'\|_{BV+L^\infty[0,1]}\}.$$

It states that for any $0<\alpha\leq 1$ (and for $t\to 0+$)

$$f\in D(\alpha) \iff K(L^1,S;f,t) = O(t^\alpha).$$

But this is only a reformulation of the problem which now asks for concrete characterizations of this abstract K-functional in terms of weighted smoothness conditions. This still remains an open question. Let us just sketch those few results known.

General estimates in terms of the ordinary first modulus of continuity were given by Bojanic-Shisha [9], who proved

$$\| \sqrt{\varphi}\,[K_n f - f] \|_1 \leq C \sup_{0<|h|\leq n^{-1/2}} \| f(\cdot+h) - f(\cdot) \|_1 .$$

Then Grundmann [14] observed that the weight $\sqrt{\varphi}$ on the left (which even smoothes the problem) is superfluous (for the corresponding result in L^p, $1<p<\infty$, see Müller [26]). In fact, Berens-DeVore [6] in the same year established quantitative Korovkin theorems in L^p and derived all these estimates as an immediate consequence of their general theorems, even involving second moduli of continuity. However, all these results are not best possible since they do not incorporate correct endpoint weights.

In this connection we do have some more information in the particular situation of $\alpha=1/2$. Extending a first result of Levi [18] for step-functions, Hoeffding [16] proved

Theorem H. One has

$$\| K_n f - f \|_1 \leq C\, n^{-1/2} \int_0^1 \sqrt{\varphi(u)}\, |df(u)| ,$$

thus $H := \{f \in L^1;\ f \in BV_{loc}(0,1) \text{ and } \int_0^1 \sqrt{\varphi(u)}\, |df(u)| < \infty\} \subset D(1/2)$.

So far this seems to be the most significant contribution to the subject, apart from the saturation results. In a certain sense the results of [16] are sharp, even asymptotically with respect to constants. But H does not provide the most effective way to exhaust the approximation class D(1/2). Indeed,

Lemma 1. For any $0<\alpha<1$ one has $f_\alpha \in D(\alpha)$, where $f_\alpha(x) := x^{\alpha-1}$. In fact, this rate of approximation is best possible for f_α.

The proof being rather intricate (see [5]), let us only mention that one may hope that f_α for $0<\alpha<1$ may serve as a testing function as $\log x$ does for $\alpha=1$. As a consequence of La.1 one has $x^{-1/2} \in D(1/2)$ but $x^{-1/2} \notin H$. Introducing

$$V := \{f \in L^1;\ f \in BV_{loc}(0,1) \text{ and } \sqrt{\varphi}\, f \in BV[0,1]\} ,$$

then $x^{-1/2} \in V$, and H is a (proper) subset of V. Nevertheless,

Theorem 2. One has $V \subset D(1/2)$, in particular

$$\| K_n f - f \|_1 \leq 6\,(n+1)^{-1/2} \| \sqrt{\varphi}\, f \|_{BV+L^\infty[0,1]} .$$

In fact this estimate is sharp (with respect to the rate) as the counterexample $x^{-1/2}$ shows. The method of proof of Thm.2 is essentially that employed by Riemenschneider [27] in the saturation case, but for the details we refer to [5].

In concluding let us recall (cf.[4])

Conjecture. For $0<\alpha\leqslant 1$ one has $f\in D(\alpha)$ iff

$$\|\varphi^{\alpha}\Delta_h^2 F\|_{BV+C[h,1-h]} = \mathcal{O}(h^{2\alpha}).$$

This would be quite parallel to Result II. That the conjecture might be true is supported by Thm.1. Moreover, it is consistent with Thm.2 since we have (cf.[5])

Lemma 2. If $f\in V$, then

$$\|\sqrt{\varphi}\,\Delta_h^2 F\|_{BV+C[h,1-h]} \leqslant C\,h\,\|\sqrt{\varphi}\,f\|_{BV+L^{\infty}[0,1]}.$$

REFERENCES

1. Becker, M., An elementary proof of the inverse theorem for Bernstein polynomials, Aequationes Math. 19(1979), 145-150.
2. Becker, M., and Nessel, R.J., Iteration von Operatoren und Saturation in lokal konvexen Räumen, Forsch.-Ber.Nordrhein Westfalen Nr.2470, Westd.Verlag, Opladen 1975, 27-49.
3. Becker, M., and Nessel, R.J., Inverse results via smoothing, in:"Constructive Function Theory"(Proc.Conf., Blagoevgrad, 1977), Sofia, to appear.
4. Becker, M., and Nessel, R.J., On global saturation for Kantorovitch polynomials, in:"Approximation and Function Spaces"(Proc.Conf., Gdańsk, 1979), to appear.
5. Becker, M., and Nessel, R.J., Some global direct estimates for Kantorovitch polynomials, to appear.
6. Berens, H., and DeVore, R.A., Quantitative Korovkin theorems for L_p-spaces, in:"Approximation Theory II"(Proc.Conf., Austin, 1976), Academic Press, New York 1976, 289-298.
7. Berens, H., and Lorentz, G.G., Inverse theorems for Bernstein polynomials, Indiana Univ. Math. J. 21(1972), 693-708.
8. Bernstein, S., Démonstration du théorème de Weierstrass, fondée sur le calcul des probabilités, Commun. Soc. Math. Kharkow 13(1913), 1-2.
9. Bojanic, R., and Shisha, O., Degree of L_1 approximation to integrable functions by modified Bernstein polynomials, J. Approx. Theory 13(1975), 66-72.

10. Butzer, P.L., On Bernstein polynomials, Thesis, Toronto 1951.
11. Butzer, P.L., Dominated convergence of Kantorovitch polynomials in L^p, Trans. Royal Soc. Canada 46(1952), 23-27.
12. Ditzian, Z., A global inverse theorem for combinations of Bernstein polynomials, J. Approx. Theory 26(1979), 277-292.
13. Ditzian, Z., Interpolation theorems and the rate of convergence of Bernstein polynomials, (these Proc.) to appear.
14. Grundmann, A., Güteabschätzungen für den Kantorovic-Operator in der L_1-Norm, Math. Z. 150(1976), 45-47.
15. Grundmann, A., Inverse theorems for Kantorovic-polynomials, in: "Fourier Analysis and Approximation Theory" (Proc. Conf., Budapest, 1976), North-Holland, Amsterdam 1978, 395-401.
16. Hoeffding, W., The L_1 norm of the approximation error for Bernstein-type pol., J. Approx. Theory 4(1971), 347-356.
17. Kantorovitch, L.V., Sur certaines développements suivant les polynomes de la forme de S. Bernstein, I, II, C.R. Acad. Sci. URSS. A(1930), 563-568, 595-600.
18. Levi, E., Sopra un'applicazione dei polinomi di Bernstein all'approssimazione in media delle funzioni sommabili, Atti. Accad. Naz. Lincei Rend. Cl. Sci. Fis. Mat. Nat. 9(1950), 242-246.
19. Lorentz, G.G., Zur Theorie der Polynome von S. Bernstein, Mat. Sb. 2(1937), 543-556.
20. Lorentz, G.G., Bernstein Polynomials, Toronto 1953.
21. Lorentz, G.G., Inequalities and the saturation classes of Bernstein polynomials, in: "On Approximation Theory" (Proc. Conf., Oberwolfach, 1963), Birkhäuser, Basel 1963, 200-207.
22. Lorentz, G.G., and Schumaker, L.L., Saturation of positive operators, J. Approx. Theory 5(1972), 413-424.
23. Maier, V., Güte- und Saturationsaussagen für die L_1-Approximation durch spezielle Folgen linearer positiver Operatoren, Dissertation, Dortmund 1976.
24. Maier, V., The L_1 saturation class of the Kantorovic operator, J. Approx. Theory 22(1978), 223-232.
25. Maier, V., L_p-approximation by Kantorovic operators, Anal. Math. 4(1978), 289-295.
26. Müller, M.W., Die Güte der L_p-Approximation durch Kantorovic-Polynome, Math. Z. 152(1976), 243-247.
27. Riemenschneider, S.D., The L_p-saturation of the Kantorovic-Bernstein polynomials, J. Approx. Theory 23(1978), 158-162.

A CHARACTERIZATION OF BERNSTEIN POLYNOMIALS

H. Berens

Mathematisches Institut
der Universität Erlangen-Nürnberg
Erlangen, W. Germany

R. DeVore

Department of Mathematics
University of South Carolina
Columbia, South Carolina

The Bernstein polynomials have received considerable attention in approximation theory, in part due to their shape preserving properties. On the other hand, their rate of approximation is not as good as that for other methods of approximation. In this note, we want to show that this slow rate of approximation is actually a consequence of their shape preserving, and in a certain sense they have the best rate of approximation among all operators with the same shape preserving properties.

For a function $f \in C[0,1]$, the Bernstein polynomial is defined by

$$B_n(f) := \sum_0^n f(k/n) p_{n,k}; \quad p_{n,k}(x) := \binom{n}{k} x^k (1-x)^{n-k}.$$

From the differentiation formula [2]

Copyright © 1980 by Academic Press, Inc.
All rights of reproduction in any form reserved.
ISBN: 0-12-171050-5

$$(1)\quad \Big[\sum_0^n a_k\, p_{n,k}\Big]^{(j)} = n(n-1)\ldots(n-j+1)\sum_{k=0}^{n-j} \Delta^j a_k\, p_{n-j,k},$$

it follows that B_n preserves convexity of all orders, i.e. $f^{(j)} \geq 0$ on $[0,1]$ implies $[B_n(f)]^{(j)} \geq 0$ on $[0,1]$, $j=0,1,\ldots$. Thus B_n is an operator in the class $\mathcal{L}_n$ of all those operators L_n with

$$(2)\quad \begin{array}{ll} \text{i)} & L_n f \in \mathbb{P}_n, \text{ for all } f \in C[0,1] \\ \text{ii)} & L_n(\ell) = \ell, \text{ for all } \ell \in \mathbb{P}_1, \\ \text{iii)} & [L_n(f)]^{(j)} \geq 0, \text{ if } f^{(j)} \geq 0,\ j=0,1,\ldots,n. \end{array}$$

Here $\mathbb{P}_k$ is the space of polynomials of degree k.

Each $L_n \in \mathcal{L}_n$ is a positive operator, so its degree of approximation is controlled by $L_n((\cdot-x)^2,x)$. More precisely, if $B = \{f\colon f' \text{ is a.c. and } \|f'\|_\infty = 1\}$ then [1]

$$\sup_{f\in B} \|f-L_n(f)\|_\infty = \frac{1}{2}\|L_n((\cdot-x)^2,x)\|_\infty$$

with $\|\cdot\|_\infty$ the sup norm on $[0,1]$. The main result of this note is:

$$(3)\quad \frac{x(1-x)}{n} = B_n\,((\cdot-x)^2,x) = \inf_{L_n\in\mathcal{L}_n} L_n((\cdot-x)^2,x),\quad 0 \leq x \leq 1.$$

Thus, at least in this sense, B_n has the best rate of approximation among the operators in $\mathcal{L}_n$.

The result (3) can also be formulated in terms of eigenvalues. From (2)iii), it follows that $L_n\colon \mathbb{P}_j \to \mathbb{P}_j, j=0,1,\ldots,n.$ from which it is easy to conclude that for each j, L_n has an eigenfunction $Q_j \in \mathbb{P}_j$, $Q_j = x^j + \ldots$ with corresponding eigenvalue $\lambda_j = \lambda_j(L_n)$. The eigenvalues of B_n are $\lambda_j(B_n) = 1$, $j=0,1$ and $\lambda_j(B_n) = (1-\frac{1}{2})\ldots(1-\frac{j-1}{n})$, $j=2,\ldots,n$. The inequality (3) is equivalent to the statement

$$(4)\quad \lambda_2(L_n) \leq \lambda_2(B_n),\ L_n \in \mathcal{L}_n,\ n=1,2,\ldots\ .$$

We will show that equality holds in (4) if and only if $L_n = B_n$ and therefore this characterizes the Bernstein polynomials.

We begin with some simple properties of the eigenvalues $\lambda_j(L_n)$ for $L_n \in \mathcal{L}_n$.

Lemma. <u>For any</u> $L_n \in \mathcal{L}_n$ <u>we have</u>

$$\lambda_0(L_n) = \lambda_1(L_n) = 1 \geq \lambda_2(L_n) \geq \ldots \geq \lambda_n(L_n) \geq 0$$

<u>Proof</u>. Since L_n preserves linear functions, $\lambda_0(L_n) = \lambda_1(L_n) = 1$. For the function $e_j(x) = x^j/j!$, we have $L_n(e_j) = \lambda_j e_j + a_{j-1,j}\, e_{j-1} + \ldots$. Since $e_j^{(i)} \geq 0$, $i=0,1,\ldots,n$, the polynomial $L_n(e_j)$ must be completely monotonic on [0,1] (property (2)iii) and so all the coefficients of $L_n(e_j)$ are non-negative. In particular

$$\lambda_j(L_n) = [L_n(e_j)]^{(j)} \geq 0.$$

In addition, $e_{j-1} - e_j$ has a non-negative $(j-1)^{\underline{st}}$ derivative and so (2)iii) also implies that

$$0 \leq [L_n(e_{j-1}-e_j)]^{(j-1)} = \lambda_{j-1} - \lambda_j x - a_{j-1,j}, \quad 0 \leq x \leq 1.$$

When $x=1$, we find $\lambda_{j-1} - \lambda_j \geq a_{j-1,j} \geq 0$, as desired.

Theorem. For any $L_n \in \mathcal{L}_n$.

$$(5) \quad \lambda_2(L_n) \leq \lambda_2(B_n) = 1 - \frac{1}{n}$$

<u>or equivalently</u>

$$(6) \quad \frac{x(1-x)}{n} = B_n((\cdot-x)^2, x) \leq L_n((\cdot-x)^2, x)$$

<u>Furthermore, equality can hold in</u> (5) <u>or</u> (6) <u>if and only if</u> $L_n = B_n$.

<u>Proof</u>. The operator L_n can be represented as

$$L_n(f,x) = \sum_{k=0}^{n} a_k(f)\, p_{n,k}(x); \qquad a_k(f) = \int_0^1 f\, d\alpha_k$$

with $d\alpha_k$ a Borel measure. Here we are using the fact that

$(p_{n,k})_0^n$ is a basis for $\mathbb{P}_n$. Now, L_n preserves linear functions and so

$$\int_0^1 d\alpha_k = 1; \quad \int_0^1 t\, d\alpha_k = k/n,\ k=0,1,\ldots,n,$$

where we used the fact that $1 = \sum_0^n p_{n,k}$; $x = \sum_0^n k/n\ p_{n,k}(x)$, uniquely. The two measures $d\alpha_0$ and $d\alpha_n$ are easily seen to be positive. Indeed if $f \geq 0$ then $a_0(f) = L_n(f,0) \geq 0$, similarly for $d\alpha_n$. In addition $0 \leq L_n(t^2,0) \leq L_n(t,0) = 0$, because $t^2 \leq t$ on $[0,1]$. This means that $\int_0^1 t^2\, d\alpha_0 = 0$. Since $d\alpha_0$ is a positive measure we must have $d\alpha_0 = d\rho_0$ with $d\rho_t$ denoting the Dirac measure with unit mass at t. In a similar way, we find $d\alpha_n = d\rho_1$.

We now want to prove that $d\alpha_1$ and $d\alpha_{n-1}$ are positive measures. It is enough to do this for $d\alpha_{n-1}$ since a similar argument (or symmetry) handles the case $d\alpha_1$. If f is any non-negative function on $[0,1]$ with $f(1) = 0$ then

$$0 \leq \lim_{x\to 1} \frac{L_n(f,x)}{p_{n,n-1}(x)} = \lim_{x\to 1} \sum_0^{n-1} a_k(f) \frac{p_{n,k}(x)}{p_{n,n-1}(x)} = a_{n-1}(f)$$

Thus $d\alpha_{n-1} = d\mu_{n-1} + c\,d\rho_1$ with $d\mu_{n-1} \geq 0$ and supported on $[0,1)$ and c a constant. We want to see that $c \geq 0$. To this end, consider the function $f_\varepsilon(x) \equiv (x-\varepsilon)_+^n$, $0 < \varepsilon < 1$. Since $f_\varepsilon^{(k)} \geq 0$, $k=0,1,\ldots,n$, we have from (1) and (2)iii)

$$\Delta^i a_0(f_\varepsilon) = \lim_{x\to 0} [L_n(f_\varepsilon)]^{(i)}(x) \geq 0,\ i=0,1,\ldots,n.$$

Here $\Delta^i a_k = \sum_{\nu=0}^{i} \binom{i}{\nu}(-1)^{\nu+i} a_{k+\nu}$

It follows easily that $(a_i(f_\varepsilon))$ is a monotone sequence:

$$(7)\quad 0 = a_0(f_\varepsilon) \leq \ldots \leq a_{n-1}(f_\varepsilon) \leq a_n(f_\varepsilon) = (1-\varepsilon)_+^n .$$

In particular $0 \leq a_{n-1}(f_\varepsilon) = \int_\varepsilon^1 f_\varepsilon\, d\mu_{n-1} + c(1-\varepsilon)^n$. Dividing by $(1-\varepsilon)^n$ and letting $\varepsilon \to 1$ easily establishes that $c \geq 0$, as desired. Note $(1-\varepsilon)^{-n} f_\varepsilon \leq 1$ for all $0 < \varepsilon < 1$.

Now we can estimate the eigenvalue $\lambda_2(L_n)$. Let ℓ be the linear function which satisfies $\ell(\frac{n-1}{n}) = (\frac{n-1}{n})^2$, $\ell'(\frac{n-1}{n}) = 2(\frac{n-1}{n})$. Then $\ell(t) \leq t^2$ for all $0 \leq t \leq 1$ and so

$$(\frac{n-1}{n})^2 = a_{n-1}(\ell) \leq a_{n-1}(t^2).$$

But

$$L_n(t^2,x) = \sum_{i=0}^{2} c_i x^i = \sum_0^n Q(k/n)\, p_{n,k}(x)$$

for some quadratic Q. Here we use the fact that $x^2 = \sum_0^n Q_2(k/n)p_{n,k}(x)$ for some quadratic Q_2. Now we know that $Q(0) = 0$, $Q(1) = 1$, $Q(\frac{n-1}{n}) \geq (\frac{n-1}{n})^2$. Thus $Q(t) \geq t^2$ for all $0 \leq t \leq 1$ and therefore

$$L_n(t^2,x) = \sum_0^n Q(k/n)\, p_{n,k}(x) \geq \sum_0^n (k/n)^2 p_{n,k}(x) = B_n(t^2,x).$$

Since both L_n and B_n preserve linear functions

$$(9) \quad L_n((\cdot-x)^2,x) \geq B_n((\cdot-x)^2,x) = \frac{x(1-x)}{n}, \quad 0 \leq x \leq 1$$

which establishes (6).

To prove (5), note that

$$L_n((t-x)^2,x) = L_n(t^2,x) - 2xL_n(t,x) + x^2 = (\lambda_2(L_n)-1)x^2+ax+b$$

Comparing terms with (8) shows that

$$[1-\lambda_2(L_n)]\; x\,(1-x) \geq \frac{x(1-x)}{n} \qquad 0 \leq x \leq 1$$

from which we get $\lambda_2(L_n) \leq 1 - \frac{1}{n} = \lambda_2(B_n)$ and so we have proved (5) and (6).

We now want to discuss when equality can hold in (5) or (6). We will show that equality in (5) or (6) implies $L_n = B_n$. This will be done by showing that $d\alpha_k = d\rho_{k/n}$, $k=0,1,\dots,n$.

For this purpose, it is enough to show that the measures $d\alpha_k$ are non-negative since by (2)ii) we already know that

(10) $$\int_0^1 t^i \, d\alpha_k = \left(\frac{k}{n}\right)^i$$

$i = 0,1$, and equality in (5) means that this holds for $i = 2$ as well. It is well known that any positive measure satisfying (10) for $i = 0,1,2$ must equal $d\rho_{k/n}$.

We now show that all the measures $d\alpha_0,\ldots,d\alpha_n$ are positive. Suppose this is not the case and let μ be the largest integer $\le n$ such that $d\alpha_\mu$ is not positive. Then $\mu \le n-2$ because we have shown earlier that $d\alpha_{n-1}$ and $d\alpha_n$ are positive. We then know that $d\alpha_j = d\rho_{j/n}$, $j = \mu+1,\ldots,n$. If $f \ge 0$ vanishes at $\frac{j}{n}$, $j = \mu+1,\ldots,n$, then $a_j(f) = 0$ so that

$$a_\mu(f) = \lim_{x\to 1} \sum_{k=0}^{\mu} a_k(f) \frac{p_{n,k}(x)}{p_{n,\mu}(x)} = \lim_{x\to 1} \frac{L_n(f,x)}{p_{n,\mu}(x)} \ge 0$$

Hence $d\alpha_\mu = d\beta + d\gamma$ with $d\beta \ge 0$ and $d\gamma = \sum_{\mu+1}^{n} c_j \, d\rho_{j/n}$.

We will show that $c_j \ge 0$, $j = \mu+1,\ldots,n$ which will complete the proof. First we want to see that $c_j = 0$, $j = \mu+2,\ldots,n$. Assume this were not the case and let ν be the largest integer $\nu \ge \mu+2$ for which $c_\nu \ne 0$.

Consider the function $f_\varepsilon(x) = (x-\varepsilon)_+^n$ which we used earlier. We have

$$0 = a_0(f_\varepsilon) \le \ldots \le a_\mu(f_\varepsilon) \le f_\varepsilon\left(\frac{\mu+1}{n}\right) \le \ldots \le f_\varepsilon(1).$$

Now, when $\nu \ne n$, take $\varepsilon > \frac{\nu}{n}$ so that $f_\varepsilon\left(\frac{\mu+1}{n}\right) = 0$ and hence $a_\mu(f_\varepsilon) = 0$, thus

$$\int_0^1 f_\varepsilon \, d\alpha_\mu = 0$$

Since $d\alpha_\mu \ge 0$ on $(\frac{\nu}{n},1]$, we have (taking $\varepsilon \sim \frac{\nu}{n}$) $d\alpha_\mu \equiv 0$ on $(\frac{\nu}{n},1]$.

This holds when $\nu = n$ as well. Now, take $\frac{\mu+1}{n} < \varepsilon < \frac{\nu}{n}$. Then

$$(11) \qquad 0 \le a_\mu(f_\varepsilon) = \int_\varepsilon^{\frac{\nu}{n}} f_\varepsilon d\beta + c_\nu f_\varepsilon(\tfrac{\nu}{n}) \le f_\varepsilon(\tfrac{\mu+1}{n}) = 0$$

Dividing by $f_\varepsilon(\frac{\nu}{n})$ and letting $\varepsilon \to \frac{\nu}{n}$ shows that $c_\nu = 0$ since the integral term will tend to 0. This gives a contradiction and shows that each $c_j = 0$, $j = \mu+2,\ldots,n$.

The same argument that arrived at the lefthand inequality in (11) applies to the case $\nu = \mu+1$. Dividing by $f_\varepsilon(\frac{\nu}{n})$ and taking a limit in this case shows that $c_{\mu+1} \ge 0$. Thus, we have shown that $c_j \ge 0$, $j = \mu+1,\ldots,n$ and as observed above this proved that $d\alpha_\mu \ge 0$ as desired.

REFERENCES

[1] R. DeVore, Approximation of Continuous Functions by Positive Linear Operators, Springer Lecture Notes in Mathematics, #293, Berlin, 1972, 289 pp.

[2] G. G. Lorentz, Bernstein Polynomials, Toronto U. Press, Toronto, 1953, 130 pp.

The first author acknowledges National Science Foundation support in Grant MCS 77-22982.

ON STRONG UNIQUENESS IN LINEAR AND COMPLEX CHEBYSHEV APPROXIMATION

Hans-Peter Blatt

Fakultät für Mathematik und Informatik
Universität Mannheim
Mannheim

Let Q be a compact subset of the complex plane C. We denote by C(Q) the set of continuous complex valued functions endowed with the uniform norm $||\cdot||$. A given function $f \in C(Q)$ is to be approximated by elements of a Haar subspace V of C(Q), $V = \text{span}(v_1, v_2, \ldots, v_n)$, dim V = n. If $\tilde{v}$ denotes the best approximation of f with respect to V, let

$$E(\tilde{v}) := \{x \in Q \mid \ ||f - \tilde{v}|| = |f(x) - \tilde{v}(x)|\}.$$

Then we define a <u>primitive extremal point set</u> to be any set $A \subset E(\tilde{v})$, such that $\tilde{v}$ is a best approximation of f on A but no more on any proper subset of A (Rivlin and Shapiro (4), Brosowski (1)).

It is well known that for the number $|A|$ of points of A the inequalities $n+1 \leq |A| \leq 2n+1$ hold. In the real case we always have $|A| = n+1$.

The best approximation $\tilde{v}$ of f is called <u>strong unique</u> iff there exists a constant $\gamma > 0$ such that

$$||f - v|| \geq ||f - \tilde{v}|| + \gamma||v - \tilde{v}|| \quad \text{for all } v \in V.$$

Contrary to the real case, in general such an inequality doesn't hold for the complex case. However Newman and Shapiro (3) proved:

For any neighbourhood U of the best approximation $\tilde{v}$ of f

Copyright © 1980 by Academic Press, Inc.
All rights of reproduction in any form reserved.
ISBN: 0-12-171050-5

there exists a constant $\gamma > 0$ such that

$$||f - v|| \geq ||f - \tilde{v}|| + \gamma ||v - \tilde{v}||^2 \quad \text{for all } v \in U.$$

The question of strong uniqueness is now closely related to the length $|A|$ of a primitive extremal point set. Gutknecht (2) showed that there exists no primitive extremal point set of length 2n+1, if $\tilde{v}$ is the best approximation of f, but not a strong unique one.

For abbreviation we define for each $f \in C(Q)$ the number

$$m(f) := \min \{|A| \mid A \text{ primitive extremal point set to } f\},$$

and for each natural number k with $n+1 \leq k \leq 2n+1$ the subset

$$T_k := \{f \in C(Q) \mid m(f) = k\}.$$

Our main result is the following

<u>Theorem</u>: The subset T_{2n+1} of $C(Q)$ is open and dense if and only if Q has at most n isolated points.

Using this theorem we get the

<u>Corollary</u>: The property "f has a strong unique best approximation" is a generic property in $C(Q)$ if and only if Q has at most n isolated points.

REFERENCES

1. Brosowski,B., Nicht-lineare Tschebyscheff-Approximation, Bibliographisches Institut, Mannheim, 1968.
2. Gutknecht, M. H., Non-strong uniqueness in real and complex Chebyshev approximation, J. Approximation Theory 23 (1978), 204-213.
3. Newman, D. J., and Shapiro, H. S., Some theorems on Cebysev approximation, Duke Math. J. 30 (1963), 673-681.
4. Rivlin, T. J., and Shapiro, H. S., A unified approach to certain problems of approximation and minimization, J. Soc. Indust. and Appl. Math. 9 (1961), 670-699.

POINTWISE-LIPSCHITZ-CONTINUOUS SELECTIONS FOR THE METRIC PROJECTION

Hans-Peter Blatt

Fakultät für Mathematik und Informatik
Universität Mannheim
Mannheim, West Germany

Günther Nürnberger
Manfred Sommer

Institut für Angewandte Mathematik
Universität Erlangen-Nürnberg
Erlangen, West Germany

We study the problem of existence of pointwise-Lipschitz-continuous selections for the metric projection in the following two cases: (i) We approximate by finite dimensional subspaces of C[a,b]. (ii) We approximate by special finite dimensional subspaces of C(X), where X is chosen as the union of finitely many compact real intervals.

I. INTRODUCTION

Let X be a compact metric space, and let C(X) be the space of all real-valued, continuous functions f on X under the uniform norm $||f||: = \sup\{|f(x)|: x\in X\}$. If G is an n-dimensional subspace of C(X), then for each $f\in C(X)$ we define $P_G(f): = \{g_o\in G: ||f-g_o|| = \inf\{||f-g||: g\in G\}\}$ which is called the <u>set of best approximations</u> of f from G. This defines a set-valued mapping P_G from C(X) into 2^G which is called the <u>metric projection</u> onto G. A mapping s from C(X) onto G is called <u>selection</u> for P_G if $s(f)\in P_G(f)$ for each $f\in C(X)$. Furthermore, a selection s for P_G is called <u>pointwise-Lipschitz-</u>

Copyright © 1980 by Academic Press, Inc.
All rights of reproduction in any form reserved.
ISBN: 0-12-171050-5

continuous (plc) if for each $f \in C(X)$ there exists a constant $K_f > 0$ such that for each $\tilde{f} \in C(X)$ $||s(f)-s(\tilde{f})|| \leq K_f||f-\tilde{f}||$ (this clearly implies that s is continuous). A selection s is called quasilinear (ql) if for each $f \in C(X)$, for each $g \in G$ and for all constants c,d $s(cf+dg) = cs(f) + dg$.

In the following we study the problem of existence of plc., ql. selections for P_G. At first we give a characterization of the spaces which admit plc., ql. selections from among the finite dimensional subspaces of C[a,b] [1] (Theorem 2.3). Finally we give a recent result on existence of plc., ql. selections for P_G, in case X is chosen as union of finitely many compact intervals. This result has been given by Sommer [11].

II. THE CASE X = [a,b]

Nürnberger and Sommer ([5],[6],[7],[8],[9],[10]) have completely characterized those finite dimensional subspaces of C[a,b] which admit continuous selections for P_G. This solves a problem posed by Lazar-Morris-Wulbert [4] for C(X), X compact, in the case X = [a,b]. These authors have given a characterization of those one-dimensional subspaces of C(X), X compact, which admit continuous selections. Nürnberger and Sommer have studied this problem with new methods - by using the theory of the weak Chebyshev spaces.

DEFINITION 1. An n-dimensional subspace G of $C(X), X \subseteq \mathbb{R}$, is called weak Chebyshev if each $g \in G$ has at most n-1 sign changes i.e. there do not exist n+1 points $x_o < x_1 < \dots < x_n$ in X such that $g(x_i)g(x_{i+1}) < 0$ for $i = 0,\dots,n-1$.

To give the exact statements of the characterization of Nürnberger and Sommer we would have to define a certain set of knots for each weak Chebyshev subspace of C[a,b] (see Sommer [10]). This would be a long and complicated process. Therefore, we only summarize these results in a short way in the following theorem.

THEOREM 2 (Nürnberger, Sommer). Let G be an n-dimensional subspace of C[a,b]. Then the following conditions are equivalent:

(2.1) There exists a continuous selection for P_G

(2.2) (i) G is weak Chebyshev

(ii) No $g \in G$ vanishes on more than one interval

(iii) "The numbers of the boundary zeros of the elements $g \in G$ are bounded in a certain sense" .

Condition (2.2) implies for each $f \in C[a,b]$ the existence of a $g_o \in P_G(f)$ such that $f-g_o$ has a certain number of global and local alternating extreme points and g_o is the unique best approximation of f satisfying these alternation properties. Then $s(f) := g_o$ defines a continuous selection.

Recently, we have shown [1] that each continuous selection constructed by Nürnberger and Sommer is even plc., which is a strong property of the metric projection, and also ql. .

THEOREM 3. Let G be an n-dimensional subspace of C[a,b]. Then the following conditions are equivalent:

(2.3) There exists a plc., ql. selection for P_G

(2.4) There exists a continuous selection for P_G .

This result is - to our knowledge - the first proof of existence of plc. selections for P_G, in case G is a non-Haar space. Freud [3] has proved the pointwise-Lipschitz-continuity of P_G, in case G is a Haar space.

We now want to apply Theorem 2 and Theorem 3 to the spline spaces, the most interesting subclass of the class of the weak Chebyshev spaces.

COROLLARY 4. Let $S_{m,k}$ be a spline space of degree m with k fixed knots $x_1,\dots,x_k$. Then the following conditions are equivalent:

(2.5) There exists a plc., ql. selection for $P_{S_{m,k}}$

(2.6) There exists a continuous selection for $P_{S_{m,k}}$

(2.7) $k \leq m + 1$.

The equivalence of (2.6) and (2.7) has been proved by Nürnberger-Sommer [7].

III. THE CASE $X = \bigcup_{j=1}^{m} I_j$

Let $m \in \mathbb{N}$ and for each $j = 1,\dots,m$ let I_j be a non-degenerate compact real interval. We set $X = \bigcup_{j=1}^{m} I_j$.

DEFINITION 1. An n-dimensional subspace G of C(X) is called a Z-space if no nonzero $g \in G$ vanishes on all points of some non-empty open subset of X.

Furthermore, a zero $\tilde{x}$ of a function $f \in C(X)$ is called a zero with a sign change in X if for each neighborhood U of $\tilde{x}$ there are two points $x_1, x_2 \in U$ such that $f(x_1)f(x_2) < 0$.

Brown [2] has shown that a continuous selection for a Z-subspace of C(X) can be only in very limited situations:

THEOREM 2 (Brown [2]). Let G be an n-dimensional Z-subspace of C(X). If there exists a continuous selection for P_G, then each nonzero $g \in G$ has at most n distinct zeros and at most n-1 <u>zeros</u> with <u>sign</u> <u>changes</u>.

Brown has proved this result even in the case that X is an arbitrary compact space.

Recently, Sommer [11] has given a partial converse to the preceding theorem by constructing continuous selections for P_G, in case there is a $\tilde{x} \in X$ such that $G|_{X \setminus \{\tilde{x}\}}$ satisfies the Haar condition and each $g \in G$ has at most n-1 zeros with sign changes. In addition to this he has shown that these selections are even plc. .

THEOREM 3 (Sommer [11]). Let G be an n-dimensional Z-subspace of C(X). If each $g \in G$ has at most n-1 <u>zeros</u> with <u>sign</u> <u>changes</u> and if there is a $\tilde{x} \in X$ such that each $g \in G|_{X \setminus \{\tilde{x}\}}$ has at most n-1 zeros, then there exists a plc., ql. selection for P_G.

EXAMPLE. Let $G = \text{span}\{g_1, \ldots, g_n\}$ be a Haar subspace of C(X). Let $g_o \in C(X)$, $g_o \geq 0$ and g_o have exactly one zero in X. Then by the preceding theorem the space $\tilde{G}$, defined by $\tilde{G} := \text{span}\{g_o g_1, \ldots, g_o g_n\}$ admits a plc., ql. selection for $P_{\tilde{G}}$.

Theorem 3 yields not only the first result on existence of non-weak-Chebyshev spaces which admit plc. selections, but also the first result on existence of non-weak-Chebyshev spaces with dimension > 1 which admit continuous selections. This easily can be shown by the preceding example:

Let $X = I_1 \cup I_2 \cup I_3$ where $I_1 = [-1,1]$, $I_2 = [2,3]$ and $I_3 = [4,5]$. Let g_o be defined by

$$g_o(x) := \begin{cases} |x| & \text{if } x \in I_1 \\ -1 & \text{if } x \in I_2 \\ 1 & \text{if } x \in I_3 \end{cases} .$$

Then the space $\tilde{G}$, defined by

$\tilde{G}:= \text{span}\{g_o(x), xg_o(x), \ldots, x^{n-1}g_o(x)\}$ is a non-Haar space and even a non-weak-Chebyshev space. But by the preceding example $\tilde{G}$ admits a plc., ql. selection.

This situation is quite different from the case $X = [a,b]$, because in this case by the Theorems 2.2 and 2.3 weak Chebyshev is not only necessary for existence of plc. selections, but also necessary for existence of continuous selections.

REFERENCES

1 Blatt, H. P., G. Nürnberger and M. Sommer, A characterization of pointwise-Lipschitz-continuous selections for the metric projection, preprint.

2 Brown, A. L., An extension to Mairhuber's Theorem. On metric projections and discontinuity of multivariate best uniform approximation, preprint.

3 Freud, G., Eine Ungleichung für Tschebyscheffsche Approximationspolynome, Acta Sci. Math. (Szeged) 19 (1958), 162-164.

4 Lazar, A. J., P. D. Morris and D. E. Wulbert, Continuous selections for metric projections, J. Functional Analysis, 3 (1969), 193-216.

5 Nürnberger, G., Nonexistence of continuous selections for the metric projection, SIAM J. Math. Anal., to appear.

6 Nürnberger, G. and M. Sommer, Weak Chebyshev subspaces and continuous selections for the metric projection, Trans. Amer. Math. Soc., 238 (1978), 129-138.

7 Nürnberger, G. and M. Sommer, Characterization of continuous selections of the metric projection for spline functions, J. Approximation Theory, 22 (1978), 320-330.

8 Sommer, M., Nonexistence of continuous selections of the metric projection for a class of weak Chebyshev spaces, Trans. Amer. Math. Soc., to appear.

9 Sommer, M., Characterization of continuous selections for the metric projection for generalized splines, SIAM J. Math. Anal., to appear.

10 Sommer, M., Characterization of continuous selections of the metric projection for a class of weak Chebyshev spaces, preprint.

11 Sommer, M., Existence of pointwise-Lipschitz-continuous selections of the metric projection for a class of Z-spaces, preprint.

THE GELFAND SPACE OF THE BANACH ALGEBRA OF RIEMANN INTEGRABLE FUNCTIONS

Jörg Blatter[1,2]

Instituto de Matemática e Estatística
Universidade de São Paulo
São Paulo, Brasil

In the course of his investigation of the Riemann integral, Professor Samuel Hönig of the Universidade de São Paulo encountered the problem of representing the Gelfand space of the Banach algebra of Riemann integrable functions on [0;1]. I have been unable to obtain a satisfactory representation. In hopes of stimulating interest in this problem, I shall present here (1) a review of that part of the Gelfand theory which is pertinent to the problem at hand (so as to make the problem accessible to a broader audience), (2) a satisfactory representation of a much simpler algebra (so as to let that broader audience know what constitutes a satisfactory representation), and (3) some of my results and some of the approaches I have tried.

1. The Gelfand Theory. For A a real, commutative Banach algebra, the Gelfand space of A is the set Γ_A of non-zero, multiplicative, linear functionals on A and is topologized as a subspace of the product space $\mathbb{R}^A$. Γ_A is a compact Hausdorff space (A is always assumed to have an identity). For $x \in A$, the Gelfand transform of x is the function $\hat{x} : \Gamma_A \to \mathbb{R}$ defined by

$$\hat{x}(\gamma) = \gamma(x), \quad \gamma \in \Gamma_A.$$

$x \mapsto \hat{x}$ is always a multiplicative, linear contraction of A to $C(\Gamma_A)$. If A satisfies the Arens conditions

(1) $\|x^2\| = \|x\|^2$ for all $x \in A$,

(2) $1 + x^2$ is invertible for all $x \in A$,

then $x \mapsto \hat{x}$ is an isometry onto $C(\Gamma_A)$.

In the case that A is a uniformly closed algebra of bounded functions on a set S the Arens conditions are obviously satisfied and so A is isomorphic with $C(\Gamma_A)$. If for $s \in S$, $\delta_s : A \to \mathbb{R}$ is defined by

$$\delta_s(f) = f(s), \quad f \in A,$$

then $s \mapsto \delta_s$ maps S onto a dense subset of Γ_A, and if A separates the points of S, $s \mapsto \delta_s$ is a homeomorphism of S to $\{ \delta_s : s \in S \}$ when S is given the smallest topology which renders each member of A continuous.

---

[1] Supported by an FAPESP-Grant.
[2] Present address: Instituto de Matemática e Estatística, Universidade de São Paulo, Caixa Postal 20570, São Paulo, Brazil.

Copyright © 1980 by Academic Press, Inc.
All rights of reproduction in any form reserved.
ISBN: 0-12-171050-5

Note that when S is so topologized it is completely regular, and $(\Gamma_A, s \mapsto \delta_s)$ is a compactification of S. Thus when S = [0;1] and A is the algebra of Riemann integrable functions, Γ_A can be regarded as a compactification of [0;1] when the latter has the discrete topology.

Let A be an arbitrary real, commutative Banach algebra. Let B be the set of idempotents in A, that is, those $e \in A$ such that $e = e^2$. When one defines the operations $\wedge$ and $\vee$ on B by $e \wedge f = ef$ and $e \vee f = e + f - ef$ B becomes a Boolean algebra. In the case that A is an algebra of functions on a set the idempotents are just the characteristic functions which belong to A, and for $1_E, 1_F \in A$, $1_E \vee 1_F = 1_{E \cup F}$ and $1_E \wedge 1_F = 1_{E \cap F}$. For B any Boolean algebra, the <u>Stone space</u> of B is the set Σ_B of non-zero Boolean algebra homomorphisms of B to {0,1}, and for $e \in B$, $\tilde{e} = \{ \beta \in \Sigma_B : \beta(e) = 1 \}$. When Σ_B is given the topology generated by $\{ \tilde{e} : e \in B \}$, it becomes a compact Hausdorff space, and the mapping $e \mapsto \tilde{e}$ is a Boolean algebra isomorphism of B to the Boolean algebra of open, closed subsets of Σ_B. If A is a commutative Banach algebra, and if the linear span of its set B of idempotents is dense in A, then the mapping $\gamma \mapsto \gamma|B$ is a homeomorphism of Γ_A onto Σ_B. In the case that A is the algebra of Riemann integrable functions this will be the case. It is also the case in the example considered in the next section.

<u>2. Berberian's Theorem.</u> In this section R_o is the algebra of <u>regulated functions</u> on [0;1], that is, those $f : [0;1] \to \mathbb{R}$ such that f(0+), f(1-), and f(t+), and f(t-) exist for all $0 < t < 1$. Berberian (1978) has proved that Γ_{R_o} consists of the functions

$$\delta_t : f \mapsto f(t), \quad 0 \leq t \leq 1,$$

$$\delta_t^+ : f \mapsto f(t+), \quad 0 \leq t < 1,$$

$$\delta_t^- : f \mapsto f(t-), \quad 0 < t \leq 1.$$

It is well known that the regulated functions form the (uniform) closure of the step functions. It follows that the linear span of the idempotents in R_o is dense in R_o. Berberian gives a straightforward description of the topology of Γ_{R_o}, but it doesn't seem likely that his procedure will help a great deal in determining the Gelfand space of the space of Riemann integrable functions. Here we describe the topology of Γ_{R_o} with the aid of results from (Blatter, J., and Seever, G., 1976) and (Blatter, J., 1975). Let K be the lattice cone (the terminology if from (Blatter, Seever)) of non-decreasing regulated functions. K - K is dense in R_o and so by results in (Blatter, Seever) K consists of f such that $\hat{f}$ is increasing with respect to the order $\leq$ on Γ_A defined by

$$\delta \leq \gamma \Leftrightarrow \hat{f}(\delta) \leq \hat{f}(\gamma) \text{ for all } f \in K.$$

One easily sees that the order $\leq$ on Γ_A is <u>total</u>, that is, for $\delta, \gamma \in \Gamma_A$, either $\delta \leq \gamma$ or $\gamma \leq \delta$. Since $_A$ is compact, and since the order is total, it follows from (Blatter) that the topology of Γ_A is the order topology, viz., the topology generated by the sets of the form $(\alpha;\beta) = \{ \gamma \in \Gamma_A :$

$\alpha \leq \gamma \leq \beta,\ \gamma \neq \alpha,\beta\ \}$. Berberian's description of the topology of Γ_{R_o} boils down to the following: for $0 < t < 1$, $\{\ (\delta_t;\delta_u^+) : t < u \leq 1\}$ is a neighborhood base at δ_t^+, $\{\{\delta_t\}\}$ is a neighborhood base at δ_t, and $\{\ (\delta_u^-;\delta_t) : 0 \leq u < t\}$ is a neighborhood base at δ_t^-.

The algebra of Riemann integrable functions is much more complicated than R_o and so one shouldn't expect a representation of its Gelfand space anywhere nearly as simple as that of R_o, but hope springs eternally in the human breast.

<u>3. Approaches to the Representation of Γ_R.</u> A pair of tools useful in studying Γ_A, A a Banach algebra which satisfies the Arens conditions, are the following consequences of the Stone-Weierstrass theorem.

<u>Theorem 1.</u> Let A be A commutative Banach algebra which satisfies the Arens conditions, and let B bs a closed subalgebra of A. Then B also satisfies the Arens conditions, and Γ_B is a quotient space of Γ_A, the quotient mapping being $\gamma \mapsto \gamma|B$.

<u>Theorem 2.</u> Let A be a commutative Banach algebra which satisfies the Arens conditions, and let I be a closed ideal in A. Then A/I satisfies the Arens conditions, and $\Gamma_{A/I}$ may be identified with $\{\ \gamma \in \Gamma_A : \gamma|I = 0\}$.

If A and B are as in Theorem 1, we say that $\gamma \in \Gamma_A$ is <u>tied to</u> $\delta \in \Gamma_B$ if $\gamma|B = \delta$. For example, if $A = R_o$ and B is the algebra of continuous functions, then for $t \in (0;1)$, δ_t, δ_t^+, and δ_t^- are tied to $\delta_t|B$.

The regulated functions, being the closed linear span of the set of characteristic functions of intervals, are all Riemann integrable. Thus R_o is a closed subalgebra of the algebra R of Riemann integrable functions. Whence each member of Γ_R is tied to a δ_t, a δ_t^+, or a δ_t^-. For $t \in [0;1]$. $\{\delta_t\}$ is an open, closed subset of Γ_{R_o} and so the set Δ_t of members of Γ_R tied to δ_t is an open, closed subset of Γ_R. Let $A_t \in [0;1]$ be such that $1_{A_t} \in R$ and $\hat{1}_{A_t} = 1_{\Delta_t}$. Obviously, $t \in A_t$, and for $s \neq t$, $A_t \quad A_s = \emptyset$. It follows that $A_t = \{t\}$, i.e., Δ_t has but one member, $f \mapsto f(t)$. I have not been able to exhibit a single γ which is tied to δ_t^+, much less describe all such γ. I do, however, have some strong candidates. For $h > 0$, define $\theta_h : R \to \mathbb{R}$ by

$$\theta_h(f) = \frac{1}{h}\int_t^{t+h} f,\ f \in R.$$

$\{\theta_h\}_{h>0}$ is a bounded net in R', the Banach space dual of R, and so has a weak*-cluster point. I conjecture that any such cluster point belongs to Γ_R and is tied to δ_t^+. Conceivably, every member of Γ_R which is tied to δ_t^+ is a cluster point of this net, but on this point I will hazard no guess.

Theorem 2 can be used to obtain a partial representation of Γ_R. The set I of all $f \in R$ such that $\int|f| = 0$ is a closed ideal in R. R/I is a closed subalgebra of $L^\infty[0;1]$ and so $\Gamma_{R/I}$ is a quotient space of Γ_{L^∞}. By Theorem 2 $\Gamma_{R/I}$ can be identified with $Z = \{\ \gamma \in \Gamma_R : \gamma|I = 0\ \}$ and so Z is a quotient space of Γ_{L^∞}. R is known to be the closed linear span of its idempotents (Frink,1933). 1_A is an idempotent in R iff the boundary of A has Lebesgue measure 0 (a bounded function on [0;1] is Riemann integrable iff it is continuous almost everywhere; 1_A is discontinuous only at its boundary points). If $\gamma \in \Gamma_{L^\infty}$, then $[\ \gamma = 1\ ]$ is an ultrafilter in the Lebesgue measure algebra. Conversely, if Ω is an ultrafilter in the Lebesgue measure algebra, then there is a unique $\gamma \in \Gamma_{L^\infty}$ such that $\Omega = [\ \gamma = 1\ ]$. Similarly, one can define the members of Γ_R in terms of ultrafilters in the Boolean algebra of idempotents in R. If $t \in [0;1)$, and if Ω is an ultrafilter which contains $\{\ 1_A \in R : \exists\, u > t \ni (t;u) \subset A\ \}$, then the member of Γ_R determined by Ω is tied to δ_t^+.

REFERENCES

Blatter, J.(1975) J. Approx. Theory. 13, 56.
Blatter, J., and Seever, G. (1976) TAMS, 222, 65.
Berberian, S., (1978) Pac. J. Math. 74, 15
Frink, O., (1933) Ann. Math. (2) 34, 518.

APPROXIMATE SOLUTIONS OF LINEAR DIFFERENTIAL SYSTEMS WITH BOUNDARY CONDITIONS

Gary Bogar
Ron Jeppson

Department of Mathematical Sciences
Montana State University
Bozeman, Montana

Consider the linear differential system with boundary condition

$$(1.1) \qquad \vec{y}' = A(x)\vec{y} + \vec{f}(x)$$
$$M\vec{y}(a) + N\vec{y}(b) = 0$$

where $A(x)$, M, and N are $n \times n$ matrices, $a_{ij}(x) \in C[a,b]$ and $\vec{f}(x)$ is an $n \times 1$ matrix with $f_i(x) \in C[a,b]$. In this paper we study closed form polynomial approximation to solutions of (1.1). Instead of using splines [3,5] or orthogonal polynomials [7] we use a uniform type norm. The work was motivated by a recent result by Schmidt and Wiggens [6], who studied similar results for second order linear differential equations with separated boundary conditions.

Let $P_k = \{\vec{p}(x)\colon \vec{p}(x) = (p_1(x),\ldots,p_n(x))^T$ such that $p_i(x)$ is a polynomial of degree k or less, $1 \le i \le n$ and $M\vec{p}(a) + N\vec{p}(b) = 0\}$.

A $\vec{p}_k(x) \in P_k$ is called a minimax approximate solution (MAS) of (1.1) if

$$(1.2) \qquad \max_{x\in[a,b]} \| \vec{p}_k'(x) - A(x)\vec{p}_k - \vec{f}(x) \| = \inf_{\vec{p}\in P_k} \max_{x\in[a,b]} \| \vec{p}'(x) - A(x)\vec{p} - \vec{f}(x) \|$$

where $\| \vec{f} \| = \max_{1\le i\le n} |f_i|$.

Copyright © 1980 by Academic Press, Inc.
All rights of reproduction in any form reserved.
ISBN: 0-12-171050-5

Let

$$|\vec{u}| = \max_{a\leq x\leq b} \max_{1\leq i\leq n} |u_i(x)|$$

then a MAS can be viewed as a best approximation to $\vec{y}$ from the linear subspace of P_k of

$$\{\vec{u}:\ u_i \in C^1[a,b],\ M\vec{u}(a) + N\vec{u}(b) = 0\}$$

with respect to the uniform norm $|\cdot|_{\mathcal{L}}$ given by

$$|\vec{u}|_{\mathcal{L}} = |\mathcal{L}[\vec{u}]|$$

where

$$\mathcal{L}[\vec{u}] = \vec{u}' - A(x)\vec{u}$$

Throughout we will assume that

$$\text{(I)}\quad \mathcal{L}[\vec{u}] = 0$$

$$M\vec{u}(a) + N\vec{u}(b) = 0$$

is incompatible, (i.e. has only the trivial solution) and the rank of the augmented matrix $[M,N]$ is n.

II. CONVERGENCE

Given $\vec{y}$ a solution of (1.1) we rewrite (1.1) as follows

$$y' = L\vec{y} + (A(x) - L)\vec{y} + \vec{f}(x)$$

where L is a nilpotent constant matrix of order n.

The uniform convergence of $\vec{p}_k$ to $\vec{y}$ and p_k' to $\vec{y}'$ is accomplished by consider the following system

$$\text{(2.1)}\quad \vec{y}' = L\vec{y}$$

$$M\vec{u}(a) + N\vec{u}(b) = 0$$

If (2.1) is incompatible we use a Green's function argument, while if (2.1) is compatible we are able to use the theory of Generalized Green's functions developed by Reid [4].

THEOREM 1: Assume condition I holds and $\vec{y}$ is the unique solution of (1.1). Suppose that for each $k \geq n+1$ a M.A.S. $\vec{p}_k$ from $\vec{P}_k$ is fixed. Then

$$\lim_{k\to\infty} \left|\vec{p}_k^{\,(i)} - \vec{y}^{(i)}\right| = 0 \qquad i = 0, 1.$$

The proof of the theorem also gives that

$$\| p_{ki}^{(i)} - y_i^{(i)} \| = \sigma(E_{k-n}(y_i')) \qquad 1 \leq i \leq n, \qquad j = 0, 1.$$

By Jackson's theorem [2], if $y_i' \in C^n[a,b]$ $1 \leq i,j \leq n$, then there is a constant m independent of k such that

$$\| p_{ki}^{(j)} - y_i^{(j)} \| \leq \frac{m}{k^n} \qquad 1 \leq i \leq n, \quad j = 0, 1.$$

Discretization results similar to Schmidt and Wiggens and examples have also been worked out. The approach taken by Schmidt and Wiggens is built for second order linear differential equations with separated boundary condition. The present method works for arbitrary differential systems and gives an algorithm for finding elements in P_k. Comparison with other types of closed form polynomial approximations are currently being studied, along with approximations to nonlinear boundary value problems.

REFERENCES

1. G. Allinger and M. Henry, "Approximate Solutions of Differential Equations with Deviating Arguments", SIAM J. Numerical Anal. Vol. 13 No. 3, 1976 412-426.

2. E.W. Cheney, Introduction to Approximation Theory, McGraw-Hill, New York, 1966.

3. C. De Boor, and B. Swartz, "Collocation at Gaussian Points", SIAM J. Numerical Anal. Vol. 10 No. 4, 1973 582-606.

4. W.T. Reid, "Generalized Green's Matrices for Compatible Systems of Differential Equations", AMJM 53(1931) 443-459.

5. R.D. Russell and L.F. Shampine, "A Collocation Method for Boundary Value Problems", Numer. Math. 19, (1972) 1-28.

6. D. Schmidt anf K. Wiggins, "Minimax Approximate Solutions of Linear Boundary Value Problems", Math. Comp. Vol. 33 No. 145, 1979 139-148.

7. M. Urabe, "Numerical Solution of Boundary Value Problems in Chebyshev Series - A Method of Computation and Error Estimation", Lecture Notes in Mathematics, Springer-Verlag 109, 1969.

ERROR ASYMPTOTICS FOR LINEAR ELLIPTIC BOUNDARY VALUE PROBLEMS ON GENERAL DOMAINS

Klaus Böhmer

Institut für Praktische Mathematik

Universität Karlsruhe

Karlsruhe, W-Deutschland

I. INTRODUCTION

In this note we discuss the numerical solution of linear elliptic boundary value problems on general domains with smooth boundary via difference methods. Since in this case there are not enough grid points on the boundary the usual five point star has to be modified, as indicated by Kreiss, to achieve asymptotic expansions. We generalize and improve known results about error asymptotics and give numerical results.

Copyright © 1980 by Academic Press, Inc.
All rights of reproduction in any form reserved.

ISBN: 0-12-171050-5

II THE BASIC DISCRETIZATION AND ITS HISTORY

For an easier presentation we confine our discussion to problems of the form

$$(1)\qquad Fz:=\begin{cases} z_{xx}+z_{yy}+f\cdot z+f^* = 0 \text{ in } \Omega\subset \mathbb{R}^2, f\leq 0,\\ z-g = 0 \text{ on } \partial\Omega,\end{cases}$$

where Ω is an open bounded and connected domain in $\mathbb{R}^2$ and f, f*, g are continuous functions. For more general results see Böhmer [2],[3]. Introducing the usual sets of equidistant grid lines parallel to the x-and y-axises, $\Gamma_{h,\ell}$, and their corresponding intersection points, the set of grid points, Γ_h, let

$$\Omega_h:=\{(x,y)\in\Gamma_h\cap\Omega \mid (x\pm h,y),(x,y\pm h)\in\Omega\},$$

$$\Omega_{h,i}:=(\Omega\cap\Gamma_h)\setminus\Omega_h,\ \partial\Omega_h:=\partial\Omega\cap\Gamma_{h,\ell}$$

be the sets of regular, irregular and boundary grid points. In regular grid points we discretize (1) with the usual five point star and obtain for $(x,y)\in\Omega_h$ and the approximate solution ζ_h the equation

$$(2)\qquad \begin{cases} (\varphi_h F)\zeta_h(x,y):=\frac{1}{h^2}\{\zeta_h(x+h,y)+\zeta_h(x-h,y)+\zeta_h(x,y+h)\\ +\zeta_h(x,y-h)-4\zeta_h(x,y)\}+f(x,y)\zeta_h(x,y)+f^*(x,y)=0\\ \text{for } (x,y)\in\Omega_h.\end{cases}$$

For $(x,y)\in\Omega_{h,i}$ we modify (2). Let h be so small, that there is at most one intersection point on each of the line segments $(x,y),(x\pm h,y),(x,y\pm h)$ for every $(x,y)\in\Omega_{h,i}$, and let there be at least k consecutive grid points in Ω opposite to the intersection point with $\partial\Omega$ on every grid line through $(x,y)\in\Omega_{h,i}$. For small h and smooth $\partial\Omega$ this is always possible. For $(x,y)\in\Omega_{h,i}$ and, e.g., $(x+h,y)\notin\Omega$ we have, with the intersection point x^*, the situation in Fig. 1:

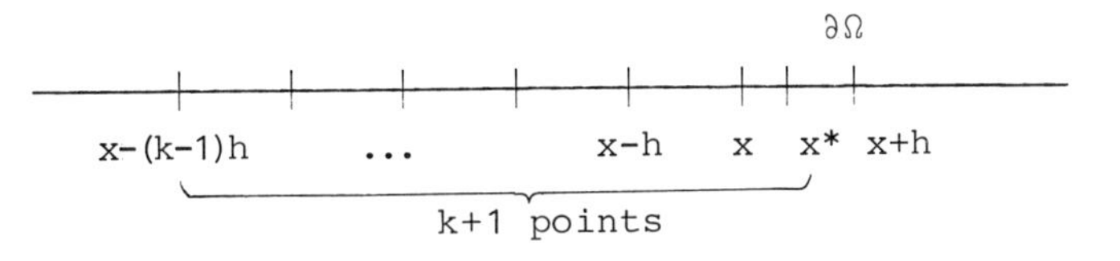

FIGURE 1: Interpolation polynomial defined by k+1 points

Let P be the polynomial of degree k which interpolates $g(x^*,y)$ in x^* and $\zeta_h(x-\nu h,y)$ in $x-\nu h, \nu=0,\ldots,k-1$. Then we replace in(2) $\zeta_h(x+h,y)$ by $P(x+h)$ to obtain $(\varphi_h F)\zeta_h(x,y)$ for $(x,y)\in\Omega_{h,i}$. Finally compute ζ_h from $(\varphi_h F)\zeta_h(x,y)=0$ for $(x,y)\in\Omega_h\cup\Omega_{h,i}$.

Pereyra-Proskurowski-Widlund [7] published a result on error asymptotics of the form

$$(3) \quad \begin{cases} \varepsilon_h(x,y):=\zeta_h(x,y)-z(x,y)=h^2e_2(x,y)+h^4e_4(x,y)+O(h^{k-1/2}) \\ \text{for } k\leq 6,\ f^*=0 \text{ in (1) and } \|\cdot\|_2 \end{cases}$$

There is a long history to this problem indicated in Table I.

TABLE I: History of boundary transfer

k	author	year	order of convergence in $\|\circ\|_{..}$	error asymptotic up to k-1/2
0	Gerschgorin	1930 [5]	1 in $\|\cdot\|_\infty$	-.5
1	Collatz	1933 [4]	2 in $\|\cdot\|_\infty$	.5
2	Shortley Weller	1938 [8]	2 in $\|\cdot\|_\infty$	1.5
	Mikeladse	1941 [6]	2 in $\|\cdot\|_\infty$	1.5
$k\leq 6$	Pereyra-Proskurowski-Widlund	1977 [7]	2 in $\|\cdot\|_2$	k-1/2

Table I immediately raises some questions:

1) Is (3) valid for (1), or even for more general linear or nonlinear elliptic equations?
2) Is $k \leq 6$ a natural restriction?
3) Is it possible to replace in (3) $O(h^{k-1/2})$ and $\|\cdot\|_2$ by $O(h^{k+1})$ and $\|\cdot\|_\infty$ to meet the results for k=0,1,2 ? Especially, for k=6 that is conjectured in [7] .

III A GENERAL RESULT AND SOME CONSEQUENCES

As an answer to the final questions in II we formulate THEOREM: For the problem (1) and its discretization in II, for $k \in \mathbb{N} \cup \{0\}$, small enough h and smooth enough $\partial\Omega, f, f^*, g$ we have

(4) $\varepsilon_h(x,y) = h^2 e_2(x,y) + h^4 e_4(x,y) + \ldots + O(h^{k+1})$ in $\|\cdot\|_\infty$.

This result as well applies to more general linear equations, see [3]. Further (3) is valid even for nonlinear equations, see [2]. There are some consequences to the result (4):

CONSEQUENCES: 1) Questions 1)-3) in II are answered essentially.
2) For k=0,1,2 the result (4) yields the correct orders of convergence known from other techniques, see Table I.
3) For k=6 (4) proves the conjecture of Pereyra-Proskurowski-Widlund [7].
4) For linear elliptic boundary value problems unrestricted error asymptotics are available, but not advisable (for computational reasons).

5) Based on (4) difference methods of arbitrarily high order may be developed.

IV A NUMERICAL EXAMPLE

We apply the above theory to the following example

$z_{xx}+z_{yy}+2\sin(x+y)=0$ in Ω

$z(x,y)-\sin(x+y)=0$ on $\partial\Omega$, exact solution $z(x,y)=\sin(x+y)$

$\Omega:=\{(x,y)\in \mathbb{R}^2 \mid (\frac{x-1}{0.5})^2+(\frac{y-1}{0.78})^2<1\}$.

With discrete Newton methods, [1],we obtain better errors $\varepsilon_{h,1},\varepsilon_{h,2},\varepsilon_{h,3}$ of orders $\leq k+1$, which is very clearly shown in Table II .

TABLE II: High order errors $\varepsilon_h,\varepsilon_{h,1},\varepsilon_{h,2},\varepsilon_{h,3}$

k	$\|\varepsilon_h\|_\infty$	$\|\varepsilon_{h,1}\|_\infty$	$\|\varepsilon_{h,2}\|_\infty$	$\|\varepsilon_{h,3}\|_\infty$	max. order
0	3.1,-2	3.3,0	2.4,+2	1.7,+4	1
1	1.7,-4	1.3,-4	1.5,-4	1.8,-4	2
2	2.3,-5	2.4,-6	1.7,-6	8.6,-7	3
3	2.4,-5	9.2,-8	4.1,-8	3.1,-8	4
4	2.4,-5	4.2,-9	5.0,-10	2.5,-10	5
5	2.4,-5	9.9,-9	8.7,-12	3.6,-12	6
6	2.4,-5	9.9,-9	4.7,-12	5.0,-14	7
7	2.4,-5	9.9,-9	4.8,-12	1.1,-14	8
8	2.4,-5	9.9,-9	4.8,-12	1.1,-14	9

REFERENCES

1. Böhmer, K: Discrete Newton methods and iterated defect corrections, I General theory, II Proofs and application to initial and boundary value problems, submitted to Numer.Math.
2. Böhmer, K: High order difference methods for quasilinear elliptic boundary value problems on general regions, University of Wisconsin-Madison, MRC, Technical Summary Report 1980.
3. Böhmer, K: Asymptotic expansion for the discretization error in linear elliptic boundary value problems on general domains, to appear
4. Collatz, L.: Bemerkungen zur Fehlerabschätzung für das Differenzenverfahren bei partieller Differentialgleichungen, Z.Angew.Math.Mech. 13, 56-57(1933).
5. Gerschgorin, S.: Fehlerabschätzung für das Differenzenverfahren zur Lösung partieller Differentialgleichungen, Z.Angew. Math. Mech. 10, 373-382 (1930).
6. Mikeladse, S.: Neue Methoden der Integration von elliptischen und parabolischen Differentialgleichungen, Izv.Akad. Nauk SSSR, Seria Mat. 5, 57-74 (1941)(russian).
7. Pereyra, V., W. Proskurowski and O. Widlund: High order fast Laplace solvers for the Dirichlet problem on general regions, Math. Comp. 31, 1-16 (1977).
8. Shortley, G., R. Weller: The numerical solutions of Laplace's equation, J. Appl. Phys. 9, 334-348 (1938).

AN ESTIMATE OF THE RATE OF CONVERGENCE OF THE NÖRLUND-VORONOI MEANS OF THE FOURIER SERIES OF FUNCTIONS OF BOUNDED VARIATION

R. Bojanic

Department of Mathematics
The Ohio State University
Columbus, Ohio

S. M. Mazhar

Department of Mathematics
Kuwait University
Kuwait

Let (p_n) be a non-increasing sequence of positive numbers. Without loss of generality we can assume that $p_o = 1$. Let

$$P_n = \sum_{k=o}^{n} p_k, \; n = 0,1,2,\ldots$$

We shall assume that $P_n \to \infty$ as $n \to \infty$.

We consider here the Nörlund-Voronoi means w_n^p of a sequence (s_n), defined by

$$w_n^p = \frac{1}{P_n} \sum_{k=o}^{n} p_{n-k} s_k \; , \; n = 0,1,2,\ldots$$

From the hypothesis that (p_n) is a non-increasing sequence of positive numbers, it follows that $P_n \geq (n+1)p_n$ and so $p_n/P_n \to 0$ as $n \to \infty$. This condition is the

Copyright © 1980 by Academic Press, Inc.
All rights of reproduction in any form reserved.
ISBN: 0-12-171050-5

regularity condition for the Nörlund-Voronoi sequence-to-sequence transformation, which means that for every sequence (s_n) for which $\lim_{n\to\infty} s_n = \ell$ exists, we have $\lim_{n\to\infty} w_n^p = \ell$.

We shall consider here the Nörlund-Voronoi means $W_n(f,x)$ of the sequence $(S_n(f,x))$ of the partial sums of the Fourier series of a 2π-periodic function f of bounded variation on $[-\pi,\pi]$.

We have

$$W_n(f,x) = \frac{1}{P_n}\sum_{k=o}^{n} p_{n-k}S_k(f,x).$$

Since

$$S_k(f,x) - \frac{1}{2}(f(x+0) + f(x-0)) = \frac{1}{\pi}\int_o^{\pi} g_x(t)\,\frac{\sin(k+1/2)t}{2\sin(t/2)}\,dt$$

where $g_x(t) = f(x+t) + f(x-t) - f(x+0) - f(x-0)$, $t \neq x$ and $g_x(t) = 0, t=x$, we have

$$W_n(f,x) - \frac{1}{2}(f(x+0) + f(x-0)) = \frac{1}{\pi}\int_o^{\pi} g_x(t)K_n(t)\,dt$$

where

$$K_n(t) = \frac{1}{P_n}\sum_{k=o}^{n} p_{n-k}\,\frac{\sin(k+1/2)t}{2\sin(t/2)}.$$

In view of Dirichlet-Jordan's theorem which states that

$$\lim_{n\to\infty} S_n(f,x) = \frac{1}{2}(f(x+0) + f(x-0))$$

and the regularity of Nörlund-Voronoi transformation, we have

$$\lim_{n\to\infty} W_n(f,x) = \frac{1}{2}(f(x+0) + f(x-0)).$$

The main result of this paper can be stated as follows:

THEOREM. *If (p_n) is a non-increasing sequence of positive numbers and $P_n = \Sigma_{k=o}^{n} p_k \to \infty$ $(n \to \infty)$, then for every function f of bounded variation on $[-\pi,\pi]$ we have*

$$(1) \quad |W_n(f,x) - \frac{1}{2}(f(x+0) + f(x-0))| \le \frac{6}{P_n} \sum_{k=o}^{n} p_k V_o^{\pi/P_k}(g_x),$$

where $V_a^b(f)$ is the total variation of the function f on $[a,b]$.

If f is a continuous function of bounded variation on $[-\pi,\pi]$, inequality (1) becomes

$$(2) \qquad |W_n(f,x) - f(x)| \le \frac{6}{P_n} \sum_{k=o}^{n} p_k \, V_{x-\pi/P_k}^{x+\pi/P_k}(f).$$

Inequalities similar to (1) and (2) for the partial sums of the Fourier series of 2π-periodic functions of bounded variation on $[-\pi,\pi]$ and 2π-periodic continuous functions of bounded variation on $[-\pi,\pi]$, respectively, were obtained recently by R. Bojanic.*

The proof of our theorem is based on two lemmas which give two different evaluations of the integral of the function $K_n(t)$.

Lemma 1. Let $\lambda_n(y) = \int_o^y K_n(t)dt$. *Then for* $0 \le y \le \pi$

$$(3) \qquad |\lambda_n(y)| \le 2\pi.$$

Proof of Lemma 1. We have

$$\lambda_n(y) = \int_o^y \frac{1}{P_n} \sum_{k=o}^{n} p_{n-k} \frac{\sin(k+1/2)t}{2\sin(t/2)} dt$$

---

*R. Bojanic, An estimate for the rate of convergence for Fourier series of functions of bounded variation, Publ. de l´Inst. Math. (Belgrade), to appear.

or

$$\lambda_n(y) = \frac{1}{2}y + \frac{1}{P_n}\sum_{k=1}^{n} p_{n-k}\left(\sum_{m=1}^{k} \frac{\sin my}{m}\right)$$

and (3) follows since

$$\left|\sum_{m=1}^{k} \frac{\sin my}{m}\right| \leq 2\sqrt{\pi}$$

for all k and y.

Lemma 2. Let (p_n) be a non-increasing sequence with $p_o = 1$ and let

$$\Lambda_n(y) = \int_y^{\pi} K_n(t)\, dt$$

Then for $0 < y \leq \pi$

$$|\Lambda_n(y)| \leq \frac{\pi^2}{P_n y}. \tag{4}$$

Proof of Lemma 2. We have

$$K_n(t) = \frac{1}{2P_n \sin(t/2)} \text{Im}\left(e^{i(n+1/2)t}\sum_{k=o}^{n} p_k e^{-ikt}\right)$$

Since (p_n) is a non-increasing sequence of positive numbers with $p_o = 1$, we have by partial summation

$$\left|\sum_{k=o}^{n} p_k e^{-ikt}\right| \leq 2 \max_{0\leq k\leq n}\left|\sum_{r=o}^{k} e^{-irt}\right| \leq \frac{2}{\sin(t/2)}.$$

Consequently

$$|K_n(t)| \leq \frac{1}{P_n(\sin(t/2))^2} \leq \frac{\pi^2}{P_n t^2}.$$

Thus

$$|\Lambda_n(y)| \leq \int_y^{\pi} |K_n(t)|dt \leq \frac{\pi^2}{P_n} \int_y^{\pi} \frac{dt}{t^2} \leq \frac{\pi^2}{P_n y}.$$

<u>Proof of the Theorem</u>. To estimate $W_n(f,x) - \frac{1}{2}(f(x+0) + f(x-0)) = \Delta_n(f,x)$, observe that

$$\Delta_n(f,x) = \frac{1}{\pi} \left(\int_0^{\delta} + \int_{\delta}^{\pi}\right) g_x(t) K_n(t)\, dt \tag{5}$$

$$= A_n(g_x) + B_n(g_x).$$

We have first by partial integration

$$A_n(g_x) = \frac{1}{\pi} \lambda_n(\delta) g_x(\delta) - \frac{1}{\pi} \int_0^{\delta} \lambda_n(t) dg_x(t)$$

and so by Lemma 1

$$|A_n(g_x)| \leq 4V_0^{\delta}(g_x). \tag{6}$$

Next, we have

$$B_n(g_x) = \frac{1}{\pi} g_x(\delta) \Lambda_n(\delta) + \frac{1}{\pi} \int_{\delta}^{\pi} \Lambda_n(t) dg_x(t).$$

Hence by (4)

$$|B_n(g_x)| \leq \frac{\pi}{P_n \delta} V_0^{\delta}(g_x) + \frac{\pi}{P_n} \int_{\delta}^{\pi} \frac{1}{t}\, dV_0^t(g_x).$$

Since

$$\int_{\delta}^{\pi} \frac{1}{t}\, dV_0^t(g_x) = \frac{1}{\pi} V_0^{\pi}(g_x) - \frac{1}{\delta} V_0^{\delta}(g_x) + \int_{\delta}^{\pi} V_0^t(g_x) \frac{dt}{t^2},$$

it follows that

$$|B_n(g_x)| \leq \frac{1}{P_n} V_0^{\pi}(g_x) + \frac{\pi}{P_n} \int_{\delta}^{\pi} V_0^t(g_x) \frac{dt}{t^2}. \tag{7}$$

From (5), (6) and (7) follows that

$$|\Delta_n(f,x)| \leq 4\ V_o^{\delta}(g_x) + \frac{1}{P_n} V_o^{\pi}(g_x) + \frac{\pi}{P_n}\int_{\delta}^{\pi} V_o^{t}(g_x)\ \frac{dt}{t^2}.$$

Choosing $\delta = \pi/P_n$ and replacing t by π/t in the last integral, we find that

$$|\Delta_n(f,x)| \leq 4\ V_o^{\pi/P_n}(g_x) + \frac{1}{P_n} V_o^{\pi}(g_x) + \frac{1}{P_n}\int_{1}^{P_n} V_o^{\pi/t}(g_x)\ dt.$$

Since

$$\begin{aligned}
\int_{1}^{P_n} V_o^{\pi/t}(g_x)\ dt &= \sum_{k=o}^{n-1} \int_{P_k}^{P_{k+1}} V_o^{\pi/t}(g_x)\ dt \\
&\leq \sum_{k=o}^{n-1} p_{k+1}\ V_o^{\pi/P_k}(g_x) \\
&\leq \sum_{k=o}^{n} p_k\ V_o^{\pi/P_k}(g_x)
\end{aligned}$$

we have

$$\begin{aligned}
|\Delta_n(f,x)| &\leq 4\ V_o^{\pi/P_n}(g_x) + \frac{1}{P_n} V_o^{\pi}(g_x) + \frac{1}{P_n}\sum_{k=o}^{n} p_k\ V_o^{\pi/P_k}(g_x) \\
&\leq 4\ V_o^{\pi/P_n}(g_x) + \frac{2}{P_n}\sum_{k=o}^{n} p_k\ V_o^{\pi/P_k}(g_x)
\end{aligned}$$

and the theorem is proved since

$$P_n V_o^{\pi/P_n}(g_x) \leq \sum_{k=o}^{n} p_k\ V_o^{\pi/P_k}(g_x).$$

A PROBLEM OF LORENTZ ON APPROXIMATION BY INCOMPLETE POLYNOMIALS

I. Borosh and C. K. Chui[1]

Department of Mathematics
Texas A&M University
College Station, Texas 77843

I. INTRODUCTION

Let $f \in C[-1,1]$ and denote by $p_n(f,\cdot)$ the best uniform approximant of f from π_n, the space of all polynomials of degree not exceeding n. In [6] and [7] Lorentz stated the following two conjectures:

CONJECTURE 1. *If $p_n(f;0) = 0$ for $n = 0, 1, 2, \ldots,$ then f is an odd function.*

CONJECTURE 2. *If there exists an $\alpha \neq 0$ such that $p_n(f;\alpha) = 0$ for $n = 0, 1, 2, \ldots,$ then f is identically zero in $[-1,1]$.*

The following simple lemma due to Lorentz [1] answers some simpler cases of the above conjectures.

LEMMA 1. *If $p_{n+1}(f;\cdot) \not\equiv p_n(f;\cdot)$ then $p_{n+1}(f;\cdot) - p_n(f;\cdot)$ has $(n+1)$ simple zeros in $(-1,1)$.*

As a corollary of Lemma 1, we see that if $p_{n+1} - p_n$ has a double zero then $p_{n+1} \equiv p_n$. In particular, if $p_n(f;\alpha) = p_n'(f;\alpha)$ for all n then f is identically zero. Also, if $p_n(f;\alpha) = 0$ for $n = 0, 1, 2, \ldots$ and α is one of the end points -1 or 1, then f is identically zero.

Recently Saff and Varga [8] proved the following special case of Lorentz's conjecture 1:

---

[1]Supported partially by the U.S. Army Research Office under Grant Number DAAG 29-78-0097.

Copyright © 1980 by Academic Press, Inc.
All rights of reproduction in any form reserved.
ISBN: 0-12-171050-5

PROPOSITION 1. *Let* $f \in C[-1,1]$. *Define* F *in* $[0,1]$ *by the equation* $F(x^2) = f(x) + f(-x)$. *If* F *can be extended to an entire function of exponential type* $\tau \le \pi/2$, *and if* $p_{2k+1}(f;0) = 0$ *for* $k = 0, 1, \ldots,$ *then* f *is an odd function.*

Note that the Lorentz condition that says that the best approximant vanishes at the origin is needed only for half of the approximants.

In this article, we will first prove a generalization of Saff and Varga's result, and then turn our attention to the case where $f \in L_2[-1,1]$ and $p_n(f,x)$ is the best L_2-approximant of f in $[-1,1]$. In this case we will see that Conjecture 1 is trivial and Conjecture 2 will lead to some interesting problems on Legendre polynomials. We will conclude with some open problems.

II. A GENERALIZATION OF SAFF AND VARGA'S RESULTS

Write $p_n(f;x) = \sum_{k=0}^{n} a_k^{(n)} x^k$. Then Lorentz's Conjecture 1 can be stated in the following way: *If* $a_0^{(n)} = 0$ *for* $n = 0, 1, \ldots$ *then* f *is an odd function.* Now, if f is an even function, it is well-known that all its best approximants from π_n are even and in particular $a_1^{(n)} = 0$ for $n = 0, 1, \ldots$. We can ask about the converse. Does the condition $a_1^{(n)} = 0$ $n = 0, 1, \ldots$ imply that f is even? We could ask a similar question about any one of the coefficients $a_k^{(n)}$, $n = 0, 1, \ldots$. It turns out that the methods of Saff and Varga in [8] can be adapted to answer these questions at least for the same class of very smooth functions for which they proved Lorentz's Conjecture 1. Our results in this direction are summarized in the following:

PROPOSITION 2: *Let* $f \in C[-1,1]$ *and assume that* f *has an entire extension of exponential type* τ *with* $0 \le \tau \le \frac{\pi}{2}$. *Let* $p_n(f;x) = \sum_{i=1}^{k} a_i^{(n)} x^i$ *denote the nth best uniform approximant of* f *on* $[-1,1]$.

i) *If* $a_{2\ell+1}^{(2k)} = 0$ *for* $k > \ell$, *then the odd part of* f *is a polynomial of degree at most* $2\ell-1$.

ii) *If* $a_{2\ell}^{(2k+1)} = 0$ *for* $k \ge \ell$, *then the even part of* f *is a polynomial of degree at most* $2\ell-2$.

The proofs of parts i) and ii) of Proposition 2 are very similar and are along the lines of [8]. We will therefore present only the proof of i). It can be shown from this proof that the condition in the statement of Proposition 2 is too strong and a condition similar to that in Proposition 1 is sufficient. However, these conditions would be too technical and are different for parts i) and ii), and besides it wouldn't add much to the content. We begin with some preliminary notation and remarks.

Let $e_k(\cdot) = f(\cdot) - p_k(f;\cdot)$. Then e_k has at least $k+2$ alternation points $-1 < \xi_1^{(k)} < \ldots < \xi_{k+2}^{(k)} \leq 1$ such that $|e_k(x)|$ attains its maximum at these points and $e_k(\xi_i^{(k)})e_k(\xi_{i+1}^{(k)}) < 0$ for $i = 1,\ldots,k+1$. Let $E_k(x) = e_k(x) - e_k(-x)$. Then we have the following

LEMMA 2. *E_k has at least $k+1$ zeros, where each zero is counted with multiplicity at most 2 and any zero counted with multiplicity 2 is an alternation point for e_k.*

The only difference with Lemma 1 in [8] is the sign in the definition of E_k. However, the proof is exactly the same.

As a consequence of Lemma 2, the function E_{2k} has at least $(2k+1)$ zeros in $[-1,1]$, each counted with multiplicity at most 2, and by Rolle's theorem the 2ℓ-th derivative $E_{2k}^{(2\ell)}$ has at least $2k - 2\ell + 1$ zeros, all counted with multiplicity at most 2. Since f has an entire function extension, we have $p_k^{(2\ell+1)} \to f^{(2\ell+1)}$ and therefore $f^{(2\ell+1)}(0) = 0$ and $E_{2k}^{(2\ell+1)}(0) = 0$. Also $E_{2k}^{(2\ell)}$ is an odd function since E_{2k} is an odd function; hence $E_{2k}^{(2\ell)}(0) = 0$ and $E_{2k}^{(2\ell)}(x)$ has at least a zero of order two at 0. But since it is odd, it has at least a triple zero at zero. The total number of zeros of $E_{2k}^{(2\ell)}$ in $[-1,1]$ is at least $2k - 2\ell + 2$; and again since $E_{2k}^{(2\ell)}$ is odd it has at least $2k - 2\ell + 3$ zeros in $[-1,1]$. We have proved the following

LEMMA 3. *Under the hypotheses of Proposition 2, the function $E_{2k}^{(2\ell)}$ has at least $2k - 2\ell + 3$ zeros in $[-1,1]$.*

Now, $E_{2k}^{(2\ell)}(x) = \left(f^{(2\ell)}(x) - f^{(2\ell)}(-x)\right) - [p_{2k}^{(2\ell)}(f;x) - p_{2k}^{(2\ell)}(f;x)]$, and the difference $q(x) = p_{2k}^{(2\ell)}(f;x) - p_{2k}^{(2\ell)}(f;-x)$ is a polynomial of degree at most $2k - 2\ell - 1$. Hence, we can write the previous equation in the form

$$E_{2k}^{(2\ell)}(x) = x\left(G(x^2) - S(x^2)\right),$$

where G is an entire function and S is a polynomial of degree at most $k-\ell-1$. From Lemma 3, we see that $H(t) = G(t) - S(t)$ has at least $[(2k-2\ell+3)-1]/2 = k-\ell+1$ zeros in $[0,1]$. By Rolle's theorem, the function $H^{(k-\ell)} = G^{(k-\ell)}$ has at least one zero t_k in $[0,1]$. As in [8] we conclude, using a result of Schoenberg [9], that G is identically zero. Therefore, $f^{(2\ell)}(x) = f^{(2\ell)}(-x)$, or $f^{(2\ell)}$ is an even function and the odd part of f is a polynomial of degree at most $2\ell-1$. This concludes the proof of Proposition 2.

Concerning Conjecture 2, Saff and Varga [8] pointed out that the function $f = \sum_{n=N}^{\infty} b_j T_{3^j} \not\equiv 0$ where $T_n \in \pi_n$ denote the Chebyshev polynomials, $b_j \geq 0$ and $0 < \sum_N^{\infty} b_j < \infty$, satisfies $p_n(f, \sin(\pi/3^N)) = 0$ for all n, showing that Conjecture 2 is false in general. However, it might still be true for certain choices of α. For example, we saw from Lemma 1 that it is true for $\alpha = \pm 1$. To gain some insight into this problem, we study the L_2 case in the next section.

III. THE L_2 CASE

Let $f \in L_2[-1,1]$ and let P_n denote the Legendre polynomial of degree n in $[-1,1]$. Then

$$f = \sum_{k=0}^{\infty} \beta_k P_k \quad \text{where} \quad \beta_k = \int_{-1}^{1} f(x) P_k(x)\, dx$$

Let us denote in this section the best L_2 approximant to f in $[-1,1]$ by $q_n(f;\cdot)$. Then

$$(*) \qquad q_n(f;x) = \sum_{k=0}^{n} \beta_k P_k(x) .$$

The L_2-version of Conjecture 1 follows immediately:

PROPOSITION 3. Let $f \in L_2[-1,1]$ such that $q_{2k+1}(f;0) = 0$ for $k = 0, 1, \ldots$ then f is an odd function.

Let $\alpha \in [-1,1]$ and assume $q_n(f;\alpha) = q_{n+1}(f;\alpha) = q_{n+2}(f;\alpha)$. Then from (*), we have

$$\beta_{n+1} P_{n+1}(\alpha) = \beta_{n+2} P_{n+2}(\alpha) = 0,$$

and since the polynomials P_{n+1} and P_{n+2} do not have a common zero, it follows that $\beta_{n+1}\beta_{n+2} = 0$, i.e. either $q_n \equiv q_{n+1}$ or $q_{n+1} \equiv q_{n+2}$. We have proved the following

LEMMA 4. If three successive best L_2 approximants of f pass through the same point, then two of them must be equal.

The following version of Conjecture 2 follows immediately from Lemma 4.

PROPOSITION 4. Let $\beta \in [-1,1]$ and $f \in L_2[-1,1]$ such that $q_n(f;\beta) = 0$ for $n = 1, 2, \ldots$.

1) If $\beta = 0$, f is an odd function

2) If $\beta \neq 0$ and is not a zero of any Legendre polynomial, then $f \equiv 0$.

3) If $\beta \neq 0$ is a zero of the Legendre polynomial P_N but $P_n(\beta) \neq 0$ for $n \neq N$, then $f \equiv P_N$.

Remark: Part 1) repeats Proposition 3 and is included for completeness. Part 2) shows that the L_2 version of Conjecture 2 is true except for a countable set of points. Part 3) would settle completely the question raised in Conjecture 2 for the L_2 case, if the conjecture that two distinct Legendre polynomials do not have a common zero (except for the trivial $x = 0$) holds. It has been conjectured long ago that this is the case. This conjecture appears for example in the Hermite and Stieltjes correspondence [3], but it is still left open. This conjecture would be implied by the stronger conjecture that P_n for n even or $P_n(x)/x$ for n odd is irreducible. Several partial results are known about this irreducibility [2], [4], [5]. Most of them are obtained by applying Eisenstein's criterion or some less well known criteria like Dumas [1] to the Legendre polynomials or to the polynomials obtained from them by some transformations. Most of these results are summarized in the following: Let

$$L_n(x) = \begin{cases} P_n(x) & \text{if } n \text{ even} \\ \frac{1}{x} P_n(x) & \text{if } n \text{ odd} \end{cases},$$

and p denote any odd prime. Then L_n is irreducible for $n = p, p\pm1, p-2, p-4, 2p-1, 2p-2, 2^m, 2^m \pm 1$. In particular, this implies that L_n is irreducible for $n \leq 25$.

IV. SOME QUESTIONS

Of course, Lorentz's Conjecture 1 is still open for the general case $f \in C[-1,1]$. In view of Saff and Varga's example, we know that Conjecture 2 is false but we could ask the question: Characterize the points α for which Conjecture 2 holds. In view of the results in the L_2 case, we make the following

CONJECTURE A. *The set of points* α *for which a function* $f \in C[-1,1]$, $f \not\equiv 0$ *exists such that* $p_n(f;\alpha) = 0$ *for* $n = 0, 1, 2, \ldots$ *is countable.*

Another question is to prove an analog to Lemma 4 for the uniform norm:

CONJECTURE B. *If* $f \in C[-1,1]$ *and* $p_n(f;\alpha) = p_{n+1}(f;\alpha) = p_{n+2}(f;\alpha)$, *then either* $p_n \equiv p_{n+1}$ *or* $p_{n+1} \equiv p_{n+2}$.

This would be equivalent to the statement that $p_{n+1} - p_n$ and $p_{n+2} - p_{n+1}$ have no common zeros.

Finally, since the conjecture about the non-existence of common zeros to two Legendre polynomials seems hopeless, we could make the following simpler conjecture C which would still imply in the case that α is the zero of a Legendre polynomial that f is a polynomial:

CONJECTURE C. *There does not exist a sequence* $n_1 < n_2 < \ldots$ *and a point* $\beta \in [-1,1]$ *such that* $P_{n_1}(\beta) = P_{n_2}(\beta) = \ldots = 0$.

REFERENCES

1. Dumas, M. G., Sur quelques cas d'irreducibilité des polynomes à coefficients rationels, Journal des Mathematiques (Liouville), 6 eine Serie vol. 2 1906 pp. 191-258.
2. Grosswald, E., Contribution to the problem of irreducibility of Legendre's Polynomials, (unpublished manuscript).
3. Hermite and Stieltjes: Correspondence, (Letter # 275).
4. Holt, J. B., The irreducibility of Legendre's polynomials, Proceedings London Math. Soc. Series 2, Vol. 11 (1913) pp. 351-356.
5. Ille, H., On the Irreducibility of Legendre Polynomials, University of Berlin, unpublished thesis (1924).
6. Lorentz, G. G., Approximation by incomplete polynomials, in *Padé and Rational Approximation*, Saff & Varga, eds., Academic Press, N.Y. 1977, pp. 289-302.
7. Lorentz, G. G., New and unsolved problems, in *Linear Spaces and Approximation*, P. L. Bützer and B. Sz Nagy eds., Birkhauser Verlag, Stuttgart 1978, p. 665.
8. Saff, E. B. and Varga, R. S., Remarks on some conjectures of G. G. Lorentz, to appear in J. Approximation Theory.
9. Schoenberg, I. J., On the zeros of successive derivatives of integral functions, Trans. Amer. Math. Soc., vol. 40 (1936), pp. 12-23.

A KOROVKIN-TYPE THEOREM FOR DIFFERENTIABLE FUNCTIONS

Bruno Brosowski

Johann Wolfgang Goethe-Universität
Fachbereich Mathematik
Frankfurt(M), Germany

In an earlier paper [1] we gewe an approach to Korovkin's theorem which used the Dedekind-completion of a partially ordered vector space. Korovkin's theorem can the be considered as an extension of the convergence behaviour of an operator sequence on a partially ordered space to its Dedekind-completion. The special Korovkin-theorem is obtained if one takes into account that the Dedekind-completion of the polynomials of degree at most 2 (endowed with the pointwise ordering) contains C[0,1]. In this paper we apply this approach to the case of differentiable functions.

Let X be a partially ordered real vector space which is Archimedean and denote by δX the Dedekind-completion of X. The elements of δX can be considered as cuts, i.e. as pairs (L,U) of nonempty subsets of X with the following properties:

(a) $\forall_{\ell \in L} \ \forall_{u \in U} \ \ell \leq u$

(b) If $c \in X$ and $\forall_{\ell \in L} \ \ell \leq c$ then $c \in U$.

(c) If $c \in X$ and $\forall_{u \in U} \ c \leq u$ then $c \in L$.

Copyright © 1980 by Academic Press, Inc.
All rights of reproduction in any form reserved.
ISBN: 0-12-171050-5

Assume now that in X a mode of convergence is given, i.e. to certain sequenced (x_n) in X a limitpoint x in X is assigned, in this case we write "$x_n \longrightarrow x$". A mode of convergence in X can be extended to δX by the following steps:

(1) A sequence of subsets $A_n \subset X$ converges to a subset $A \subset X$ if and only if

$$A = \lim A_n := \{x \in X \mid \forall_{n \in \mathbb{N}} \ \exists_{a_n \in A_n} \ a_n \longrightarrow x\}.$$

(2) A sequence (L_n, U_n) of cuts in X converges to a cut (L,U) in X if and only if

$$L = \mathbb{L}(\lim U_n) \quad \& \quad U = \mathbb{U}(\lim L_n) \ ,$$

where for any subset $W \subset X$

$$\mathbb{L}(W) := \{x \in X \mid x \leq W\} \quad \text{and}$$

$$\mathbb{U}(W) := \{x \in X \mid W \leq x\}.$$

Then we have the following

THEOREM 1. Condition (2) extends the mode of convergence in X to δX if and only if the mode of convergence in X satisfies the condition

(*) If $x_n \leq y_n$, $x_n \longrightarrow x$, $y_n \longrightarrow y$, then $x \leq y$.

Now we state a generalized Korovkin-theorem

THEOREM 2. Let X,Y be partially ordered vector spaces with $X \subset Y \subset \delta X$, let $T_n : Y \longrightarrow \delta X$, $n \in \mathbb{N}$, be monotonic operators, and let "$\xrightarrow[X]{}$" be a mode of convergence in X satisfying (*).

If for each x in X we have $T_n(x) \xrightarrow[\delta X]{} x$, then we have also $T_n(y) \xrightarrow[\delta X]{} y$ for each y in Y, where "$\xrightarrow[\delta X]{}$" denotes the extension of "$\xrightarrow[X]{}$" to δX.

We apply now this theorem to the special case of the approximation of the identity in $\delta C^1[a,b]$. Let $[a,b]$, $a < b$, be a compact interval and denote by $C^1[a,b]$ the real vectorspace of all continuously differentiable functions $x : [a,b] \longrightarrow \mathbb{R}$

partially ordered by

$$x \geqq 0 \quad :\Longleftrightarrow \quad \forall_{t\in[a,b]} \quad [x(t) \geqq 0 \quad \& \quad x'(t) \geqq 0].$$

For applying theorem 2 to this case we have to find subspaces V of $C^1[a,b]$ with $\delta V \supset C^1[a,b]$ and we have to determine the convergence in $\delta C^1[a,b]$ when in V a mode of convergence is given.

We have then:

<u>LEMMA 3.</u> Let V be a subspace of $C^1[a,b]$ such that for each x in $C^1[a,b]$ one has

(B1) $\quad x(a) = \sup\limits_{\ell\in L_x} \ell(a) = \inf\limits_{u\in U_x} u(a)$,

(B2) $\quad \forall_{t\in[a,b]} \quad x'(t) = \sup \ell'(t) = \inf u'(t).$

Then $\delta V \supset C^1[a,b]$.

<u>THEOREM 4.</u> The conditions (B1) and (B2) are equivalent to:

(A1) For each $t \in [a,b]$ there exists one and only one positive linear functional $x^* : C^1[a,b] \longrightarrow \mathbb{R}$ with

(*) $\quad \forall_{v\in V} \; x^*(v) = v'(t).$

(A2) There are v_1, v_2 in V such that $v_1(a) > 0$, $v_2(a) > 0$, and

$$\forall_{t\in[a,b]} \quad [v_1'(t) > 0 \quad \& \quad v_2'(t) < 0].$$

<u>PROOF.</u> We prove only that (A1) & (A2) implies (B1) & (B2). Assume that there exists an element x in $C^1[a,b]$ such that (B1) is not true or that (B2) is not true. We consider first the case that (B1) is not true, i.e.

$$x(a) > \sup \ell(a) \quad \text{or} \quad x(a) < \inf u(a).$$

It suffices to consider only the first case, since the second one can be proved similar. Since V contains an element v_2 with (A2) we can find an $\varepsilon > 0$ such that

$$x(a) \geqq \ell(a) - \varepsilon v_2(a) > \sup \ell(a)$$

and

$$\forall_{t \in [a,b]} \quad x'(t) \geq \ell'(t) \geq \ell'(t) - \varepsilon\, v_2'(t) \,,$$

i.e. the element $\ell - \varepsilon v_2$ is contained in L_x and we have

$$\ell(a) - \varepsilon v_2(a) > \sup \ell(a) \,,$$

which is a contradiction.

Now assume that (B2) is not true, i.e. there exist an element x in $C^1[a,b]$ and a point t_o in [a,b] such that

$$s' := \sup \ell'(t_o) < x^*(x).$$

Then we define on the linear space

$$V_1 := \text{span}\,(x,V)$$

a monotonic linear functional x_1^* as follows:

$$x_1^*(\alpha x + v) := \alpha \cdot \frac{s' + x'(t_o)}{2} + v'(t_o), \alpha \in \mathbb{R} \text{ and } v \in V.$$

Further we define the convex and absorbent subset

$$U := \{z \in C^1[a,b] \mid z \leq v_1\}.$$

Now let p be the Minkowski-functional of U. For any $w \in V_1 \cap U$ we have the estimate $\alpha x + v = u \leq v_1$. If we apply the linear functional x_1^* to this inequality we receive

$$x_1^*(\alpha x + v) \leq x_1^*(v_1) = v_1'(t_o).$$

Since we can assume without loss of generality $v_1'(t_o) = 1$ it follows that

$$x_1^* \leq 1 \quad \text{on} \quad V_1 \cap U.$$

For any w in V_1 we have then the estimate $w = \lambda u \leq \lambda v_1$ with $\lambda > 0$ and $u \in U$. From the last inequality we conclude

$$x_1^*(w) \leq x_1^*(\lambda v_1) = \lambda v_1'(t_o) = \lambda$$

for any $\lambda > 0$ satisfying $w \in \lambda U$. Hence we have $x_1^*(w) \leqq p(w)$ for each $w \in V_1$. By the Hahn-Banach-theorem x_1^* has an extension x_2^* defined on $C^1[a,b]$, which is dominated by p. If $y \leqq 0$, then $x_2^*(y) \leqq p(y) = 0$. Consequently if $y \geqq 0$ we have

$$x_2^*(y) = - x_2^*(-y) \geqq 0.$$

Since $x_1^* \neq x_2^*$ and since both functionals satisfy the condition (*), we have a contradiction.

For the extension of the mode of convergence we have the following results: Let X be such that $X \subset C^1[a,b] \subset \delta X$.

<u>THEOREM 5.</u> If for a sequence (x_n) in X one has

$$x_n \longrightarrow x \quad \& \quad x_n' \longrightarrow x' \quad \text{uniformly}$$

then $L_x = \lim L_{x_n}$ & $U_x = \lim U_{x_n}$.

<u>THEOREM 6.</u> Let V be a subspace of $C^1[a,b]$ with (B1) and (B2). If for a sequence (x_n) in $C^1[a,b]$ on has $L_x = \lim L_{x_n}$ and $U_x = \lim U_{x_n}$ then

$$x_n \longrightarrow x \quad \& \quad x_n' \longrightarrow x' \quad \text{uniformly.}$$

With the aid of these results we can derive the

<u>THEOREM 7.</u> Let V be a subspace of $C^1[a,b]$ with (B1) and (B2). Assume that

$$T_n : C^1[a,b] \longrightarrow C^1[a,b]$$

is a sequence of monotonic operators such that

$$\forall_{x \in X} \quad (T_n(x) \longrightarrow x \quad \& \quad T_n'(x) \longrightarrow x') \quad \text{uniformly}$$

then

$$\forall_{y \in C^1[a,b]} \quad (T_n(y) \longrightarrow y \quad \& \quad T_n'(y) \longrightarrow y') \quad \text{uniformly.}$$

REFERENCES

[1] BROSOWSKI,B.: The completion of partially ordered vector spaces and Korovkin's theorem.
Appoximation Theory and Functional Analysis (ed.J.Prolla) p. 63 - 69, North-Holland, 1979.

ON TWO CONJECTURES ON THE ZEROS OF GENERALIZED BESSEL POLYNOMIALS

M. G. de Bruin

Department of Mathematics
University of Amsterdam
Amsterdam, Netherlands

E. B. Saff[1]

Department of Mathematics
University of South Florida
Tampa, Florida

R. S. Varga[2]

Department of Mathematics
Kent State University
Kent, Ohio

I. INTRODUCTION

In a path-finding paper, Krall and Frink [3] studied properties of the so-called <u>Bessel polynomials</u> (BP), and they defined there a generalization which has become known under the name <u>generalized Bessel polynomials</u> (GBP).

In his recent monograph, Grosswald [2] has given a systematic treatment of the GBP, including a chapter on the location of their zeros. Because of the still growing interest in GBP's for their many applications, it is important to study the location of the zeros of GBP's even more closely than has been done up to now. Using techniques developed in Saff

[1] Research supported in part by AFOSR.

[2] Research supported in part by AFOSR and by the Dept. of Energy.

Copyright © 1980 by Academic Press, Inc.
All rights of reproduction in any form reserved.
ISBN: 0-12-171050-5

and Varga [7]-[11], it is possible to improve many previous results on the location of the zeros of GBP. The full statements of these results, along with their proofs, will appear elsewhere (cf. de Bruin, Saff, and Varga [1]).

The purpose of this note is to focus on two outstanding conjectures associated with the zeros of GBP. The first, to be answered affirmatively in Section II, concerns a conjecture of Grosswald [2, p. 163, no. 6] on the stability of the zeros of GBP. The second conjecture, due to Luke [4, p. 194], concerns the asymptotic behavior of the unique (negative) real zero of odd-degree GBP. With our asymptotic results, given in Section III, this conjecture is shown to be incorrect, and the true asymptotic formula is obtained.

We now give the definition of GBP by means of an explicit formula.

Definition 1.1. For any real number a and for any nonnegative integer n, the GBP $y_n(z;\, a)$ is given by

$$y_n(z;\, a) := \sum_{k=0}^{n} \binom{n}{k} (n + a - 1)_k \left(\frac{z}{2}\right)^k, \tag{1.1}$$

where, for any real number x and for any positive integer k,

$$(x)_0 := 1, \text{ and } (x)_k := x(x+1)\cdots(x+k-1). \tag{1.2}$$

It is immediately clear that the GBP $y_n(z;\, a)$ is of exact degree n iff $a \notin \{-2n+2,\ -2n+3,\ \cdots,\ -n+1\}$.

II. STABILITY AND THE GROSSWALD CONJECTURE

First, we give the new result of

Theorem 2.1. For any $n \geq 2$ with $n+a-1>0$, all zeros of the GBP $y_n(z;\, a)$ lie in the sector

$$S(n,a) := \{z = re^{i\theta} \in \mathbb{C} : |\theta| > \cos^{-1}(-a/(2n+a-2)),\ -\pi < \theta \leq \pi\}. \tag{2.1}$$

This result immediately implies the following known stability result (Martinez [5]) for the zeros of GBP.

Corollary 2.2. For any $n \geq 2$ and any $a \geq 0$, all zeros of the GBP $y_n(z; a)$ lie in the open left-half plane.

Using the methods of Saff and Varga [7] and [8], further improvements on Corollary 2.2 can be made:

Theorem 2.3. For each real number a, there exists a positive integer $n_0(a)$ such that all zeros of the GBP $y_n(z; a)$ lie in the open left-half plane for all $n \geq n_0(a)$. For $a \geq -1$, and for $a \geq -2$, one can take $n_0(a) = 2$ and $n_0(a) = 4$, respectively, while for $a < -2$, one can take $n_0(a) = 1 + [\![2^{3-a}]\!]$, where $[\![\cdot]\!]$ denotes the greatest integer function.

This last result then establishes a conjecture of Grosswald [2, p. 162, no. 6] on the stability of zeros of GBP's. Concerning the sharpness of Theorem 2.3 for the cases $a \geq -1$ and $a \geq -2$, the reader is referred to the more detailed results of [1].

III. ASYMPTOTICS AND THE LUKE CONJECTURE

Olver [6] proved that the zeros of the normalized ordinary BP $y_n(z/n; 2)$ tend, as $n \to \infty$, to a curve Γ in the closed left-hand plane, defined by

$$\Gamma := \{z \in \mathbb{C} : |\omega(z)| = 1 \text{ and } \operatorname{Re} z \leq 0\}, \tag{3.1}$$

where

$$\omega(z) := \frac{e^{\sqrt{1+z^{-2}}}}{z\{1+\sqrt{1+z^{-2}}\}}, \tag{3.2}$$

so that $\{\pm i\}$ are endpoints of Γ. Using the asymptotic methods of Saff and Varga [11], this can be substantially sharpened for the case $a = 2$, as well as generalized to any real a.

Theorem 3.1. For any fixed real a, $\hat{z}$ is a limit point of zeros of the normalized GBP $y_n(2z/(2n+a-2); a)$, as $n \to \infty$, iff $\hat{z} \in \Gamma$. If γ is a closed arc of $\Gamma \setminus \{\pm i\}$ with endpoints μ_1 and μ_2 (with $\frac{\pi}{2} < \arg \mu_1 \leq \arg \mu_2 < \frac{3\pi}{2}$), where $\omega(\mu_j) = e^{i\phi_j}$, $j = 1, 2$, $(\frac{\pi}{2} < \phi_2 \leq \phi_1 < \frac{3\pi}{2})$, let $\tau_n(\gamma)$ denote the number of zeros of $y_n(z; a)$ which satisfy $\arg \mu_1 \leq \arg z \leq \arg \mu_2$. Then,

$$\lim_{n\to\infty} \tau_n(\gamma)/n = (\phi_1 - \phi_2)/\pi. \tag{3.3}$$

Moreover, for any fixed $a \geq 2$, there exists an integer $n_1(a)$ such that for each $n \geq n_1(a)$, every zero z of the normalized GBP $y_n(2z/(2n+a-2); a)$ satisfies $|\omega(z)| > 1$ and $\operatorname{Re} z < 0$.

The last statement of the above theorem has the following geometrical interpretation. Letting $\gamma := \{z = iy \in \mathbb{C} : -1 \leq y \leq 1\}$, then, under the conditions given in Theorem 3.1, these zeros all lie inside the closed curve $\Gamma \cup \gamma$.

Finally, we turn to the behavior of the unique (negative) real zero of an odd-degree GBP $y_n(z; a)$, denoted by $\alpha_n(a)$. Our new result is

Theorem 3.2. For any fixed real number a,

$$\frac{2}{\alpha_n(a)} = (2n+a-2)\hat{r} + K(\hat{r}; a) + \mathcal{O}\left(\frac{1}{2n+a-2}\right), \text{ as } n \to \infty, \tag{3.4}$$

where $\hat{r}$ is the unique negative root of

$$-\hat{r}\, e^{\sqrt{1+\hat{r}^2}} = 1 + \sqrt{1+\hat{r}^2} \quad (\hat{r} \doteq -0.662\ 743\ 419), \tag{3.5}$$

and where

$$K(\hat{r}; a) := \hat{r}\left[\sqrt{1+\hat{r}^2} + (2-a)\ln(\hat{r} + \sqrt{1+\hat{r}^2})\right]/\sqrt{1+\hat{r}^2}. \tag{3.6}$$

With the approximate value of $\hat{r}$ from (3.5), we have

Corollary 3.3. For any fixed real number a,

$$\frac{2}{\alpha_n(a)} \doteq 2n\,\hat{r} - 1.006\ 289\ 950\ a + 1.349\ 836\ 480 + \mathcal{O}\left(\frac{1}{2n+a-2}\right), \quad \text{as } n \to \infty. \tag{3.7}$$

Recalling now the conjecture of Luke [4, p. 194] that

$$\frac{2}{\alpha_n(a)} \sim 2n\,\hat{r} - a + (\pi+1)/\pi, \ a > 0, \text{ as } n \to \infty, \tag{3.8}$$

we see that, since $(\pi+1)/\pi \doteq 1.318\ 309\ 886$, neither of the constant terms in this conjecture is correct, although these conjectured constants had a maximum relative error of only 2%.

REFERENCES

1. de Bruin, M. G., E. B. Saff, and R. S. Varga, On the Zeros of Generalized Bessel Polynomials, Indag. Math. (to appear).

2. Grosswald, E., Bessel Polynomials, Lecture Notes in Math. #698, Springer-Verlag, New York, 1978.

3. Krall, H. L. and Frink, O., A new class of orthogonal polynomials: the Bessel polynomials, Trans. Amer. Math. Soc. 65, (1949), 100-115.

4. Luke, Y. L., Special Functions and Their Approximations, Vol. II, Academic Press, N.Y., 1975.

5. Martinez, J. R., Transfer functions of generalized Bessel polynomials, IEEE CAS 24 (1977), 325-328.

6. Olver, F. W. J., The asymptotic expansions of Bessel functions of large order, Phil Trans. Roy. Soc. London Ser. A 247 (1954), 338-368.

7. Saff, E. B. and Varga, R. S., On the zeros and poles of Padé approximants to e^z, Numer. Math. 25 (1975), 1-14.

8. Saff, E. B. and Varga, R. S., Zero-free parabolic regions for sequences of polynomials, SIAM J. Math. Anal. 7 (1976 a), 344-357.

9. Saff, E. B. and Varga, R. S., On the sharpness of theorems concerning zero-free regions for certain sequences of polynomials, Numer. Math. 26(1976 b), 345-354.

10. Saff, E. B. and Varga, R. S., in Padé and Rational Approximations: Theory and Applications (E. B. Saff and R. S. Varga, eds.), pp. 195-213, Academic Press, New York, 1977.

11. Saff, E. B. and Varga, R. S., On the zeros and poles of Padé approximants. III, Numer. Math. 30(1978), 241-266.

STABILITY OF NUMERICAL METHODS FOR COMPUTING PADÉ APPROXIMANTS

A. Bultheel
Applied Mathematics Division
University of Leuven
Heverlee, Belgium

L. Wuytack[1]
Department of Mathematics
University of Antwerp
Wilrijk, Belgium

1. NOTATIONS AND DEFINITIONS.

Let $f(x) = \sum_{i=o}^{\infty} c_i . x^i$ be a given power series, with $c_o \neq o$. Let R_n^m be the class of (ordinary) rational functions $r = \frac{p}{q}$ where p and q are polynomials of degree at most m and n respectively such that $\frac{p}{q}$ is irreducible. The Padé approximation problem for f of order (m,n) is to find an element $r = \frac{p}{q}$ in R_n^m satisfying $f.q-p = O(x^{m+n+1+j})$, where j is an integer which is as high as possible. This problem has a unique solution for all values of m and n, denoted by $r_{m,n}$. The elements $r_{m,n}$ can be ordered in a 2-dimensional array, called the Padé table, as follows

$$\begin{matrix} r_{00} & r_{01} & r_{02} & r_{03} & \cdots \\ r_{10} & r_{11} & r_{12} & r_{13} & \cdots \\ r_{20} & r_{21} & r_{22} & r_{23} & \cdots \\ r_{30} & r_{31} & r_{32} & r_{33} & \cdots \\ \cdot\cdot & \cdot\cdot & \cdot\cdot & \cdot\cdot & \end{matrix}$$

This table has a "block structure" which means that equal elements appear in square blocks. An element is called normal if it appears only once in the table. Further in this paper we will consider only the case of a normal Padé table which means that every element in the table is assumed to

1 Supported in part by the FKFO under grant number 2.0021.75

Copyright © 1980 by Academic Press, Inc.
All rights of reproduction in any form reserved.
ISBN: 0-12-171050-5

be normal.

If $r_{m,n} = \frac{p_{m,n}}{q_{m,n}}$, with $p_{m,n} = \sum_{i=o}^{m} a_i^{(m,n)}.x^i$ and $q_{m,n} = \sum_{i=o}^{n} b_i^{(m,n)}.x^i$ is normal then $a_o^{(m,n)} \neq o$, $a_m^{(m,n)} \neq o$, $b_o^{(m,n)} \neq o$, $b_n^{(m,n)} \neq o$ and

$$f(x).q_{m,n}(x) - p_{m,n}(x) = \sum_{i=o}^{\infty} d_i^{(m,n)}.x^{m+n+1+i} \text{ with } d_o^{(m,n)} \neq o. \quad (1)$$

This implies that $a_i^{(m,n)}$, $b_i^{(m,n)}$ satisfy

$$\sum_{i=o}^{n} c_{k-i}.b_i^{(m,n)} = a_k^{(m,n)} \quad \text{for } k = 0,1,2,\dots,m \quad (2.a)$$

$$\sum_{i=o}^{n} c_{m+k-i}.b_i^{(m,n)} = o \quad \text{for } k = 1,2,3,\dots,n \quad (2.b)$$

where $c_i = o$ if $i < o$. Remark that $a_i^{(m,n)}$ can be computed easily from (2.a) as soon as a solution of (2.b) is known. Before describing the basis of some recursive methods for solving (2.b) we consider the conditioning of the Padé approximation problem.

2. THE CONDITIONING OF THE PADE APPROXIMATION PROBLEM

Let m and n be fixed and $r = \frac{p}{q}$ be the Padé approximant of order (m,n) for f. What is the effect of "small" changes on the coefficients c_i of f on the coefficients of p and q. Since r only depends on $c_o, c_1, \dots, c_{n+m}$ we will only consider changes in these coefficients. Let $c = [c_o, c_1, \dots, c_{n+m}]^T$ and $\|c\| = \sup_{o \leq i \leq n+m} |c_i|$. A neighbourhood U_δ of f, with $\delta > o$, be defined as the set of power series $f' = \sum_{i=o}^{\infty} c_i'.x^i$ such that $\|c-c'\| < \delta$. Using a characterization theorem for normal Padé approximants ([14],p.34) it is not difficult to prove the following result.

<u>THEOREM</u>. Let f have a normal Padé approximant of order (m,n), then there exists a neighbourhood U_δ of f such that every f' in U_δ has a normal Padé approximant of order (m,n).

This result guarantees that the system (2.a), (2.b) has a unique solution if the coefficients c_i are changed "slightly". This system will therefore be well-conditioned if the condition number of its matrix of coefficients is not too large ([10],p.38). There are however examples ([9],p. 242) for which this condition number is very large. Therefore the problem of finding a Padé approximant of a given order for f can be ill-conditioned. This fact is illustrated numerically in [12].

Let T be the operator which associates $r_{m,n}$ to f. We conjecture that, for certain classes of functions f, the operator T satisfies a local Lipschitz condition, or $\|Tf-Tf'\| \leqslant K.\|f-f'\|$ for all f' in a neighbourhood" of f and where the norm must be chosen suitably.

3. FORMULATION OF RECURSIVE ALGORITHMS FOR COMPUTING PADÉ APPROXIMANTS

It has been remarked in ([8], [16]) that several algorithms for computing a Padé approximant can be interpreted as algorithms for factorizing a Toeplitz or Hankel matrix. A more extensive study of this fact is given in [3]. Here we will only give some of the basic ideas of these algorithms and their relations with other algorithms in the field, of which the exact references can be found in [15].

3A. Algorithms for constructing a row or column in the Padé table.

Let m be fixed, then (1) and (2b) for different values of n can be combined to give : T.U = L where T is a Toeplitz matrix with $t_{ij} = c_{m+1+i-j}$, U is upper triangular with $u_{ij} = b_{i-1}^{(m,j-1)}$, and L is lower triangular with $\ell_{ij} = d_{i-j}^{(m,j-1)}$. Using (2.a) and (2.b) we get T'.L' = U' where $t'_{ij} = c_{m+i-j}$, L' is lower triangular with $\ell'_{ij} = u_{i-j+1'n-j+2}$, and U' is upper triangular with $u'_{ij} = a_{m-j+i}^{(m,n-j+1)}$.
The equations T.U. = L and T'.L'=U' can be combined to give a three-term recurrence relation between the denominators of the row $\{r_{m,j}\}_{j\geqslant o}$ in the Padé table. It is also possible to derive relations between the elements on neighbouring rows and columns in the Padé table. Similar relations (of Frobenius-type) can be obtained for the elements of the column $\{r_{i,n}\}_{i\geqslant o}$.
The algorithms of W.F. Trench (1964), J. Rissanen (1974), O. Bussonais (1978), A. Bultheel (1978) are special cases of these techniques, which are in fact algorithms to factorize a Toeplitz matrix or its inverse.

3B. Algorithms for constructing a diagonal or staircase in the Padé table.

The equations (1), (2.b), corresponding to the elements on a diagonal in the Padé table, can be written in the form H.V = L where H is a Hankel matrix with $h_{i,j} = c_{m-n+i+j-1}$, V is upper triangular with $v_{ij} = b_{j-i}^{(m+j-1,j-1)}$ and L is lower triangular with $\ell_{ij} = d_{i-j}^{(m+j-1,j-1)}$. Similar relations as above can now be derived between the elements on a diagonal (or anti-diagonal) in the Padé table. The algorithms of Massey J.L. (1969), Rissanen J. (1971), Mills W.H. (1975), C. Brezinski (1976) are special

cases of these techniques.
By combining the relations which correspond to neighbouring diagonals (or antidiagonals) we get algorithms for computing the elements on a descending (or ascending) staircase in the Padé table. The algorithms of Thacher H.C. (1960), Watson P.J. (1973), Murphy J.A. and O'Donohoe M.R. (1977) for computing the elements on a descending staircase and of Baker G.A. (1970), Claessens G. (1975) and McEliece R.J. and Shearer J.B. (1978) for computing the elements on an ascending staircase are special cases of these techniques. They are in fact based on factorizing a Hankel matrix or its inverse. Also Bussonais O. (1978) has this Hankel-type recursions. Remark that it is also possible to use a classical factorization algorithm (Gauss, Crout,...) to solve (2.b). These algorithms use however $O(n^3)$ operations, instead of $O(n^2)$ in case the special structure of a Hankel or Toeplitz matrix is exploited. The disadvantage (with respect to stability) of exploiting this structure is however that no pivoting is possible.

4. STABILITY OF RECURSIVE ALGORITHMS

A detailed analysis of the stability of the algorithms mentioned in the preceding section would require a detailed description of these algorithms. Since this is not possible here we only mention some of the results obtained in this field and refer to the literature for further details.
In [6] it is shown that the algorithms of Rissanen and Massey-Berlekamp (and of a variant of it) are unstable. A stable variant of the algorithm of Rissanen is proposed, but it uses $O(n^3)$ operations.
In [5] the algorithm of Durbin for factorizing a real symmetric positive definite Toeplitz matrix is considered. This algorithm uses $O(n^2)$ operations and shows similar stability properties as the Cholesky algorithm, which uses $O(n^3)$ operations.
A more general result was obtained in [4], for several recursive techniques to factorize Toeplitz matrices or their inverses. It is proved that these algorithms are unstable when applied to the whole class of Toeplitz matrices. However if these algorithms are restricted to the class of symmetric positive definite Toeplitz matrices, bounded away from singularity, then they are stable. The Levinson algorithm or its continued fraction form is mentioned as a special case from this class of stable algorithms.
Remark that the system (2.a), (2.b) is based on a representation of the numerator and denominator of a rational function using powers of x as base functions. It is an open problem whether another choice of base functions

would improve the general situation.

5. EVALUATION OF THE VALUE OF A PADE APPROXIMANT

Most of the algorithms mentioned in section 3 can also be interpreted as algorithms for computing the coefficients in continued fractions whose convergents form a sequence of neighbouring elements in the Padé table. For example the algorithms of Watson and Brezinski can be used to find the coefficients e_i in the continued fraction $e_o + \sum_{i=1}^{\infty} \frac{e_i . x|}{|1}$, whose convergents are the elements $r_{o,o}, r_{1,o}, r_{1,1}, r_{2,1}, r_{2,2}, \ldots$ on a staircase in the Padé table. If the value of a Padé approximant, from this sequence, must be computed for several values of x then the continued fraction representation of this element can be used.

The problem of finding the value of a finite part of this continued fraction is directly connected with the problem of evaluating three-term recurrence relations. Two basic techniques can be used to do this (see [7]): forward and backward recursion. It is known that forward recursion can be unstable. The use of backward recursion seems to be more appropriate [11]. Certain variants of these basic techniques do exist [13] and can be seen as algorithms for solving a special tridiagonal system of equations.

REFERENCES

1. BREZINSKI C. : Computation of Padé approximants and continued fractions. Journal of Computational and Applied Mathematics 2 (1976), pp. 113-123.
2. BULTHEEL A. : Recursive algorithms for the Padé table : two approaches. In "Padé Approximation and its Applications" (ed. L. Wuytack, Springer-Verlag, Berlin, 1979), pp.211-230.
3. BULTHEEL A. : Fast algorithms for the factorization of Hankel and Toeplitz matrices and the Padé approximation problem. Report TW 42, University of Leuven, 1978.
4. BULTHEEL A. : Towards an error analysis of fast Toeplitz factorization. Report TW 44, University of Leuven, 1979.
5. CYBENKO G. : Error analysis of Durbin's algorithm. Submitted for publication. 1979.
6. DE JONG L.S. : Numerical aspects of realization algorithms in linear systems theory. Ph. D. Thesis, T.H. Eindhoven, 1975.
7. GAUTSCHI W. : Computational aspects of three-term recurrence relations. SIAM Review 9 (1967), pp.24-82.
8. GRAGG W.B. : Matrix interpretations and applications of the continued fraction algorithm. Rocky Mountain Journal of Mathematics 4 (1974), pp.213-225.

9. GRAVES-MORRIS P.R. : The numerical calculation of Padé Approximants. In "Padé Approximation and its Applications" (ed. L. Wuytack, Springer Verlag, Berlin, 1979), pp.231-245.

10. ISAACSON E., KELLER H.B.: Analysis of Numerical methods. Wiley, N.Y. 1966.

11. JONES W.B. and THRON W.J.: Numerical stability in evaluating continued fractions. Mathematics of Computation 28 (1974), pp. 795-810.

12. LUKE Y.L. : Computations of coefficients in the polynomials of Padé approximants by solving systems of linear equations. To appear in the Journal of Computational and Applied Mathematics.

13. MIKLOSKO J. : Investigation of algorithms for numerical computation of continued fractions. USSR Computational Mathematics and Mathematical Physics 16 (1976, nr. 4), pp.1-12.

14. PADE H. : Sur la representation approchée d'une fonction par des fractions rationnelles. Annales Scientifiques de l'Ecole Normale Supérieure de Paris 9 (1892), pp. 1-93.

15. WUYTACK L. : Commented bibliography on techniques for computing Padé Approximants. In "Padé Approximation and Its Applications" (Lecture Notes in Mathematics 765, Springer-Verlag, Berlin, 1979), pp.375-392.

16. WYNN P. : The rational approximation of functions which are formally defined by a power series expansion. Mathematics of Computation 14 (1960), pp.147-186.

PROPERTIES OF FUNCTIONS OF GENERALIZED BOUNDED VARIATION

H. G. Burchard and K. Höllig

We prove regularity properties of approximation by quasiinterpolants for functions of generalized bounded variation. The quasiinterpolant can be used to give another natural definition of the derivative D^k introduced in [1,9]. We study the properties of D^k and show in particular that the map D^k: $AC \to L^\sigma$ is onto.

1. INTRODUCTION

The spaces $V^p_{k,\lambda}$ have been studied by Bergh and Peetre, Brudnyi, Burchard and Höllig [1 - 4, 6, 7, 9, 10] and the spaces $AC^p_{k,\lambda}$ first appear in [1,7]. They arise in a natural way in the study of spline approximation with free knots [1, 3 - 7, 9, 10, 12]. We need some basic notation.

$P^r_{k,\pi}$ ≙ piecewise (r - 1)-times continuously differentiable polynomials of degree k - 1 on a partition π of [0,1].

$S^r_{k,\pi}$: $L_p \to P^r_{k,\pi}$ may be any bounded projection locally defined via quasiinterpolation.

$E_k(f;I)_p = \inf \{\|f - s\|_p : s \in P_{k,I}\}$.

$\lambda(\pi) := \max \{|I| : I \in \pi\}$, $\mu(\pi,f) = \max \{E_k(f;I)_p : I \in \pi\}$.

For $0 < p \leq \infty$, $\sigma(k,p) = (k + 1/p)^{-1} \leq \lambda < p$ we let

$$\|f\|_{V^p_{k,\lambda}} = \sup \{\sum_{I \in \pi} E_k(f;I)^\lambda_p : \pi \text{ any partition of } [0,1]\}$$

Copyright © 1980 by Academic Press, Inc.
All rights of reproduction in any form reserved.
ISBN: 0-12-171050-5

and define the F-spaces $V^p_{k,\lambda}$ by the condition

$$|||f|||_{V^p_{k,\lambda}} = ||f||_p^\lambda + ||f||_{V^p_{k,\lambda}} < \infty.$$

$AC^p_{k,\lambda}$ is the closure of C^∞ in $V^p_{k,\lambda}$.

Indices on norms and spaces are mostly dropped in the text below.

2. APPROXIMATION IN V BY QUASIINTERPOLANTS

In this section we prove that the quasiinterpolants provide regular approximation processes in the space V.

THEOREM 1. *The quasiinterpolants $S_{j,\pi}$ are uniformly bounded projections on V_k with respect to π for every positive integer j.*

The next result is a corollary to Theorem 1.

THEOREM 2. *V is the relative completion of AC in L^p, i.e., $f \in V$ iff there is a sequence $f_n \in AC$,*

$$\sup_n |||f_n|||_V < \infty \text{ and } \lim_{n\to\infty} ||f - f_n||_p = 0.$$

Also,

$$|||f|||_V = \inf \{\sup_n |||f_n|||_V : f_n \to f \text{ in } L^p\}.$$

THEOREM 3. *For $f \in AC_k$ and $j > k$ we have*

$$\lim_{\mu(\pi;f)\to 0} ||f - S_{j,\pi} f||_{V_k} = 0.$$

3. GENERALIZED DERIVATIVES

We may use the quasiinterpolants to give another natural definiton of the derivative introduced in [7,9] extending it from AC to the larger space V. A similar definition has been employed in [9].

<u>Definition.</u> For $f \in V^p_{k,\sigma}$, $1 \leq p \leq \infty$, $\sigma = \sigma(k,p)$, we define

$$D^k f = \lim_{\mu(\pi,f)\to 0} D^k S_{k+1,\pi} f$$

where the limit is understood in the sense of pointwise convergence a.e. on [0,1]. For non-smooth quasiinterpolation $D^k S_{k+1,\pi} f$ is defined only in the interior of mesh intervals. For other values of p and λ the derivative must be defined by different methods [9].

The proof of the existence of the limit on the right hand side for $f \in V$ in the above definition when only $\mu(f;\pi)\to 0$ (rather than $\lambda(\pi)\to 0$) is based on the nontrivial fact [3, 4, 8, 9] that $f \in V$ possesses a Taylor polynomial p_x of degree k for a.a. $x \in (0,1)$,

$$\| f - p_x \|_{p,I} = o(|I|^{1/\sigma}); \quad x \in I, \ |I| \to 0.$$

For such x $(D^k f)(x)$ equals the coefficient of $(k!)^{-1}(\cdot - x)^k$ in the Taylor expansion of p_x.

THEOREM 4. <u>For</u> $1 \leq p \leq \infty$ <u>the map</u>

$$D^k : V^p_{k,\sigma} \to L^\sigma$$

<u>is linear and continuous.</u>

For $f \in AC$ the continuity of D^k is a consequence of the original definition given in [7].

By Theorem 3 we may of course replace the pointwise convergence in the definition of D^k by norm-convergence in L^σ as already observed in [7].

Let $|f|_{V^p_{k,\sigma}} = \overline{\lim}_{\lambda(\pi)\to o} \sum_{I\in\pi} E_k(f,I)^\sigma_p$.

For $f\in AC^p_{k,\sigma}$ we have $|f|_{V^p_{k,\sigma}} = c\|D^k f\|^\sigma_\sigma$ [6, 7].

THEOREM 5. a) $\ker\{D^k\colon\ AC^p_{k,\sigma} \to L^\sigma\} = \operatorname{clos}_{\|\ \|_{V^p_{k,\sigma}}} \bigcup_\pi P_{k,\pi}$.

b) _If_ $f\in AC^p_{k,\sigma}$ _then_ $\|D^k f\|^\sigma_\sigma \sim \inf_\pi \|f - S_{k,\pi}f\|_{V^p_{k,\sigma}}$

$$\sim \overline{\lim}_{\mu(\pi,f)\to o} \|f - S_{k,\pi}f\|_{V^p_{k,\sigma}} \sim |f|_{V^p_{k,\sigma}},$$

where "~" indicates estimates in both directions with constants depending on k _and_ p _only_.

From this it is clear that the kernel of D^k has some unusual properties. But, the definition of the derivative D^k fits nicely into the scale of V spaces [9].

THEOREM 6. _We have the commutative diagram_

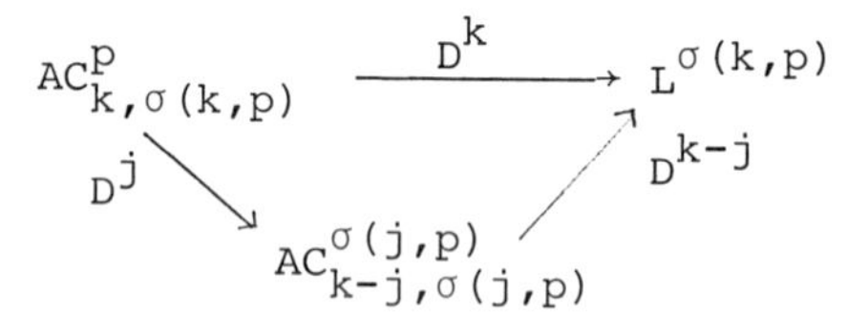

The proof is a consequence of the embedding $V^p_{k,\sigma(k,p)} \hookrightarrow AC^p_{j,\sigma(j,p)}$ and the recurrence relation $\sigma(k-j,\ \sigma(j,p)) = \sigma(k,p)$. This relation is also reflected in the following result [9].

THEOREM 7. *Denote by* $E_k(f;N)_p$ *the error of the best* L^p *approximation with splines of degree* $k - 1$ *with* N *free knots. Then for* $j < \alpha$, $j < k$

$$E_k(f;N)_p = O(N^{-\alpha})$$

implies

$$E_{k-j}(D^j f;N)_{\sigma(j,p)} = O(N^{-\alpha+j}).$$

We obtain the following result as an application of Theorem 5(b). It states that a function in L^σ may be integrated in a generalized sense, even for $0 < \sigma < 1$. As remarked in [7] this cannot be done, however, by any linear continuous right inverse of D^k. For $1 \leq p \leq \infty$ this associates a distribution with each element of L^σ.

THEOREM 8. D^k *maps* $AC^p_{k,\sigma}$ *onto* L^σ.

It follows that for $\sigma < 1$ AC is not locally convex, a fact already proved in [7] by a different method.
Moreover by Theorem 4 it follows now that

$$V^p_{k,\sigma} = AC^p_{k,\sigma} + \ker \{D^k : V^p_{k,\sigma} \to L^\sigma\}.$$

J. Peetre [11] has shown that under the classical definition the Sobolev spaces W^k_σ are isomorphic to L^σ for $o < \sigma < 1$. According to the above results, the spaces $AC^p_{k,\sigma}$ possess at least some of the expected properties. The precise relationship of the $V^p_{k,\lambda}$-spaces with Peetre's work via H^p and Besov spaces [12] remains to be worked out.

REFERENCES

1. Bergh, J., and Peetre, J., On the spaces V_p (0<p<∞). Bolletino U.M.I. (4) 10 (1974), 632-648.

2. Brudnyi, Ju. A., Spaces defined by local approximation, Trudy Moskov. Mat. Obsc. 24 (1971), 69-132 = Trans. Moskow Math. Soc. 24 (1971), 73-139 (Engl. Transl.).

3. Brudnyi, Ju. A., Spline Approx. and functions of bounded variation. Dokl. Akad. Nauk USSR 215 (1974), 511-513 = Soviet Math. Dokl. 15 (1974), 518-521) (Engl. Transl.).

4. Brudnyi, Ju. A., Piecewise polynomial approximation, embedding theorem and rational approximation, in: Approximation Theory, Bonn, R. Schaback and K. Scherer, eds., Springer Lecture Notes 556 (1976), 73-99.

5. Boor, C. de, Good approximation with variable knots. II, Conf. on the Numerical Solution of Differential Equations, Dundee, Springer Lecture Notes 363 (1973), 12-20.

6. Burchard, H.G. and Hale, D. F., Piecewise polynomial approximation on optimal meshes, J. Approximation Theory 14 (1975), 128-147.

7. Burchard, H. G., On the degree of convergence of piecewise polynomial approximation on optimal meshes, Trans. Amer. Math. Soc. 234 (1977), 531-559.

8. Calderon, A. P. and Zygmund, A., Local properties of solutions of elliptic partial differential equations, Studia Math. 20 (1961), 171-225.

9. Höllig, K., Spline-Approximation auf freien Knoten und $V^p_{k,\lambda}$-Räume, Diplomarbeit, Inst.f.Angew.Math., Universität Bonn (1977).

10. Kahane, J. P., Teoria constructiva de funciones, Cursos y Sem. Mat., Fasc. 5, Univ. Buenos Aires, Buenos Aires (1961).

11. Peetre, J., A remark on Sobolev spaces. The case 0<p<1, J. Approximation Theory 13 (1975), 218-228.

12. Peetre, J., New Thoughts on Besov spaces, Duke University Math. Ser. 1, Duke University, Durham N.C. (1976).

H.G. Burchard
Oklahoma State University
Stillwater, Oklahoma 74074,USA
u. Institut f. Angew.Mathematik
der Universität Bonn/SFB 72
Wegelerstraße 6
5300 Bonn 1, FRG

K. Höllig
Institut für Angew. Mathematik
der Universität Bonn/SFB 72
Wegelerstraße 6
5300 Bonn 1
Federal Republic of Germany

THE ORDER OF MAGNITUDE OF FUNCTIONS OF A POSITIVE VARIABLE AND THEIR DEGREE OF APPROXIMATION BY PIECEWISE INTERPOLATING POLYNOMIALS

J. S. Byrnes
Department of Mathematics
University of Massachusetts
Boston, Massachusetts 02125

and

O. Shisha
Department of Mathematics
University of Rhode Island
Kingston, Rhode Island 02881

1. Our purpose is to relate the order of magnitude of real functions $f(x)$ as $x \to \infty$ to their degree of approximation by piecewise polynomials interpolating them on some simple denumerable sets of points. A similar relation, for unbounded functions on $(0, 1]$, is given in [1].

2. Let f be a real function on $[1, \infty)$, let k be a positive integer and let $r = (r_n)_{n=1}^{\infty}$ be a strictly monotone sequence of numbers with $r_1 \geq 1$, $r_n \to \infty$ and $r_n - r_{n-1} \to 0$. For every $a \in [0, 1]$ we denote by $P_{a,k}(f,r,x) \equiv P_{a,k}(f)$ the function with domain $[a+r_1, \infty)$ which in each

(1) $I_{a,n} = [a+r_{n-1}, a+r_n), \quad n = 2,3,\cdots,$

coincides with the polynomial of degree $\leq k$ interpolating f at the $k+1$ equally spaced points

(2) $x_j = a+r_{n-1} + (d_n/k)j, \quad j = 0,1,\cdots,k,$

where $d_n = r_n - r_{n-1}$ is the length of $I_{a,n}$. In particular, $P_{a,1}(f)$ is a polygonal function, interpolating f at $a+r_n$, $n = 1,2,\cdots$. In the following theorem we relate the order of magnitude of $f(x)$ as $x \to \infty$ to that of our "degree of approximation"

Copyright © 1980 by Academic Press, Inc.
All rights of reproduction in any form reserved.
ISBN: 0-12-171050-5

$$\langle f\rangle_{k,\gamma} \equiv \sup_{0\le a\le 1} \sup_{x\ge\gamma} |f(x) - P_{a,k}(f,r,x)|$$

as $\gamma \to \infty$, for a particularly simple choice of r.

Had we defined the "degree of approximation" as $\sup_{x\ge\gamma} |f(x) - P_{a,k}(f,r,x)|$ for some fixed a, say $a = 0$, we would have given undue weight to the behavior of f at the points r_n. For example, if f is linear on each $[r_{r-1}, r_n]$, $n = 2,3,\cdots$, the last sup is $\equiv 0$ while f can be of an arbitrarily large order of magnitude as $x \to \infty$. It is to avoid such a state of affairs that we define $\langle f\rangle_{k,\gamma}$ as we do.

Finally, in the last section we show that, in our theorem (in one direction), $P_{a,k}$ can be replaced by <u>any</u> piecewise polynomial of degree $\le k$ whose knots are $a+r_n$, $n = 1,2,\cdots$, not necessarily one arising from interpolation.

3. <u>Theorem</u>. <u>Let</u> $r_n = \sum_{j=1}^{n} 1/j$, $n = 1,2,\cdots$, <u>let</u> $0<\alpha\le k+1$ <u>and let</u> $f^{(k+1)}$ <u>exist and be nondecreasing or nonincreasing in</u> $[1, \infty)$. <u>Then</u> $f(x) = O(e^{\alpha x})$ <u>as</u> $x \to \infty$ <u>iff</u> $\langle f\rangle_{k,\gamma} = O(e^{(\alpha-k-1)\gamma})$ <u>as</u> $\gamma \to \infty$.

<u>Proof</u>. Assume that $f^{(k+1)}$ is nondecreasing in $[1, \infty)$ (otherwise, consider $-f$). Let

$$g(x) \equiv f(x) - \sum_{j=0}^{k+1} \frac{f^{(j)}(1)}{j!} (x-1)^j$$

so that $g(1) = g'(1) = \cdots = g^{(k+1)}(1) = 0$ and $g^{(k+1)}(x) \equiv f^{(k+1)}(x) - f^{(k+1)}(1)$. Also $g(x) - f(x) \equiv \psi(x) - [f^{k+1}(1)/(k+1)!](x-1)^{k+1}$ where $\psi(x)$ is a polynomial of degree $\le k$ so that

$$P_{a,k}(g-f) = \psi - [f^{(k+1)}(1)/(k+1)!]P_{a,k}((x-1)^{k+1}) .$$

Let $t\ge\gamma\ge 2$ and let $a \in [0, 1]$. Then t belongs to some $I_{a,n}(n\ge 2)$ and

$$(3) \quad \gamma < a+r_n \le 2 + \sum_{j=2}^{n} 1/j < 2 + \sum_{j=2}^{n} \int_{j-1}^{j} dx/x = 2 + \log n .$$

Since, clearly, $P_{a,k}((x-1)^{k+1},r,t) = (t-1)^{k+1} - \Pi_{j=0}^{k}(t-x_j)$ where the x_j are given by (2), we have

$$|(t-1)^{k+1} - P_{a,k}((x-1)^{k+1},r,t)| < n^{-k-1} < e^{(2-\gamma)(k+1)} .$$

Hence $h(t) = g(t) - f(t) - P_{a,k}(g-f,r,t)$ satisfies

$$|h(t)| \leq |f^{(k+1)}(1)/(k+1)!|e^{(2-\gamma)(k+1)} .$$

Observe that $g(t) - P_{a,k}(g,r,t) = f(t) - P_{a,k}(f,r,t) + h(t)$ which clearly implies that $\langle f\rangle_{k,\gamma} = O(e^{(\alpha-k-1)\gamma})$ as $\gamma \to \infty$ iff $\langle g\rangle_{k,\gamma} = O(e^{(\alpha-k-1)\gamma})$ as $\gamma \to \infty$. Also $f(x) = O(e^{\alpha x})$ as $x \to \infty$ iff $g(x) = O(e^{\alpha x})$ as $x \to \infty$. Therefore we may assume without loss of generality that

$$f^{(j)}(1) = 0,\ j = 0,1,\cdots,k+1, \text{ and hence } f^{(j)}(x) \text{ is } \geq 0 \tag{4}$$
$$\text{and nondecreasing in } [1,\infty) \text{ for } j = 0,1,\cdots,k+1 .$$

Suppose now that M is a number such that

$$\langle f\rangle_{k,\gamma} \leq Me^{(\alpha-k-1)\gamma} \text{ for all } \gamma \geq \text{ some } \gamma_0 \geq 2 . \tag{5}$$

Let $x \geq \gamma_0$. Define the integer $n(\geq 2)$ and the numbers a, γ and $\tilde{x}$ by

$$r_{n-1} \leq x < r_n , \tag{6}$$

$$a = x - r_{n-1}, \quad \gamma = x, \quad \tilde{x} = x + (2kn)^{-1} .$$

By the remainder theorem for Lagrange interpolation [2, p.56], using again the notation (2), for some $\xi \in (a+r_{n-1}, a+r_n)$,

$$|f(\tilde{x}) - P_{a,k}(f,r,\tilde{x})| = \frac{f^{(k+1)}(\xi)}{(k+1)!}\prod_{j=0}^{k}|\tilde{x}-x_j| = \frac{f^{(k+1)}(\xi)}{(k+1)!}\,\frac{1\cdot 3\cdots(2k-1)}{(2kn)^{k+1}}$$

so that

$$0 \leq f^{(k+1)}(x) \leq f^{(k+1)}(\xi) \leq M_k n^{k+1} e^{(\alpha-k-1)x} \leq M_k e^{\alpha x}$$

where $M_k = M(k+1)!(2k)^{k+1}[1\cdot 3\cdots(2k-1)]^{-1}$; the first two inequalities are from (4), the third from (5) and the fourth from (6) as $\log n = \sum_{j=2}^{n} \int_{j-1}^{j} dx/x < \sum_{j=2}^{n}(j-1)^{-1} = r_{n-1} \leq x$. Thus, for some constant $\mu_k \geq M_k$,

$$0 \leq f^{(k+1)}(x) \leq \mu_k e^{\alpha x} \quad \text{throughout} \quad [1, \infty).$$

Successive integrations, using (4), yield

$$0 \leq f(x) \leq \mu_k \alpha^{-k-1} e^{\alpha x} \quad \text{throughout} \quad [1, \infty)$$

as required.

For the converse, suppose that, for some constant J,

$$0 \leq f(x) \leq J e^{\alpha x} \quad \text{throughout} \quad [1, \infty).$$

Then, for $j = 0,1,\cdots,k+1$,

$$(7) \quad 0 \leq f^{(j)}(x) \leq J(\alpha e)^j e^{\alpha x} \quad \text{throughout} \quad [1, \infty).$$

This is true for $j = 0$ and assuming its truth for some j, $0 \leq j \leq k$, we have, for every $x \in [1, \infty)$ and a proper $y > x$,

$$Je(\alpha e)^j e^{\alpha x} \geq f^{(j)}(x+\alpha^{-1}) \geq f^{(j)}(x+\alpha^{-1}) - f^{(j)}(x) = \alpha^{-1} f^{(j+1)}(y)$$
$$\geq \alpha^{-1} f^{(j+1)}(x) \quad \text{so that} \quad 0 \leq f^{(j+1)}(x) \leq J(\alpha e)^{j+1} e^{\alpha x}.$$

Let $0 \leq a \leq 1$, $2 \leq \gamma \leq x$. For a proper $n \geq 2$, $a+r_{n-1} \leq x < a+r_n$.

Using again (2) and the above remainder theorem, we have, for some $\eta \in (a+r_{n-1}, a+r_n)$,

$$|f(x) - P_{a,k}(f,r,x)| = [f^{(k+1)}(\eta)/(k+1)!] \, \Pi_{j=0}^{k} |x-x_j|.$$

Since, by (3), $a+r_n$ is smaller than $2 + \log n$, so are η and γ. Hence, by (7) with $j = k+1$,

$|f(x) - P_{a,k}(f,r,x)| \leq J_k e^{\alpha \eta_n} {}^{-k-1} \leq J_k e^{2\alpha} n^{\alpha-k-1}$
$\leq J_k e^{2k+2} e^{(\alpha-k-1)\gamma}$ where $J_k = J(\alpha e)^{k+1}/(k+1)!$. This completes the proof.

4. <u>Corollary.</u> <u>Assume the hypotheses of the Theorem. A necessary and sufficient condition for</u> $f(x)$ <u>to be</u> $O(e^{\alpha x})$ <u>as</u> $x \to \infty$ <u>is the existence, for each</u> $a \in [0, 1]$, <u>of a function</u> $Q_a(x)$ <u>with domain</u> $[a+1, \infty)$, <u>continuous there, which in each</u> $I_{a,n}$ <u>of</u> (1) <u>coincides with some polynomial of degree</u> $\leq k$ <u>such that</u>

$$\sup_{0 \leq a \leq 1} \sup_{x \geq \gamma} |f(x) - Q_a(x)| = O(e^{(\alpha-k-1)\gamma})$$

as $\gamma \to \infty$.

<u>Proof.</u> Only sufficiency needs proof. Let μ be a number such that

$$\sup_{0 \leq a \leq 1} \sup_{x \geq \gamma} |f(x) - Q_a(x)| \leq \mu e^{(\alpha-k-1)\gamma} \quad \text{for all } \gamma \geq \text{ some } \gamma_1 \geq 2.$$

Let $\gamma \geq \gamma_1 + \frac{1}{2}$, $t \geq \gamma$, $a \in [0, 1]$ and set

$$R_a(x) \equiv P_{a,k}(f,r,x) - Q_a(x).$$

For some $n \geq 3$, $t \in I_{a,n}$ and, using (2),

$$R_a(t) = \sum_{j=0}^{k} R_a(x_j) \prod_{\substack{s=0 \\ s \neq j}}^{k} (t-x_s)/(x_j-x_s).$$

For $j = 0,1,\cdots,k$, $x_j \geq a+r_{n-1} \geq a+r_n - \frac{1}{3} > \gamma - \frac{1}{2} \geq \gamma_1$ and hence

$$|R_a(x_j)| \leq \mu_k e^{(\alpha-k-1)\gamma}$$

where $\mu_k = \mu e^{-(\alpha-k-1)/2}$.

Therefore $|R_a(t)| \leq (k+1)k^k \mu_k e^{(\alpha-k-1)\gamma}$ and hence
$|f(t) - P_{a,k}(f,r,t)| \leq [\mu+(k+1)k^k\mu_k]e^{(\alpha-k-1)\gamma}$. Thus

$$\langle f \rangle_{k,\gamma} \leq [\mu+(k+1)k^k\mu_k]e^{(\alpha-k-1)\gamma} \quad \text{if } \gamma \geq \gamma_1 + \frac{1}{2}$$

and, by our Theorem, $f(x) = O(e^{\alpha x})$ as $x \to \infty$.

References

1. Byrnes, J.S. and O. Shisha, The order of magnitude of unbounded functions and their degree of approximation by piecewise interpolating polynomials, J. Approximation Theory (to appear).

2. Davis, P.J., "Interpolation and Approximation", Blaisdell, New York, 1963; Dover, New York, 1975.

MULTIPLICATIVE VARIATIONS LEAD TO THE VARIATIONAL EQUATIONS FOR MINIMAL PROJECTIONS

B. L. Chalmers and F. T. Metcalf

Department of Mathematics
University of California
Riverside, California

I. INTRODUCTION

Let X denote an arbitrary Banach space (with norm $\| \ \|$) and let $V = [v_1, \ldots, v_n] = [v]$ denote an n-dimensional subspace of X. A projection P from X onto V is a linear map such that $Py = y$, for all $y \in V$. For $x \in X$ consider Px as an approximation to x from V. The error $\|x - Px\|$ is estimated as follows: for any $y \in V$,

$$\|x - Px\| = \|(I-P)(x-y)\| \leq (1 + \|P\|)\text{dist}(x,V) .$$

Thus a "minimal projection" P_{min} yielding $\min\|P\|$ is sought.

Note that a bounded projection P can be identified with an n-dimensional subspace $P = [g_1, \ldots, g_n] = [g]$ of the dual space X^*; i.e., $Px = \langle x,g \rangle \cdot v = \sum_{i=1}^{n} \langle x,g_i \rangle v_i$, where v is chosen dual to g (i.e., $\langle v_i, g_j \rangle = \delta_{ij}$).

For compactness of notation we will in the following write $X \subset L^\infty$ for $X \subset C$, and $X^* \subset L^1$ for $X^* \subset C^*$, where C denotes the continuous functions. As usual let $\frac{1}{q} + \frac{1}{p} = 1$ where $1 \leq p \leq \infty$; and if $X \subset L^p(Q)$, identify X^* with a subspace of $L^q(Q)$. Also in the following, Q will denote an arbitrary compact separable T_1-space and "$'$" will denote 2-sided differentiation along any continuously differentiable (C^1) vector field on Q. Further, for $v = (v_1, \ldots, v_n) \in V^n$ we write $|v|^2 = \Sigma v_i^2 = v \cdot v$. Finally "c" will denote "constant" and

Copyright © 1980 by Academic Press, Inc.
All rights of reproduction in any form reserved.
ISBN: 0-12-171050-5

"pcw." will abbreviate "piecewise."

II. KNOWN EXAMPLES OF P_{min}

There are very few known examples of P_{min} which are structurally nontrivial. From these few examples we can, however, gain insight into formulae ((VE) in Section III below) completely determining the linear structure of minimal projections.

Example 1. $n = 1$.

$L^p(Q) = X \supset V = [v]$.

Then $L^q(Q) = X^* \supset P_{min} = [g]$ where $\underline{g = |v|^{p-2}v}$ by the Hölder (in)equality. Note $g' = (p-1)|v|^{p-2}v'$; thus

$$(*) \quad \frac{1}{p}g'v = \frac{1}{q}gv' , \qquad \text{if } v \in C^1_{pcw.}(Q) .$$

Note 1. Even in the case $p = \infty$, $g = (\operatorname{sgn} v)|v|^{\infty}$ has an appropriate interpretation as a limit, but in any event (*) is unambiguous.

Note 2. In the case $p = \infty$, nonuniqueness occurs if $|v|$ has more than one maximum ($v' = 0$ at interior maxima). Likewise, in general ($n \geq 1$), the possibility of nonuniqueness depends on the "local" dependency of the (*)-equation (on the $n-1$ (o)-equations (see below)).

Example 2. $n = 2$ (Cheney-Franchetti [6]).

$L^{p=1}([-1,1]) = X \supset V = [1,t]$.

Then $L^{q=\infty}([-1,1]) = X^* \supset P_{min} = [g_1,g_2] = [g]$ where $\underline{\underline{g = (1+\delta)|v|^{p-2}v + \vec{c}}}$, $v = (1,\alpha t) \in V^2$, $\vec{c} = (0,\pm c)$,

$\delta = (p-1)c\alpha \int_0^{|t|} |v|^{-p}$. Note $g' = \pm c(p-1)\alpha|v|^{-2}v +$ $(1+\delta)|v|^{p-4}[(v\cdot v)v' + (p-2)(v'\cdot v)v]$ and $v' = (0,\alpha)$; thus

$$(*)\quad \frac{1}{p}g'\cdot v = \frac{1}{q}g\cdot v'$$

$$(\circ)\quad g\cdot v_{(2)} = \pm c\ , \qquad v_{(2)} = (-\alpha t, 1)\in V^2\ .$$

Note 3. This example extends to all p , $1 \le p \le \infty$, as will be discussed in Section III. For the specific values of α and c in the case $p = 1$, see [6].

Example 3. $n = 3$ (Chalmers-Metcalf [4]).

$$L^{p=\infty}([-1,1]) = X \supset V = [1,t,t^2]\ .$$

Then $L^{q=1}([-1,1]) = X^* \supset P_{min} = [g_1,g_2,g_3] = [g]$ where, off $\{-1,0,1\}$, $g = w(1,t,\pm t)$, $w = \frac{c_\pm}{(b\pm|t|)^3}$, $c_\pm$ is symmetric and $= c_-$ on $[t_1,t_2]$, $= c_+$ on $[t_3,t_4]$, $= 0$ elsewhere, where $0 < t_1 < t_2 < t_3 < t_4 < 1$. Thus

$$(*)\quad \frac{1}{p}g'\cdot v_{(1)} = \frac{1}{q}g\cdot v'_{(1)}\ , \qquad v_{(1)} = (\tfrac{t^2}{2}, -t, 0)\in V^3$$

$$(\circ)\quad g\cdot v_{(2)} = c_\pm\ , \qquad v_{(2)} = (b^3+3bt^2, 0, 3b^2+t^2)\in V^3$$

$$(\circ)\quad g\cdot v_{(3)} = 0\ , \qquad v_{(3)} = (0,1,-1)\in V^3\ .$$

Note 4. If an antisymmetric function and a symmetric function occur together in a $(\circ)$-equation (as in the second $(\circ)$-equation above), the g_i are understood to be modulo sign.

III. VARIATIONAL EQUATIONS AND FURTHER EXAMPLES

Theorem (Variational Equations (VE)).

$$L^p(Q) \supset X \supset V = [v_1,\ldots,v_n] = [v] , \qquad 1 \le p \le \infty .$$

Then $L^q(Q) \supset X^* \supset P_{min} = [g_1,\ldots,g_n] = [g]$ where $\exists$ nonzero $v_{(k)} \in V^n$, $1 \le k \le n$, such that

$$\text{(VE)} \quad \begin{aligned} &(*) \quad \frac{1}{p} g' \cdot v_{(1)} = \frac{1}{q} g \cdot v'_{(1)} && \text{if } V \subset C^1_{pcw.}(Q) \\ &(o) \quad g \cdot v_{(k)} = c_{pcw.} , && 2 \le k \le n \quad \text{(independent).} \end{aligned}$$

Note 5. The theorem applies to any separable Banach space X , identified as usual with a subspace of $L^\infty(Q)$, where Q is the set of extreme points of the ball of the dual.

As specific applications of the theorem, we have the following further examples.

Example 4. Example 2 extends to all $1 \le p \le \infty$. In particular for $p = 2$, $c = 0$, $\alpha \ne 0$; while for $p = \infty$, $c = 0$, $\alpha > 0$, yielding the interpolating projection at $\{-1,1\}$.

Example 5. The quadratic L^1-example ($L^{p=1}([-1,1]) = X \supset V = [1,t,t^2]$) is currently being calculated ([2]).

Other examples. One can check that the (VE) hold in any example where P_{min} is known, e.g. Hilbert space $(p = 2)$ and the Fourier projection in $L^p[-\pi,\pi]$, $1 \le p \le \infty$ (see the following note).

Note 6. For any orthogonal (see e.g. [5]) projection $P_\perp = [g] = [wv]$, $\langle v_i, wv_j \rangle = \delta_{ij}$, we have $n-1$ (o)-equations, namely (o) $g_{k-1}v_k - g_k v_{k-1} = 0$ $(2 \le k \le n)$. If we assume (*)

holds with $v_{(1)} = v$, we obtain $w = |v|^{p-2} = \text{ext}(|v|)/|v|$, the "natural" projections P_p (see [1],[7]) where $\|P_p\| \leq n^{|\frac{1}{2}-\frac{1}{p}|}$, $1 \leq p \leq \infty$. (The Fourier projections are examples of natural projections P_p , $1 \leq p \leq \infty$, which are minimal.)

IV. MULTIPLICATIVE VARIATIONS LEAD TO (VE)

Proof of (VE)- Theorem (Sketch (for details see [3])). Since operators under composition are norm-submultiplicative, the advantage of multiplicative variations over additive variations for our minimum norm problem is illustrated below.

Additive Variations:

$$\|P_{min}\| \leq \|\underbrace{P_{min} + 0_\varepsilon}_{\text{projection}}\| \leq \|P_{min}\| + \|0_\varepsilon\| \Rightarrow \|0_\varepsilon\| \not< 0$$

which gives no information.

Multiplicative Variations:

$$\|P_{min}\| \leq \|\underbrace{P_{min} \circ I_\varepsilon}_{\text{projection}}\| \leq \|P_{min}\| \cdot \|I_\varepsilon\| \Rightarrow \|I_\varepsilon\| \not< 1$$

which gives much information as seen in the following.

For simplicity and purposes of illustration take $Q = [a,b]$ (with Lebesgue measure) and partition Q into N equal disjoint intervals $Q = \bigcup_{r=1}^{N} Q_r$. Consider on $L^p(Q)$ the integral operator $(I_\varepsilon x)(t) = \int_a^b x(s)k(s,t)ds$ with kernel

$$k(s,t) = \mu(t)^{\frac{1}{p}} \mu(s)^{\frac{1}{q}} \sum_{r=1}^{N} \chi_{Q_r}(s) \chi_{Q_r}(t)/mQ_r ,$$

where χ_A denotes the characteristic function of A and mA denotes the measure (length) of A . Note that I_ε is close to the identity operator I if μ is close to 1 and N is large. But the kernel $k(s,t)$ is supported and separable on disjoint squares $Q_r \times Q_r$ in $Q \times Q$ and so we easily compute (using Hölder's inequality) that $\|I_\varepsilon\| = \max_{1 \le r \le N} \int_{Q_r} |\mu| / mQ_r$ for all $1 \le p \le \infty$.

For further simplicity consider $p = 1$. Then $(I_\varepsilon x)(t) = x^N(t)\mu(t)$, where $x^N(t)$ is the average value of x on Q_r if $t \in Q_r$. Now I_ε is an admissible multiplicative variation of P_{min} if $P_{min} \circ I_\varepsilon$ is a projection, i.e., $P_{min} \circ I_\varepsilon$ has the same action on V as does P_{min}:

$$\text{(i)} \quad \int_a^b g_i v_j^N \mu = \int_a^b g_i v_j \qquad (1 \le i,j \le n).$$

Further $\|P_{min} \circ I_\varepsilon\| < \|P_{min}\|$ if $\|I_\varepsilon\| < 1$, in particular if $\mu \ge 0$ and

$$\text{(ii)} \quad \int_{Q_r} 1 \cdot \mu < \int_{Q_r} 1 \qquad (1 \le r \le N).$$

But now (i) and (ii) together form an interpolation problem for μ. It is then seen that (i) and (ii) can be solved for $\mu \ge 0$ (contradicting the minimality of P_{min}), unless as $N \to \infty$ the $g_i v_j^N$ and 1 are becoming dependent (locally (on each Q_r), yielding the (*)-equation, and globally (on Q), yielding the $n-1$ (o)-equations).

The above procedure extends easily to all compact subsets Q of $\mathbb{R}^m$. The theorem then extends to an arbitrary compact separable T_1-space Q by embedding Q in the countable cube $[0,1]^\omega$ with metric $d(x,y) = \sum_{k=1}^{\infty} 2^{-k} |x_k - y_k|$, approximating Q by compact subsets of $\mathbb{R}^m$, applying the theorem, and then taking limits. ∎

Note 7. When Q is a compact subset of $\mathbb{R}^m$, then the proof of the (o)-equations also requires initially that $V \subset C^1_{pcw.}$. Next, if $V \not\subset C^1_{pcw.}$, we approximate V by $\tilde{V} \subset C^1_{pcw.}$, obtain (o)-equations for $\tilde{V}$, and then take limits to obtain (o)-equations for V. In fact, this procedure can also be applied with the (*)-equation to obtain P_{min} even if $V \not\subset C^1_{pcw.}$. That is, approximate V by $\tilde{V} \subset C^1_{pcw.}$, solve for $\tilde{P}_{min}$ via the n (VE)-equations, and take $P_{min} = \lim \tilde{P}_{min}$. (Think of the $n = 1$ case.)

REFERENCES

1. B. L. Chalmers, "A natural simple projection with norm $\le \sqrt{n}$," submitted for publication.
2. ______, "A minimal projection from $L^1[a,b]$ onto $\mathcal{P}_2$," in preparation.
3. B. L. Chalmers and F. T. Metcalf, "The variational equations for minimal projections," submitted for publication.
4. ______, "A minimal projection from $C[a,b]$ onto $\mathcal{P}_2$," in preparation.
5. E. W. Cheney and K. H. Price, "Minimal projections," in: Approximation Theory, A. Talbot, ed., Acad. Press, London, 1970, 261-289. MR 42#751.
6. C. Franchetti and E. W. Cheney, "The problem of minimal projections in L_1-spaces," in: Approximation Theory II, G. G. Lorentz, et al, ed., Acad. Press, New York, 1976, 365-368. (Also CNA Report 106, Univ. of Texas, Austin, 1975.)
7. D. R. Lewis, "Finite dimensional subspaces of L^p," Studia Math. 63(1978), 207-212.

ON THE EXISTENCE OF STRONG UNICITY OF ARBITRARILY SMALL ORDER

B. L. Chalmers[1]

Department of Mathematics
University of California
Riverside, California

G. D. Taylor[2]

Department of Mathematics
Colorado State University
Fort Collins, Colorado

The strong unicity theorem, first given by Newman and Shapiro (4), may be described as follows: Given C[a,b] and W an n-dimensional Haar subspace of C[a,b]. Let $f \in C[a,b]$ and $p_f \in W$ be the best approximation to f from W. Then there exists a positive constant γ, depending only on f, such that

$$\|f - p\| \geq \|f - p_f\| + \gamma\|p - p_f\| \tag{1.1}$$

for all $p \in W$ where $\|h\| = \max\{|h(t)|: t \in [a,b]\}$, $h \in C[a,b]$. The extension of this theorem to the setting of monotone approximation has recently been studied by Fletcher and Roulier (3) and Schmidt (5). Specifically, fix an interval [a,b], integers $1 \leq r_0 < \ldots < r_k$, signs $\varepsilon_i = \pm 1$, i=0,...,k and define $K = K(r_0, \ldots, r_k; \varepsilon_0, \ldots, \varepsilon_k)$ by

$$K = \{p \in \Pi_n: \varepsilon_j p^{(r_j)}(x) \geq 0,\ a \leq x \leq b,\ j=0,1,\ldots,k \text{ with } k \leq n\} \tag{1.2}$$

where Π_n denotes the class of all real algebraic polynomials of degree $\leq$ n. The study of approximation of C[a,b] by K is called the monotone

[1]Research supported in part by the National Science Foundation, under grant MCS-76-08518.
[2]Research supported in part by the Air Force Office of Scientific Research, Air Force Systems Command, USAF, under contract F-49620-79-C-0124 and by the National Science Foundation, under grant MCS-78-05847.

Copyright © 1980 by Academic Press, Inc.
All rights of reproduction in any form reserved.
ISBN: 0-12-171050-5

approximation problem. Professor G.G.Lorentz has played a major role in the development of the theory for this problem. See (2) for a brief expository treatment of this problem and an extensive bibliography.

In (3), Fletcher and Roulier constructed an example in $K=\{p\in\Pi_3 : p'(x)\geq 0$ on $[-1,1]\}$ which shows that the best result of form (1.1) that could hold in this setting would be where $\|p-p_f\|$ is replaced by $\|p-p_f\|^2$. Also, some positive results were given that were extended by Schmidt (5). In (5) it is proved that given $f\in C[a,b]$, K as defined in (1.2), $p_f\in K$ the best monotone approximation to f and a positive constant M, there exists $\gamma>0$ depending only on f and M such that

$$\|f - p\| \geq \|f - p_f\| + \gamma\|p - p_f\|^2 \tag{1.3}$$

for all $p \in K$ satisfying $\|p\| \leq M$.

In (5) one has the following definition: If p_f is the best uniform approximation to $f\in C[a,b]$ from W a subset of C[a,b], we say that p_f is <u>strongly unique of order</u> α $(0<\alpha\leq 1)$ if for each $M>0$ there is a constant $\gamma>0$ such that

$$\|f - p\| \geq \|f - p_f\| + \gamma\|p - p_f\|^{1/\alpha}$$

for all $p\in W$ satisfying $\|p\|\leq M$. Thus, these two papers taken together show that in monotone approximation strong unicity of order 1/2 holds and this is a best possible result.

In this paper we shall show that by taking an appropriate combination of interpolatory constraints with a monotone constraint one obtains an approximation problem in which strong unicity of order $\frac{1}{2m}$, m a positive integer, holds and that this is also a best possible result.

Thus, fix m a positive integer and define $K\subset \Pi_n$ by

$$K = \{p\in \Pi_n : p^{(1)}(x)\geq 0, a\leq x\leq b \text{ and } p^{(2)}(x_0)=\dots=p^{(2m-1)}(x_0)=0 \text{ for } x_0\in(a,b) \text{ fixed, } n\geq 2m+1\}. \tag{1.4}$$

Now, by referring to the general theory of (1), one can prove that corresponding to each $f\in C[a,b]$, there exists a unique best approximation,

p_f from K to f. The basic tools of this theory are extreme linear functionals (extremals) of the dual of Π_n corresponding to f and a given $p \in K$. In this particular setting the extremals are as follows. Given $f \in C[a,b]$ and $p \in K$, define for $x \in [a,b]$, e_x^0 on $C[a,b]$ by $e_x^0(g)=g(x)$ for all $g \in C[a,b]$ (point evaluation) and for $x \in [a,b]$, and $1 \leq j \leq 2m$, e_x^j on Π_n by $e_x^j(q)=q^{(j)}(x)$ for all $q \in \Pi_n$. The linear functional e_x^0, $x \in [a,b]$, is said to be an extremal for f and p provided $|e_x^0(f-p)| = \|f-p\|$. The linear functional e_x^1, $x \in [a,b]$ is said to be extremal for f and p provided $e_x^1(p)=0$. Whenever e_x^1 is an extremal for f and p and $x \notin \{a,x_0,b\}$ then an additional extremal called an <u>augmented extremal</u> is also present; namely, the extremal e_x^2 for which $e_x^2(p)=0$ must also hold (since $p^{(1)}(x) \geq 0$). If $e_{x_0}^1$ is an extremal for f and p, then the linear functional $e_{x_0}^{2m}$ is an <u>augmented extremal</u> for f and p with $e_{x_0}^{2m}(p)=0$ holding (since $p^{(1)}(x) \geq 0$). If one starts with an extremal set for f and p (which contains $e_{x_0}^2, \ldots, e_{x_0}^{2m-1}$) and adds all possible augmented extremals (as described above) to this set, then one has the augmented set of extremals for f and p corresponding to the original extremal set. Observing that these augmented extremal sets always correspond to Hermite-Birkhoff interpolation problems in which every supported block is even, it is relatively straightforward to prove that the maximal augmented extremal set for f and its best approximation, p_f, from K must have n+2 elements which span the dual of Π_n. Thus, K is generalized Haar and uniqueness of best approximations holds (1). In addition, suppose p_f is the best approximation to f from K. Then there exists $k \leq n+2$ extremals (e.g. (2)), $E=\{e_i\}_{i=1}^k$, none of which are augmented extremals, for which $\underset{\sim}{0}$ belongs to the convex hull of $\{\sigma(e)e\colon e \in E\}$ where $\sigma(e)=\operatorname{sgn}(f(x)-p_f(x))$ if $e=e_x^0$ for some $x \in [a,b]$, $\sigma(e)=1$ if $e=e_y^1$ for some $y \in [a,b]$ and $\sigma(e_{x_0}^j)=1$, $j=2,\ldots,2m-1$. Then, by adjoining to E the set E^a={all augmented extremals corresponding to elements of E} we must have that the set $E^{aug}=E \cup E^a$ contains at least n+2 elements of Π_n^* which will necessarily span Π_n^* by the fact that every supported block in the corresponding Hermite-Birkhoff problem is even. Likewise, we must have that there exists $e \in E$ for which $e=e_x^0$ some $x \in [a,b]$ as otherwise E is also an extremal set for f and p_f+c, c any constant, for which $\underset{\sim}{0}$ is in the convex hull of $\{\sigma(e)e: e \in E\}$ violating uniqueness of best approximation. Using these observations we can now prove

<u>THEOREM</u>. Let $f \in C[a,b]$ and $p_f \in K$ be the best approximation to f from K. Given $M>0$ there exists $\gamma=\gamma(f,M)>0$ such that for $p \in K$ satisfying $\|p\| \leq M$,

$$\|f - p\| \geq \|f - p_f\| + \gamma\|p - p_f\|^{2m}$$

(i.e. strong unicity of order $\frac{1}{2m}$) and this inequality is best possible.

<u>Proof</u>: The proof is an extension of the techniques of Fletcher and Roulier and Schmidt. If $f \in K$ then $\gamma=(2M)^{1-2m}$ suffices. Thus, assume $f \notin K$. Let $E=\{e_i\}_{i=1}^k$ be a set of k extremals, which contains $\{e_{x_0}^j\}_{j=2}^{2m-1}$ but contains no augmented extremals, for which $\underset{\sim}{0}$ is in the convex hull of $\{\sigma(e)e: e \in E\}$. Set $E^{aug}=E \cup E^a$. Further, define E^0, $E^1 \subset E$, where $e \in E$ is in E^0 if $e=e_x^0$ for some $x \in [a,b]$ and $e \in E^1$ if $e=e_y^1$ for some $y \in [a,b]$. Define the semi-norm $\|\cdot\|'$ on Π_n by $\|q\|'=\max\{|e(q)|: e \in E\}$. Set $Q=\{q=\frac{p_f-p}{\|p_f-p\|'}: \|p_f-p\|' \neq 0 \text{ and } p \in K\}$. We claim that $\inf_{q \in Q} \max_{e \in E^0} \sigma(e)e(q)=\tau > 0$. Indeed, if there exist $q \in Q$ with $\max_{e \in E^0} \sigma(e)e(q) \leq 0$. Then from $q=\frac{p_f-p}{\|p_f-p\|'}$ with $\|p_f-p\|' \neq 0$ and $p \in K$ we see that $e(q) \neq 0$ for some $e \in E$ and $e(q) \leq 0$ for all $e \in E^1$. Thus, $\sigma(e)e(q) \leq 0$ for all $e \in E$ with strict inequality holding at least once. This violates the fact that $\underset{\sim}{0}$ belongs to the convex hull of $\{\sigma(e)e: e \in E\}$. Using this lower bound, we have for $p \in K$ with $\|p_f-p\|' \neq 0$ that there exists $e \in E^0$ for which $\sigma(e)e(p_f-p) \geq \tau \|p_f-p\|'$. Now observe that (as usual)

$$\|f - p\| \leq \|f - p_f\| + \gamma\|p_f - p\|' .$$

As this inequality holds for $\|p_f-p\|'=0$, we have a strong uniqueness-type result for the seminorm $\|\cdot\|'$. Next, the norm, $\|p\|^*=\max\{|e(p)|: e \in E^{aug}\}$, is introduced. Thus, there exists a constant $\lambda>0$ such that $\|p\|^* \geq \lambda\|p\| \;\forall\; p \in \Pi_h$. Finally, we claim that there exists $A>0$ for which $\|p_f-p\|' \geq A\,(\|p_f-p\|^*)^{2m}$, $\forall\; p \in K$ satisfying $\|p\| \leq M$. First observe that $\|p_f-p\|'=0$ with $p \in K$ implies

$e(p_f-p)=0 \ \forall$ a E^{aug} so that $\|p_f-p\|^*=0$. Now, for $e\in E$, there exists a constant K_1 for which $|e(p_f-p)|\geq K_1|e(p_f-p)|^{2m}$ as $\|p\|\leq M$. Let $e\in E^{aug}\sim E$ and assume that $e=e^{2m}_{x_0}$ (the augmented extremal corresponding to $e^1_{x_0}$). We claim that there exists $K_2>0$ for which $|e^1_{x_0}(p_f-p)|\geq K_2|e^{2m}_{x_0}(p_f-p)|^{2m} \ \forall \ p\in K$ satisfying $\|p\|\leq M$. If this is not the case, then corresponding to each integer $\nu>0$ there exists $q_\nu \in K$ with $\|q_\nu\|\leq M$ for which $|q'_\nu(x_0)|<\frac{1}{\nu}|q^{(2m)}_\nu(x_0)|^{2m}$. Now we may assume that q_ν converges uniformly to $q\in K$. Clearly, $q'(x_0)=0$. We can write $q'_\nu(x)=q'_\nu(x_0)+\frac{q^{(2m)}_\nu(x_0)}{(2m-1)!}(x-x_0)^{2m-1}+s_\nu(x)(x-x_0)^{2m}=\beta_\nu+\alpha_\nu(x-x_0)^{2m-1}$ $+s_\nu(x)(x-x_0)^{2m}$ where $\beta_\nu\to 0,\alpha_\nu\to 0$ (as $q^{(2m)}(x_0)=0$ since $q\in K$), $|s_\nu(x)|\leq M_1$ for all $x\in[a,b]$, some M_1 independent of ν and $q'_\nu(x)\geq 0 \ \forall \ x\in[a,b]$. Thus, $0\leq\beta_\nu+\alpha_\nu(x-x_0)^{2m-1}+M_1(x-x_0)^{2m}$ for $x\in[a,b]$. For ν sufficiently large (so that $x\in(a,b)$), set $x-x_0=-\frac{\alpha_\nu(2m-1)}{2mM_1}$. This gives $\frac{(2m-1)}{M_1}\left(\frac{2mM_1}{2m-1}\right)^{2m}\beta_\nu\leq\alpha_\nu^{2m}$ or that there exists a constant K_1 independent of ν (sufficiently large) such that $|q'_\nu(x_0)|\geq K_1|q^{(2m)}_\nu(x_0)|^{2m}$ which is our desired contradiction. Finally, if $e\in E^{aug}\sim E$ is of the form $e=e^2_y$ some $y\in(a,b)\sim\{x_0\}$, the above argument (modified) shows that there exists K_3 for which $|e^1_y(p_f-p)|\geq K_2|e^2_y(p_f-p)|^2$ $\geq K_3|e^2_y(p_f-p)|^{2m} \ \forall \ p\in K$ satisfying $\|p\|\leq M$ where K_3 is independent of p. By taking A to be the smallest of the constants produced above, we have that $\|p_f-p\|'\geq A(\|p_f-p\|^*)^{2m}$ implying $\|f-p\|\geq\|f-p_f\|+\gamma\|p_f-p\|^{2m} \ \forall \ p\in K$ satisfying $\|p\|\leq M$ with $\gamma=\gamma(M,f)>0$ independent of p.

To show this result is best possible we construct an example. Fix m a positive integer and let r_1,r_2,r_3 denote the three roots of $p_0(x)=x^{2m+1}$ $+2x^{2m}-1$ (note $-2<r_1<-1$, $r_2=-1$, $0<r_3<1$). Define $K=\{p\in\Pi_{2m+1}: p'(x)\geq 0,$ $x\in[r_1,r_3],\ 0=p^{(2)}(0)=\dots=p^{(2m-1)}(0)\}=\{p(x)=a_0x^{2m+1}+a_1x^{2m}+a_2x+a_3: p'(x)\geq 0$ on $[r_1,r_3]\}$. Define $g\in C[r_1,r_3]$ by $g(r_1)=\frac{1}{2}$, $g(-1)=\frac{1}{2}$, $g(r_3)=\frac{1}{2}$ and extend g linearly to all $[r_1,r_3]$. Set $f=g+2x^{2m+1}$ and $p_f(x)=2x^{2m+1}$. Note that $\{-e^0_{r_1},$ $e^0_{-1},-e^0_{r_3},e^1_0\}$ is an extremal set for f and p_f whose convex hull contains the zero of V^*, $V=\{a_0x^{2m+1}+a_1x^{2m}+a_2x+a_3\}$. (Coefficients are: $\alpha_1=1$, $\alpha_2=1+\alpha_3$,

$\alpha_3 = \frac{-r_1^{2m+1}-1}{r_3^{2m+1}+1}$, $\alpha_4 = r_1 + \alpha_2 + \alpha_3 r_3$, respectively.) Thus, p_f is the desired best approximation to f from $K^{(2)}$. Next, define $p_\alpha(x) = p_f(x) + \alpha p_0(x) + 4m\alpha^{2m}x$, for $0 < \alpha \leq \alpha_0$ where α_0 is chosen so small that $|f-p_\alpha| = |g-\alpha[p_0+4m\alpha^{2m-1}x]|$ decreases as x moves away from r_i in a neighborhood of $\{r_1, r_2, r_3\}$ for all α $(0<\alpha\leq\alpha_0)$. This can be done since $|g|$ decreases linearly as x moves away from r_i. Hence α_0 can be chosen so small that $\|f-p_\alpha\| = \max_{i=1,2,3} |(f-p_\alpha)(r_i P|$, $0<\alpha\leq\alpha_0$ $= f(-1)-p_\alpha(-1) = \frac{1}{2}+4m\alpha^{2m}$. Also, $\|f-p_f\| = \frac{1}{2}$, $\|p_f-p_\alpha\| \geq |p_f(0)-p_\alpha(0)| = \alpha$ and $p'_\alpha(x) = 2(2m+1)x^{2m} + \alpha((2m+1)x^{2m}+4mx^{2m-1}) + 4m\alpha^{2m}$. Now, for $x>0$, $p'_\alpha(x)>0$; for $x\in[r_1,-\alpha]$, the term $2(2m+1)x^{2m}$ dominates showing that $p'_\alpha(x)>0$ here; and for $x\in[-\alpha,0]$ the term $4m\alpha^{2m} \geq |4\alpha m x^{2m-1}|$ again implying that $p'_\alpha(x)\geq 0$. Thus $p_\alpha \in K$ and $(\|f-p_\alpha\| - \|f-p_f\|)/\|p_f-p_\alpha\|^\beta \leq \frac{4m\alpha^{2m}}{\alpha^\beta}$. This implies that we must have $\beta \geq 2m$ in order for the strong unicity theorem to hold for this f and p_f. ∎

By suitably selecting g, it can be shown that this weaker strong uniqueness result holds for an f which also satisfies all the constraints of K. Additional results on this topic will appear elsewhere.

REFERENCES

1. B.L.Chalmers, A unified approach to uniform real approximation by polynomials with linear restrictions, Trans.Amer.Math.Soc., 166(1972), 309-316.

2. B.L.Chalmers and G.D.Taylor, Uniform approximation with constraints, Iber.d.Dt.Math.-Verein., 81(1979),49-86.

3. Y.Fletcher and J.A.Roulier, A counterexample to strong unicity in monotone approximation, preprint.

4. D.J.Newman and H.S.Shapiro, Some theorems on Cebysev approximation, Duke Math.J., 30(1963),673-682.

5. D.Schmidt, Strong unicity and Lipschitz conditions of order 1/2 for monotone approximations, preprint.

PROBLEMS AND RESULTS ON BEST INVERSE APPROXIMATION

Charles K. Chui[1]

Department of Mathematics
Texas A&M University
College Station, Texas 77843

DEDICATED TO PROFESSOR G. G. LORENTZ
ON THE OCCASION OF HIS SEVENTIETH BIRTHDAY

The purpose of this paper is to discuss some recent results and related problems on best inverse approximation by polynomials.

1. INTRODUCTION

The method of least-squares inverses is a well known and useful technique in geophysical studies and recursive digital filter design. It was introduced by E. A. Robinson [11] to obtain a minimum-delay finite-length wavelet whose convolution with a given finite-length wavelet deviates least from the unit spike in the ℓ^2 sequence norm. The interested reader should refer to Robinson's book [10] for a detailed study of this subject. In terms of polynomial approximation in the complex plane $\mathbb{C}$, Robinson's result can be stated in the following way. Let H^2 denote the usual Hardy space of analytic functions in the open unit disc of the complex plane $\mathbb{C}$ with finite H^2 norm. If f is a given polynomial satisfying $f(0) \neq 0$, and P_n is the polynomial with degree no greater than n so chosen that the product fP_n is closest to unity in the H^2 norm, then P_n does not vanish on the closed unit disc. A simple proof of this statement via orthogonal polynomials is contained in [3]. In fact, the proof in [3] applies to any $f \in H^2$.

[1]Supported by the U.S. Army Research Office under Grant Number DAAG 29-78-0097.

Copyright © 1980 by Academic Press, Inc.
All rights of reproduction in any form reserved.
ISBN: 0-12-171050-5

The analogous problem for real-valued functions on a compact interval of the real line $\mathbb{R}$ was posed in [3]. Let π_n denote the space of all polynomials with real coefficients and with degrees no greater than n. If $f \in L_p[0,1]$ is given we propose to study the extremal problem:

$$(1.1) \qquad D_{n,p}(f) = \inf\{\|1 - fg\|_p : g \in \pi_n\}.$$

This problem will be called the problem of <u>best L_p inverse approximation</u> by polynomials. If $P_n^* \in \pi_n$ satisfies $\|1 - fP_n^*\|_p = D_{n,p}(f)$, then P_n^* is a <u>best L_p inverse approximant</u> of f from π_n. If f vanishes on a set of positive measure, then $D_{n,p}(f) \not\to 0$ for any p, $1 \le p \le \infty$. Similarly, even if f vanishes at a single point in $[0,1]$, we have $D_{n,\infty}(f) = 1$. In studying the order of $D_{n,p}(f)$, these trivial cases should be ruled out. A moment of reflection also tells us that if f is zero-free on $[0,1]$, then the problem reduces to the usual best L_p approximation problem that has been studied very extensively. Hence, we only consider the case where $1 \le p < \infty$ and where the zero set of f is a non-empty subset of $[0,1]$ with zero measure. In this case, it turns out that the order of best L_p inverse approximation depends not only on the smoothness of the function f to be approximated, but also on the structure of the zero set of f. If f is a real analytic function on $[0,1]$, then the order of $D_{n,p}(f)$ has been completely determined in [2].

2. INVERSE APPROXIMATION IN H^2

In this section, we denote by $\| \ \|_2$ the norm of the Hardy space H^2. Let $f \in H^2$ with $f(0) \ne 0$, and $f^*(\theta)$ be the almost everywhere limit of $f(re^{i\theta})$ as $r \to 1^-$. Let ϕ_k, $k = 0, 1, \ldots$, be polynomials of degree k, orthonormal on the unit circle $|z| = 1$ with respect to the measure $|f^*(\theta)|^2 d\theta/2\pi$. If P_n is in π_n^c, the space of all (complex-valued) polynomials with degree no greater than n, so chosen that fP_n deviates least from unity in the norm $\| \ \|_2$ then P_n is called the least-squares inverse (LSI) of f in π_n^c. It can be shown as in [3] that $P_n(z) = cz^n\overline{\phi_n(1/\overline{z})}$, where c is a constant satisfying $|c|^2 = P_n(0)f(0) > 0$. Since ϕ_n has all its zeros in $|z| < 1$ (cf. [14]), we conclude that P_n is zero-free in $|z| \le 1$. Also, $P_n(z) = a_0 + \ldots + a_n z^n$ can be determined by solving the linear system $C[a_0, \ldots, a_n]^T = [\overline{f}(0), 0, \ldots, 0]^T$ where $C = [c_{\ell,j}]$ is an $n \times n$ positive definite Toeplitz matrix with $c_{\ell,j}$ being the $(\ell - j)$th

Fourier coefficient of the function $|f^*(\theta)|^2$, and the superscript T denoting the transpose of a matrix. If, in addition, f is zero free in $|z| < 1$, then

$$(2.1) \qquad \|1 - P_n f\|_2^2 = |f(0)|^2 \sum_{k=n+1}^{\infty} |\phi_k(0)|^2$$

which tends to zero as $n \to \infty$ (cf. [3, 14]). For each n, let $Q_{k,n} \in \pi_k^c$ be chosen such that $\|1 - P_n Q_{k,n}\|_2$ is minimum over all polynomials in π_k^c. $Q_{k,n}$ is called the generalized double least-squares inverse (GDLSI) of f in π_k^c through π_n^c. If $1/f$ is a bounded analytic function in $|z| < 1$, it is shown in [3] that $Q_{k,n} \to Q_k$ as $n \to \infty$, where Q_k is the LSI of $1/f$ in π_k. We conjecture that the same conclusion holds for any f in H^2 such that $1/f$ is also in H^2. If f is a polynomial with m zeros (counted according to their multiplicities) on $|z| = 1$ and is zero-free in $|z| < 1$, we also conjecture that $Q_{k,n} \to 0$ as $n \to \infty$ for $k = 0, \ldots, m-1$. Let us now consider the case when $f = f_k$ is a polynomial in π_k^c such that $f_k(0) \neq 0$, and let $Q_{k,n}$ be its double least-squares inverse (DLSI) in π_k through π_n. The following result is obtained in [3].

THEOREM 2.1. *$Q_{k,n} \to f_k$ as $n \to \infty$ if and only if f_k is zero free in $|z| < 1$. If, however,*

$$f_k(z) = (\alpha_1 - z) \ldots (\alpha_\ell - z)g(z)$$

where $0 < |\alpha_1|, \ldots, |\alpha_\ell| < 1$ and g is zero-free in $|z| < 1$, then

$$Q_{k,n}(z) \to (1/\bar{\alpha}_1 - z) \ldots (1/\bar{\alpha}_\ell - z)g(z),$$

as $n \to \infty$.

As a consequence, we can conclude immediately that $|Q_{k,n}(e^{i\theta})| \to c|f_k(e^{i\theta})|$ for all real θ as $n \to \infty$, where $c \geq 1$ is some constant. Furthermore, $c = 1$ if and only if f_k is zero-free in $|z| < 1$. This result shows that the procedure of DLSI in recursive digital filter design needs modification. For more detail in DLSI, see [9, 12, 13].

3. BEST L_p INVERSE APPROXIMATION

We now consider the inverse approximation problem in the setting which is of particular interest to the approximation theorist. Throughout the remaining portion of this article, all functions to be considered will be real-valued functions of a real variable. Hence, π_n will denote the space of all polynomials with real coefficients and with degrees no greater than n. We are interested in the study of the order of the error $D_{n,p}(f)$ of best L_p inverse approximation by polynomials as defined in (1.1). As discussed in section 1, we are only interested in $1 \le p < \infty$, and that the zero set of f is a non-empty subset of $[0,1]$ with zero measure. In this case, that is, if $f \in L_p[0,1]$ and $f \neq 0$ a.e. on $[0,1]$, it is easy to see that $D_{n,p}(f) \to 0$ as $n \to \infty$. The order of $D_{n,p}(f)$ will depend not only on the smoothness of f on $[0,1]$, but also on the zero structure of f there. For real analytic functions f, this order has been determined in [2]. The main results in [2] can be summarized as follows:

THEOREM 3.1. *Let f be a real analytic function on $[0,1]$ such that f is not identically zero, and let $1 \le p < \infty$. If $f(x) = 0$ for some $x \in (0,1)$, then there exist positive constants C_1 and C_2 such that*

$$C_1 n^{-1/p} \le D_{n,p}(f) \le C_2 n^{-1/p} . \tag{3.1}$$

If $f(x) \neq 0$ for all $x \in (0,1)$, but $f(0)f(1) = 0$ then there exist positive constants C_3 and C_4 such that

$$C_3 n^{-2/p} \le D_{n,p}(f) \le C_4 n^{-2/p} . \tag{3.2}$$

If f is a zero-free real analytic function on $[0,1]$, it is clear that $D_{n,p}(f) = o(1/n^{\alpha})$ for any $\alpha > 0$, $1 \le p \le \infty$. It should be mentioned that the above theorem holds for a much larger class of functions (cf. [2]). The proof in [2], however, depends very heavily on the hypothesis that f has a finite number of zeros of "finite order". In this direction, Beatson, Hasson, and the author have also proved that

$$D_{n,2}(x^{\alpha}) = \frac{\alpha}{\alpha + n + 1} \tag{3.3}$$

for any real number $\alpha > 0$. This example seems to indicate that the order of $D_{n,p}(f)$ would depend heavily on the zero structure of f (namely, the order of the zero at $x = 0$ of the function $f(x) = x^{\alpha}$ when $p = 2$). It would be interesting to study the order of $D_{n,p}(f_j)$, $1 \le p < \infty$ and $j = 1, 2$, where

$$f_1(x) = e^{-1/x} \tag{3.4}$$

(with $x = 0$ being a zero of "infinite order") and

$$(3.5) \qquad f_2(x) = 1/\log\left(\frac{2}{x}\right)$$

(with $x = 0$ being a zero of "zeroth order"). One is lead to believe that for each p, $1 \le p < \infty$, $D_{n,p}(f_1)$ would tend to zero very slowly, while $D_{n,p}(f_2)$ would converge to zero very rapidly. Their exact orders should be determined.

4. SOME PROBLEMS IN BEST INVERSE APPROXIMATION

We now list several problems in best inverse approximation that would be of interest to the approximation theorist. We will only concentrate on the order of best inverse approximation, although uniqueness and characterization questions are also of interest.

The first obvious extension is the problem of simultaneous approximation. Let $f_1, \ldots, f_m$ be functions in $L_p[0,1]$ which do not vanish on a set of positive measure, where $1 \le p < \infty$. Study the order of

$$(4.1) \qquad D_{n,p}(f_1, \ldots, f_m) = \inf\{\|1 - (f_1g_1 + \ldots + f_mg_m)\|_p : g_1, \ldots, g_m \in \pi_n\}.$$

Also, if $f_1, \ldots, f_m$ are continuous on $[0,1]$ with $f_1^2(x)+\ldots+f_m^2(x) > 0$ for all $x \in [0,1]$, study the order of $D_{n,\infty}(f_1,\ldots,f_m)$. It is not difficult to show that $D_{n,p}(f_1,\ldots,f_m) \to 0$ as $n \to \infty$ for any p, $1 \le p \le \infty$. Knowing the smoothness and the zero structure of $f_1,\ldots,f_m$, we should be able to know the order of $D_{n,p}(f_1,\ldots,f_m)$.

Next, let $\pi_{k,0}$ be the class of all polynomials $q \in \pi_k$ satisfying $q(0) = 1$. For each $f \in L_p[0,1]$ for $1 \le p < \infty$ and $f \in C[0,1]$ for $p = \infty$ not vanishing on a subset of positive measure, study the order of

$$(4.2) \qquad D_{k,n,p}(f) = \inf\{\|q_k - fp_n\|_p : q_k \in \pi_{k,0},\ p_n \in \pi_n\}.$$

If $k = 0$, then $D_{0,n,p}(f) = D_{n,p}(f)$ which has been discussed previously. A similar problem is that of quasi-rational approximation (cf. [4]), namely

$$(4.3) \qquad E_{k,n,p}(f) = \inf\{\|q_k - fp_n\|_p : q_k \in \pi_k,\ p_n \in \pi_{n,0}\}.$$

If f is a "normal" function in $C^{n+k}[0,\delta]$, $\delta > 0$, and $q_\varepsilon \in \pi_k$, $p_\varepsilon \in \pi_{n,0}$ are chosen such that the quantity $\|q_\varepsilon - fp_\varepsilon\|_{[0,\varepsilon]}$ is minimized, it was shown in [4] that the net $\{q_\varepsilon/p_\varepsilon\}$ of rational functions converges to the $[k/n]$ Padé approximant of f as $\varepsilon \to 0^+$.

The best L_p inverse approximation problem is also related to the problem of best approximation by incomplete polynomials (cf. [5,6,7,8]). Let $0 \le \alpha < \beta$ or $0 < \alpha \le \beta$, and $f \in L_p[0,1]$ if $1 \le p < \infty$ and $f \in C[0,1]$ if $p = \infty$. Study the order of

$$E_{n,p}(f;\alpha,\beta) = \inf\{\|f - g\|_p : g \in S_n(\alpha,\beta)\} \tag{4.4}$$

where

$$S_n(\alpha,\beta) = \mathrm{span}[x^{\alpha}, x^{\alpha+\beta}, \ldots, x^{\alpha+\beta n}]. \tag{4.5}$$

Since there is an Arithmetic structure in the exponents of the polynomial space $S_n(\alpha,\beta)$, a direct application of the estimates of Bak and Newman [1] is not desirable. For instance, if k is a positive integer, then we have

$$E_{n,2}(x^k;\ k+1,\ 2k+1) = (2k+1)^{-1/2}\ D_{n,2}\left(x^{1/(2k+1)}\right) = O\left(\frac{1}{n}\right).$$

REFERENCES

1. Bak, J. and D. J. Newman, Müntz-Jackson theorems in $L^p[0,1]$ and $C[0,1]$, Amer. J. Math., 94 (1972), 437-457.
2. Beatson, R. K., C. K. Chui, and M. Hasson, Degree of best inverse approximation by polynomials, (to appear).
3. Chui, C. K., Approximation by double least-squares inverses, J. Math. Analysis and Appl. (to appear).
4. Chui, C. K., O. Shisha, and P. W. Smith, Best local approximation, J. Approx. Theory, 15 (1975), 371-381.
5. Lorentz, G. G., Incomplete polynomials of best approximation, Israel J. Math. (to appear).
6. __________, Problems in Approximation Theory. The University of Arkansas Lecture Notes in Mathematics, Vol. 1, 1977.
7. __________, Approximation by incomplete polynomials (problems and results), in Padé and Rational Approximation, Ed. E. B. Saff and R. S. Varga, Academic Press, Inc., N.Y., 1977.
8. __________, Properties of incomplete polynomials, this volume.
9. Oppenheim, A. V. and R. W. Schafer, Digital Signal Processing. Prentice-Hall, Inc., Englewood Cliffs, N.J., 1975.
10. Robinson, E. A., Statistical Communication and Detection. Hafner Publ. Co., N.Y. 1967.
11. __________, Structural properties of stationary stochastic process with applications, in Time Series, Ed. M. Rosenblatt, Wiley Inc., N.Y. 1963.
12. Shanks, J. L., Recursion for digital processing, Geophysics, 32 (1967), 33-51.
13. __________, The design of stable two-dimensional recursive filters, in Proc. UMR - M.J. Kelley Commun. Conf., Univ. of Missouri, Rolla, 1970.
14. Szegö, G., Orthogonal Polynomials, Amer. Math. Soc. Colloq. Publ. Vol. 23, Providence, R.I., 1939.

APPROXIMATION BY M-IDEALS IN THE DISC ALGEBRA[1]

Charles K. Chui
Philip W. Smith
Joseph D. Ward

Department of Mathematics
Texas A&M University
College Station, Texas 77843

DEDICATED TO PROFESSOR G. G. LORENTZ
ON THE OCCASION OF HIS SEVENTIETH BIRTHDAY

1. INTRODUCTION

In approximation theory, it is sometimes of interest and important to consider the problem of approximation from subspaces of finite or small codimension. Let X be a normed linear function space and M_N an ideal corresponding to functions which vanish on a set E_N, which is usually small. Let $m_N(f)$ denote a best approximant, whenever it exists, to f from M_N. Hence $s_N(f) \equiv f - m_N(f)$ interpolates f on E_N. It is therefore of interest to consider the convergence of the error $m_N(f) = f - s_N(f)$. If $E_1 \subset E_2 \subset \dots$, the study of the convergence of $\|f - s_N(f)\| = \|m_N(f)\|$ is related to the problem of convergence of interpolants of f with minimal norm. Such questions have previously been considered in the Sobolev space setting (cf. [3], [6], [8], [9]).

As an example of the above considerations, let $X = C[0,1]$ and $M_N = \{f \in C[0,1]: f(E_N) = 0\}$ where E_N are finite point sets in $[0,1]$ with $E_1 \subset E_2 \subset \dots$. It is well known that M_N is a proximinal subspace and that $\mathrm{dist}(f,M_N) = \|f|_{E_N}\|$. It is also easily seen that for each $f \in C[0,1]$ there exists an $m_N \in M_N$ so that $\|f - s_N(f)\| = \|m_N\| = \|f|_{[0,1]\setminus E_N}\| - \|f|_{E_N}\| \equiv \gamma_n \geq 0$. Furthermore, if $T_N(f)$ is any other

---

[1] Supported in part by an NSF grant.

Copyright © 1980 by Academic Press, Inc.
All rights of reproduction in any form reserved.
ISBN: 0-12-171050-5

interpolant of f on E_N of minimal norm, then $\|f - T_N(f)\| \geq \gamma_N$.

Less obvious is the case of the disc algebra A. For this situation, the non-trivial ideals considered correspond to the subspaces of functions which vanish on a closed set E_N of measure zero on the boundary T of the unit disc D in the complex plane. Unlike the case $C[0,1]$ the proximinality of the ideal $M_N = \{g \in A: g|_{E_N} = 0\}$ is no longer clear although it is assured since from [7] and [10] such ideals are also M-ideals which are known to be proximinal [12].

A closed subspace N of a real Banach space X is said to be an L-ideal if there exists a closed subspace N' such that $X = N \oplus N'$ and

$$\|p + q\| = \|p\| + \|q\|$$

for all $p \in N$ and $q \in N'$. N is said to be an M-ideal if $N^{\perp}$ is an L-ideal in X^*. Also, N is said to be an M-summand if there exists a closed subspace M so that $X = N + M$ and

$$\|r + s\| = \max(\|r\|, \|s\|).$$

It was shown in [12] that every M-ideal is proximinal in its ambient space X and that the linear span of the set of best approximants of any element not in the M-ideal spans the M-ideal.

2. THE MAIN RESULTS

In what follows, E_N will denote a closed subset of measure zero of the unit circle T in the complex plane, A the disc algebra, M_N the subspace $\{g \in A: g|_{E_N} \equiv 0\}$, $m_N = m_N(f)$ a best approximant to a given $f \in A$ from M_N, and $\Delta(E_N, T)$ will designate the Hausdorff distance between E_N and T. The following result guarantees convergence of certain interpolants to f.

<u>Proposition 2.1.</u> <u>Let $E_1, E_2, \ldots$ be sets as defined above such that $\lim_{N\to\infty} \Delta(E_N, T) = 0$. Then for each N there exists an $m_N \in M_N$ so that $s_N \equiv f - m_N$ converges to f in A.</u>

<u>Remark</u>: It should be noted that care must be exercised in selecting the best approximants m_N. To see this, let E_N denote the 2^Nth roots of unity and $M_N = \{f \in A: f(E_N) = 0\}$. It is easily checked that $m_N = 1 - z^{2^N}$ is a best approximant to $f(z) \equiv 1$; however, $s_N \equiv f - m_N = z^{2^N}$ does not converge to f in A.

Proof: Since $\Delta(E_N,T) \to 0$ we have $\mathrm{dist}(f,M_N) \to \|f\|$. Pick $\varepsilon_N > 0$ and $\varepsilon_N \to 0$, such that

$$\mathrm{dist}(f,M_N) > \|f\| - \varepsilon_N \quad \text{for all } N.$$

Since

i) $\mathrm{int}\, B(0,2\varepsilon_N) \cap B\big(f, \mathrm{dist}(f,M_N)\big) \neq \phi$, and

ii) $B(0,2\varepsilon_N) \cap M_N \neq \phi$ and $B\big(f, \mathrm{dist}(f,M_N)\big) \cap M_N \neq \phi$,

we conclude by [1, Thm. 5.8] that there exists an $m_N \in M_N$ that lies in $B(0,2\varepsilon_N) \cap B\big(f, \mathrm{dist}(f,M_N)\big)$. Thus, $\|s_N - f\| = \|m_N\| \le \varepsilon_N \to 0$. This completes the proof.

The above method of proof does not give the rate of convergence. The main result in [5] was to establish this rate, and the main result in [5] is the following.

Theorem 2.1. Let $f \in A$ and E_N, $N = 1, 2, \ldots,$ be closed subsets of T of measure zero such that

$$\gamma_N = \log\big(\|f\|_T / \|f\|_{E_N}\big) \to 0 .$$

Then there exist minimal norm interpolants s_N of f satisfying $f - s_N \in M_N \equiv \{g \in A\colon g(E_N) = 0\}$ such that

$$\|f - s_N\| = O(\gamma_N) .$$

Moreover, this rate of convergence is sharp in the sense that $O(\gamma_N)$ cannot be replaced by $o(\gamma_N)$.

3. M-IDEALS IN A

In this section, we strengthen a known result concerning M-ideals in the disc algebra A.

It has been shown in [10, Thm 3.1] that for a function algebra A, the following are equivalent:

i) J is an M-ideal, and

ii) $J = \{a \in A \mid a \equiv 0 \text{ on } E\}$ where E is a peak set for A.

Since it is well known that the peak sets in A correspond to closed sets on T having zero (one-dimensional) Lebesgue measure, this yields a result of Fakhoury.

Proposition 3.1. The M-ideals in A correspond to $J = \{f\colon f(E) \equiv 0\}$ where E is a closed subset of T of Lebesgue measure zero.

On the other hand, R. R. Smith [15] has proved the following.

Theorem 3.1. *Let* A *be a function algebra contained in* $C(\Omega)$ *where* Ω *is a compact Hausdorff space. Then the M-ideals in* A *are precisely the closed ideals with a bounded approximate identity.*

These two facts may be reconciled directly using complex analysis. These arguments in turn will yield a strengthening of Smith's theorem in the case of A. Let $J = \{f \in A: f(E) \equiv 0,\ E$ is a closed set, $E \subset T,\ m(E) = 0\}$. Then J has a bounded approximate identity. To see this note that by the Rudin-Carleson theorem [4], E must be a peak set. By a theorem of Bishop [4], there exists $f \in A$ so that $f(E) \equiv 1$, $|f| < 1$ elsewhere. Then it is easily checked that $g_N = 1 - f^N$ is a bounded approximate unit corresponding to the ideal J. In fact, a norm one approximate unit for J may be constructed. To prove the converse, assume J is an ideal in A with bounded approximate unit e_N. It must be established that $J \equiv I_E$ where I_E is an ideal with corresponding ideal set E, and $E \subset T$ with $m(E) = 0$. Let $E = \bigcap_N Z(e_N)$ where $Z(e_N)$ denotes the zero set of e_N. Clearly $J \subset I_E$ where E is as defined above. Since $\bigcap_N Z(e_N) = E$, there exists $f_x \in J$ so that $f_x(x) > 1/2$ for all $x \in E^C$ where E^C denotes the complement of E relative to T. Let O be any open set such that $O \cap E = \emptyset$. Then $O_x = \{y \in T: f_x(y) > 1/2\}$ forms an open cover for $\overline{O}$. Thus, there exists a finite subcover $\{O_{x_1}, \ldots, O_{x_N}\}$. Evidently $\{e_N\}$ converges uniformly on $\left(\bigcup_{i=1}^N O_{x_i}\right) \cap \overline{O} = \overline{O}$. Since the e_N's are uniformly bounded, for each $y \in I_E$, $\|e_N y - y\| \to 0$ as $n \to \infty$. Since J is a closed ideal, $e_N y \in J$ for all N and thus $y \in J$. Hence $I_E \subset J$. Now E cannot contain an interior point. For suppose $z_0 \in \text{int } D$, and let $f_0 \in J$ have a zero of minimal order n_0 at z_0. By the Cohen factorization theorem [2], $f_0 = g_0 h_0$ where $g_0, h_0 \in J$. Thus, f_0 has a zero of order at least $2n_0$ at z_0 which is impossible. Finally since each e_N must in particular be in $H_1(T)$, then by [11, p. 52], E must have measure zero. This completes the argument.

Corollary 3.1. *The M-ideals of* A *are exactly the two-sided ideals with norm one approximate identity.*

REFERENCES

1. Alfsen, E. M. and E. Effros, Structure in real Banach spaces, Ann. of Math., 96 (1972), 98-173.
2. Bonsall, F. F. and J. Duncan, Complete normed algebras, Ergebnisse der Math. 80, Springer-Verlag, New York, 1973.

3. de Boor, C., On "best" interpolation, J. Approximation Theory, 15 (1976), 28-42.

4. Browder, A., Introduction to Function Algebras, W. A. Benjamin, Inc., New York, 1969.

5. Chui, C. K., P. W. Smith and J. D. Ward, Approximation by minimum norm interpolants in the disc algebra, J. Approximation Theory, to appear.

6. ________, Favard's solution is the limit of $W^{k,p}$-splines, Trans. of AMS, 220 (1976), 299-305.

7. Fakhoury, H., Projections de meilleure approximation continues dans certains espaces de Banach, C. R. Acad. Sci., 276 (1973), 45-48.

8. Fisher, S. D., and J. W. Jerome, Minimum norm extremals in function spaces, Lecture notes in mathematics # 479, Springer-Verlag, New York, 1975.

9. Golomb, M., and H. F. Weinberger, Optimal approximation and error bounds, in On Numerical Approximation, (R. E. Langer, ed.) Univ. of Wisconsin Press, Madison, Wis. 1959, pp. 117-190.

10. Hirsberg, B., M-ideals in complex function spaces and algebras, Israel J. Math., 12 (1972), 133-146.

11. Hoffman, K., Banach Spaces of Analytic Functions. Modern Analysis Series, Prentice-Hall, Inc., N. J., 1962.

12. Holmes, R., B. Scranton, and J. Ward, Approximation from the space of compact operators and other M-ideals, Duke Math. J., 42 (1975), 259-269.

13. Macintyre, A. J., and W. W. Rogosinski, Extremum problems in the theory of analytic functions, Acta Math., 82 (1950), 275-325.

14. Rogosinski, W. W., and H. S. Shapiro, On certain extremum problems for analytic functions, Acta Math. 90 (1953), 287-318.

15. Smith, R. R., An addendum to "M-ideal structure in Banach algebras", J. Functional Analysis, (to appear).

Numerical treatment of some boundary value problems with hidden singularities

Lothar Collatz
Institute for Applied Mathematics
University of Hamburg
Germany

SUMMARY

Many numerical calculations deal with singularities which are often clear but also often hidden; sometimes the location of the singularity is unknown, sometimes even the type; the success of the numerical calculation depends strongly on the careful treatment of the singularities. This is illustrated in the following survey lecture by several examples from the applications.

1. INTRODUCTION

It is for numerical calculation very improtant to take care of singularities (abbreviated by "Sing") of the considered problem, otherwise the numerical results will be unsatisfactory. This will be illustrated here in the case of boundary value problems, expecially with partial differential equations. There are different types of singularities of the wanted solution:

Case I: The location and the type of the Sing. are known a priori

Ia) Singularities of functions occuring in the differential equation (Analytic Sing)

Copyright © 1980 by Academic Press, Inc.
All rights of reproduction in any form reserved.
ISBN: 0-12-171050-5

b) geometric singularities of the domain, corners, unbounded domains a.o., compare f.i. Whiteman [79].

Case II: "Hidden" singularities. The location or the type of the Sing. or both are not known immediately;

IIa) the location P of the singularity is unknown

α) P lies in the domain B, in which the problem is stated,

β) P lies outside of the domain B.

IIb) the type of the singularity at the place P is unknown; Here we have again the cases α), β) as in a).

2. UNKNOWN LOCATION OF THE SINGULARITIES

Movable Sing. have been studied extensively for ordinary differential equations in the complex plane. Recently H. Werner [79] has made numerical tests with aid of rational Approximation. For instance: for the initial value problem

$$\frac{dy}{dx} = 1+xy^2, \quad y(o)=1 \tag{2.1}$$

there is a pole at an unknown abscissa $x=x_o$. Here it is better, to approximate $y(x)$ by

$$y(x) \approx w(x) = \sum_{v=o}^{p} a_v x^v + \frac{b}{x-c}$$

with unknown constants a_v, b, c as to approximate $y(x)$ by polynomials only. $x=c$ is the unknown location of the pole.

Examples for the other cases are given below; often occur different types of singularities in the same problem.

The phenomenon of "hidden" poles is also very well known since long time in other areas, for instance the function $f(x)=\frac{1}{1+x^2+x^{1o}}$ is infinitely often differentiable for all real x, Fig. 1, but the power series $f(x)= \sum_{v=o}^{\infty} a_v x^v$ converges only for $|x| \leq x_o$ with $x_o < \infty$, because there are poles of $f(x)$ in the complex plane. Therefore the fact, whether a singularity is

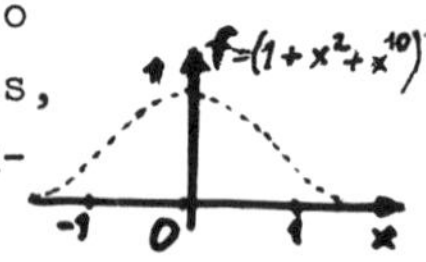

Fig.1

"hidden", depends of the knowledge of the viewer. Other hidden singularities can occur in the numerical quadrature, cubature a.o.
One has also to mention the wide area of research on moving free boundary value problems, compare Baiocchi [72], Hoffmann [78], Wilson-Solomon-Boggs [78], Collatz [78], Ockendon [78] a.o.

3. SINGULARITIES AT CORNERS

Examples of this type were treated numerically f.i. by Whiteman [79]. The ideal flow of a liquid around a corner P in the x-y-plane, fig. 2, was considered by Collatz,

The ideal flow through a channel with the coasts, Fig. 2, $\pm\, y = \frac{2+x^2}{3+x^2}$ is numerically calculated, compare Collatz, 1979, Proc. Symp. Bad Honnef, to appear. (Other examples Collatz [78])

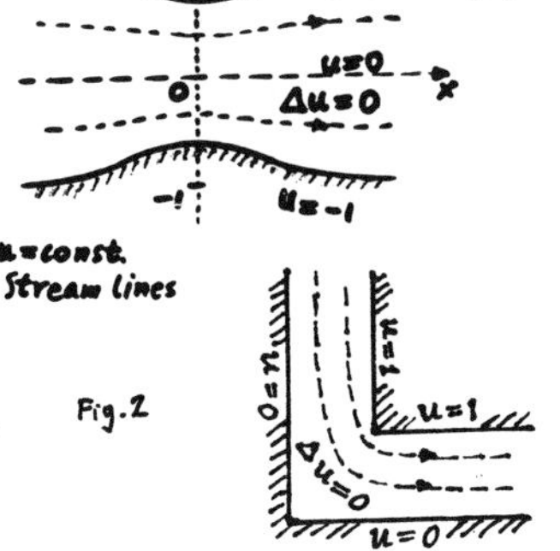

Fig.2

But there are at corners also harmless singularities, which may be neglected. For instance we ask for the temperature u(x,t) in a wall (space coordinate x, time t):

$$Lu = \frac{\partial u}{\partial t} - \frac{\partial^2 u}{\partial x^2} = o \tag{3.1}$$

$$\text{in } B = \{(x,t),\ o<x<1,\ t>o\}$$

with the mixed boundary conditions, fig. 3.

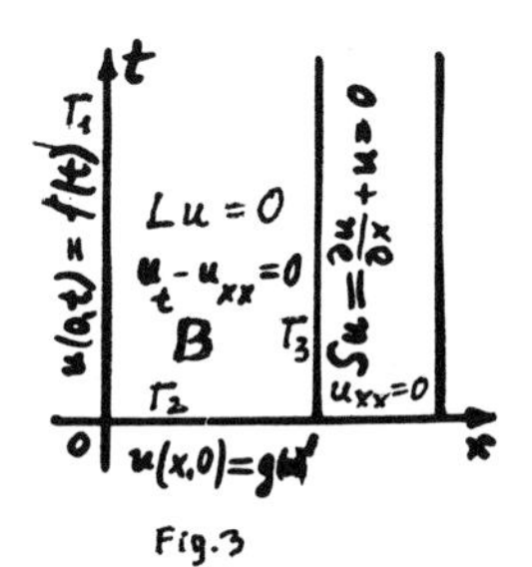

Fig.3

$$(3.2)\quad \begin{cases} u(o,t) = f(t) \text{ on } \Gamma_1 \ (x=o,\ t>o) \\ u(x,o) = g(x) \text{ on } \Gamma_2 \ (o<x<1,\ t=o), \\ Su = \frac{\partial u}{\partial x} + u = o \text{ on } \Gamma_3 \ (x=1,\ t>o) \end{cases}$$

The following results are calculated for $g(x)=2-x$, $f(t)=2+t$. The differential equation (3.1) is not satisfied in the origin, because there is

$$\frac{\partial u}{\partial t} = 1, \quad \frac{\partial^2 u}{\partial x^2} = o.$$

But we do not look on this singularity and take as approximate solution

$$(3.3)\qquad u(x,y) \approx v(x,y) = \sum_{i=1}^{p} a_i v_i(x,t) + \sum_{j=1}^{q} b_j w_j(x,t)$$

with $Lv_i = Lw_j = o$ for all i,j.

The v_i are the well known polynomials satisfying the homogeneous heat-conduction-equation.

ν	1	2	3	4	5	6	7	..
v_ν	1	x	x^2+2t	x^3+6xt	$x^4+12x^2t+12t^2$	$x^5+20x^3t+60xt^2$	$x^6+30x^4t+180x^2t^2+120t^3$	..

j					...
w_j	$e^{-t}\cos x$	$e^{-t}\sin x$	$e^{-4t}\cos(2x)$	$e^{-4t}\sin(2x)$	...

We use the monotonicity (compare f.i. Redheffer [67], Collatz [68], Walter [7o], Meyn-Werner [79], a.o.) of vector $T\phi = \{L\phi \text{ in } B,\ \phi \text{ on } \Gamma_1 \text{ and } \Gamma_2,\ S\phi \text{ on } \Gamma_3\}$

$$(3.4)\qquad T\phi \leq T\psi \quad \text{implies} \quad \phi \leq \psi \text{ in } B$$

The sign $\leq$ means the classical ordering of real numbers, for every component of the vector and pointwise in B and on Γ_k $(k=1,2,3)$. In $S\phi$ we have to use the outer normal n; on Γ_3 we have $\frac{\partial u}{\partial n} = \frac{\partial u}{\partial x}$.

In a wellknown manner we can get lower bounds $\underline{v}$ and upper bound $\bar{v}$ for u, using onesided Chebychev-A(see f.i. Cheney [66], Meinardus [67], Hoffmann [69], Collatz [67]-[71], Collatz-Wetterling [75], a.o.) and simultaneous (see E. Bredendiek [7o], [76]) approximation.

The tables gives values for ε in the sense, that

$$\operatorname*{Max}_{B}\ (\bar{v}-\underline{v})=2\varepsilon, \qquad \underline{v} \le u \le \bar{v}.$$

Table for ε

p \ q	0	1	2	3	4
1	1.5	1.144 9	0.827 6	0.436 9	0.436 8
2	0.5	0.134 5	0.132 4	0.118 4	0.043 03
3	0.25	0.126 4	0.051 93	0.024 23	0.011 55
4	0.074 97	0.037 49	0.027 76	0.024 02	0.008 274
..	..	..	..	..	..
6	0.036 06	0.013 04	0.009778	0.007 528	0.006 653
..	..	..	..	..	..
8	0.012 01	0.007 796	0.006 913	0.006 164	0.004 903

4. GEOMETRIC SINGULARITY OF TYPE I b)

The classical torsion-problem for the torsion of a beam with the simply connected cross-section B in a x-y-plane asks for a function $u(x,y)$ with

$$(4.1)\quad \Delta u = \frac{\partial^2 u}{\partial x^2} + \frac{\partial^2 u}{\partial y^2} = o \text{ in } B$$

and the boundary condition

$$(4.2)\quad u = r^2 = x^2+y^2 \quad \text{on } \partial B.$$

B may be the rectangle $|x|<2$, $|y|<1$, fig. 4.

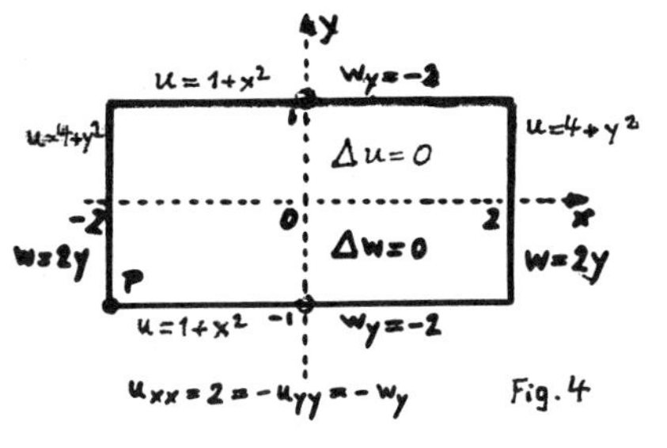

Fig. 4

The tensions can be calculated from the partial derivatives of u; we expect the maximal tension at the points $x=o$, $y=\pm 1$ and ask for the value of $w=\frac{\partial u}{\partial y}$ at these points.

We have for w the mixed boundary value problem

$$(4.3)\quad \begin{cases} \Delta w = o \text{ in } B \\ w = 2y \text{ for } |x| = 2 \\ \dfrac{\partial w}{\partial y} = -2 \text{ for } |y| = 1. \end{cases}$$

Again we transform the problem into another problem for a function $z(x,y)$:

$$w = -2y + 4z$$

$$(4.4) \quad \begin{cases} \Delta z = o \text{ in } B \\ z = y \text{ for } |x| = 2 \\ \dfrac{\partial z}{\partial y} = o \quad \text{for } |y| = 1 \end{cases}$$

z has singularities at the 4 corners of the rectangle, because $\zeta = \frac{\partial z}{\partial y}$ is jumping there, for instance in the point P: $x=-2$, $y=-1$: $\zeta = o$ for $y=-1$ and $\zeta = 1$ for $x=-2$. We introduce new coordinates ξ, η and polarcoordinates r, φ at this corner with $\varphi = \arctan \frac{\eta}{\xi}$, $r^2 = \xi^2 + \eta^2$, fig. 5.

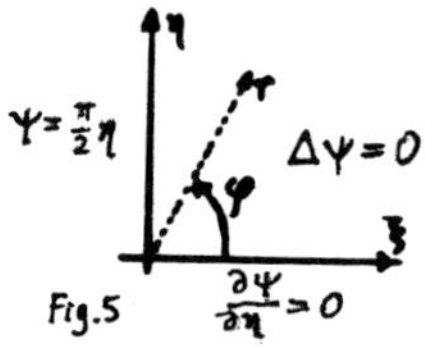

Then the following type of a singularity, which is used not so often:

$$(4.5) \qquad \psi = \eta \arctan \frac{\eta}{\xi} - \xi \ln \sqrt{\xi^2 + \eta^2}$$

Fig.5

is useful for our purpose. The function $\psi^* = \frac{2}{\pi}\psi$ has the same singularity at P as z.

5. TYPE AND LOCATION OF THE SINGULARITY ARE NOT KNOWN A PRIORI (Type IIb)

A tube has the cross-section B of an annulus in a x-y-plane

$$B = \{(x,y),\ 1 < r^2 = x^2 + y^2 < 2\}$$

and the unknown temperature $u(x,y)$. We have at the inner zylinder the prescribed temperature

$$u = a \quad \text{for} \quad r = 1. \quad (\text{boundary } \Gamma_1).$$

In B we assume $\Delta u = o$. The tube may lie half in the earth (boundary part Γ_3) of temperature 1 and half in the air (boundary Γ_2) of temperature 0, fig. 6:

$$\begin{cases} u = o & \text{on } \Gamma_2 \ (r = 2,\ x > o) \\ u = 1 & \text{on } \Gamma_3 \ (r = 2,\ x < o) \end{cases}$$

Here we have known singularities at the points P_1, P_2 ($x=o$, $y=\mp 2$) of the form φ_1 resp. φ_2 (polarcoordinates at P_1, resp. P_2, Fig. 6). We put

$$(5.1) \qquad S(x,y) = \frac{1}{\pi}(\varphi_1 + \varphi_2) = \begin{cases} 1/2 & \text{on } \Gamma_2 \\ 3/2 & \text{on } \Gamma_3 \end{cases} .$$

Subtracting $(\varphi_1+\varphi_2)/\pi$ from u we get a function z(x,y), which is regular at P_1, P_2; but we can approximate z(x,y) by polynomials only very bad. We guess that u(x,y) has a logarithmic singularity and probably also poles inside the circle r<1. Using only the logarithmic singularity and no poles, gives numerical results which are unsatisfactory. Therefore we introduce

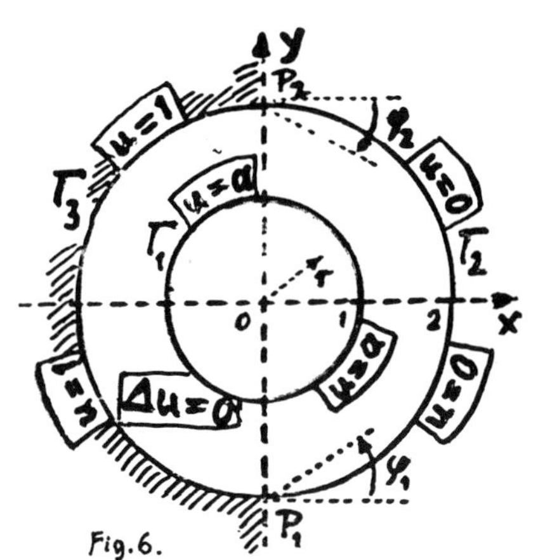

Fig.6.

$$(5.2) \qquad v(x,y) = S(x,y) + a - 1 + \left(\frac{1}{2} - a\right)\frac{\ln r}{\ln 2}$$

and we have for $w=u-\hat{v}$ the boundary value problem

$$\begin{cases} \Delta w = 0 \text{ in } B \\ w = \begin{cases} 0 \text{ on } \Gamma_2 \text{ and } \Gamma_3, & r=2 \\ 1-S \text{ on } \Gamma_1, & r=1 \end{cases} \end{cases}$$

We take as approximate solution

$$(5.4) \qquad w(x,y) \approx v(x,y) = \sum_{\nu=0}^{p} a_\nu v_\nu(x,y)$$

and as v_ν (x,y) the antisymmetric harmonic functions with $v_\nu(x,y)=-v_\nu(-x,y)$,

$$(5.5) \qquad v_0 = \frac{x-\gamma}{(x-\gamma)^2+y^2} + \frac{x+\gamma}{(x+\gamma)^2+y^2}, \quad v_1 = x,$$

$$v_2 = x^3 - 3xy^2, \quad v_\nu = \text{Re}\,(x+iy)^{2\nu-1} \quad (\nu=1,\dots,p)$$

This takes care of 3 types of singularities: the angles φ_j, the logarithms and poles.

Furthermore we have taken as approximate solutions:

$$w \approx \hat{v}(x,y) = \sum_{\nu=0}^{p} \hat{a}_\nu \hat{v}_\nu(x,y)$$

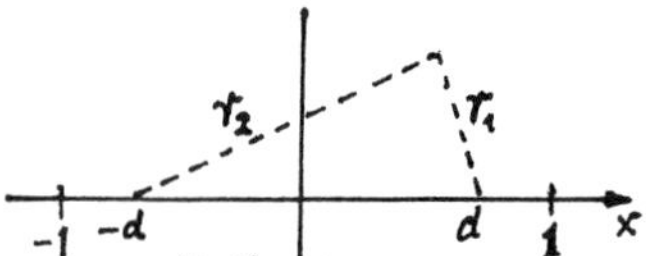

$$\hat{v}_o = \frac{x}{x^2+y^2},\quad \hat{v}_1 = v_1 = x,\quad \hat{v}_2 = v_2 = x^3 - 3xy^2$$

Fig.7

$\hat{v}_k = \ln \frac{r_2}{r_1}$ with w_1, w_2 as distances from the points $x = \pm d_k$,

$y = o$, fig. 7, $d_k = \frac{(k-2)}{1o}$, $k \geq 3$.

The following table gives some numerical results

for δ, where

$$|\varepsilon| = |w - v| \leq \delta \quad \text{in } B$$

Table for δ

P	using v_o with $\gamma=0$	with $\hat{v}_\nu$
o	o.1oo1	o.1oo1
1	o.o279	o.o2713
2	o.o242	o.o2419
3	o.o242	o.oo4268
4	o.o242	o.ooo8839
5	o.o238	o.ooo231o

One sees, that it is important, to use a good trial. Perhaps it is interesting, that even with the simple function

$$w \approx v^*(x,y) = a_o^* x + a_1^* \frac{x}{x^2+y^2} + a_3^* \left[\frac{x - o.1}{(x-o.1)^2+y^2} + \frac{x+o.1}{(x+o.1)^2+y^2} \right]$$

one gets the result

$$|\varepsilon| = |w - v^*| \leq o.oo4374$$

for $a_1^* = -o.1o6748$, $a_2^* = 2\ 1.575559$, $a_3^* = -1.363o21$

I thank Mr. Uwe Grothkopf for numerical calculations.

Lothar Collatz
Institut für Angewandte Mathematik
Bundesstrasse 55
2ooo Hamburg 13
Germany

REFERENCES

1. Baiocchi, C. [72] Su un problema frontiera libera connesso a questioni di idraulica; Ann. Pura Appl. (4) 92 (1972) 1o7-127.
2. Bredendiek, E. [7o] Charakterisierung und Eindeutigkeit bei Simultanapproximationen. Z.Ange.Math.Mech. 5o, 4o3-41o (197o).
3 Bredendiek, E. - L. Collatz [76] Simultan Approximation bei Randwertaufgaben, Internat.Ser.Num.Math. 3o (1976), 147-174.
4. Cheney, E-W. [66] Introduction to Approximation Theory, New York (1966) 259 S.
5. Collatz, L. [56] Approsimation von Funktionen bei einer und bei mehreren unabhängigen Veränderlichen, Z.Angew. Math.Mech. 36 (1956) 198-211.
6. Collatz, L. [67] Nichtlineare Optimierungsaufgaben Wiss.Z.Hochsch.Archit.Bauwesen Weimar 1967, 33-44.
7. Collatz, L. [68] Funktional Analysis und Numerische Mathematik, Springer 1968, 371 S.
8. Collatz, L. [69] Nichtlineare Approximationen bei Randwertaufgaben, V. IKM. Weimar, 169-182 (1969).
9. Collatz, L. [71] Some applications of Functional Analysis to analysis particularly to nonlinear integral equations, Proc. Sympos. Nonlinear functional analysis and applications, edited by L.B. Rall, Academic Press 1971,S.1-43.

1o. Collatz, L. [78] Application of approximation to some singular boundary value problems, Proc. Conference Num. Anal. Dundee, Springer Ledt.Not.Math. Bd. 63o (1978) 41-5o.

11. Collatz, L. - W. Wetterling [75] Optimization problems, Springer 1975, 356 p.

12. Hoffmann, K.H. [69] Zur Theorie der nichtlinearen Tschebyscheff Approximation mit Nebenbedingungen, Numer.Math.14, 24-41 (1969).

13. Hoffmann, K.H. [78] Monotonie bei nichtlinearen Stefan Problemen, Internat. Ser.Num.Math. vol. 39 (1978) 162-19o.

14. Meinardus, G. [67] Approximation of functions, Theory and numerical methods, Springer 1967, 198 p.

15. Meyn, K.H. - B. Werner [79] Maximum and Monotonicity Principles for elliptic boundary value problems in partitioned domains, to appear.

16. Ockendon, J.R. [78] Numerical and Analytic Solutions of Moving Boundary Problems, (In the book of Wilson a.o., see below, p. 129-145.

17. Redheffer, R.M. [67] Differentialungleichungen unter schwachen Voraussetzungen, Abhandl. Math.Sem.Univ. Hamburg, 31 (1967) 33-5o.

18. Walter, W. [7o] Differential and Integral Inequalities Springer 197o, 352 S.

19. Werner, H. [79] Extrapolationsmethoden zur Bestimmung der beweglichen Singularitäten von Lösungen gewöhnlicher Differentialgleichungen, Intern.Ser.Num.Math. 49 (1979) 159-176.

2o. Whiteman, J.R. [79] Two dimensional biharmonic problems. Proc. Manchester/Liverpool Sommer school on Numerical Solut. of P.D.E's. Oxford Univ. Press 1979, to appear.

ON ABSTRACT BOOLEAN INTERPOLATION

F.J. Delvos

Fachbereich Mathematik
University of Siegen
Siegen, W. Germany

H. Posdorf

Rechenzentrum
University of Bochum
Bochum, W. Germany

I. INTRODUCTION

Boolean methods in multivariate approximation were first introduced by GORDON [9]. Since then Boolean sum interpolation has been applied successfully in surface fitting and finite element analysis.
It is the purpose of this paper to discuss some features of Boolean interpolation in the general setting of linear spaces (CHENEY-GORDON [3], DAVIS [4]).

II. THE INTERPOLATION PROBLEM

Let V be a (real) vector space (of infinite dimension). Suppose that

$$L_{ij} \in V^* \quad (i,j=o,1,2,\ldots)$$

is a set of linear independent functionals on V. Let

$$V_{m,n} \subset V \quad (m,n=o,1,2,\ldots)$$

be a family of vector subspaces such that

$$\dim(V_{m,n}) = (m+1)(n+1).$$

Furthermore, we assume that the system

$$(V_{m,n}\ ;\ \{L_{ij}:\ i=o,\ldots,m\ ;\ j=o,\ldots,n\})$$

Copyright © 1980 by Academic Press, Inc.
All rights of reproduction in any form reserved.
ISBN: 0-12-171050-5

possesses the interpolation property in the sense of DAVIS [4]: For any set of real numbers

$$\{a_{ij}: i=o,\dots,m \; ; \; j=o,\dots,n\}$$

there is a unique element $f\in V_{m,n}$ such that

$$L_{ij}(f)=a_{ij} \quad (i=o,\dots,m \; ; \; j=o,\dots,n).$$

In particular, there is a unique $f_{ij}^{mn}\in V_{m,n}$ satisfying

$$L_{kl}(f_{ij}^{mn})=\delta_{ik}\delta_{jl} \quad (i,k=o,\dots,m \; ; \; j,l=o,\dots,n). \tag{1}$$

$\{f_{ij}^{mn}: i\leq m, j\leq n\}$ is the dual basis of $\{L_{ij}: i\leq m, j\leq n\}$ in $V_{m,n}$.

The linear operator $P_{m,n}\in L(V)$ defined by

$$P_{m,n}[f]=\sum_{i\leq m}\sum_{j\leq n} L_{ij}(f)f_{ij}^{mn}, \quad (f\in V) \tag{2}$$

is the interpolation projector associated with the interpolatory system $(V_{m,n} \; ; \; \{L_{ij}: i\leq m, j\leq n\})$.

The object of this paper is to show that the system

$$(\sum_{m+n=t} V_{m,n} \; ; \; \{L_{ij}: i+j\leq t\}) \tag{3}$$

possesses the interpolation property when the relations

$$P_{m,n}P_{r,s}=P_{r,s}P_{m,n}=P_{u,v}$$
$$(u=\inf\{m,r\}, \; v=\inf\{n,s\}, \; m,n,r,s=o,1,\dots) \tag{4}$$

are satisfied.

Our method of proof is Boolean sum interpolation:

Let

$$\mathcal{P}=\{P_j: j\in J\}\subset L(V)$$

be a set of commuting projectors:

$$P_j^2=P_j \; , \quad P_{j_1}P_{j_2}=P_{j_2}P_{j_1} \; , \quad (j,j_1,j_2\in J).$$

Then the set of projectors

$$K(\mathcal{P})=\{P\in L(V): P^2=P, \; PP_j=P_jP \; (j\in J)\}$$

contains $\mathcal{P}$: $\mathcal{P}\subset K(\mathcal{P})$.

We introduce the subset $B(\mathcal{P})$ of commuting projectors of $K(\mathcal{P})$:

$$B(\mathcal{P})=\{Q\in K(\mathcal{P}): QP=PQ, \; P\in K(\mathcal{P})\}. \tag{5}$$

It can be shown that $B(\mathcal{P})$ is a Boolean algebra with respect to the order relation

$$Q_1\leq Q_2 \Leftrightarrow Q_2Q_1=Q_1 \quad (Q_1,Q_2\in B(\mathcal{P})).$$

The following relations hold:

$$Q_1,Q_2 \in \mathcal{B}(P) \Rightarrow Q_1Q_2 \in \mathcal{B}(P),$$
$$Q_1,Q_2 \in \mathcal{B}(P) \Rightarrow Q_1 \oplus Q_2 = Q_1+Q_2-Q_1Q_2 \in \mathcal{B}(P),$$
$$P \subset \mathcal{B}(P).$$

Since the elements of $\mathcal{B}(P)$ commute we have

$$\mathrm{Im}(Q_1 \oplus Q_2) = \mathrm{Im}(Q_1)+\mathrm{Im}(Q_2). \tag{6}$$

The Boolean sum of $Q_1,\ldots,Q_k \in \mathcal{B}(P)$ is defined inductively:

$$Q_1 \oplus \ldots \oplus Q_k = (Q_1 \oplus \ldots \oplus Q_{k-1}) \oplus Q_k \in \mathcal{B}(P), \quad (Q_1,\ldots,Q_k \in \mathcal{B}(P)).$$

For our interpolation problem (3) we consider the set of commuting projectors ($t \in \mathbb{Z}_+$)

$$P = \{P_{o,t},\ P_{1,t-1},\ldots,\ P_{t,o}\}.$$

Then

$$B_t := P_{o,t} \oplus \ldots \oplus P_{t,o} \in \mathcal{B}(P). \tag{7}$$

It follows from (6) that

$$\mathrm{Im}(B_t) = \sum_{m+n=t} \mathrm{Im}(P_{m,n}) = \sum_{m+n=t} V_{m,n} \ . \tag{8}$$

Furthermore, we have

LEMMA 1: For any $f \in V$ the following relation is valid:

$$L_{ij}(B_t[f]) = L_{ij}(f) \ , \quad (i+j \leq t). \tag{9}$$

Proof: The construction of $P_{m,n}$ yields

$$L_{ij}(f) = L_{ij}(P_{m,n}[f]) \ , \quad (i \leq m,\ j \leq n,\ m+n=t). \tag{1o}$$

Since

$$P_{m,n}B_t = B_tP_{m,n} = P_{m,n} \ , \quad (m+n=t)$$

we can conclude from (1o)

$$L_{ij}(B_t[f]) = L_{ij}(P_{m,n}B_t[f]) = L_{ij}(P_{m,n}[f]) = L_{ij}(f)$$

which proves (9).¤

Let $\{f_{ij}^{tt}: i \leq t,\ j \leq t\}$ be the dual basis of $\{L_{ij}: i \leq t,\ j \leq t\}$ in $V_{t,t}$.
Then we put

$$f_{ij}^{t} := B_t[f_{ij}^{tt}] \ , \quad (i \leq t,\ j \leq t). \tag{11}$$

It follows from Lemma 1 that

$$L_{kl}(f^t_{ij})=\delta_{ik}\delta_{jl}\ , \quad (i+j\leqslant t,\ k+l\leqslant t). \tag{12}$$

We will show now that $\{f^t_{ij}: i+j\leqslant t\}$ is the dual basis of $\{L_{ij}: i+j\leqslant t\}$ in $\sum\limits_{m+n=t} V_{m,n}$.

LEMMA 2: Suppose that

$$S_t=P_{o,t}+P_{1,t-1}+\ldots+P_{t,o}\ , \quad (t\in\mathbb{Z}_+).$$

Then

$$B_t=S_t-S_{t-1}.$$

Proof: Using the relation (4) we obtain

$$\begin{aligned}B_t&=(P_{o,t}\oplus\ldots\oplus P_{t-1,1})\oplus P_{t,o}\\ &=-(P_{o,t}\oplus\ldots\oplus P_{t-1,1})P_{t,o}+P_{t,o}+P_{o,t}\oplus\ldots\oplus P_{t-1,1}\\ &=-(P_{o,t}P_{t,o})\oplus\ldots\oplus(P_{t-1,1}P_{t,o})+P_{t,o}+P_{o,t}\oplus\ldots\oplus P_{t-1,1}\\ &=-P_{o,o}\oplus\ldots\oplus P_{t-1,o}+P_{t,o}+P_{o,t}\oplus\ldots\oplus P_{t-1,1}\\ &=-P_{t-1,o}+P_{t,o}+P_{o,t}\oplus\ldots\oplus P_{t-1,1}\\ &\ \ \vdots\\ &=-S_{t-1}+S_t.\ \square\end{aligned}$$

An immediate consequence of Lemma 2 is

$$f^t_{ij}=B_t[f^{tt}_{ij}]=0\ , \quad (i\leqslant t,\ j\leqslant t,\ i+j>t). \tag{13}$$

We are now able to state our main result.

THEOREM 1: For any set of real numbers

$\{a_{ij}: i+j\leqslant t\}$,

$f=\sum\limits_{i+j\leqslant t} a_{ij}f^t_{ij}$ is the unique element in $\sum\limits_{m+n=t} V_{m,n}$

such that

$$L_{ij}(f)=a_{ij}\ , \quad (i+j\leqslant t);$$

i.e. the system

$$\Big(\sum_{m+n=t} V_{m,n}\ ;\ \{L_{ij}: i+j\leqslant t\}\Big)$$

is interpolatory.

Proof: It follows from (12) and (13) that f possesses the interpolation property. To show the uniqueness of f it is sufficient to prove that

$$\{f_{ij}^{t}: i+j\leq t\} \text{ generates } \sum_{m+n=t} V_{m,n}.$$

Since

$$V_{m,n}=\langle\{f_{ij}^{mn}: i\leq m,\ j\leq n\}\rangle$$

we have

$$\sum_{m+n=t} V_{m,n}=\langle\{f_{ij}^{mn}: i\leq m,\ j\leq n,\ m+n=t\}\rangle.$$

It follows from (4) that $V_{m,n}\subset V_{t,t}$ $(m+n=t)$. Thus, we have

$$f_{ij}^{mn}=\sum_{k\leq t}\ \sum_{l\leq t} c_{kl} f_{kl}^{tt}\ , \quad (m+n=t).$$

On the other hand $f_{ij}^{mn}=B_t[\,f_{ij}^{mn}]$ yields

$$f_{ij}^{mn}=\sum_{k+l\leq t} c_{kl} f_{kl}^{t}.$$

This proves

$$\sum_{m+n=t} V_{m,n}=\langle\{f_{kl}^{t}: k+l\leq t\}\rangle.\square$$

Lemma 2 yields an explicit expression for the elements f_{ij}^{t} of the dual basis.

THEOREM 2: For any $i+j\leq t$

$$f_{ij}^{t}=\sum_{\substack{m+n=t\\ m\geq i\\ n\geq j}} f_{ij}^{mn} - \sum_{\substack{m+n=t-1\\ m\geq i\\ n\geq j}} f_{ij}^{mn}\ .$$

REFERENCES

1. Barnhill, R.E., "Blending Function Interpolation: A Survey and Some New Results". ISNM 3o, 43-89 (1976).
2. Baszenski, G., Delvos, F.J., Posdorf, H., "Boolean Methods in Bivariate Reduced Hermite Interpolation". ISNM 51, 11-29 (1979).
3. Cheney, E.W., Gordon, W.J., "Bivariate and Multivariate Interpolation with Noncommutative Projectors". ISNM 4o, 381-387 (1977).
4. Davis, Ph.J., Interpolation and Approximation. Blaisdell 1963.
5. Delvos, F.J., Posdorf, H., "N-th Order Blending". Lecture Notes in Mathematics 571, 53-64 (1977).
6. Delvos, F.J., Posdorf, H., "A Representation Formula for Reduced Hermite Interpolation". ISNM 42, 124-137 (1978).
7. Delvos, F.J., Posdorf, H., "Boolesche zweidimensionale Lagrange Interpolation". Computing 22, 311-323 (1979).

8. Delvos, F.J., Posdorf, H., Schempp, W., "Serendipity Type Bivariate Interpolation". Multivariate Approximation, 47-56. Academic Press 1978.
9. Gordon, W.J., "Distributive Lattices and the Approximation of Multivariate Functions". Proc.Symp. Madison, Wisc. 1969, 223-277.
1o. Gordon, W.J., "Blending Function Methods of Bivariate and Multivariate Interpolation and Approximation". SIAM, J.Num.Anal. 8, 158-177 (1971).
11. Gordon, W.J., Hall, C.A., "Transfinite Element Methods: Blending-Function Interpolation over Arbitrary Curved Element Domains". Num.Math. 21, 1o9-129 (1973).
12. Lancaster, P., Watkins, D.S., "Interpolation in the Plane and Rectangular Finite Elements". ISNM 31, 125-145 (1976).

WHEN DOES THE METRIC PROJECTION ADMIT A CONTINUOUS SELECTION?

Frank Deutsch

Department of Mathematics
The Pennsylvania State University
University Park, Pennsylvania

Petar Kenderov

Mathematics Institute
Sofia, Bulgaria

Let M be a proximinal subset of the normed linear space X with metric projection P_M. Suppose that either M is a one dimensional subspace or that the set $\{x \in X \mid P_M(x)$ is a singleton$\}$ is dense in X. Then P_M has a continuous selection if and only if P_M is "2-continuous". Whether this characterization is valid for arbitrary proximinal sets is not known.

I. INTRODUCTION

Let M be a nonempty subset of the normed linear space X and for each $x \in X$ let $P_M(x)$ denote the (possibly empty) set of all best approximations to x from M. That is,

$$P_M(x) = \{y \in M \mid \|x-y\| = d(x,M)\},$$

where $d(x,M) = \inf\{\|x-y\| \mid y \in M\}$. The mapping $P_M : X \to 2^M$ thus defined is called the <u>metric projection</u> onto M. We call

Copyright © 1980 by Academic Press, Inc.
All rights of reproduction in any form reserved.
ISBN: 0-12-171050-5

M proximinal if $P_M(x) \neq \phi$ for every $x \in X$. A continuous selection for P_M is a continuous function $s : X \to M$ such that $s(x) \in P_M(x)$ for every $x \in X$.

In the particular case when M is a finite dimensional subspace of X, there has been substantial interest in determining precise conditions on M and/or P_M which guarantee that P_M admits a continuous selection (see [1]-[3], [6]-[16]). Perhaps the best known sufficient condition is the following (special case of the) celebrated selection theorem of Michael [8]:

1 THEOREM. If M is a finite dimensional subspace of X and P_M is lower semicontinuous (ℓ.s.c.) on X, then P_M has a continuous selection.

Recall that P_M is ℓ.s.c. at a point x_0 if for each open set V with $P_M(x_0) \cap V \neq \phi$, there is a neighborhood U of x_0 such that $P_M(x) \cap V \neq \phi$ for every $x \in U$. Spaces X in which the metric projection onto every finite dimensional subspace of X is ℓ.s.c were characterized geometrically by Brown [1]. These spaces include all strictly convex spaces, polyhedral spaces (e.g. $\ell_1^{(n)}$ and $\ell_\infty^{(n)}$), and $c_0(\Gamma)$ for any set Γ. (They do not, however, include the spaces $C(T)$ or $L_1(\mu)$ in general.)

It is not hard to give examples which show that lower semicontinuity of P_M is not necessary in order that P_M admit a continuous selection. Thus it is natural to ask whether there is any kind of "continuity" property of P_M which is equivalent to the existence of a continuous selection for P_M. In the next section we observe that the answer is affirmative if either M is a one dimensional subspace or if the set $\{x \in X \mid P_M(x)$ is a singleton$\}$ is dense in X. Detailed proofs, along with related material, will appear elsewhere.

II. MAIN RESULTS

3 DEFINITION. Let M be a proximinal subset of the normed linear space X and $x \in X$. P_M is said to be <u>n-continuous</u> at x if for every $\varepsilon > 0$ there exists a neighborhood U of x such that

$$\bigcap_{i=1}^{n} B_\varepsilon(P_M(x_i)) \neq \phi$$

for each choice of n points $x_i \in U$ $(i=1,\ldots,n)$. (Here $B_\varepsilon(A)$ denotes the closed ε-neighborhood of A in M : $B_\varepsilon(A) = [A + \{y \in X \mid \|y\| \leq \varepsilon\}] \cap M$.)

Obviously, every metric projection is 1-continuous, and if P_M is n-continuous, then P_M is also k-continuous for every $k \leq n$. Also, it is not difficult to show that if either P_M is lower semicontinuous or P_M admits a continuous selection, then P_M is n-continuous for every n. Finally, if P_M is singleton-valued, then n-continuity and ordinary continuity are the same. Of special interest to us is 2-continuity which may be equivalently defined as: P_M is 2-continuous at x if and only if for each $\varepsilon > 0$ there exists a neighborhood U of x such that for each x_1, x_2 in U,

$$\inf \{\|y_1 - y_2\| \mid y_i \in P_M(x_i) \quad (i=1,2)\} < \varepsilon.$$

Our first main result is:

4 THEOREM. Let M be a proximinal subset of X and suppose that the set of points having unique best approximations,

$$U(M) \equiv \{x \in X \mid P_M(x) \text{ is a singleton}\},$$

is dense in X. Then P_M has a continuous selection if and only if P_M is 2-continuous. Moreover, if P_M has a continuous selection, it is unique.

It is easy to verify that if M is a proximinal subset of a _strictly convex_ space X, then $U(M)$ is dense in X and P_M admits a continuous selection only if M is Chebyshev. Thus, in strictly convex spaces, P_M has a continuous selection if and only if M is Chebyshev and P_M is continuous.

A subset M of X is called _almost Chebyshev_ if $X \setminus U(M)$ is a set of first category in X. Garkavi [4], [5] studied almost Chebyshev subspaces and showed that they exist in abundance. For example, in every separable Banach space [dual space], there exist almost Chebyshev subspaces of every finite dimension [finite codimension]. Since the complement of a first category set in a Banach space is dense, we have:

5 COROLLARY. Let M be an almost Chebyshev subset of a Banach space X. Then P_M has a continuous selection if and only if P_M is 2-continuous. Moreover, if P_M has a continuous selection, it is unique.

A finite dimensional subspace M of $C(T)$ (T compact Hausdorff) is called a _Z-subspace_ [7] if the only element of M which vanishes on an open subset of T is the zero function. As a consequence of the main result of Garkavi [5], Z-subspaces are almost Chebyshev. This fact and Corollary 5 implies:

6 COROLLARY. Let M be a Z-subspace of $C(T)$. Then P_M has a continuous selection if and only if P_M is 2-continuous. If P_M has a continuous selection, it is unique.

The uniqueness statement in this corollary was also proved by Brown [2] without appealing to Garkavi's theorem.

In related work, Brown [3] proved a "Mairhuber-type" theorem when he showed that if $C(T)$ contains a Z-subspace M of dimension at least two and P_M admits a continuous selection, then T is essentially an "interval". Nürnberger [10], building on earlier work in [12] and [16], gave an <u>intrinsic</u> characterization of those (finite dimensional) Z-subspaces of $C[a,b]$ whose metric projections admit continuous selections. More generally, Sommer [15] filled the remaining gap in [10], [12], and [16] by giving an <u>intrinsic</u> characterization of those finite dimensional subspaces of $C[a,b]$ whose metric projections admit continuous selections.

Our second main result is:

7 THEOREM. Let M be a one dimensional subspace of the normed linear space X. Then P_M has a continuous selection if and only if P_M is 2-continuous.

The proof of this result uses machinery similar to that developed by Michael [8] along with applications of Helly's theorem concerning the intersection of convex sets.

Related to this theorem, we note that Lazar, Morris, and Wulbert [7] gave an <u>intrinsic</u> characterization of those one dimensional subspaces of $C(T)$ whose metric projections admit continuous selections. Also, Lazar [6] gave an <u>intrinsic</u> characterization of those one dimensional subspaces of ℓ_1 whose metric projections admit continuous selections.

We do not know whether Theorem 7 is valid for n dimensional subspaces M, when $n > 1$. However, there is some evidence which supports the following

CONJECTURE. Let M be an n dimensional subspace. Then P_M has a continuous selection if (and only if) P_M is $n + 1$-continuous.

Note that Theorem 7 is the statement that the conjecture is true in case $n = 1$.

REFERENCES

1. A. L. Brown, Best n-dimensional approximation to sets of functions, Proc. London Math. Soc., 14 (1964), 577-594.
2. A. L. Brown, On continuous selections for metric projections in spaces of continuous functions, J. Functional Anal., 8 (1971), 431-449.
3. A. L. Brown, An extension to Mairhuber's theorem. On metric projections and discontinuity of multivariate best uniform approximation, 1979 (preprint).
4. A. L. Garkavi, On Čebyšev and almost Čebyšev subspaces, Izv. Akad. Nauk SSSR, Ser. Mat. 28 (1964), 799-818 [Translation in: Amer. Math. Soc. Transl., 96 (1970), 153-175.]
5. A. L. Garkavi, Almost Čebyšev systems of continuous functions, Izv. Vysš. Učebn. Zaved. Matematika, 45 (1965), 36-44. [Translation in: Amer. Math. Soc. Transl., 96 (1970), 177-187.]
6. A. J. Lazar, Spaces of affine continuous functions on simplexes, Trans. Amer. Math. Soc., 134 (1968), 503-525.
7. A. J. Lazar, P. D. Morris, and D. E. Wulbert, Continuous selections for metric projections, J. Functional Anal., 3 (1969), 193-216.
8. E. Michael, Selected selection theorems, Amer. Math. Monthly, 63 (1956), 233-237.
9. G. Nürnberger, Continuous selections for the metric projection and alternation, J. Approximation Theory (to appear).
10. G. Nürnberger, Nonexistence of continuous selections for the metric projection and weak Chebyshev subspaces, 1978 (preprint).
11. G. Nürnberger, Schnitte für die metrische Projektion, J. Approximation Theory, 20 (1977), 196-219.

12. G. Nürnberger and M. Sommer, Weak Chebyshev subspaces and continuous selections for the metric projection, Trans. Amer. Math. Soc., 238 (1978), 129-138.
13. G. Nürnberger and M. Sommer, Characterization of continuous selections of the metric projection for spline functions, J. Approx. Theory, 22 (1978), 320-330.
14. M. Sommer, Characterization of continuous selections for the metric projection for generalized splines, SIAM J. Math. Anal. (to appear).
15. M. Sommer, Continuous selections for the metric projection, SIAM J. Math. Anal. (to appear).
16. M. Sommer, Nonexistence of continuous selections of the metric projection for a class of weak Chebyshev spaces, preprint.

CHARACTERIZATIONS OF GOOD AND BEST APPROXIMATIONS IN AN ARBITRARY SUBSET OF A NORMED LINEAR SPACE

Claude Dierieck

Philips Research Laboratory

Brussels , Belgium

Based on a simple separation principle, necessary and sufficient characterization conditions are obtained for good approximations in non-linear problems. This result is further particularized to obtain a complete characterization of best approximations in arbitrary sets.

I. GOOD APPROXIMATION IN A NORMED LINEAR SPACE

In this first part we analyse the problem of characterizing good (or ε-) approximations. We consider approximating functions which are elements of an arbitrary subset G of a normed linear space E, together with a target function $f \in E \setminus \bar{G}$. For some strict positive scalar ε we want thus to characterize the set of ε-approximations :

$$P_G^{\varepsilon}(f) = \{g \in G \mid \|g-f\| \leqslant d(f,G) + \varepsilon\}, \tag{1}$$

where $d(.,G)$ is the distance functional of the set G. The set $P_G^{\varepsilon}(f)$ is

Copyright © 1980 by Academic Press, Inc.
All rights of reproduction in any form reserved.
ISBN: 0-12-171050-5

equally described as the intersection of the set G and the ball $B(f,d(f,G)+\varepsilon)$. We can thus immediately formulate the following primal characterization : an element $g_0 \in G$ is an ε-approximation if and only if no approximating function is situated in the open ball $B_0(f,\|g_0-f\|-\varepsilon)$. In order to obtain a characterization in terms of the existence of some linear continuous functionals, we will apply the separation theorem. Consequently we obtain that an element $g_0 \in G$, will be in $P_G^\varepsilon(f)$ if and only if for every approximating function $g \in G$, there exists some real closed hyperplane H, which separates this element g from the ball $B(f,\|g_0-f\|-\varepsilon)$. A dual formulation of the characterization is thus : for every approximating function g exists some linear continuous functional L in the dual unit sphere (S_{E^*}) and a real scalar α such that $H = H[\mathrm{Re}L,\alpha] \triangleq \{x \in E \mid \mathrm{Re}L(x)=\alpha\}$ with $\alpha \in [\mathrm{Re}L(f) + \|g_0-f\| -\varepsilon, \mathrm{Re}L(g)]$. In fact, as soon as for a fixed $L \in S_{E^*}$, the preceding interval of $\mathbb{R}$ is nonvoid, then for any scalar α of the interval, the hyperplane $H[\mathrm{Re}L,\alpha]$ satisfies the requirements of separation. For the upper bound ($\alpha=\mathrm{Re}L(g)$) the hyperplane H passes through the element g, while for the lower bound ($\alpha=\mathrm{Re}L(f)+\|g_0-f\|-\varepsilon$) , H is a supporting hyperplane of the ball $B(f, \|g_0-f\|-\varepsilon)$ (in the sense that H supports the ball B if $d(B,H)=0$ and $H \cap \mathrm{int}\, B=\phi$).
We should also remark that whenever exists a separating hyperplane, there also exists an extremal one (i.e. $L \in \mathcal{E}(B_{E^*})$) as results from the following main characterization theorem.

THEOREM 1. <u>Let</u> E <u>be a normed linear space</u>, G <u>a subset of</u> E, $g_0 \in G$, $f \in E\setminus\bar{G}$ <u>and</u> $\varepsilon \in [0,\|g_0-f\|[$. <u>We have</u> $g_0 \in P_G^\varepsilon(f)$ <u>if and only if for every</u> $g \in G$ <u>exists a linear continuous functional</u> L^g <u>extremal point of the dual unit ball</u> B_{E^*} <u>such that</u>:

$$\mathrm{Re}L^g(g) \geqslant \mathrm{Re}L^g(f) + \|g_0-f\|-\varepsilon \ . \tag{2}$$

Proof. Suppose the condition to hold for every approximating function. Thus for every $g \in G$ exists a l.c.f. $L^g \in \mathcal{E}(B_{E^*})$ satisfying (2). Consequently for all $g \in G$ we have

$$\| g_0 - f\| - \varepsilon \leq \mathrm{Re} L^g(g-f) \leq |L^g(g-f)| \leq \| g-f\| ,$$

which proves that $g_0 \in P_G^{\varepsilon}(f)$.

To prove the converse, we first show that whenever $g_0 \in P_G^{\varepsilon}(f)$ then for every approximating function $g \in G$ exists a l.c.f. $L^g \in B_{E^*}$ such that (2) is satisfied. Suppose there exists some element $g' \in G$ such that for every $L \in B_{E^*}$ we have : $\mathrm{Re}L(g'-f) < \| g_0-f\| - \varepsilon$. By the Hahn-Banach extension theorem, there exists a l.c.f. $L \in E^*$ such that $\| L\| = 1$ and $\mathrm{Re}L(g'-f) = \| g'-f\|$ (indeed $f \neq g'$). We can thus conclude that there exists some element $g' \in G$ for which $\| g'-f\| < \| g_0-f\| - \varepsilon$, which shows that whenever the condition is not verified, g_0 cannot be a good approximation. Finally, to prove that there necessarily exists an extremal point of B_{E^*} satisfying (2) (whenever g_0 is a good approximation) we consider, for every $g \in G$, the map

$$\phi_{g-f} : L \in (E^*)_{w-*} \longrightarrow \mathrm{Re}L(g-f) \in \mathbb{R} .$$

This map is weak-* continuous from the dual space endowed with the weak-* topology into $\mathbb{R}$. Consequently if for every $g \in G$ exists a l.c.f. $L^g \in B_{E^*}$ satisfying (2), then since B_{E^*} is weak-* compact, we have that for every $g \in G$:

$$\max\{\phi_{g-f}(L) \mid L \in B_{E^*}\} \geq \| g_0-f\| - \varepsilon .$$

Moreover this maximum is attained at an extremal point of B_{E^*} [3,P.74]. This proves that the condition is also necessary, which completes the proof.

It is interesting to emphasize the fact that an equivalent characterization theorem can be proved where the condition is then replaced by : for every $g \in G$ exists a l.c.f. $L^g \in B_{E^*}$ satisfying (2).

For the particular case G=C, a convex set, we can repeat the preceding geometrical reasoning and obtain sharper results. The set $P_C^\varepsilon(f)$ is again described as the intersection of the set C and the ball $B(f,d(f,C)+\varepsilon)$. In this particular case the separation theorem can be applied on the whole of the geometrical setting. We mean that the set C and the ball $B_0(f, \|g_0-f\|-\varepsilon)$ are disjoint if and only if there exists a real closed hyperplane which separates the whole of C from the ball $B(f, \|g_0-f\|-\varepsilon)$. The corresponding dual formulation is thus : an element g_0 is in $P_C^\varepsilon(f)$ iff there exists a l.c.f. L in S_{E^*} and a real scalar α such that $\mathcal{H}= H[\mathrm{Re}L,\alpha]$ where $\alpha \in [\mathrm{Re}L(f)+\|g_0-f\|-\varepsilon,\ t_C(L)]$, with $t_C : L \in E^* \to \inf\{\mathrm{Re}L(g) \mid g \in C\}$ the support functional of the set C. Here again, as soon as the interval for α is nonvoid, there exists a separating hyperplane as required by the characterization. When $\alpha=t_C(L)$, $\mathcal{H}$ can be considered as a supporting hyperplane of the convex set while for $\alpha = \mathrm{Re}L(f)+\|g_0-f\|-\varepsilon$, $\mathcal{H}$ supports the ball $B(f, \|g_0-f\|-\varepsilon)$.

By a similar reasoning, we obtain the following analytical form of a simultaneous characterization.

THEOREM 2. Let E be a normed linear space, C a convex subset of E, M a nonvoid convex subset of C, $f \in E\setminus\bar{C}$ and $\varepsilon \in [0, \sup\{\|m-f\| \mid m \in M\}[$. We have $M \subset P_C^\varepsilon(f)$ if and only if there exists a linear continuous functional L of the dual unit ball B_{E^*} such that:

$$t_C(L) \geq \mathrm{Re}L(f) + \sup\{\|m-f\| \mid m \in M\} - \varepsilon. \tag{3}$$

Proof. If $M \subset P_C^\varepsilon(f)$ then for every $m \in M$, we have $\|m-f\| \leq d(f,C)+\varepsilon$. Denoting $\beta = \sup\{\|m-f\| \mid m \in M\}$ we have $\beta \leq d(f,C)+\varepsilon$. Consequently we also have that $B_0(f, \beta-\varepsilon) \cap C = \phi$. We can apply the separation theorem and obtain that there exists a real closed hyperplane separating the ball

$B(f,\beta-\varepsilon)$ from the set C. Thus there exists a l.c.f. $L \in S_{E^*}$ such that $\text{ReL}(f)+\beta-\varepsilon \leqslant t_C(L)$.

Conversely if there exists a l.c.f. $L \in B_{E^*}$ such that (3) is valid, then for some $m \in M$ we have that : $\text{ReL}(g) \geqslant \text{ReL}(f) + \|m-f\| - \varepsilon$, for every element $g \in C$. Consequently for every element $g \in C$ we obtain the following inequality

$$\|m-f\| - \varepsilon \leqslant \text{ReL}(g-f) \leqslant |L(g-f)| \leqslant \|g-f\| ,$$

which proves that $m \in P_C^{\varepsilon}(f)$. This can be repeated for every $m \in M$ and the proof is thus complete.

In case theorem 2. is applied for the characterization of a single element $g_0 \in C$, then condition (3) states that the l.c.f. L is necessarily an element of the set

$$\partial_{\varepsilon}\|g_0-f\| = \{L \in E^* \mid \text{ReL}(g_0-f) \geqslant \|g_0-f\| - \varepsilon , \|L\| \leqslant 1\} , \tag{4}$$

which is known as the <u>ε-subdifferential</u> of the norm at the point (g_0-f). [3]. This set is nonvoid, convex and is a weak-* compact subset of the dual unit ball B_{E^*}. Consequently for $M=\{g_0\}$, the characterization condition of theorem 2. can also be stated as : there exists a l.c.f. $L \in E^*$ such that $L \in \partial_{\varepsilon}\|g_0-f\|$ and which satisfies (3),

II. BEST APPROXIMATION IN A NORMED LINEAR SPACE

In this second part we consider best approximation problems. We suppose thus given an, arbitrary subset G of a normed linear space E and a function $f \in E \setminus \bar{G}$. Now we want to characterize elements of the set

$$P_G(f) = \{g \in G \mid \|g-f\| = d(f,G)\}.$$

As is well known, this set containing the best approximations to f in G is in fact a limit case, for $\varepsilon \to 0$, of the set of good approximations. Consequently main results of the first part, can now easily be particularized. For example, from Theorem 1. we obtain:

THEOREM 3. <u>Let</u> E <u>be a normed linear space</u>, G <u>a subset of</u> E, $g_0 \in \dot{G}$, $f \in E \setminus \bar{G}$. <u>We have</u> $g_0 \in P_G(f)$ <u>if and only if for every</u> $g \in G$ <u>exists a linear continuous functional</u> L^g <u>extremal point of the dual unit ball</u> B_{E^*}, <u>such that</u> : $\mathrm{Re}L^g(g) \geq \mathrm{Re}L^g(f) + \|g_0 - f\|$.

It is thus also evident now that l.c.f. L^g of the preceding condition is not necessarily in the subdifferential of the norm at $(g_0 - f)$:

$$\partial\|g_0-f\| = \{L \in E^* \mid \|L\| = 1 \ , \ \mathrm{Re}L(g_0-f) = \|g_0-f\|\} \ ,$$

as is requested by the so called <u>extended Kolmogoroff condition</u>. (see e.g. [1]) . It is thus a not surprising fact that the ext. Kolm. cond. can only be sufficient when g_0 is a best nonlinear approximation.

However when best convex approximations are to be characterized, then the l.c.f. involved in the characterization is necessarily an element of $\partial\|g_0-f\|$, since the separating hyperplane necessarily supports the ball $B(f,\|g_0-f\|)$ at the point g_0. Particularizing theorem 2.we obtain thus also the well-known characterization condition for convex problems, that $g_0 \in P_C(f)$ iff there exists a l.c.f. $L \in \partial\|g_0-f\|$ which satisfies $t_C(L) = \mathrm{Re}L(f) + \|g_0-f\|$.

REFERENCES

1. Dierieck, C., J. Approximation Theory 14, 163-187 (1975).
2. Ekeland, I., and Temam, R., "Analyse Convexe et Problèmes Variationnels", Dunod, Paris, (1974).
3. Holmes, R., "Geometric Functional Analysis and its Applications", Springer Verlag, Berlin, (1975).

INTERPOLATION THEOREMS AND THE RATE OF CONVERGENCE OF BERNSTEIN POLYNOMIALS

Z. Ditzian[1]

Department of Mathematics
University of Alberta
Edmonton, Alberta
Canada

We recall that the Bernstein polynomials are given by

$$B_n(f,x) \equiv \sum_{k=0}^{n} f\left(\frac{k}{n}\right)\binom{n}{k} x^k (1-x)^{n-k} . \tag{1.1}$$

Berens and Lorentz [1,p.706] have shown that for $0 \le \alpha \le 2$, $0 < \beta < 2$ and $X \equiv x(1-x)$ and $f \in C[0,1]$,

$$X^{-\alpha/2}|B_n(f,x) - f(x)| \le \frac{M(f)}{n^{\beta/2}} \tag{1.2}$$

implies

$$X^{-\alpha/2}|f(x-t) - 2f(x) + f(x+t)| \le M_1(f)\left(\frac{t^2}{X}\right)^{\beta/2} \tag{1.3}$$

for $0 < x < 1$ and $t \le \min(x,1-x)$.

It is our goal here to show that (1.3) implies (1.2) and, therefore, that (1.3) and (1.2) are equivalent. This result was proved by Berens and Lorentz [1] for $\alpha = \beta$ and by the author for $\alpha = 0$ and all β [2] and will be shown here for $0 \le \alpha \le \beta$.

The method uses the Peetre K interpolation theory. It was proved [1] that f satisfies (1.2) if and only if $f_1(x) = f(x) - Ax - B$ such that $f_1(0) = f_1(1) = 0$ belongs

[1]Supported by NRC grant A 4816.

Copyright © 1980 by Academic Press, Inc.
All rights of reproduction in any form reserved.
ISBN: 0-12-171050-5

to $(C_\alpha, C_\alpha^2)_\beta$. We recall that $C_0 = \{f \in C[0,1]; f(0)=f(1)=0\}$ and C_α, C_α^2 are given by:

$$C_\alpha = \{f \in C_0; X^{-\alpha/2} f \text{ C.B. } (0,1)\} \quad ; \qquad \|f\|_{C_\alpha} = \sup |X^{-\alpha/2} f(x)| \tag{1.4}$$

and

$$C_\alpha^2 = \{f \in C_0; X^{1-\alpha/2} f'' \in \text{C.B. } (0,1) f' \in \text{A.C. locally}\} \quad , \qquad \|f\|_{C_\alpha^2} = \sup |X^{1-\alpha/2} f''(x)| \tag{1.5}$$

The interpolation space $(C_\alpha, C_\alpha^2)_\beta$ is given by its norm $\|f\|_{\alpha,\beta}$ and the K functional as:

$$\|f\|_{\alpha,\beta} = \sup_{t>0} \frac{K(t^2,f)}{t^\beta} \quad \text{and} \qquad K(t^2,f) = \left(\inf_{g \in C_\alpha^2} \|f-g\|_{C_\alpha} + t^2 \|g\|_{C_\alpha^2} \right) . \tag{1.6}$$

We will show that (1.2), (1.3) and $(C_\alpha, C_\alpha^2)_\beta$ are all equivalent to $(C_0, C_\gamma^2)_\beta$ where $\gamma = \frac{2\alpha}{\beta}$, (and therefore for $0 \le \alpha \le \beta$ $0 \le \gamma \le 2$).

2. THE MAIN RESULT

The result of this paper can be summarized in the following theorem:

Theorem. *For* $0 \le \alpha \le \beta$, $0 < \beta < 2$ *the conditions*

a. $X^{-\alpha/2} |B_n(f,x) - f(x)| \le \dfrac{M(f)}{n^{\beta/2}}$;

b. $X^{-\alpha/2} |f(x-h) - 2f(x) + f(x+h)| \le M_1(f) \left(\dfrac{h^2}{X} \right)^{\beta/2}$

for $0 < x < 1$ *and* $h \le \min(x, 1-x)$;

c. $f_1(x) - f(x) = Ax + B$, $f_1(x) \in C_0$ *and* $f_1 \in (C_\alpha, C_\alpha^2)_\beta$

and

d. $f_1(x) - f(x) = Ax + B$, $f_1(x) \in C_0$ *and* $f_1 \in (C_0, C^2_{2\alpha/\beta})_\beta$;

are equivalent

Proof. It was shown in [1] that (a) and (c) are equivalent and imply (b). We will show that (d) implies (a) and (b) implies (d). To show that (d) implies (a), we recall by definition of $(C_0, C^2_{2\alpha/\beta})_\beta$ that for every $t > 0$ there exists a function g_t such that $\|f-g_t\|_C \le Mt^\beta$ and $t^2\|g_t\|_{C^2_{2\alpha/\beta}} \le Mt^\beta$ where M is independent of t. (But it is possible that for different t we may have to choose different g_t). This implies

$$\begin{aligned} |B_n(f,x) - f(x)| &\le |B_n((f-g_t),x) - f(x)-g_t(x))| \\ &\quad + |B_n(g_t,x) - g_t(x)| \qquad (2.1) \\ &\le 2Mt^\beta + |B_n(g_t,x) - g_t(x)| \end{aligned}$$

following [1,p.704], we have for $g_t \in C^2_\gamma$ where $\gamma = \frac{2\alpha}{\beta}$,

$$\begin{aligned} |B_n(g_t,x) - g_t(x)| &\le \left(\frac{7}{n}\right)^{1-\gamma/2}\left(\frac{X}{2n}\right)^{\gamma/2} \|g_t\|_{C^2_\gamma} \qquad (2.2) \\ &\le 7\frac{1}{n}X^{\alpha/\beta}Mt^{\beta-2} \end{aligned}$$

In order to prove our theorem, we choose $t = \frac{X^{\alpha/2\beta}}{\sqrt{n}}$ for a fixed x and n, and the estimate follows substituting this t in (2.1) and (2.2). One could prove directly that (d) implies (b), but this is not necessary.

To prove that (b) implies (d), we will construct for f satisfying (b) a function $g_t(x)$ such that $\|f-g_t\|_C \le Mt^\beta$ and $\|g_t\|_{C^2_{2\alpha/\beta}} \le Mt^\beta$. We shall follow earlier papers of the author [2] and [3] with some modifications and some necessary changes. We first observe that it is enough to find $g_{2^{-\ell}}$ for all ℓ and write $g_t = g_{2^{-\ell}}$ for $2^{-\ell-1} < t \le 2^{-\ell}$. This will only change the constant M in the estimates above. We now define a decreasing function $\psi(x) \in C^2$, $\psi(x) = 1$ on $[0,\frac{1}{4}]$ and $\mathrm{Supp}\psi \subset [0,\frac{3}{4}]$, and then find $g_{t,1}$ and $g_{t,2}$ such that $\|g_{t,1}-f\cdot\psi\| \le Mt^{\beta}$, $\|g_{t,2}-f(1-\psi)\| \le Mt^{\beta}$ and $\|g_{t,i}\|_{C^2_{2\alpha/\beta}} \le Mt^{\beta-2}$; and since finding $g_{t,1}$ and $g_{t,2}$ constitute symmetric problems, we will construct in detail only $g_{t,1}$. We can define $f_h(x)$ by

$$f_h(x) = \left(\frac{2}{h}\right)^2 \int_0^{h/2}\int_0^{h/2} [2f(x+(u_1+u_2)) - f(x+2(u_1+u_2))]du_1du_2 \,. \tag{2.3}$$

Using (b), we have for $x < \frac{3}{4}$ and $h < \frac{1}{8}$ (and therefore $x + h < \frac{7}{8}$)

$$\begin{aligned}|f(x) - f_h(x)| &\le M_1(f)8^{\frac{\beta-\alpha}{2}}\left(\frac{2}{h}\right)^2\int_0^{h/2}\int_0^{h/2}\frac{(u_1+u_2)^{\beta}du_1du_2}{(x+u_1+u_2)^{(\beta-\alpha)/2}} \\ &\le M\min\left(\frac{h}{x^{(\beta-\alpha)/2}}, h^{(\alpha+\beta)/2}\right) \,.\end{aligned} \tag{2.4}$$

Moreover denoting $\Delta_h^2 f(x+a) = f(x+a-h) - 2f(x+a) + f(x+a+h)$, we have

$$\begin{aligned}|f_h''(x)| &\le \left|\left(\frac{2}{h}\right)^2\left[2\Delta_{h/2}^2 f(x+\frac{h}{2}) - \Delta_h^2 f(x+h)\right]\right| \\ &\le Mh^{-2}\min\left(\frac{h}{x^{(\beta-\alpha)/2}}, h^{(\alpha+\beta)/2}\right) \,.\end{aligned} \tag{2.5}$$

We now define $\psi_k(x) \equiv \psi(4^k\cdot x)$, and since $\mathrm{Supp}\psi_{k+1}(x) \subset \{x: \psi_k(x) = 1\}$, we have

$$f(x)\psi(x) = f(x)\psi_0(x) = \sum_{k=0}^{r} f(x)\psi_k(x)(1-\psi_{k+1}(x)) + f(x)\psi_r(x) \quad . \tag{2.6}$$

Define for $2^{-\ell-1} < t \le 2^{-\ell}$, $g_{t,1} \equiv g_{2^{-\ell},1}$ by

$$g_{2^{-\ell},1}(x) = \sum_{k=0}^{r-1} f_{k,\ell}(x)\psi_k(x)(1-\psi_{k+1}(x)) + f_{r,\ell}(x)\psi_r(x) \tag{2.7}$$

where $f_{m,\ell}(x)$ is $f_h(x)$ with $h = 2^{-\ell}2^{-m(1-\frac{\alpha}{\beta})}$ and $r = \left[\ell/(1+\frac{\alpha}{\beta})\right] + 1$. For every x at most two terms of the sum in either (2.6) or (2.7) are different from 0. In particular in $4^{-m-1} < x < 3\cdot 4^{-m-1}$, $\psi_m(x)(1-\psi_{m+1}(x)) = \psi_m(x)$ and $\psi_{m-1}(x)(1-\psi_m(x)) = 1-\psi_m(x)$, and in $3\cdot 4^{-m-1} < x < 4^{-m}$ and $x < 4^{-\ell-1}$, $(1-\psi_m(x))\psi_{m-1}(x) = 1$ and $\psi_\ell(x) = 1$ respectively. Therefore we can estimate $|g_{1,2^{-\ell}}(x)-f(x)\psi(x)|$

$$\le M\left\{\frac{\left(2^{-\ell-m(1-\alpha/\beta)}\right)^\beta}{4^{-m(\beta-\alpha)/2}} + \frac{\left(2^{-\ell-(m-1)(1-\alpha/\beta)}\right)^\beta}{4^{-m(\beta-\alpha)/2}}\right\} \le M_1 2^{-\ell\beta} \quad \text{for}$$

$4^{-m-1} < x < 4^{-m}$; and since $\beta \ge \alpha$ and $r \sim \ell(1+\frac{\alpha}{\beta})^{-1}$,

$$|g_{1,2^{-\ell}}(x) - f(x)\psi(x)| \le M\left(\left(2^{-\ell}\, 2^{-r(1-\frac{\alpha}{\beta})}\right)^{\frac{\alpha+\beta}{2}}\right) \le M_2 2^{-\ell\beta} \quad \text{for}$$

$x < 4^{-r}$. To estimate $X^{1-\frac{\alpha}{\beta}}g''_{t,1}(x)$ we use (2.5) to obtain

for $3\cdot 4^{-m-1} < x \le 4^{-m}$ $\quad |X^{1-\frac{\alpha}{\beta}}g_{t,1}(x)| \le MX^{1-\frac{\alpha}{\beta}}h^{-2}\dfrac{h^\beta}{x^{(\alpha-\beta)/2}}$

$$\le M_1 4^{-m(1-\frac{\alpha}{\beta})}\left[\frac{\left(2^{-\ell}2^{-m(1-\frac{\alpha}{\beta})}\right)^\beta}{4^{-m(\beta-\alpha)/2}}\right]\left[2^{-\ell}2^{-m(1-\frac{\alpha}{\beta})}\right]^{-2} \le M_1 2^{2\ell}2^{-\ell\beta} \quad ; \text{ and}$$

for $x < 4^{-r}$ $\quad |X^{1-\frac{\alpha}{\beta}}g''_{t,1}(x)| \le MX^{1-\frac{\alpha}{\beta}}h^{-2}\cdot h^{\frac{\alpha+\beta}{2}}$

$\le M4^{-r(1-\frac{\alpha}{\beta})}\left(2^{-\ell}2^{-r(1-\frac{\alpha}{\beta})}\right)^{\frac{\alpha+\beta}{2}-2} \le M(2^{-\ell})^{\beta-2}$. For $4^{-m-1} < x < 3\cdot 4^{-m-1}$ we write $g_{t,1} = f_{2^{-\ell-m(1-\alpha/\beta)}}(x) + (1-\psi_m(x))\left[f_{2^{-\ell-(m-1)(1-\alpha/\beta)}}(x) - f_{2^{-\ell-m(1-\alpha/\beta)}}(x)\right] \equiv I_1(x) = I_2(x)$. The estimate of $X^{1-\frac{\alpha}{\beta}}I_1''(x)$ is similar to the above. Let $I_2(x) \equiv (1-\psi_m(x))\phi_m(x)$, $|X^{1-\frac{\alpha}{\beta})}I_2''(x)| \le (1-\psi_m(x)|X^{1-\frac{\alpha}{\beta}}\phi_m''(x)| + 2|\psi_m'(x)||\phi_m'(x)X^{1-\frac{\alpha}{\beta}}| + |\psi_m''(x)||X^{1-\frac{\alpha}{\beta}}\phi_m(x)| \equiv J_1 + J_2 + J_3$. The estimate of J_1 follows the above pattern. Recalling $|\psi_m^{(i)}(x)| \le M_3 4^{mi}$, and since $m \le r \sim \ell(1+\frac{\alpha}{\beta})^{-1}$, $|J_3| \le M_3 4^{2m}\cdot(4^{-m})^{(1-\frac{\alpha}{\beta})}2^{-\ell\beta} \le M_4(2^{-\ell})^{\beta-2}$. To estimate J_2, the following Lemma will be sufficient. The Lemmas is probably well-known and in this form it is shown in [3] (for a more general set up).

Lemma. *For* $f \in C(a,b)$ *and* $f' \in A.C.$ *we have*

$$\|f'\|_{C[a,b]} \le M\left(\frac{\|f\|_{C[a,b]}}{(b-a)} + (b-a)\|f''\|_{C[a,b]}\right)$$

where M *is independent of* f,b *and* a.

3. REMARKS

Remark 1. In [5] Strukov and Timan show $|x^\gamma - B_n(t^\gamma,x)| \sim \frac{1}{n^\gamma}$ for $\gamma < 1$. This follows easily, (b) implies (a) for $\alpha = 0$ and $\beta = 2\gamma$ since x^γ satisfies (b).

Remark 2. As a result of [3] the K fucntional of the pair $(C_0, C^2_{2\alpha/\beta})$ satisfies $A\omega_*(f,t) \le K(t^2,f) \le B\omega_*(f,t)$ where

$$\omega_*(f,t) = \sup_{|h|\le t} \left| f\left(x-hX^{(1-\frac{\alpha}{\beta})/2}\right) - 2f(x) + f\left(x+hX^{(1-\frac{\alpha}{\beta})/2}\right)\right|$$

where x and $x \pm hX^{(1-\frac{\alpha}{\beta})/2}$ belong to $[0,1]$. Therefore also $\omega_*(f,t)$ Mt^β is equivalent to (a),(b),(c) and (d).

The saturation result given by:

Theorem. For $f \in C[a,b]$ $\|X^{-\alpha/2}(B_n(f,x)-f(x)\| \le \frac{M}{n}$ and $X^{1-\frac{\alpha}{2}}f'' \in L_\infty[0,1]$ are equivalent.

This result follows (2.1) and (2.2) in one direction and the local case, using the method of section 5 of [4], in the other direction

REFERENCES

(1) Berens, H. and Lorentz, G.G. (1972). Inverse theorems for Bernstein polynomials, *Indiana Univ. Math. Journal, Vol. 21, 693-708.*

(2) Diztian, Z. (1979). Global inverse thoeerems for combinations of Bernstein polynomials, *Jour. of Approx., 26, 277-292.*

(3) Ditzian, Z. On interpolation of $L_p[a,b]$ and weighted Sobolev spaces, *Pacific Jour. of Math. (to appear).*

(4) Ditzian, Z. (1979). On global inverse theorems for Szasz and Baskakov operators, *Can. Math. J. (2), 255-263.*

(5) Strukov, L.I. and Timan, A.F. (1978). Mathematical expectation of continuous functions of random variables. Smoothness and Variance, *Siberian Math. J., 18, 469-479.*

FINITE-DIFFERENCE COLLOCATION METHODS FOR SINGULAR BOUNDARY VALUE PROBLEMS

E. J. Doedel[1]

Computer Science Department
Concordia University
Montreal, Quebec

G. W. Reddien[1]

Mathematics Department
Southern Methodist University
Dallas, Texas

I. INTRODUCTION

We consider numerical methods for problems of the form

$$Lu = \frac{d^2u}{dx^2} + \frac{\sigma}{x}\frac{du}{dx} - q(x)u = f(x) \tag{1.1}$$

on (0,1) with boundary conditions depending on σ. If $0<\sigma<1$, we require $u(0) = u(1) = 0$. If $\sigma \geq 1$, we add $u'(0) = u(1) = 0$. These problems have been considered recently by several authors. We cite here only [1,2,5-8]. We give a finite difference method and a related collocation scheme in this paper.

Introduce a mesh $0 = x_0 < x_1 < \cdots < x_J = 1$ with $h_j = x_j - x_{j-1}$ and $h = \max h_j$. Our difference equation at the point x_j has the form

$$L_h u = \sum_{i=-1}^{1} d_{j,i} u_{j+i} = \sum_{i=1}^{m} e_{j,i} f(z_{j,i}), \tag{1.2}$$

where the points $z_{j,i}$ are given in $[x_{j-1}, x_{j+1}]$. These

[1]Supported in part by a grant from the U. S. Army Research Office

Copyright © 1980 by Academic Press, Inc.
All rights of reproduction in any form reserved.
ISBN: 0-12-171050-5

methods are constructed as follows. Let $\{\phi_{j,l}\}_{\ell=1}^{m+2}$ be a basis for some set of functions defined on $[x_{j-1}, x_{j+1}]$, $j=1,\cdots, J-1$. These functions should be chosen to match the behavior of the solutions to (1.1). The coefficients $\{d_i\}$ and $\{e_i\}$ are then defined by the equations

$$\sum_{i=-1}^{1} d_{j,i}\phi_{j,\ell}(x_{j+1}) = \sum_{i=1}^{m} e_{j,i}L\phi_{j,\ell}(z_{j,i}), \quad 1 \le \ell \le m+2, \tag{1.3}$$

$$\sum_{i=1}^{m} e_{j,i} = 1.$$

Once (1.3) is solved, then the finite difference method is given by (1.2) plus appropriate boundary conditions. These methods have been analyzed for smooth problems in [3].

Our collocation method has the following form: Find $\{p_j\}_{j=1}^{J-1}$, where $p_j(x) = \sum_{\ell=1}^{m+2} \alpha_{j,\ell}\phi_{j,\ell}(x)$, such that

$$p_j(x_{j+i}) = v_{j+1}, \quad -1 \le i \le 1 \tag{1.4a}$$

$$Lp_j(z_{j,i}) = f(z_{j,i}), \quad 1 \le i \le m, \quad 1 \le j \le J-1 \tag{1.4b}$$

plus the appropriate boundary conditions. Eq(1.4)(a) gives matching conditions for the overlapping functions p_j, and (1.4)(b) represents collocation equations. With matching boundary conditions, it follows directly that (1.2)-(1.3) and (1.4) are equivalent with $v_j = u_j$, provided the finite difference method is defined, i.e. (1.3) is uniquely solvable. Thus two approaches are available for both the computation and analysis of (1.2).

We next give a specific example of the equations (1.3). Let $Lu = (x^{\sigma}u')'$ and let the ϕ's be $1, x^{1-\sigma}, x^{2-\sigma}, x^2, x^{3-\sigma}, \cdots, x^{n-\sigma}, x^n$. Then $m = 2n-3$ and (1.3) gives the matrix

$$A_j = \left[\begin{array}{ccc|ccc}
1 & 1 & 1 & 0 & \cdots & 0 \\
x_{j-1}^{1-\sigma} & x_j^{1-\sigma} & x_{j+1}^{1-\sigma} & 0 & \cdots & 0 \\
x_{J-1}^{2-\sigma} & x_j^{2-\sigma} & x_{j+1}^{2-\sigma} & 2-\sigma & \cdots & 2-\sigma \\
\hline
x & x & x & & (L\phi_{j,\ell})(z_{j,i}) & \\
\cdot & \cdot & \cdot & & & \\
\cdot & \cdot & \cdot & & & \\
\cdot & \cdot & \cdot & & & \\
x & x & x & & & \\
0 & 0 & 0 & 1\ 1 & \cdots & 1
\end{array}\right]$$

where x indicates a nonzero entry whose exact value is not needed here.

Lemma. Let $z_{j,i}$ be distinct points in $[x_{j-1}, x_{j+1}]$. Then A_j^{-1} exists for each j.

Proof. Adding $-(2-\sigma)$ times row 2n to row 3, we have that A_j is row equivalent to a matrix having the form $\left[\begin{array}{c|c} A & 0 \\ \hline B & C \end{array}\right]$. Thus A_j is invertible if A and C are. But this follows from the fact that both $\{1, x^{1-\sigma}, x^{2-\sigma}\}$ and $\{1;\ L\phi_{j,i}, i \geq 4\}$ are Tchebychev systems.

II. CONVERGENCE

We next describe a convergence result for (1.1) in the case that $0<\sigma<1$. We first need to define certain local projections. Let $\{\phi_{i,j}\}$ be chosen consecutively from the list $1, x^{1-\sigma}, x^{2-\sigma}, x^2, x^{3-\sigma}, x^3, \cdots$. A consideration of the Frobenius expansions for the solution u in (1.1) with q and f smooth leads one to this choice for the $\phi_{i,j}$'s. See also [4]. Define $\psi_{i,j} = (x^{\sigma}\phi'_{ij})'$ and let $Y_j = \text{span}\{\psi_{i,j}\}$ over $[x_{j-1}, x_{j+1}]$. Let m+2 $\phi_{i,j}$'s be chosen so that $\dim(Y_j) = m$. A projection Q_j is now defined as a mapping from $C[x_{j-1}, x_{j+1}]$ into Y_j (taken as a subspace of $C[x_{j-1}, x_{j+1}]$) so that $Q_j f = s$ if and only if $f(z_{j,i}) = s(z_{j,i})$, $i = 1, \cdots, m$, and s is in Y_j.

Theorem. Let the homogeneous problem associated with (1.1) and the boundary conditions admit only the trivial solution. For h sufficiently small, let the local projections Q_j be well-defined and satisfy $||Q_j||_{\infty} \leq c$ for some constant c independent of j and the partition. The norm is the supremum norm over $[x_{j-1}, x_{j+1}]$ for each j. Then for all partitions with h sufficiently small, a solution $\{p_j\}$ to (1.4) with boundary conditions $p_1(0) = p_{J-1}(0) = 0$ exists and is unique. Moreover, there is a constant c>0 so that

$$||u - p_j||_{\infty} \leq c \cdot \max ||(1-Q_j)(x^{\sigma}u')'||_{\infty}, \qquad (2.1)$$

where u solves (1.1). The norms in (2.1) are taken over $[x_{j-1}, x_{j+1}]$.

The proof of this theorem will appear elsewhere. It is a modification of the usual projection method type proof using some ideas of Kreiss. See also [4]. An analogous result holds for the case $\sigma \geq 1$.

The boundedness condition on the Q_j's is in general difficult to establish since the ψ's come from the set $\{1,x^{\beta},x,x^{1+\beta},x^2,\cdots\}$, $\beta=1-\sigma$. In certain special cases however, the projections $\{Q_j\}$ represent local Lagrange interpolation and bounding their norms is straightforward. For example, in the case $\sigma=1$, the ϕ's can be chosen from 1, ℓnx, x^2,x^3, $\cdots$ and the ψ's become $1,x,x^2,x^3,\cdots$. If $\sigma=2$, the ϕ's can be chosen from $1,x^{-1},x^2,x^3,\cdots$ and the ψ's again become $1,x,x^2,x^3,\cdots$.

As an example, we considered the problem

$$(x^{1/2}y') - x^{1/2}y = - (3 + x^{-1/2})e^{-x}/2$$

with boundary conditions $y(0) = 1$, $y(1) = 2e^{-1}$. In this case $y(x) = (x^{1/2}+ 1)e^{-x}$. The ϕ's were chosen to be 1, $x^{1/2},x,x^{3/2},x^2,x^{5/2},x^3,x^{7/2}$ $(m=6)$ over $[0,.25]$. We let $z_{j,i} = x_j + s_i h$, $s_i = -.8, -.4, -.2, .2, .4, .8$. On $[.25,1]$ we used an $O(h^4)$ compact scheme with the ϕ's polynomials [3]. Assuming the error behaves like $e(h)\approx ch^{\beta}$, an observed β can be computed. We obtained $e(1/8) = .261\cdot 10^{-3}$, $e(1/16) = .864\cdot 10^{-5}$, $e(1/32) = .400\cdot 10^{-6}$ and $e(1/64) = .218\cdot 10^{-7}$. The β for the last two numbers is $\beta = 4.19$. The methods are easy to program and high order accuracy is possible.

REFERENCES

1. Brabston, D.C., and Keller, H.B., A numerical method for singular two-point boundary value problems, SIAM J. Numer. Anal. 14, 779-791(1977).
2. DeHoog, F.R., and Weiss, Richard, Collocation methods for singular boundary value problems, SIAM J. Numer. Anal. 15, 198-217(1978).

3. Doedel, E.J., The construction of finite difference approximations to ordinary differential equations, SIAM J. Numer. Anal. 15, 450-465(1978).
4. Doedel, E.J., On the stability of certain discretizations on nonuniform meshes, to appear.
5. Jamet, P., On the convergence of finite difference approximations to one-dimensional singular boundary value problems, Numer. Math. 14, 87-99(1970).
6. Jesperson, D., Ritz-Galerkin methods for singular boundary value problems, SIAM J. Numer. Anal. 15, 813-834 (1978).
7. Nitsche, J.A., Der Einfluß von Randsingularität beim Ritschen Verfahren, Numer. Math. 25, 263-278(1976).
8. Reddien, G.W., and Schumaker L.L., On a collocation method for singular two-point boundary value problems, Numer. Math. 25, 427-432(1976).

KOROVKIN THEOREMS FOR L^p-SPACES

Klaus Donner

Mathematisches Institut
Universität Erlangen-Nürnberg
Erlangen, W.-Germany

We state a characterization of Korovkin systems in L^p-spaces. Furthermore, it is shown that there exist Korovkin systems in L^p such that the H-affine functions are not dense in L^p.

Let X be a second countable locally compact space and let μ be a positive Radon measure on X. Consider the space $\mathcal{L}^p$ of all μ-measurable, real-valued functions f on X for which $|f|^p$ is μ-integrable, $p \in [1,\infty[$. A linear subspace $\mathcal{H}$ of $\mathcal{L}^p$ will be called a <u>Korovkin</u> <u>space</u> in $\mathcal{L}^p$ whenever the following condition holds:

> If $(T_i)_{i\in I}$ is an equicontinuous net of positive linear operators from $\mathcal{L}^p$ into itself such that $(T_i h)_{i\in I}$ converges to h with respect to the $\mathcal{L}^p$-seminorm for all $h \in \mathcal{H}$, then $(T_i f)_{i\in I}$ converges to f for <u>all</u> $f \in \mathcal{L}^p$ in the $\mathcal{L}^p$-topology.

Every linearly generating subset of $\mathcal{H}$ is called a <u>Korovkin system</u>.

Several authors tried to determine Korovkin spaces in $\mathcal{L}^p$ ([1],[3],[5]), but only sufficient conditions were given.

One of these conditions starts from the so-called shadow of $\mathcal{H}$ (for nets of positive linear operators) stating that $\mathcal{H}$ is a Korovkin space in $\mathcal{L}^p$ if the $\mathcal{H}$-affine functions are dense in $\mathcal{L}^p$.

Copyright © 1980 by Academic Press, Inc.
All rights of reproduction in any form reserved.
ISBN: 0-12-171050-5

Here a function $f \in \mathcal{L}^p$ is called $\mathcal{H}$-affine iff

$$H_f := \{[h] \in L^p : h \in \mathcal{H}, [h] \geq [f]\} \neq \emptyset,$$

$$H^f := \{[h] \in L^p : h \in \mathcal{H}, [h] \leq [f]\} \neq \emptyset \quad \text{and}$$

$$\inf_{[h] \in H_f} [h] = [f] = \sup_{[h] \in H^f} [h],$$

where $[h],[f]$ denote the equivalence classes of h, f in L^p. Since the $\mathcal{H}$-affine functions are easy to describe in many cases of practical interest, our first question should be

1. Is the denseness of the set $\mathcal{A}$ of all $\mathcal{H}$-affine functions also a necessary condition for $\mathcal{H}$ to be a Korovkin space?

We shall give a counterexample showing that this conjecture is false even if $\mathcal{H}$ is finite-dimensional. Thus the second problem is

2. Give necessary and sufficient conditions characterizing the Korovkin subspaces of $\mathcal{L}^p$ which are easy to handle.

For simplicity, let us confine this problem to the most interesting case of finite-dimensional Korovkin subspaces of $\mathcal{L}^p$, only.

For a μ-negligible set N and $x \in X$ let $\mathcal{M}_x^N(\mathcal{H})$ denote the set of all linear functionals $\sum_{i=1}^{n+1} \lambda_i \varepsilon_{x_i}$ on $\mathcal{L}^p$ such that

$$\sum_{i=1}^{n+1} \lambda_i h(x_i) = h(x) \quad \text{for all } h \in \mathcal{H},$$

where $n := \dim(\mathcal{H})$, $x_1,\ldots,x_{n+1} \in X \setminus N$, $\lambda_1,\ldots,\lambda_{n+1} \geq 0$ and ε_{x_i} denotes the evaluation functional at x_i.

The following theorem now solves the second problem:

<u>Theorem</u>: $\mathcal{H}$ is a Korovkin space in $\mathcal{L}^p$ iff there exists a μ-negligible set $N \subset X$ such that

$$\mathcal{M}_x^N(\mathcal{H}) = \{\varepsilon_x\}$$

for all $x \in X \setminus N$.

As an application of this theorem let us first give a counterexample for conjecture 1:

<u>Counterexample</u>: Consider the functions $h_1, h_2, h_3, h_4 : \mathbb{R} \to \mathbb{R}$ defined by

$$h_1(x) = 1,$$

$$h_2(x) = x,$$

$$h_3(x) = \begin{cases} (x+1)^2 & \text{for} \quad x < -1 \\ 0 & \text{for} \quad x \in [-1,1] \\ (x-1)^2 & \text{for} \quad x > 1 \end{cases}$$

$$h_4(x) = (x+2)^3,$$

and let μ denote the Lebesgue measure on $\mathbb{R}$ multiplied by the density $x \to e^{-x^2}$. We claim that $\{h_1, h_2, h_3, h_4\}$ is a Korovkin system in $\mathcal{L}^p$ but

$$\mathcal{A} \subset M := \{f \in \mathcal{L}^p : \exists\ \lambda_1, \lambda_2, \lambda_4 \in \mathbb{R}\ f(x) =$$

$$= (\lambda_1 h_1 + \lambda_2 h_2 + \lambda_4 h_4)(x) \text{ for } \mu\text{-a.e. } x \in [-1,1]\}.$$

In particular, $\bar{\mathcal{A}} \subset M \neq \mathcal{L}^p$.

In order to prove that $\{h_1, \ldots, h_4\}$ is a Korovkin system in $\mathcal{L}^p$, let $N := \emptyset$ and let $x_o \in \mathbb{R}$ be arbitrary. By the theorem stated above it suffices to show that $\mathcal{M}_{x_o}^{\emptyset}(\mathcal{H}) = \{\varepsilon_{x_o}\}$, where $\mathcal{H}$ is the linear subspace of $\mathcal{L}^p$ generated by $\{h_1, \ldots, h_4\}$.

Given $x_1, \ldots, x_5 \in \mathbb{R}$, $\lambda_1, \ldots, \lambda_5 \geq 0$ such that

$$\sum_{i=1}^{5} \lambda_i h(x_i) = h(x_o) \quad \text{for all } h \in \mathcal{H}. \tag{*}$$

We shall construct a function $h_o \in \mathcal{H}$ satisfying $h_o(x_o) = 0$ and $h_o(x_i) > 0$ for all $i \in \{1,...,5\}$ with $x_i \neq x_o$. Inserting h_o in (*) yields $\lambda_i = 0$ for all $i \in \{1,...,5\}$ such that $x_i \neq x_o$. Moreover, we have

$$\sum_{i=1}^{5} \lambda_i = \sum_{i=1}^{5} \lambda_i h_1(x_i) = h_1(x_o) = 1,$$

which shows that $\sum_{i=1}^{5} \lambda_i \varepsilon_{x_i} = \varepsilon_{x_o}$.

To obtain the function h_o suppose first that $x_o < -1$. Then

$$h_o = h_3 - 2(x_o + 1)h_2 + (x_o^2 - 1)h_1$$

is strictly positive on $\mathbb{R} \setminus \{x_o\}$ and $h_o(x_o) = 0$.
The same holds for $h_o = h_3 - 2(x_o - 1)h_2 + (x_o^2 - 1)h_1$, if $x_o > 1$. Finally, if $x_o \in [-1,1]$, then

$$h := h_4 - 3(x_o + 2)^2 h_2 + 2(x_o + 2)^2(x_o - 1)h_1$$

has a double zero and a local minimum at x_o the second zero $x_1 = -2(x_o + 3)$ being smaller then -4. Hence there exists $n \in \mathbb{N}$ such that

$$h_o := h + n h_3$$

satisfies $h_o(x_i) > 0$ for all $i \in \{1,...,5\}$ with $x_i \neq x_o$.

In order to show that $\mathcal{A} \subset M$ let $f \in \mathcal{A}$ be given. If $h = \sum_{i=1}^{4} \sigma_i h_i$, $h' = \sum_{i=1}^{4} \sigma_i' h_i$ are such that $[h] \leq [f] \leq [h']$, then $\sum_{i=1}^{4} (\sigma_i' - \sigma_i)[h_i] = [h' - h] \geq 0$, which is only possible if $\sigma_4' - \sigma_4 = 0$. Indeed there exists a fixed number $\sigma \in \mathbb{R}$ such that $h - \sigma h_4$, $h' - \sigma h_4$ are contained in the three-dimensional subspace $\mathcal{H}_o$ of $\mathcal{H}$ generated by $\{h_1,h_2,h_3\}$ for all $h,h' \in \mathcal{H}$ satisfying $[h] \leq [f] \leq [h']$. Since f is $\mathcal{H}$-affine, we conclude

$$\inf_{\substack{h''\in\mathcal{H}_o \\ [h'']\geq[f-\sigma h_4]}}[h''] = [f]-\sigma[h_4] = \sup_{\substack{h''\in\mathcal{H}_o \\ [h'']\leq[f-\sigma h_4]}}[h''] .$$

Infima and suprema of functions in $\mathcal{H}_o$ being concave and convex on $[-1,1]$, respectively, we obtain that $[f]-\sigma[h_4]$ is the equivalence class of an affine-linear function on $[-1,1]$. Hence there exist $\lambda_1,\lambda_2 \in \mathbb{R}$ such that

$$[f]-\sigma[h_4] = [\lambda_1 h_1 + \lambda_2 h_2] \quad \text{on } [-1,1],$$

which yields

$$f(x) = (\lambda_1 h_1 + \lambda_2 h_2 + \sigma h_4)(x) \quad \text{for } \mu\text{-a.e. } x \in [-1,1].$$

Application to Fourier analysis. Consider a sequence (μ_n) of probability measures on some locally compact abelian group G. For each character χ on G we have $\mu_n * \chi = \hat{\mu}_n(\chi)\cdot\chi$, where $\hat{\mu}_n$ denotes the Fourier transform of μ_n. Hence $(\mu_n * \chi)_{n\in\mathbb{N}}$ converges uniformly to χ iff $\lim_{n\to\infty}\hat{\mu}_n(\chi) = 1$.

If G is the unit circle $\{z\in\mathbb{C} : |z| = 1\}$ it follows from Korovkin's classical theorems (cf.[4]) that

$$\lim_{n\to\infty}\mu_n * f = f \quad \text{uniformly for all} \quad f\in\mathcal{C}(G)$$

whenever $\lim_{n\to\infty}\dot{\mu}_n(1) = 1$, where $\dot{\mu}_n(1) = \hat{\mu}_n(e^{-ix})$ denotes the first Fourier coefficient of μ_n.

If $G = \mathbb{R}$ it can be shown that for any countable subset $A\subset\mathbb{R}$ there exists a sequence (μ_n) of probability measures on $\mathbb{R}$ such that the Fourier transforms of μ_n converge to the constant 1 on A but not on $\mathbb{R}$. This shows that the sequence of positive convolution operators induced by (μ_n) on the space

$L^p (p \in [1,\infty[\,)$ of all equivalence classes of p-times Lebesgue integrable functions on $\mathbb{R}$ does not converge to the identity operator.

However, the following result can be proved using the theorem stated above:

The sequence $(\mu_n * f)_{n \in \mathbb{N}}$ converges to f for all $f \in L^p$ (with respect to the L^p-norm) iff the following conditions hold

(i) there exist two real numbers $t_1, t_2 \in \mathbb{R}$ independent over the rationals such that $\lim_{n\to\infty} \dot{\mu}_n(t_j) = 1$ for $j = 1,2$.

(ii) for every $\varepsilon > 0$ there exists a compact subset $K \subset \mathbb{R}$ such that $\mu_n(\mathbb{R} \setminus K) < \varepsilon$ for all $n \in \mathbb{N}$.

REFERENCES

[1] Berens, H., and Lorentz, G. G.: Theorems of Korovkin type for positive linear operators on Banach lattices. In: Approximation Theory (Ed.: G. G. Lorentz), p. 1-30. Academic Press, New York - London 1973.

[2] Donner, K.: Fortsetzung positiver Operatoren und Korovkinsätze (unpublished).

[3] Kitto, W., and Wulbert, D. E.: Korovkin approximation in L^p-spaces. Pacific J. Math. 63, 153-167 (1976).

[4] Korovkin, P. P.: Linear Operators and Approximation Theory. Hind. Publ. Comp., Delhi 1960.

[5] Wolff, M.: Über die Korovkinhülle von Teilmengen in lokalkonvexen Vektorverbänden. Math. Ann. 213, 97-108 (1975).

ON A PROBLEM OF DEVORE, MEIR AND SHARMA

Wolfgang Drols

Department of Mathematics
University of Duisburg
Duisburg, West Germany

I. INTRODUCTION

In 1973 DeVore, Meir and Sharma [1] studied a special Birkhoff interpolation problem for (algebraic) polynomials $Q \in \Pi_n$, for a set of three nodes $x_1 < x_2 < x_3$ and for a $3 \times (n+1)$-incidence matrix $E = (e_{k,m})_{1 \le k \le 3, 0 \le m \le n}$ with p Hermite data in the first row, q Hermite data in the third row and with two non-zero entries in the middle row in the positions i and j and with the condition $n = q+p+1$. Without loss of generality we assume $p \le q$. We denote such an incidence matrix by $E_{q,p;i,j}$.

Since this interpolation problem is a linear problem for the coefficients of $Q \in \Pi_n$, $E_{q,p;i,j}$ is called regular, if the

Copyright © 1980 by Academic Press, Inc.
All rights of reproduction in any form reserved.
ISBN: 0-12-171050-5

determinant of the corresponding linear problem is regular and called singular otherwise.

Although a number of results are contained in [1],[2] and [3], the question of singularity is completely solved for the case $i < p$ or $q+1 < j$ only. For the case $p \le i < j \le q+1$ there is no solution except for $p = 1$. In this case DeVore, Meir and Sharma [1] proved

Theorem A

$E_{q,1;i,j}$ with $1 = p \le i < j \le q+1$ is singular, if

$$(q+2)\cdot(i+j-1)^2 - 4ij\cdot(q+1) \ge 0 .$$

One of the aims of this paper is to show that this theorem holds true for all p.

II. PRELIMINARIES

Without loss of generality we choose the special nodes $x_1 = -1$, $x_2 = 0$ and $x_3 = u$. All polynomials $Q \in \Pi_n$ fulfilling the equations $e_{k,m} \cdot Q^{(m)}(x_k) = 0$ for all entries $e_{k,m}$ of $E_{q,p;i,j}$ necessarily can be written in the form

$$Q(x) = (x+1)^q \cdot (x-u)^p \cdot (a\cdot x+b) = \sum_{r=0}^{n} x^r \cdot (a\cdot P_r^{q,p}(u) + b\cdot P_{r-1}^{q,p}(u)),$$

where a and b are real numbers and $P_r^{q,p}$ is a so-called "coefficient polynomial" with the representation

$$P_r^{q,p}(u) = \sum_{\mu+\nu=r} \binom{q}{\mu}\cdot\binom{p}{\nu}\cdot(-u)^{p-\nu} \quad ; \quad \mu,\nu,r \in \mathbb{Z} .$$

Some properties of the coefficient polynomials are proved in [2] and listed in the following lemma.

Lemma A

For all r with $p-1 \le r \le q$ the following assertions are true:

(1) All roots of $P_r^{q,p}$ are real and simple.

(2) The roots of $P_r^{q,p}$ and $P_{r+1}^{q,p}$ alternate.

(3) The primitive function of $(-p-1)\cdot P_r^{q,p}$ is $P_r^{q,p+1}$.

If $p \le i < j \le q+1$ then:

(4) $E_{q,p;i,j}$ is singular, iff there exists an u_o with

$$(P_i^{q,p}\cdot P_{j-1}^{q,p} - P_j^{q,p}\cdot P_{i-1}^{q,p})(u_o) \le 0 .$$

(5) $E_{q,p;i,j}$ is singular, iff there exist an u_o and a pair of real numbers $(a,b) \neq (0,0)$ with

$$(a\cdot P_i^{q,p} + b\cdot P_{i-1}^{q,p})(u_o) = (a\cdot P_j^{q,p} + b\cdot P_{j-1}^{q,p})(u_o) = 0 .$$

(6) $E_{q,p;i,j}$ is singular, if the roots of $P_i^{q,p}$ and $P_j^{q,p}$ or the roots of $P_{i-1}^{q,p}$ and $P_{j-1}^{q,p}$ don't alternate.

III. MAIN RESULT

Lemma 1

(1) For all pairs of real numbers (a,b) with $a\cdot b \neq 0$ and for all r with $p \le r \le q+1$ the roots of $a\cdot P_r^{q,p} + b\cdot P_{r-1}^{q,p}$ and $P_r^{q,p}$ alternate.

(2) $E_{q,p;i,j}$ with $p \le i < j \le q+1$ is singular, iff there exists a pair of numbers $(a,b) \neq (0,0)$, such that the roots of $a\cdot P_i^{q,p} + b\cdot P_{i-1}^{q,p}$ and $a\cdot P_j^{q,p} + b\cdot P_{j-1}^{q,p}$ don't alternate.

Proof:

The assertion (1) follows immediately from Lemma A.

Lemma A shows that the condition in (2) is necessary, so that only the sufficiency must be proved.

Because of Lemma A we can assume $a \cdot b \neq 0$. If the roots of $a \cdot P_i^{q,p} + b \cdot P_{i-1}^{q,p}$ and $a \cdot P_j^{q,p} + b \cdot P_{j-1}^{q,p}$ don't alternate without possessing a common root, we can assume that there exist two neighbouring roots $z_1 < z_2$ of $a \cdot P_i^{q,p} + b \cdot P_{i-1}^{q,p}$, so that no root of $a \cdot P_j^{q,p} + b \cdot P_{j-1}^{q,p}$ lies in the closed interval $[z_1 , z_2]$.
Because of (1) there is exactly one (simple) root of $P_j^{q,p}$ in $[z_1 , z_2]$. Altogether we get

$$(a \cdot P_j^{q,p} + b \cdot P_{j-1}^{q,p})(z_1) \cdot (a \cdot P_j^{q,p} + b \cdot P_{j-1}^{q,p})(z_2) \geq 0 ,$$

$$a \cdot P_i^{q,p}(z_k) = -b \cdot P_{i-1}^{q,p}(z_k) \quad k = 1 , 2 ,$$

$$a \cdot P_j^{q,p}(z_1) = b \cdot P_{j-1}^{q,p}(z_2) .$$

This leads to

$$(P_i^{q,p} \cdot P_{j-1}^{q,p} - P_j^{q,p} \cdot P_{i-1}^{q,p})(z_1) \cdot (P_i^{q,p} \cdot P_{j-1}^{q,p} - P_j^{q,p} \cdot P_{i-1}^{q,p})(z_2) \leq 0 ,$$

i.e. $E_{q,p;i,j}$ is singular. □

Now the main result can be proved.

Theorem 1

If $E_{q,p;i,j}$ with $p < i < j \leq q+1$ is singular, then $E_{q,p+1;i,j}$ is singular too.

Proof:

Because of the singularity of $E_{q,p;i,j}$ there exists a pair of real numbers $(a,b) \neq (0,0)$, so that the roots of $a \cdot P_i^{q,p} + b \cdot P_{i-1}^{q,p}$

and $a \cdot P_j^{q,p} + b \cdot P_{j-1}^{q,p}$ don't alternate. By a theorem of Markoff [4] the same is true for their primitive functions. This means that the roots of $a \cdot P_i^{q,p+1} + b \cdot P_{i-1}^{q,p+1}$ and $a \cdot P_j^{q,p+1} + b \cdot P_{j-1}^{q,p+1}$ don't alternate. By Lemma 1 the assertion is proved. □

IV. APPLICATIONS

<u>Theorem 2</u>

$E_{q,p;i,j}$ with $p \le i < j \le q+1$ is singular, if

$$(q+2) \cdot (i+j-1)^2 - 4ij \cdot (q+1) \ge 0 .$$

Proof:

By Theorem A the incidence matrix $E_{q,1;i,j}$ is singular. If we apply Theorem 1 (p-1)-times we get the assertion. □

To get some more sufficient conditions for singularity we mention a special case of a theorem which is proved in [2].

<u>Theorem B</u>

$E_{q,p;i,j}$ is singular, iff $E_{q,p;q+p+1-j,q+p+1-i}$ is singular.

The next results are corollaries of Theorem B and Theorem 1.

<u>Theorem 3</u>

If $E_{q,p;i,j}$ with $p \le i < j \le q$ is singular, then $E_{q,p+1;i+1,j+1}$ is singular too.

Proof:

If $E_{q,p;i,j}$ is singular, then so is $E_{q,p;q+p+1-j,q+p+1-i}$ by Theorem B and $E_{q,p+1;q+p+1-j,q+p+1-i}$ by Theorem 1. Theorem B applied once more gives the singularity of the incidence matrix $E_{q,p+1;i+1,j+1}$. □

Theorem 4

$E_{q,p;i,j}$ with $p \le i < j \le q+1$ is singular, if

$$(q+2)\cdot(i+j-2\cdot p+1)^2 - 4\cdot(i+1-p)\cdot(j+1-p)\cdot(q+1) \ge 0 .$$

Proof:

$E_{q,1;i+1-p,j+1-p}$ is singular by Theorem A. If we apply Theorem 3 (p-1)-times we get the assertion. □

REFERENCES

1. DeVore, R., Meir, A. and Sharma, A., *Canad. J. Math.* 25, 1040 (1973).
2. Drols, W., Zur Hermite-Birkhoff-Interpolation: DMS-Inzidenzmatrizen, to appear.
3. Lorentz, G.G., Stangler, S.S. and Zeller, K.L., in "Approximation Theory II" (G.G. Lorentz, C.K. Chui and L.L. Schumaker, eds.), p. 423. Academic Press, New York (1976).
4. Markoff, A.A., *Petersb. Bull.* 18, 1 (1903).

FAILURE OF COMPLEX STRONG UNIQUENESS

Charles B. Dunham

Department of Computer Science
The University of Western Ontario
London, Ontario, Canada

Consider complex Chebyshev approximation by an approximating function F with parameter $A = (a_1, \ldots, a_n)$ from a parameter set P, a subset of complex n-space C_n, to a continuous function f on a compact Hausdorff space X.

Following Bartelt and McLaughlin (1) and the text of Cheney (2, p.80, p.165) we define

Definition: $F(A,.)$ is a strongly unique best approximation to f if there is $\gamma > 0$ such that for all parameters $B \in P$,

$$||f - F(B,.)|| \geq ||f - F(A,.)|| + \gamma||F(B,.) - F(A,.)|| .$$

We define a parameter norm $||A|| = \max\{|a_i|: 1 \leq i \leq n\}$.

Definition: Let there exist continuous partial derivatives F_k of F with respect to parameter a_k of A. Define

$$D(A,B,x) = \sum_{k=1}^{n} b_k F_k(A,x) ,$$

$$R(A,B,x) = F(A+B,x) - F(A,x) - D(A,B,x) ,$$

and let $R(A,B,x) = o(||B||)$ as $||B|| \to 0$. Let a neighbourhood of A in n-space be in P. We say F is locally linear at A.

Copyright © 1980 by Academic Press, Inc.
All rights of reproduction in any form reserved.
ISBN: 0-12-171050-5

Definition: $M(A) = \{x: |f(x) - F(A,x)| = \|f - F(A,.)\|\}$.

LEMMA: Let F be locally linear at A. Let there exist $D(A,B,.) \not\equiv 0$ in the tangent space of F at A such that

$$\mathrm{Re}\ (\overline{f(x) - F(A,x)}) \cdot D(A,B,x) = 0, \quad x \in M(A)$$

and $f \neq F(A,.)$. Then $F(A,.)$ is not strongly unique to f.

Proof: Let $\lambda \in [0,1]$.

$$\begin{aligned}\|F(A+\lambda B,.) - F(A,.)\| &= \|D(A,\lambda B,.) + R(A,\lambda B,.)\| \\ &= \|\lambda D(A,B,.) + R(A,\lambda B,.)\|.\end{aligned}$$

The first term above goes down proportionally to λ. As the second term is $o(\lambda)$, the expression above goes down roughly as λ and in particular is greater than $o(\lambda)$ for λ small.

Apply the arguments of Theorem 5 of (1) to approximation of $f - F(A,.)$ by multiples of $D(A,B,.)$ to get

$$\|f - F(A,.) - D(A,\lambda B,.)\| \leq \|f - F(A,.)\| + o(\lambda)\|D(A,B,.)\|.$$

But by the triangle inequality

$$\|f - F(A+\lambda B,.)\| \leq \|f - F(A,.) - D(A,\lambda B,.)\| + \|R(A,\lambda B,.)\|,$$

$$\|f - F(A+\lambda B,.)\| \leq \|f - F(A,.)\| + o(\lambda) + o(\lambda).$$

THEOREM 1: Let F be locally linear at A. Let $f - F(A,.) \not\equiv 0$ and be real on $M(A)$. Let there exist $D(A,C,.) \not\equiv 0$ in the tangent space of F at A which is real on $M(A)$. Then $F(A,.)$ is not strongly unique to f.

Proof: Let $B = iC$ and apply the previous lemma.

REMARK: The hypotheses on reality on $M(A)$ are always satisfied if f, $F(A,.)$, and $D(A,C,.)$ are real.

COROLLARY: Let Q be a subset of the parameter space P, such that for all $A \in Q$

i) F is locally linear at A

ii) $F(A,.)$ is real

iii) there exists a non-zero real element of the tangent space $D(A,.,.)$.

Then no element of (F,Q) is strongly unique in (F,P) to real $f \notin (F,Q)$.

Hypothesis (iii) is always satisfied when there is a parameter a_k whose only function is to multiply a real function $\phi \not\equiv 0$. The case of most interest is when $\phi = 1$.

REMARK: Let X be real. Many approximating functions F have the property that $F(A,.)$ is real on X if A is real: if F is locally linear at A this implies that $D(A,B,.)$ is real for all real B . The above Corollary may then be applied with Q the set of real parameters.

Gutknecht (3) gives a special case of the above remark for real X and ordinary rational functions.

A case of application of the above Corollary when X need not be real is when F is a sum of a_1 and an approximation (not depending on a_1) which, together with its tangent space, can vanish. An example is

$$F(A,.) = a_1 + a_2\phi_2 + \dots + a_n\phi_n .$$

If f is real and non-constant, no real constant can be strongly unique to f .

COROLLARY: Let F be a linear approximating function with real basis $\theta = \{\phi_1, \dots, \phi_n\}$. No best approximation to real f not an approximation is strongly unique.

Proof: If there is a best real Chebyshev approximation to f by real linear combinations of Θ , it is also a best complex approximation (use the zero in the convex hull results of Cheney (2, p.73)). Apply the previous Corollary.

In the most elementary examples of failure of strong uniqueness, X is real and f is real. It is therefore of interest to develop examples in which this is not the case. Of special interest are cases in which all functions are analytic and X is not necessarily a subset of an arc.

EXAMPLE: Let $X = \{x: |Re(x)| \leq 1, |Im(x)| \leq 1\}$ and $f(x) = x^2$. Approximate by first degree polynomials. Consider the zero approximation. M(0) consists of the four corners of X, namely $\{1+i, -1+i, -1-i, 1-i\}$ and f takes the values 2i, -2i, 2i, -2i respectively on these points. Let $I(x) = (1,x)$. On adding up $\overline{f(x)}\, I(x)$ for $x \in M(0)$ we get the zero vector. By the characterization theorem involving the zero in the convex hull (2, p.73), 0 is is a best approximation. As f is imaginary on M(0), real constants are orthogonal to the error on M(0) and 0 is not strongly unique.

REFERENCES

1. Bartelt, M. W., and McLaughlin, H.W., *J. Approximation Theory 9*, 255-266 (1973).
2. Cheney, E. W., "Introduction to Approximation Theory", McGraw Hill, New York (1966).
3. Gutknecht, M., *J. Approximation Theory 23*, 204-213 (1978).
4. Saff, E., and Varga, R., *Bull. Amer. Math. Soc. 83*, 375-377 (1977).

HERMITE-BIRKHOFF QUADRATURE FORMULAS OF GAUSSIAN TYPE

Nira Dyn[1]

Department of Mathematical Sciences
Tel Aviv University
Israel

INTRODUCTION

In this talk we present results concerning the existence of quadrature formulas of Gaussian type related to Hermite-Birkhoff interpolation problems. Given the mxn incidence matrix $E = \{e_{ij}\}_{i=1,j=0}^{m \quad n-1}$ with entries consisting of zeros and ones, a Hermite-Birkhoff quadrature formula of Gaussian type (HB-GQF), is defined as a formula of the form

$$\int_a^b p \, d\sigma = \sum_{e_{ij}=1} a_{ij} \, p^{(j)}(x_i), \quad p \in \Pi_{n-1}, \tag{1}$$

such that the number of parameters in the formula equals the dimension of Π_{n-1} (the space of polynomials of degree $\leq$ n-1):

$$n = \sum_{i=1}^{m} \sum_{j=0}^{n-1} e_{ij} + \#\{x_i \mid x_i \neq a,b\}. \tag{2}$$

In (1) σ is a positive measure on [a,b] and $(x_1,\ldots,x_m) \in S^m$ where $S_m = \{X = (x_1,\ldots,x_m) \mid a \leq x_1 < \ldots < x_m \leq b\}$.

This notion extends the classical notion of Gaussian

---

[1]Part of this work was accomplished at the Mathematics Research Center of the University of Wisconsin-Madison.

Copyright © 1980 by Academic Press, Inc.
All rights of reproduction in any form reserved.
ISBN: 0-12-171050-5

quadrature formulas (principal representations in [6]):

$$\int_a^b p\, d\sigma = \sum_{i=1}^{m} a_i\, p(x_i),\ p \in \Pi_{n-1},\ n = m + \#\{x_i \mid x_i \neq a,b\}$$

where $(x_1,\ldots,x_m) \in S^m$ and $a_i > 0$, $i = 1,\ldots,m$.

Two classes of incidence matrices admitting quadrature formulas of the type (1) are known in the literature:

(i) Hermite matrices consisting of odd Hermite sequences in rows corresponding to interior points of [a,b],

(ii) Quasi-Lagrange matrices consisting of Lagrange sequences in rows corresponding to interior points of [a,b], and satisfying for $\ell = 1,\ldots,n$:

$$\sum_{j=0}^{\ell-1} \sum_{i=1}^{m} e_{ij} \geq \ell - \tilde{m},\quad \sum_{j=0}^{n-1} e_{ij} = n - \tilde{m},\quad \tilde{m} = \#\{x_i \mid x_i \neq a,b\} \qquad (3)$$

The existence uniqueness and extremal properties of the first type of HB-GQF (multiple nodes Gaussian quadratures) are studied in [2], [5], [9] and those of the second type are studied in [8].

In [4] the problem of characterizing incidence matrices admitting HB-GQF is posed and the following necessary condition on such matrices is proved:

Result A. Let $E = \{e_{ij}\}_{i=1,j=0}^{m\quad n-1}$ admits a quadrature formula exact for Π_{n-1}, for some positive measure σ supported on more than m points of [a,b], and let r be the minimal number of ones which must be added to E to obtain a matrix $\tilde{E} = \{\tilde{e}_{ij}\}$ without odd sequences in rows corresponding to interior points of [a,b]. Then $\tilde{E}$ is a Pólya matrix (the number of ones in any first ℓ columns exceeds $\ell - 1$), and

$$n \leq \sum_{i=1}^{m} \sum_{j=0}^{n-1} e_{ij} + r \qquad (4)$$

In the classical Gaussian quadrature formulas (GQF), and in the above mentioned two classes of HB-GQF there is equality in (4), since r is also the number of interior points in these quadrature formulas.

In the following we present two classes of incidence matrices admitting HB-GQF with equality in (4). The first of these classes contains the Hermite and quasi-Lagrange matrices as a special case. For certain subclasses of matrices results concerning the uniqueness and extremal property of the corresponding HB-GQF are also obtained.

The results are formulated for polynomials, but are easily extendable to Extended-Complete-Chebychev systems.

MAIN RESULTS

In the analysis of HB-GQF of the form (1) the rows of E corresponding to the end points of [a,b] have a distinct role. For convenience we term hereafter the rows of E corresponding to interior points of [a,b] "interior rows".

Theorem 1. Let $E = \{e_{ij}\}_{i=1,j=0}^{m \quad n-1}$ be an incidence matrix such that each interior row of E contains an odd Hermite sequence, and all other sequences in such a row are of even length. If E satisfies the extended Pólya conditions (3), then E admits a HB-GQF defined by (1) and (2). Moreover in this HB-GQF the coefficients of the even order derivatives prescribed by the Hermite sequences in the interior rows are all positive.

The proof of Theorem 1 appears in [3]. It is a direct conclusion from the following three results: the Atkinson-Sharma sufficient condition for the regularity of an incidence matrix [1] and the related Chebychev spaces of polynomials [7], the existence of classical GQF for any Chebychev system [6], and the Brouwer Fixed-Point Theorem.

Although our method of proof does not yield uniqueness of the HB-GQF in general, nor in the special case of Hermite matrices, yet the proof of existence is simpler than previous ones. Moreover, the uniqueness and extremal property of the HB-GQF corresponding to Quasi-Lagrange matrices are obtained directly by this method. By the same construction it is also possible to extend the known uniqueness and extremal property of the HB-GQF admitted by Hermite matrices to HB-GQF admitted by Quasi-Hermite matrices (See §3 in [3]).

An additional class of incidence matrices admitting HB-GQF is investigated in Theorem 2. In these matrices again each interior row contains precisely one odd sequence ($\tilde{m} = r$), but this sequence is not necessarily Hermite.

<u>Theorem 2</u>. Let $E = \{e_{ij}\}_{i=1,j=0}^{m \quad n-1}$ be an incidence matrix which can be vertically decomposed into ℓ sub-matrices, each consisting of m rows and $n_1, n_2, \ldots, n_\ell$ columns respectively:

$$E = E(n_1) \oplus E(n_2) \oplus \ldots \oplus E(n_\ell), \quad \sum_{i=1}^{\ell} n_i = n, \tag{5}$$

such that all the non-zero entries of each interior row of E belong to one of the sub-matrices only. If each sub-matrix $E(n_i)$ admits a HB-GQF exact for Π_{n_i-1}, for any positive measure σ, and if each $E(n_i)$ satisfies Result A with equality in (4), then E admits a HB-GQF exact for Π_{n-1}, for any positive measure σ, with points $X = (x_1,\ldots,x_m) \in \Omega^m$ where

$$\Omega^m = \{X \in [a,b]^m \mid x_i < x_j \text{ if } i < j \text{ and rows } i,j \text{ are non-zero in the same sub-matrix of } E\}.$$

Moreover if each $E(n_i)$ admits a unique HB-GQF for each positive measure σ, so does E.

It should be noted that matrices of Theorem 1 can be taken as sub-matrices in Theorem 2. If in addition all the sub-matrices are Hermite then the composite matrix admits a unique HB-GQF. In the following example the matrix E admits a HB-GQF by the results of Theorems 1, 2:

$$E = \begin{pmatrix} 110000 & 1010100 \\ 101100 & 0000000 \\ 000000 & 1001100 \end{pmatrix}, \quad \tilde{E} = \begin{pmatrix} 110000 & 1010100 \\ 111100 & 0000000 \\ 000000 & 1101100 \end{pmatrix}$$

$$a = x_1, \quad a < x_2 < b, \; a < x_3 < b, \quad n_1 = 6, \quad n_2 = 7.$$

Under certain conditions the HB-GQF of Theorem 2 has an extremal property.

<u>Definition 1</u>. The HB-GQF (1) related to $E = \{e_{ij}\}_{i=1,j=0}^{m \quad n-1}$ has an extremal property, if for any f satisfying $f^{(n)} \geq 0$ on [a,b]:

$$\min_{p \in P(E,f)} [(-1)^{\mu} \int_a^b (f-p)\,d\sigma] = (-1)^{\mu} [\int_a^b f\,d\sigma - \sum_{e_{ij}=1} a_{ij} f^{(j)}(x_i)]$$

where $\mu = \sum_{j=0}^{n-1} e_{mj}$ if $x_m = b$ and $\mu = 0$ otherwise, and

$$P(E,f) = \{p \in \Pi_{n-1} \mid p^{(j)}(y_i) = f^{(j)}(y_i),\ \tilde{e}_{ij} = 1,\ Y \in S^m\}$$

with $\tilde{E} = \{\tilde{e}_{ij}\}$ the matrix of Result A related to E.

<u>Theorem 3</u>. Let E satisfy the conditions of Theorem 2. If for any positive measure σ each $E(n_k)$, $1 \leq k \leq \ell$, admits a unique HB-GQF of the form (1) with the extremal property of Definition 1, then the unique HB-GQF admitted by E, for any positive measure σ, has the extremal property of Definition 1

with S^m replaced by Ω^m in the definition of P(E,f).

The proofs of Theorems 2,3 are based on the observation that the Peano kernel K(y,t) for the error in interpolation, specified by a Pólya matrix with even sequences only in interior rows, is of constant sign on $[a,b]^2$. The details of the proofs will appear elsewhere.

REFERENCES

[1] Atkinson, K. and A. Sharma, *SIAM J. Numer. Anal.* 6 (1969), 230-235.

[2] Barrow, D.L., *Math. Comp.* 32 (1978), 431-439.

[3] Dyn, N., *Mathematics Research Center TSR* #1978, July 1979.

[4] Lorentz, G.G. and S.D. Riemenschneider, *in* "Linear Spaces and Approximation", P.L. Butzer and Sz. -Nagy (Ed.) Birkhäuser Verlag, Basel, (1978) (ISNM 40), 359-374.

[5] Karlin, S. and A. Pinkus, *in* "Studies in Spline Functions and Approximation Theory", Academic Press (1976), 113-141 and 142-162.

[6] Karlin, S. and Studden, W.J., "Tchebycheff Systems", Interscience Publishers (1966).

[7] Kimchi, E. and N. Richter-Dyn, *J. Approximation Theory,* 15 (1975), 85-100.

[8] Micchelli, C.A. and T.J. Rivlin, *Advances in Math.* 11 (1973), 93-113.

[9] Popoviciu, T., *Acad. R.P. Romine Fil. Iasi Studii Cerc. Sti.* 6 (1955), 29-57.

THE CARDINALITY OF EXTREME SETS OF RATIONAL FUNCTIONS

Norman Eggert
John Lund

Department of Mathematics
Montana State Univeristy
Bozeman, Montana

I. INTRODUCTION

Let $C[-1,1]$ be the set of continuous real valued functions on the interval $[-1,1]$. Let Π_n be the set of real algebraic polynomials of degree at most n, and $B_n \varepsilon \Pi_n$ be the best approximation to f. The extreme set for f is given by $E_n = \{x \varepsilon [-1,1]: |r_n(x)| = \| r_n \| \}$, where $r_n = f - B_n$ is the error function and $\| - \|$ is the uniform norm. Let $|E_n|$ denote the cardinality of E_n, then by the Chebyshev equioscillation theorem $|E_n| \geq n + 2$.

The strong unicity constant is defined to be the minimal value M_n such that the inequality $M_n(\| f - p\| - \| f - B_n\|) \geq \| P - B_n\|$ is satisfied for all $P \varepsilon \Pi_n$. The sequence $\{M_n\}$ has been considered in several recent papers, including some articles in the present volume. If $f \varepsilon \Pi_k$, for some k, then the sequence $\{M_n\}$ is the constant sequence of ones for $n \geq k$. Let $\mathcal{B} = \{f \varepsilon C[-1,1]: \{M_n\}$ is bounded$\}$ and notice that $\Pi \subseteq \mathcal{B}$, where Π is the set of all algebraic polynomials. It has been conjectured by M. Henry and J. Roulier [3] that $\Pi = \mathcal{B}$. M. Bartelt and D. Schmidt [1] established a criterion for a function to be in the compliment of $\mathcal{B}$. By defining $m_n = |E_n| - (n + 2)$, $A = \{f \varepsilon C[-1,1]: \overline{\lim_n}\, m_n$ is finite$\}$, their formulation can be stated as the following theorem.

Copyright © 1980 by Academic Press, Inc.
All rights of reproduction in any form reserved.
ISBN: 0-12-171050-5

THEOREM 1 (Bartelt and Schmidt [1]): Let $f \in C[-1,1]$. If there exists a positive integer N such that $m_n \leq 2$ wherever $n \geq N$ and $B_n \neq B_{n+1}$ then $f \in \mathcal{A} \setminus \mathcal{B}$.

Notice if $f \in \Pi_k$ then m_n is infinite for all $n \geq k$ and therefore $\mathcal{A} \cap \Pi$ is empty.

It can be shown that if $f \notin \Pi$ is a real analytic function on $[-1,1]$ then m_n is finite. The example in M. Henry and L. Huff [2] shows that the analyticity of a function is not a sufficient condition for the boundedness of $\{m_n\}$. Their example is in fact an entire function that does not belong to $\mathcal{A}$.

One method used to examine the sequence $\{m_n\}$ is to place certain restrictions on the derivatives of f. For example, if $f^{(n)}$ does not vanish in $(-1,1)$, then $m_n = 0$. Since the rational function $f(x) = 1/(x + 2)^q$ has non-vanishing derivatives on $(-1,1)$ it follows from Theorem 1 that $f \in \mathcal{A} \setminus \mathcal{B}$. However, it can be shown that the derivatives of the function $f(x) = 1/(x^2 + 1)$ do vanish in $(-1,1)$. In particular $f^{(2p+1)}$ has at least $(p - 1)$ zeros in $(-1,1)$, and in the next section it is shown that the sequence $\{m_n\}$ for this function is also the zero sequence and thus $f \in \mathcal{A} \setminus \mathcal{B}$.

II. A CLASS OF RATIONAL FUNCTIONS

Let the rational function $f = P/Q \in C[-1,1] \setminus \Pi$ where the polynomials P and Q have degrees p and q respectively. Let

$$\mathcal{N} = \{n \in \mathbb{N}: B_n \neq B_{n+1} \text{ and } \deg B_n \geq p - q\},$$

and for $n \in \mathcal{N}$ define $\ell = \ell_n = n + 1 - \deg B_n \geq 1$. Since

$$B_{n-\ell} \neq B_{n-\ell+1} = \cdot\cdot\cdot = B_n \neq B_{n+1}$$

ℓ_n measures the number of times that the best approximation repeats itself. The following lemma shows that an upper bound for ℓ_n is independent of both n and p.

LEMMA 1: If $f = P/Q$ and $n \in N$ then $0 \leq q - \ell$ and $1 \leq \ell \leq q$.

Proof. For $n \in N$, an alternation set for $r_n = f - B_n = (P - Q B_n)/Q = R/Q$ has exactly $n + 2$ points. By the intermediate value theorem r_n has at least $n + 1$ zeros. The polynomial R has degree at most $n + q - \ell + 1$. Thus R, and hence r_n, has at most $n + q - \ell + 1$ zeros, and the inequality $n + 1 \leq n + q - \ell + 1$ is established.

Since $f(x) = 1/(x^2 + 1)$ is an even function, the corresponding $\ell_n \geq 2$. It follows from Lemma 1 that ℓ_n is equal to 2 and the set N is $\{2k - 1: k \geq 1\}$. Actually this function is in a class of rational functions studied by T. Rivlin [4] which are defined by

$$(1) \quad f = [\beta(1 + t^2 - 2t\, T_q) + \alpha(T_b - t\, T_{|b-q|})]/[1 + t^2 - 2t\, T_q]$$

where T_a is the Chebyshev polynomial of degree a, $\alpha \neq 0$ and β are real numbers, $q \geq 1$ and $0 < |t| < 1$. From the results in [4] the set N and the length ℓ_n are easily computed for all f of the form (1). In fact, for $k \geq 1$, $n = kq + b \in N$ and $\deg B_n = kq + b$, so that $\ell_n = q$; the degree of the denominator of f. This observation and Lemma 1 imply that f is in reduced form. Furthermore, for the functions defined by (1), Rivlin proves that $|E_n| = n + 2$ and therefore $m_n = 0$. By Theorem 1, $f \in A \setminus B$.

III. THE EXTREME SET OF A RATIONAL FUNCTION

Let f be a rational function P/Q that is not a polynomial. In this section the subscript n will be omitted. Recall that if $n \in N$ then a maximal alternation set for r contains $n + 2$ points. Let the extreme set associated with $n \in N$ be given by

$E = \{x_i: i = 0,1,\ldots,n + 1 + m\}$ where $-1 \leq x_0 < x_1 < \ldots < x_{n+1+m} \leq 1$.

Let $\|r\| = \delta$, and suppose that $r(x_{i-1}) = r(x_i)$ for some i, $1 \leq i \leq n + 1 + m$ and $n \in N$. The Mean Value Theorem implies that there is a $y_i \in (x_{i-1}, x_i)$ which is a zero of $D_x r$. The definition of m implies that there are exactly m such pairs of extreme points. The subscripts of these extreme points are partitioned into the following sets:

$$S_+ = \{i: \ 1 \leq i \leq n + 1 + m \ \text{ and } \ r(x_{i-1}) = r(x_i) = \delta\}$$

$$S_- = \{i: \ 1 \leq i \leq n + 1 + m \ \text{ and } \ r(x_{i-1}) = r(x_i) = -\delta\}.$$

One of these sets contains at least $[(m + 1)/2]$ elements and without loss of generality assume that $|S_+| \geq [(m + 1)/2]$. Let $\gamma = \max_{i \in S_+} \{\frac{1}{2}[\delta + r(y_i)]\}$ and it follows that $-\delta < r(y_i) < \gamma < \delta$ for all $i \in S_+$. By considering the triples of points $\{(x_{i-1},\delta), (y_i, r(y_i)), (x_i,\delta)\}$, $i \in S_+$ and the $n + 2$ alternates of r_n, the intermediate value theorem implies that $r - \gamma$ has at least $2[(m + 1)/2] + (n + 1)$ zeros in $(-1,1)$. These observations lead to the following theorem.

THEOREM 2: If $f = P/Q \in C[-1,1] \setminus \Pi$ and $n \in N$ then $m \leq 2[(q - \ell)/2]$.

Proof. The numerator of the rational function $r - \gamma$ has degree $n + q - \ell + 1$, the same as r. Since $r - \gamma$ has at least $2[(m + 1)/2] + (n + 1)$ zeros in $(-1,1)$, $n + q - \ell + 1 \geq 2[(m + 1)/2] + (n + 1)$ or $(q - \ell) \geq 2[(m + 1)/2]$. By considering the four cases when $(m + 1)$ and $(q - \ell)$ are odd and even, the theorem follows.

If the number of distinct complex roots of Q is known, then an alternate bound on m can be obtained. This alternate bound is sharper when Q has fewer than $q - \ell$ distinct complex roots. It is obtained by considering the zeros of $g = r - \gamma$ and $D_x r = D_x g$.

LEMMA 2: The zeros of $g = (r - \gamma)$ and $D_x r$ are exactly the zeros of $D_x g^2$. Furthermore $(r - \gamma)$ has at least $2[(m + 1)/2] + (n + 1)$ zeros in $(-1,1)$ and $D_x r$ has at least $n + 2m$ zeros in $(-1,1)$.

Proof: The first statement follows since $D_x g^2 = 2(r - \gamma)(D_x r)$. The statement on the number of zeros of $(r - \gamma)$ was established in Theorem 2. With the possible exception of two points, the elements of the extreme set are roots of $D_x r$. If $i \in S_+ \cup S_-$ then y_i is also a root of $D_x r$. Thus $D_x r$ has at least $(n + m) + m$ zeros in $(-1,1)$.

Let $2s$ be the number of distinct real roots of Q.

THEOREM 3: If $f = P/Q \in C[-1,1]\setminus \Pi$ and $n \in N$ then $m \leq \frac{2}{3}(q - \ell + s)$.

Proof: Since $g = r - \gamma$, write $g = \overline{R}/Q$ where $\deg \overline{R} = n + q - \ell + 1 > q$. Let t be the number of distinct real roots of Q. Factor Q as AB where all the roots of Q are roots of A and A has simple roots. The degree of A is $2s + t$. Since B is a factor of AB', where $B' = D_x B$, write $AB' = BC$. Then:

$$D_x g^2 = 2g\, D_x(\overline{R}/AB) =$$

$$2g(\overline{R}'\, AB - \overline{R}\, A'B - \overline{R}\, AB')/(AB)^2 =$$

$$2\overline{R}(\overline{R}'A - \overline{R}A' - \overline{R}C)/A^3B^2 = S/A^3B^2.$$

The degree of the numerator S is

$(n + q - \ell + 1) + (n + q - \ell + 2s + t) = 2(n + q - \ell + s) + t + 1.$

The t real roots of Q give vertical asymptotes for g^2. Furthermore $\lim_{x\to\pm\infty} g^2(x) = +\infty$. Since g^2 is a non-negative function the Mean Value Theorem gives t zeros of $D_x g^2$ not in $(-1,1)$. Thus $D_x g^2$ has at most $[2(n + q - \ell + s) + t + 1] - t = 2(n + q - \ell + s) + 1$ zeros in $(-1,1)$. By Lemma 2, $D_x g^2$ has at least $2(n + m + [(m + 1)/2]) + 1$ zeros in $(-1,1)$. It now follows that $q - \ell + s \geq m + [(m + 1)/2]$. By considering the two cases, m odd and m even, the theorem follows.

The following theorem follows from Theorems 2 and 3 by removing the condition that $n \in N$.

THEOREM 4: Let $f = P/Q \in C[-1,1] \setminus \Pi$ and $\| E_n \| = n + 2 + m_n$. Then:

(i) $m_n \leq 2[(q - 1)/2]$

(ii) $m_n \leq \frac{2}{3}(q + s - 1)$

If $2s < q$, that is Q has a real root or a multiple root, then inequality (ii) implies inequality (i).

Since the bounds on m_n in Theorem 4 are independent of n the following corollary is immediate.

COROLLARY 1: $f = P/Q \in A$.

Finally Theorem 4 implies that certain rational functions, which cannot be written in the form (1), are not in B.

COROLLARY 2: Let $f = P/Q$. If $\deg Q < 5$ or $\deg Q = 5$ and Q has only real roots, then $r \in A \setminus B$.

Proof: It follows from Theorem 4 that $m_n \leq 2$ for all n. By Theorem 1, $f \notin B$.

REFERENCES

1. M.W. Bartelt and D. Schmidt, On Poreda's Problem for Strong Unicity Constants, Submitted.

2. M.S. Henry and L.R. Huff, On the behavior of the strong unicity constant for changing dimension, J. Approximation Theory, to appear.

3. M.S. Henry and J.A. Roulier, Lipschitz and strong unicity constants for changing dimension, J. Approximation Theory 22(1978), 85-94.

4. T.J. Rivlin, Polynomials of best uniform approximation to certain rational functions, Numerische Mathematik 4(1962), 345-349.

SOME APPLICATIONS OF APPROXIMATION THEORY IN NUMERICAL CONFORMAL MAPPING

S.W. Ellacott

Department of Mathematics
Brighton Polytechnic
Brighton, Sussex
England

There has been recent interest in methods of conformal mapping based on approximation techniques. These have emphasised the fact that the problems involved in accurate numerical conformal mapping are essentially approximation theoretic in nature. A survey is given of some applications of existing results, and some remaining problems pointed out.

Early work on numerical conformal mapping tended to concentrate on integral equation methods (see (1) for a comprehensive survey up to 1964) but recently there has been increased interest in methods based on construction of simple explicit approximations to the required mapping function. Both as regard techniques and theory, such methods naturally depend on ideas of approximation theory.

In order to fix ideas, we first describe briefly some methods.

The Bergman Kernal method (BKM) is concerned with mapping a simply connected region onto the unit disc, and can be regarded as an orthogonal expansion of the derivative of the mapping function. A good discussion of the theory is given by Nehari(2). Levin, Papamichael and Sideridis have recently shown by a large number of numerical examples that provided a good basis is chosen a very effective numerical method results (3) "Good" in this context means, of course, that the series con-

Copyright © 1980 by Academic Press, Inc.
All rights of reproduction in any form reserved.
ISBN: 0-12-171050-5

verges rapidly and we will discuss this below.

When applied to the same mapping problem the authors own "approximation method" (AM)(4) involves approximating $\ell n\ |z|$ by the real part of a polynomial or other "simple" function p(z) and then taking as the approximate mapping function

$$z \exp(-p(z)) .$$

Some early experiments along these lines were performed using interpolation (see (1)). But the use of modern best (real) minimax approximation algorithms makes this approach much more viable.

Although we shall not discuss it further, it is also worth mentioning at this point some experiments of Opfer(5) with extremal principles for mapping simply connected regions.

The corresponding mapping problem for multiply connected regions is to map the region onto an annulus (if the connectivity is 2) or an annulus with concentric slits (for connectivity > 2). Although orthogonal expansions of the mapping are known (2) they lack the elegance of the BKM and this author is not aware of any serious numerical experiments with them. Generalisation of the AM is reasonably straightforward (6).

The basic justification for any such method is the following well known result. (7)

Theorem 1

If a region Ω is bounded by a finite number of Jordan curves, then any function which is analytic on Ω and continuous on its closure can be uniformly approximated (arbitrarily closely) by a rational function. The poles of this rational function can be prescribed arbitrarily subject only to the restriction that at least one be in each region separated from Ω by the boundary curves. ■

Armed with this result, together with some properties of

families of analytic functions, it can be shown that both the BKM and the AM with polynomial bases, when applied to the problem of determining ϕ which maps a simply connected Jordan region (with 0 an interior point and such that $p(0) = 0$, $p'(0)$ real and positive), converge as the degree of the approximating polynomial tends to ∞. Moreover convergence is uniform on every closed subset of the interior (Refs (2), (4).) (That the region be a Jordan region is required only to show the completeness of the polynomial basis and can be weakened in various ways). For the AM it is also a triviality to show that the real part of the approximating polynomial converges uniformily on the whole region to its Green's function with respect to zero (4). For a multiply connected region one must take a basis of rational functions with poles chosen as described in Theorem 1: the convergence analysis of the AM then extends to the problem of mapping a multiply connected region onto a slit annulus (6).

In practice, however, these results are of limited usefulness since it is easy to find examples for which convergence is slow. One needs estimates of the rate of convergence (and hopefully clues as to the choice of a better basis than just power functions).

The situation here is reasonably satisfactory when the region is bounded by analytic Jordan curves. For a simply connected region Ω we have the following result due to Walsh. ((7) p75). Suppose the boundary curve C is analytic, and let $G(z)$ be the Green's function with respect to ∞ of the *outside* of C. We denote by C_R the curve $G(z) = \ell n\, R$ $(R \geq 1)$ so that $C = C_1$.

Theorem 2

Suppose a function f is analytic on the region inside C_R. Then there exists a constant M such that for each positive integer n with $n \geq 0$ there is a polynomial q_n of degree $\leq n$ with

$$||f - q_n||_\infty \leq M/R^n$$

where $|| \; ||_\infty$ denotes the uniform norm on Ω . ■

Now since the region Ω is being mapped onto the unit disc, the fact that C is analytic means that the mapping function ϕ itself can be extended analytically over the boundary. One can then apply Theorem 2 to the functions

$$\ell n(\phi(z)/) \text{ for the AM}$$

and

$$\phi'(t) \qquad \text{for the BKM}$$

One can then easily prove geometric uniform convergence of

$\text{Re}(p_n(z))$ to the Green's function with respect to zero of Ω as $n \to \infty$

(where p_n is the approximating polynomial of degree $\leq n$ in the AM (4).)

That a similar result holds in the least squares sense for the convergence of the Bergman Kernal series is a trivial consequence of Theorem 2, although oddly enough this author has not been able to find it explicitly stated anywhere in the literature.

Since the image region is the unit disc it is often possible (e.g. using Schwarz' Reflection Principle) to find exactly or approximately the positions of the poles (and in the case of the AM, zeros) nearest to C of the analytic extension of ϕ. By incorporating these into the approximation, a considerable improvement in rate of convergence is obtained (3), (4), as again would be predicted by Theorem 2. An analagous (although rather more complicated) version of Theorem 2 holds for multiply connected regions ((7) p252).

When the boundary curve is not analytic, (i.e. when one actually wants to perform the mapping to solve singular plane potential problems!) the situation is much less satisfactory. Obviously one needs to straighten out corners by means of a suitable fractional power but numerical experiments show that a single term of this type is not always sufficient to get good convergence (3),(4). An asymptotic expansion for the mapping function is known (8) and this was successfully exploited by Levin, Papamichael and Sideridis(3), but this author is aware of no actual expressions for the rate of convergence.

Certain known results ((7) p371) appear to suggest that some bound in terms of the order of Lipshitz continuity of derivatives of ϕ might be possible.

From a practical point of view a hopeful way forward would seem to lie in the use of non-linear approximation techniques, but there is a need for more results on degree of approximation in order to guide in a suitable choice of approximating functions. Other open problems concern questions of conditions for uniform convergence of the approximate mappings and (a related problem) conditions for the approximation to be schlicht.

REFERENCES

1. Gaier, D., "Konstruktive Methoden der Konformen Abbildung" Springer-Verlay, Berlin, (1964).
2. Nehari, Z., "Conformal Mapping" McGraw Hill, New York,(1952) (Reprinted Dover, New York (1975)).
3. Levin, D., Papamichael, N., and Sideridis A., *J. Inst. Maths. Applics. 22,* 171-187 (1978).

4. Ellacott, S., "A Technique for Approximate Conformal Mapping" in "Multivariate Approximation" (D. Handscomb, ed), Academic Press, London, (1978.)
5. Opfer, G., "New Extremal Properties for Constructing Conformal Mappings" *Numer. Math. 32*,437-476 (1979).
6. Ellacott, S., "On the Approximate Conformal Mapping of Multiply Connected Domains" *Numer. Math. 33*, 437-476 (1979).
7. Walsh, J.L., "Interpolation and Approximation by Rational Functions in the Complex Domain" *American Math. Soc. 5th ed.* (1969.)
8. Lehman, R.S., "Development of the Mapping Function at an Analytic Corner" *Pacific J. Math. 7*, 1437-1444 (1957).

APPROXIMATION BY POLYNOMIALS WITH INTEGRAL COEFFICIENTS IN SEVERAL COMPLEX VARIABLES

Le Baron O. Ferguson

Department of Mathematics
University of California
Riverside, California

By a polynomial with integral coefficients we mean a function of the form

$$\sum_{k_1,\ldots,k_n=0}^{m} a_{k_1,\ldots,k_n} z_1^{k_1} \cdots z_n^{k_n}$$

where the a's belong to a fixed but arbitrary discrete subring A of the complex numbers $\underset{\sim}{C}$ and possessing rank 2. For example, A could be the ring of Gaussian integers, $\underset{\sim}{Z} + i\underset{\sim}{Z}$, where $\underset{\sim}{Z} = \{\cdots -1, 0, 1, \cdots\}$. Let X be a compact subset of the space $\underset{\sim}{C}^n$ of n complex variables. We will study the question of what complex valued functions on X can be approximated in the uniform norm $(\|f\|_X = \sup_{x \in X} |f(x)|)$ by elements of the ring $A[z]$ of polynomials with integral coefficients. The ring of such functions will be denoted by $A[z]^-$. It is clear that such functions must be continuous on X and holomorphic on the interior X° of X. The spaces of all such functions are denoted by $C(X)$ and $H(X^\circ)$ respectively.

In any study of polynomial approximation in $\underset{\sim}{C}^n$ one must consider the following concept. See [5].

1. DEFINITION. Let X be a compact subset of $\underset{\sim}{C}^n$. Then we set

$$h(X) = \{z \in \underset{\sim}{C}^n : |p(z)| \le \|p\|_X, \text{all } p \in \underset{\sim}{C}[z]\}$$

Copyright © 1980 by Academic Press, Inc.
All rights of reproduction in any form reserved.
ISBN: 0-12-171050-5

and call it the polynomial hull of X . If $h(X) = X$ then X is said to be polynomially convex.

It is easy to see that $h(X)$ is compact and also that $h(h(X)) = h(X)$, i.e., the polynomial hull is polynomially convex.

For the connection with ordinary convexity, we have the following result. The converse is false as we will see later.

2. EXAMPLE. If X is a compact convex subset of $\mathbb{C}^n$, then X is polynomially convex.

Proof. Let z' be any element of $\mathbb{C}^n \setminus X$. Since $X \subset h(X)$ it suffices to show that there exists a polynomial $p' \in \mathbb{C}[z]$ such that $\|p'\|_X < |p'(z')|$. But X is compact and convex, hence there exist constants c and ε , $\varepsilon > 0$, and a linear functional f on $\mathbb{C}^n$ such that

$$\sup_{z \in X} \operatorname{Re} f(z) \le c - \varepsilon < c \le \operatorname{Re} f(z') , \tag{1}$$

by the Hahn-Banach theorem. In general, a linear functional on $\mathbb{C}^n$ has the form

$$f(z) = \lambda_1 z_1 + \cdots + \lambda_n z_n$$

where $\lambda_i \in \mathbb{C}$ $(1 \le i \le n)$. Thus f is continuous and $f(X \cup \{z'\})$ is compact. It follows that for some truncation $p(w) = 1 + w + \cdots + w^k/k!$ of the MacLauren series for the complex exponential function we have

$$\|p(w) - \exp(w)\|_{f(X \cup \{z'\})} < \frac{e^c - e^{c-\varepsilon}}{2} .$$

It follows that

$$\|p(f(z)) - \exp(f(z))\|_{X \cup \{z'\}} < \frac{e^c - e^{c-\varepsilon}}{2} . \tag{2}$$

From (1) we have

$$\|\exp(f(z))\|_X \le e^{c-\varepsilon} < e^c \le |\exp(f(z'))|$$

and with (2) this gives

$$\|p(f(z))\|_X < \frac{e^{c-\varepsilon}+e^c}{2} < |p(f(z'))| \ .$$

It only remains to notice that $p(f(z))$ is a polynomial in z . ▯

In the case of uniform approximation by polynomials, we can restrict our attention to polynomially convex X in view of the following.

3. PROPOSITION. Let X be a compact subset of $\underset{\sim}{C}^n$, $n \ge 1$, f a map of X into $\underset{\sim}{C}$ and P any subset of $\underset{\sim}{C}[z]$. If f is uniformly approximable by elements of P , then there exists a unique extension $\tilde{f}$ of f to $h(X)$ such that $\tilde{f}$ is uniformly approximable by elements of P .

<u>Proof</u>. Let $\{p_m\}$ be a sequence in P which converges uniformly to f on X . Then by definition of $h(X)$, $\{p_m\}$ is Cauchy in the uniform norm on all of $h(X)$, hence converges uniformly to a limit function $\tilde{f}$ on $h(X)$. It is clear that this $\tilde{f}$ is an extension of f to $h(X)$. To see uniqueness, suppose that $\tilde{f}'$ is another extension of f to $h(X)$ and that $\{g_m\}$ is a sequence in P which converges uniformly to $\tilde{f}'$ on $h(X)$. Then for any $\epsilon > 0$ there exists an integer m_ϵ such that $i,j > m_\epsilon$ implies

$$\|p_i - \tilde{f}\|_X = \|p_i - f\|_X < \frac{\varepsilon}{2}$$

and

$$\|g_j - \tilde{f}'\|_X = \|g_j - f\|_X < \frac{\varepsilon}{2} .$$

It follows that

$$\varepsilon > \|p_i - g_j\|_X = \|p_i - g_j\|_{h(X)}$$

whence both $\{p_m\}$ and $\{g_m\}$ tend to the same limit on $h(X)$, i.e., $\tilde{f}'$ and $\tilde{f}$ agree on $h(X)$. ▯

For the case $n = 1$ the polynomial hull of X has the following topological characterization. See [1, Thm. 1.3.3] for example.

4. PROPOSITION. Let X be a compact subset of $\mathbb{C}$ and $\{U_i\}_{i=1}^{\infty}$ the collection (possibly finite) of bounded connected components of $\mathbb{C}\backslash X$. Then

$$h(X) = X \cup \left(\bigcup_{i=1}^{\infty} U_i\right) .$$

Thus X is polynomially convex if and only if it has a connected complement.

Proof. Clearly, $X \subset h(X)$. If U_i is a bounded connected component of $\mathbb{C}\backslash X$, then U_i is closed in $\mathbb{C}\backslash X$. Thus $(cl(U_i)\backslash U_i) \subset X$ and by the maximum modulus principle, if $z_0 \in U_i$ and $p \in \mathbb{C}[z]$, then we have

$$|p(z_0)| \leq \sup_{z \in cl(U_i)\backslash U_i} |p(z)| \leq \|p\|_X .$$

Thus $U_i \subset h(X)$ for all i.

Let U be the unbounded component of $\mathbb{C}\backslash X$ and z_0 an element of U. We will be done once we show that z_0 is not in $h(X)$. Since U is open in $\mathbb{C}\backslash X$ and $\mathbb{C}\backslash X$ is open in $\mathbb{C}$, U is open in $\mathbb{C}$. There exists $\varepsilon > 0$ such that

$(z_0+\varepsilon D)\cap X = \phi$, where D is the closed unit disk in $\underset{\sim}{C}$. Pick z_1 in $\underset{\sim}{C}$ such that

$$|z_0-z_1| = \frac{\varepsilon}{4} , \tag{1}$$

that is

$$\frac{1}{|z_0-z_1|} = \frac{4}{\varepsilon} . \tag{2}$$

Since $|z-z_0| > \varepsilon$, for $z\in X$ we have, by (1),

$$|z-z_1| > \frac{3\varepsilon}{4} \qquad z\in X$$

and

$$\frac{1}{|z-z_1|} < \frac{4}{3\varepsilon} \qquad z\in X . \tag{3}$$

Since $1/(z-z_1)$ is a holomorphic function of z in a neighborhood of

$$X_o = X\cup(\bigcup_{i=1}^{\infty} U_i)\cup\{z_0\}$$

there exists, by Runge's theorem (cf. [4, 13.7]), a p in $\underset{\sim}{C}[z]$ such that

$$\|p(z) - \frac{1}{z-z_1}\|_{X_o} < \frac{4}{3\varepsilon} . \tag{4}$$

By (3) and (4)

$$\|p\|_X < \frac{8}{3\varepsilon}$$

whereas by (2) and (4)

$$|p(z_0)| > \frac{8}{3\varepsilon} .$$

Thus $p(z_0) > \|p\|_X$ which shows that $z_0\notin h(X)$. ▯

We conclude this introduction to polynomial convexity with two examples.

5. EXAMPLE. Let a compact subset X of $\underset{\sim}{C}^2$ be defined by

$$X = \{z \in \underset{\sim}{C}^2 : |z_1| = 1 = |z_2|\} .$$

By considering the polynomials $p_1(z) = z_1$ and $p_2(z) = z_2$ we see from the definition of $h(X)$ that

$$h(X) \subset \{z \in \underset{\sim}{C}^2 : |z_1| \leq 1\}$$

and

$$h(X) \subset \{z \in \underset{\sim}{C}^2 : |z_2| \leq 1\} .$$

It remains to show that $X_o \subset h(X)$. Let z' be an element of $\underset{\sim}{C}^2$ such that $|z_1'| \leq 1$ and $|z_2'| \leq 1$. Let p be any polynomial in $\underset{\sim}{C}[z]$. Since $z' \in X_o$ we have

$$|p(z')| \leq \sup_{z \in X_o} |p(z)| = \sup_{z \in X_o} |p(z_1,z_2)| .$$

But

$$\sup_{z \in X_o} |p(z_1,z_2)| = \sup_{|z_1| \leq 1} \left(\sup_{|z_2| \leq 1} |p(z_1,z_2)| \right)$$

and for each fixed z_1 , $p(z_1,z_2)$ is a polynomial in the single complex variable z_2 and so by the maximum modulus principle

$$\sup_{|z_1| \leq 1} \left(\sup_{|z_2| \leq 1} |p(z_1,z_2)| \right) = \sup_{|z_1| \leq 1} \left(\sup_{|z_2| = 1} |p(z_1,z_2)| \right)$$

$$= \sup_{|z_2| = 1} \left(\sup_{|z_1| \leq 1} |p(z_1,z_2)| \right)$$

$$= \sup_{|z_2|=1} \left(\sup_{|z_1|=1} |p(z_1,z_2)| \right)$$

$$= \sup_{z \in X} |p(z)| \ .$$

Thus $|p(z')| \leq \|p\|_X$ and z' is in $h(X)$, as desired.

It is clear that the idea can be applied to prove that for any positive integer n, if

$$X = \{z \in \underset{\sim}{C}^n : |z_1| = \cdots = |z_n| = 1\} ,$$

then

$$h(X) = \{z \in \underset{\sim}{C}^n : |z_i| \leq 1, 1 \leq i \leq n\} .$$

6. REMARKS. First, it is clear that X does not contain the topological boundary of $h(X)$ in these examples (for $n>1$) although this is always the case for $n = 1$ as we see by Proposition 4. Secondly, Proposition 4 shows that if $\underset{\sim}{C} \backslash X$ is connected, then $X = h(X)$. This is false for general $X \subset \underset{\sim}{C}^n$ ($n>1$) since in the example above, $\underset{\sim}{C} \backslash X$ is connected, as is easily seen.

7. EXAMPLE. Let the compact subset X of $\underset{\sim}{C}^n$ ($n>1$) be defined by

$$X = \{z \in \underset{\sim}{C}^n : |z_1| = 1, z_2 = \cdots = z_n = 0\} .$$

Then we have $h(X) = X_o$ where, by definition,

$$X_o = \{z \in \underset{\sim}{C}^n : |z_1| \leq 1, z_2 = \cdots = z_n = 0\} .$$

This is quite close to the previous example and is proved by the same technique.

We next give some necessary conditions in order that a complex valued function f on a compact subset X of $\underset{\sim}{C}^n$ be approximable by elements of $A[z]$.

8. PROPOSITION. If f is a complex valued function on a compact subset X of $\underset{\sim}{C}^n$ and f is in $A[z]^-$, then f has

a unique extension $\tilde{f}$ to $h(X)$ which is continuous on $h(X)$, holomorphic on $h(X)^\circ$ and in $A[z]^-$.

The proof is immediate from Proposition 3 since uniform limits preserve continuity and holomorphicity.

An example of the use of Proposition 8 is the following. Let $X = \{z \in \underset{\sim}{C} : \delta_1 \le |z| \le \delta_2\}$ where $0 < \delta_1 < \delta_2 < 1$ and let the function $z + 1/2$ be restricted to X. By Proposition 4, $h(X) = \delta_2 D$ where D is the closed unit disk in $\underset{\sim}{C}$. By the principle of holomorphic continuation, the only continuous extension of f to $h(X)$ which is holomorphic on $\delta_2 D^\circ$ is the function $z + 1/2$ restricted to $\delta_2 D$. But this function is not in $A[z]^-$, since any $p \in A[z]$ has an element of A as its value at $z = 0$ and every nonzero element of a discrete subring of $\underset{\sim}{C}$ has modulus at least unity, hence $1/2 \notin A^-$. Thus $f \notin A_o^-[z]$ for any discrete subring A_o of $\underset{\sim}{C}$.

In the case $X \subset \underset{\sim}{C}$ it is easy to use Proposition 8 as follows. Given a complex valued function f on X we seek an extension f' of f to $h(X)$ which is continuous on $h(X)$ and holomorphic on $h(X)^\circ$. If none such exists, then $f \notin A[z]^-$. If such an f' exists, then $f \in A[z]^-$ if and only if $f' \in A[z]^-$. This is true since if $f \in A[z]$ then by Proposition 8 there exists an extension $\tilde{f}$ of f to $h(X)$ with $\tilde{f} \in A[z]^-$ and which is continuous on $h(X)$ and holomorphic on $h(X)^\circ$. By Proposition 4 and the maximum modulus principle, $\tilde{f}$ is uniquely determined by f hence $\tilde{f} = f'$ which shows that $f' \in A[z]^-$. The converse is obvious.

This procedure is based on the fact that if $X \subset \underset{\sim}{C}$, then there is at most one extension f' of f to $h(X)$ which is continuous on $h(X)$ and holomorphic on $h(X)^\circ$. In the case $X \subset \underset{\sim}{C}^n$, $n > 1$, the extension need not be unique, as the following example shows.

Let $X = \{z: |z_1| = 1, z_2 = \cdots = z_n = 0\}$. We know from Example 7 that $h(X) = \{z: |z_1| = 1, z_2 = \cdots = z_n = 0\}$. If we define $f(z) = 1$ for all z in X , then certainly $f \in (\underset{\sim}{Z} + i\underset{\sim}{Z})(z)$. Also, the function

$$f'(z) = |z_1| \qquad z \in h(X)$$

is an extension of f to $h(X)$ which is continuous on $h(X)$ and holomorphic on $h(X)^\circ$ (indeed, $h(X)^\circ$ is empty). But it is clear that the unique extension $\tilde{f}$ of f which is in $(\underset{\sim}{Z} + i\underset{\sim}{Z})(z)$ is simply $\tilde{f}(z) = 1$, $z \in h(X)$.

In view of the foregoing we need only consider the problem for those X such that $X = h(X)$, i.e., the polynomially convex subsets of $\underset{\sim}{C}^n$.

9. PROPOSITION. Let X be a compact subset of $\underset{\sim}{C}^n$ and f a complex valued function on X . If $f \in A[z]^-$ then the coefficients of its power series expansion about each point in $A^n \cap X^\circ$ are elements of A .

Proof. The idea of the proof is simply that if a sequence of holomorphic functions converges uniformly to f in a neighborhood of a point, then so do the corresponding power series expansions. It follows from this that the corresponding coefficients coverge. Here the coefficients are in A , which is closed, hence the limits are in A .

For a detailed proof, let $f \in A[x]^-$. Since $X^\circ \subset h(X)^\circ$ we see from Proposition 8 that f has a power series expansion (with nonzero radius of convergence) about each point of X° . For z_0 in $A^n \cap X^\circ$ let $\delta > 0$ be so small that the polydisk $z_0 + \delta D^n \subset X^\circ$ and

$$f(z) = \sum_{k \in \underset{\sim}{N}^n} b_k (z-z_0)^k , \qquad z \in (z_0 + \delta D^n) .$$

Let $\{q_m\}$ be a sequence of polynomials in $A[z]$ which converge uniformly on X to f. Then $q_m(z+z_0) \to f(z+z_0)$ uniformly on δD^n and

$$f(z+z_0) = \sum_{k \in \underset{\sim}{N}^n} b_k z^k , \qquad z \in \delta D^n .$$

If, for each m,

$$q_m(z+z_0) = \sum_{k \in \underset{\sim}{N}^n} a_k^{(m)} z^k$$

where, of course, all but finitely many of the $a_k^{(m)}$, for fixed m, are 0 then we have $a_k^{(m)} \in A$ since $z_0 \in A^n$ and the coefficients of q_m are in A by hypothesis. Since the convergence is uniform on δD^n, the respective power series converge term by term. Thus for each $k \in \underset{\sim}{N}^n$, $a_k^{(m)} \to b_k$ $(m \to \infty)$. But A is a discrete subring of $\underset{\sim}{C}$, by hypothesis, hence closed. Thus $b_k \in A$ for all k. ∎

By restricting a complex valued function on $\underset{\sim}{C}^n$ to a function of one complex variable, in certain ways, we can get additional necessary conditions for approximability, as follows.

For any pair (k,a), $1 \le k \le n$ and $a \in A^{n-1}$ we define an injection $j_{k,a}: \underset{\sim}{C} \to \underset{\sim}{C}^n$ by

$$j_{k,a}(z) = (a_1, \ldots, a_{k-1}, z, a_k, \ldots, a_{n-1}) .$$

Then $\pi_k \circ j_{k,a} = id_{\underset{\sim}{C}}$ (the identity map on $\underset{\sim}{C}$) where π_k is the kth coordinate projection of $\underset{\sim}{C}^n$ and "$\circ$" denotes composition of functions. For each pair (k,a) as above let

$$X_{k,a} = \pi_k(X \cap j_{k,a}(\mathbb{C})) \text{ .}$$

If f is a complex valued function on X and $\|f-q\|_X < \varepsilon$, where $q \in A[z_1,\ldots,z_n]$, then we have

$$\|f \circ j_{k,a} - q \circ j_{k,a}\|_{X_{k,a}} < \varepsilon$$

and $q \circ j_{k,a} \in A[z]$ $(z \in \mathbb{C})$. Thus, in order that f be in $A[z]^-$ it is necessary that for all (k,a), as above, $f \circ j_{k,a}$ be in $A[z]^-$ on $X_{k,a}$, i.e., a number of one-dimensional problems arise. Note that only finitely many of the $X_{k,a}$ are nonempty. This follows from the discreteness of A^n.

We turn now to sufficient conditions for approximability. Henceforth X is assumed to be polynomially convex.

10. DEFINITION. A compact subset X of $\mathbb{C}^n$ is said to be Mergelyan if it satisfies the following condition. Any complex valued function which is continuous on X and holomorphic on X° is in $\mathbb{C}[z]^-$.

The terminology is motivated by the following result of Mergelyan [3] (see also [4]). A compact subset of $\mathbb{C}$ is Mergelyan if and only if its complement $\mathbb{C}\backslash X$ is connected.

In higher dimensions this criterion is no longer valid, as can be seen from the examples above.

A result which does extend, in a sense, is Runge's theorem. In one of its forms, this theorem states that every complex valued function holomorphic in a neighborhood of a compact set X of $\mathbb{C}$ is in $\mathbb{C}[z]$ if and only if $\mathbb{C}\backslash X$ is connected ([4]). From Proposition 4 we see that $\mathbb{C}\backslash X$ is connected if and only if X is polynomially convex. It is this last criterion which remains in force as we pass to higher dimensions.

11. THEOREM. Let X be a compact polynomially convex subset of $\mathbb{C}^n$. Then any function which is holomorphic in a neighborhood of X is in $\mathbb{C}[z]^-$.

For a proof see [1, Thm. 2.7.7].

It is easy to see from this theorem that every Mergelyan set is polynomially convex.

The following technical result will be used to exhibit a large class of Mergelyan subsets of $\mathbb{C}^n$, for any n.

12. PROPOSITION. Let X_1 and X_2 be disjoint compact subsets of $\mathbb{C}$ such that $X_1 \cup X_2$ is polynomially convex, and suppose that each X_i satisfies the following condition (*): There exists z_i' in X_i and $\delta_i > 0$ such that whenever $1 < a < 1+\delta_i$ we have

$$(aX_i - (a-1)z_i')^\circ \supset X_i .$$

Then $X_1 \cup X_2$ is Mergelyan. In addition, any polynomially convex compact subset of $\mathbb{C}^n$ which satisfies condition (*) is Mergelyan.

Condition (*) might be viewed as saying that the dilation of X_i about z_i' of magnitude a is a neighborhood of X for all a in $(1, 1+\delta_i)$. The idea of the proof is simply to dilate the domain of definition of the function to be approximated. This gives a function which is holomorphic in a neighborhood of $X_1 \cup X_2$, but still close to the original one, for dilations close to 1, by uniform continuity. We omit the details.

As a corollary of this proposition we see that every compact convex subset X of $\mathbb{C}^n$ with nonvoid interior is Mergelyan. It suffices to notice that condition (*) of the proposition is satisfied whenever $z_i' \in X^\circ$ and $a > 1$. The assumption here that the interior be void is necessary as can be seen by means of Example 7.

The following is the main result on approximation by integral polynomials in $\underset{\sim}{C}^n$ where $n > 1$.

13. THEOREM. Let A be a discrete subring of $\underset{\sim}{C}$ with rank 2 and let X be a Mergelyan subset of the open unit polydisk, i.e.,

$$X \subset \{z \in \underset{\sim}{C}^n : |z_1| < 1, \ldots, |z_n| < 1\},$$

and $0 \in X^\circ$. If f is a complex valued function on X, then the following are equivalent:

(i) $f \in A[z]^-$;

(ii) f is continuous on X, holomorphic on X° and the coefficients of its power series expansion about 0 lie in A.

<u>Proof</u>. We have already seen that (i) implies (ii). Conversely, suppose that (ii) holds and that $\epsilon > 0$. Since X is compact and the coordinate projections are continuous we have, where ρ and the ρ_i's are so defined, $\rho_i = \|z_i\|_X < 1$ so $\rho = \max_{1 \le i \le n} \rho_i < 1$. It is easy to see that $(1-\rho)^{-n} = \sum_{k \in \underset{\sim}{N}^n} \rho^{k_1 + \cdots + k_n}$. Let δ be chosen so that $z \in \underset{\sim}{C}$ implies $|z-a| < \delta$ for some $a \in A$. There exists a finite subset F of N^n such that $\sum_{k \in \underset{\sim}{N}^n \setminus F} \rho^{k_1 + \cdots + k_n} < \frac{\epsilon}{3\delta}$. Since X is Mergelyan, by hypotheses there exists a sequence of polynomials $\{p_m\}$ in $\underset{\sim}{C}[z]$ such that $p_m \to f$ uniformly on X. Let $p_m(z) =$

$\sum_{k \in \underset{\sim}{N}^n} a_k^{(m)} z^k$, where, of course, all but finitely many of the $a_k^{(m)} = 0$, for each m , and let $f(z) = \sum_{k \in \underset{\sim}{N}^n} a_k z^k$ in a neighborhood of the origin. Then, as in the proof of Proposition 9, $a_k^{(m)} \to a_k$ as $m \to \infty$. Thus there exists a positive integer N such that $m > N$ implies

$$\|p_m - f\|_X < \frac{\varepsilon}{3} \tag{*}$$

and

$$|a_k^{(m)} - a_k| < \frac{\varepsilon}{3M(\text{card } F)} \qquad k \in F$$

where $M = \max_{k \in F} \|z^k\|_X$. Thus if $m > N$ and $[p_m]$ denotes the polynomial p_m with each coefficient replaced by a nearest element of A , we have

$$\begin{aligned} \|p_m - [p_m]\| &\le \sum_{k \in F} |a_k^{(m)} - a_k| \, \|z^k\| + \sum_{k \notin F} \delta \|z^k\| \\ &\le \sum_{k \in F} \frac{\varepsilon}{3M(\text{card } F)} M + \delta \sum_{k \notin F} \rho^{k_1 + \cdots + k_n} < \frac{\varepsilon}{3} + \frac{\varepsilon}{3} \\ &= \frac{2\varepsilon}{3} . \end{aligned}$$

This estimate, together with (*), gives $\|[p_n] - f\|_X < \varepsilon$. ∎

When X does not contain the origin, a similar result holds, as follows.

14. COROLLARY. Let A be a discrete subring of $\underset{\sim}{C}$ with rank 2 and let X be a polynomially convex compact subset of the open unit polydisk, $0 \notin X$ and let X satisfy condition (*) of Proposition 12. Then every continuous complex valued function on X which is holomorphic on X° is in $A[z]$.

Proof. We first quote the "separation lemma" of [2]. This says that if X_1 and X_2 are two compact subsets of $\mathbb{C}^n$ and there exists a polynomial f such that $h(f(X_1)) \cap h(f(X_2)) = 0$, then $h(X_1 \cup X_2) = h(X_1) \cup h(X_2)$.

Now suppose we are in the situation stated in the corollary. Since $0 \notin X$ and X is polynomially convex by hypothesis, there is a p in $\mathbb{C}[z]$ with $|p(0)| > \|p\|_X$. Thus $p(0)$ is a point of the complex w-plane lying outside of the circle $|w| = \|p\|_X$ which in turn contains $p(X)$. By the continuity of p there exists a closed ball X_1 centered at $z = 0$ such that $p(X_1)$ lies outside of the circle $|w| = \|p\|_X$. By Proposition 4 we have $h(p(X_1)) \cap h(p(X)) = \phi$ and applying Kallin's [2] separation lemma we see that $X_1 \cup X$ is polynomially convex. It is clear that every dilation of X_1 about the origin of magnitude $a > 1$ is a neighborhood of X_1. Thus X_1 and X_2 satisfy the hypotheses of Proposition 12; hence $X_1 \cup X$ is Mergelyan. Let f be continuous on X and holomorphic on X°. Extend f to $X_1 \cup X$ by setting $f(X_1) = \{0\}$. Then f is continuous on $X_1 \cup X$, holomorphic on $(X_1 \cup X)^\circ$ and the coefficients of its power series expansion about 0 are all 0, hence in A. Thus $f \in A[z]^-$ by Theorem 13. ▯

REFERENCES

1. Hörmander, L., An Introduction to Complex Analysis in Several Variables, University Series in Higher Math., D. Van Nostrand Co., Inc., Princeton, New Jersey, 1966.

2. Kallin, E., Polynomial convexity: The three spheres problem, in "Proc. Conf. Complex Analysis (Minneapolis, 1964)," pp. 301-304, Springer-Verlag, New York, 1965.

3. Mergelyan, S. N., "On the representation of functions by a series of polynomials on closed sets," Dokl. Akad. Nauk SSSR 78(1951), 405-408. (Russian). Trans: Amer. Math. Soc. Transl., no. 35.
4. Rudin, W., Real and Complex Analysis, 2d ed., McGraw-Hill Series in Higher Math., McGraw-Hill Publ. Co., New York, 1974.
5. Wermer, J., "Banach algebras and analytic functions," Advances in Math. 1(1961), 51-102.

Research sponsored by the Air Force Office of Scientific Research, Air Force Systems Command, USAF, under Grant No. AFOSR-78-3599-A. The United States government is authorized to reproduce and distribute reprints for government purposes, notwithstanding any copyright notation hereon.

THE N-WIDTHS OF SOME SETS OF ANALYTIC FUNCTIONS

Stephen D. Fisher[1]

Department of Mathematics
Northwestern University
Evanston, Illinois

Charles A. Micchelli

IBM Thomas J. Watson Research Center
Yorktown Heights, New York

I. INTRODUCTION

The n-width of a subset A of a Banach space X is defined by the formula

$$d_n(A,X) = \inf_{X_n} \sup_{f \in A} \inf_{g \in X_n} ||f - g||$$

where X_n varies over all n dimensional subspaces of X. In this report we summarize several results we have recently attained on n-widths of certain sets of analytic functions. Specifically, Ω is a domain in the complex plane, K a compact subset of Ω, and A is the restriction to K of the closed unit ball of $H^\infty(\Omega)$, the space of bounded holomorphic functions on Ω. Further, ν is a non-negative measure on K of mass one and the Banach space X is either $L^q(\nu)$, $1 \leq q \leq \infty$, or C(K), the space of continuous functions on K. We seek the n-width of A in X; here n is the complex dimension.

A. The Simply-connected Case

If Ω is simply connected, then the Riemann mapping theorem allows us to assume that Ω is the open unit disc Δ_0. A Blaschke product B(z), of degree r, is an analytic function

[1]Research supported in part by a grant from National Science Foundation.

Copyright © 1980 by Academic Press, Inc.
All rights of reproduction in any form reserved.
ISBN: 0-12-171050-5

of the form

$$B(z) = \lambda \Pi_1^r (z - a_j)(1 - \bar{a}_j z)^{-1}$$

where $|\lambda| = 1$ and $a_1, \ldots, a_r$ are points of Δ_o. We let $\mathcal{B}_n$ denote the set of all Blaschke products of degree n or less.

Theorem 1. If $\Omega = \Delta_o$, then

$$d_n(A,X) = \inf \{||B||_X: B \in \mathcal{B}_n\}$$

B. General Domains

Let Ω be a domain which is regular for the Dirichlet problem; let $g(z,\zeta)$ be the Green's function for Ω with pole at ζ. A Blaschke product of degree r is a (multiple-valued) analytic function of the form

$$B(z) = \lambda \exp[-\sum_1^r \{g(z,\zeta_j) + ih(z,\zeta_j)\}]$$

where $|\lambda| = 1$, $\zeta_1, \ldots, \zeta_r$ are points of Ω, and $h(z,\zeta_j)$ is the (multiple-valued) harmonic conjugate of $g(z,\zeta_j)$. Let $\mathcal{B}_n$ denote the set of Blaschke products of degree n or less

Theorem 2. Suppose Ω has m + 1 complementary components none of which is trivial. Then

$$\inf \{||B||_X: B \in \mathcal{B}_{n+m}\} \leq d_n(A,X)$$
$$\leq \inf \{||B||_X: B \in \mathcal{B}_n\}$$

Theorem 3. For a general Ω, we have

$$d_n(A,X) \leq \inf \{||B||_X: B \in \mathcal{B}_n\}$$

Definition. The capacity of K relative to Ω, $c(K,\Omega)$, is defined to be $\sup\{\mu(K)\}$ where μ runs over all non-negative measures on K which satisfy

$$\int_K g(z,\zeta)d\mu(\zeta) \leq 1 \text{ if } z \in \Omega$$

This notion was introduced by Widom (1972). Theorem 5 below was first proved by Widom, as well as other related results.

Theorem 4. Given $\varepsilon > 0$, there is an N such that

$$\{d_n(A,X)\}^{1+\varepsilon} \leq \exp[-n/c(K,\Omega)], \quad n \geq N.$$

Theorem 5. If Ω has finitely or countably many complementary components, then

$$\lim_{n\to\infty} \{d_n(A,X)\}^{1/n} = \exp[-1/c(K,\Omega)]$$

Theorem 6. Let Ω have finitely many complementary components and suppose $K = \{z \in \Omega: |B_0(z)| \leq p\}$ for some $B_0 \in \mathcal{B}_N$ and p, $0 < p < 1$. Then there are constants a,b with

$$a \leq d_n(A,X) \exp[n/c(K,\Omega)] \leq b$$

Definition. The Gel'fand n-width of A in X is defined by

$$d^n(A,X) = \inf_{Y_n} \sup_{x \in A\cap Y_n} ||x|| \text{ where } Y_n \text{ runs over all}$$

subspaces of X of codimension n.

Theorem 7. Let $\Omega = \Delta_0$ and $K = [a,b]$ or $K = \{z: |z| \le r\}$. Then

$$d^n(A,C(K)) = d_n(A,C(K))$$

REFERENCE

Widom, H. (1972). J. of Approx. Theory 5, 343-361.

KOROVKIN CLOSURES IN AM-SPACES

Hans O. Flösser[1]

Fachbereich Mathematik-Informatik

Universität-Gesamthochschule Paderborn

I. THE PROBLEM

The meanwhile classical theorem of Korovkin states that a sequence (T_n) of positive linear operators on $C[0,1]$ converges pointwise to the identity if the three sequences $(T_n(f_i))$ converge to f_i for $f_i(x) = x^i$, $i = 0,1,2$ (9). We will use ideas of H. Berens and G.G. Lorentz outlined in (2) and (3) to generalize Korovkins theorem to quite a large class of Banach lattices.

To begin with let a class $\mathcal{T}$ of linear operators on a Banach lattice E be given. If M is a subset of E we call a net (T_α) of operators in $\mathcal{T}$ <u>M-admissible</u> if

1. (T_α) is normbounded,
2. $\lim T_\alpha(p) = p$ for all $p \in M$.

[1]Attendance of the Conference was made possible by support by DFG.

Copyright © 1980 by Academic Press, Inc.
All rights of reproduction in any form reserved.
ISBN: 0-12-171050-5

(The reason why we only consider normbounded nets is illuminated by K. Donners contributions to this volume). Next we define the T-Korovkin closure of M as

$$K_T(M) = \{ e \in E \mid \lim T_\alpha(e) = e \text{ for all } M\text{-admissible nets in } T\}.$$

Now, of course, the problem is, given $M \subset E$ and T, to determine the T-Korovkin closure of M; in particular, we are interested in characterizing those sets M the T-Korovkin closure of which is the whole space E.

But the problem stated as it is seems to be to general to obtain many non-trivial results. We shall therefore content ourselves with considering only special classes T of operators on special classes of Banach lattices. To do so we introduce the sets

T^+ of all positive linear operators on E,
T_1 of all linear contractions on E, and
$T_1^+ = T^+ \cap T_1$ of all positive linear contractions on E.

Instead of general Banach lattices we only consider AM-spaces, i.e. those Banach lattices the norm of which satisfies

$$\|e \vee f\| = \|e\| \vee \|f\| \quad \text{for all positive } e,f \in E,$$

$e \vee f$ denoting the supremum of e and f in E. The main examples of AM-spaces are

1. $E = C(X)$, the set of all continuous realvalued functions on the compact space X, endowed with its supremum-norm and canonical order;

2. $E = C_0(X) = \{f \in C(X) \mid f \text{ vanishes at infinity}\}$, X locally compact, again with its supremum-norm and canonical order;

3. $E = Cv_0(X) = \{f \in C(X) \mid vf \text{ vanishes at infinity}\}$, X completely regular and v an u.s.c. positive function on X, provided $\|f\| = \sup_{x \in X} v(x) \mid f(x) \mid$ is a norm on E which makes E complete.

While the T-Korovkin closure of a subset M of an AM-space E is meanwhile well known for $T = T^+$ ((5) and (12), Theorem 4.3), this is not true for $T = T_1$ or T_1^+. Indeed, up to now, only the problem for which M we have the equality $K_T(M) = E$ has been studied and this only in the special case $E = C(X)$ (see (3)).

In this paper we present a unified treatment of Korovkin approximation in AM-spaces, the operators approximating the identity belonging either to the class T^+, T_1 or T_1^+, and we shall give descriptions of the respective closures at least for a large subclass of the class of all AM-spaces. The proofs for the case $T = T^+$ can be found in (5), (7), (12), the ones for $T = T_1^+$ in (6) while the ones for $T = T_1$ are either trivial or - modulo some minor modifications - the same as in the case $T = T_1^+$.

II. UNIQUENESS CLOSURES

Here again let E be an arbitrary Banach lattice. Denote by L one of the following sets of continuous linear functionals on E:

L^+ : the set of positive linear functionals on E,

L_1 : the unit ball of E', and

$L_1^+ = L^+ \cap L_1$: the positive part of the unit ball of E'.

While L^+ is a weak*-closed convex cone in E', L_1 and L_1^+ are weak*-compact convex sets. Let ex L denote the set of extremal elements in L which means that ex L^+ is the set of all functionals in L^+ lying on some extreme ray of L^+ and is the set of extreme points of L in the two other cases. Finally, let $\text{ex}^* L = \text{ex } L \setminus \{0\}$.

For a subset M of E define the L-<u>uniqueness</u> <u>closure</u> by

$$E_L(M) = \{e \in E \mid \forall\, \delta \in \overline{\text{ex}^* L} \quad \forall\, \mu \in L \quad \delta \underset{M}{=} \mu \Rightarrow \delta(e) = \mu(e)\};$$

the closure of $\text{ex}^* L$ is taken in the weak*-topology.

To find an alternative description of the respective L-uniqueness closure we introduce one more notation: For a non-empty subset A of a vector lattice let $\hat{A}$ denote the set of all finite infima of elements of A, i.e.

$$\hat{A} = \{\wedge A' \mid \emptyset \neq A' \subset A, \quad A' \quad \text{finite}\}$$

and, dually,

the set of all finite suprema of elements of A is denoted by

$$\overset{\vee}{A} = \{\vee A' \mid \emptyset \neq A' \subset A,\ A' \text{ finite}\}.$$

Recall that if E is an AM-space its bidual E" is one, too, containing a strong order unit u.

<u>Theorem 1. Let</u> E <u>be an AM-space,</u> M <u>a subset and</u> H <u>its linear hull. Then</u>

(i) $E_{L^+}(M) = \overline{\hat{H}} \cap \overline{\overset{\vee}{H}}$;

(ii) $E_{L_1^+}(M) = \overline{(H + \mathbb{R}_+ u)^{\wedge}} \cap \overline{(H - \mathbb{R}_+ u)^{\vee}}$;

(iii) $E_{L_1}(M) = \overline{H}$ <u>if</u> $0 \in \overline{\text{ex } L_1}$.

The proof of (i) can be found in (5); statement (ii) has been proved in (6) and, essentially, follows from (i), while (iii) is trivial. Observe that in (ii) we consider H as a subspace of E" and that the lattice operations $\wedge$ and $\vee$ as well as the norm closure are taken in E".

<u>Example 1.</u> For $E = C_0(X)$ it is easily seen that the order unit u of E" in statement (ii) of Theorem 1 can be replaced by the constant function 1 on X. Then $H \pm \mathbb{R}_+ 1$ is contained in the space $C_b(X)$ of all continuous bounded realvalued functions on X and both the lattice operations and the norm closure can be taken in $C_b(X)$.

<u>Example 2.</u> Again let $E = C_0(X)$. By the description of ex L as ex $L^+ = \{ r\delta_x \mid r \geq 0, x \in X \}$ and ex $L_1^+ = \{ \delta_x \mid x \in X \} \cup \{0\}$ one

easily derives peak-point criteria for $E_L(M) = C_0(X)$. Indeed, if we assume that H contains a strictly positive function and if every point of X is a (P^+) resp. (P_1^+) peak-point in the sense of 1.5 in (3) then $E_{L^+}(M) = C_0(X)$ or $E_{L_1^+}(M) = C_0(X)$. In contrast to this for not compact X we always have $E_{L_1}(M) = \overline{H}$. Only if X is compact the peak-point criterion (P_1) in 1.5 of (3) implies $E_{L_1}(M) = C(X)$, too.

III. KOROVKIN CLOSURES

An often used method in Korovkin approximation essentially due to P.P. Korovkin himself ((9), p. 16) yields the following theorem.

Theorem 2. *Let E be an AM-space, M a subset and $T = T^+$ or T_1 or T_1^+. Then* $E_L(M) \subset K_T(M)$.

Of course, in this theorem you have to take $L = L^+$ in the case of $T = T^+$ ... . The hypothesis E being an AM-space is used only for $T = T^+$. A proof can be found in (12); alternatively, you may use Theorem 1 and prove $\hat{H} \cap \check{H} \subset K_{T^+}(M)$ which is true in any topological vector lattice (11). The statements for $T = T_1$ and $T = T_1^+$ are true in any Banach lattice (10).

Example 3. Combining Theorem 2 and Example 2 one obtains peak-point criteria ensuring $K_T(M) = C_0(X)$. Borrowing an example from (1) consider $M = \{f_1, f_2, f_3\} \subset C_0[0,\infty)$ where

$$f_i(x) = \exp(-t_i x), \quad 0 < t_1 < t_2 < t_3, \quad x \in [0,\infty).$$

Then $K_{T^+}(M) = C_0[0,\infty)$. For $M = \{f_1, f_2\}$ one obtains $K_{T_1^+}(M) = C_0[0,\infty)$. On the other hand, for not compact X the theorems 1 and 2 yield only the trivial result $\overline{H} \subset K_{T_1}(M)$.

Now, one is interested whether the lower estimate for $K_T(M)$ given in Theorem 2 in best possible. In the case $T = T^+$ this is indeed true in all AM-spaces as can be proved using the metric approximation property of E". Instead of doing so we propose to use the abstract counterparts of the multiplication operators on C(X), namely the operators belonging to the ideal center Z(E) of E. The advantage of this procedure is that one can treat all three cases T^+, T_1 and T_1^+ in a similar way, the disadvantage that we have to impose some restrictions on the AM-space under consideration to ensure to existence of sufficiently many of those operators. This can be done using the isomorphism between the ideal center Z(E) and the AM-space $C_b(S)$ of all continuous bounded realvalued functions on the structure space S of E (4). Recall that this <u>structure space</u> is the set $ex^* L_1^+$ endowed with the facial topology the closed sets of which are the traces of the weak*-closed faces of L_+. It is well known that this space can be very nasty (8).

<u>Theorem 3.</u> <u>Suppose the structure space of the AM-space E is completely regular. Then</u> $K_T(M) \subset E_L(M)$.

Under a slightly stronger hypothesis this theorem was proved in (7) for $T = T^+$. The proof for $T = T_1$ is essentially the same as the one for $T = T_1^+$ which can be found in (6).

Taking into account all our theorems we may summarize our results as follows.

Corollary. Under the assumptions of Theorem 3 the following descriptions of the T-Korovkin closures of M *hold true:*

(i) $K_{T^+}(M) = E_{L^+}(M) = \overline{\hat{H}} \cap \overline{\overset{\vee}{H}}$;

(ii) $K_{T_1^+}(M) = E_{L_1^+}(M) = \overline{(H + \mathbb{R}_+ u)^{\wedge}} \cap \overline{(H - \mathbb{R}_+ u)^{\vee}}$;

(iii) $K_{T_1}(M) = E_{L_1}(M) = \overline{H}$,

the last equality under the condition $0 \in \overline{\text{ex}\, L_1}$. *Here again* H *denotes the linear hull of* M, u *the order unit of* E".

As the structure space of $E = C_0(X)$ is homomorphic to X our Corollary applies to this situation. Using the descriptions of ex L in Example 2 and Example 1 statement (i) reduces to a theorem proved by H. Bauer and K. Donner in (1), while even for compact X (ii) and (iii) generalize results of H. Berens and G.G. Lorentz (3).

REFERENCES

1. Bauer, H. and Donner, K., Korovkin approximation in $C_0(X)$. *Math. Ann. 236*, 225 (1978).
2. Berens, H. and Lorentz, G.G., Theorem of Korovkin type for positive operators on Banach lattices, in "Approximation Theory"(G.G. Lorentz, ed.), p. 1. Academic Press, New York (1973).

3. Berens, H. and Lorentz, G.G., Geometric theory of Korovkin sets. *J. Approx. Theory 15,* 161 (1975).

4. Flösser, H.O., Das Zentrum archimedischer Vektorverbände. *Mitt. Math. Sem. Gießen 137* (1979).

5. Flösser, H.O., A Korovkin type theorem for locally convex M-spaces. *Proc. Amer. Math. Soc. 72,* 456 (1978).

6. Flösser, H.O., Sequences of positive contractions on AM-spaces. Preprint (1979).

7. Flösser, H.O., Irmisch, R. and Roth, W., Infimum-stable convex cones and approximation. To appear in *Proc. London Math. Soc.*

8. Goullet de Rugy, A., La structure idéale des M-espaces. *J. Math. pures et appl. 51,* 331 (1972).

9. Korovkin, P.P., Linear Operators and Approximation Theory. Hindustan Publ. Corp., Delhi (1960).

10. Scheffold, E., Über die Konvergenz linearer Operatoren. *Mathematica (Cluj) 20* (1978).

11. Wolff, M., Über die Korovkinhülle von Teilmengen in lokalkonvexen Vektorverbänden. *Math. Ann. 213,* 97 (1975).

12. Wolff, M., On the theory of approximation by positive linear operators in vector lattices, in "Functional Analysis: Surveys and Recent Results" (K.-D. Bierstedt and B. Fuchssteiner, eds.), p. 73, North-Holland Mathematics Studies 27 (1977).

MULTIVARIATE INTERPOLATION TO SCATTERED DATA USING DELTA ITERATION

Thomas A. Foley
California Polytechnic State University
San Luis Obispo, California

Gregory M. Nielson[1]
Arizona State University
Tempe, Arizona

I. INTRODUCTION

We present an iterative procedure that solves the multivariate interpolation problem: Given N distinct points $v_i \varepsilon R^n$ and N values $Z_i \varepsilon R$, construct a function $F(v)$ that satisfies $F(v_i) = Z_i$ for $i = 1, \ldots, N$.

The first stage of this iterative procedure is a process called delta sum, which is defined to be $S\Delta B = S[I-BS] + BS$. The delta sum of Shepard's interpolant with a bicubic spline or a Bernstein polynomial approximation yields an interpolant to the scattered data that is visually smoother and more accurate than the original. This process is repeated to form a procedure called delta iteration. Dramatic improvements are observed when Shepard's method is iterated with bicubic splines to form smooth C^2 interpolants to the scattered data that are globally defined. In certain instances, the iterates converge to a bicubic spline defined on a rectangular grid that interpolates the scattered data.

Although the examples are for bivariate functions using Shepard's method and bicubic splines, the analysis is presented in the multivariate case for general interpolation schemes.

---

[1]Supported by Office of Naval Research. ONR-044-043.

Copyright © 1980 by Academic Press, Inc.
All rights of reproduction in any form reserved.
ISBN: 0-12-171050-5

2. TWO-STAGE APPROXIMATIONS AND DELTA ITERATION

In (2), Schumaker introduced the ideas of a two-stage process for approximating scattered data. Part of the motivation for these methods is based upon the fact that many of the methods that apply directly to scattered data do not yield interpolants which are smooth or have desirable fitting characteristics. On the other hand, many of the methods which do yield smooth interpolants only apply to regularly spaced data.

An easily implemented method for interpolating scattered data is Shepard's method (3):

$$S[f](v) = \frac{\sum_{i=1}^{N} \frac{f_i}{\rho_i(v)}}{\sum_{i=1}^{N} \frac{1}{\rho_i(v)}}, \text{ where } \rho_i(v) = ||v - v_i||_2^2 .$$

Even though this is a C^∞ interpolant, without significant modifications, Shepard's method does not yield a very desirable approximation to the data. An example based upon $N = 100$ evaluations of the function

$$f(x,y) = 9 \{.75 \text{ EXP } (-\frac{(x-3)^2 + (y-3)^2}{4})$$
$$+ .75 \text{ EXP } (-\frac{x}{49} - \frac{y}{100}) + .5 \text{ EXP } (-\frac{(x-8)^2+(y-4)^2}{4})$$
$$- .2 \text{ EXP } (-(x-5)^2 - (y-8)^2))\}, \; 1 \le x \le 10 \text{ and } 1 \le y \le 10,$$

is shown in Figure 1.

An example of a method which generally yields smooth approximations and is computationally efficient is the bicubic spline. This approximation only applies to gridded data

$$U = \{u_j=(x_i,y_k), \; i=1,\ldots,m; \; k=1,\ldots,n; \; j=m(k-1)+i; \; m\cdot n = M\}.$$

A two stage approximation can be formed from these two interpolants by first applying S to the scattered data and then evaluating S at u_j, j=1,...,M to obtain data which is interpolated by a bicubic spline. An example of this approximation is shown in Figure 2a where the bicubic spline interpolates S[f] of Figure 1a at the 8 x 8 rectangular grid illustrated in Figure 1b. While the two-stage approximation BS[f] generally yields a desirable approximation, it does not necessarily interpolate the original data. This can be remedied by the addition of a correction term S[f-BS[f]] which leads to the definition of

$$\Delta_1[f] = S[f-BS[f]] + BS[f] = S \oplus BS[f].$$

This interpolant is illustrated in Figure 2b. Since this is quite an improvement over the original interpolant of Shepard's method, it seems reasonable to continue this process by smoothing the new approximation with the bicubic spline and then correcting it back to an interpolant. That is,

$$\Delta_2[f] = S[f-B\Delta_1[f]] + B\Delta_1[f] = S \oplus B\Delta_1[f]. \qquad (2.1)$$

In general we have

$$\Delta_m[f] = S \oplus B\Delta_{m-1}[f], \ m = 1,2,\ldots \text{ with } \Delta_0 = S.$$

Examples of these C^2 interpolants are shown in Figures 2 and 3. Another way to continue this improvement scheme is to replace the original interpolant S of equation (2.1) with the new interpolant $\Delta_1[f]$. This leads to the iteration:

$$\hat{\Delta}_m[f] = \hat{\Delta}_{m-1} \oplus B\hat{\Delta}_{m-1}[f], \ m = 1,2,\ldots \text{ with } \hat{\Delta}_0 = S.$$

However, this really leads to nothing new in that it can be shown that $\hat{\Delta}_m = \Delta_{2^m-1}$.

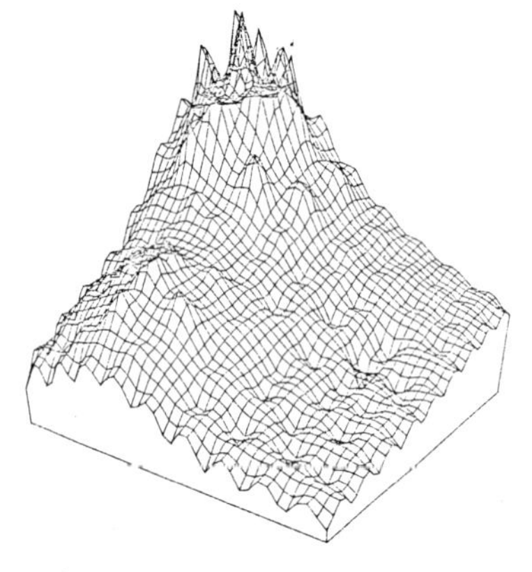

Fig. 1a. Shepard's Method

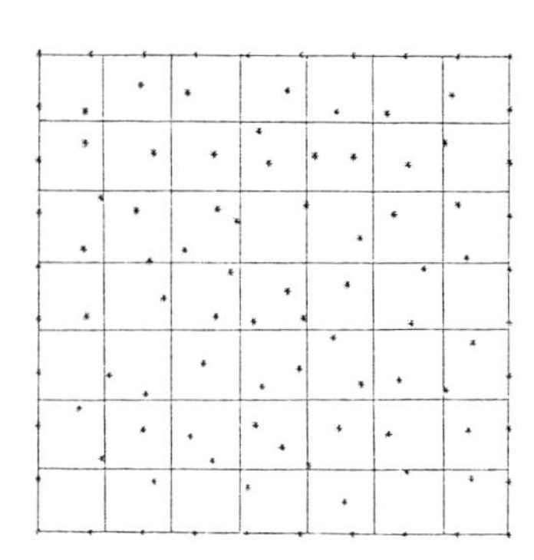

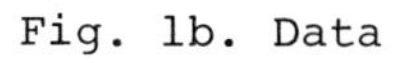

Fig. 1b. Data

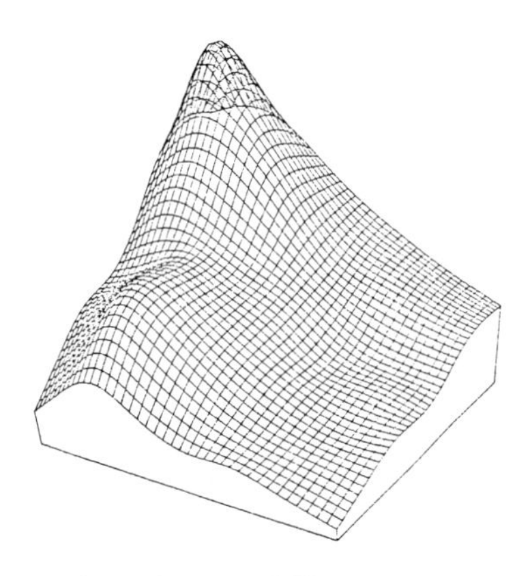

Fig. 2a. BS[f]

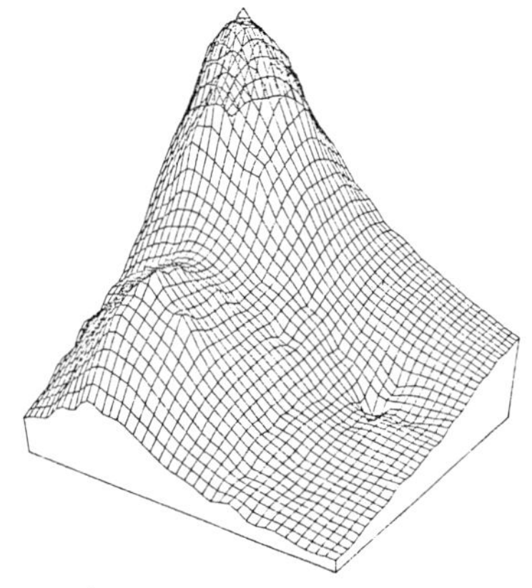

Fig. 2b. $\Delta_1[f]$

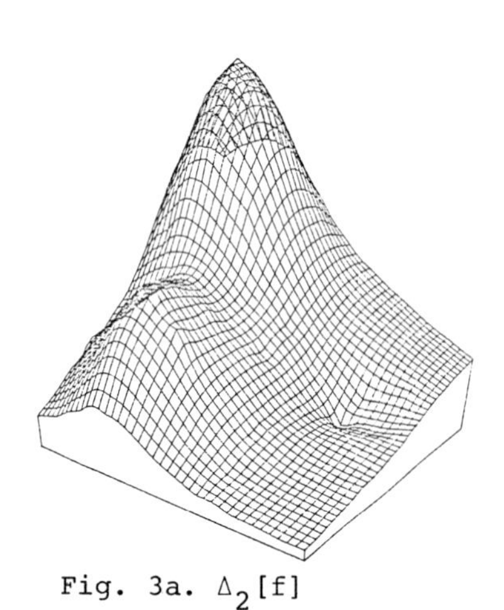

Fig. 3a. $\Delta_2[f]$

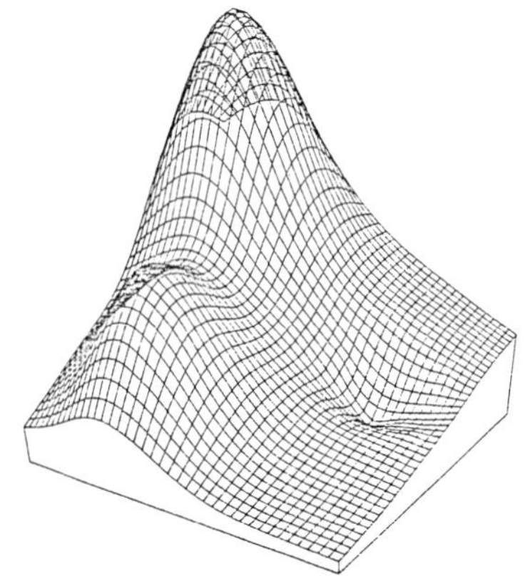

Fig. 3b. $\Delta_{10}[f]$

3. CONVERGENCE OF DELTA ITERATION

We now consider the question of convergence and show that in some cases the interpolants converge to a bicubic spline defined on a rectangular grid that interpolates the scattered data.

It will be convenient to represent B and S in terms of their cardinal bases. Let

$$S[f](v) = \sum_{i=1}^{N} f(v_i)\phi_i(v), \text{ where } \phi_i(v_j) = \delta_{ij} ,$$

and

$$B[f](v) = \sum_{i=1}^{M} f(u_i)\psi_i(v), \text{ where } \psi_i(u_j) = \delta_{ij}.$$

We also define two vectors and a matrix: $F_m = (\Delta_m[f](u_i))$, $C = (S[f](u_i))$, $A = (S[\psi_j](u_i))$, $i = 1,\ldots,M;\ j = 1,\ldots,M.$ The proofs of the following Theorems can be found in (1).

Theorem 1. If $||I - A|| < 1$, then $\lim_{m\to\infty} F_m = A^{-1}C.$

The condition $||I - A|| < 1$ depends only on the distribution of the u_j's and v_i's but not on the function f. This condition will be satisfied if there is at least one data point v_i near each grid point u_j. In the case of convergence, let $B^*[f]$ denote the bicubic spline defined from the grid values $A^{-1}C$. The limiting interpolant is defined to be $\Delta^*[f] = S[f-B^*[f]] + B^*[f]$. Δ^* has the same interpolation, continuity, and precision properties of each Δ_i. The rate of convergence is given in the following theorem.

Theorem 2. If $||I - A|| < 1$ then

$$||\Delta_m[f] - \Delta^*[f]||_D \le K\cdot\overline{f}\cdot||I - A||^m ,$$

where K is independent of f, $\overline{f} = \max|f(v_i)|$, D is a bounded subset of R^n, and $||g||_D = \sup_{v\in D}|g(v)|$.

The delta iterates shown in Figure 3 appear to resemble a bicubic spline more than they do Shepard's interpolant. Since $\Delta_m[f] = S[f-B\Delta_{m-1}[f]] + B\Delta_{m-1}[f]$, it would be of interest to know which dominates, the bicubic part $B\Delta_{m-1}[f]$, or the correction term $S[f-B\Delta_{m-1}[f]]$. In many examples, the correction term is very small and in some cases it approaches zero. This means the limit is a bicubic spline defined over gridded data that also interpolates the scattered data.

Theorem 3. Let $R = (\phi_j(u_i))$. If $\lim_{m\to\infty} F_m$ exists and R has a left inverse, then $\Delta^*[f] = B^*[f]$.

Corollary 1. Let $G = (B[\phi_j](v_i))$. If $\lim_{m\to\infty} F_m$ exists and $||I - G|| < 1$, then R has a left inverse and $\Delta^*[f] = B^*[f]$.

Note that G is analogous to A. If there is a grid point u_i near each data point v_j, then $||I - G|| < 1$. The u_i's are free to be chosen, while the v_i's are generally fixed. A grid selection algorithm which attempts to satisfy both $||I - A|| < 1$ and $||I - G|| < 1$ is presented in (1).

REFERENCES

1. Foley, Thomas A., Smooth multivariate interpolation to scattered data, Ph.D. dissertation, Arizona State University, 1979.
2. Schumaker, L. L., Fitting surfaces to scattered data, "Approximation Theory II," edited by G. G. Lorentz, C. K. Chui, L. L. Schumaker, Academic Press, 1976.
3. Shepard, D., A two dimensional interpolation function for irregularly spaced data, Proc. 23rd Nat. Conf. ACM, 1968.

ON THE RADIAL PROJECTION IN BANACH SPACES

Carlo Franchetti

Department of Mathematics
University of Genoa
Genoa, Italy

I. INTRODUCTION

Let U be the closed unit ball of a real Banach space X. The radial projection $R: X \to U$ is defined by

$$Rx = \begin{cases} x \ , & \text{if } x \in U \\ x/\|x\| \ , & \text{if } x \notin U \ . \end{cases}$$

Elementary computations show that:

$\|Rx-Ry\| \leq 2\|x-y\|$ in any space X ;

$\|Rx-Ry\| \leq \|x-y\|$ when X is a Hilbert space .

In 1967 de Figueiredo and Karlovitz proved in [2] that the second inequality characterizes the Hilbert spaces whose dimension is greater than two.

The best Lipschitz constant for R, i.e. the number

$$k(X) = \sup \left\{ \frac{\|Rx-Ry\|}{\|x-y\|} \ : \ x \neq y \right\}$$

is a parameter of the space X which we shall call the radial constant of X. $k(X)$ satisfies the inequality

$$1 \leq k(X) \leq 2 \ .$$

Copyright © 1980 by Academic Press, Inc.
All rights of reproduction in any form reserved.
ISBN: 0-12-171050-5

A further study of $k(X)$ was given in 1974 by Thele [4]. Using the results obtained in [4] we will prove in Section III that the equality $k(X) = k(X^*)$ holds for every Banach space X (here X^* denotes the norm dual of X).

II. KNOWN FACTS

We recall here some definitions and results which will be needed in the sequel.

A. Duality Mapping

The (multivalued) duality mapping $J: X \to 2^{X^*}$ is defined by

$$J(x) = \{f \in X^*: f(x) = \|x\|^2 , \|f\| = \|x\|\} .$$

B. James Orthogonality

If x, y are elements in a real normed space X, then x is orthogonal to y $(x \perp y)$ if the inequality $\|x\| \leq \|x + \lambda y\|$ holds for every real λ.

The following characterization of the orthogonality in X is essentially (although not explicitly) contained in the literature.

Proposition. We have

$$x \perp y \Leftrightarrow \exists\, f \in J(x): f(y) = 0 . \tag{1}$$

C. Spaces Which Are Uniformly Non-square

The following definition was given by James in [3]: a normed space X is called uniformly non-square (u.n.s.) if

$$\exists\, \delta > 0: \|x\| = \|y\| = 1 \Rightarrow \min\|x \pm y\| \leq 2 - \delta .$$

James also proved in [3] the following:

$$\text{u.n.s. Banach spaces are reflexive .} \tag{2}$$

D. Thele's Formula

Thele provided in [4] the following formula:

$$k(X) = \sup\left\{ \frac{1}{\|y+\lambda x\|} : \|x\| = \|y\| = 1, \quad x \perp y , \quad \lambda \in R \right\} ; \tag{3}$$

from which he obtained:

$$k(X) = 1 \Leftrightarrow \text{the orthogonality is symmetric in } X$$

(this is the de Figueiredo-Karlovitz result);

$$k(X) = 2 \Leftrightarrow X \text{ is not u.n.s.} \tag{4}$$

III. $k(X) = k(X^*)$

Let X^* denote the norm dual of the space X. We now want to prove the following:

<u>Theorem</u>. For a real Banach space X we have

$$k(X) = k(X^*) .$$

<u>Proof</u>. First note that λ_0 minimizes the function $\lambda \to \|y+\lambda x\|$ if and only if $\lambda_0 x + y \perp x$ (the set of such λ_0 is a compact interval, it reduces to a point for every nonzero x and every y if and only if X is strictly convex, see for ex. [1]). It is therefore clear by (3) that

$$k(X) = \sup\left\{ \frac{1}{\|y+\lambda x\|} : \|x\| = \|y\| = 1 , \quad x \perp y , \quad \lambda x + y \perp x \right\} .$$

Let λ_0, x_0, y_0 be such that $\|x_0\| = \|y_0\| = 1$, $x_0 \perp y_0$, $\lambda_0 x_0 + y_0 \perp x_0$. By (1) we can select f, f_{x_0} such that $f \in J(\lambda_0 x_0 + y_0)$, $f_{x_0} \in J(x_0)$ and $f(x_0) = f_{x_0}(y_0) = 0$, let also f_{y_0} be any element in $J(y_0)$. The functionals f_{x_0} and f_{y_0} are independent: we try to "approximate" f with f_{x_0} and f_{y_0}. Indeed let $\varphi = \alpha f_{x_0} + \beta f_{y_0}$, we require that $\varphi(x_0) = f(x_0) = 0$ and $\varphi(\lambda_0 x_0 + y_0) = f(\lambda_0 x_0 + y_0) = \|\lambda_0 x_0 + y_0\|^2$.

An easy computation shows that

$$\varphi = \|\lambda_0 x_0 + y_0\|^2 (f_{y_0} - f_{y_0}(x_0) f_{x_0}) \; ; \; \|\varphi\| \geq \|\lambda_0 x_0 + y_0\| \; .$$

The functional φ is "near" to belong to $J(\lambda_0 x_0 + y_0)$, remark also that when $\dim X = 2$ we actually have $f = \varphi$ since f is in fact of the form $\alpha f_{x_0} + \beta f_{y_0}$. This last remark is sometimes useful for the actual computation of the radial constant for two dimensional spaces.

We now have

$$\|f_{y_0}(x_0) f_{x_0} - f_{y_0}\| = \frac{\|\varphi\|}{\|\lambda_0 x_0 + y_0\|^2} \geq \frac{1}{\|\lambda_0 x_0 + y_0\|} \; .$$

Since $\|f_{y_0}\| = 1$ this can also be written:

$$\frac{\|f_{y_0}(x_0) f_{x_0} - f_{y_0}\|}{\|-f_{y_0}(x_0) f_{x_0} + (f_{y_0}(x_0) f_{x_0} - f_{y_0})\|} \geq \frac{1}{\|\lambda_0 x_0 + y_0\|} \; .$$

Now observe that f_{x_0} (hence also $-f_{y_0}(x_0) f_{x_0}$) is orthogonal to $f_{y_0}(x_0) f_{x_0} - f_{y_0}$: indeed for any $h \in R$ we have $\|f_{x_0} + h(f_{y_0}(x_0) f_{x_0} - f_{y_0})\| \geq |(f_{x_0} + h(f_{y_0}(x_0) f_{x_0} - f_{y_0}))(x_0)| = 1 = \|f_{x_0}\|$. Therefore

$$\frac{1}{\|\lambda_0 x_0 + y_0\|} \leq \sup \left\{ \frac{\|g\|}{\|g+f\|} \; : \; f \perp g \, , \;\; g \neq 0 \right\} = k(X^*) \; .$$

We conclude that $k(X) \leq k(X^*)$ and, of course, the equality holds if X is reflexive. If X is not reflexive then $k(X) = 2$, indeed if $k(X) < 2$ by (4) we have that X is u.n.s. and by (2) that X is reflexive. The proof is therefore completed since $2 \geq k(X^*) \geq k(X) = 2$.

REFERENCES

1. Diestel, J., "Geometry of Banach spaces. Selected topics", Lecture Notes in Mathematics, 485, Springer-Verlag.
2. de Figueiredo, D.G., and Karlovitz, L.A., "On the radial projection in normed spaces", Bull. Amer. Math. Soc., 73(1967), 364-368.
3. James, R.C., "Uniformly non-square Banach spaces", Annals of Math., 80(1964), 542-550.
4. Thele, R.L., "Some results on the radial projection in Banach spaces", Proc. Amer. Math. Soc., 42(1974), 483-486.

REMARKS ON FUNCTIONAL REPRESENTATION

Manfred v. Golitschek

Institut für Angewandte Mathematik und Statistik
der Universität Würzburg
Bundesrepublik Deutschland

There has been some recent interest (see Professor Buck [1 ; 2]) in the following problem, indirectly related to the 13th Hilbert Problem to which Professor Lorentz has contributed many important results. (See [3].)

PROBLEM. Characterize the subclasses $\dot{G}^p$ of $C^p(\mathbb{R}^3)$ consisting of functions F of form

$$F(x,y,z) = f(\phi(x,y) , \psi(y,z))$$

where f, ϕ, ψ belong to $C^p(\mathbb{R}^2)$ and $p \in \{0;1;2;\ldots\}$.

In [4], Polya and Szego proved by a simple argument that the trial function $F^*(x,y,z) = xy + yz + zx$ does not belong to G^3, on all $\mathbb{R}^3$. The following result of Professor Buck [2] shows that F^* is not locally in G^1 anywhere.

THEOREM (Buck [2]). Let A,B,C,D be real constants. The function $F(x,y,z) = Axy + Byz + Czx + Dy^2$ is nowhere locally in G^1, unless $AB = CD$ or $C = 0$.

Copyright © 1980 by Academic Press, Inc.
All rights of reproduction in any form reserved.
ISBN: 0-12-171050-5

The method of proof of this theorem depends on the nature of certain differential equations. As Professor Buck points out, his method does not seem capable of confirming the conjecture that F* is nowhere locally in G^0 or that, for instance, $F^{**}(x,y,z) := xy^2 + yz^2 + zx^2$ is nowhere locally in either G^1 or in G^0.

The main result of my talk is the MAIN THEOREM below. It states that whenever a function $F \in C(\mathbb{R}^3)$ is locally in G^0 and has certain mild monotonicity properties, then F can locally be represented in the simple form (2). This Main Theorem is a powerful mean to decide if a given function F is or is not (locally) in G^0. For instance, we can confirm Professor Buck's conjecture that the above functions F* and F** are nowhere locally in G^0.

MAIN THEOREM. Let $P_0 = (x_0, y_0, z_0) \in \mathbb{R}^3$, $\delta > 0$, and

$$U_\delta(P_0) := \{(x,y,z) \in \mathbb{R}^3 \mid -\delta \le x-x_0, y-y_0, z-z_0 \le \delta\}.$$

Suppose that F is a continuous real-valued function defined in $U_\delta(P_0)$, for which $F(x,y_0,z_0)$ and $F(x_0,y_0,z)$ are strictly monotone functions in x and z, respectively.

1. If F can be represented by

$$F(x,y,z) = f(\phi(x,y), \psi(y,z)), \quad (x,y,z) \in U_\delta(P_0), \tag{1}$$

where f, ϕ, ψ are continuous and real-valued, then there exist a number δ_0, $0 < \delta_0 \le \delta$, and continuous real-valued functions g and h such that $g(x_0,y_0) = x_0$, $h(y_0,z_0) = z_0$ and

$$F(x,y,z) = F(g(x,y), y_0, h(y,z)), \tag{2}$$

$(x,y,z) \in U_{\delta_0}(P_0)$, holds.

2. Conversely, if F satisfies (2) then F is of form (1) in $U_{\delta_o}(P_o)$ for $f(u,v) := F(u,y_o,v)$, $\phi(x,y) := g(x,y)$, $\psi(y,z) := h(y,z)$.

PROOF. 1. Suppose that (1) holds. We may assume without loss of generality that $x_o = y_o = z_o = 0$ and $\phi(0,0) = \psi(0,0) = 0$. Since $F(x,0,0) = f(\phi(x,0),\ 0)$ and $F(0,0,z) = f(0,\ \psi(0,z))$ are strictly monotone in x and z, respectively, the functions

$$G(x) := \phi(x,0) \quad , \quad H(z) := \psi(0,z) \quad , \quad -\delta \le x,\ z \le \delta \ ,$$

are also strictly monotone, say monotone increasing. We set

$$u_o := \phi(-\delta,0) \ , \ u_1 := \phi(\delta,0) \ , \ v_o := \psi(0,-\delta) \ , \ v_1 := \psi(0,\delta).$$

Then we obtain the equation

$$f(u,v) = F(G^{-1}(u) \ , \ 0 \ , \ H^{-1}(v)) \tag{3}$$

for any $u \in [u_o,u_1]$, $v \in [v_o,v_1]$, where G^{-1} and H^{-1} denote the inverse functions of G and H. Since $u_o < 0 < u_1$, $v_o < 0 < v_1$, it follows by continuity arguments that there exists a number δ_o, $0 < \delta_o \le \delta$, such that

$$u_o \le \phi(x,y) \le u_1 \quad , \quad v_o \le \psi(y,z) \le v_1 \tag{4}$$

holds for all $(x,y,z) \in U_{\delta_o}(P_o)$. Therefore, by (3) and (4),

$$f(\phi(x,y),\ \psi(y,z)) = F(G^{-1}(\phi(x,y)),\ 0 \ , \ H^{-1}(\psi(y,z)))$$

is valid for all $(x,y,z) \in U_{\delta_o}(P_o)$. Hence equality (2) is valid for

$$g(x,y) := G^{-1}(\phi(x,y)) \quad , \quad h(y,z) := H^{-1}(\psi(y,z)) \ .$$

Statement 2. of our theorem is trivial.

APPLICATIONS

If F is of form (2) then differentiability properties of g and h can be derived from corresponding properties of F by standard analysis arguments. In this short note I shall not proceed systematically, but only want to consider the above mentioned trial functions F* and F**.

Lemma 1. If F can be represented by (2) and has the special form

$$F(x,y,z) = F_o(x,y) + F_1(y,z) + (\alpha x+\beta)\, F_2(z) \tag{5}$$

where $\alpha \neq 0$ and β are real numbers, F_o, F_1, F_2 are continuous functions, F_2 is strictly monotone, then the function g in the representation (2) has the form

$$g(x,y) = g_o(y) + x\, g_1(y) \tag{6}$$

with continuous functions g_o and g_1.

Proof. It follows by (2) and (5) that

$$F(x,y,z) - F(x,y,z_o) - F(x_o,y,z) + F(x_o,y,z_o)$$

$$= \alpha(x-x_o)(F_2(z) - F_2(z_o))$$

$$= \alpha(g(x,y) - g(x_o,y))(F_2(h(y,z)) - F_2(h(y,z_o)))$$

and, hence,

$$\frac{g(x,y) - g(x_o,y)}{x - x_o} = \frac{F_2(z) - F_2(z_o)}{F_2(h(y,z)) - F_2(h(y,z_o))} =: g_1(y)$$

That concludes the proof of Lemma 1.

<u>Theorem 2.</u> The function

$$F(x,y,z) = Axy + Byz + Cxz + D(y) \ , \tag{7}$$

where A, B, C are real constants, $C \neq 0$, and D is a continuous function, is nowhere locally in G^o, unless

$$D(y) = \text{const} + \frac{AB}{C} y^2 \ .$$

<u>Proof.</u> Suppose that F is locally in $P_0 = (x_o, y_o, z_o) \in \mathbb{R}^3$ in subclass G^0. We may assume without loss of generality that the first partial derivatives $F_x(x_o, y_o, z_o) \neq 0$ and $F_z(x_o, y_o, z_o) \neq 0$. Then, by the Main Theorem and Lemma 1, F can be represented by

$$F(x,y,z) = Ay_o(g_o(y) + xg_1(y)) + By_o(h_o(y) + zh_1(y))$$

$$+ C(g_o(y)+xg_1(y))(h_o(y)+zh_1(y)) + D(y_o) \tag{8}$$

with continuous functions g_o, g_1, h_o, h_1.

Comparing in (7) and (8) the coefficients (i.e. functions in y) of xz, x, z, and $x^o z^o$ leads immediately to the above stated form of D(y).

<u>Theorem 3.</u> The function $F(x,y,z) = xy^2 + yz^2 + zx^2$ is nowhere locally in G^o.

<u>Proof.</u> It suffices to consider any point $P_o = (x_o, y_o, z_o)$ in $\mathbb{R}^3$ for which $y_o \neq 0$, $x_o \neq 0$, say $x_o > 0$, and $F_x(x_o, y_o, z_o) \neq 0$, $F_z(x_o, y_o, z_o) \neq 0$. Suppose that F is locally in P_o in subclass G^o. Then the function $K(x,y,z) := F(\sqrt{x}, y, z)$ is locally in $P_1 := (x_o^2, y_o, z_o)$ in subclass G^o, Hence, by our Main Theorem and Lemma 1, it follows that K can be represented in a neighborhood of P_1 by

$$K(x,y,z) = \sqrt{x}\, y^2 + yz^2 + zx$$

$$= \sqrt{g_o(y) + xg_1(y)}\; y_o^2 + y_o(h_o(y) + zh_1(y))^2$$

$$+ (h_o(y) + zh_1(y))(g_o(y) + xg_1(y)) \qquad (9)$$

with continuous functions g_o, g_1, h_o, h_1. Comparing in (9) the coefficients of xz, z^2, and $\sqrt{x}$ we find the equations

$$g_1(y)h_1(y) = 1 \quad , \quad y_o(h_1(y))^2 = y \quad , \qquad (10)$$

$$g_o(y) = 0 \quad , \quad y_o^2 \sqrt{g_1(y)} = y^2 \qquad (11)$$

for any y in a sufficiently small neighborhood of y_o. But (10) and (11) lead to a contradiction.

Final Remark. A more detailed investigation of the subclasses $G^p(p = 0,1,\ldots)$ and related subclasses by the aid of our Main Theorem will be published elsewhere.

REFERENCES

1. Buck, R.C., Approximate complexity and functional representation. J.Math.Analysis and Appl.70,280-298 (1979).

2. Buck, R.C., Characterization of classes of functions, preprint 1979.

3. Lorentz, G.G., The 13th Problem of Hilbert, in "Mathematical Developments arising from Hilbert's Problems" Proc.Symp.Pure Math., v.28, Amer.Math.Soc. Providence, RI, 1976.

4. Polya, G., and Szego, G., "Aufgaben und Lehrsätze", v. 1, Part II, No. 119a, Dover Publ. NY 1945.

LINEARIZED ELASTICA AND CLASSICAL CUBIC SPLINE INTERPOLATION

Michael Golomb

Department of Mathematics
Purdue University
Lafayette, Indiana

Joseph Jerome

Department of Mathematics
Northwestern University
Evanston, Illinois

I. INTRODUCTION

Let $P = \{p_0, p_1, \ldots, p_m\}$ be a prescribed set of ordered points in $\mathbb{R}^2$, with $p_i \neq p_{i-1}$, $i = 1,\ldots,m$. A nonlinear spline function x, interpolating P, is a smooth curve with continuous curvature κ_x vanishing at the terminal points such that, if $x(s_{i-1}) = p_{i-1}$, $x(s_i) = p_i$, then on (s_{i-1}, s_i) the differential equation

$$\ddot{\kappa}_x + \frac{1}{2}\kappa_x^3 = 0 \tag{1.1}$$

holds and characterizes x (cf. §2). Such functions are designated extremal P-interpolants in the sequel and are realized as critical points of the curvature functional $\int \kappa^2\, ds$. To date, no general existence theory has been assembled, though many properties have been described (cf. Lee and Forsythe (1973), Golomb and Jerome (1979), Golomb (1978) and Golomb (1979)). A satisfactory existence theory has been established, however, if the differential equation (1.1) is replaced by

Copyright © 1980 by Academic Press, Inc.
All rights of reproduction in any form reserved.
ISBN: 0-12-171050-5

Copyright © 1980 by Academic Press, Inc.
All rights of reproduction in any form reserved.
ISBN: 0-12-171050-5

$$\ddot{\kappa}_x + \frac{1}{2}\kappa_x^3 - (\lambda/2)\kappa_x = 0 \tag{1.2}$$

where λ is an appropriate nonnegative Lagrange multiplier associated with the global length-constrained minimization of $\int \kappa^2\, ds$ (cf. Jerome (1975) and Fisher and Jerome (1976)). We shall not consider such interpolants in this note, however.

It is trivial to see that, if P lies on a ray segment, then this segment is the graph of an extremal P-interpolant. Failing a general existence theory, we inquire whether existence holds in an $\mathbb{R}^{m+1}$ neighborhood of the ray configuration $P = \overline{P}$. Using a standard implicit function theorem, we are able to provide an affirmative answer. We find, not surprisingly, that the associated linearization condition reduces to the assertion that natural cubic spline interpolation is uniquely possible.

II. EXTREMAL CHARACTERIZATIONS

Set

$$H^2[0,1] = \{x = (x^1,x^2)\colon [0,1] \to \mathbb{R}^2\colon x,\dot{x},\ddot{x} \in L^2(0,1)\},$$

$$H^2_{reg} = \{x \in H^2\colon |\dot{x}| > 0 \text{ on } [0,1]\},$$

and note that H^2_{reg} is open. For $x \in H^2_{reg}$, set

$$\kappa_x = [\dot{x},\ddot{x}]\,|\dot{x}|^{-3}, \quad [p,q] = p^1q^2 - q^1p^2$$

and define the curvature functional

$$U(x) = \int \kappa_x^2\, ds = \int_0^1 [\dot{x},\ddot{x}]^2|\dot{x}|^{-5} = \int_0^{\overline{s}} |\ddot{\overline{x}}|^2\, ds, \tag{2.1}$$

where

$$\overline{x} = x \circ s_x^{-1}, \quad s_x(t) = \int_0^t |\dot{x}|, \quad t \in [0,1].$$

Remark 2.1. For $x \in H^2_{reg}$, $U'(x)$ exists and is given by

$$U'(x)[y] = \int_0^{\bar{s}} (2\ddot{\bar{x}}\ddot{\bar{y}} - 3\kappa_x^2 \dot{\bar{x}}\dot{\bar{y}})\,ds \tag{2.2}$$

for $y \in H^2[0,1]$ and $s = s_x$.

Definition 2.1. $x \in H^2_{reg}$ is an admissible P-interpolant if $\bar{x}(s_i) = p_i$, $i = 0,\ldots,m$ for some points $0 \leq s_0 < \ldots < s_m \leq \bar{s}$. x is extremal if

$$U'(x)[z] = 0 \tag{2.3}$$

for all $z \in H^2[0,1]$ such that $\bar{z}(s_i) = 0$, $i = 0,\ldots,m$.

It will be convenient to use the so called normal representation rather than the Cartesian representation. Thus, let θ denote the angle between the tangent line and a reference line. Then $\kappa_x = \dot{\theta}$ and the Cartesian representation is retrieved by

$$\begin{aligned} x^1(s) &= x^1(0) + \int_0^s \cos\theta_x, \\ x^2(s) &= x^2(0) + \int_0^s \sin\theta_x. \end{aligned} \tag{2.4}$$

The following result characterizes the extremal P-interpolants (cf. Golomb and Jerome (1979), p. 31).

Proposition 2.1. θ is the normal representation of an extremal x which interpolates the configuration P $= \{0,p_1,\ldots,p_m\}$ at the nodes $0 = \bar{s}_0 < \bar{s}_1 < \ldots < \bar{s}_m < \bar{s}$ if and only if $\dot{\theta}(0+) = 0$ and

$$\text{(i)}\quad \begin{aligned} &\ddot{\theta}(s) + \frac{1}{2}\dot{\theta}^3(s) = 0, \quad \bar{s}_{i-1} < s < \bar{s}_i, \\ &\qquad i = 1,\ldots,m, \\ &\dddot{\theta}(s) = 0, \quad \bar{s}_m < s < \bar{s}. \end{aligned}$$

$$(2.5)\qquad \text{(ii)}\quad \dot{\theta}(\bar{s}_i - 0) - \dot{\theta}(\bar{s}_i + 0) = 0, \quad i = 1,\ldots,m.$$

$$\text{(iii)}\quad \int_0^{\bar{s}_i} \cos\theta = p_i^1, \quad \int_0^{\bar{s}_i} \sin\theta = p_i^2, \quad i = 1,\ldots,m.$$

Remark 2.2. Note that we have set $p_0 = 0$ and have linearly extended the extremal to $[0,\bar{s}]$, $\bar{s} > s_m$.

III. THE EXTREMAL MAPPING

Suppose an extremal P-interpolant is given in normal form for some $P = \bar{P}$ (cf. Proposition 2.1). We introduce two spaces of functions:

$$\begin{aligned} \text{NBV} = \{\kappa\colon [0,\bar{s}] \to \mathbb{R}, &\ \text{of bounded variation } V(\kappa), \text{ and} \\ &\text{right continuous with } \kappa(0) = \kappa(\bar{s} - 0) = 0 \text{ and} \\ &\text{norm } V(\kappa)\}, \end{aligned}$$

$$\begin{aligned} \text{NBV}_1 = \{\theta\colon [0,\bar{s}] \to T^1, &\ \theta \text{ locally absolutely continuous,} \\ &\text{with } \dot{\theta} \in \text{NBV and norm } \sup|\theta| + V(\dot{\theta})\}. \end{aligned}$$

Here T^1 is a real copy of the torus $\{e^{i\phi}\colon -\infty < \phi < \infty\}$. Setting

$$(3.1)\qquad \dot{\theta}(\bar{s}_i) = a_i, \quad \ddot{\theta}(\bar{s}_i + 0) = b_i \quad (a_0 = b_m = 0),$$

and integrating (2.5i) twice, we are led to formulate a mapping $G = G(\Theta,P)$ whose zeros correspond to extremal P-interpolants; Θ is defined below. We are primarily interested here in deciding whether $\bar{P}$ is an interior point in the set of configurations P for which extremal P-interpolants exist. Specifically, we define a mapping $G = (g, r_i, q_i)$ with components $g \in \text{NBV}$, $r_i \in \mathbb{R}^1$, $q_i \in \mathbb{R}^2$ as follows.

Let

$$\varepsilon = \min\left(\frac{\bar{s}_i - \bar{s}_{i-1}}{2}, \bar{s} - \bar{s}_m\right), \quad i = 1, \ldots, m.$$

Given (Θ, P),

$$P = (0, p_1, \ldots, p_m),$$

$$\Theta = (\theta;\ s_1, \ldots, s_m;\ a_1, \ldots, a_m;\ b_0, \ldots, b_{m-1}),$$

where $\theta \in \text{NBV}_1$, $s_i \in (\bar{s}_i - \varepsilon, s_i + \varepsilon)$, $a_i \in \mathbb{R}^1$, $b_i \in \mathbb{R}^2$, then $G(\Theta, P)$ is defined as follows.

$$\text{(i)} \quad g(s) = \dot{\theta}(s) + \frac{1}{2} \int_{s_{i-1}}^{s} (s - t)\dot{\theta}^3(t)\,dt - a_{i-1} - b_{i-1}(s - s_{i-1}), \quad s_{i-1} \le s < s_i.$$

$$g(s) = \dot{\theta}(s), \quad s_m \le s \le \bar{s}.$$

$$(3.2) \qquad \text{(ii)} \quad r_i = a_i - a_{i-1} - b_{i-1}(s_i - s_{i-1}) + \frac{1}{2} \int_{s_{i-1}}^{s_i} (s_i - t)\theta^3(t)\,dt.$$

$$\text{(iii)} \quad q_i^1 = \int_0^{s_i} \cos\theta - p_i^1, \quad q_i^2 = \int_0^{s_i} \sin\theta - p_i^2.$$

<u>Remark 3.1</u>. It is an elementary fact that

$$G(\bar{\Theta}, \bar{P}) = 0 \tag{3.3}$$

characterizes $\bar{P}$-extremal interpolants.

IV. THE LINEARIZATION

The Frechét derivative $G'_\Theta(\overline{\Theta},\overline{P})$ can be computed without difficulty. We quote here a result from (Golomb and Jerome) (1979), p. 33).

Theorem 4.1. $G'_\Theta(\Theta,P)$ is an isomorphism of

$$NBV_1 \times \mathbb{R}^m \times \mathbb{R}^m \times \mathbb{R}^m \text{ onto } NBV \times \mathbb{R}^m \times (\mathbb{R}^2)^m$$

if and only if $(\overline{\Theta},\overline{P})$ satisfies the following hypothesis:

(A) The system for the unknown $\psi \in NBV_1$:

$$\text{(i)} \quad \ddot{\psi} + \frac{3}{2}\overline{\kappa}^{-2}\dot{\psi} = 0 \text{ on } (\overline{s}_{i-1}, \overline{s}_i), \quad i = 1,\ldots,m, \qquad \dot{\psi} = 0 \text{ on } (\overline{s}_m, \overline{s}).$$

$$(4.1) \qquad \text{(ii)} \quad \Delta_i\dot{\psi} + \Delta_i\dot{\overline{\kappa}} \int_0^{\overline{s}_i} \psi \sin(\overline{\theta} - \overline{\theta}_i) = 0, \quad i = 1,\ldots,m,$$

$$\text{(iii)} \quad \int_0^{\overline{s}_i} \psi \cos(\overline{\theta} - \overline{\theta}_i) = 0, \quad i = 1,\ldots,m,$$

has only the trivial solution $\psi = 0$. Here, $\overline{\theta}_i = \theta(\overline{s}_i)$ and $\Delta_i\phi = \phi(\overline{s}_i + 0) - \phi(\overline{s}_i - 0)$, $i = 1,\ldots,m$.

The implicit function theorem (cf. Nirenberg (1974)) and Theorem 4.1 yield immediately the following

Corollary 4.2. If hypothesis (A) holds for an extremal P-interpolant defined by $\overline{\Theta}$, then there exist neighborhoods $N_{\overline{\Theta}}$ and $N_{\overline{P}}$ of $\overline{\Theta}$ and $\overline{P}$ and a diffeomorphism Π,

$$\Pi: N_{\overline{\Theta}} \xrightarrow{\text{onto}} N_{\overline{P}}$$

such that $\Theta \in N_{\overline{\Theta}}$ defines an extremal P-interpolant for P $= \Pi(\Theta)$.

V. THE RAY CONFIGURATION

We close this note with the result alluded to in the title (cf. Golomb and Jerome (1979), p. 37).

Theorem 5.1. Suppose $P = \{0, p_1, \ldots, p_m\}$ is the ray configuration $\overline{p}_i = (\overline{s}_i, 0)$ $(i = 1, \ldots, m)$ with the trivial interpolant. The hypothesis (A) is satisfied, hence the conclusion of Corollary 4.2 holds.

Remark 5.1. With the substitutions

$$s = x, \ \overline{s}_i = x_i, \ \overline{s} = \overline{x}, \ y(x) = \int_0^x \psi,$$

then $\dot{y}(0) = 0$, $\ddot{y}(0) = 0$ and equations (4.1) become

$$\text{(5.1)} \quad \begin{array}{ll} \text{(i)} & \begin{array}{l} y^{(4)} = 0 \quad \text{on } (x_{i-1}, x_i), \quad i = 1, \ldots, m, \\ y'' = 0 \quad \text{on } (x_m, \overline{x}), \end{array} \\ \text{(ii)} & y''(x_i + 0) - y''(x_i - 0) = 0, \quad i = 1, \ldots, m, \\ \text{(iii)} & y(x_i) = 0, \quad i = 1, \ldots, m. \end{array}$$

These are exactly the equations for a natural cubic spline which interpolates the points $(x_i, 0)$ $(i = 0, \ldots, m)$. It follows that $y \equiv 0$.

REFERENCES

Fisher, S., and Jerome, J. (1976). J. Math. Anal. Appl. 53, 367-376.
Golomb, M. (1978). "Stability of interpolating elastica," MRC Technical Summary Report 1852, Madison, Wis.
Golomb, M. (1979). "Stability of Interpolating Elastica," Trans. 24th Conf. Army Mathematicians. ARO-Report 1, pp. 301-350.
Golomb, M., and Jerome, J. (1979). "Equilibria of the curvature functional and manifolds of nonlinear interpolating spline curves," MRC Technical Summary Report 2024, Madison, Wis.

Jerome, J. (1975). "Smooth interpolating curves of prescribed length and minimum curvature," Proc. Amer. Math. Soc. 51, 62-66.

Lee, E. and Forsythe, G. (1973). "Variational study of nonlinear spline curves," SIAM Review 15, 120-133.

Nirenberg, L. (1974). "Topics in Nonlinear Functional Analysis," Courant Institute of Mathematical Sciences, New York University, New York.

ON MAMEDOV ESTIMATES FOR THE APPROXIMATION OF FINITELY DEFINED OPERATORS

Heinz H. Gonska

Department of Mathematics
University of Duisburg
Duisburg, West Germany

I. INTRODUCTION

In the present paper we will give Mamedov (or quantitative Korovkin-type) estimates for the difference $|(L - A)(f,y)|$, where L is positive and linear and A is finitely defined, both mapping the space $C[a,b]$ of real-valued continous functions on an interval $[a,b]$ to the space $B(Y)$ of real-valued bounded functions on a non-empty set Y. The estimates given below extend recent quantitative results of Nagel [6]; even for the approximation of the injection from $C[a,b]$ to $B[a,b]$ they lead to rather sharp error bounds.

Copyright © 1980 by Academic Press, Inc.
All rights of reproduction in any form reserved.
ISBN: 0-12-171050-5

II. GENERAL ESTIMATES AND APPLICATIONS

In this section we will first outline the proof of a general estimate, then show how the number of test differences in the majorant may be reduced for special A's and give two applications of the theorems formulated below. The general estimate is contained in

THEOREM 1. Let $[a,b]$ be a compact interval of the real axis and let Y be a non-empty set. If $A \neq 0$ is of the form

$$A(f,y) = \sum_{i=1}^{n} \psi_i(y)\cdot f(\varphi_i(y))$$

($f \in C[a,b]$, $y \in Y$, $n \in \mathbb{N}$, $\psi_i \in B(Y)$ and $\varphi_i \in [a,b]^Y$), and if L is a positive linear operator mapping $C[a,b]$ to $B(Y)$, then there are constants c_1 and c_2, depending only upon $[a,b]$ and n, such that with $\Delta_1 := \sup_{0\le j\le 2n-1} |(L-A)(id^j,y)|$ and

$$\Delta_2 := \left(\sup_{0\le j\le 2n} \frac{|(L-A)(id^j,y)|}{\|L\| + \|A\|}\right)^{1/2n}$$

for all $f \in C[a,b]$ and every $y \in Y$ the estimate

$$|(L-A)(f,y)| \le c_1\cdot\Delta_1\cdot\|f\| + c_2\cdot(\|L\| + \|A\|)\cdot\omega_{2n}(f,\Delta_2)$$

holds, where ω_{2n} denotes the 2n-th order modulus of continuity and where id^j is the j-th monomial.

For the proof of this theorem we need the following three lemmas, the first of which is comparable to a result due to Butzer-Scherer-Westphal [1].

LEMMA 1. Let E denote a real vector space, U a subspace of E, and p and $\bar{p}$ seminorms on E and U respectively. We define the functional $\tilde{K} : \mathbb{R}_+^2 \times E \to \mathbb{R}_+$ by

$$\tilde{K}(t_1,t_2,f;E,U)_{p,\bar{p}} := \inf\{p(f-g)+t_1\cdot p(g)+t_2\cdot\bar{p}(g) : g \in U\} .$$

Let F be another vector space with seminorm q. Suppose $L,A : E \to F$ are linear mappings. Then for a constant $\eta > 0$ and real numbers $t_1, t_2 \geq 0$ the following are equivalent:

(i) $q((L-A)(f)) \leq \eta \cdot \tilde{K}(\frac{1}{\eta} \cdot t_1, \frac{1}{\eta} \cdot t_2, f; E, U)_{p,\bar{p}}$ for all $f \in E$.

(ii) $q((L-A)(f)) \leq \eta \cdot p(f)$ for all $f \in E$ and
$q((L-A)(g)) \leq t_1 \cdot p(g) + t_2 \cdot \bar{p}(g)$ for all $g \in U$.

Proof. The implication '(ii) $\Rightarrow$ (i)' which will be used later is a consequence of the inequality

$$q((L-A)(f)) \leq \eta \cdot [p(f-g) + \frac{1}{\eta} \cdot t_1 \cdot p(g) + \frac{1}{\eta} \cdot t_2 \cdot \bar{p}(g)]$$

which holds for an arbitrary $g \in U$.

The following lemma will be used in the proof of Lemma 3.

LEMMA 2. Let $[a,b]$ be a compact interval, $m \in \mathbb{N}_0$ and $x_0,..,x_m$ a family of points in $[a,b]$. Let $n \in \mathbb{N}$ be given with $m \leq n-1$. Given an arbitrary Hermite interpolation problem for $x_0,..,x_m$ with exactly $2n$ conditions and a function g in $C^{2n-1}[a,b]$, there exist constants $c^{(l)}$, $0 \leq l \leq 2n-1$, none of which depend upon $x_0,..,x_m$ or on the form of the special incidence matrix, such that the coefficients a_k of the interpolating polynomial $\sum_{k=0}^{2n-1} a_k \cdot id^k \in \Pi_{2n-1}$ satisfy the inequalities

$$|a_k| \leq \sum_{l=k}^{2n-1} c^{(l)} \cdot \| g^{(l)} \| \quad , \ 0 \leq k \leq 2n-1 .$$

Here $\| \cdot \|$ denotes the usual sup-norm over $[a,b]$.

The proof of Lemma 2 utilizes the Newton formula for the interpolating polynomial.

The next lemma gives the Jackson-type inequality needed for the use of Lemma 1.

LEMMA 3. Let $[a,b]$, L and A be given as in Theorem 1. Then for a function $g \in C^{2n}[a,b]$ and a point $y \in Y$ the following estimate holds true, where c_3 and c_4 are constants depending on $[a,b]$ and n only:

$$|(L-A)(g,y)| \leq c_3 \cdot \Delta_1 \cdot \|g\| + c_4 \cdot (\|L\| + \|A\|) \cdot \Delta_2^{2n} \cdot \|g^{(2n)}\| \ .$$

Proof. Useful tools for the proof are the interpolation process considered in Lemma 2, the error estimate for $g - p$ (where p denotes a suitable interpolation polynomial), the information obtained from Lemma 2, the equivalence of the norms $\sum_{i=0}^{2n} \|\cdot^{(i)}\|$ and $\|\cdot\| + \|\cdot^{(2n)}\|$ on $C^{2n}[a,b]$ and the fact that the coefficients of polynomials of the type $\prod_{i=1}^{n} (id - x_i)^2$, $x_i \in [a,b]$, have bounds depending on $[a,b]$ and n only.

Proof of Theorem 1. We first note that $|(L-A)(f,y)|$ is less or equal to $(\|L\| + \|A\|) \cdot \|f\|$. Lemma 3 gives the Jackson-type estimate for $C^{2n}[a,b]$-functions g. Thus we conclude from Lemma 1 that for an arbitrary $f \in C[a,b]$ and every $y \in Y$ the inequality

$$|(L-A)(f,y)| \leq (\|L\| + \|A\|) \cdot \tilde{K}\left(\frac{c_3 \cdot \Delta_1}{\|L\| + \|A\|}, c_4 \cdot \Delta_2^{2n}, f; C, C^{2n}\right)$$

holds. Moreover,

$$\tilde{K}(t_1, t_2, f; E, U)_{p,\bar{p}} \leq t_1 \cdot p(f) + 2 \cdot K(t_2, f; E, U)_{p,\bar{p}} \ ,$$

where $K(t_2, f; E, U)_{p,\bar{p}} := \inf\{p(f-g) + t_2 \cdot \bar{p}(g) : g \in U\}$.

For the latter functional and the case $(E,p) = (C[a,b], \|\cdot\|)$, $(U,\bar{p}) = (C^{2n}[a,b], \|\cdot^{(2n)}\|)$ the inequality

$$K(t^{2n}, f; E, U)_{p,\bar{p}} \leq c_5 \cdot \omega_{2n}(f,t) \qquad (c_5 \text{ a real constant})$$

is well known. This yields the assertion of Theorem 1.

While Theorem 1 deals with the case of arbitrary A's of the form given above, we now treat the case where the points

$\varphi_i(y)$ do not depend on y. As has been shown by Nagel [6] among others, the number of test differences which have to be taken into consideration in this case may be reduced if endpoints of [a,b] occur among the $\varphi_i(y)$'s. The analogue of Theorem 1 is

THEOREM 2. Let [a,b] and Y be given as above. Let $A \neq 0$ be an operator of the form $A(f,y) = \sum_{i=1}^{n} \psi_i(y)\cdot f(x_i)$, where $\psi_i \in B(Y)$ and w.l.o.g. $x_i \neq x_j$ for $i \neq j$, and let L be a positive linear operator mapping C[a,b] to B(Y). We define $w(x_i)$ by

$$w(x_i) := \begin{Bmatrix} 1 & \text{if } x_i \in \{a,b\} \\ 2 & \text{if } x_i \in [a,b] \end{Bmatrix}$$

and $m+1 := \sum_{i=1}^{n} w(x_i)$. Then for $f \in C[a,b]$ and $y \in Y$ the estimate

$$|(L-A)(f,y)| \leq c_1\cdot\Delta_1\cdot\|f\| + c_2\cdot(\|L\| + \|A\|)\cdot\omega_{m+1}(f,\Delta_2)$$

holds, where $\Delta_1 := \sup_{0\leq j\leq m} |(L-A)(\mathrm{id}^j,y)|$,

$$\Delta_2 := \left(\sup_{0\leq j\leq m+1} \frac{|(L-A)(\mathrm{id}^j,y)|}{\|L\| + \|A\|}\right)^{1/m+1} ,$$

and where c_1 and c_2 are constants depending on [a,b] and n only.

With certain modifications, the proof follows the pattern of that of Theorem 1.

We will now discuss briefly some applications of the estimations given above. In view of Wolff's representation theorem [7], the case n = 1 [i.e. $A(f,y) = \psi(y)\cdot f(\varphi(y))$] yields a quantitative result for the approximation of certain lattice homomorphisms. A special case thereof, namely $Y = [a,b]$, $\psi \equiv 1$, $\varphi = \mathrm{id}$, is of major interest. Theorem 1 gives for $f \in C[a,b]$, $x \in [a,b]$ the inequality

$$|L(f,x) - f(x)| \leq c_1\cdot \sup_{0\leq j\leq 1} |L(\mathrm{id}^j,x) - x^j|\cdot\|f\| + \ldots$$

$$\ldots + c_2 \cdot (\|L\| + 1) \cdot \omega_2\left(f, \left(\sup_{0 \le j \le 2} \frac{|L(id^j, x) - x^j|}{\|L\| + 1}\right)^{1/2}\right),$$

which is a pointwise improvement of a result due to Freud [3] and implies the main result of a paper by Esser [2].

As an application of Theorem 2 we mention the following quantitative version of an approximation theorem for iterates $B_n^{k_n}$ of Bernstein operators (see e.g. Kelisky-Rivlin [5]). If 'too much' iteration takes place (i.e. $\frac{k_n}{n} \to \infty$ for $n \to \infty$), $B_n^{k_n}(f,x) \to B_1(f,x)$ uniformly for $x \in [0,1]$; this can be derived from Theorem 2 giving

$$|B_n^{k_n}(f,x) - B_1(f,x)| \le c \cdot \omega_2\left(f, \left(\frac{(1-1/n)^{k_n} \cdot x \cdot (1-x)}{2}\right)^{1/2}\right).$$

Further material on the subject is contained in [4].

REFERENCES

1. Butzer, P. L., Scherer, K., and Westphal, U., *Acta Sci. Math.* (*Szeged*) *34*, 25 (1973).
2. Esser, H., *Indag. Math. 38*, 189 (1976).
3. Freud, G., *Studia Sci. Math. Hungar. 3*, 365 (1968).
4. Gonska, H. H., Dissertation, Universität Duisburg, (1979).
5. Kelisky, R. P., and Rivlin, T. J., *Pacific J. Math. 21*, 511 (1967).
6. Nagel, J., Dissertation, Universität Essen, (1978).
7. Wolff, M., *Math. Ann. 182*, 161 (1969).

ON THE CONTINUITY OF CONVEXLY CONSTRAINED INTERPOLATION

Rudolf Gorenflo
Martin Hilpert

Institut für Mathematik III
Freie Universität Berlin
Berlin, Germany

1. Introduction

Interpolation under shape constraints is an important area of investigation ([3],[8]). A naturally arising question is that of continuous dependence of the interpolate on the data. We treat this problem in the abstract setting of E-spaces. For general background on functional and convex analysis we refer to [6], [7], [9].

From [6], 1.46, we quote a characterization of E-spaces: A real Banach space X is an E-space iff it is reflexive, rotund, and (*) is valid.

$$x_n, x \in \{y \in X \mid \|y\| = 1\},\ x_n \rightharpoonup x \Rightarrow x_n \to x. \tag{*}$$

X is rotund iff X is strictly normed, and (*) is equivalent to

$$x_n,\ x \in X,\ x_n \rightharpoonup x,\ \|x_n\| \to \|x\| \quad \Rightarrow \quad x_n \to x. \tag{1}$$

Every Hilbert space is an E-space.

(P1) <u>Minimum norm interpolation:</u> Let U be an E-space, $m \in \mathbb{N}$, $Z = \mathbb{R}^m$ normed, $A \in L(U,Z)$ (meaning "linear and bounded"), $S \subset U$ closed and convex, $T = A(S)$ (hence also convex). For $z \in T$ to determine the norm-minimal element $\tilde{u} \in S$ with $A\tilde{u} = z$.

Copyright © 1980 by Academic Press, Inc.
All rights of reproduction in any form reserved.

ISBN: 0-12-171050-5

(P2) Minimum norm fitting: Under the additional assumptions that Z is strictly normed and T is closed for $a \in Z$ to determine the norm-minimal $\tilde{u} \in S$ with $A\tilde{u} = z := P_T a$ where P_T is the metric projector from Z on T.

In both cases exactly one solution $\tilde{u}$ exists. It is the metric projection of the zero element $\Theta \in U$ on the closed convex set $S \cap \{u \in U \mid Au = z\}$. With the notation

$$\tilde{u} =: A_S^+ z \text{ in (P1)}, \quad \tilde{u} =: A_S^+ a \text{ in (P2)}$$

the question is that of continuity of the "constrained pseudo-inverse" $A_S^+ : T \to S$ or $Z \to S$. Because P_T in (P2) is continuous it suffices to investigate $A_S^+|T$. The answer is affirmative for (P1) if $z \in \text{icr}(T)$ or if T is a polyhedron, for (P2) if T is a polyhedron. By $\text{icr}(T)$ the "intrinsic core" of T is denoted ([6], 7-8), T is a polyhedron if there exist $B \in L(Z, \mathbb{R}^k), k \in \mathbb{N}, c \in \mathbb{R}^k$, so that $z \in T \Leftrightarrow Bz \geq c$.

2. Minimum Norm Interpolation

For (P1) $\tilde{u} = A_S^+(z)$ with $z \in T$ is defined by

$$\tilde{u} \in S, \quad A\tilde{u} = z, \quad \|\tilde{u}\| \text{ minimal}. \tag{2}$$

For abbreviation we put $\varphi(z) := \|A_S^+(z)\|$.

Theorem 1. For any $z \in T$ we have: A_S^+ continuous at $z \Leftrightarrow \varphi$ continuous at z.

Proof: $\Rightarrow$ is obvious. To prove $\Leftarrow$, let $\{z_n\}$ be a sequence in T, $z_n \to z \in T$. Then $\varphi(z_n) \to \varphi(z)$, $\{\varphi(z_n)\}$ is bounded, and there exists a subsequence $\{z_{n_j}\}$ and $u \in S$ with $A_S^+ z_{n_j} \rightharpoonup u$ (S closed convex $\Rightarrow$ S weakly closed). A also being weakly continuous yields $Au = z$. Using the weak lower semi-continuity of $\|.\|$ and the continuity of φ at z we get

$$\|u\| \leq \liminf_{j\to\infty} \varphi(z_{n_j}) = \lim_{j\to\infty} \varphi(z_{n_j}) = \varphi(z),$$

hence $u = A_S^+ z$ from the uniqueness of the norm-minimizing $\tilde{u}$ in (2).

By (1) follows $A_S^+ z_{n_j} \to A_S^+ z$, and by deducing a contradiction from assuming the contrary we obtain $A_S^+ z_n \to A_S^+ z$.

Theorem 2: If T is a polyhedron then $A_S^+ : T \to S$ is continuous.

Proof: We shall show: (a) φ is convex, (b) φ is lower semi-continuous. [9], Thoerem 10.2, then gives continuity of φ, hence of A_S^+ by Theorem 1.

(a) $z_1, z_2 \in T$, $\lambda + \mu = 1$, $0 \leq \lambda \leq 1$ imply $t := \lambda A_S^+ z_1 + \mu A_S^+ z_2 \in S$, $At = \lambda z_1 + \mu z_2 \in T$, and the norm-minimizing property of A_S^+ yields

$$\varphi(\lambda z_1 + \mu z_2) \leq \|t\| \leq \lambda\varphi(z_1) + \mu\varphi(z_2).$$

(b) If φ is not lower semi-continuous there exists a sequence $\{z_n\}$ in T with $z_n \to z \in T$ and (α) $\lim_{n\to\infty} \varphi(z_n) < \varphi(z)$ so that $\{\varphi(z_n)\}$ is bounded. As in the proof of Theorem 1 there follows the existence of $u \in S$ and of a subsequence $\{z_{n_j}\}$ with $Au = z$ and (β) $\liminf_{j\to\infty} \varphi(z_{n_j}) \geq \|u\|$.

The minimum norm property gives $\|u\| \geq \varphi(z)$, hence from (β) and (α) a contradiction.

Theorem 3: In problem (P1) $A_S^+ | \text{icr}(T)$ is continuous; if T is a polyhedron even $A_S^+ : T \to S$ is continuous. In problem (P2) if T is a polyhedron $A_S^+ : Z \to S$ is continuous.

Proof: For (P1) surround $z \in \text{icr}(T)$ or $z \in T$ by a polyhedral relative neighborhood ("relative" referring to the relative topology of the affine hull of T). For (P2) take account of the continuity of P_T. In both cases then apply Theorem 2.

3. Specialization to Hilbert Space

If U is a Hilbert space and $N = N(A)$ the null-space of A we have $U = N \oplus N^{\perp}$, and in case (P1) for $z \in T$ we can uniquely decompose $A_S^+ z = x + v(x)$ where $x = A^+ z \in M = P_{N^{\perp}}(S)$, $v(x) \in N$ norm-minimal so that $x + v(x) \in S$.

Here A^+ is the pseudo-inverse, here continuous because A has closed range (see [5]). The question of continuity is now reduced to that of $v : M \to N$. From discussions one of us (R. Gorenflo) had with Thomas Seidman in October 1979 at the International Symposium on Ill-Posed Problems (University of Delaware) the following theorem arose which now is deductible from Theorem 3: assume A surjective and take account of the linear isomorphism $A^+ : Z \to N^{\perp}$.

<u>Theorem 4:</u> Let $U = N \oplus N^{\perp}$ be a Hilbert space, $\dim N^{\perp} = m \in \mathbb{N}$, $S \subset U$ closed and convex, $M = P_{N^{\perp}}(S)$, M_o the interior of M relative to $N^{\perp}$. For $x \in M$ let $x + v(x) \in S$, $v(x) \in N$ norm-minimal, $f(x) = \|v(x)\|$. Then $f : M \to \mathbb{R}$ is convex, and $f|M_o$ and $v|M_o$ are continuous.

Engl and Kirsch in [4] state a variant of this theorem. Not requiring $\dim N^{\perp}$ finite but assuming S also bounded they obtain continuity of $v|P_{N^{\perp}}(\text{int } S)$.

Seidman has supplied an example for discontinuity of v at a point of ∂M. Let $U = \mathbb{R}^3$ be equipped with the natural scalar product, $N = \{(0,0,x_3) | x_3 \in \mathbb{R}\}$, and S the closed convex hull of

$$\{\Theta\} \cap \{x \in \mathbb{R}^3 \;|x_1 > 0,\; x_2 \geq x_1^2,\; x_3 \geq e^{-x_1},\; \|x\|_\infty \leq 2\}.$$

Then, as long as $t > 0$ and $(t,t^2,0) \in M$, we have $v(t,t^2,0) = (0,0,e^{-t})$, but $v(\Theta) = \Theta$. Thus v is discontinuous at the boundary point Θ.

<u>Application</u> of Theorem 3 is possible to convexly constrained spline interpolation or fitting. As in [1] let $U,Y,Z = \mathbb{R}^m$ be Hilbert spaces, $A \in L(U,Z)$ and $B \in L(U,Y)$ surjective, $N(A) \cap N(B) = \{\Theta\}$, $S \subset U$ closed and convex, $z \in A(S)$. To determine $\tilde{u} \in S$ so that $\|B\tilde{u}\|_Y$ is minimal and $A\tilde{u} = z$.

Because $\|Bu\|_Y$ generally is only a semi-norm on U renorm U by

$$|||u|||^2 = \|Au\|_Z^2 + \|Bu\|_Y^2 .$$

Then $|||.|||$ and $\|.\|_U$ are equivalent norms ([2], 99-100), and $\tilde{u} = A_S^+ z$ or $A_S^+ a$, respectively, A_S^+ taken with regard to $|||.|||$.

References

1. Anselone, P.M., and Laurent, P.J., A general method for the construction of interpolating and smoothing spline functions, Numer. Math. 12, 66-82 (1968).
2. Böhmer, K., Spline-Funktionen, Teubner, Stuttgart (1974).
3. Copley, P. and Schumaker, L.L., On plg-splines, J. Approximation Theory 23, 1-28 (1978).
4. Engl, H.W., and Kirsch, A., On the continuity of the metric projection onto a convex set subject to an additional constraint, Institutsbericht 153, Institut für Mathematik, Universität Linz, Austria (1979).
5. Groetsch, C.W., Generalized Inverses of Linear Operators, Marcel Dekker, Inc., New York and Basel (1977).
6. Holmes, R.B., A Course on Optimization and Best Approximation, Springer-Verlag, Berlin (1972).
7. Holmes, R.B., Geometric Functional Analysis and its Applications, Springer-Verlag, New York (1975).
8. Hornung, U., Interpolation by smooth functions under restrictions for the derivatives, J. Approximation Theory (to appear).
9. Rockafellar, R., Convex Analysis, Princeton University Press, Princeton (1972).

TRUNCATION ERROR BOUNDS FOR T-FRACTIONS

William B. Gragg

Department of Mathematics
University of Kentucky
Lexington, Kentucky

We describe best possible inclusion lunes for approximants of positive definite T-fractions, and give several bounds for their diameters. As a corollary we obtain an analog of the Carleman condition, for determinateness of the strong Stieltjes moment problem.

I. INTRODUCTION

The connections between strong Stieltjes moment problems,

$$\gamma_n = \int_0^\infty t^n d\mu(t) \quad , \quad n = 0, \pm 1, \pm 2, \cdots ,$$

(μ isotone with infinitely many points of increase), double asymptotic expansions

$$\Phi(z) = \int_0^\infty \frac{d\mu(t)}{1+tz} , \qquad |\arg z| < \pi ,$$

$$\sim \textstyle\sum_0^\infty \gamma_n(-z)^n, \quad z \to 0 ,$$

$$\sim -\textstyle\sum_1^\infty \gamma_{-n}/(-z)^n, \quad z \to \infty ,$$

and positive definite T-fractions,

$$F(z) = a_1/1 + b_2 z + a_2 z/1 + b_2 z + \cdots ,$$

all $a_n > 0$, $b_n > 0$, have recently been given in [4]. A solution μ exists if and only if, for some k, $\gamma_n^{(k-n)} > 0$ and $\gamma_n^{(k-n+1)} > 0$ for $n \geq 0$. Here, $\gamma_n^{(k)} \equiv \det\,(\gamma_{k+i+j})_{i,j=0}^{n-1}$,

Copyright © 1980 by Academic Press, Inc.
All rights of reproduction in any form reserved.
ISBN: 0-12-171050-5

$-\infty < k < +\infty$, $n \geq 0$, are the Hankel determinants of $\{\gamma_n\}_{-\infty}^{\infty}$. This is shown in [4] for $k = 0$. But, it follows from Jacobi's identity,

$$\gamma_n^{(k-1)}\gamma_n^{(k+1)} - \gamma_{n-1}^{(k+1)}\gamma_{n+1}^{(k-1)} = (\gamma_n^{(k)})^2 ,$$

that either implies $\gamma_n^{(m)} > 0$ for *all* m and n. That is, the doubly infinite Hankel matrix (γ_{i+j}) is strictly totally positive (definite). This justifies use of the quotient-difference algorithm to compute the coefficients of the T-fraction from the moments. See [5], where some of our results below are given for real z. We have

$$a_1 a_2 \cdots a_{n+1} = \frac{\gamma_{n+1}^{(-n)}}{\gamma_n^{(-n)}} , \quad b_1 b_2 \cdots b_n = \frac{\gamma_n^{(1-n)}}{\gamma_n^{(-n)}}$$

It is also known that the nth approximant, $w_n(z)$, of $F(z)$ is of the same form as $\Phi(z)$, but with μ having n points of increase (partial fraction decomposition), and that μ is essentially unique if and only if $F(z)$ converges for some, and hence all, z with $|\arg z| < \pi$.

II. INCLUSION LUNES

Fix $z = \zeta^2$, $\xi = \mathrm{Re}\,\zeta > 0$, and consider

$$\zeta F(z) = a_1/\beta_1(\zeta) + a_2/\beta_2(\zeta) + \cdots ,$$

with $\beta_n(\zeta) = b_n\zeta + 1/\zeta$, and approximants $\omega_n(\zeta) = \zeta w_n(z)$. The set of admissible first approximants is

$$\begin{aligned} K_0(\zeta) &= \{a/b\zeta + 1/\zeta : a > 0,\ b > 0\} \\ &= \{\omega : |\arg \omega| < |\arg \zeta|\} \text{ if } \zeta \neq \bar{\zeta} , \\ &= \mathbb{R}^+ \text{ if } \zeta = \bar{\zeta} , \\ &= \overline{K_0(\zeta)} \text{ in general.} \end{aligned}$$

Clearly, $\omega_n(\zeta) \in K_0(\zeta)$ for all $n \geq 1$. The boundary rays of the sector $K_0(\zeta)$, $\mathbb{R}^+\zeta$ and $\mathbb{R}^+/\zeta$, are (generalized) circular arcs which meet at $\omega = \omega_0(\zeta) \equiv 0$ and $\omega = \omega_{-1}(\zeta) \equiv \infty$ with angle $|\arg z|$.

Put $t_n(\omega) = a_n/\beta_n(\zeta) + \omega$ $(n \geq 1)$, $t_0(\omega) = \omega$, and $T_n(\omega) = t_0 \circ t_1 \circ \cdots \circ t_n(\omega)$. These are nonsingular linear fractional transformations. We have $\omega_n(\zeta) = T_n(0)$, $\omega_{n-1}(\zeta) = T_n(\infty)$, $\omega_{n+1}(\zeta) = T_n(a_{n+1}/\beta_{n+1}(\zeta)) \in T_n(L_0(\zeta)) \equiv K_n(\zeta) \subset K_{n-1}(\zeta)$, and $K_n(\zeta)$ is the best possible inclusion lune for all (n+1)th, and later, approximants of $\zeta F(z)$ under the assumption that $\{a_k\}_1^n$ and $\{b_k\}_1^n$, that is $\{\gamma_k\}_{-n}^{n-1}$, are known. For $n \geq 1$, $K_n(\zeta)$ is the intersection of two bounded circular disks, and is hence convex. (Convexity can also be seen from the partial fraction decomposition.) $K_n(\zeta)$ has vertices $\omega_n(\zeta)$, $\omega_{n-1}(\zeta)$, and angle $\theta = |\arg z|$. Thus, diam $K_n(\zeta) \leq |\omega_n(\zeta) - \omega_{n-1}(\zeta)|/\kappa(\theta)$, with $\kappa(\theta) \equiv 1$ if $\theta \leq \pi/2$ and $\kappa(\theta) \equiv \sin\theta$ if $\pi/2 \leq \theta < \pi$.

Suppose now that $\{a_k\}_1^{n+1}$ and $\{b_k\}_1^n$, that is $\{\gamma_k\}_{-n}^n$, are known. Let $K_n'(\zeta)$ be the set of all possible (constrained) (n+2)th approximants, so that

$$K_n'(\zeta) = T_n(a_{n+1}/\mathbb{R}^+\zeta + K_0(z) + 1/\zeta)$$
$$= T_n(a_{n+1}/K_0(\zeta) + 1/\zeta).$$

Clearly, $K_n'(\zeta) \subset K_n(\zeta) \subset K_{n-1}'(\zeta)$. $K_n'(\zeta)$ is also the convex intersection of circular disks, with vertices $\omega_n(\zeta)$, $\omega_n'(\zeta) \equiv T_n(a_{n+1}\zeta)$, and angle $\theta = |\arg z|$. Hence, diam $K_n'(\zeta) \leq |\omega_n(\zeta) - \omega_n'(\zeta)|/\kappa(\theta)$.

We could similarly derive best possible inclusion lunes $K_n''(\zeta)$ when the moments $\{\gamma_k\}_{-n-1}^{n-1}$ are known, and we would then have $K_{n+1}(\zeta) = K_n'(\zeta) \cap K_n''(\zeta)$. We also omit the detailed description of the boundary arcs of the lunes, and of their circular extensions.

Finally, these ζ-lunes translate into the corresponding inclusion lunes $L_n(z) = K_n(\zeta)/\zeta$, $L_n'(z) = K_n'(\zeta)/\zeta$, for the approximants $w_n(z)$ of the T-fraction $F(z)$, and the associated values $w_n'(z) \equiv \omega_n'(\zeta)/\zeta$. In fact the latter are the approximants of

$$\gamma_0 - a_1'z/1 + b_1'z + a_2'z/1 + b_2'z + \cdots ,$$

in which $a_1' = a_1 b_1' = \gamma_1$, $b_1' = b_1 + a_2$, and for $n \geq 2$, $a_n' = a_n r_n$, $b_n' = b_{n-1} r_n$, with $r_n = (b_n + a_{n+1})/(b_{n-1} + a_n)$. The T-fraction (or proper) part of this continued fraction is associated with Stieltjes transforms $[\gamma_0 - \Phi(z)]/z = \int_0^\infty t d\mu(t)/(1 + tz)$. When $z = x > 0$ our lunes become intervals of $\mathbb{R}^+$:

$$0 = w_0(x) < w_1'(x) < w_2(x) < \cdots < w_2'(x) < w_1(x) < w_0'(x)$$

As the moments enter in the order $\gamma_0, \gamma_{-1}, \gamma_1, \gamma_{-2}, \cdots$, the approximants enter in the order $w_0'(z)$, $w_1(z)$, $w_1'(z)$, $w_2(z)$, $w_2'(z)$, $\cdots$. Likewise for the corresponding inclusion lunes.

III. BOUNDS FOR $|w_n'(z) - w_n(z)|$

From the very elementary theory of continued fractions,

$$|\omega_{n-1}(\zeta) - \omega_n(\zeta)| = \frac{a_1 a_2 \cdots a_n}{|\psi_n \psi_{n-1}|} \leq \frac{a_1 a_2 \cdots a_n}{\text{Re } \psi_n \bar{\psi}_{n-1}} \equiv \frac{1}{d_n}$$

and

$$|\omega_n'(\zeta) - \omega_n(\zeta)| = \frac{a_1 a_2 \cdots a_{n+1}}{|\psi_n' \psi_n|} \leq \frac{a_1 a_2 \cdots a_{n+1}}{\text{Re } \psi_n' \bar{\psi}_n} \equiv \frac{1}{d_n'} ,$$

with $\psi_{-1} = 0$, $\psi_0 = 1$, $\psi_{n+1} = \psi_n \beta_{n+1}(\zeta) + \psi_{n-1} a_{n+1}$, and $\psi_n' \equiv \psi_{n-1} a_{n+1} + \psi_n/\zeta$. From the three term recursion formula,

$$d_n' = d_n + \xi\sigma_n/|\zeta|^2 , \quad d_{n+1} = d_n' + \xi b_{n+1}\sigma_n ,$$

with $d_0 = 0$ and $\sigma_n \equiv |\psi_n|^2/a_1a_2\cdots a_{n+1}$. By the arithmetic-geometric mean inequality, together with $|\psi_n\psi_{n-1}| > \text{Re}\ \psi_n\bar{\psi}_{n-1}$ and $d_n \geq d'_{n-1}$,

$$\begin{aligned} d'_n &= d'_{n-1} + \xi(b_n\sigma_{n-1} + \sigma_n/|\zeta|^2) \\ &\geq d'_{n-1} + \frac{2\xi}{|\zeta|} b_n^{\frac{1}{2}} (\sigma_{n-1}\sigma_n)^{\frac{1}{2}} \\ &\geq d'_{n-1} + \frac{2\xi}{|\zeta|}\left(\frac{b_n}{a_{n+1}}\right)^{\frac{1}{2}} d_n \\ &\geq d'_0\ \Pi_1^n(1 + 2\alpha_k) \quad , \qquad \alpha_k \equiv \left(\frac{b_k}{a_{k+1}}\right)^{\frac{1}{2}} \frac{\xi}{|\zeta|} \end{aligned}$$

This gives our first bound,

$$|w'_n(z) - w_n(z)| \leq \frac{a_1|\zeta|}{\xi}\prod_1^n\left[1 + \frac{2\xi}{|\zeta|}\left(\frac{b_k}{a_{k+1}}\right)^{\frac{1}{2}}\right]^{-1}.$$

By Chrystal's inequality, [1, p. 61], and the relations in the introduction,

$$\begin{aligned} \Pi_1^n(1 + 2\alpha_k) &\geq \left[1 + 2(\alpha_1\alpha_2\cdots\alpha_n)^{1/n}\right]^n \\ &= \left[1 + \frac{2\xi}{|\zeta|}\left(\gamma_0\ \frac{\gamma_n^{(-n+1)}}{\gamma_{n+1}^{(-n)}}\right)^{1/2n}\right]^n . \end{aligned}$$

Jacobi's identity, and the positivity of all $\gamma_n^{(m)}$, now lead to

$$\left(\frac{\gamma_n^{(-n+1)}}{\gamma_{n+1}^{(-n)}}\right)^2 > \frac{\gamma_n^{(-n+1)}}{\gamma_{n+1}^{(-n-1)}}\ \frac{\gamma_n^{(-n+1)}}{\gamma_{n+1}^{(-n+1)}} > \frac{1}{\gamma_{-n-1}\gamma_{n+1}} ,$$

in which the second inequality expresses a basic fact about positive definite matrices. Thus, our second bound is

$$|w'_n(z) - w_n(z)| \leq \frac{a_1|\zeta|}{\xi}\left[1 + \frac{2\xi}{|\zeta|}\left(\frac{\gamma_0^2}{\gamma_{-n-1}\gamma_{n+1}}\right)^{1/4n}\right]^{-n} .$$

Finally, we have

$$\begin{aligned} \Pi_1^n(1 + 2\alpha_k) &\geq 1 + 2\ \textstyle\sum_1^n \alpha_k \\ &> 1 + \frac{2}{e}\ \textstyle\sum_1^n(\alpha_1\alpha_2\cdots\alpha_k)^{1/k} , \end{aligned}$$

by Carleman's inequality, [1, p. 249]. This gives our final

$$|w_n'(z) - w_n(z)| \le \frac{a_1|\zeta|}{\xi}\left[1 + \frac{2\xi}{e|\zeta|}\sum_1^n\left(\frac{\gamma_0^2}{\gamma_{-k-1}\gamma_{k+1}}\right)^{1/4k}\right]^{-1}.$$

Hence, the T-fractions converge for $|\arg z| < \pi$, and the moment problem is determinate, if

$$\sum_2^\infty(\gamma_{-n}\gamma_n)^{-1/4(n-1)} = +\infty .$$

Our treatment has been inspired by that of the classical Stieltjes case in [2,3].

REFERENCES

1. Hardy, G. H., J. E. Littlewood, and G. Pólya, "Inequalities." Cambridge University Press, 1959.
2. Henrici, Peter, "Applied and Computational Complex Analysis," Volume Two. Wiley, New York, 1977.
3. Henrici, Peter, and Pia Pfluger, Truncation error estimates for Stieltjes fractions. *Numer. Math.* 9 (1966) 120-138.
4. Jones, William B., W. J. Thron, and Haakon Waadeland, A strong Stieltjes moment problem. *Trans. Amer. Math. Soc.*, to appear.
5. McCabe, J. H., and J. A. Murphy, Continued fractions which correspond to power series expansions at two points. *J. Inst. Math. Appl.* 17 (1976) 233-247.

SMOOTHEST LOCAL INTERPOLATION FORMULAS FOR EQUALLY SPACED DATA

T. N. E. Greville[1]

Mathematics Research Center
University of Wisconsin
Madison, Wisconsin

Hubert Vaughan[2]

Sydney, Australia

1. INTRODUCTION

Schoenberg pointed out in 1946 [3] that a large class of local interpolation formulas for equally spaced data can be expressed in the form

$$v_x = \sum_{\nu=-\infty}^{\infty} L(x - \nu) y_\nu \ , \tag{1}$$

where y_ν denotes a given ordinate, v_x is an interpolated value, and $L(x)$ is a given function of finite support called by Schoenberg the basic function of the interpolation formula. There is an extensive literature on such formulas, but it is mostly in actuarial journals and not well known to mathematicians and approximation theorists; some references are given in [1,3].

---

[1]Sponsored by the United States Army under Contract No. DAAG29-75-C-0024.

[2]Deceased.

Copyright © 1980 by Academic Press, Inc.
All rights of reproduction in any form reserved.
ISBN: 0-12-171050-5

Formula (1) is called <u>exact for the degree</u> r if $\sum_{\nu=-\infty}^{\infty} L(x-\nu)p(\nu) = p(x)$ for all x for $p \in \pi_r$, but this is not true, in general, for $p \in \pi_{r+1}$. It is called a <u>reproducing formula</u> when $L(x)$ is such that $v_\nu = y_\nu$ for every integer ν, whatever may be the values of the quantities y_ν. It is clear that (1) is reproducing if and only if

$$L(\nu) = \delta_{0\nu} \qquad (\nu = \ldots, -1, 0, 1, \ldots),$$

where $\delta_{0\nu}$ is a Kronecker symbol.

II. DEFINITIONS

Let $I = (\alpha,\beta)$ be a finite open interval on the real line and let $\mu = \beta - \alpha$. Let F_{Irm} denote the set of interpolation formulas (1) such that:

1. The support of L is contained in I.
2. The formula is exact for the degree r.
3. $L \in C^{m-1}$.
4. $L^{(m)}$ is piecewise continuous: that is, all its discontinuities are jumps and the number of these is finite.

Let F_{Irm}^{rep} denote the subset of F_{Irm} that contains only reproducing formulas.

Certain consequences follow easily from these definitions. F_{Irm}^{rep} is empty unless $0 \in I$. When $0 \in I$, F_{Irm} and F_{Irm}^{rep} are identical when the number of integers contained in I is less than or equal to $r + 1$; otherwise F_{Irm}^{rep} is a proper subset.

F_{Irm} is empty for $\mu < r + 1$, and it contains a single formula (of Newtonian-Lagrangian interpolation) when $\mu = r + 1$. In the latter case, L is discontinuous at arguments $\alpha + \nu$ and $\beta - \nu$ (ν being any integer) within its support, unless α and β are integers and $0 \in I$. For $\mu > r + 1$, F_{Irm} is an infinite set.

III. THE MAIN RESULT

Let a quantity J be defined by:

$$J = \int_{\alpha}^{\beta} [L^{(m)}(x)]^2 dx .$$

In a previous paper [1] we called a formula (1) a <u>minimized derivative formula</u> (m. d. f.) of its class if, for that formula, J has its minimum value for the class. The following two theorems are the main results; proofs will be given in the complete paper [2], which will appear in 1980 as a Mathematics Research Center technical report.

<u>Theorem 1.</u> For any nonnegative integers r and m, and any finite interval $I = (\alpha,\beta)$ such that $\mu > r + 1$, the class F_{Irm} contains a single formula whose basic function L satisfies the following three conditions:

(i) L is a piecewise polynomial function of degree at least r and at most $d = \max(r, 2m - 1)$.

(ii) Each knot of L is an argument that differs by an integer from α or β (or both).

(iii) The piecewise polynomial function $\delta^{r+1} L$ is given in $(\alpha + \rho, \beta - \rho)$ by a simple polynomial (i.e. without knots) of degree at most $2m - 1$. (Here δ denotes the central finite difference and $\rho = \frac{1}{2}(r + 1)$.)

The formula uniquely determined by these three conditions is the m. d. f. of F_{Irm}.

This theorem requires interpretation for $m = 0$. In that case we interpret a polynomial degree -1 (in condition (iii)) to mean one that is identically zero.

Theorem 2. For any nonnegative integer r and positive integer m, and any finite interval $I = (\alpha, \beta)$ containing 0 and such that $\mu > r + 1$, the class F_{Irm}^{rep} contains a single formula satisfying the following three conditions:

(i) L is a piecewise function of degree at least r and at most $d = \max(r, 2m - 1)$.

(ii) Each knot of L is either an integer or an argument that differs by an integer from α or β (or both).

(iii) The piecewise polynomial function $\delta^{r+1} L$ is given in $(\alpha + \rho, \beta - \rho)$ by a spline function of degree $2m - 1$ with simple knots. These knots are at the integers when r is odd and at integers + ½ when r is even.

IV. COMPACT EXPRESSION FOR BASIC FUNCTION

The key to the proofs which will appear in [2] is the possibility of expressing the basic function of an m. d. f. in a compact form that involves few parameters. Let p_{rn} denote the unique element of π_r such that

$$\sum_{j=1}^{n} j^k \, p_{rn}(j) = \delta_{0k} \qquad (k = 0, 1, \ldots, r).$$

Let the spline functions of degree r, $S_{rn}(x)$ and $S^*_{rn}(x)$ be defined by

$$S_{rn}(x) = x_+^r - \sum_{j=1}^{n} p_{rn}(j)(x - j)_+^r ,$$

$$S^*_{rn}(x) = x_+^r - \sum_{j=1}^{n} p_{rn}(j)(x + j)_+^r ,$$

where $x_+ = \frac{1}{2}(x + |x|)$. We note that the support of $S_{rn}(x)$ is $(0,n)$ and that of $S^*_{rn}(x)$ is $(-n,0)$, and $S^*_{rn}(x) = (-1)^{r+1} S_{rn}(-x)$. It can be shown that the basic function of a formula (1) satisfying conditions (i)-(iii) of Theorem 1 can be expressed in the form

$$L(x) = \sum_{i=m}^{d} [c_i S_{rn}^{(r-i)}(x - \alpha) + g_i S_{rn}^{*(r-i)}(x - \beta)]. \qquad (2)$$

In case $i > r$, the derivative of negative order is interpreted to mean that particular integral which vanishes for $x < \alpha$ or for $x > \beta$, as the case may be. The proof is completed by showing that the coefficients c_i and g_i are uniquely determined and that J assumes its minimum value for the resulting interpolation formula.

For a formula (1) satisfying conditions (i)-(iii) of Theorem 2, there must be added to the right member of (2) the expression

$$\sum_{j \in E} h_j (x - j)_+^{2m-1} ,$$

where E is the set of integers in I .

V. SOME PUBLISHED FORMULAS ARE m. d. f.'S

In a few instances, minimized derivative formulas are formulas that were already known. Table I lists, for several such formulas, the class F_{Irm} or F_{Irm}^{rep} involved, the name by which the formula is commonly known, and the year of publication. Publication citations will be found in [1].

TABLE I. Previously Published Formulas That Are m. d. f.'s

I	r	m	Rep or Nonrep	Originator	Publication Year
(-2,2)	2	2	Both	Karup	1898
(-2,2)	1	2	Nonrep	Jenkins	1927
(-3,3)	4	2	Both	Shovelton	1913
(-3,3)	4	3	Both	Sprague	1880
(-3,3)	3	3	Nonrep	Vaughan "C"	1946
(-3,3)	3	2	Rep	Henderson	1906

REFERENCES

1. Greville, T. N. E., and Vaughan, H., Polynomial interpolation in terms of symbolic operators, Trans. Soc. Actuar., 6 (1954), 413-476.
2. Greville, T. N. E., and Vaughan, H., Smoothest local interpolation formulas for equally spaced data, to appear in 1980 as a Mathematics Research Center technical report.
3. Schoenberg, I. J., Contributions to the problem of approximation of equidistant data by analytic functions. Part A. On the problem of smoothing or graduation. A first class of analytic approximation formulae, Quart. Appl. Math., 4 (1946), 45-99.

TWO APPLICATIONS OF PERIODIC SPLINES

Martin H. Gutknecht[1]
Seminar für angewandte Mathematik
Eidgenössische Technische Hochschule
Zürich, Switzerland

Periodic splines with equidistant knots are a class of functions whose Fourier coefficients can be computed by multiplying discrete Fourier transform coefficients by known attenuation factors [1, 3, 9]. Therefore, exact Fourier series of periodic splines can be utilized in a number of applications, where up to now the Fourier series has usually been approximated by an interpolating trigonometric polynomial. We present two such applications: The evaluation of the conjugate periodic function on a uniform N-point mesh and the computation (on a N^d-point mesh) of the solution to Poisson's equation on a cube in $\mathbb{R}^d$. In both cases we end up with fast algorithms requiring $O(N \log N)$ and $O(N^d \log N)$ operations, respectively, plus preliminary computations that do not depend on the data, but only on N and the order of the spline.

I. NOTATION; ATTENUATION FACTORS

We denote by L_2 the space of square integrable real 2π-periodic functions, by Π_{2N} the space of complex 2N-periodic sequences $\underline{f} := (f_k)_{k\in\mathbb{Z}}$, by $X := \{x_k := \pi k/N;\ \forall k\in\mathbb{Z}\}$ a uniform mesh on $\mathbb{R}$, and by $S_{2N,\ell}$ the space of periodic splines of order ℓ (degree $\ell-1$) with knots x_k if ℓ is even, or $(x_k + x_{k+1})/2$ if ℓ is odd, respectively. Let $\phi := Ff$ be the sequence of complex Fourier coefficients of $f \in L_2$, i.e. $\phi = (\phi_k)_{k\in\mathbb{Z}}$ where

$$\phi_k := \frac{1}{2\pi} \int_0^{2\pi} f(x)\, e^{-ikx}\, dx .$$

[1]Presently on leave at the Computer Science Department of Stanford University, Stanford, California. This work has been supported by the Swiss National Science Foundation fellowship 82.723.0.79.

Copyright © 1980 by Academic Press, Inc.
All rights of reproduction in any form reserved.
ISBN: 0-12-171050-5

The discrete Fourier transform (DFT) $\underline{F}_{2N}: \Pi_{2N} \to \Pi_{2N}$, $\underline{f} \mapsto \underline{\hat{f}}$, which can be implemented with the fast Fourier transform (FFT), is defined by

$$\hat{f}_n := \frac{1}{2N} \sum_{k=0}^{2N-1} f_k \, e^{-ikn\pi/N}, \quad \forall n \in \mathbb{Z}.$$

The <u>attenuation factor</u> theory [1, 3, 9] states: If $s \in S_{2N,\ell}$, then $\sigma := Fs$ and $\underline{\hat{s}} := \underline{F}_{2N}\underline{s} := \underline{F}_{2N}s(X)$ are related by

$$\sigma_k = \hat{s}_k \, \tau_k^{(\ell)} \qquad (\forall k \in \mathbb{Z}), \tag{1}$$

where $\tau_0^{(\ell)} = 1$, $\tau_{2Nk}^{(\ell)} = 0$ $(k \neq 0)$, and for $k \not\equiv 0 \pmod{2N}$

$$\tau_k^{(\ell)} = \left(\frac{\sin(x_k/2)}{x_k/2}\right)^{\ell} \Bigg/ \begin{cases} q_{\ell-2}(\cos(x_k/2)) & \text{if } \ell \text{ is even,} \\ \tilde{q}_{\ell-1}(\cos(x_k/2)) & \text{if } \ell \text{ is odd,} \end{cases} \tag{2}$$

with

$$q_0(t) \equiv 1, \quad q_m(t) = tq_{m-1}(t) + \frac{1-t^2}{m+1} q'_{m-1}(t), \quad m = 1, 2, \ldots, \tag{3a}$$

$$\tilde{q}_0(t) \equiv 1, \quad \tilde{q}_m(t) = t\tilde{q}_{m-1}(t) + \frac{1-t^2}{m} \tilde{q}'_{m-1}(t), \quad m = 1, 2, \ldots \tag{3b}$$

II. THE EVALUATION OF THE FUNCTION CONJUGATE TO A PERIODIC SPLINE

The <u>conjugation operator</u> $K: L_2 \to L_2$ (also called <u>discrete Hilbert transform</u> [8]) is defined by

$$K: f \in L_2 \overset{F}{\mapsto} \phi = (\phi_k) \overset{\hat{K}}{\mapsto} (-i \,\mathrm{sign}(k)\, \phi_k) \overset{F^{-1}}{\mapsto} Kf \in L_2 . \tag{4}$$

K is skew-symmetric, and $\|K\| = 1$. Kf is called the <u>conjugate periodic function</u> of f [6].

The classical discretization of K is based on trigonometric interpolation of the data $(x_k, f(x_k))_{k \in \mathbb{Z}}$ [2]:

$$\begin{array}{ccccc} \underline{f} := f(X) \in \Pi_{2N} & \overset{\underline{F}_{2N}}{\longmapsto} & \hat{\underline{f}} \in \Pi_{2N} \\ & & \downarrow \hat{\underline{K}}_N \\ (Kf)(X) \approx \underline{g} \in \Pi_{2N} & \overset{\underline{F}_{2N}^{-1}}{\longleftarrow} & \hat{\underline{g}} \in \Pi_{2N} \end{array} \tag{5}$$

where

$$\hat{g}_k := (\hat{\underline{K}}_N \hat{\underline{f}})_k := \begin{cases} -i \operatorname{sign}(k)\ \hat{f}_k & \text{for } |k| < N, \\ 0 & \text{for } |k| = N. \end{cases}$$

$\underline{K}_N := \underline{F}_{2N}^{-1} \hat{\underline{K}}_N \underline{F}_{2N}$ is a skew-symmetric banded Toeplitz matrix (essentially Wittich's matrix [2]) which before the rediscovery of the FFT was used to compute $\underline{g} = \underline{K}_N \underline{f}$ in $O(N^2)$ operations. Note that $\underline{g} = (KT)(X)$ if T denotes the normalized trigonometric interpolant to $\underline{f}$. Hence, $\underline{g} = (Kf)(X)$ if f is a trigonometric polynomial of degree at most N with vanishing sin(Nt) term.

However, if f is not very smooth, trigonometric interpolation is known to be inaccurate, and it is better to use other approximants. E.g., assume that we interpolate with $s \in S_{2N,\ell}$. Unfortunately, $r := Ks$ is not a spline function. But often we only want to evaluate r on X. Since $\rho := Fr$ and $\hat{\underline{r}} := \underline{F}_{2N}\underline{r} := \underline{F}_{2N} r(X)$ are related by the aliasing formula if $\ell \geq 2$, i.e.

$$\hat{r}_n = \sum_{k=-\infty}^{\infty} \rho_{n+2Nk} \quad (\forall n \in \mathbb{Z}) \tag{6}$$

(see, e.g., [6]), $\underline{r}$ can be computed according to the scheme

$$\begin{array}{ccccc} \underline{f} = \underline{s} \in \Pi_{2N} & \overset{\underline{F}_{2N}}{\longmapsto} & \hat{\underline{f}} = \hat{\underline{s}} \in \Pi_{2N} & \overset{(1)}{\longmapsto} & \sigma \\ & & \downarrow \hat{\underline{L}} & & \downarrow \hat{K} \\ \underline{r} \in \Pi_{2N} & \overset{\underline{F}_{2N}^{-1}}{\longleftarrow} & \hat{\underline{r}} \in \Pi_{2N} & \overset{(6)}{\longleftarrow} & \rho = (-i \operatorname{sign}(k)\, \phi_k)_{k \in \mathbb{Z}} \end{array} \tag{7}$$

Now, the main point is that $\hat{\underline{r}}$ may be obtained directly from $\hat{\underline{f}}$ by multiplication with a diagonal matrix $\hat{\underline{L}}$ which does not depend on $\underline{f}$. In fact, $\hat{r}_k = \hat{L}_{kk} \hat{f}_k$ $(\forall k \in \mathbb{Z})$, where for ℓ even

$$\hat{L}_{kk} = \overline{\hat{L}_{-k,-k}} = -i\, \tau_k^{(\ell)} [1 + \psi_{\ell-1}(\tfrac{k}{2N}) - \psi_{\ell-1}(\tfrac{-k}{2N})], \quad 1 \leq k < N,$$

$$\psi_m(t) := \frac{1}{m!} (-t)^{m+1} \left(\frac{d}{dt}\right)^{m+1} \log \Gamma(t+1) - 1, \quad m = 1, 2, \ldots$$

(See [5] for details.) Efficient polynomial approximations to the polygamma functions $(d/dt)^{m+1} \log \Gamma(t)$ have been given by Luke [7]. Once $\underline{L}$ is computed, we only need two FFTs and 2N real multiplications to evaluate $\underline{r}$ for a given $\underline{f}$.

It can be shown that $\underline{r}$ is an optimal approximation of $(Kf)(X)$ in the sense of Sard and Schoenberg [5]. Moreover, the approach generalizes to other approximants with known attenuation factors, including, e.g., deficient spline functions and smoothing splines [5]. Several types of approximants have been used with great success in conformal mapping [4].

III. FAST POISSON SOLVERS BASED ON THE FFT AND PERIODIC SPLINES

The idea for this application of attenuation factors emerged from discussions with Prof. P. Henrici and is heavily based on his new view of fast Poisson solvers using the FFT [6]. For simplicity we consider here only the one-dimensional case, which is rather trivial. However, our approach is easily generalized to the Poisson problem with homogeneous or periodic boundary conditions on a cube in $\mathbb{R}^d$. We first note that the given problem

$$\begin{aligned} -\Delta u \equiv -u''(x) &= f(x), \quad 0 < x < \pi, \\ u(0) = u(\pi) &= 0, \end{aligned} \tag{8}$$

can be solved via an expansion into the orthonormal eigenfunctions $u_n(x) = \sqrt{2/\pi}\ \sin(nx)$ $(n = 1, 2, \ldots)$ of the corresponding eigenvalue problem $-\Delta u = \lambda u$, $u(0) = u(\pi) = 0$. The eigenvalues are $\lambda_n = n^2$. If we extend f to be an odd 2π-periodic function and set $\lambda_{-n} := \lambda_n$, we may use our previous complex notation:

$$f \overset{F}{\mapsto} \phi = (\phi_k)_{k\neq 0} \mapsto (\phi_k / \lambda_k)_{k\neq 0} \overset{F^{-1}}{\mapsto} u \tag{9}$$

A simple discretization is again obtained by trigonometric interpolation (as proposed by Prof. P. Henrici in a lecture):

$$\begin{array}{lcl} \underline{f} = f(X) & \overset{\underline{F}_{2N}}{\mapsto} & \hat{\underline{f}} \\ & & \downarrow \\ u(X) \approx \underline{g} & \overset{\underline{F}_{2N}^{-1}}{\leftarrow} & \hat{\underline{g}} := (\hat{f}_k / \lambda_k)_{0<|k|<N} \end{array} \tag{10}$$

(Of course, $\hat{g}_0 := \hat{g}_N := 0$ and $\hat{\underline{g}} \in \Pi_{2N}$.) Note again that $\underline{g} = u(X)$ is the true solution if f is itself a real trigonometric polynomial of degree at most N-1. Otherwise, there is an approximation error, which, e.g., is of order $O(e^{-N\alpha})$ if f is analytic and bounded in a strip $|\mathrm{Im}\, z| < \alpha$. Hence, with respect to order this very simple fast Poisson solver is far superior to the usual ones based on difference methods, which have inherent discretization errors of order $O(N^{-p})$ for some small p. In fact, as Henrici [6] pointed out, all those methods, including the Mehrstellenverfahren of Collatz, use difference equations of the form $\underline{d} * \underline{g} = \frac{1}{N} \underline{c} * \underline{f}$ for $\underline{g} \approx u(X)$ with suitable $\underline{c}, \underline{d} \in \Pi_{2N}$. (The star denotes convolution.) In frequency space, $\hat{d}_n \hat{g}_n = \frac{1}{N} \hat{c}_n \hat{f}_n$ $(\forall n \in \mathbb{Z})$ is immediately solved for $\hat{\underline{g}}$. Comparing with (10) we see that $1/\lambda_n$ is approximated by $\frac{1}{N} \hat{c}_n / \hat{d}_n$, and that is where the $O(N^{-p})$ error comes from. However, as mentioned before, trigonometric interpolation is inaccurate if f is not smooth enough. Then, an interpolating spline may be appropriate:

$$
\begin{array}{ccccc}
\underline{f} = f(X) & \overset{\underline{F}_{2N}}{\mapsto} & \underline{\hat{f}} = \underline{\hat{s}} & \overset{(1)}{\mapsto} & \sigma = (\sigma_k)_{k \neq 0} \\
 & \underline{F}_{2N}^{-1} & \downarrow \underline{\hat{L}} & & \downarrow \\
u(X) \approx \underline{r} & \leftarrow & \underline{\hat{r}} & \leftarrow & \rho = (\sigma_k / \lambda_k)_{k \neq 0}
\end{array}
\tag{11}
$$

To make this approach feasible we only need the diagonal matrix $\underline{\hat{L}}$. Using (6) one gets after some elementary calculations: $\hat{L}_{kk} = k^{-2}\, \tau_k^{(\ell)} / \tau_k^{(\ell+2)}$ $(|k| \leq N)$. However, this is trivial in view of $u = r \in S_{2N,\ell+2}$ if $f = s \in S_{2N,\ell}$. But if we apply tensor product splines to higher dimensional Poisson problems on a cube, this is no longer true: While there is still a diagonal operator $\underline{\hat{L}}$, it cannot be of this simple form. But it still only depends on ℓ and N, and not on the data $\underline{f}$.

REFERENCES

1. Eagle, A., On the relations between the Fourier constants of a periodic function and the coefficients determined by harmonic analysis, Philos. Mag. 5, 113-132 (1928).

2. Gaier, D., Konstruktive Methoden der konformen Abbildung, Springer-Verlag, Berlin, 1964.

3. Gautschi, W., Attenuation factors in practical Fourier analysis, Numer. Math. 18, 373-400 (1972).

4. Gutknecht, M.H., Numerical experiments on solving Theodorsen's integral equation for conformal maps with the fast Fourier transform and various nonlinear iterative methods, to appear.

5. Gutknecht, M.H., The evaluation of the conjugate function of a periodic spline on a uniform mesh, to appear.

6. Henrici, P., Fast Fourier methods in computational complex analysis, SIAM Review 21, 481-527 (1979).

7. Luke, Y.L., The Special Functions and Their Approximation, Vol. I and II, Academic Press, New York, 1969.

8. Oppenheim, A.V., and Schafer, R.W., Digital Signal Processing, Prentice-Hall, Englewood Cliffs, N.J., 1975.

9. Quade, W., and Collatz, L., Zur Interpolationstheorie der reellen periodischen Funktionen, Sitzungs-Berichte Preuss. Akad. Wiss., phys.-math. Kl. 30, 383-429 (1938).

Martin H. Gutknecht
Seminar für angewandte Mathematik
ETH-Zentrum
CH-8092 Zürich, Switzerland

A NUMBER-THEORETIC PROBLEM IN NUMERICAL APPROXIMATION OF INTEGRALS

Seymour Haber

Mathematical Analysis Division
Center for Applied Mathematics
National Bureau of Standards
Washington, D.C.

We shall take as our point of departure for the developments to be surveyed in this paper, an old observation in numerical quadrature: the trapezoid rule is remarkably accurate in integrating periodic functions over a full period. The classical formula

$$\int_0^1 f(x)dx = \frac{1}{N}\left(\frac{1}{2}f(0) + f\left(\frac{1}{N}\right) + f\left(\frac{2}{N}\right) + \cdots + f\left(\frac{N-1}{N}\right) + \frac{1}{2}f(1)\right) + \frac{f''(\zeta)}{12N^2}$$

indicates $1/N^2$-type convergence; but when f has period 1, much more rapid convergence is usual. Fourier series provide an explanation. Supposing that f is fairly smooth, we may write

$$f(x) = \sum_{m=-\infty}^{\infty} c_m e^{2\pi imx}$$

 ISBN: 0-12-171050-5

where the series converges as nicely as we wish. Then--writing "$T_N f$" for the trapezoid-rule mean--

$$T_N f = \sum_{m=-\infty}^{\infty} c_m T_N(e^{2\pi imx}) = \sum_{N|m} c_m$$

(the last sum is over all integers m that are divisible by N). Since c_o is the integral of f, we have

$$\int_0^1 f - T_N f = -\sum_{N|m}' c_m \quad , \tag{1}$$

the prime indicating that the m=o term is omitted. If $f \varepsilon C^k$, k>2, then we know that $|c_m| = 0(|m|^{-k})$; and so (1) implies that the trapezoid rule's convergence will be at least as fast as $1/N^k$, rather than only $1/N^2$.

This is very good; but when we go on to multiple integrals, problems appear. Let E_s^α be the class of functions of s variables that are of period 1 in each variable, and whose Fourier series (notation: $\underline{x} = (x_1, x_2, \ldots, x_s)$)

$$f(\underline{x}) = \sum_{m_1,\ldots,m_s=-\infty}^{\infty} c_{\underline{m}} e^{2\pi i \underline{m}\cdot\underline{x}}$$

satisfy the condition

$$|c_{\underline{m}}| \leq C(f)\left(\prod_{r=1}^{s} \max\{1, |m_r|\}\right)^{-\alpha} . \tag{2}$$

Suppose $\alpha>1$, $f \varepsilon E_s^\alpha$, and we wish to integrate f over a period interval--specifically, over the unit s-cube, which we will denote "G_s". In the case s=1 the trapezoid rule will give $N^{-\alpha}$-type convergence, as we have seen. For s>1 we could do the integral numerically as an iterated integral, using T_M (we will reserve "N" for the total number of quadrature nodes) for each variable. The error would be

$$\sum_{M|m_1,\ldots,M|m_s}{}' c_{\underline{m}} = 0(M^{-\alpha}),$$

But now N--which is a good measure of the computational effort involved--is M^s, so that the error goes to zero only as $N^{-\alpha/s}$, which is not good at all.

A better s-dimensional extension of the trapezoid rule was devised by N.M. Korobov [1], [2], E. Hlawka [3], and H. Conroy [4]: Let $\underline{a} = (a_1,\ldots,a_s)$ be a vector of integers and approximate the integral of f by

$$Q_{N,\underline{a}}\, f = \frac{1}{N} \sum_{r=1}^{N} f\left(\frac{r}{N}\,\underline{a}\right) \tag{3}$$

The error turns out to be

$$\sum_{N|\underline{a}\cdot\underline{m}}{}' c_{\underline{m}}\ , \tag{4}$$

the prime indicating that the $\underline{m}=(0,\ldots,0)$ term is omitted. For $f\varepsilon E_s^\alpha$, (2) and (4) imply that

$$\int_{G_s} f - Q_{N,\underline{a}}f \leq C(f) \sum_{N|\underline{a}\cdot\underline{m}}{}' \left(\prod_{r=1}^{s} \max\,\{1,|m_r|\}\right)^{-\alpha}. \tag{5}$$

Korobov and Hlawka showed that (at least when N is a prime) one can find $\underline{a} = \underline{a}(N)$ for which the last sum is bounded by

$$C_0 \frac{(\log N)^{s\alpha}}{N^\alpha} \tag{6}$$

for some absolute constant C_0. Thus the convergence rate is almost $N^{-\alpha}$; a great improvement over the iterated trapezoid rule.

But it is necessary to find the very special $\underline{a}$'s--Korobov calls them "optimal coefficients"; Hlawka "good lattice points"--which afford us the bound (6). The proofs of their existence are constructive, but the calculations involved are impracticable for the most important range of values of s and N. The proofs, based on averaging over all possible $\underline{a}$'s, give no information on the nature of the good lattice points. More understanding is needed, to allow us to construct good lattice points efficiently.

Let us look at the quadrature nodes they define. Formula (3) gives the nodes as $(r/N)\underline{a}$; we can replace these by nodes inside G_s--since our integrands have period 1--by replacing each component by its fractional part. Our system of quadrature nodes is then

$$\left(\left\{\frac{ra_1}{N}\right\}, \ldots, \left\{\frac{ra_s}{N}\right\}\right) : r=1,2,\ldots,N \tag{7}$$

In two dimensions we have some useful information ([5], [6], [7]): we know that when N is a Fibonacci number F_n, a "best" $\underline{a}$ is $(1,F_{n-1})$. The resulting nodes are pictured in Figure 1, for N=21, 34, and 55.

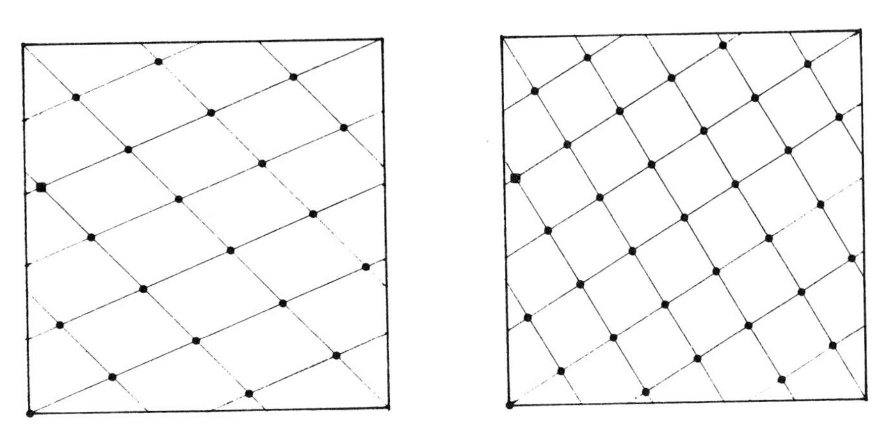

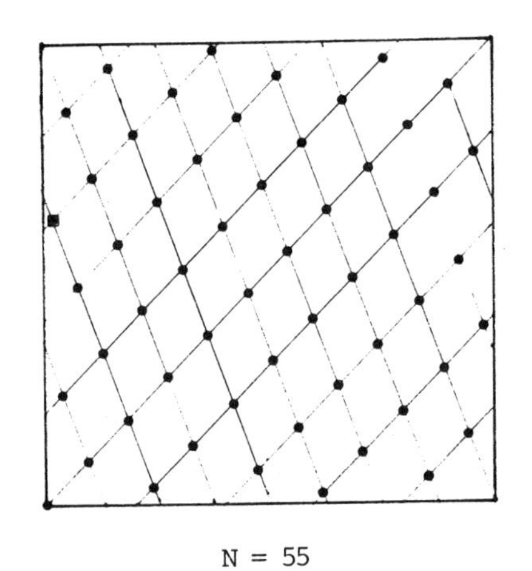

FIGURE 1. Sets of quadrature nodes defined by good lattice points.

These are strange slanted lattices! Why are they better than the simple lattice

$$\left(\frac{r_1}{M}, \frac{r_2}{M}, \ldots, \frac{r_s}{M}\right) \quad : \quad \text{each } r_i = 1, 2, \ldots, M \tag{8}$$

defined by the iterated trapezoid rule? An important clue is that for a given number of nodes (i.e. $M^s=N$) the lattice (7) is *more uniformly distributed* in G_s than is (8).

The sense in which we measure uniformity of distribution is this: First, for any point $\underline{t} = (t_1, \ldots, t_s)$ in G_s, we define the box $B(\underline{t})$ to be the set of all points $\underline{x}$ in G_s with

$$0 \leq x_i < t_i, \quad i=1,2,\ldots,s;$$

its volume is $t_1 \cdot t_2 \cdot \ldots \cdot t_s$. If P is a set of N points in G_s, its departure from uniformity is defined as its "discrepancy": the supremum, over all $\underline{t}$ in G_s, of

$$\left| \frac{\text{the number of points in } P \cap B(\underline{t})}{\text{the number of points in } P} - t_1 \cdot t_2 \cdot \ldots \cdot t_s \right| .$$

(This is a much-studied concept. See, e.g., [8].) The discrepancy of the lattice (8), with $N=M^s$, is of the order of $N^{-1/s}$. The slanted lattice (7) has far lower discrepancy when $\underline{a}(N)$ is a good lattice point; it is $O(N^{-1}\log^A N)$, for some constant A ([2], p. 141).

There is a more direct connection between quadrature and discrepancy. If we integrate a function f over G_s by simply averaging its values over some set P, of N points, then the error is bounded by

$$C(f) \cdot (\text{The discrepancy of } P)$$

as long as f is of bounded variation in the sense of Hardy and Krause. (For a definition see, e.g., [9], pp. 965-966; this class of functions includes all the E_s^α, $\alpha>1$.) The best-distributed sets of points that are known, in G_s, have discrepancies that are

$$O\left(\frac{\log^{s-1} N}{N}\right) \tag{9}$$

[10] and so yield (9) also as a quadrature error estimate. That is rather like (6), with $\alpha=1$ (and in fact [5] the s in (6) can be replaced by s-1). However these non-lattice sets of points do not afford the stronger estimate (6) for the more restricted function classes E_s^α with $\alpha>1$.

The mysterious looking log N factors in (6) and (9) are not accidents of analysis; it is known that they cannot be entirely eliminated. It is conjectured that there are no sets with lower discrepancy than that given by (9). This was shown for s=2 in a remarkably elementary proof by W. M. Schmidt [11]. For general s, K. F. Roth [12] showed that the exponent s-1 cannot be lowered past (s-1)/2. As regards (6), I. F. Šarygin [13] showed that no quadrature rule whatever can give faster convergence, for the function class E_s^α, than $N^{-\alpha}\log^{s-1}N$. This can be attained for s=2 by good lattice points [7]; whether it can be attained at all for s>2 is unknown.

We note a number of unresolved issues: How good are good lattice points--i.e. how far can we lower the exponent in (6)? What enables the sets of quadrature nodes derived from good lattice points to afford us the error estimate (6) rather than only a weaker one like (9)--is it the rational lattice character of those sets? And above all: how can we find good lattice points efficiently?

There is some extra moment to these questions. Chemists and physicists have been giving considerable attention to the good-lattice-point method of quadrature (e.g. [4], [14], [15]; they see it as promising, for practical application.

REFERENCES

1. Korobov, N. M., *On Approximate Calculation of Multiple Integrals*, Dokl. Akad Nauk SSSR 124 (1959), pp. 1207-1210. (Russian)
2. ----------, *Number-theoretic Methods of Approximate Analysis*, Fizmatgiz, Moscow, 1963. (Russian)
3. Hlawka, E., *Zur Angenaherten Berechnung Mehrfacher Integrale*, Monatsh. Mat. 66 (1962), pp. 140-151.
4. Conroy, H., *Molecular Schrodinger Equation VIII: A New Method for the Evaluation of Multidimensional Integrals*, J. Chemical Phys. 47, (1967), pp. 5307-5318.
5. Bahvalov, N. S., *Approximate Computation of Multiple Integrals*, Vestnik Moskov. Univ. Ser. Mat. Meh. Astr. Fiz. Him. 1959, no. 4, pp. 3-18. (Russian)
6. Hua, L. K. and Wang, Y., *Remarks Concerning Numerical Integration*, Sci. Record (N.S.) 4 (1960), pp. 8-11.
7. Zaremba, S.K., *Good Lattice Points, Discrepancy, and Numerical Integration*, Ann. Mat. Pura Appl. 73, 1966, no. 4, pp. 293-317.
8. Kuipers, L. and Niederreiter, H., Uniform Distribution of Sequences, Wiley, N.Y., 1974.
9. Niederreiter, H., *Quasi-Monte Carlo Methods and Pseudo-random Numbers*, Bull. AMS 84, 1978, pp. 957-1041.
10. Halton, J. H., *On the Efficiency of Certain Quasi-random Sequences of Points in Evaluating Multi-dimensional Integrals*, Numer. Math. 2 (1960), pp. 84-90; Correction, ibid., 196.
11. Schmidt, W. M., *Irregularities of Distribution VII*, Acta Arith. 21 (1972), pp. 45-50.
12. Roth, K. F., *On Irregularities of Distribution*, Mathematika 1 (1954), pp. 73-79.
13. Šarygin, I. F., *A Lower Estimate for the Error of Quadrature Formulae for Certain Classes of Functions*, Zh. Vychisl. Mat. i Mat. Fiz. 3 (1963), pp. 370-376 = U.S.S.R. Comput. Math. and Math. Phys. 3 (1963), pp. 489-497.
14. Daudey, J. P., Diner, S., and Savinelli, R., *Numerical Integration Techniques for Quantum Chemistry*, Theoret. Chim. Acta 37 (1975), pp. 275-283.
15. Zakrzewska, K., Dudek, J., and Nazarewicz, N., *A Numerical Calculation of Mathdimensional Integrals*, Comput. Phys. Commun. 14 (1978), pp. 299-309.

TURNING CORNERS WITH FOUR-SIDED POLYNOMIAL PATCHES

D.C. Handscomb
Computing Laboratory
University of Oxford

A common practice in computer-aided design of curved three-dimensional objects is to represent a surface by a vector-valued function $\underline{r}(u,v)$ of two real parameters. The surface will have continuous slope and continuous curvature if $\underline{r}$ is a C^2 function of u and v, provided that the vector product $\underline{r}_u \times \underline{r}_v$ nowhere vanishes. The first of these conditions is met if each component of $\underline{r}$ is a bicubic spline of u and v, of the form

$$p_{33}(u,v) + \Sigma\,\Sigma\, a_{ij}(u-u_i)_+^3 \quad (v-v_j)_+^3 \quad .$$

This has the effect of breaking the surface up into four-sided patches, separated by the families of curves $u = u_i$ and $v = v_j$, with $\underline{r}$ as a bicubic polynomial on each patch, such that $\underline{r}$, $\underline{r}_u$ and $\underline{r}_{uu}$ are continuous across $u = u_i$, and $\underline{r}$, $\underline{r}_v$ and $\underline{r}_{vv}$ across $v = v_j$.

Often, however, we can not conveniently use the same two parameters over the whole surface. If the object is essentially box-shaped, for instance, we can without difficulty represent the top and one side, the top and the front, or the front and one side in this manner, but there is no natural way of representing all three. In fact if we set out to cover such an object with four-sided patches in an obvious fashion, we shall find when we reach the corner of the box

Copyright © 1980 by Academic Press, Inc.
All rights of reproduction in any form reserved.
ISBN: 0-12-171050-5

that we need either to insert a three-sided patch or to make three instead of four patches meet at a common vertex. Yet there is a need to be able to represent smooth surfaces of this kind, instances being the roof of a car or the toe of a boot.

I wish to consider the possibility of taking three bicubic spline surfaces and joining them edge to edge, along curves of constant parameter, to produce a single surface of continuous curvature. As we shall see, this turns out not to be possible, but the desired effect can be achieved by very slight local modification near the common vertex.

Suppose that we have three vector-valued functions $\underline{r}(u,v)$, $\underline{s}(v,w)$ and $\underline{t}(w,u)$, such that

$$\underline{r}(0,v) = \underline{s}(v,0), \ \underline{s}(0,w) = \underline{t}(w,0), \ \underline{t}(0,u) = \underline{r}(u,0).$$

These functions then map the positive quadrants of the (u,v), (v,w) and (w,u) planes into surfaces meeting at the edges, and the value of the common parameter is continuous across each edge.

The straightforward way to ensure C^2 continuity between $\underline{r}(u,v)$ and $\underline{s}(v,w)$ across their common edge is to identify $\underline{s}(v,w)$ with $\underline{r}(-w,v)$ for small w, giving us the conditions

$$\underline{r}_u(0,v) + \underline{s}_w(v,0) = 0, \ \underline{r}_{uu}(0,v) = \underline{s}_{ww}(v,0). \qquad (1)$$

If (1) holds, however, and corresponding conditions hold across the other two common edges, then we find by considering what happens where the three edges meet that we must have $\underline{r}_u(0,0) = \underline{r}_v(0,0) = 0$, so that this point is a singular point of the surface. Conditions (1) are therefore too strict to permit the three functions to combine to form a smooth surface.

Suppose that we instead identify $\underline{s}(v,w)$ with $\underline{r}(-p(v,w),$

$q(v,w))$, where p and q are non-negative functions such that $p(v,0) = 0$, $p_w(v,0) > 0$ and $q(v,0) = v$. Conditions for C^2 continuity are now

$$\underline{s}_w(v,0) = - p_w(v,0)\underline{r}_u(0,v) + q_w(v,0)\underline{r}_v(0,v)$$

and

$$\begin{aligned}\underline{s}_{ww}(v,0) &= p_w(v,0)^2\underline{r}_{uu}(0,v) - 2p_w(v,0)q_w(v,0)\underline{r}_{uv}(0,v)\\ &+ q_w(v,0)^2\underline{r}_{vv}(0,v) - p_{ww}(v,0)\underline{r}_u(0,v)\\ &+ q_{ww}(v,0)\underline{r}_v(0,v).\end{aligned}$$

There is therefore C^2 continuity if there exist functions $A(v)$, $B(v)$, $C(v)$ and $D(v)$, with $A(v) > 0$, such that

$$\underline{s}_w + A(v)\underline{r}_u + B(v)\underline{r}_v = 0$$

and (2)

$$\underline{s}_{ww} = A(v)^2\underline{r}_{uu} + 2A(v)B(v)\underline{r}_{uv} + B(v)^2\underline{r}_{vv} + C(v)\underline{r}_u + D(v)\underline{r}_v$$

everywhere on the edge. Provided that $\underline{r}_u \times \underline{r}_v$ does not vanish, conditions (2) do ensure continuity of slope and curvature. We can make (2) symmetric in $\underline{r}$ and $\underline{s}$ by taking $A(v) = 1$ and $D(v) = B(v)\{B'(v) + C(v)/2\}$, when the conditions take the simpler form

$$\underline{r}_u + \underline{s}_w + B(v)\underline{r}_v = 0 \tag{3}$$

and

$$\underline{r}_{uu} + B(v)\underline{r}_{uv} + C(v)\underline{r}_u/2 = \underline{s}_{ww} + B(v)\underline{s}_{vw} + C(v)\underline{s}_w/2.$$

If a condition of the form (3), with the same functions B and C, holds on each of the common edges, then the point where the edges meet need not be a singular point of the surface provided that

$$B(0) = 1, \quad C(0) = 2B'(0). \tag{4}$$

Consider now the possibility that $\underline{r}$, $\underline{s}$ and $\underline{t}$ are bicubic splines. For simplicity, assume that the knots occur at integer values of the parameters. I shall require (1) to hold for $v > 1$ and (3) to hold for $0 \leq v \leq 1$, and likewise on the other common edges. Then in order that there should not be a singular point when $v = 1$, we find that we need

$$B(1) = B'(1) = B''(1) = C(1) = C'(1) = C''(1) = 0. \qquad (5)$$

Since B and C should clearly be polynomials, the simplest solution to (4) and (5) is to take

$$B(v) = (1-v)^3, \quad C(v) = -6(1-v)^3$$

when (3) becomes

$$\underline{r}_u + \underline{s}_w + (1-v)^3 \underline{r}_v = 0$$

and (6)

$$\underline{r}_{uu} + (1-v)^3(\underline{r}_{uv} - 3\underline{r}_u) = \underline{s}_{ww} + (1-v)^3(\underline{s}_{vw} - 3\underline{s}_w).$$

There are eight linearly independent sets of bicubic functions satisfying the three pairs of conditions of form **(6).** In the three patches meeting at the corner, therefore, the function must be a linear combination of these eight **sets,** with vector coefficients. Unfortunately, this causes $\underline{r}(0,v)$ to degenerate to a single point.

I propose that we remedy this by increasing the order of the polynomial in these three patches alone; since we must still have C^2 continuity with the neighbouring patches, we must continue to insist that $\underline{r}(1,v)$, $\underline{r}_u(1,v)$ and $\underline{r}_{uu}(1,v)$ are cubics in v, etc. If we allow the degrees of $\underline{r}(u,v)$, $\underline{s}(v,w)$ and $\underline{t}(w,u)$ in $u < 1$, $v < 1$ to rise to 4, 5 or 6, then the number of independent solutions rises from 8 to 9, 13 or 24, respectively, and the edge no longer degenerates to a point. At degrees 4 and 5, however, we still have one undesirable property, in that we always have $\underline{r}(0,1) = \underline{s}(0,1) = \underline{t}(0,1)$, so that the surface is bound to intersect itself. Degree 6, on the other hand, appears capable of giving useful results.

ON USING CURVED KNOT LINES

Peter Hartley

Department of Mathematics

Coventry (Lanchester) Polytechnic

Coventry, England

I. INTRODUCTION

Hayes (1974) introduced the idea of a bicubic spline with variable knots, as a means for obtaining surface shapes with features not tied to the essentially rectangular structure implied by constant knots. Here we discuss some of the problems associated with the use of varying knots and the automatic construction of the associated 'curved knot lines', giving some partial solutions.

This work is of relevance to (at least) two practical problems. First it is being applied in research into surface representations for computer aided geometric design. This is reported in Hartley and Judd (1980) and Judd (1980) and some similar considerations can be found in Nielson and Wixom (1977). Second it may be applied to the surface fitting of data given at the vertices of a rectangular grid. An outline of a method for this second problem is given here. Consider the function b defined by

$$b(x,y) = \sum_{s=1}^{n_1+k_1} \sum_{t=1}^{n_2+k_2} a_{st} N_{s,k_1}(x;\underline{\xi}(y))\, N_{t,k_2}(y;\underline{\eta}(x)), \tag{1.1}$$

Copyright © 1980 by Academic Press, Inc.
All rights of reproduction in any form reserved.

ISBN: 0-12-171050-5

where $(x,y) \in [\xi_o, \xi_{n_1+1}] \times [\eta_o, \eta_{n_2+1}] \subset \mathbb{R}^2$. $N_{s,k_1}(\cdot;\underline{\xi}(y))$ for fixed y are the normalised B-splines of order k_1 (degree k_1-1) with knots $\{\xi_u\}_{u=1-k_1}^{n_1+k_1}$ defined by

$$\xi_{1-k_1} = \ldots = \xi_{-1} = \xi_o < \xi_1 \leq \xi_2 \leq \ldots \leq \xi_{n_1} < \xi_{n_1+1}$$
$$= \xi_{n_1+2} = \ldots = \xi_{n_1+k_1} \qquad (1.2)$$

$$\xi_u = \xi_u(y) \qquad u = 1(1)n_1 \qquad (1.3)$$

where $(\xi_1(y), \ldots, \xi_{n_1}(y)) = \underline{\xi}(y) \in \{C^r[\eta_o, \eta_{n_2+1}]\}^{n_1}$ for suitable r (to be discussed later). $N_{t,k_2}(\cdot;\underline{\eta}(x))$ are defined correspondingly.

If the functions $\xi_u(y)$ and $\eta_v(x)$ are constants then (1.1) is the usual formulation of a bivariate spline in terms of B-splines.

II. SURFACE FITTING

Assume data of the form $(x_i, y_j, z_{ij}) \in \mathbb{R}^3$, $i = 1(1)m_1$, $j = 1(1)m_2$, with $x_i < x_{i+1}$, $y_j < y_{j+1}$, all i and j. It is well established that such data can be represented by splines of the form (1.1) when the knots are constant, with $\xi_o = x_1$, $\xi_{n_1+1} = x_{m_1}$, and similarly for η and y. Interpolation depends on the knots and data in each coordinate direction satisfying the Schoenberg-Whitney conditions for non-singularity; the properties of the least-squares fit will also depend on the knot/data relationships (see Hartley (1976) for instance). Moreover, in such cases the required function can be obtained by separating the two-variable problem into two univariate problems as follows.

<u>Stage 1</u> Find c_{sj} such that

$$f_j(x) = \sum_{s=1}^{n_1+k_1} c_{sj} N_{s,k_1}(x; \underline{\xi}) \qquad j = 1(1)m_2 \qquad (2.1)$$

are the interpolants or least-squares fits to the data sets $\{(x_i, z_{ij}) \mid i = 1(1)m_1\}$.

Stage 2 Find a_{st} such that

$$c_s(y) = \sum_{t=1}^{n_2+k_2} a_{st} N_{t,k_2}(y;\underline{\eta}) \qquad s = 1(1)n_1+k_1 \tag{2.2}$$

are the corresponding functions for the sets of data $\{(y_j, c_{sj}) \mid j = 1(1)m_2\}$.

Then a_{st} are the required coefficients for b(x,y). (Here $\underline{\xi}$ stands for $\underline{\xi}(y)$ when the $\xi_u(y)$ are constant functions; similarly $\underline{\eta}$.)

As a starting point for variable knots consider adapting the above approach by allowing each of the functions (2.1) to have different knots:

$$f_j^*(x) = \sum_{s=1}^{n_1+k_1} c_{sj}^* N_{s,k_1}(x;\underline{\xi}^{(j)}) \qquad j = 1(1)m_2 \tag{2.3}$$

Note that although we allow different knots along each 'section' $y = y_j$ we constrain the *number* of knots n_1 to be the same in each case.

Carrying on formally to the second stage (ignoring the fact that the c_{sj}^* are in general associated with splines from different spaces):

$$c_s^*(y) = \sum_{t=1}^{n_2+k_2} a_{st}^* N_{t,k_2}(y;\underline{\eta}^{(s)}) \qquad s = 1(1)n_1+k_1 \tag{2.4}$$

and then assuming for the moment that we can construct a 'reasonable' vector of knot functions $\underline{\xi}(y)$ satisfying $\underline{\xi}(y_j) = \underline{\xi}^{(j)}$, we obtain

$$b^*(x,y) = \sum_{s=1}^{n_1+k_1} \sum_{t=1}^{n_2+k_2} a_{st}^* N_{s,k_1}(x;\underline{\xi}(y)) N_{t,k_2}(y;\underline{\eta}^{(s)}) \tag{2.5}$$

Note that we have not and cannot achieve the form (1.1) because we cannot infer knot functions $\underline{\eta}(x)$ from the $\underline{\eta}^{(s)}$. (There are no values x_s for which the equation $\underline{\eta}(x_s) = \underline{\eta}^{(s)}$ has any meaning.) Rather (2.5) represents a 'lofted' or 'blended' surface, in which the 'sections' $f_j^*(x)$ are blended into a surface by the functions $c_s^*(y)$.

This bias to one of the coordinate directions is unnatural for general purpose surface fitting. We suggest, therefore, the following (symmetric) strategy.

1) Fit curves $f_j(x)$ with knots $\underline{\xi}^{(j)}$ for each data set $\{(x_i, z_{ij})\}$, constructing associated knot functions $\underline{\xi}(y)$ satisfying $\underline{\xi}(y_j) = \underline{\xi}^{(j)}$.

2) Fit curves $g_i(y)$ with knots $\underline{\eta}^{(i)}$ for each data set $\{(y_j, z_{ij})\}$, constructing associated knot functions $\underline{\eta}(x)$ satisfying $\underline{\eta}(x_i) = \underline{\eta}^{(i)}$.

3) Find (without separation of variables) the surface of the form (1.1) fitting all of the data.

Note that, as a simple extension to the constant knots case, if the Schoenberg-Whitney conditions for interpolation are satisfied along each mesh line of the data grid in 1) and 2) above, then 3) will produce a unique interpolant. Similar remarks apply to the least-squares case.

It remains to be judged whether this obviously costly process is worthwhile for its gain in flexibility.

III. GENERATING THE KNOT LINES

There are two stages in the generation of the curved knot lines. First the discrete knots are chosen along each mesh line in the data grid, and, second, corresponding knots are connected by interpolating functions. The first stage will not concern us here, but is clearly non-trivial.

Suppose then that discrete knots $\underline{\xi}^{(j)} = (\xi_{1j}, \ldots, \xi_{n_1 j})$ have been obtained for each mesh line $y = y_j$, $j = 1(1)m_2$. We need functions $\xi_i(y)$ satisfying the following conditions.

1) $\xi_i(y_j) = \xi_{ij} \qquad i = 1(1)n_1, \quad j = 1(1)m_1$ (3.1)

i.e. interpolation of corresponding knots;

2) $\xi_{i+1}(y) \geq \xi_i(y) \qquad i = 0(1)n_1$ (3.2)

where $\xi_o(y) = x_1$, $\xi_{n_1+1}(y) = x_{m_1}$, i.e. the knots in x must remain ordered for all y;

3) $\xi_i(y) \in C^r[\eta_o, \eta_{n_2+1}] \qquad i = 1(1)n_1$ (3.3)

where r would normally be $k_2 - 2$, i.e. the knot functions would have the same order of continuity as the surface.

Lastly we would prefer an efficient method for constructing the $\xi_i(y)$ and an efficient representation for evaluating them. (All the above applies correspondingly to the $\eta_j(x)$.)

If we assume that the knot lines should be 'fair', that is, have a shape which looks natural (relative to the interpolation points) to the eye, then this appears to be a difficult problem. Despite some effective recent work on producing fair curves automatically, there is, as far as we are aware, only one known class of curves which guarantees that two adjacent interpolants do not overlap.

If $\{\phi_j\}$ denotes the canonical basis for a space of interpolants on the data grid $\{y_j\}$, i.e. $\phi_j(y_k) = \delta_{jk}$, then we require

$$\xi_{i+1}(y) - \xi_i(y) = \sum_{j=1}^{m_2} (\xi_{i+1,j} - \xi_{i,j})\, \phi_j(y)$$

$$\geqslant 0 \qquad y \in [\eta_o, \eta_{n_2+1}] \tag{3.4}$$

Since $\xi_{i+1,j} \geqslant \xi_{i,j}$ (see 1.2)) a sufficient condition for no overlap is $\phi_j(y) \geqslant 0$, all j. The obvious example of such functions is the 'hat' functions of piecewise linear interpolation, such that

$$\xi_i(y) = (1 - s)\, \xi_{i,j} + s\, \xi_{i,j+1} \qquad 0 \leqslant s \leqslant 1 \tag{3.5}$$

where $s = (y - y_j)/(y_{j+1} - y_j)$. From at least a theoretical point of view we would obviously prefer, e.g., C^2 knot functions for cubic splines.

In order to maintain the positivity of the canonical functions while raising the order of continuity it appears to be necessary to maintain the compactness of their support, viz. two intervals only. To do this requires the imposition of stringent end conditions. For example, the piecewise cubics

$$\xi_i(y) = (1 - F(s))\, \xi_{i,j} + F(s)\, \xi_{i+1,j} \qquad 0 \leqslant s \leqslant 1 \tag{3.6}$$

$$F(s) = s^2\,(3 - 2s) \tag{3.7}$$

with the same change of variable as before, are C^1 functions, they do not overlap, but are forced to satisfy $\xi_i'(y_j) = 0$, all j. Similarly the piecewise quintics defined by (3.6) with

$$F(s) = s^3 (6s^2 - 15s + 10)$$

are C^2 and non-overlapping, but satisfy $\xi_i'(y_j) = \xi_i''(y_j) = 0$, all j.

Although these functions satisfy all the mathematical requirements they are not 'fair'. The zero derivatives at the interpolation points produce a step-like appearance. (Fritsch and Carlson (1978) give some graphs of the cubic variety of equations (3.6), (3.7.) Many people believe that the human concept of a fair curve is best achieved by human interaction with a computing machine. This may be the best solution to the problem of constructing non-overlapping curves.

REFERENCES

Fritsch, F.N. and Carlson, R.E. (1978). "Monotone piecewise cubic interpolation." UCRL-82453, Lawrence Livermore Laboratory, Livermore.

Hartley, P.J. (1976). *Comp. J. 19*, 348.

Hartley, P.J. and Judd, C.J. (1980). "Curve and surface representations for Bézier B-spline systems." To appear in the Proceedings of CAD80.

Hayes, J.G. (1974). "New shapes from bicubic splines." NAC58, National Physical Laboratory, Teddington, England.

Judd, C.J. (1980). "A lofted surface representation using B-splines." In preparation.

Nielson, G.M. and Wixom, J.A. (1977). *In* SCHAD, 1977, Society of Naval Architects and Marine Engineers, New York.

FUNCTIONS f FOR WHICH $E_n(f)$ IS EXACTLY OF THE ORDER $\frac{1}{n}$

Maurice Hasson

Department of Mathematics
Texas A&M University
College Station, Texas 77843

I. INTRODUCTION AND NOTATION

Let $C[-1,1]$ denote the space of continuous real valued functions defined on $[-1,1]$, endowed with the supremum norm denoted by $\| \ \|$. If $f \in C[-1,1]$, let $E_n(f)$ be defined by

$$E_n(f) = \inf_{P_n \in \pi_n} \|P_n - f\|$$

where π_n is the space of algebraic polynomials of degree at most n.

The purpose of this paper is to find sufficient conditions on f in order for $E_n(f)$ to be exactly of the order $\frac{1}{n}$, which means that there exist positive constants K_1 and K_2 such that

$$\frac{K_1}{n} \le E_n(f) \le \frac{K_2}{n}, \quad n = 1, 2, \ldots \tag{1.1}$$

An example of such functions is given by $f(x) = |x|$. In fact, Bernstein proved that $nE_n(|x|)$ has a limit as n goes to infinity (see [1]). The theorem of Bernstein has been generalized by Nikol'skii (see [4]):

If $f \in C[-1,1]$ is absolutely continuous on $[-1,1]$, if $f'(x)$ has only discontinuities of the first kind on $[-1,1]$ and if there exists $x_0 \in (-1,1)$ such that $f'(x_0+0) \neq f'(x_0-0)$, then $\lim_{n\to\infty} nE_n(f)$ exists.

We will weaken the hypothesis of the theorem of Nikol'skii but will not be any more able to prove that $nE_n(f)$ has a limit but only that $E_n(f)$ satisfies (1.1).

We say that f belongs to the Zygmund class of $[-1,1]$ if f is continuous there and if $|f(x+h) + f(x-h) - 2f(x)| \le Kh,\ h > 0,$ $x+h,\ x-h \in [-1,1]$.

The Dini derivatives $D_+f(x)$ and $D^-f(x)$ of a function f at a point x are defined by

$$D_+f(x) = \underline{\lim}_{h\to 0^+} \frac{f(x+h) - f(x)}{h}$$

$$D^-f(x) = \overline{\lim}_{h\to 0^+} \frac{f(x+h) - f(x)}{h}$$

$D^+f(x)$ and $D_-f(x)$ are similarly defined.

In this work, P_n denotes algebraic polynomials of degree at most n.

If $f \in C[-1,1]$ and $-1 < a < b < 1$,

$$\|f\|_{[a,b]} = \sup_{x\in[a,b]} |f(x)| .$$

II. THE MAIN THEOREM

Theorem 2.1. Suppose that f belongs to the Zygmund class of $[-1,1]$ and that there exists $a \in (-1,1)$ with one of the following properties

$$D_+f(a) > D^-f(a) \tag{2.1}$$

$$D_-f(a) > D^+f(a) \tag{2.2}$$

Then $E_n(f)$ is exactly of the order $\frac{1}{n}$, i.e. satisfies (1.1).

Proof. We prove the theorem in the case that (2.1) is satisfied. (For 2.2, it suffices to replace f by $-f$). We suppose first that the two Dini derivatives in (2.1) are finite. It is easy to see that, without loss of generality, we may suppose $a = f(a) = 0$ and $D_+f(0) = 1$, $D^-f(0) = -1$.

Because f belongs to the Zygmund class of $[01,1]$, we have (see [5]):

$$E_n(f) \le \frac{K_2}{n}, \quad n = 1, 2, \ldots \tag{2.3}$$

It remains, therefore, to prove the reverse inequality. Let P_n be the polynomial of best approximaiton to f on $[-1,1]$. We have, because of (2.3) (see [3])

$$\|P_n''\|_{[-1/2,1/2]} \leq Mn, \; n = 1, 2, \ldots \tag{2.4}$$

The mean value theorem and (2.4) yield

$$|P_n'(x) - P_n'(y)| \leq \frac{1}{2}, \quad x \in [-\frac{1}{4Mn}, \frac{1}{4Mn}] \tag{2.5}$$

Suppose that

$$P_n'(x) \leq \frac{1}{2}, \quad x \in [0, \frac{1}{4Mn}] \tag{2.6}$$

Because $D_+f(0) = 1$ and $f(0) = 0$, we have, for n big enough:

$$f(x) \geq \frac{3}{4}x, \; x \in [0, \frac{1}{4Mn}] \; . \tag{2.7}$$

If

$$|P_n(0) - f(0)| = |P_n(0)| \leq \frac{1}{32Mn} ,$$

then, because of (2.6),

$$P_n(\frac{1}{4Mn}) \leq \frac{1}{32Mn} + \frac{1}{8Mn} \; . \tag{2.8}$$

But, from (2.7), we have

$$f(\frac{1}{4Mn}) \geq \frac{3}{16Mn} \tag{2.9}$$

for n big enough. (2.8) and (2.9) yield:

$$f(\frac{1}{4Mn}) - P_n(\frac{1}{4Mn}) \geq \frac{6}{32Mn} - \frac{5}{32Mn} = \frac{1}{32Mn} \; .$$

We have proved that $E_n(f) \geq \frac{1}{32Mn}$ (for n big enough) if $P_n'(x) \leq \frac{1}{2}$, $x \in [0, \frac{1}{4Mn}]$.

Suppose now that there exists $x \in [0, \frac{1}{4Mn}]$ such that $P_n'(x) > \frac{1}{2}$. It follows from (2.5) that

$$P_n'(x) > 0, \quad x \in [-\frac{1}{4Mn}, 0] \tag{2.10}$$

On the other hand

$$f(x) > -\frac{3}{4}x, \quad x \in [-\frac{1}{4Mn}, 0] \tag{2.11}$$

because $D^-f(0) = -1$ and $f(0) = 0$. Suppose that

$$|P_n(0) - f(0)| = |P_n(0)| \leq \frac{3}{32Mn} \; .$$

Then, because of (2.10),

$$P_n(-\frac{1}{4Mn}) > \frac{3}{32Mn} \; .$$

But (2.11) implies that

$$f(-\frac{1}{4Mn}) > \frac{3}{16Mn}$$

It follows that

$$f\left(-\frac{1}{4Mn}\right) - P_n\left(-\frac{1}{4Mn}\right) > \frac{3}{32Mn} .$$

We have proved that $E_n(f) \geq \frac{K_1}{n}$ in the case of the Dini derivatives in (2.1) are finite. If one of the Dini derivatives takes infinite value then (2.7) and (2.11) remain true and the rest of the proof remains unchanged. Theorem 2.1 is proved.

Theorem 2.2. Suppose that f is a periodic function of period 2π belonging to the Zygmund class (of $[-\pi,\pi]$). Suppose that there exists a satisfying (2.1) or (2.2). Then $E_n^*(f)$ is exactly of the order $\frac{1}{n}$, when $E_n^*(f)$ is the degree of approximation by trigonometric polynomials.

The proof of this result is omitted.

We now give an application of Theorem 2.2.

Theorem 2.3. Let $f(x) = \sum_{n=0}^{\infty} a^{-n} \cos a^n x$, $a > 1$, be the Weierstrass nowhere differentiable function. Then $E_n^*(f)$ is exactly of the order $\frac{1}{n}$.

Proof. It suffices to remark that f belongs to the Zygmund class of $[-\pi,\pi]$ (see [5]) and satisfies (2.2) (see [2]). Theorem 2.3 follows from Theorem 2.2.

Theorem 2.4. Suppose that $f \in C^k[-1,1]$ and that $f^{(k)}$ satisfies the hypothesis of Theorem 2.1. Then there exist positive constants K_k, L_k such that

$$\frac{K_k}{kn} \leq E_n(f) \leq \frac{L_k}{kn}, \quad n = 1, 2, \ldots$$

The second inequality is a consequence of a theorem of Zygmund [5]. We omit the proof of the first inequality.

BIBLIOGRAPHY

1. S. Bernstein, Sur la meilleure approximation de $|x|$ par des polynomes de degrés donnés, Acta Mathematica, 37, (1914), 1-57.
2. G. Hardy, Weierstrass's non-differentiable function, Trans. Amer. Math. Soc., 17, (1916), 301-325.
3. M. Hasson, Derivatives of the algebraic polynomials of best approximation, to appear.
4. S. M. Nikol'skii, On the best approximation of functions, the s-th derivative of which has discontinuities of the first kind. Dolk. Akad. Nauk SSSR 55, No. 2, (1947), 99-102.
5. A. Zygmund, Smooth functions, Duke Math Journal 12, (1945), 47-76.

MULTIVARIATE APPROXIMATION AND TENSOR PRODUCTS

Werner Haussmann

Department of Mathematics
University of Duisburg
Duisburg, West Germany

Karl Zeller

Department of Mathematics
University of Tübingen
Tübingen, West Germany

I. INTRODUCTION

It is the purpose of the present note to present an approach to multivariate approximation which utilizes tensor product methods. This approach leads to computationally simple proofs and rather sharp error bounds. We use two types of estimates, one depending on univariate operator norms, the other on properties of corresponding univariate functions. There are direct applications to Jackson-Favard theorems. We formulate the results for the two-dimensional case; the n-dimensional case can be treated in an analogous manner.

Copyright © 1980 by Academic Press, Inc.
All rights of reproduction in any form reserved.
ISBN: 0-12-171050-5

II. CONTINUOUSLY DIFFERENTIABLE FUNCTIONS

Let $I \subset \mathbb{R}$ be a compact interval, $p \in \mathbb{N}_o$. We consider the vector space $C^p(I)$ endowed with the norm

$$\|f\|_p := \max_{o \leq \kappa \leq p} \|f^{(\kappa)}\|_\infty .$$

Let $C^p_{2\pi}$ be the subspace of $C^p([o,2\pi])$ given by $f(o) = f(2\pi)$. The algebraic tensor product $C^p(I) \otimes C^q(J)$ will be endowed with the ε-norm

$$\varepsilon(h) := \sup_{\|\phi\| \leq 1} \sup_{\|\psi\| \leq 1} \left| \sum_{\mu=1}^{m} \phi(f_\mu) \cdot \psi(g_\mu) \right| ,$$

where $h := \sum_{\mu=1}^{m} f_\mu \otimes g_\mu$, and ϕ, ψ are in the respective duals. Completion under the ε-norm yields the isometries

$$C^p(I) \hat{\otimes}_\varepsilon C^q(J) \cong C^{p,q}(I \times J), \quad C^p_{2\pi} \hat{\otimes}_\varepsilon C^q_{2\pi} \cong C^{p,q}_{2\pi,2\pi} .$$

Here $C^{p,q}(I \times J)$ consists of all $h \in C(I \times J)$ for which all derivatives appearing in

$$\|h\|_{p,q} := \sup_{o \leq \kappa \leq p, o \leq \lambda \leq q} \|D^{\kappa,\lambda} h\|_\infty$$

exist and are continuous. And $C^{p,q}_{2\pi,2\pi}$ is the subspace defined by $h(o,y) = h(2\pi,y)$ and $h(x,o) = h(x,2\pi)$.

We note that the ε-norm is a uniform cross norm (with respect to $(C^{p,q}(I \times J), C^{r,s}(I \times J))$, $p,q,r,s \in \mathbb{N}_o$, cf. [4]). Hence for any tensor product operator $S \hat{\otimes} T$ mapping $C^{p,q}$ into $C^{r,s}$ we have

$$\|S \hat{\otimes} T\|_{[p,q;r,s]} = \|S\|_{[p,r]} \cdot \|T\|_{[q,s]} ,$$

i. e. the norm of the product operator is the product of the corresponding one-dimensional operator norms. The indices show which spaces and element norms (see above) are involved.

For further details about tensor products we refer to Haussmann-Pottinger [4] and Treves [12].

III. BIVARIATE APPROXIMATION THEOREMS

For $m,n,p,q,r,s \in \mathbb{N}_o$, we consider continuous linear operators $K, K_m : C^p(I) \longrightarrow C^r(I)$, and $L, L_n : C^q(J) \longrightarrow C^s(J)$. Given $h \in C^{p,q}(I\times J)$, we approximate $(K \hat{\otimes} L)h$ by $(K_m \hat{\otimes} L_n)h$, estimating the difference in terms of one-dimensional data (partial mappings h_x, h_y and/or operator norms). For $p=q=r=s=o$ we get

THEOREM 1.

Let $h \in C(I\times J)$. Then the following estimate holds:

$$\|(K\hat{\otimes}L)h - (K_m\hat{\otimes}L_n)h\|_{o,o}$$

$$\leq \|K\|_{[o,o]} \cdot \max_{x\in I}\|Lh_x - L_n h_x\|_\infty + \|L_n\|_{[o,o]} \cdot \max_{y\in J}\|Kh_y - K_m h_y\|_\infty,$$

where $h_x(y) := h(x,y)$, x fixed, and $h_y(x) := h(x,y)$, y fixed.

Proof. Using broken line interpolation in one direction we verify (with id := identity mapping)

$$\|(K\hat{\otimes}\mathrm{id})h - (K_m\hat{\otimes}\mathrm{id})h\|_{o,o} = \sup_{y\in J}\|Kh_y - K_m h_y\|_\infty,$$

$$K\hat{\otimes}L - K_m\hat{\otimes}L = (\mathrm{id}\hat{\otimes}L) \circ (K\hat{\otimes}\mathrm{id} - K_m\hat{\otimes}\mathrm{id}).$$

From these and two corresponding formulas our estimate follows (we have a uniform cross norm).- For any $p,q,r,s \in \mathbb{N}_o$, we get

THEOREM 2.

Let $h \in C^{p,q}(I\times J)$. Then the following estimate holds:

$$\|(K\hat{\otimes}L)h - (K_m\hat{\otimes}L_n)h\|_{r,s}$$

$$\leq \|h\|_{p,q} \cdot (\|K\|_{[p,r]} \cdot \|L-L_n\|_{[q,s]} + \|L_n\|_{[q,s]} \cdot \|K-K_m\|_{[p,r]}).$$

Indeed, we have

$$\|(K\hat{\otimes}L)h - (K_m\hat{\otimes}L_n)h\|_{r,s} \leq \|h\|_{p,q} \cdot \|K\hat{\otimes}L - K_m\hat{\otimes}L_n\|_{[p,q;r,s]}$$

$$\leq \|h\|_{p,q} \cdot (\|K\|_{[p,r]} \cdot \|L-L_n\|_{[q,s]} + \|L_n\|_{[q,s]} \cdot \|K-K_m\|_{[p,r]}),$$

where we again used the uniform cross norm property.

IV. JACKSON THEOREMS IN TWO VARIABLES

Bivariate Jackson estimates (for the rectangle) can be obtained by treating the variables successively (cf. Lorentz [5], Nikol'skiĭ [8], Timan [11]). The tensor product approach shown below avoids most computations and leads to rather sharp constants, thus providing more insight.

Here we approximate by (real) trigonometric polynomials

$$T_{m,n}(x,y) = \sum_{\mu=-m}^{m} \sum_{\nu=-n}^{n} a_{\mu\nu} \cdot e^{i\mu x + i\nu y} .$$

We use the univariate considerations due to Favard, Achieser and Krein (see Achieser [1], Lorentz [5], Müller [6]). They apply linear trigonometric operators $K_m = K_{m,p}$ (defined as L^1-approximations of convolution kernels acting on $f^{(p)}$) with

$$\| f - K_m f \|_\infty \le c_p (m+1)^{-p} \| f^{(p)} \|_\infty \; ; \quad c_p := \frac{4}{\pi} \sum_{\kappa=0}^{\infty} \left(\frac{(-1)^\kappa}{2\kappa+1} \right)^{p+1}$$

(the constants are optimal for the $C^p_{2\pi}$-problem). One knows $c_p \in [1, \frac{\pi}{2}]$. As a mapping from $C^p_{2\pi}$ to $C_{2\pi}$ (both normed as above) the operator K_m has a norm

$$\| K_m \|_{[p,o]} \le 1 + c_p (m+1)^{-p} \quad (m \in \mathbb{N}_o) \quad \text{(and } =1 \text{ for } m=o).$$

Applying Theorem 2 with this K_m and a corresponding L_n we obtain (where $E_{m,n}(h)$ is the approximation constant of h):

THEOREM 3.

For each $h \in C^{p,q}_{2\pi,2\pi}$, $m,n \in \mathbb{N}_o$, $p,q \ge 1$, we have

$$E_{m,n}(h) \le \| h - (K_m \hat{\otimes} L_n) h \|_{o,o}$$

$$\le ((1+c_p(m+1)^{-p}) c_q (n+1)^{-q} + (1+c_q(n+1)^{-q}) c_p (m+1)^{-p}) \cdot \| h \|_{p,q}$$

$$\le (1+\tfrac{\pi}{2}) \cdot (c_p (m+1)^{-p} + c_q (n+1)^{-q}) \cdot \| h \|_{p,q} .$$

In order to treat the case where $h \in C_{2\pi,2\pi}$ we start from the Favard-Achieser-Krein estimate for $f \in C^1_{2\pi}$:

$$E_m(f) \le \|f - K_m f\|_\infty \le \frac{\pi}{2(m+1)} \|f'\|_\infty.$$

With the aid of the Steklov mean operator S_d we obtain for all $f \in C_{2\pi}$ (with $f_d := S_d f$, $d>o$, then $d := \pi(m+1)^{-1}$)

$$E_m(f) \le \|f - f_d\|_\infty + \|f_d - K_m f_d\|_\infty \le \frac{3}{2}\omega(f, \frac{\pi}{m+1})$$

(cf. Cheney [2], Müller [6]). In order to estimate

$$\|K_m \circ S_d\|_{[o,o]} \le \frac{3}{2} \quad \text{for } d := \pi(m+1)^{-1}$$

we consider (with inj := injection operator)

$$\|S_d\|_{[o,1]} \le \max(1, \frac{1}{d}) \quad (\text{for } d>o), \quad \|\text{inj} - K_m\|_{[1,o]} \le \frac{\pi}{2}(m+1)^{-1}$$

(the latter by Favard-Achieser-Krein) and get for $d = \pi(m+1)^{-1}$

$$\|(\text{inj} - K_m) \circ S_d\|_{[o,o]} \le \|\text{inj} - K_m\|_{[1,o]} \cdot \|S_d\|_{[o,1]} \le \frac{1}{2},$$

hence the bound $\frac{3}{2}$ above. Now we put

$$P_m := K_m \circ S_d \quad \text{and} \quad Q_n := L_n \circ S_{d'},$$

with $d := \pi(m+1)^{-1}$ and $d' := \pi(n+1)^{-1}$ and apply Theorem 1 (symmetrizing for the last estimate) to get

THEOREM 4.

For each $h \in C_{2\pi,2\pi}$, $m,n \in \mathbb{N}_o$, we have

$$E_{m,n}(h) \le \|h - (P_m \hat{\otimes} Q_n)h\|_{o,o}$$

$$\le \max_x \|h_x - Q_n h_x\|_\infty + \|Q_n\|_{[o,o]} \cdot \max_y \|h_y - P_m h_y\|_\infty$$

$$\le \frac{3}{2} \max_x \omega(h_x, \frac{\pi}{n+1}) + \frac{9}{4} \max_y \omega(h_y, \frac{\pi}{m+1}),$$

and

$$E_{m,n}(h) \le \frac{15}{8} \left(\max_x \omega(h_x, \frac{\pi}{n+1}) + \max_y \omega(h_y, \frac{\pi}{m+1}) \right).$$

V. CONCLUDING REMARKS

One can consider other tensor product spaces (see [4]) and problems like numerical integration (cf. Scherer [10]), also further Jackson estimates (combining Theorems 3 and 4 or using algebraic polynomials). Non-rectangular regions provide more difficulties (cf. Newman and Shapiro [7,9]); for some purposes mapping or imbedding methods are useful. Another goal is to obtain best constants and sharper estimates for individual functions.

REFERENCES

1. Achieser, N. I., Vorlesungen über Approximationstheorie, Akademie-Verlag, Berlin 1967.
2. Cheney, E. W., Introduction to Approximation Theory, McGraw Hill, New York-St. Louis-San Francisco-Toronto-London-Sydney 1966.
3. de Boor, C., A Practical Guide to Splines, Springer, Berlin-Heidelberg-New York 1978.
4. Haussmann, W., and Pottinger, P., On the Construction and Convergence of Multivariate Interpolation Operators, J. Approximation Theory 19 (1977), 205-221.
5. Lorentz, G. G., Approximation of Functions, Holt, Rinehart and Winston, New York-Chicago-San Francisco-Toronto-London 1966.
6. Müller, M. W., Approximationstheorie, Akademische Verlagsgesellschaft, Wiesbaden 1978.
7. Newman, D. J., and Shapiro, H. S., Jackson's Theorem in Higher Dimensions, in "On Approximation Theory", Internat. Ser. Numer. Math. 5, Birkhäuser, Basel-Stuttgart 1972.
8. Nikol'skiǐ, S. M., Approximation of Functions of Several Variables and Imbedding Theorems, Springer, Berlin-Heidelberg-New York 1975.
9. Shapiro, H. S., Approximation by Trigonometric Polynomials to Periodic Functions of Several Variables, in "Abstract Spaces and Approximation", Internat. Ser. Numer. Math. 10, Birkhäuser, Basel-Stuttgart 1969.
10. Scherer, R., Über Fehlerschranken bei Produkt-Kubatur, to appear in Z. Angew. Math. Mech.
11. Timan, A. F., Theory of Approximation of Functions of a Real Variable, MacMillan, New York 1963.
12. Treves, F., Topological Vector Spaces, Distributions and Kernels, Academic Press, London-New York 1967.

GROWTH RATES FOR STRONG UNICITY CONSTANTS

Myron S. Henry[1,2]

Department of Mathematical Sciences
Montana State University
Bozeman, Montana

John J. Swetits

Department of Mathematical Sciences
Old Dominion University
Norfolk, Virginia

I. INTRODUCTION

Let C(I) denote the set of real valued, continuous functions on I = [-1, 1], and let π_n be the set of polynomials of degree at most n. Denote the uniform norm on C(I) by $||\cdot||$.

For each $f \in C(I)$ with the best approximation $B_n(f) \in \pi_n$ on I, there is a constant $r \geq 1$ such that for any $p \in \pi_n$,

$$||p - B_n(f)|| \leq r(||f - p|| - ||f - B_n(f)||). \quad (1.1)$$

Inequality (1.1) is the classical strong unicity theorem (2, p. 80). The smallest $r \geq 1$ such that (1.1) is valid for all $p \in \pi_n$ is defined to be the strong unicity constant $M_n(f)$.

---

[1]Research for this paper was effected while this author was on leave from Montana State University to Old Dominion University, August, 1979 - July, 1980.

[2]Present address: School of Arts & Sciences, Central Michigan University, Mt. Pleasant, Michigan.

Copyright © 1980 by Academic Press, Inc.
All rights of reproduction in any form reserved.

ISBN: 0-12-171050-5

The behavior of the sequence $\{M_n(f)\}_{n=0}^{\infty}$ has been the subject of much recent research (1, 3, 4-7, 9). In particular, Henry and Roulier (5) have conjectured that $\{M_n(f)\}_{n=0}^{\infty}$ is bounded if and only if f is a polynomial. Evidence supporting this conjecture appears in (1, 5, 9). The order of growth of $M_n(f)$ has been examined in (4). The present work extends the results of this latter reference.

II. PRELIMINARIES

For $n \geq 1$, let $f \in C^{(n+2)}(I)$ and assume

$$f^{(n+1)}(x) \cdot f^{(n+2)}(x) \neq 0, \quad x \in I. \tag{2.1}$$

Let $e_n(f) = f - B_n(f)$ and $E_{n+1}(f) = \{x \in I: |e_n(f)(x)| = ||e_n(f)||\}$. It is known that if $f^{(n+1)}(x) \neq 0$ on I, then $E_{n+1}(f)$ contains exactly n+2 points, $-1 = x_0 < x_1 \ldots < x_n < x_{n+1} = 1$, and these points are separated (8) by the extreme points of the Chebyshev polynomials of degree n and n+1, respectively.

For each i = 0, 1, ..., n+1, let q_{in} be the polynomial of degree n defined by

$$q_{in}(x_k) = \text{sgn } e_n(f)(x_k), \quad k = 0, \ldots, n+1; \; k \neq i.$$

Assuming $E_{n+1}(f)$ has exactly n+2 points, Henry and Roulier (5), using a result of Cline (3), have shown that

$$M_n(f) = \max_{0 \leq i \leq n+1} ||q_{in}||. \tag{2.2}$$

Let Q_{n+1} be the polynomial of degree n+1 defined by

$$Q_{n+1}(x_k) = \text{sgn } e_n(f)(x_k), \quad k = 0, \ldots, n+1.$$

If a_{n+1} is the leading coefficient of Q_{n+1}, then for each $i = 0, \ldots, n+1$,

$$q_{in}(x) = Q_{n+1}(x) - a_{n+1} \prod_{\substack{j=0 \\ j \neq i}}^{n+1} (x - x_j), \quad x \in I. \tag{2.3}$$

III. ORDER OF GROWTH OF $M_n(f)$

The following lemma is of interest in its own right.

LEMMA. Let $f \in C^{(n+2)}(I)$ *and satisfy* (2.1). *Then*

$$2n+1 < M_n(f). \tag{3.1}$$

The proof of the lemma depends on the separation property alluded to in section II, equation (2.3), and the inequality

$$|a_{n+1}| \geq 2^n.$$

This latter inequality may be established by utilizing the theorem of de LaVallee Poussin (6).

Henry and Huff (4) have shown for $a \geq 2$ that

$$c_1 n \leq M_n\left(\frac{1}{a-x}\right) \leq c_2 n, \quad x \in I,$$

where c_1 and c_2 are positive constants that are independent of n. Thus (3.1) is best possible in an asymptotic sense.

The next theorem is the main result of the present paper.

THEOREM 1. Let $f \in C^{\infty}(I)$ *and satisfy* (2.1) *for all n sufficiently large. Suppose there is a constant* $\alpha > 0$, *independent of* n, *such that*

$$\frac{f^{(n+1)}(\varepsilon)}{f^{(n+1)}(\eta)} \leq \alpha \tag{3.2}$$

for all $\varepsilon, \eta \in I$ *and for all n sufficiently large. Then there are positive constants* k_1 *and* k_2, *independent of* n, *such that*

$$k_1 n \le M_n(f) \le k_2 n^2. \tag{3.3}$$

Examples to which Theorem 1 applies include $f(x) = e^x$ and $g(x) = \sin(\frac{x}{2}) + \cos(\frac{x}{2})$.

The lower bound in (3.3) follows from the Lemma. Establishing the upper bound requires verification that $||q_{on}||$, $||q_{n+1,n}||$, and $||Q_{n+1}||$ are all $0(n)$. The equation

$$q_{in}(x) = \frac{(x^2-1)}{2} \frac{q_{n+1,n}(x) - q_{on}(x)}{x-x_i} + Q_{n+1}(x),\ i = 1, \ldots, n,$$

which can be obtained (6) from (2.3), is used to establish that $||q_{in}|| = 0(n^2)$ $i = 1, \ldots, n$. An application of (2.2) then completes the proof.

The final theorem provides a bound on $M_n(f)$ which depends on $||Q_{n+1}||$.

THEOREM 2. *Let* $f \in C^{(n+2)}(I)$ *and satisfy* (2.1). *Define* α_n by

$$\alpha_n = \sup_{\eta, \varepsilon \in I} \left| \frac{f^{(n+2)}(\eta)}{f^{(n+1)}(\varepsilon)} \right| \tag{3.4}$$

Then

$$M_n(f) \ge \frac{(n+1)(n+2)}{2\pi\ \alpha_n} (||Q_{n+1}||-1) - ||Q_{n+1}||.$$

If $f \in C^{\infty}(I)$ and $\{\alpha_n\}_{n=1}^{\infty}$ is bounded, then it can be shown that (3.2) is satisfied. Thus, if there exists a $\beta > 0$ independent of n, such that $||Q_{n+1}|| - 1 > \beta$, then it follows from Theorems 1 and 2 that

$$\mu_1 n^2 \le M_n(f) \le \mu_2 n^2,$$

where μ_1 and μ_2 are positive constants independent of n. A detailed analysis of Theorems 1 and 2 appears in (6).

REFERENCES

1. Bartelt, M. W. and Schmidt, D., On Poreda's Problem for Strong Unicity Constants, J. Approximation Theory (to appear).
2. Cheney, E. W., "Introduction to Approximation Theory," McGraw-Hill, New York, 1966.
3. Cline, A. K., Lipschitz conditions on uniform approximation operators, J. Approximation Theory, 8(1973), 160-172.
4. Henry, M. S. and Huff, L. R., On the behavior of the strong unicity constant for changing dimension, J. Approximation Theory (to appear).
5. Henry, M. S. and Roulier, J. A., Lipschitz and strong unicity constants for changing dimension, J. Approximation Theory, 22(1978), 85-94.
6. Henry, M. S., Swetits, J. J. and Weinstein, S., Order of Strong Unicity Constants, (submitted).
7. Poreda, S. J., Counter examples in best approximation, Proc. Amer. Math. Soc. 56(1976). 167-171.
8. Rowland, J. H., On the location of the deviation points in Chebyshev approximation by polynomials, SIAM J. Numer. Anal., 6(1969), 118-126.
9. Schmidt, D., On an unboundedness conjecture for strong unicity constants, J. Approximation Theory 24(1978), 216-223.

LEBESGUE AND STRONG UNICITY CONSTANTS

Myron S. Henry[1,2]

Department of Mathematical Sciences
Montana State University
Bozeman, Montana

John J. Swetits
Stanley E. Weinstein

Department of Mathematical Sciences
Old Dominion University
Norfolk, Virginia

I. INTRODUCTION

Let C(I) denote the space of real valued continuous functions on the interval I = [-1, 1], and let $\pi_n \subseteq C(I)$ be the subspace of polynomials of degree at most n. For each $f \in C(I)$ with best uniform approximation $B_n(f) \in \pi_n$ on I, there is a constant r > 0 such that for any $p \in \pi_n$,

$$||p - B_n(f)|| \leq r(||f - p|| - ||f - B_n(f)||). \quad (1.1)$$

This inequality is the strong unicity theorem (2, p. 80), and

---

[1]Research for this paper was effected while this author was on leave from Montana State University to Old Dominion University, August, 1979 - July 1980.

[2]Present address: School of Arts and Sciences, Central Michigan University, Mount Pleasant, Michigan.

Copyright © 1980 by Academic Press, Inc.
All rights of reproduction in any form reserved.
ISBN: 0-12-171050-5

the smallest constant for which (1.1) is valid for all $p \in \pi_n$ is defined to be the strong unicity constant, designated $M_n(f)$.

Given any n + 1 distinct points $\{x_i\}_{i=0}^n$ the corresponding Lebesgue function $\lambda_n(x)$ is defined by $\lambda_n(x) = \sum_{i=0}^{n} |L_i(x)|$, $x \in I$, where $L_i(x) = \prod_{\substack{j=0 \\ j\neq i}}^{n} \frac{x - x_j}{x_i - x_j}$, $i = 0, \ldots, n$, are the classical Lagrange polynomials determined by $\{x_i\}_{i=0}^n$, (8). The Lebesgue constant λ_n is now defined to be

$$\lambda_n = ||\lambda_n(\cdot)|| = \max_{-1\leq x\leq 1} |\lambda_n(x)|.$$

For $f \in C(I)$, $e_n(f)(x) = f(x) - B_n(f)(x)$, $x \in I$, is the standard error function, and

$$E_{n+1}(f) = \{x \in I: |e_n(f)(x)| = ||e_n(f)||\}$$

is the set of extreme points of $e_n(f)$.

II. MAIN RESULTS

The following theorem shows that the strong unicity constant for any function whose error curve $e_n(f)$ contains exactly n + 2 extreme points equals the largest of the Lebesgue constants determined by omitting one point at a time from $E_{n+1}(f)$.

THEOREM 1. *For* $f \in C(I)$, *suppose that* E_{n+1} *contains exactly* n + 2 *points* $\{x_i\}_{i=0}^{n+1}$. *Let* λ_{n+1}^j *denote the Lebesgue*

<u>constant determined by</u> $E^j_{n+1} = E_{n+1}(f) - \{x_j\}, j = 0, \ldots, n+1$.
<u>Then</u>

$$M_n(f) = \max_{0 \le j \le n+1} \lambda^j_{n+1}.$$

The keys to the proof of Theorem 1 are an observation of Henry and Roulier (see (2.4) below) and the following theorem, (1).

THEOREM 2. (Bartelt and Schmidt). <u>If</u> $f \in C(I) - \Pi_n$, <u>then</u>

$$M_n(f) = \max\{||p|| : p \in \Pi_n \text{ and } \operatorname{sgn} e_n(f)(x)p(x) \le 1 \text{ for all } x \in E_{n+1}(f)\}.$$

The next theorem relates the strong unicity constant $M_n(f)$ to λ_{n+1}, the Lebesgue constant determined by all of $E_{n+1}(f) = \{x_0, \ldots, x_{n+1}\}$. First, define $Q_{n+1} \in \pi_{n+1}$ by

$$Q_{n+1}(x_i) = \operatorname{sgn} e_n(f)(x_i), \; i = 0, \ldots, n+1. \tag{2.1}$$

THEOREM 3. <u>For</u> $f \in C(I)$, <u>suppose that</u> $E_{n+1}(f)$ <u>contains exactly</u> $n + 2$ <u>points</u> $\{x_i\}_{i=0}^{n+1}$, <u>with</u> $x_0 = -1$ <u>and</u> $x_{n+1} = 1$. <u>Then</u>

(i) $||Q_{n+1}|| \le M_n(f)$,

<u>and</u>

(ii) $\lambda_{n+1} \le 2M_n(f)$,

<u>where</u> Q_{n+1} <u>is defined in</u> (2.1) <u>and</u> λ_{n+1} <u>is the Lebesgue constant determined by all of</u> E_{n+1}.

Proof: It can be shown (7) that

$$Q_{n+1}(x) = q_{in}(x) + a_{n+1} \prod_{\substack{j=0 \\ j\neq i}}^{n+1} (x - x_j) \qquad (2.2)$$

where $q_{in} \in \Pi_n$ is defined by $q_{in}(x_j) = \text{sgn } e_n(f)(x_j)$, $j = 0, \ldots, n+1$; $j \neq i$, and where a_{n+1} is the coefficient of the highest power of Q_{n+1}. It follows from (2.2) that

$$2Q_{n+1}(x) = (x + 1)q_{0n}(x) - (x - 1)q_{n+1,n}(x). \qquad (2.3)$$

But under the hypotheses of this theorem, (see (6)),

$$M_n(f) = \max_{0\leq i\leq n+1} \{||q_{in}||\}. \qquad (2.4)$$

Therefore (2.3) implies that $||Q_{n+1}|| \leq M_n(f)$. Thus part (i) is established. From section I, if $\lambda_n(x) = ||\lambda_n(\cdot)||$, then

$$\lambda_{n+1} = \sum_{i=0}^{n+1} \prod_{\substack{j=0 \\ j\neq i}}^{n+1} |x - x_j| \Big/ \prod_{\substack{j=0 \\ j\neq i}}^{n+1} |x_i - x_j|$$

$$\leq \max_{0\leq i\leq n+1} \prod_{\substack{j=0 \\ j\neq i}}^{n+1} |x - x_j| \sum_{i=0}^{n+1} 1 \Big/ \prod_{\substack{j=0 \\ j\neq i}}^{n+1} |x_i - x_j|$$

$$= |a_{n+1}| \max_{0\leq i\leq n+1} \prod_{\substack{j=0 \\ j\neq i}}^{n+1} |x - x_j|.$$

But (2.2) implies that

$$|a_{n+1}| \max_{0\leq i\leq n+1} \prod_{\substack{j=0 \\ j\neq i}}^{n+1} |x - x_j| \leq |Q_{n+1}(x)| + \max_{0\leq i\leq n+1} |q_{in}(x)|.$$

Conclusion (ii) now follows from (i) and (2.4).

A comparison of Theorems 1 and 2 reveals the following interesting observation: for functions satisfying the hypotheses of Theorem 2, the maximum of the Lebesgue constants obtained by removing one point at a time from the extremal set $E_{n+1}(f)$ grows at least as fast as the Lebesgue constant determined by all of $E_{n+1}(f)$ as n tends to infinity. The following example further illustrates this observation.

III. EXAMPLE

Let $P(x) = x^{n+1}$, $x \in I$. Then $M_n(P) = 2n + 1$, see (3,5). Consequently Theorem 1 implies that $\max_{0 \leq j \leq n+1} \lambda_{n+1}^{j} = 2n + 1$.

On the other hand, $E_{n+1}(P)$ consists of precisely the n + 2 extreme points of the Chebyshev polynomial of degree n + 1, and consequently $\lambda_{n+1} = 0(\log(n + 1))$,(4). Therefore the orders of $M_n(f) = \max_{0 \leq j \leq n+1} \lambda_{n+1}^{j}$ and λ_{n+1} may indeed differ significantly.

REFERENCES

1. M. W. Bartelt and D. P. Schmidt, On Poreda's problem for strong unicity constants, J. Approximation Theory, to appear.
2. E. W. Cheney, Introduction to Approximation Theory, McGraw-Hill, New York, 1966.
3. A. K. Cline, Lipschitz conditions on uniform approximation operators, J. Approximation Theory, 8(1973), 160-172.
4. H. Ehlich and K. Zeller, Auswertung der Normen Von Interpolations operatoren, Math. Annalen, 164(1966), 105-112.
5. M. S. Henry and L. R. Huff, On the behavior of the strong unicity constant for changing dimension, J. Approximation Theory, to appear.
6. M. S. Henry and J. A. Roulier, Lipschitz and strong unicity constants for changing dimension, J. Approximation Theory, 22(1978), 85-94.

7. M. S. Henry, J. J. Swetits, and S. E. Weinstein, Orders of Strong Unicity constants, submitted.
8. T. J. Rivlin, An Introduction to the Approximation of Functions, Blaisdell Publishing Co., Waltham, Mass., 1969.

GENERALIZATION OF A THEOREM OF BERNSTEIN*

A. S. B. Holland

Department of Mathematics
University of Calgary
Alberta

G. M. Phillips

Mathematical Institute
University of St Andrews
Scotland

P. J. Taylor

Department of Mathematics
University of Stirling
Scotland

I. PROPERTY B

We shall be concerned in this paper with semi-norms on $C^{n+1}[a,b]$ which satisfy the following property.

DEFINITION A semi-norm E on $C^{n+1}[a,b]$ is said to satisfy Property B of order n if

$$|f^{(n+1)}(x)| \leq g^{(n+1)}(x), \qquad a \leq x \leq b, \tag{1}$$

implies that $E(f) \leq E(g)$.

*This work was supported by grant No.A9225 from the National Research Council of Canada.

Copyright © 1980 by Academic Press, Inc.
All rights of reproduction in any form reserved.
ISBN: 0-12-171050-5

We have used the term 'Property B' to mark its connection with S. N. Bernstein, 1926, in the special case of

$$E(f) = \inf_{q \in P_n} \|f - q\|_\infty . \tag{2}$$

(See Lorentz, 1966 or Meinardus, 1967.) If a semi-norm E satisfies Property B, then

$$E(q) = 0, \quad \forall q \in P_n; \tag{3}$$

$$E(f + q) = E(f), \quad \forall f \in C^{n+1}[a,b], \quad q \in P_n. \tag{4}$$

To verify (3) we note that (1) holds with $f(x) = q(x) \in P_n$ and $g(x) = \varepsilon x^{n+1}$, where $\varepsilon > 0$ is arbitrary. Then

$$E(q) \leqslant |\varepsilon| E(x^{n+1})$$

which may be made arbitrarily small. To verify (4) we have

$$E(f + q) \leqslant E(f) + E(q) = E(f)$$

and

$$E(f) = E(f + q - q) \leqslant E(f + q).$$

II. THEOREM OF BERNSTEIN

We now generalize a theorem which Bernstein, 1926, gave for the special case of the error of minimax approximation (2).

THEOREM 1. If the semi-norm E satisfies Property B, then for each $f \in C^{n+1}[a,b]$ there exists $\xi \in (a,b)$ such that

$$E(f) = \frac{|f^{(n+1)}(\xi)|}{(n+1)!} E(x^{n+1}). \tag{5}$$

PROOF The proof is similar to that in Meinardus, 1967 (Theorem 60), for the special case of minimax approximation. Let

$$g_c(x) = \frac{cx^{n+1}}{(n+1)!}$$

where $c > 0$. Then

$$g_c^{(n+1)}(x) = c \quad \text{and} \quad E(g_c) = \frac{c}{(n+1)!} E(x^{n+1}),$$

since E is a semi-norm. If

$$0 < m \leqslant f^{(n+1)}(x) \leqslant M$$

or

$$0 < m \leqslant -f^{(n+1)}(x) \leqslant M$$

on [a,b], it follows from Property B that

$$\frac{m}{(n+1)!} E(x^{n+1}) = E(g_m) \leqslant E(f) \leqslant E(g_M) = \frac{M}{(n+1)!} E(x^{n+1}).$$

Alternatively, if $f^{(n+1)}$ has a zero on [a,b] and

$$f^{(n+1)}(x) \leqslant M,$$

we have

$$0 \leqslant E(f) \leqslant E(g_M) = \frac{M}{(n+1)!} E(x^{n+1}).$$

In either case the derived result follows from the continuity of $f^{(n+1)}$.

As we have already remarked, Bernstein has shown that the semi-norm (2) satisfies Property B. As a second example, we cite

$$E(f) = |f(\alpha) - q(\alpha)|, \tag{6}$$

where $\alpha \in [a,b]$ is some fixed point and q is the interpolating polynomial for f at certain fixed points $x_0, x_1, \ldots, x_n$ of [a,b]. We now prove:

THEOREM 2. The interpolation error operator (6) satisfies Property B.

PROOF We first exclude the trivial case where α coincides with one of the x_i. Next we suppose that f and g satisfy (1). We can add an amount εx^{n+1} to g, as in Meinardus, 1967 (Theorem 59), and so need consider only the case

$$|f^{(n+1)}(x)| < g^{(n+1)}(x) \quad \text{on} \quad [a,b],$$

when

$$g^{(n+1)}(x) > 0 \text{ on } [a,b].$$

Now let q_f and q_g denote the interpolating polynomials for f and g respectively, constructed at $x_0, x_1, \ldots, x_n$ and define

$$h(x) = E(g)(f(x) - q_f(x))\operatorname{sign}(f(\alpha) - q_f(\alpha))$$
$$-E(f)(g(x) - q_g(x))\operatorname{sign}(g(\alpha) - q_g(\alpha)).$$

Then $h(x_i) = 0$, $i = 0,1, \ldots, n$ and $h(\alpha) = 0$. By repeated application of Rolle's theorem, there exists $\xi \in (a,b)$ such that

$$h^{(n+1)}(\xi) = 0 = E(g)f^{(n+1)}(\xi)\operatorname{sign}(f(\alpha) - q_f(\alpha))$$
$$-E(f)g^{(n+1)}(\xi)\operatorname{sign}(g(\alpha) - q_g(\alpha)).$$

Thus

$$E(f) = \frac{f^{(n+1)}(\xi)}{g^{(n+1)}(\xi)}. \operatorname{sign}(f(\alpha) - q_f(\alpha))\operatorname{sign}(g(\alpha) - q_g(\alpha)).E(g)$$

and Property B follows.

REFERENCES

Bernstein, S.N. (1926). "Leçons sur les propriétés extrémales et la meilleure approximation des fonctions analytiques d'une variable réele." Gauthier-Villars, Paris.

Lorentz, G.G. (1966). "Approximation of Functions". Holt, Rinehart and Winston, New York.

Meinardus, G. (1967). "Approximation of Functions: Theory and Numerical Methods." Springer-Verlag, New York.

AN IDENTITY FOR THE CHARACTERISTIC POLYNOMIALS OF THE L_2-PROJECTIONS ON SPLINES FOR A GEOMETRIC MESH

K. Höllig and K. Scherer

Institute für Angewandte Mathematik
der Universität Bonn
Federal Republic of Germany

1. INTRODUCTION

Given a biinfinite sequence of knots $\vec{x} = \{x_\nu\}$, $\Delta_\nu := x_{\nu+k} - x_\nu > 0$ we denote by $N_\nu^{(k)}$, $\operatorname{supp} N_\nu^{(k)} = [x_\nu, x_{\nu+k}]$ the corresponding B-splines normalized by $\sum N_\nu^{(k)} = 1$ and by $S_{k,\vec{x}} = \operatorname{span}\{N_\nu^{(k)}\}$ the space of splines of degree k-1 with knots $\vec{x}$ [2,3].

The L_2-projection $P_{k,\vec{x}} : L_2 \to S_{k,\vec{x}} \cap L_2$ is defined by the Gramian system

$$P(f) = \sum a_\nu(f) N_\nu \in S_{k,\vec{x}} \cap L_2$$

$$\sum (N_\nu, N_\mu) a_\mu(f) = (N_\nu, f)$$

C. de Boor [1] conjectured that P is also a bounded projection on L_∞, independently of the underlying mesh, in particular

$$C_k(x) := ||P_{k,\vec{x}}||_\infty \leq C_k < \infty$$

In this note we will investigate the case of a geometric mesh $\vec{x}_t$ with j coalescing knots

$$\vec{x}_t = \{x_\nu\};\ x_{\nu+\mu j} = t^\mu,\ 0 \leq \nu < j,\ t \in (1,\infty) \tag{1}$$

which was also studied recently by B. Mityagin in the case k = 2j.

Copyright © 1980 by Academic Press, Inc.
All rights of reproduction in any form reserved.
ISBN: 0-12-171050-5

2. GENERAL CONSIDERATIONS

Using the basis properties of B-splines, in particular the well known isomorphism

$$||\sum a_\nu (\Delta_\nu^{-1/p} N_\nu)||_p \sim ||\{a_\nu\}||_{l_p}$$

of C. de Boor [2] we may easily estimate the L_∞-norm of P.

LEMMA 1. *Writing* $G_{\nu\mu} = \Delta_\nu^{-1/2}(N_\nu, N_\mu)\Delta_\mu^{-1/2}$ *for the Gramian matrix, we have*

$$C_k(\vec{x}) \sim \sup_\nu \sum_\mu \Delta_\nu^{-1/2} |(G^{-1})_{\nu\mu}| \Delta_\mu^{1/2}$$

where "$\sim$" denotes inequalities in both directions with constants only depending on k.

Since $|(G^{-1})_{\nu\mu}| \le C\,\lambda^{|\nu-\mu|}$, $\lambda \in (0,1)$ [4] the norm of $P_{k,\vec{x}}$ can obviously be bounded in terms of the global mesh ratio. Therefore a geometric mesh (1) is a reasonable test for C. de Boor's conjecture. In this case one verifies by direct computation that G has constant block band structure

$$G = \begin{pmatrix} \ddots & & & & \\ \cdots A_{-1} & A_0 & A_1 \cdots & & \\ & \cdots A_{-1} & A_0 & A_1 \cdots & \\ & & & \ddots & \end{pmatrix}$$

where $(A_r)_{\nu\mu} = \Delta_\nu^{-1/2}(N_\nu, N_{\mu+rj})\Delta_{\mu+rj}^{-1/2}$, $0 \le \nu, \mu < j$ and that

$$G^{-1} = \begin{pmatrix} \ddots & & & & \\ \cdots B_{-1} & B_0 & B_1 \cdots & & \\ & \cdots B_{-1} & B_0 & B_1 \cdots & \\ & & & \ddots & \end{pmatrix}$$

The special structure of G^{-1} leads to the following simplification of lemma 1.

LEMMA 2. *For a geometric mesh* $\vec{x}_t$ *we have*

$$C(\vec{x}_t) \sim \sum_r ||B_r||\, t^{r/2}$$

where $||\ \ ||$ *may be any matrix norm.*

Using lemma 2 we can derive by the extension of the classical Töplitz theory to constant block band matrices a criterion for the boundedness of P with respect to the L_∞-norm.

THEOREM 1. $C(\vec{x}_t)$ <u>is finite for any</u> $t\varepsilon(1,\infty)$ <u>iff</u>

$$Q(t;z) := \det \{\sum_r A_r(t^{1/2}z)^r\} \tag{2}$$

<u>has no roots on the unit circle</u> $\{z\varepsilon\mathbb{C}: |z| = 1\}$ <u>for every</u> $t\varepsilon(1,\infty)$.

In the case k = 2j a similar characterization has been obtained by B. Mityagin who used a more abstract approach leading to different matrices A_r.

3. THE CASE OF SIMPLE KNOTS

Following a suggestion of C. de Boor we consider more generally the projections P_r, $0\leq r<2k$, defined by

$$\begin{aligned} P_r(f) &= \sum a_\nu(f)N_\nu^{(2k-r)} \\ \sum (N_\nu^{(r)},N_\mu^{(2k-r)})a_\mu(f) &= (N_\nu^{(r)},f) \end{aligned} \tag{3}$$

where $N_\nu^{(o)} := \delta(\cdot - t^\nu)$. Hence P_o is the interpolation projection on S_{2k} and P_k the L_2-projection on S_k. Integration by parts shows that for smooth f with compact support

$$s = P_o(f) \Rightarrow s^{(r)} = P_r(f^{(r)}) \ . \tag{4}$$

It is well known, but may also be derived from theorem 1, that (3) defines a bounded projection on L_∞ iff the characteristic polynomial $Q_r(t;z)$ does not vanish on the unit circle where

$$Q_r(t;z) = \sum q_\nu^{(r)} z^\nu \quad ; \ q_\nu^{(r)} = (N_o^{(r)},N_\nu^{(2k-r)}) \ . \tag{5}$$

We recall that the B-splines $N_\nu^{(r)}$ and therefore the coefficients $q_\nu^{(r)}$ depend on the mesh parameter t. We prove a relationship between the polynomials $Q_r(t;z)$ by formulating a discrete analogon of the implication (4). To this end we define some operations on sequences

$$(D^+a)_\nu = a_{\nu+1}-a_\nu \quad ; \ (D^-a)_\nu = a_\nu - a_{\nu-1}$$
$$(\mathrm{diag}(\lambda_\nu)a)_\mu = \lambda_\mu \ a_\mu$$

LEMMA 3. *Define the matrices* $(A^{(r)})_{\nu\mu} = q^{(r)}_{\mu-\nu} = t^{-\nu}(N^{(r)}_{\nu}, N^{(2k-r)}_{\mu})$. *For all sequences with compact support*

$$A^{(o)} a = F \tag{6}$$

implies

$$A^{(r)}\{\mathrm{diag}(t^{-\nu})D^{-}\}^{r} a = C(t)\ \{\mathrm{diag}(t^{-\nu})D^{+}\}^{r} F \tag{7}$$

where here and in the following $C(t)$ *stands for various constants depending on* k,r *and* t.

Proof. Set $s = \sum a_\nu N^{(2k)}_\nu$, $F_\nu = s(t^\nu)$ and consider (4) with $f = s$. We have $s^{(r)} = \sum b_\nu N^{(2k-r)}_\nu$ where in case of a geometric mesh

$$b = C(t)\ \{\mathrm{diag}(t^{-\nu})D^{-}\}^{r} a\ ,$$

on the other hand by (3) and the Peano-kernel theorem

$$(A^{(r)}b)_\nu = t^{-\nu}(N^{(r)}_\nu, s^{(r)}) = C(t)\ (\{\mathrm{diag}(t^{-\mu})D^{+}\}^{r} F)_\nu\ .$$

LEMMA 4. *The coefficients of the polynomials* $Q_r(t;z)$ *are related by*

$$(N^{(r)}_o, N^{(2k-r)}_\nu) = C(t)\ t^{r\nu}\ N^{(2k)}_o(t^{r-\nu})\ . \tag{8}$$

Proof. Restricting ourselves to sequences with compact support we may invert D^+ by

$$((D^+)^{-1}\ b)_\nu = \sum_{\mu=-\infty}^{\nu-1} b_\mu\ .$$

We apply the operation $(D^+)^{-1}\mathrm{diag}(t^\nu)$ to both sides of the equation (7) and obtain by direct computation

$$A^{(r),1}\{\mathrm{diag}(t^{-\nu})D^{-}\}^{r-1} a = C(t)\ \{\mathrm{diag}(t^{-\nu})D^{+}\}^{r-1} F$$

where $(A^{(r),1})_{\nu\mu} = t^{\nu-\mu-1}q^{(r)}_{\mu-\nu+1}$.

Iterating this process it follows that

$$A^{(r),r}\ a = C(t)\ F \tag{9}$$

where $(A^{(r),r})_{\nu\mu} = C(t)\ t^{r(\nu-\mu)}\ q^{(r)}_{\mu-\nu+r}$.

Comparing (9) with (6) we obtain the desired identity (8).

We now can formulate our main result.

COROLLARY. <u>The characteristic polynomials for the projections P_r are related by</u>

(10) $Q_r(t;z) = C(t)\, Q_o(t;t^r z)$.

This is an immediate consequence of lemma 4 and the definition of Q_r.

By C. de Boor and C. A. Micchelli it was pointed out to us that in [5] (formula (23)) the following explicit representation of $Q_o(t;t^r z)$ has been given

$$(11)\quad Q_o(t;t^r z) = C(t) \sum_{j\in Z} \prod_{\nu=0}^{2k-1} \frac{1}{i(\phi+2\pi j) + (r-\nu)\ln t}$$

where $z = e^{i\phi}$. This then allows to estimate the norm of P_r. More precisely one has for $r = k,k-1$

$$(12)\quad \|P_r\|_\infty^{-1} = C(t) \sum_{j\in Z} \prod_{\nu=1}^{k} \frac{1}{(\pi+2\pi j)^2 + (\nu \ln t)^2}$$

which implies that P_r is a bounded projection on L_∞ for any $t \in [1,\infty)$. Also one has

$$(13)\quad \lim_{t\to\infty} \|P_r\|_\infty \leq 2k-1 ,$$

i.e. the norms are uniformly bounded with respect to t.

REMARK. Before being informed by C. de Boor and C. A. Micchelli about this way of proving the boundedness of $\|P_r\|_\infty$ we had the idea to use a direct method working with $\mathcal{L}$- splines and a suitable convolution structure analogous to the classical polynomial case. This leads to a systematic approach using only elementary Fourier-analysis which is the content of a forthcoming paper [6]. It does not involve the interpolation problem (but instead gives a second proof of lemma 4) and gives an independent simple proof of (12). Moreover it is shown in [6] that for all other values of r in (3) $\|P_r\|_\infty$ becomes unbounded for at least one $t \in (1,\infty)$.

REFERENCES

1. Boor, C. de, The quasi-interpolant as a tool in elementary polynomial spline theory, in "Approximation Theory", G.G. Lorentz ed., Academic Press (1973), 269-276.

2. Boor, C. de, Splines as linear combinations of B-splines, in "Approximation Theory II", G.G. Lorentz, C.K. Chui, L.L. Schumaker, eds., Academic Press (1976), 1-47.

3. Boor, C. de, A practical guide to splines, Springer 1978.

4. Demko, S., Inverses of band matrices and local convergence of spline projections, Siam J. Numer. Anal. 14 (1977), 616-619.

5. Micchelli, C. A., Cardinal $\mathcal{L}$-splines, in "Studies in spline functions and approximation theory", S. Karlin, C.A. Micchelli, A. Pinkus, I.J. Schoenberg, eds., Academic Press 1976.

6. Höllig, K., L_∞-boundedness for the L_2-projection on splines for a geometric mesh, to appear.

K. Höllig K. Scherer
Institut für Angewandte Mathematik
der Universität Bonn
Wegelerstr. 6
53 Bonn
Federal Republic of Germany

APPLIED APPROXIMATION THEORY

Richard B. Holmes[1]

M.I.T. Lincoln Laboratory
Lexington, Massachusetts

A brief description is given of some problems and procedures which are important in practice, and which have some approximation theoretic content.

I. INTRODUCTION

Instances abound in the statistical and engineering sciences where one makes measurements or otherwise observes data, and then must construct an estimate of some quantity of interest. Very frequently the procedure used is one of orthogonal projection, perhaps in a disguised form. Mathematically, the reasons for this widespread usage included the natural occurrence of various Hilbert spaces in the modeling process, and the (characteristic) properties enjoyed by the metric projectors (on closed convex sets) in Hilbert space, namely that they contract distances and are linear whenever possible.

We propose a rough taxonomy of approximation theory: classical, abstract, and applied. Most readers will easily be capable of delineating the first two areas. Generally, "abstract" is distinguished from "classical" by the substantial use of functional analysis. Here we shall concentrate on defining and illustrating applied approximation theory.

[1]This work was supported by the Department of the Army.

Copyright © 1980 by Academic Press, Inc.
All rights of reproduction in any form reserved.
ISBN: 0-12-171050-5

Let us say that "applied" means playing a serious role in the construction of devices (software or hardware), or in the evaluation of systems. In practice, applied approximation theory involves the identification of an appropriate Hilbert space and metric projector thereon, and specification of an effective means of computating this projector.

We shall now consider five cogent examples of applied approximation theory and make a few comments about their application and technical content. They, along with many others (such as signal detection, pdf estimation, splines, etc.), are discussed in greater detail in the forthcoming book (1).

II. OPTIMAL FILTERING

This is a subject with a long and venerable history; associated with its development are the names of Gauss, Legendre, Wiener, Kolmogorov, Kalman, Bucy, Kailath, etc. It is concerned with the very general problem of separating a desired signal from noisy observations thereof. Over the last two decades, various filtering algorithms have had a decisive impact in areas ranging from aerospace engineering and navigation to econometric modeling.

The essential ingredients of a modern analysis of this problem are the Gram-Schmidt orthogonalization procedure (which gives the eventual algorithm its all-important recursive character), orthogonal projections (viewed either as strict or wide-sense conditional expectations) in a Hilbert space of second order random variables, and a dynamic system (state-space) model.

At the core of this analysis is the formula

$$E(Y|X) := E(Y|\mathfrak{I}) = P_M(Y);$$

here, X,Y are random variables in $L^2(\Omega,\mathcal{S},P)$, $\mathfrak{I}$ is the sub-σ-algebra of $\mathcal{S}$ generated by X, M is the subspace $\{f \circ X\colon f$ a Borel function$\}$, and P_M denotes orthogonal projection on M. Thus, the conditional expectation $E(Y|X)$ appears as the best (mean square) approximation to Y by Borel functions of X. If, in fact, $E(Y|X) = g(X)$, then the regression function $E(Y|X = x)$, so useful for prediction, is simply $g(x)$, and conversely. Extending this to the case of random vectors, invoking a joint normality hypothesis, and developing a recursive updating procedure à la Gram-Schmidt, provides the basic machinery for the Kalman filter algorithm (2,3).

III. SIGNAL AND OBJECT RECONSTRUCTION

Here we have another broad class of problems of estimation from partial information. There is now no state space model, nor is the sequential presentation of data significant. Typical applications are to the extrapolation of band-limited signals, and to the reconstruction of density functions from a finite number of projections, as in electron microscopy and radiology.

A general approach is based on the Alternating Projections Lemma due to von Neumann and Halperin: if $M_1,\ldots,M_r$ are closed subspaces of a Hilbert space, $M := \bigcap M_i$ and $P_i := P_{M_i}$, then

$$P_M = \lim_{n\to\infty}(P_1P_2 \cdots P_r)^n,$$

with convergence in the strong operator topology (4). It is

also known that such a formula can only be valid in an inner product space.

Now to approximate a particular element f in a Hilbert space (e.g., in x-ray tomography f might represent the tissue density of a body slice), assume that $\cap_{i=1}^{\infty} M_i = \{\theta\}$, whence $\lim P_i = 0$, strongly. Then for sufficiently large r, $f \approx g$:= projection of an initial guess (often, θ) on the flat f + M. It remains to apply the Lemma to approximate g. In the tomography example each M_i is the nullspace of the radiograph operator in a direction α_i, the necessary projectors can be determined from these operators, and g will have the same radiographs as the unknown f.

The resultant reconstruction algorithm is often called ART (Algebraic Reconstruction Technique), and formed the basis for early CAT scanner software. Its rate of convergence (wrt n) has recently been studied as a function of the angles between the subspaces M_i (which, in turn, can be related to the known directions α_i) by Hamaker and Solomon (5).

IV. CRAMÉR-RAO BOUND

This is a lower bound on the variance of a (usually unbiased) estimator of a real or vector parameter. The estimator will typically be some function of a sample, which may be chosen by any of several popular methods such as moments, least squares, or max. likelihood. It is important to emphasize that the C-R bound is independent of the estimation procedure used and of simulation techniques. It is widely used to assess the relative value of estimation procedures. A particular application is to the comparison of trajectory tracking and impact point prediction algorithms. Here the parameter is the 6-dim. vehicle state vector at some time of interest.

Let $(\Omega, \mathcal{S}, P)$ be a probability space and, for each admissable $\alpha \in R^n$, let P_α be a probability measure on $\mathcal{S}$. Then for any subspace $M := \text{span}\,\{q_1, \text{---}, q_n\}$ and T in $L^2(P_\alpha)$,

$$\text{Var}_\alpha(T) := ||T - E_\alpha(T)||^2 \geq ||P_M(T - E_\alpha(T))||^2$$

$$:= ||\lambda_1 q_1 + \cdots + \lambda_n q_n||^2 = \lambda^* \Phi_\alpha \lambda = \lambda^* \nabla E_\alpha(T);$$

here E_α denotes integration wrt P_α, Φ_α is the Gram matrix of $\{q_j\}$, and the q's must be specially chosen so that, *inter alia*, $\langle q_j, f \rangle = \nabla E_\alpha(f)_j$, $1 \leq j \leq n$, $f \in L^2(P_\alpha)$ (6). Applying this inequality to linear functionals of a given unbiased estimator $\underline{T}$ of α eventually yields the C-R bound:

$$\text{cov}_\alpha(\underline{T}) \geq \Phi_\alpha^{-1},$$

and exhibits its essential geometric interpretation.

V. MULTIVARIATE INTERPOLATION

Given a finite set of values of an unknown function f of two or more variables, it is desired to reconstruct f. Again, this is a widely investigated problem because of the numerous potential applications. Many of these are suggested in the survey of Schumaker (7). We note, in particular, applications to the design of real-time graphics software for microprocessors, and to the implementation of digital reconstruction algorithms for bivariate signals. For example, a frequency domain approach to the "reconstruction from projections" problem attempts to approximate the Fourier transform of the unknown signal. Transforming the given projections yields "slices" of the desired transform. Because of the digital computation this transform is thus known only on a polar raster, and interpolation is needed to define it on a rectangular raster, whence an inverse discrete Fourier transform leads to an approximation to the actual signal.

In general, let X be a RKHS of functions on a set Q, and let there be given values $x(q_i) = c_i$, $1 \leq i \leq n$, $\{q_i\} \subseteq Q$. The Golomb-Weinberger estimate $\hat{x}$ may then be thought of as a "smoothest" interpolant of the data and is sometimes called an abstract interpolating spline. It is obtained by orthogonal projection onto the span of the representers of the evaluation functionals for $\{q_i\}$. Cases of particular practical importance occur when X is one of the Sard-type spaces of bivariate functions, whose reproducing kernel has been determined by Barnhill and Nielson (8), and Mansfield (9). Both local and global interpolation schemes ensue and seem to perform favorably relative to other methods (10).

VI. GAUSS-MARKOV ESTIMATION

For our final example we consider the general linear model of statistics:

$$y = Z\alpha + u,$$

where Z is a known m x n matrix, α an unknown parameter vector to be estimated, and u (hence y) is random. We shall assume that $E(u) = \theta$, $\text{cov}(u) := D$ (known and positive definite), and that rank $(Z) = n$ (absence of multicollinearity). Then the interest is in a optimal unbiased estimate $\hat{\alpha}$ of α. According to the classical Gauss-Markov theorem, this can be obtained by projecting y orthogonally onto the column space of Z, considered as a subspace of R^n normed by $||x||^2 :=$ $x^T D^{-1}x$. (Here, "optimal" means that $\mathrm{var}(a^T \cdot \hat{\alpha}) \leq \mathrm{var}(a^T \cdot \tilde{\alpha})$, for all $a \in R^n$ and any other unbiased estimate $\tilde{\alpha}$ of α; a more intrinsic geometric definition of optimality, involving the notion of concentration ellipsoid, is also possible (11).)

Computationally this operation is carried out by recalling that, in R^n so normed, the matrix of any orthogonal projector must have the form

$$P_M = A(A^T D^{-1} A)^{-1} A^T D^{-1},$$

where M := col space(A), and A has full rank. Thus $\hat{\alpha} :=$ $(Z^T D^{-1} Z)^{-1} Z^T D^{-1}(y)$, $E(\hat{\alpha}) = \alpha$, and $\mathrm{cov}(\hat{\alpha}) = (Z^T D^{-1} Z)^{-1}$. Extensions to the case of singular D and to a more general constraint on the location of E(y) are available (11).

In practice this technique is used to estimate the parameters in a multiple regresssion: The data vector of dependent variable observations is projected onto the space of the explanatory variable observations, and the D matrix is estimated from the data. The resulting formula is notationally virtually identical to that of the wide sense conditional expectation of Section II; the essential distinctions between the two are that here covariances are obtained empirically, and the relevant Hilbert spaces are inherently of finite dimension.

REFERENCES

1. Holmes, R. B., "Hilbert Spaces, Operators, and Applications, Vol. 1. Geometric Theory", to appear.
2. McGarty, T. P., "Stochastic Systems and State Estimation", McGraw Hill, New York, (1974).
3. Rhodes, I. B., *IEEE Trans. Auto. Control 16*, 688 (1971).

4. Halperin, I., *Acta Sci. Math. (Szeged) 23,* 96 (1962).

5. Hamaker, C., and Solomon, D., *J. Math. Anal. Appl. 62,* 1 (1968).

6. Fabian, V., and Hannan, J., *Ann. Stat. 5,* 197 (1977).

7. Schumaker, L. L., *in,* "Approximation Theory II" (Lorentz, Chui, Schumaker, ed's.), p. 203, Academic Press, (1976).

8. Barnhill, R., and Nielson, G., *SIAM J. Num. Anal. 11,* 37 (1974).

9. Mansfield, L., *J. Approx. Theory 5,* 77 (1972).

10. Franke, R., *J. Inst. Maths Applics 19,* 471 (1977).

11. Drygas, H., "The Coordinate-Free Approach to Gauss-Markov Estimation," Springer, Berlin, (1971).

The views and conclusions contained in this document are those of the contractor and should not be interpreted as necessarily representing the official policies, either expressed or implied, of the United States Government.

INTERPOLATION BY FUNCTIONS WITH m-th DERIVATIVE IN PRE-ASSIGNED SPACES

A. Jakimovski
Tel-Aviv University
Tel-Aviv, Israel

D. C. Russell
York University
Toronto, Canada

0. INTRODUCTION

Let ω denote the space of all doubly-infinite complex-valued sequences $y = (y_k)_{k \in Z}$. Throughout this paper, $x = (x_k)_{k \in Z}$ will be a fixed strictly increasing sequence with $\lim_{k \to -\infty} x_k = -\infty$, $\lim_{k \to +\infty} x_k = +\infty$ (a sequence of simple knots); and, for any $k \in Z$, $e^k := (\ldots,0,1,0,\ldots)$ (with 1 in the k-th place). Let $\mathcal{L}$ be a fixed linear space of functions from R into C. Given a prescribed sequence $y \in \omega$, the problem of finding a function f such that $f(x_k) = y_k$ $(\forall k \in Z)$ is called the <u>interpolation problem</u> $IP(y; \mathcal{L}, x)$. This symbol will also denote the set of its solutions. If a semi-norm $\|\cdot\|_{\mathcal{L}}$ is defined on $\mathcal{L}$ we say that f_* is an <u>optimal</u> (or <u>extremal</u>) <u>solution</u> of $IP(y; \mathcal{L}, x)$ if $f_* \in IP(y; \mathcal{L}, x)$ and $\|f_*\|_{\mathcal{L}} \leqslant \|f\|_{\mathcal{L}}$ for each $f \in IP(y; \mathcal{L}, x)$. The object in this paper is to consider the existence of solutions and optimal solutions of $IP(y; \mathcal{L}, x)$ for certain spaces $\mathcal{L} = \Lambda^m$ introduced below.

0.1 <u>The Space D.</u> $(D, \|\cdot\|_D)$, $D \subset \omega$, denotes, throughout, a Banach sequence space which is:

(i) <u>solid</u>: that is, whenever $u \in D$ and $v \in \omega$, with $|v_k| \leqslant |u_k|$ $(\forall k \in Z)$, we have $v \in D$ and $\|v\|_D \leqslant \|u\|_D$;

(ii) <u>AK</u>: that is, $\{e^k\}_{k \in Z} \subset D$ and, for each $u \in D$,

$$\lim_{m,n \to +\infty} \|u - (\ldots,0,u_{-m},\ldots,u_n,0,\ldots)\|_D = 0 ;$$

Copyright © 1980 by Academic Press, Inc.
All rights of reproduction in any form reserved.
ISBN: 0-12-171050-5

(iii) <u>with bounded shift operators</u>: that is, $\exists K$, $0<K<+\infty$, such that for each $u \in D$, $\|(u_{k+1})_{k\in Z}\|_D \leqslant K\|u\|_D$, $\|(u_{k-1})_{k\in Z}\|_D \leqslant K\|u\|_D$.

0.2 <u>The α-dual of a Sequence Space</u>. If $(E, \|\cdot\|_E)$ is a normed sequence space, $\{e^k\}_{k\in Z} \subset E \in \omega$, then its <u>$\alpha$-dual</u> is a Banach space defined by

$$E^\alpha := \{v \mid v \in \omega,\ \|v\|_{E^\alpha} := \sup_{\|u\|_E = 1} \sum_{k\in Z} |u_k v_k| < +\infty\}$$

0.21 <u>Example</u>. Suppose $1 \leqslant p \leqslant +\infty$ and $E = \ell_p$. Then ℓ_p is solid, AK (except for $p = +\infty$), has bounded shift operators, and $(\ell_p)^\alpha = \ell_q$, where $p^{-1} + q^{-1} = 1$.

0.22 <u>Example</u>. Suppose $E = c^o := \{u \mid \lim_{|k|\to\infty} u_k = 0\}$, $\|u\|_{c^o} := \|u\|_{\ell_\infty}$. Then c^o is solid, AK, has bounded shift operators, and $(c^o)^\alpha = \ell_1$.

0.23 <u>Example</u>. If E is a normed sequence space and $\rho = (\rho_k)_{k\in Z}$ is a fixed sequence of positive real numbers, the <u>weighted space</u> E_ρ is defined by $u \in E_\rho \Leftrightarrow (\rho_k u_k) \in E$; $\|u\|_{E_\rho} := \|(\rho_k u_k)_{k\in Z}\|_E$. Then $(E_\rho)^\alpha = (E^\alpha)_{\rho^{-1}}$, where $\rho^{-1} := (\rho_k^{-1})_{k\in Z}$. Also if E is solid then so is E_ρ; and if E is AK then so is E_ρ. But if E has bounded shift operators, we need to impose the condition

$$0 < m \leqslant \rho_{k+1}/\rho_k \leqslant M < +\infty \quad (\forall k \in Z)$$

to ensure that E_ρ should have the same property.

0.3 <u>The Sequence Space</u> $E_{m,p,x}$. If $(E, \|\cdot\|_E)$ is a normed sequence space and $\sigma_k := \frac{1}{m}(x_{k+m} - x_k)$, $\sigma^{1/p} := (\sigma_k^{1/p})_{k\in Z}$, where $m \in Z^+$ and $1 \leqslant p \leqslant +\infty$, we denote by $E_{m,p,x}$ the weighted space $E_{\sigma^{1/p}}$.

0.4 <u>The Function Space</u> $EL_{p,x}$. Suppose $(E, \|\cdot\|_E)$ is a normed sequence space and $1 \leqslant p \leqslant +\infty$. Define

$$EL_{p,x} := \{ f \mid f: R \to C,\ (\|f\|_{L_p(x_k, x_{k+1}]})_{k\in Z} \in E \},$$

$$\|f\|_{EL_{p,x}} := \|(\|f\|_{L_p(x_k, x_{k+1}]})_{k\in Z}\|_E .$$

0.5 <u>The Function Space</u> $EL_{(p),x}$. This is a weighted form of the space $EL_{p,x}$ in which, given $(E, \|\cdot\|_E)$ and $1 \leq p \leq +\infty$, we write

$$EL_{(p),x} := \{f \mid f:R\to C,\ ((x_{k+1}-x_k)^{1-1/p}\|f\|_{L_p(x_k,x_{k+1}]})_{k\in Z} \in E\},$$

$$\|f\|_{EL_{(p),x}} := \|((x_{k+1}-x_k)^{1-1/p}\|f\|_{L_p(x_k,x_{k+1}]})_{k\in Z}\|_E .$$

0.6 <u>The Function Space</u> Λ^m. Suppose Λ is a linear space of functions from R into C, and $m \in Z^+$. Then we denote by Λ^m the space of functions with m-th derivative in Λ, namely

$$\Lambda^m := \{f \mid f^{(m-1)} \in AC(R),\ f^{(m)} \in \Lambda\} .$$

If a semi-norm $\|\cdot\|_\Lambda$ is defined on Λ, we denote $\|f\|_{\Lambda^m} := \|f^{(m)}\|_\Lambda$.

1. THEOREM 1

<u>Let</u> D <u>be a sequence space with the properties in 0.1, and</u> $E = D^\alpha$. <u>For a sequence</u> $y \in \omega$ <u>and a fixed</u> $m \in Z^+$, <u>let</u> $z = (z_k)_{k\in Z}$ <u>denote its sequence of divided differences</u> $z_k := y[x_k,\ldots,x_{k+m}]$.

(i) <u>Let</u> $1 < p \leq +\infty$. <u>With the notation of 0.3,0.4,0.6, we have</u>

$$IP(y;(EL_{p,x})^m, x) \neq \emptyset \iff z \in E_{m,p,x} .$$

(ii) <u>Let</u> $1 < p \leq +\infty$. <u>If</u> $z \in E_{m,p,x}$ <u>then</u> $IP(y;(EL_{p,x})^m, x)$ <u>has an optimal solution, and</u> $\exists K = K_m < +\infty$ <u>such that each optimal solution</u> f_* <u>satisfies</u>

$$\|f_*^{(m)}\|_{EL_{p,x}} \leq K\|z\|_{E_{m,p,x}} .$$

(iii) <u>Let</u> $1 < p < +\infty$ <u>and</u> E <u>be strictly convex</u>. <u>If</u> $z \in E_{m,p,x}$ <u>then</u> $IP(y;(EL_{p,x})^m, x)$ <u>has exactly one optimal solution</u>.

1.1 <u>Example</u>. For the same p as in Theorem 1, and $p^{-1}+q^{-1}=1$, choose $D = \ell_q$. Then $E = D^\alpha = \ell_p$ and $EL_{p,x} = L_p(R) = L_p$. The problem $IP(y;\ L_p^m, x)$ has been extensively considered. See, for example, de Boor (1976) and Schoenberg (1969).

1.2 <u>Example</u>. Choose $D = c^0$ in Theorem 1. Here

$$EL_{p,x} = \{g \mid g:R\to C,\ \|g\| := \sum_{k\in Z} \|g\|_{L_p(x_k,x_{k+1}]} < +\infty\}.$$

In the cardinal case $x_k = k$, and for $p = +\infty$, this is the space $L_{[1]}$ of Schoenberg (1969), who considers $CIP(y;\ L_{[1]}^m)$.

1.3 Example. Suppose $1 < p < +\infty$, $p^{-1} + q^{-1} = 1$. Let $\rho(\cdot) \in C(R)$, $\rho(t) > 0$ $(t \in R)$, and $\max\limits_{[x_k, x_{k+1}]} \rho(t) \,/ \min\limits_{[x_k, x_{k+1}]} \rho(t) \leqslant M < +\infty$.

If $D := \{u \mid u \in \omega, \|u\|_D := (\sum\limits_{k \in Z} [\rho(x_k)]^{1-q} |u_k|^q)^{1/q} < +\infty\}$,

$$\Lambda_p := \{g \mid g: R \to C,\ \|g\|_{\Lambda_p} := (\int_R \rho(t) |g(t)|^p \, dt)^{1/p} < +\infty\},$$

and if $E = D^\alpha$, it is easy to verify that $EL_{p,x}$ and Λ_p are isomorphic Banach spaces. Hence, by Theorem 1, with

$z_k = y[x_k, \ldots, x_{k+m}]$ $(\forall k \in Z)$,

$$IP(y;\ \Lambda_p^m,\ x) \neq \emptyset \iff IP(y; (EL_{p,x})^m,\ x) \neq \emptyset$$

$$\iff \sum_{k \in Z} (x_{k+m} - x_k)\, \rho(x_k)\, |z_k|^p < +\infty .$$

1.4 Example. In Example 1.3 when $x_k = k$, we may choose $\rho(t) = e^t$, $\rho(t) = e^{-t}$, $\rho(t) = (t^2 + 1)^\delta$ $(\delta \in R)$, $\rho(t) = \ln(2 + |t|)$.

2. THEOREM 2

Let the initial hypotheses and notation of Theorem 1 hold, and let $1 < p \leqslant +\infty$. Then, with the notation of 0.3, 0.5, 0.6, we have

(i) $$IP(y; (EL_{(p),x})^m,\ x) \neq \emptyset \iff z \in E_{m,1,x} ;$$

(ii) if $z \in E_{m,1,x}$ then $IP(y; (EL_{(p),x})^m,\ x)$ has an optimal solution, and $\exists K = K_m < +\infty$ such that each optimal solution f_* satisfies

$$\|f_*^{(m)}\|_{EL_{(p),x}} \leqslant K \|z\|_{E_{m,1,x}} .$$

2.1 Example. Choose $D = c^o$, $E = (c^o)^\alpha = \ell_1$. The solution of the interpolation problem corresponding to this choice in Theorem 2 was given by Jakimovski and Russell (1980), Theorem 8(b).

2.2 Example. Choose, in Theorem 2, $D = \ell_1$, $E = (\ell_1)^\alpha = \ell_\infty$. This gives conditions for a solution of $IP(y;\ \Lambda^m,\ x)$, where

$$\Lambda := \{g \mid g: R \to C,\ \|g\|_\Lambda := \sup_{k \in Z} (x_{k+1} - x_k)^{1-1/p} \|g\|_{L_p(x_k, x_{k+1}]} < +\infty\} .$$

3. AN AUXILIARY RESULT

Let $\mathcal{S}_{m,x}$ denote the linear space of all spline functions of degree $m-1$ with simple knots at x, and let $M_{k,m}(t;\, x)$, $N_{k,m}(t;\, x)$ $(k \in Z,\ t \in R)$, be respectively the B-splines, and normalized B-splines (e.g., see de Boor (1976), Jakimovski and Russell (1980)). The following result, of independent interest, is used in the proof of Theorems 1 and 2.

THEOREM 3. Let D be a sequence space with the properties in 0.1. If $a \in \omega$ and $S(\cdot) \in \mathcal{S}_{m,x}$ are related by $S(t) = \sum_{k \in Z} a_k N_{k,m}(t;\, x)$ then for all q, $1 \leqslant q \leqslant +\infty$,

$$S(\cdot) \in \mathcal{S}_{m,x} \cap DL_{q,x} \Leftrightarrow a \in D_{m,q,x} .$$

3.1 Corollary $(q = +\infty)$. $\quad S(\cdot) \in \mathcal{S}_{m,x} \cap DL_{\infty,x} \Leftrightarrow a \in D$.

4. THE METHOD OF PROOF OF THEOREMS 1 AND 2

4.1 Let Γ be a Banach space of functions from R to C such that

(i) $\quad N_{k,m}(\cdot;\, x) = \sigma_k M_k(\cdot;\, x) \in \Gamma$, where $\sigma_k := \frac{1}{m}(x_{k+m} - x_k)$;

(ii) $\quad \mathcal{S}_{m,x} \cap \Gamma$ is a closed subspace of Γ ;

(iii) $\quad (N_{k,m}(\cdot;\, x))_{k \in Z}$ is a Schauder basis for $\mathcal{S}_{m,x} \cap \Gamma$.

4.2 Suppose the functional dual Γ^* of Γ is isomorphic to a Banach space Λ of functions from R to C such that, for each $g \in \Lambda$,

$$F(f) := \int_R gf \quad (\forall f \in \Gamma) \quad \text{satisfies} \quad F \in \Gamma^* ,$$

and that each $F \in \Gamma^*$ is of this form, with $\|F\| = \|g\|_\Lambda$.

4.3 Let $z_k := y[x_k, \ldots, x_{k+m}]$. Then

$$IP(y;\, \Lambda^m,\, x) \neq \emptyset \Leftrightarrow \exists f \in \Lambda^m,\ y_k = f(x_k) \qquad (\forall k \in Z)$$

(i) $$\Leftrightarrow \exists g \in \Lambda,\ z_k = \frac{1}{m!} \int_R g\, M_{k,m}(\cdot;\, x) \qquad (\forall k \in Z)$$

(ii) $$\Leftrightarrow \exists F \in \Gamma^*,\ z_k = F(M_{k,m}(\cdot;\, x)) \qquad (\forall k \in Z)$$

(iii) $$\Leftrightarrow \exists F \in (\mathcal{S}_{m,x} \cap \Gamma)^*,\ \sigma_k z_k = F(N_{k,m}(\cdot;\, x)) \quad (\forall k \in Z).$$

Implication (i) comes from Peano's formula

$$f[x_k,\ldots,x_{k+m}] = \frac{1}{m!}\int_R f^{(m)}\, M_{k,m}(\cdot;\, x)\ ;$$

implication (ii) from 4.2 and 4.1(i); and (iii) from 4.1(ii) with the use of the Hahn-Banach theorem in one direction. Also, by 4.1(iii), we have

(iv) $F \in (\mathcal{S}_{m,x} \cap \Gamma)^*$ is uniquely determined by the sequence z .

4.4 Suppose that $\mathcal{S}_{m,x} \cap \Gamma$ is a Banach space, isomorphic under some mapping to a solid AK sequence space μ , and that the element in μ corresponding to $N_{k,m}(\cdot;\, x)$ in $\mathcal{S}_{m,x} \cap \Gamma$ is e^k. Then

$$\exists F \in (\mathcal{S}_{m,x} \cap \Gamma)^*,\ \sigma_k z_k = F(N_{k,m}(\cdot;\, x)) \quad (\forall k \in Z) \quad \Leftrightarrow$$

$$\Leftrightarrow \exists F_1 \in \mu^*,\quad \sigma_k z_k = F_1(e^k)\ (\forall k \in Z)$$

$$\Leftrightarrow \quad (\sigma_k z_k)_{k\in Z} \in \mu^\alpha\ .$$

4.5 By taking D to be a sequence space satisfying 0.1, $\Gamma = DL_{q,x}$ $(1 \leqslant q < +\infty)$, and using Theorem 3, the hypotheses of 4.4 hold with $\mu = D_{m,q,x} = D_{\sigma^{1/q}}$ (in the notation of 0.3). Writing $E = D^\alpha$ we then have $\mu^\alpha = (D_{\sigma^{1/q}})^\alpha = (D^\alpha)_{\sigma^{-1/q}} = E_{\sigma^{-1/q}}$, and so

$$(\sigma_k z_k)_{k\in Z} \in \mu^\alpha = E_{\sigma^{-1/q}} \Leftrightarrow z \in E_{\sigma^{1/p}} \quad (p^{-1} + q^{-1} = 1)\ .$$

Since the choice $\Gamma = DL_{q,x}$ gives $\Lambda = D^\alpha L_{p,x} = EL_{p,x}$, Theorem 1(i) follows. An alternative choice of Γ gives Theorem 2(i).

REFERENCES

de Boor, C. (1976). Splines as linear combinations of B-splines: a survey. Approximation Theory II (Symposium Proceedings, Texas 1976, ed. G.G.Lorentz, C.K.Chui, L.L.Schumaker), pp.1-47. Academic Press, New York/London.

Jakimovski, A., and Russell, D.C. (1980). On an interpolation problem and spline functions. General Inequalities II (Conference Proceedings, Oberwolfach 1978, ed. E.F.Beckenbach). Birkhäuser Verlag, Basel/Stuttgart.

Schoenberg, I.J. (1969). Cardinal interpolation and spline functions. J. Approx. Theory 2, 167-206.

SOME IDENTITIES FOR B-SPLINES WITH MULTIPLE KNOTS, AND APPLICATIONS TO INTERPOLATION PROBLEMS

Amnon Jakimovski[1]
Department of Mathematical Sciences
Tel-Aviv University
Tel-Aviv, Israel

Michael Stieglitz
Mathematisches Institut I der
Universität Karlsruhe
Karlsruhe, Bundesrepublik Deutschland

I. INTRODUCTION

After having defined the divided difference of a complex sequence $y \equiv (y_j)$ $(j \in Z)$ with respect to a fixed doubly-infinite increasing sequence $x \equiv (x_j)$ $(j \in Z)$ of (possibly multiple) knots in such a manner that it is consistent with the definition of that of a function defined at the points $(x_j)_{j \in Z}$ we consider Spline functions of order m with respect to x. Some identities involving Spline functions, B-splines and truncated

[1]Supported by the Israel Commission for Basic Research

Copyright © 1980 by Academic Press, Inc.
All rights of reproduction in any form reserved.
ISBN: 0-12-171050-5

power functions are deduced. Applications will be made to interpolation problems with multiple knots which in the case of simple knots were treated before by de Boor (1) Schoenberg (4), Subbotin (5) and Jakimovski and Russell (2). No proofs are given here.

II. DIVIDED DIFFERENCES, SPLINE FUNCTIONS AND TRUNCATED POWER FUNCTIONS

Suppose $m \in N$. Let $x \equiv (x_j)$ $(j \in Z)$ be a fixed sequence of knots satisfying $x_j \to \pm\infty$ as $j \to \pm\infty$, $x_j \leqslant x_{j+1}$ and $x_j < x_{j+m} (\forall j \in Z)$. Denote for each $i \in Z$ $i^- := \min\{j : x_j = x_i\}$ and $i^+ := \max\{j : x_j = x_i\}$. Given $k \in Z$ the relative left and right multiplicities and the total relative multiplicity of i with respect to k are, respectively, $[i]_k^- := \min(i+1-i^-, i+1-k)$, $[i]_k^+ := \min(i^+ + 1 - i, k+m+1-i)$ and $\rho_{ki} := [i]_k^+ + [i]_k^- - 1$. The divided difference of a complex sequence $y \equiv (y_i)$ $(i \in Z)$ with respect to x is defined by

$$(1) \qquad y[x_k, \ldots, x_{k+m}] := \det(b_{io}, a_{i1}, \ldots, a_{im})_{k \leqslant i \leqslant k+m} / \\ / m! \cdot \det(a_{io}, a_{i1}, \ldots, a_{im})_{k \leqslant i \leqslant k+m}$$

where $b_{i,o} := y_{i^- + [i]_k^- - 1}$ and $a_{ir} := x_i^{m+1-r-[i]_k^-} / (m+1-r-[i]_k^-)!$ $(o \leqslant r \leqslant m)$. We may express the divided difference in the following form

$$y[x_k, \ldots, x_{k+m}] := \sum_{i=k}^{k+m} \{([i]_k^+ - 1)!([i]_k^+ - 1)!\}^{-1}.$$

$$\cdot\left(\frac{d}{dz}\right)^{[i]_k^+-1}\left(\frac{(z-x_i)^{\rho_{ki}}}{(z-x_k)\cdots(z-x_{k+m})}\right)\Bigg|_{z=x_i}\cdot y_{i^-+[i]_k^--1}$$

The definition (1) of the divided difference $y[x_k,\ldots,x_{k+m}]$ of a numerical sequence y is consistent with the ordinary definition of a divided difference $F[x_k,\ldots,x_{k+m}]$ of a function $F:R^1\to\mathbb{C}$ when $y_i:=F^{(i-i^-)}(x_i)$ $(\forall i\in Z)$. When $F^{(m-1)}\in AC(R^1)$ we have by Peano's theorem $F[x_k,\ldots,x_{k+m}]= =\{(m-1)!(x_{k+m}-x_k)\}^{-1}\int_{R^1}F^{(m)}(t)N_{k,m}(t)dt$ $(\forall k\in Z)$, where the so called normalized B-splines $N_{k,m}(\cdot;x)$ are defined by $N_{k,m}(t;x):=(x_{k+m}-x_k)(\cdot-t)_+^{m-1}[x_k,\ldots,x_{k+m}]$ $(\forall k\in Z, t\in R^1)$. The space $S_{m,x}$ of Spline functions of order m and the knots x is given by

$$S_{m,x}:=\{S(\cdot)\,|\,S:R^1\to\mathbb{C}, S|_{(z_i,z_{i+1}]}\in\pi_{m-1}, S(\cdot)\in C^{m-1-\mu_i}(z_{i-1},z_{i+1})$$

$$(\forall i\in Z)\},$$

were z_i are the different knots of the sequence x and μ_i denotes the multiplicity of z_i, i.e. $\cdots<z_{-1}:=x_{o^--\mu_{-1}}=\ldots= =x_{o^--1}<z_o:=x_{o^-}=\ldots=x_o=\ldots=x_{o^+}\equiv x_{o^-+\mu_o-1}<x_{o^-+\mu_o}=\ldots=x_{o^-+\mu_o+\mu_1-1}:= =z_1<\cdots$. As is well-known there exists a one-to-one correspondence T between the space ω of all double-infinite, complex valued, sequences $\alpha\equiv(\alpha_k)$ $(k\in Z)$ and the space $S_{m,x}$ given by $T(\alpha):=\sum_{k\in Z}\alpha_k N_{k,m}(\cdot;x)$.

The following results can be proved

THEOREM 1. *Let* $S\in S_{m,x}$, $n\in Z$ *and* $S(t)=0$ *for* $t<x_n$, *so that there is a unique* $\alpha\in\omega$, *with* $\alpha_k=0$ *for* $t<x_n$, *such that* $S(t)=\sum_{k=n}^{\infty}\alpha_k N_{k,m}(t;x)$ $(\forall t\in R^1)$. *Let* (z_r) $(r\geq 1)$ *be the sequence of different knots within* (x_k) $(k\geq n)$ *and* (λ_r) $(r\geq 1)$ *their corresponding multiplicities*, i.e. $z_1:=x_n=\ldots=x_{n+\lambda-1}<$

$<z_2:=x_{n+\lambda_1}=\ldots=x_{n+\lambda_1+\lambda_2-1}<\ldots<z_r:=x_{n+\lambda_1+\ldots+\lambda_{r-1}}=\ldots<\ldots$.

Define the sequence $(\beta_r)(r\geq n)$ *by*

$$\alpha_k:=\sum_{r=0}^{\rho(k)}\sum_{j=0}^{\lambda_{r+1}-1}\beta_{n+\lambda_1+\ldots+\lambda_{r+1}-1-j}(-1)^j\left(\binom{m-1}{j}j!\right)^{-1}\cdot$$

$$\cdot\frac{d^j}{dz^j}\prod_{i=1}^{m-1}(x_{k+i}-z)\Big|_{z=z_{r+1}}$$

when $k\geq n$, *where* $\rho(k)$ *is defined by* $n+\lambda_1+\ldots+\lambda_{\rho(k)}\leq k<n+\lambda_1+\ldots+\lambda_{\rho(k)+1}$. *Then* $S(t)=\sum_{r=n}^{\infty}\beta_r(t-x_r)_+^{m-r^+}$ $(\forall t\in R^1\setminus(x_r)_{r\geq n})$.

The above result when x has only simple knots is given in Jakimovski and Russell (2).

THEOREM 2. *For each* $r\in Z$ *and each* j, $0\leq j\leq r^+-r+1$

$$(t-x_r)_+^{m-1-j}=(-1)^j\left[\binom{m-1}{j}j!\right]^{-1}\sum_{k=r}^{\infty}\frac{d^j}{dz^j}\prod_{i=1}^{m-1}(x_{k+i}-z)\Big|_{z=x_r}\cdot$$

$$\cdot N_{k,m}(t,x)$$

for $t\in R^1\setminus(x_r)$ *when* $j=m-1$ *and for* $t\in R^1$ *when* $0\leq j\leq m-2$.

The last result is known when $j=0$ (see Marsden (3)).

By applying the ideas of the proofs in Jakimovski and Russell (2) it is possible to prove, for example, the following results which extend thosed of Jakimovski and Russell (2) to the case when x has multiple knots, too.

THEOREM 3. <u>The map</u> $T:\omega\to S_{m,x}$ <u>is bijective and defines an isomorphism between the following pairs of Banach spaces:</u>

(i) $T:(\ell_{m,p,x}, \|\cdot\|_{m,p,x})\to(S_{m,x}\cap L_p(R^1), \|\cdot\|_{p,R^1}), 1\leq p\leq+\infty;$

(ii) $T:(c^o, \|\cdot\|_\infty) \to(S_{m,x}\cap C^o_{p,x}, \|\cdot\|_{C^o_{p,x}}), 1\leq p\leq+\infty;$

(iii) $T:(c^o, \|\cdot\|_\infty) \to(S_{m,x}\cap C^o, \|\cdot\|_{+\infty,R^1}),$

<u>where</u> $\ell_{m,p,x}:= \{u\in\omega: \|u\|_{m,p,x}:= =(\sum_{k\in Z}(x_{k+m}-x_k)m^{-1}|u_k|^p)^{\frac{1}{p}}<+\infty\}$ $(1\leq p<+\infty)$, $\ell_{m,+\infty,x}:=\ell_{+\infty}:=\{u\in\omega: : \|u\|_{m,+\infty,x}:=\sup_{k\in Z}|u_k|<+\infty\}$, $c^o:=\{u\in\omega:\lim_{|k|\to\infty}u_k=o\}$,

$C^o:=\{f:R^1\to¢, f\in C(R^1), \lim_{|t|\to\infty}f(t)=o\}$, $C^o_{p,x}:=\{f:R^1\to¢, f|_{Z_k}\in L_p(Z_k)$ $(\forall k\in Z)$, $\lim_{|k|\to\infty}|Z_k|^{-1/p}\|f\|_{p,Z_k}=o\}$ $(1\leq p\leq+\infty)$ <u>where</u> $Z_k:= =(z_k,z_{k+1}]$, $|Z_k|:= z_{k+1}-z_k$, <u>and for a set</u> $E\subset R^1$, $\|f\|_{p,E}:= =(\int_E|f|^p)^{1/p}$ $(1\leq p<+\infty)$, $\|f\|_{+\infty,E}:=\text{ess.}\sup_{t\in E}|f(t)|$ <u>and</u> $\|f\|_{C^o_{p,x}}:=\sup_{k\in Z}|Z_k|^{-1/p}\|f\|_{p,Z_k}$.

III. <u>AN INTERPOLATION PROBLEM</u>

Given a sequence $y\in\omega$, the problem of finding a function $F:(a,b)\to¢$ belonging to a preassigned linear space L of function and such that $F(x_i)=y_i$ for every $i\in Z$ is called the

<u>interpolation problem</u> $IP(y;L,x)$. The symbol $IP(y;L,x)$ will also denote the set of all its solutions. If a semi-norm $\|\cdot\|_L$ is defined for all functions in L, then F_* is an <u>optimal solution</u> when $F_* \in IP(y;L,x)$ and $|F_*|_L \leq \|F\|_L$ for all $F \in IP(y;L,x)$. The following result, for example, can be proved.

THEOREM 4. <u>Suppose</u> $L \equiv L_2^m := \{F : F^{(m-1)} \in AC(R^1), F^{(m)} \in L_2(R^1)\}$ <u>and</u> $\|F\|_{L_2^m} := \|F^{(m)}\|_{2,R^1}$. <u>Then:</u> (i) $y \in \ell^m_{2,x} \Longleftrightarrow IP(y;L_2^m,x) \neq \phi$, <u>where</u> $\ell^m_{p,x} := \{y \in \omega : (y[x_k,\ldots,x_{k+m}]_{k \in Z} \in \ell_{m,p,x}\}$ <u>and</u> $\|y\|_{\ell^m_{p,x}} := = \|(y[x_k,\ldots,x_{k+m}])_{k \in Z}\|_{m,p,x}$ $\{1 \leq p < +\infty\}$. (ii) <u>If</u> $y \in \ell^m_{2,x}$ <u>then</u> $IP(y;L_2^m,x)$ <u>has a unique optimal solution</u> S_* <u>which is a Spline function in</u> $S_{2m,x}$. <u>Moreover</u>, S_* <u>is the only solution of</u> $IP(y;L_2^m,x)$ <u>in</u> $S_{2m,x}$. <u>We have</u> $\|S_*^{(m)}\|_{2,R^1} \leq (m!)^2 \cdot (2m^2)^{\frac{1}{2}m(m-1)+1} \cdot \|y\|_{\ell^m_{2,x}}$.

REFERENCES

1. de Boor C., in "Approximation Theory II" (G.G. Lorentz, C.K. Chui and L.L. Schumaker,ed.),p.1. Academic Press,N.Y.(1976).

2. Jakimovski A., and Russell D.C., in "General Inequalities 2" (E.F. Beckenbach,ed.)Birkhäuser-Verlag, Basel (1979).

3. Marsden M.J., J. Approx. Theory 3,7 (1970).

4. Schoenberg I.J., J. Approx. Theory 2,167 (1969).

5. Subbotin Ju.N., Proc. Steklov Inst. Math. 88,30 (1967).

SPECTRAL REPRESENTATION AND APPROXIMATION OF THE GENERALIZED INVERSE OF A CLOSED LINEAR OPERATOR

Virginia V. Jory

School of Mathematics
Georgia Institute of Technology
Atlanta, Georgia

I. INTRODUCTION

Let H_1 and H_2 be Hilbert spaces and let B denote a closed, linear operator mapping a dense subset of H_1 into H_2. Solutions in the sense of generalized inverses are constructed for the operator equation

$$Bx = y. \tag{1}$$

It is well known that in the Hilbert space setting the Moore-Penrose or orthogonal generalized inverse B^+ exists and is the unique solution to the following four equations

$$BB^+B = B \quad , \quad B^+B = 1 - P, \quad \text{on} \quad D(B) \quad ,$$

$$B^+BB^+ = B^+, \quad BB^+ = Q \quad , \quad \text{on } D(B^+) \quad ,$$

where P and Q are the orthogonal projections onto N(B) and $\overline{R(B)}$ respectively [6]. Furthermore, if y is in the domain of B^+, then B^+y is the least squares solution of minimal norm of Equation (1).

Spectral methods associated with the operational calculus [8] may be used to construct representations for B^+. Groetsch has shown, for the bounded operators B, that the operational calculus is an excellent tool for obtaining the convergence

Copyright © 1980 by Academic Press, Inc.
All rights of reproduction in any form reserved.
ISBN: 0-12-171050-5

of approximation schemes to the generalized inverse B^+ [1], [2]. He established the following result.

Theorem 1 (Groetsch [1]). Suppose T is bounded linear, $T : H_1 \to H_2$. Let S be an unbounded subset of $(0,\infty)$. Let $\{U_\beta(t) : \beta \in S\}$ be a family of real valued functions such that each U_β is continuous on $[0,||T||^2]$ and such that there exists $M > 0$ with

$$|tU_\beta(t)| \le M$$

for all t and β and

$$U_\beta(t) \to t^{-1}$$

as $\beta \to \infty$ for each $t \neq 0$. Then

$$\lim_{\beta\to\infty} U_\beta(T^*T)T^*b = T^+b$$

for all $b \in D(T^+)$.

Examples of families $\{U_\beta : \beta \in S\}$ satisfying the assumptions of Theorem 1 are shown by Groetsch [1] to include

(i) $U_\beta(t) = \int_0^\beta e^{-\xi t} d\xi$, for $t \in S = (0,\infty)$,

(ii) $U_\beta(t) = [1+\beta t]^{-1}$, for $t \in S = (0,\infty)$,

(iii) $U_\beta(t) = \alpha \sum_{k=0}^{\beta} [1+\alpha t]^k$, for $\alpha > 0$ and $t \in Z^+$, the positive integers,

(iv) $U_\beta(t) = \sum_{k=0}^{\beta} \frac{1}{k+1} \prod_{j=1}^{k} (1 - \frac{1}{1+j} t)$, for $t \in Z^+$.

In a 1975 paper Lardy [4] utilized the operational calculus for unbounded self adjoint operators to obtain the following result.

Theorem 2 (Lardy [4]). Let A be a densely defined, closed linear operator mapping H_1 to H_2. If y is in $D(A^+)$, then

$$A^+y = \sum_{k=1}^{\infty} A^*(1+AA^*)^{-k}y.$$

Moreover, if $R(A)$ is closed then

$$A^+ = \sum_{k=1}^{\infty} A^*(1+AA^*)^{-k}.$$

II. GENERAL REPRESENTATION FOR B^+

In all that follows, let E_{B^*B} denote the spectral family for B^*B (see for example [6]), where B^* is the Hilbert space adjoint of B, and let S be an unbounded subset of $(0,\infty)$. We obtain two general representation theorems for the generalized inverse B^+.

Theorem 3. Suppose for each β in S, f_β is a continuous function on $[0,\infty]$ such that for $\lambda \neq 0$, $\lim_{\beta\to\infty} f_\beta(\lambda) = \frac{1}{\lambda}$. Suppose there exist a real-valued function h and a number γ such that

a) $\int_0^\infty |h(\lambda)|^2 d||E_{B^*B}(\lambda)x||$ exists for all x in $D(B)$, and

b) if $\beta \geq \gamma$, then $||h_\beta|| \leq ||h||$.

Suppose, finally, that

$$B^*f_\beta(BB^*) = f_\beta(B^*B)B^* \text{ on } D(B^*) \text{ and}$$

$$Bf_\beta(B^*B) = f_\beta(BB^*)B \text{ on } D(B).$$

Then

$$s - \lim_{\beta\to\infty} B^*f_\beta(BB^*) = B^+.$$

The proof of Theorem 3 utilizes results from the operational calculus and the Lebesgue-Fatou Lemma. Details will appear elsewhere.

III. SPECTRAL REPRESENTATION

The next theorem gives a representation on $D(B^*) \cap D(B^+)$ for the generalized inverse B^+ in terms of the spectral family for B^*B.

Theorem 4. Under the hypotheses of Theorem 3, if y is in $D(B^+) \cap D(B^*)$, then

$$B^+y = \lim_{\beta\to\infty} \int_0^\infty f_\beta(\lambda)\,d(E_{B^*B}(\lambda)B^*y).$$

IV. REGULARIZATION ALGORITHMS

In this section we note that families $\{f_\beta : \beta \in S\}$ satisfying the hypotheses of Theorem 3 yield regularization algorithms for the possibly ill-posed problem (1). The definitions and first theorem are from [7].

Definition 1. Equation (1) is said to be *well-posed* in (H_1,H_2) if for each y in H_2, (1) has a unique pseudo-solution $x = B^+y$; otherwise the equation is *ill-posed*.

As the next theorem indicates, if the range of B is not closed, then (1) is an ill-posed problem. (See [5] for examples).

Theorem 5 (Nashed and Wahba [7]). The following statements are equivalent:

(i) Equation (1) is well-posed in (H_1,H_2);

(ii) B has closed range in H_2;

(iii) B^+ is a bounded operator on H_2 into H_1.

Definition 2. Let I be an indexing set. A *regularization algorithm* for (1) is a family of bounded linear operators $\{F_\alpha : \alpha \in I\}$ from H_2 into H_1 such that for every y_0 in $D(B^+)$, there is an element x in the set of all least squares solutions of (1) such that for every $\varepsilon > 0$, there exists $\delta > 0$ and α in I with the property that $|F_\alpha y - x| \le \varepsilon$ if $|y - y_0| \le \delta$.

Theorem 6. Under the hypotheses of Theorem 3, the family of operators $\{B^*f_\beta(BB^* : \beta \in S\}$ is a regularization algorithm for Equation (1).

V. APPLICATION

The following three lemmas are useful in establishing families $\{f_\beta : \beta \in S\}$ which satisfy the hypotheses of Theorem 3.

<u>Lemma 1</u> (see for example [3] or [4]). Let $\lambda > 0$, z be in $D(B^*)$, x be in $D(B)$. Then

(i) $(1+\lambda B^*B)^{-1}B^*z = B^*(1+\lambda BB^*)^{-1}z$ and

(ii) $(1+\lambda BB^*)^{-1}Bx = B(1+\lambda B^*B)^{-1}x$.

<u>Lemma 2</u> [3]. Let z be in $D(B^*)$; let T_{-B^*B}, T_{-BB^*} be the C_0 semigroups generated by $-B^*B$, $-BB^*$ respectively. Then for all $t > 0$,

$$T_{-B^*B}(t)B^*z = B^*T_{-BB^*}(t)z.$$

<u>Lemma 3</u> [3]. Let z be in $D(B^*)$ and $\beta > 0$. Then

$$\frac{1}{\beta}\int_0^\beta \xi T_{-B^*B}(\beta-\xi)B^*z\,d\xi = B^*\frac{1}{\beta}\int_0^\beta \xi T_{-BB^*}(\beta-\xi)z\,d\xi.$$

Examples of families of real-valued functions satisfying the hypotheses of Theorem 3 include the families $\{f_\beta : \beta \in S\}$ defined by

(i) $f_\beta(\lambda) = \frac{1}{\beta}\sum_{j=1}^{\beta}\sum_{i=1}^{j}(1+\lambda)^{-i}$, $\beta \in S = Z^+$,

(ii) $f_\beta(\lambda) = \int_0^\beta e^{-\xi\lambda}d\xi$, $\beta \in S = R^+$,

(iii) $f_\beta(\lambda) = \beta(1+\beta\lambda)^{-1}$, $\beta \in S = R^+$,

(iv) $f_\beta(\lambda) = \sum_{p=1}^{\beta}\lambda_p\sum_{j=1}^{p}(1+\lambda_j\lambda)^{-1}$, $\beta \in S = Z^+$, where $\{\lambda_p\}$ is a positive number sequence such that $\sum\lambda_p = +\infty$.

In case B is bounded, the family in (ii) yields the Showalter integral formula for B^+ and the family in (iii) gives the Tikhonov regularization algorithm of order zero. The family in (iv) includes the result of Lardy as a special case with $\lambda_p = 1$ for all p.

If B has closed range, the convergence in Theorem 3 is uniform. Furthermore for the approximators to B^+ constructed from the families in Examples (i)-(iii), the convergence is of order $\frac{1}{\beta}$. In Example (iv) if $\sum \lambda_p^2 = +\infty$, then the convergence is of order $\left(\sum_{p=1}^{\beta} \lambda_p^2\right)^{-1}$.

Full details will appear elsewhere.

REFERENCES

1. Groetsch, C. W., *Generalized Inverses of Linear Operators: Representation and Approximation*, Dekker, New York (1977).

2. Groetsch, C. W., On rates of convergence for approximations to the generalized inverse, *Numer. Funct. Anal. and Optimiz.*, 1, 195-201(1979).

3. Jory, V. V., Approximators to the generalized inverse B^+ as regularizers of the ill-posed problem $Bx = y$, (to appear).

4. Lardy, L. J., A series representation for the generalized inverse of a closed linear operator, *Atti Acad. Naz. Lincei Rend. Cl. Sci. Fis. Mat. Natur.*, Ser. VIII, 58, 152-156(1975).

5. Lavrentiev, M. M., *Some Improperly Posed Problems of Mathematical Physics*, Springer-Verlag, Berlin, (1967).

6. Nashed, M. Z. and Votruba, G. F., A unified operator theory of generalized inverses, in *Generalized Inverses and Applications,* (M. Z. Nashed, ed.), Academic Press, New York, (1976).

7. Nashed, M. Z., and Wahba, G., Generalized inverses in reproducing kernel spaces: An approach to regularization of linear operator equations, MCR Technical Summary Report #1200, the University of Wisconsin Mathematics Research Center, Madison, Wisconsin, (1972).

8. Yosida, K., *Functional Analysis,* Springer-Verlag, New York, (1978).

DIFFERENTIAL APPROXIMATION OF COMPLETELY MONOTONIC FUNCTIONS

David W. Kammler[1]

Department of Mathematics
Southern Illinois University
Carbondale, Illinois

I. GENERALIZED DIFFERENTIAL APPROXIMATION

We consider the following scheme for numerically finding the parameters a_i, λ_i, $i = 1, \ldots, n$ of an exponential sum

$$Y: = a_1 E_{\lambda_1} + \ldots + a_n E_{\lambda_n} \tag{1}$$

where

$$E_\lambda(t): = \exp(-\lambda t),\ 0 \leq t < \infty,\ 0 \leq \lambda$$

with which to approximate a given completely monotonic function

$$F = \int_0^\infty E_\lambda \, df(\lambda),$$

where $df(\lambda)$ is a non-negative measure on $(0, \infty)$ with at least n points of increase. Using a suitable semi-definite bilinear form $\langle , \rangle$ and basis functions $U_1, \ldots, U_n, V_1, \ldots, V_n$ we:

---

[1]Supported in part by AFOSR Grant 74-2653

Copyright © 1980 by Academic Press, Inc.
All rights of reproduction in any form reserved.
ISBN: 0-12-171050-5

(i) Solve the system of linear equations

$$\langle (D^n + b_1 D^{n-1} + \dots + b_n)F,\ U_i \rangle = 0,\ i = 1, \dots, n$$

to find the auxiliary parameters $b_1, \dots, b_n$. (Here $D = d/dt$ is the differential operator.)

(ii) Effect the factorization

$$\lambda^n + b_1 \lambda^{n-1} + \dots + b_n = (\lambda + \lambda_1) \dots (\lambda + \lambda_n)$$

to obtain the exponents $\lambda_1, \dots, \lambda_n$.

(iii) Solve the system of linear equations

$$\langle F - a_1 E_{\lambda_1} - \dots - a_n E_{\lambda_n},\ V_i \rangle = 0,\ i = 1, \dots, n$$

to obtain the coefficients $a_1, \dots, a_n$.

In addition to the hypothesis that F is completely monotonic, we shall naturally assume that the functions $D^i F$, $i = 0, 1, \dots, n$ and U_i, V_i, E_{λ_i}, $i = 1, \dots, n$ are all finite with respect to the seminorm induced by $\langle , \rangle$. We shall further assume that each of the two sets of transforms

$$\mathcal{U}_i(\lambda) := \langle E_\lambda, U_i \rangle,\ i = 1, \dots, n$$

$$\mathcal{V}_i(\lambda) := \langle E_\lambda, V_i \rangle,\ i = 1, \dots, n$$

satisfies the Haar condition on $(0, \infty)$. We then obtain the following:

<u>Theorem 1</u>. A unique exponential sum Y of the form (1) with $0 < \lambda_1 < \dots < \lambda_n$ is determined by (i), (ii), and (iii). If F is itself of the form (1), then $Y \equiv F$. If $Y \not\equiv F$, the parameters $\lambda_1, \dots, \lambda_n$ interlace the support of $df(\lambda)$.

II. SOME EXAMPLES

The following choices of the bilinear form and basis functions fit within the above framework.

Example 1. $\langle G, H\rangle : = \int_0^b G(t)H(t)\, dt,\ 0 < b \leq \infty$

$U_i : = D^{i-1}F,\ i = 1, \ldots, n$

$V_i : = E_{\lambda_i}$ with $\lambda_1 < \ldots < \lambda_n$ obtained from (ii).

This choice results in the well known differential approximation scheme of R. Bellman (1). The proof that $U_1, \ldots, U_n$ satisfy the Haar condition is based on a suitable extension of Lemma 1 from (2).

Example 2. $\langle G, H\rangle : = \sum_{k=1}^{N} w_k G(t_k)H(t_k),\ w_k > 0,\ N \geq 2n$, and $0 \leq t_1 < \ldots < t_N$

U_i, V_i as in Example 1.

This choice results from one natural discretization of Bellman's scheme.

Example 3. $\langle G, H\rangle : = \sum_{k=1}^{N} w_k G^{(k-1)}(t_0)\ H^{(k-1)}(t_0),\ w_k > 0,\ N \geq 2n,$ $0 \leq t_0 < \infty$

U_i, V_i as in Example 1.

This choice results in a good agreement of Y with F in a neighborhood of the point t_0.

Example 4. $\langle G, H\rangle := \int_0^\infty G(t)H(t)\,dt$

$$U_i = V_i := E_{\mu_i} \text{ where } 0 < \mu_1 < \dots < \mu_n.$$

This choice results in a Galerkin rather than a least squares type approximation in step (i) of the algorithm.

III. THE SPECIAL CASE OF EXAMPLE 4

We now focus on Example 4 where

$$\langle F, E_\lambda\rangle = \int_0^\infty F(t)\, e^{-\lambda t}\, dt =: \mathcal{F}(\lambda)$$

is the ordinary Laplace transform of F so that the matrix elements

$$\langle D^k F, E_{\mu_i}\rangle = \mu_i^k\, \mathcal{F}(\mu_i) - \mu_i^{k-1} F(0) - \dots - F^{(k-1)}(0)$$

needed in (i) and the matrix elements

$$\langle E_{\lambda_j}, E_{\mu_i}\rangle = 1/(\lambda_j + \mu_i),\ i,j = 1, \dots, n$$

needed in (iii) are easy to evaluate. Within this context we obtain:

Theorem 2. The coefficient $a_1, \dots, a_n$ produced in (iii) are positive so that Y is completely monotonic.

(Whether or not this is the case with Examples 1, 2, 3 is not known.)

Although any choice $0 < \mu_1 < \dots < \mu_n$ will serve to determine a_i, λ_i, $i = 1, \dots, n$, it would seem advantageous to somehow use the underlying function F to determine reasonable values for these parameters. In practice, we have found that a particularly satisfying choice can be obtained by simply iterating steps (i) and (ii) of the algorithm, i.e., we arbitrarily select $0 < \lambda_{10} < \dots < \lambda_{n0}$ and for $\nu = 1, 2, \dots$ determine $0 < \lambda_{1\nu} < \dots < \lambda_{n\nu}$ such that

$$\langle [(D + \lambda_{1\nu}) \dots (D + \lambda_{n\nu})] F, E_{\lambda_{i,\nu-1}} \rangle = 0, \; i = 1, \dots, n.$$

Numerical experimentation suggests that the resulting sequence rapidly converges to yield a unique set of exponents, but we have only succeeded in proving:

Theorem 3. There exist exponents $0 < \lambda_1 < \dots < \lambda_n$ such that

$$\langle [(D + \lambda_1) \dots (D + \lambda_n)] F, E_{\lambda_i} \rangle = 0, \; i = 1, \dots, n.$$

Usually 3-4 iterations suffice to determine each of the exponents $\lambda_1, \dots, \lambda_n$ to within a few percent, and a few dozen iterations suffice to reach the point where the relative error in the exponents is approximately Cu where C is the condition number of the $n \times n$ matrix from (i) and u is the machine epsilon.

A more complete discussion of the above which includes numerical results and extensions to the cases where the differential operator D of (i) is replaced by the shift operator

$$E\,F(t) = F(t + h), \; h > 0,$$

as used in variations of Prony's method or by the integral

operator

$$S\,F(t) = \int_0^t F(\tau)\,d\tau$$

as used by W. B. Gearhart will appear elsewhere.

REFERENCES

1. Bellman, R., Methods of Nonlinear Analysis I, Academic Press, 1970.
2. Kammler, D. W., SIAM J. Numer. Anal 16, 801 (1979).

RESTRICTED - DENOMINATOR RATIONAL APPROXIMATION

Edwin H. Kaufman, Jr.

Department of Mathematics
Central Michigan University
Mount Pleasant, Michigan

Gerald D. Taylor[1]

Department of Mathematics
Colorado State University
Fort Collins, Colorado

[1]Supported in part by the Air Force Office of Scientific Research, Air Force System Command, USAF, under grant AFOSR-76-2878C and by the National Science Foundation under grant MCS-78-05847.

Copyright © 1980 by Academic Press, Inc.
All rights of reproduction in any form reserved.
ISBN: 0-12-171050-5

I. INTRODUCTION

In this paper we consider the uniform approximation of continuous functions by generalized rational functions whose denominators are bounded away from zero and have bounded coefficients. Proofs of theorems and further references and examples can be found in (2). C.B. Dunham (preprint) has independently considered the similar situation where the denominators are bounded above and below.

There are several reasons for considering such denominator restrictions rather than the weaker standard restriction that the denominator merely be positive. For one thing, best approximations will always exist. From a practical standpoint, keeping the denominator away from zero avoids some numerical difficulties in both finding and using good approximations. The lower bound on the denominator may also have some physical interpretation; for example, if one is approximating the magnitude squared response of a digital filter, the lower bound is related to the feedback gain of the filter. M.T. McCallig (Honeywell Electro-Optics; preprint) has considered the restricted-denominator filter problem in more detail and has done some computations.

Formally, we let X be a compact metric space, let $\mathbb{P}$ and $\mathbb{Q}$ be subspaces of C[X] with bases $\{\theta_1, \cdots, \theta_m\}$ and $\{\psi_1, \cdots, \psi_n\}$ respectively, and let L be a continuous strictly positive function on X satisfying $\max_{x\varepsilon X} L(x) < \min_{x\varepsilon X} \psi_1(x)$ (this last assumption, which is normally satisfied in practice, is needed to make the theorems go through). Our approximating family is then

$$\mathbb{R}_L = \{P/Q : P = p_1\,\theta_1 + \cdots + p_m\,\theta_m,\ Q = q_1\,\psi_1 + \cdots + q_n\,\psi_n,\ Q \geq L \text{ on } X,\ |q_j| \leq 1 \text{ For } j = 1, \cdots, n\}.$$

Given $f \in C[X]$, we wish to choose $R^* \in \mathbb{R}_L$ to minimize $||f-R^*|| = \max_{x \in X} |f(x) - R^*(x)|$.

We will need some notation. Given any $f \in C[X]$ and any $R^* = P^*/Q^* = (p_1^*\,\theta_1 + \cdots + p_m^*\,\theta_m)/(q_1^*\,\psi_1 + \cdots + q_n^*\,\psi_n)$, we define

$$\sigma(x) = \operatorname{sgn}(f(x) - R^*(x))\ \forall x \in X,$$

$$\mathbb{P} + R^*\mathbb{Q} = \{P + R^*\,Q : P \in \mathbb{P},\ Q \in \mathbb{Q}\},$$

$$X_0 = \{x \in X : |f(x) - R^*(x)| = ||f - R^*||\},$$

$$Y_0 = \{x \in X : Q^*(x) = L(x)\},$$

$$I_0 = \{j \in \{1, \cdots, n\} : |q_j^*| = 1\},$$

$$S = \{\sigma(x)\ [\theta_1(x), \cdots, \theta_m(x), R^*(x)\psi_1(x), \cdots, R^*(x)\psi_n(x)]^T : x \in X_0\} \cup \{[0, \cdots, 0, -\psi_1(x), \cdots, -\psi_n(x)]^T : x \in Y_0\} \cup \{[0, \cdots, 0, 0, \cdots, q_j^*, \cdots, 0] : j \in I_0, \text{ with the } q_j^* \text{ in position } m + j\},$$

$$H(S) = \text{convex hull of } S = \{\sum_{i=1}^{k} \lambda_i\, s_i : k \text{ is a positive integer, } s_i \in S,\ \lambda_i \geq 0,\ \sum_{i=1}^{k} \lambda_i = 1\},$$

$$\text{int } H(S) = \text{the interior of } H(S).$$

II. THEORETICAL RESULTS

The following two theorems and example concern characterization of best approximations.

Theorem 1. Suppose $f \in C[X] - \mathbb{R}_L$. Then $R^* = P^*/Q^* \in \mathbb{R}_L$ is a best approximation to f iff $\not\exists$ $\overline{P} \in \mathbb{P}$, $\overline{Q} \in \mathbb{Q}$ satisfying

(i) $\sigma(x)\ (\overline{P} + R^*\ \overline{Q})(x) > 0\ \forall x \in X_0$,

(ii) $\overline{Q}(x) < 0\ \forall x \in Y_0$,

(iii) $\overline{q}_j > 0$ if $q_j^* = 1$,

(iv) $\overline{q}_j < 0$ if $q_j^* = -1$.

Theorem 2. Suppose $f \in C[X] - \mathbb{R}_L$. Then $R^* = P^*/Q^* \in \mathbb{R}_L$ is a best approximation to f iff $\overline{0} \in H(S)$, where $\overline{0}$ is the origin of $(m + n)$-space.

Example 1. Let $X = \{0,1\}$, $f(x) = x$, $\mathbb{P} = \pi_0$ = the set of all polynomials of degree ≤ 0, $\mathbb{Q} = \pi_1$, $L(x) \equiv .1$, $R^*(x) = \frac{1}{11}/(1 - .9x)$. We have $X_0 = \{0,1\}$, $Y_0 = \{1\}$, $I_0 = \{1\}$, $\sigma(0) = -1$, $\sigma(1) = 1$. Thus

$$\sigma(x)\begin{bmatrix}\theta_1(x)\\(R^*\psi_1)(x)\\(R^*\psi_2)(x)\end{bmatrix} = \sigma(x)\begin{bmatrix}1\\\frac{1}{11}/(1 - .9x)\\\frac{x}{11}/(1 - .9x)\end{bmatrix}, \quad S = \left\{\begin{bmatrix}-1\\-\frac{1}{11}\\0\end{bmatrix}, \begin{bmatrix}1\\\frac{10}{11}\\\frac{10}{11}\end{bmatrix}, \begin{bmatrix}0\\-1\\-1\end{bmatrix}, \begin{bmatrix}0\\1\\0\end{bmatrix}\right\}.$$

We have $\overline{0} \in H(S)$ (the coefficients which show this are $\lambda_1 = \frac{1}{3}$, $\lambda_2 = \frac{1}{3}$, $\lambda_3 = \frac{10}{33}$, $\lambda_4 = \frac{1}{33}$), so R^* is a best approximation.

If in Example 1 $\mathcal{Q}$ is taken to be π_{1980} instead of π_1, then similar arguments can be used to show that the given R* is still a best approximation, but it is not unique (e.g. $\frac{1}{11}/(1 - .9x^2)$ is also a best approximation). It turns out that although $\overline{0} \in H(S)$ is not sufficient for uniqueness of best approximations, $\overline{0} \in \text{int } H(S)$ is actually sufficient for strong uniqueness.

Theorem 3. Suppose $f \in C[X] - \mathbb{R}_L$ and $R^* = P^*/Q^* \in \mathbb{R}_L$. If $\overline{0} \in \text{int } H(S)$, then there is a constant $\gamma > 0$ such that for any $R \in \mathbb{R}_L$, $||f - R|| \geq ||f - R^*|| + \gamma\, ||R - R^*||$.

III. COMPUTATION OF BEST APPROXIMATIONS

The differential correction algorithm (e.g. see (1)) can be modified to compute approximations from $\mathbb{R}_L$ by inserting extra constraints to force $Q \geq L$. We have

Algorithm (Restricted-denominator differential correction - RDDC)

(i) Choose $P_0/Q_0 \in \mathbb{R}_L$;

(ii) Having found $P_k/Q_k \in \mathbb{R}_L$ with $||f - R_k|| = \Delta_k$, choose P_{k+1}, Q_{k+1} as a solution to the problem

$$\text{minimize} \max_{x \in X} \frac{|f(x)Q(x) - P(x)| - \Delta_k Q(x)}{Q_k(x)}$$

subject to $|q_j| \leq 1\ \forall j$, and $Q(x) \geq L(x)\ \forall x \in X$;

(iii) Continue until some stopping criterion is met.

Step (ii) can be done by linear programming if X is finite. The following theorem shows that the algorithm has some desirable convergence properties; an example using this algorithm can be found in (2).

Theorem 4. The RDDC algorithm converges monotonically and at least linearly; furthermore, if X contains at least $m + n + 1$ points, $R^* = P^*/Q^*$ is a best approximation to $f \in C[X] - R_L$ such that the space spanned by $\{\theta_1, \cdots, \theta_m, R^*\psi_1, \cdots, R^*\psi_n\}$ is a Haar subspace of dimension $m + n - 1$, and either $Y_0 = \emptyset$ or $\overline{0} \in \text{int } H(S)$, then the convergence is at least quadratic.

We conclude by remarking that one can prove de La Vallée Poussin estimates, and this theory can be extended to include a positive multiplicative weight function and restricted range restrictions on the approximating functions, as well as complete restricted range restrictions on the denominator.

REFERENCES

1. Barrodale, I., Powell, M. J. D., and Roberts, F.D.K., The differential correction algorithm for rational ℓ_∞ approximation, SIAM J. Numer. Anal. 9, 493 (1972).
2. Kaufman, E. H. Jr., and Taylor, G. D., Uniform approximation by rational functions having restricted denominators, J. Approx, Theory, submitted.

SMOOTH FUNCTIONS AND GAUSSIAN PROCESSES

Robert Kaufman

Department of Mathematics
University of Illinois
Urbana, Illinois

0. Let $X(B)$ be a Gaussian process indexed by the Borel subsets $B \subseteq R^k$ of Lebesgue measure $m(B) < \infty$, of mean 0 and covariance $E(X(A)X(B)) \equiv m(A \cap B)$. A family F of subsets is called G.B. (Gaussian-bounded) if $\sup\{|X(B)| : B \in F\} < +\infty$ almost surely; when F is uncountable we replace F by a countable subset of F, dense in F in the L^2-metric. When $p \geq 1$ is a real number, we can study open sets U with boundary ∂U of class C^p, subject to a certain technical boundedness condition [2]. These constitute families F_p and the best result known [1, 2] on the classes F_p is this: F_p is a G.B. set when $p > k-1$, but is not when $p < k-1$.

In this paper we prove the negative result for $p = k-1$; because our sets are much less general than the sets described in [2], we have not reproduced the definition; it is somewhat complicated.

1. To construct open sets of class C^{k-1} in R^k $(k \geq 2)$, we fix once and for all a bounded open subset W contained in the half-space $x_k < 0$, whose boundary is of class C^∞ and contains the rectangle $Q : 0 \leq x_1 \leq 1, \cdots, 0 \leq x_{k-1} \leq 1$ in the plane $x_k = 0$. Let $f(x_1, \cdots, x_{k-1}) \geq 0$ be defined on Q and vanish on its edges. The region $W(f) = W \cup 0 \leq x_k < f(x_1, \cdots, x_{k-1}\}$ has the same smoothness as f, and the set $0 \leq x_k < f(x_1, \cdots, x_{k-1})$ is the difference $W(f) - W$. To obtain a negative result on classes of sets, it will be sufficient to

Copyright © 1980 by Academic Press, Inc.
All rights of reproduction in any form reserved.
ISBN: 0-12-171050-5

obtain lower bounds for the random variables $\Gamma(f)$, corresponding to the various sets $0 \leq x_k < f(x_1,\ldots,x_{k-1})$, or the variables $\Gamma(f) - \Gamma(f/2)$.

2. To generate large collections of functions let us consider functions $f_1,\cdots,f_N$ of class $C^k(R^{k-1})$, whose supports are disjoint (or meet only at their boundaries); then any sum $f = \Sigma a_j f_j$ with $0 \leq a \leq 1$, has partial derivatives Df no larger than the maximum of $|Df_j|$. When D has order $k-1$ we can estimate $|Df(x) - Df(x')|$ by the modulus of continuity of the individual terms Df_j. This is clear if x and x' belong to the support of a single function f_j, and in the opposite case we draw the line $\overline{xx'}$ and mark on it the zeroes of Df_j nearest x and x'.

We define a specific function G of class $C^k(R^1)$: set $G(t) = (t-t^2)^{k+1}$ on $(0,1)$ and $g = 0$ outside, and then $g(x) = G(x_1)\cdots G(x_{k-1})$. Let now N be a large integer and let us divide the cube Q into N^{k-1} similar cubes of side N^{-1}, and consider the total measure of those subcubes on which $g(x) \geq N^{1-k}$. This will certainly be true if $G(x_j) \geq N^{-1}$ $(1 \leq j \leq k-1)$, or $(x_j - x_j^2)^{k+1} \geq N^{-1}$ $(1 \leq j \leq k-1)$. Hence the measure of the cubes exceeds $1 - N^{-\delta}$ for a certain $\delta > 0$, depending on k.

3. We divide the cube Q into cubes of side N_1^{-1} and draw on each of these cubes a function similar to g, but with height $a_1 N_1^{1-k}$ $(0 < a_1 < 1)$. At the j-th step we have cubes of side $(N_1 N_2 \cdots N_j)^{-1}$ and a function, say g_j, of height

$$a_1 a_2 \cdots a_j \; (N_1 N_2 \cdots N_j)^{1-k} \quad (0 < a_j < 1).$$

In passing from the j-th step to the next one, we keep only cubes on which $g_j > 2g_{j+1}$, and this is certainly true if $g_j > 2N_j^{1-k}$. If, then, $N_j \geq j^L$ for a certain L, and N_1 is suitably large, the total measure retained always exceeds 1/2.

The random variables $\Gamma(g_j) - \Gamma(g_j/2)$ are independent when the cube of step j is varied, but since we have the inequality $g_j > 2g_{j+1}$, the sample functions at step j will be independent of all previous steps as well.

Let $Y > 0$ be fixed, and then let us test, at each step, the inequality $\{\Gamma(g_j) - \Gamma(g_j/2) \geq Y(N_1 \cdots N_j)^{1-k}\}$. The variable $\Gamma(g_j) - \Gamma(g_j/2)$ has variance $c(a_1 a_2 \cdots a_j)(N_1 N_2 \cdots N_j)^{2(1-k)}$, and therefore the inequality has probability equal to $P\{X_1 > c^{-1/2} Y(a_1 \cdots a_j)^{-1/2}\}$, where $\sigma^2(X_1) = 1$. Let us now add the requirement

$$(a_1 a_2 \cdots a_j) \log j \to +\infty.$$

Then $\Sigma P\{X_1 > c^{-1/2} Y (a_1 \cdots a_j)^{-1/2}\} = +\infty$ for any $Y > 0$.

4. For each x contained in a cube of every generation, it is almost sure that the inequality $\Gamma(g_j) - \Gamma(g_j/2) > Y(N_1 \cdots N_j)^{1-k}$ is fulfilled on some cube (of some step j) containing x. Hence the successful experiments cover a set of measure $> 1/2$, and we can find functions $f_1, \cdots, f_N$, among the various g_j, with disjoint supports, so that $\Sigma\Gamma(f_i) - \Gamma(f_i/2) > Y/2$. Now one of the inequalities $\Gamma(\Sigma f_i) > Y/4$ or $\Gamma(\Sigma f_i/2) > Y/4$ is fulfilled, almost surely.

For definiteness we choose $0 < \alpha < 1$ and $a_1 \cdots a_j = (\log(3+j))^{-\alpha}$, and recall that $(N_1 N_2 \cdots N_j)^{-1} \geq C(j!)^{-L}$ for some $L > 0$. To estimate $|D\, g_j(x) - D\, g_j(x')|$ when x and x' belong to the support of g_j and D has order $K-1$, we use the inequality

$$|D\, g_j(x) - D\, g_j(x')| \leq C\, a_1 a_2 \cdots a_j\, N_1 N_2 \cdots N_j\, |x - x'|.$$

Distinguishing two cases, $|x - x'| \leq |N_1 \cdots N_j|^{-1/2}$ or $|x - x'| > |N_1 \cdots N_j|^{-1/2}$, we find the upper bound $C(\log^+(\log|x - x'|^{-1}))^{-\alpha}$. We believe the exponent $\alpha = 1$ could be attained, at the expense of considerable detail.

References

[1] R. M. Dudley, Sample functions of the Gaussian process, Annals of Probability 1 (1973), 66-103.

[2] R. M. Dudley, Metric entropy of some classes of sets with differentiable boundaries. J. Approx Theory 10 (1974), 227-236.

On Characterizing Generalized Convex Functions by Their Best Approximations

Esther Kimchi

Department of Mathematics
Fordham University
New York

I. INTRODUCTION

Let $\{u_0, u_1, \ldots, u_k\}$ be a positive Tchebycheff system on $[a,b]$, namely

$$\Delta\begin{pmatrix} u_0, u_1, \ldots & u_k \\ x_0, x_1, \ldots & x_k \end{pmatrix} = \begin{vmatrix} u_0(x_0) \ldots & u_0(x_k) \\ u_1(x_0) \ldots & u_1(x_k) \\ \vdots & \\ u_k(x_0) \ldots & u_k(x_k) \end{vmatrix} > 0 \qquad (1.1)$$

for all $\underline{a} < x_0 < x_1 < \ldots < x_n \leq b$, and let $\Lambda_k = \Lambda_k[a,b]$ denote its linear span. We denote by C_{n-1} the cone of functions "generalized convex" with respect to Λ_{n-1}, i.e. functions for which

$$\Delta(f, x_0, x_1, \ldots x_{n-1}) = \Delta\begin{pmatrix} u_0, u_1, \ldots u_{n-1}, f \\ x_0, x_1, \ldots x_{n-1}, x_n \end{pmatrix} \geq 0 \qquad (1.2)$$

for all $a \leq x_0 < x_1 < \ldots < x_n \leq b$.

As introduced in [3], norms defined on $C[a,b]$ for which

$$f(x) \cdot g(x) \geq 0, \ |f(x)| \leq |g(x)|, \ a \leq x \leq b \ \text{ implies } \|f\| \leq \|g\| \qquad (1.3)$$

are called "sign-monotone norms".

We denote by $T_k^* = T_k^*(f, I, \|\cdot\|)$ a polynomial of best approximation (p.b.a) to f from Λ_k with respect to $\|\cdot\|$ on

Copyright © 1980 by Academic Press, Inc.
All rights of reproduction in any form reserved.
ISBN: 0-12-171050-5

the interval I, by $a_k = a_k(f,I,\|\cdot\|)$ its leading coefficient, and by $E_k = E_k(f,I,\|\cdot\|)$ the degree of approximation of f by such polynomials.

In [3] it is proved that if $f\in C[a,b]$ is generalized convex with respect to Λ_{n-1}, and $\|\cdot\|$ is a sign monotone norm then $a_n(f,[a,b],\|\cdot\|) \geq 0$. On the other hand this condition is sufficient for generalized convexity only if satisfied for all subintervals of [a,b] and under more restrictive assumptions. More specifically, it is proved in [3]:

Theorem 1.1 Let $\{u_0,u_1,\ldots,u_n\}$ be an extended complete Tchebycheff system and $f\in C^{(n)}[a,b]$. If for each $I\subset[a,b]$ there exists a sign-monotone norm $\|\cdot\|_I$ and a p.b.a. $T_n^*(f,I,\|\cdot\|_I)$ such that $a_n(f,I,\|\cdot\|_I) \geq 0$, then $f\in C_n$.

The purpose of this work is twofold: to see whether Theorem 1.1 may hold under less restrictive assumptions (as is proved for the sup-norm in [1]) and to find some sufficient conditions for generalized convexity which do not involve all subintervals of [a,b].

In §2 we deal with the first question by presenting an example which shows that Theorem 1.1 does not hold for $f\notin C^{(n)}[a,b]$. In §3 we prove sufficient conditions for generalized convexity which involve either approximations in different norms, or approximations by subspaces of increasing degrees.

II. A COUNTER EXAMPLE

In this section we present an example which shows that Theorem 1.1 does not hold for $f\notin C^{(n)}[a,b]$. To this end we use sign-monotone norms defined on C[a,b] of the following form

$$\|f\|_L = \|f\|_\infty + A\sum_{i=0}^{n}|f(y_i)| \qquad (2.1)$$

where $A \geq 0$ is a constant and $a \leq y_0 < y_1 < \dots < y_n \leq b$ are fixed. First we prove a lemma concerning best approximation in such norms:

<u>Lemma 2.1</u> Let $\{y_0, y_1, \dots y_n\}$ be a set of n+1 fixed points in [a,b]. Then, there exists a constant A such that for every $f \in C[a,b]$ the unique polynomial of interpolation to f from Λ_n at the above fixed points is also the p.b.a. to f from Λ_n with respect to $\|\cdot\|_L$.

<u>Proof</u>. Let p_L denote the unique polynomial of interpolation to f from Λ_n at the above points. We'll show that there exists $A > 0$ such that for every $p \in \Lambda_n$, $p \neq p_L$ it is possible to find a nontrivial polynomial $q \in \Lambda_n$ and $\lambda > 0$ such that $\|f-p-\lambda q\|_L < \|f-p\|_L$. Indeed, let $q \in \Lambda_n$ be the polynomial satisfying

$$q(y_i) = \operatorname{sgn}(f(y_i)-p(y_i)) \qquad i=0,1,\dots,n \tag{2.2}$$

where sgn(x) is 1, -1, or 0 for $x > 0$, $x < 0$ or $x = 0$ respectively. It is clear that the above conditions define q uniquely and moreover, since $p \neq p_L$ q is nontrivial. Now, for $\lambda > 0$ small enough:

$$\|f-p-\lambda q\|_L = \|f-p-\lambda q\|_\infty + A\sum_{i=0}^{n} |(f-p-\lambda q)(y_i)| \leq$$
$$\|f-p\|_\infty + \lambda\|q\|_\infty + A\sum_{i=0}^{n} |(f-p)(y_i)| - \lambda A \sum |q(y_i)| =$$
$$= \|f-p\|_L + \lambda\left(\|q\|_\infty - A\sum_{i=0}^{n} |q(y_i)|\right) < \|f-p\|_L.$$

The last inequality holds, provided $A > \dfrac{\|q\|_\infty}{\sum_{i=0}^{n} |q(y_i)|}$.

But for a fixed set $\{y_0, y_1, \dots y_n\}$ we may take

$$A > \max_{q \in Q} \frac{\|q\|_\infty}{\sum_{i=0}^{n} |q(y_i)|}$$

where Q is the set of all nontrivial polynomials satisfying $q(y_i) = 0, 1$ or -1 for $i=0,1,\dots,n$.

In order to present our example, let us construct a function g continuous on [a,b] but nowhere monotonic there. (see for example [2 p. 29]). With each subinterval $I \subset [a,b]$ we associate the norm:

$$\|f\|_I = \|f\|_\infty + A(|f(x_1)| + |f(x_2)|) \tag{2.3}$$

where $x_1, x_2 \in I$ are chosen to be such that $g(x_1) < g(x_2)$, and A is large enough so that according to Lemma 2.1 $p_L = a_1 x + a_0$, which interpolates g at x_1 and x_2 is the p.b.a. to g, from the set of polynomials of degree ≤ 1, in the norm $\|\cdot\|_I$ on I. But $p_L(x_1) < p_L(x_2)$ means that $a_1(g, I, \|\cdot\|_I) > 0$ for all $I \subset [a,b]$, while g is not increasing.

It is still unclear whether the result of Theorem 1.1 holds if the assumptions $f \in C^{(n)}[a,b]$ and $\{u_0, u_1, \ldots u_n\}$ is extended and complete are replaced by the assumption that $\{\|\cdot\|_I, I \subset [a,b]\}$ is "continuous" - (see [3]) - as is implied by the case of the sup-norm studied in [1].

III. SUFFICIENT CONDITIONS FOR GENERALIZED CONVEXITY

In this section we present two types of sufficient conditions for generalized convexity. First we show that the known necessary conditions are also sufficient provided they are satisfied with respect to the class of all sign-monotone norms of the form (2.1).

Theorem 3.1 Let $f \in C[a,b]$, and suppose $a_n(f, [a,b], \|\cdot\|_L) \geq 0$ for all sign monotone norms $\|\cdot\|_L$ of the form (2.1). Then $f \in C_n$.

Proof. Suppose $f \notin C_n$. Then there exist points $a \leq y_0 < y_1, < \ldots < y_n \leq b$ such that

$$\Delta(f, y_0, y_1, \ldots, y_n) < 0 \tag{3.1}$$

Now, for A large enough, the p.b.a. to f in the norm (2.1) based on the above points is the polynomial which interpolates f at these points. That means

$$\Delta(f,y_0,y_1,\ldots y_n) = \Delta(p,y_0,y_1,\ldots,y_n) = a_n \cdot \Delta\begin{pmatrix} u_0,u_1,\ldots u_n \\ y_0,y_1,\ldots y_n \end{pmatrix}$$

which by (3.1) leads to $a_n < 0$ - a contradiction.

In a similar way, one can prove that either the property $E_{n-1}(f,[a,b],\|\cdot\|_L) > E_n(f,[a,b],\|\cdot\|_L)$ or the property $\tilde{Z}(f-T_n) = n$ when satisfied for all $\|\cdot\|_L$ of the type (2.1) are also sufficient for the generalized convexity of f or -f. ($\tilde{Z}(g)$ is number of zeroes of g, counting twice zeroes with no sign change).

The second type of sufficient conditions for generalized convexity deal with approximation by polynomials of increasing degrees. It is well known ([4],[6]) that there are functions which satisfy $f^{(k)}(x) \geq 0$ for $a \leq x \leq b$ while all their polynomials of best approximation from $\pi_n (n \geq N)$ do not share the same property. Some further assumptions (which concern the smoothness of f(x)) do guarantee that for n large enough the polynomials of best approximation from π_n satisfy the same restrictions on their derivatives as f, [4]. We will show, in the general context of approximation in sign-monotone norms by elements of Tchebycheff Systems, that the converse to the above result does hold. This result relies on the following Theorem generalizing different aspects of results from [5] and from [7]

<u>Theorem 3.2</u> Let $f \in C[a,b]$, let $\|\cdot\|$ be a sign monotone norm, and let $\{f_k\}_{k=1}^{\infty}$ be a sequence of functions satisfying

$$f_k \in C_n \qquad k=1,2,\ldots \tag{3.2}$$

$$\lim_{k\to\infty} \|f-f_k\| = 0 \tag{3.3}$$

Then $f \in C_n$.

We omit the proof which uses arguments similar to those in [5]. As a direct result one gets:

Theorem 3.3 Let $\{\Lambda_k, k=1,2,\ldots\}$ be dense in $C[a,b]$, in the topology induced by a given sign monotone norm $\|\cdot\|$, and let T_k be the p.b.a.'s to $f\epsilon C[a,b]$ from Λ_k $(k=1,2,\ldots)$ in $\|\cdot\|$. If there exists a subsequence $\{T_{k_j}, j=1,2,\ldots\}$ such that $T_{k_j}\epsilon C_n$ for all j, then $f\epsilon C_n$.

REFERENCES

[1] D. Amir and Z. Ziegler, "Functions with strictly decreasing distances from increasing Tchebycheff subspaces" J. Approximation Theory 6(1972) 332-344.

[2] B. Gelbaum and J. Olmsted, "Counter Examples in Analysis", Holden-Day Inc. 1964.

[3] E. Kimchi "Characterization of Generalized Convex functions by their best Approximation in Sign-Monotone Norms" J. Approximation Theory 24(1978) 350-360.

[4] E. Kimchi and D. Leviatan, "On restricted best approximation to functions with restricted derivatives" SIAM J. Numer. Anal. 13(1976) 51-53.

[5] E. Kimchi and N. Richter-Dyn, "Convergence properties of sequences of functions with application to restricted derivatives approximation" J. Approximation theory 22 (1978) 289-303.

[6] J. Roulier, "Polynomials of best approximation which are monotone" J. Approximation Theory 9(1973) 212-217.

[7] A. Roberts and D. Varberg " Convex Functions" Academic Press, 1973.

THEORY OF NEAREST DEVIATORS

Joseph M. Lambert

Department of Mathematics
The Pennsylvania State University
University Park, Pennsylvania

Let E be a normed linear space. The deviation of a set A from a set B, both contained in the space E, is denoted by $\beta(A,B) = \sup_{a\in A} \inf_{b\in B} \|a-b\| = \sup_{a\in A} \text{dist}\,(a,B)$. One can consider the following approximation problem: Given a set A contained in E and a collection of sets $K = \{B \mid B \subset E\}$, find $\inf_{B\subset K} \beta(A,B)$, which we call the least deviation of K from A. An element $B_0 \subset K$ such that $\beta(A,B_0) = \inf_{B\subset K} \beta(A,B)$ is called the nearest deviator of K from A.

In this short note it is shown that a theory of least deviation of K from A can be stated in terms of the Hausdorff metric and a variant of a fundamental result in the theory of approximation in metric spaces. In particular, one can recover some of the results on n-dimensional diameters of sets in Banach spaces by analytic methods as contrasted with the geometrical methods used to prove the original results.

We denote the unit ball of a space X by $U(X)$ and its boundary by $S(X)$.

Recall that if (E,d) is a metric space one defines the Hausdorff metric α on the subsets of E as follows:

Copyright © 1980 by Academic Press, Inc.
All rights of reproduction in any form reserved.

ISBN: 0-12-171050-5

$$d(a,B) = \inf_{b \in B} d(a,b) \quad \text{where} \quad B \subset E,\ B \neq \phi$$
$$\beta(A,B) = \sup_{a \in A} d(A,B) \quad \text{where} \quad A \subset E,\ A \neq \phi$$
$$\alpha(A,B) = \max\ [\beta(A,B), \beta(B,A)]$$

If E is a Banach space, then the closed bounded subsets of E with the Hausdorff metric induced by the norm is a complete metric space.

A fundamental theorem on the existence of a best approximation in a metric space states that if K is a compact subset of a metric space, then to each point p in the space, there exists a point in K of minimum distance from p, (1, p.6).

In our discussion we consider the closed bounded convex subsets of a Banach space E and associate the Hausdorff metric topology with these sets. We denote the space (X,α).

THEOREM 1. Let A be a bounded subset of a Banach space E. If K is a compact subset of (X,α) then there exists a nearest deviator of K from A.

The proof is an obvious modification of (1).

Before exhibiting a non trivial class of compact sets K in (X,α), we need a technical lemma to show their compactness.

LEMMA 2. Let G_1, G_2 be n-dimensional subspaces of E. Set $F_i = G_i \cap rU(E)$, $r > 0$. Then given z in $G_1 \setminus G_2$, $\|z-g_0\| = \inf_{g \in G_2} \|z-g\| \leq \max\{1, (2\|z\|/r)\} \sup_{f \in F_1} \inf_{g \in F_2} \|f-g\|$.

The proof is a case by case analysis of whether the norms of z and g_0 are less than or equal to r or are greater than r and by using the fundamental result in approximation theory that $\|g_0\| \leq 2\|z\|$.

THEOREM 3. Let E_{n+1} be a fixed $(n+1)$ dimensional subspace of a Banach space E. The class of bounded subsets of E_{n+1} of the form $G \cap rU(E_{n+1})$, G an n-dimensional subspace of E_{n+1}, $r > 0$, is a compact set in (X,α).

PROOF. We first show that K is closed. Let $F_k = G_k \cap rU(E_{n+1})$ be elements of K with $\lim \alpha(F_k,F_0) = 0$. We must show that F_0 is in K. To each G_k associate a functional λ_k in $S(E_{n+1})$ such that $G_k = \{z \in (E_{n+1}) \mid \lambda_k(z) = 0\}$. Geometrically λ_k corresponds to the normal vector to the hyperplane G_k. Thus $F_k = \{z \in rU(E_{n+1}) \mid \lambda_k(z) = 0\}$. Let λ_0 be a limit point of the sequence $\{\lambda_k\}$. Define $H_0 = \{z \in rU(E_{n+1}) \mid \lambda_0(z) = 0\}$. We show that F_k converges to H_0 and since induces a Hausdorff topology on X, $H_0 = F_0$.

The functional λ_0 is unique in the following sense. If λ is a limit point of λ_k then $\lambda = \lambda_0$ or $\lambda = -\lambda_0$. To see this let z be an element of $S(E_{n+1})$. Then

$$-\mathrm{dist}(g_{0m},G_k) \leq \mathrm{dist}(z,G_m) - \mathrm{dist}(z,G_k) \leq \mathrm{dist}(g_{0k},G_m),$$

where g_{0j} is a best approximation of z from G_j.
By applying lemma 2 we see that

$$|\mathrm{dist}(z,G_m) - \mathrm{dist}(z,G_k)| \leq \max\{(2\|z\|/r),1\}\alpha(F_k,F_m).$$

However, using elementary results of approximation theory $\mathrm{dist}(z,G_j) = |\lambda_j(z)|$. Since F_k is Cauchy in the Hausdorff metric topology, for a given $\varepsilon > 0$ there exists an $N(\varepsilon)$ such that for all m,k greater than N,

$\| |\lambda_k(z)| - |\lambda_m(z)| \| \le \max\{(2/r)\varepsilon, \varepsilon\}$. Thus $|\lambda_k(z)|$ is Cauchy in $\mathbb{R}$ and is bounded by 2, and converges to say $\theta(z)$. But $\lambda_k(z)$ has a convergent subsequence to $\lambda_0(z)$ for each z in $S(E_{n+1})$. Since convergence is unique in a Hausdorff topology $\theta(z) = \lambda_0(z)$. Since λ_k is a sequence in $S(E^*_{n+1})$ pointwise convergence implies norm convergence. Thus λ_k converges to either λ_0 or $-\lambda_0$. Hence H_0 as we have defined it, is unique. To show that F_k actually converges to H_0 in (X,α), fix f in F_k. Then $\inf_{g \in \text{span } H_0} \|f-g\| = \|f-g_0\| = |\lambda_0(f)-\lambda_k(f)|$ where $g_0 = f-\lambda_0(f)\lambda_0$. If

$\|g_0\| \le r$ then g_0 is in H_0. Otherwise

$$\inf_{g \text{ span } H_0} \|f-g\| = \|f-g_0\| \le \|f-(r/\|g_0\|)g_0\| \le \|f-g_0\| + \|g_0-(r/\|g_0\|)g_0\| = \|f-g_0\| + (\|g_0\|-r).$$

However, $\|g_0\| \le \|f\| + |\lambda_0(f)-\lambda_k(f)| \le \|f\|(1+\|\lambda_0-\lambda_k\|)$. Thus $\|f-g_0\| \le \|f\|(2\|\lambda_0-\lambda_k\|+1) - r \le r(2\|\lambda_0-\lambda_k\|)$. But this is true for all f in F_k. Thus $\lim \beta(F_k,H_0) = 0$ since λ_k converges to λ_0. A similar argument holds for $\beta(H_0,F_k)$. Thus F_k converges to H_0 in (X,α) and $H_0 = F_0$.

To show that K is compact, take any sequence F_k in K. To each F_k associate a linear functional λ_k as above. Denote a limit point of λ_k as λ_0. The previous analysis now shows that there is a convergent subsequence of the F_k converging to the set $F_0 = \{z \in rU(E_{n+1}) | \lambda_0(z) = 0\}$ qed.

Thus such classes K have a nearest deviator from a bounded set A in E. Sets of the form x + K, x in E, and finite unions of such translates are also compact sets in (X,α).

N DIMENSIONAL DIAMETERS

Following the notation of (3, p. 268), we let E be a normed linear space, A a set in E and n an integer with $0 \leq n < \infty$. The number

$$d_n(A,E) = \inf_{\dim G=n} \sup_{x \in A} \inf_{g \in G} \|x-g\|,$$

where the infimum is taken over all n-dimensional subspaces of E, is called the n-dimensional diameter of A. A. L. Brown (2) calls this term the n-dimensional radius of A. Singer calls a subspace G_0 such that $d_n(A,E) = \sup_{x \in A} \inf_{g \in G_0} \|x-g\|$ a best n-dimensional secant of A with respect to E. Tikomorov (4) calls the term an extremal n-dimensional subspace of A with respect to E.

If one considers the case where A is a bounded subset of E, one sets $G_A = \{g \in G \mid \|g\| \leq 2\sup_{a \in A} \|a\|\}$ and notes that $\inf_{g \in G} \|x-g\| = \inf_{g \in G_A} \|x-g\|$ for all x in. A. G_A is a closed bounded set with span $G_A = G$. Thus $d_n(A,E) = \inf_{\dim G=n} \beta(A,G)$ $= \inf_{G_A} \beta(A,G_A)$

We can now recover a theorem of Brown concerning the existence of extremal n-dimensional subspaces for A with respect to E using the results of existence of nearest deviators (2).

THEOREM . Let E be a normed linear space, n an integer $0 \leq n < \infty$ and E_{n+1} an (n+1) -dimensional subspace of E. Then for every bounded set A contained in E_{n+1} there exists an n-dimensional subspace G contained in E which is extremal for A with respect to E.

PROOF. Brown shows that $d_n(A,E) = d_n(A,E_{n+1})$ and that $d_n(A,E) = d_n(ccA,E)$, where ccA denotes the closed convex hull of A(2). Thus we can consider the problem in the space E_{n+1} and set (X,) to be the metric space of closed convex bounded sets of E_{n+1} with the induced Hausdorff metric. Denote by K the class of sets $G_A = \{G \cap 2 \sup_{a \in A} \|a\| \, U(E_{n+1})\}$, where G is an n-dimensional subspace of E_{n+1}. We apply theorem 3 to show that K is compact in (X,α). Thus there exists an element H in K which is a nearest deviator to A from K. The fact that $\inf_{h \in H} \|x-h\| = \inf_{h \in \text{span } H} \|x-h\|$ shows that span H is an extremal n-dimensional subspace of A with respect to E_{n+1}.qed.

REFERENCES

1. Cheney, E. W., "Introduction to Approximation Theory, McGraw Hill, New York 1966.
2. Brown, A. L., *Proc. Lon. Math. Soc.* 14, 577-594 (1964).
3. Singer, I. Best Approximation in Normal Linear Spaces by Elements of Linear Subspaces, Springer-Verlag, New York 1970.
4. Tikomorov, Ushpehk, Math. Nauk 15, 81-120 (1960) (English Translation), *Russ. Math. Surveys* 15, 75-111 (1960).

A PROBABILISTIC APPROACH TO THE ESTIMATION OF FUNCTIONALS

F. M. Larkin

Department of Computing and Information Science
Queen's University
Kingston, Ontario

I. INTRODUCTION

We are concerned here with making inferences about a function $h(.)$, defined over a real or complex domain D and regarded as an element of a Hilbert space H, on the basis of limited information of the type commonly encountered in practice, i.e. numerical values of continuous, linear (and possibly certain non-linear) functionals. However, we specifically abandon the requirement, central to the technique of Optimal Approximation (eg. 3), for a bound R on the norm, or a semi-norm, of h; for example,

$$||h|| \leq R \ . \qquad [1]$$

Instead of requiring assumption [1], we introduce the necessary localizing information by endowing H with a canonical weak Gaussian distribution (2). Roughly speaking, this associates with every $g \epsilon H$ a prior likelihood $\exp(-\lambda||g||^2)$, where λ is an undetermined positive real number, and formalizes the common intuitive notion that a "smooth" function (i.e. one with small norm or semi-norm) would be more likely to have given rise to the observed functional values than would a less

Copyright © 1980 by Academic Press, Inc.
All rights of reproduction in any form reserved.
ISBN: 0-12-171050-5

smooth function. This assumption is similar to [1] in intent, which effectively ensures that $||h|| \leq R$ with probability 1.

As we shall see below, the assumption of a weak Gaussian distribution on H in principle permits us to

(a) obtain joint distributions, induced by the distribution on H, between the known and the required unknown quantities, and

(b) find averages, with respect to the weak distribution $\mu(.)$, of error functionals over H, which can then be minimized with respect to any parameters involved.

Standard statistical techniques can then be used in order to

(c) show that the approach essentially agrees with Optimal Approximation when the only available measurements are exact linear functionals of h, and then obtain computable confidence limits on estimated quantities <u>without requiring further non-linear information</u>,

(d) indicate how to filter signal from noise when the measured information is inexact, thereby providing another viewpoint on Tikhonov regularization, and

(e) find reasonable, if not "optimal", solutions to non-linear problems, such as zero-finding or estimation of quadratic functionals.

In the interests of brevity, we confine ourselves to (a), (b), (c), and (e) in this presentation; a more extensive review can be found in (5).

We shall also suppose that H possesses a reproducing kernel function $K(.,\div)$ (1), which greatly facilitates the task of numerical evaluation of certain integrals of functionals over H.

II. LINEAR FUNCTIONALS

Let $\{L_j;j=1,2,...,n+1\}$ denote bounded linear functionals on H with corresponding Riesz representers $\{g_j;j=1,2,...,n+1\}$. Then, presuming H to be real, it may be shown (6) that the weak Gaussian distribution on H induces a proper probability density p(.) on the quantities

$$y_j \overset{\text{def}}{=} L_j g = (g_j,g) \quad ; \quad j = 1,2,...,n+1, \qquad [2]$$

given by

$$p(\underline{y}) = (\tfrac{\lambda}{\pi})^{n/2} \cdot |G|^{-1/2} \cdot \exp(-\lambda \underline{y}' G^{-1} \underline{y})$$

$$= (\tfrac{\lambda}{\pi})^{n/2} \cdot |G|^{-1/2} \cdot \exp(-\lambda ||\hat{g}_{n+1}||^2),$$

where G is the Gram matrix of the $\{g_j;j=1,2,...,n+1\}$ and $\hat{g}_{n+1}$ is that element of H of least norm, subject to [2]. The conditional density function of y_{n+1}, given $\{y_j;j=1,2,...,n\}$ is then found to be

$$p(y_{n+1}|y_1,y_2,...,y_n) = (\tfrac{\lambda q}{\pi})^{1/2} \cdot \exp\{-\lambda q(y_{n+1}-\hat{y}_{n+1})^2\},$$

where $\hat{y}_{n+1} = L_{n+1}\hat{g}_n$ (the optimal approximant of y_{n+1}) and q is the squared norm of that element $h_o \epsilon H$ of least norm satisfying

$$L_j\hat{h}_o = 0 \ ; \ j = 1,2,...,n \text{ and } L_{n+1}\hat{h}_o = 1 \ .$$

Also, q^{-1} is the squared norm of the Riesz representer of the error functional in the usual optimal rule for estimation of y_{n+1} from $\{y_j;j=1,2,...,n\}$.

Remarkably, it turns out (7) that the quantity

$$t = \frac{(nq)^{1/2}}{||\hat{g}_n||} \cdot |y_{n+1} - \hat{y}_{n+1}|$$

is then distributed as Student's-t with n degrees of freedom, permitting the determination of confidence intervals on the optimal approximant $\hat{y}_{n+1}$ in the usual fashion (eg. 4). In other words, a useful assessment of the error in $\hat{y}_{n+1}$ is obtained, even in the absence of any non-linear information specific to the function h.

III. ESTIMATION OF A QUADRATIC FUNCTIONAL

As an illustration of the many possibilities arising in connection with non-linear functionals (8) we show how to construct a rule for estimating the value of

$$F_o(h) \stackrel{\text{def}}{=} \int_a^b |h(x)|^2 .dx \quad ,$$

given that $h(x_j) = y_j$; $j = 1,2,..,n$.

Since $F_o(h)$ is quadratic in h we define an error functional E(.) by the relation

$$E(g) = \int_a^b |g(x)|^2 dx - \sum_{j,k=1}^{n} w_{jk} g(x_j) g(x_k) \quad , \forall g \in H \ .$$

For any weight vector $\underline{w}$, the mean-(over H)-squared-error $U(\underline{w})$ of E(.) can be expressed as

$$U(\underline{w}) = \int_H |E(g)|^2 \mu(dg) = B - 2. \sum_{j,k=1}^{n} w_{jk} C_{jk} + \sum_{j,k,r,s=1}^{n} w_{jk} w_{rs} D_{jkrs} \quad ,$$

and the coefficients $\{B, C_{jk}, D_{jkrs}, \forall \text{ relevant } j,k,r,s\}$ can be found without difficulty in terms of the reproducing kernel function K(.,.) of H. The weights $\{w^*_{jk}\}$ satisfying

$$U(\underline{w}^*) \le U(\underline{w}) \ \forall \underline{w} \in \mathbb{R}^{n^2}$$

are then evaluated by solving the linear equations

$$\sum_{r,s=1}^{n} D_{jkrs} w^*_{rs} = C_{jk} \ ; \ j,k = 1,2,...,n \quad ,$$

and the mean-square-error in the "optimal" estimation rule

$$\int_a^b |h(x)|^2 \, dx \simeq \sum_{j,k=1}^{n} w_{jk}^{*} \, h(x_j).h(x_k)$$

is simply $U(\underline{w}^*)$.

IV. ZEROS OF ANALYTIC FUNCTIONS

One way of applying the foregoing probabilistic ideas in estimating a zero of an analytic function whose ordinate values can be computed is described in (9) and summarized below.

If a complex function g(.) is analytic (possibly except for poles) and has a single (unknown) zero x_o within a simple domain $D \subset \mathbb{C}$, it may be expressed as

$$\frac{1}{g(z)} = \frac{h(z)}{z-x_o} \overset{\text{def}}{=} f(z) \quad \forall z \in D,$$

where h is a member of a suitably chosen Hilbert space H of functions analytic in D. Then, for fixed points $\{x_j \in D; j=1,2,...,n\}$ the weak Gaussian distribution on H induces on the values $\{f_j = f(x_j); j=1,2,...,n\}$ the proper probability density function

$$p(\underline{f}) = (\tfrac{\lambda}{\pi})^n |K|^{-1} \prod_{j=1}^{n} |x_j - x_o|^2 . \exp\{-\lambda \underline{f}' (X-\bar{x}_o I) K^{-1} (X-x_o I) \underline{f}\}, \quad [3]$$

where

$$X = \mathrm{diag}[x_1, x_2, ..., x_n] \text{ and } K_{jk} = K(x_j, \bar{x}_k); j,k=1,2,...,n.$$

The maximum likelihood estimator $\hat{x}_o$ of x_o, determined from [3] by the usual method (4) then satisfies the equation

$$\frac{1}{n} \cdot \sum_{j=1}^{n} \frac{1}{x_j - \hat{x}_o} = \frac{\underline{f}' (X' - \hat{\bar{x}}_o I) K^{-1} \underline{f}}{\underline{f}' (X' - \hat{\bar{x}}_o I) K^{-1} (X - \hat{x}_o I) \underline{f}} \quad [4]$$

It may also be shown that, under reasonable conditions on the $\{x_j\}$, [4] simplifies to the asymptotic form

$$\hat{x}_o \sim \frac{\underline{f}'K^{-1}X\underline{f}}{\underline{f}'K^{-1}\underline{f}} \overset{\text{def}}{=} s \text{ , say,}$$

for large n. For suitable choices of H, the estimator s of x_o can be evaluated quite efficiently and made the basis for an iterative process converging to x_o at least as fast as Newton's method, without the necessity for computing derivatives.

REFERENCES

1. Aronszajn, N., *Trans. Amer. Math. Soc. 68*, 337 (1950).
2. Gross, L., *Trans. Amer. Math. Soc. 105*, 372 (1962).
3. Handscomb, D.C., "Methods of Numerical Approximation", Pergamon Press, Oxford, (1965).
4. Kendall, M.G. and Stewart, A., "The Advanced Theory of Statistics", vol. 2, Griffin, London, (1967).
5. Larkin, F.M., Technical Report 79-90, Dept. of Computing and Information Science, Queen's University, Kingston, Ontario.
6. Larkin, F.M., *Rocky Mountain J. Math. 2, #3*, 379 (1972).
7. Larkin, F.M, in "Information Processing 74", p. 605, North-Holland, Amsterdam, (1974).
8. Larkin, F.M., in "Theory of Approximation with Applications", (Law, A.G., and Sahney, B.N., eds.), p. 43, Academic Press, New York, (1976).
9. Larkin, F.M., to appear in J. *Approximation Theory*.

ON THE RATE OF APPROXIMATION BY GENERALIZED POLYNOMIALS WITH RESTRICTED COEFFICIENTS

D. Leviatan

Department of Mathematics
Tel Aviv University
Tel Aviv, Israel

I. INTRODUCTION

Given a sequence $A = \{A_k\}(k\geq 1)$ of nonnegative constants v. Golitschek, Leviatan and Roulier ([5], [6]) proved that the set of polynomials $P_A = \{p(x): p(x) = \sum_{k=1}^{n} a_k x^k$, $n = 1,2,\ldots, |a_k| \leq A_k^k\}$ is dense in $C_0[0,1]$ if and only if there exists a subsequence of the positive integers $\{k_i\}(i\geq 1)$ such that $\sum_{i=1}^{\infty} 1/k_i = \infty$ for which $\lim_{i\to\infty} A_{k_i} = \infty$. Recently Bak, v. Golitschek and the author [1] have shown that if for all $k\geq k_0$, $A_k \geq \delta k^2$ for some $\delta > 0$, then the rate of approximation of functions by polynomials in P_A is at least that guaranteed by Jackson's theorem. This result however does not take advantage of the full strength of the previous result, namely, that we may force many of the coefficients to vanish (by demanding that $A_k = 0$ for $k \neq k_i$) and still have a dense P_A. Also we have information on the rate of approximation of functions by Müntz polynomials (see [2]) and we will apply that here.

Copyright © 1980 by Academic Press, Inc.
All rights of reproduction in any form reserved.
ISBN: 0-12-171050-5

II. MAIN RESULTS

Let $\{k_i\}$ be a subsequence of the positive integers and assume that it satisfies

$$k_i \geq 2i \quad i = 1,2,\dots . \tag{1}$$

Put

$$\varepsilon_n = \exp[-2 \sum_{i=1}^{n} 1/k_i]$$

then we prove

<u>THEOREM 1</u>. There exists an absolute constant $C>0$ (independent of the sequence $\{k_i\}$ satisfying (1)) such that for any $\varepsilon > 0$ and any $f \in C_0[0,1]$ there are Müntz polynomials $p_n(x) = \sum_{i=1}^{n} a_{in} x^{k_i}$ with the properties that for all sufficiently large n

$$\|f-p_n\| \leq C\omega(f, \varepsilon_n^{1-\varepsilon}) \tag{2}$$

and

$$|a_{in}| \leq 2\|f\| e^{2k_i} \varepsilon_i^{-k_i(1-\varepsilon)/\varepsilon}, \quad 1 \leq i \leq n. \tag{3}$$

Here $\omega(f,\cdot)$ is the ordinary modulus of continuity of f.

<u>REMARKS</u>. By the Müntz-Jackson theorem [2, Thms. 1 and 3 and Remark A on p. 337] inequality (2) is almost best possible and can be improved only to the extent of dropping the ε, but this requires lifting the restriction (3) altogether. Although we would like to be able to drop the ε in (2) while still keeping a restriction of the type (3) this is not the only case in this theory where we meet restriction of the type (3) with small ε in the denominator of the exponent. For instance v. Golitschek [4] showed that if we replace (1) by the assumption $k_i \leq 2(1-\varepsilon)i, i = 1,2,\dots$ where $\varepsilon > 0$, then there are Müntz polynomials $p_n(x) = \sum_{i=1}^{n} a_i x^{k_i}$ such that (for large n) $\|f-p_n\| \leq C\omega(f,\frac{1}{n})$ and $|a_{in}| \leq \|f\| \, i^{2k_i(1-\varepsilon)/\varepsilon}$. Also it was shown in [1] that for $f(x) = x^{1/2}$ for any $\varepsilon > 0$ there are

polynomials $p_n(x) = \sum_{k=1}^{n} a_{kn}x^k$ such that $\|f-p_n\| \leq Cn^{-(1-\varepsilon)}$ and

$$|a_{kn}| \leq k^{2k(1-\varepsilon)/\varepsilon}, \quad 1 \leq k \leq n. \tag{4}$$

However $\|f-p_n\| \leq Cn^{-1}$ cannot be obtained if (4) is satisfied no matter how small ε is.

PROOF OF THEOREM 1. Let $0<\varepsilon<1$ be given and let $1 \leq r = r(\varepsilon,n) \leq n$ be chosen such that

$$\exp[(1-\varepsilon)\sum_{i=1}^{r-1} 1/k_i - \varepsilon \sum_{i=r}^{n} 1/k_i] \leq 1 \tag{5}$$

$$\exp[(1-\varepsilon)\sum_{i=1}^{r} 1/k_i - \varepsilon \sum_{l=r+1}^{n} 1/k_i] > 1 \tag{6}$$

This is possible since for $r = 1$ the first inequality holds and for $r = n$ the second one does. Now we apply the Müntz-Jackson theorem [2, Thm. 1] to the set of exponents $\{k_r,\dots,k_n\}$ and conclude that there exists a constant $C>0$ independent of the set of exponents and of r,n and ε such that for any $f \in C_0[0,1]$ there is a Müntz polynomial $p_n(x) = \sum_{i=r}^{n} a_n x^{k_i}$ satisfying

$$\|f-p_n\| \leq C\omega(f,\eta_n) \tag{7}$$

where $\eta_n = \max_{\operatorname{Re} z=1} |\frac{1}{z} \prod_{i=r}^{n} \frac{z-k_i}{z+k_i}|$.

Now if the maximum is achieved for $|z| = R$ we have since $k_i \geq 2i$, $i \geq 1$,

$$\eta_n \leq \frac{1}{R}\exp[-2\sum_{i=r}^{n} \frac{k_i}{R^2+k_i^2}]$$

$$= \frac{1}{R}\exp[-2\sum_{i=r}^{n} 1/k_i]\exp[2\sum_{i=r}^{n} \frac{R^2}{k_i(R^2+k_i^2)}]$$

$$\leq \frac{1}{R}\exp[-2\sum_{i=r} 1/k_i]\exp[\tfrac{1}{2}\log(1+R^2)]$$

$$\leq 2 \exp[-2 \sum_{i=r}^{n} 1/k_i]$$

(and by (5))

$$\leq \exp[-2(1-\varepsilon) \sum_{i=1}^{n} 1/k_i]$$

$$= 2\varepsilon_n^{1-\varepsilon}$$

This together with (7) establishes (2) and shows that $\lim_{n\to\infty} \| f-p_n \| = 0$. Hence for sufficiently large n, $\| p_n \| \leq 2\| f \|$. A slight improvement of [5, Lemma 2] yields

$$|a_{in}| \leq (1+k_i)^{1/2} e^{3k_i/2} \varphi(r,n)^{1+2k_i} 2 \| f \| , \qquad i = r,\ldots,n, \tag{8}$$

where $\varphi(r,n) = \exp[\sum_{j=r+1}^{n} 1/k_j]$. By (6)

$$\varphi(r,n) \leq \exp[\frac{1-\varepsilon}{\varepsilon} \sum_{j=1}^{r} 1/k_j]$$

and as $k_i \geq 2i$ we have

$$\varphi(r,n)^{1/k_i} \leq \exp[\frac{1-\varepsilon}{\varepsilon r} \sum_{i=1}^{r} 1/j]$$

$$\leq \exp[\frac{1-\varepsilon}{\varepsilon r} \log r].$$

By virtue of (1) and (6) $r(n,\varepsilon) \to \infty$ as $n \to \infty$ so that it follows that for sufficiently large n

$$\varphi(r,n)^{1/k_i} \leq e^{1/4}, \quad r \leq i \leq n.$$

Combining (8), (9) and (10) we obtain (3) for $i = r,\ldots,n$ while for $i = 1,\ldots,r-1$ $a_{in} = 0$ so that (3) is trivially satisfied. This completes the proof.

One can use [2, Thm. 2] to extend our Theorem 1 to differentiable functions, namely,

THEOREM 2. Given $k \geq 0$ there exists a constant $C_k > 0$ (independent of the sequence $\{k_i\}$ satisfying (1)) such that for any $\varepsilon > 0$ and any $f \in C^{(k)}[0,1]$ with $f^{(q)}(0) = 0$, $q=0,\ldots,k$ there are Müntz polynomials $p_n(x) = \sum_{n=1}^{n} a_{in} x^{k_i}$ the coefficients of which satisfy (3) and such that

$$\| f - p_n \| \leq A_k \varepsilon_n^{(1-\varepsilon)k} \omega(f^{(k)}, \varepsilon_n^{1-\varepsilon})$$

for all sufficiently large n.

PROOF. The proof is similar to that of Theorem 1. One should notice that by (1) and (6) $r = r(n,\varepsilon) \to \infty$ as $n \to \infty$. Since $f^{(q)}(0) = 0$, $q = 0,\ldots,k$ and the first $r-1$ coefficients of $p_n(x)$ vanish we do not need the condition in [2, Thm. 2] that $x^0,\ldots,x^k$ belong to the set of exponents.

REFERENCES

1. J. Bak, R.v. Golitschek and D. Leviatan, The rate of approximation by means of polynomials with restricted coefficients, *Israel J. Math. 26* (1977), 265-275.

2. J. Bak, D. Leviatan, D.J. Newman and J. Tzimbalario, Generalized polynomial approximation, *Israel J. Math. 15* (1973), 337-349.

3. M.v. Golitschek, Permissible bounds on the coefficients of generalized polynomials, *Approx. Theory, Proceedings of a Conference on Approx. Theory, Austin,* Texas 1973 G.G. Lorentz Ed. Academic Press N.Y. 1973.

4. M.v. Golitschek, Approximation Durch Komplexe Exponentialsummen und Zülassige Koeffiziente Restriktionen, *Proceed-*

ings of a Conference on the Theory of Approx. of Functions, Kaluga, U.S.S.R. 1976.

5. M.v. Golitschek and D. Leviatan, Permissible bounds on the coefficients of approximating polynomials with real or complex exponents, *J. Math. Anal. and Appl. 60* (1977), 123-138.

6. J.A. Roulier, Restrictions on the coefficients of approximating polynomials, *J. Approx. Theory 6* (1972), 276-282.

CENTRAL PROXIMITY MAPS

William A. Light

Department of Mathematics
University of Lancaster
Lancaster, England

I. INTRODUCTION

In [2] Golomb defined the concept of a <u>central proximity map</u> (for which he used the term central extremal). P is said to be a central proximity map from a Banach space X into a linear subspace Y if firstly $||x - Px|| = \text{dist }(x,Y)$, i.e. P is a proximity map. Secondly, P must have the following centrality property : $||x - Px+y||=||x - Px - y||$ for all $y \in Y$. This concept allowed him to generalise the work of Diliberto and Straus [1] under the following hypotheses:

(I) X is a Banach space with disjoint subspaces Y_1, Y_2 such that X can be projected onto $Y = Y_1 + Y_2$.

(II) There exist central proximity maps from X to Y_1 and Y_2 respectively. The generalisation was then

<u>Theorem 1.1</u>

The sequence $\{x_n\}$ defined by $x_o = 0$, $x_n = x_{n-1} - P_i x_{n-1}$, $i = n \pmod 2$ has the property $||x_n|| \to \text{dist }(x,Y)$.

Our intention in this paper is to show that central proximity maps are not commonly occuring phenomena. However, we begin by citing an existence result.

Copyright © 1980 by Academic Press, Inc.
All rights of reproduction in any form reserved.
ISBN: 0-12-171050-5

Theorem 1.2

Let $X = C(T)$, T a compact topological space.

Then $(P_x)(t) = \frac{1}{2}\{\sup_{t \in T} x(t) + \inf_{t \in T} x(t)\}$ is a central proximity map from X to the subspace of constants on T, using the usual supremum norm.

The following result, which is hardly more than a remark, often proves useful in cases where the closest point is not unique.

Theorem 1.3

Let Y be a subspace of X. If there exists a central proximity map $P : X \to Y$, then for each $x \in X$ the set $Y_x = \{y \in Y : ||x-y|| = \text{dist}(x,Y)\}$ must be the translate of a balanced set in Y. If this is the case, then Px is the image of 0 under the translation, and so is necessarily unique.

Proof

If Y_x consists of a single point then there is nothing to prove. Suppose, therefore Y_x is not a singleton, and neither is it the translate of a balanced set with Px as the image of zero. Suppose $Px = y_o \in Y_x$. Then by the assumptions on Y_x there exists $y \in Y_x$ such that $2Px-y \notin Y_x$. Hence

$$||x - y|| < ||x - 2Px + y||$$

or $$||x - Px - (y + Px)|| < ||x - Px + (y + Px)||.$$

Since $Px + y \in Y$ this contradicts the centrality property.

Example

Let $X = L_1[0,1]$; $Y = P_1[0,1]$, the polynomials of degree one on $[0,1]$; $||x|| = \int_0^1 |x(t)|\,dt$. Then there does not exist a central proximity map from X to Y. To see this, let $x \in X$ be defined by

$$x(t) = \begin{cases} 1 & t \in [0,\frac{1}{4}) \\ -1 & t \in [\frac{1}{4},\frac{3}{4}) \\ 0 & t \in [\frac{3}{4},1] \end{cases}$$

Furthermore let x be approximated by $a_o + a_1 t$. Then the set of best approximations is as indicated in Fig. 1.

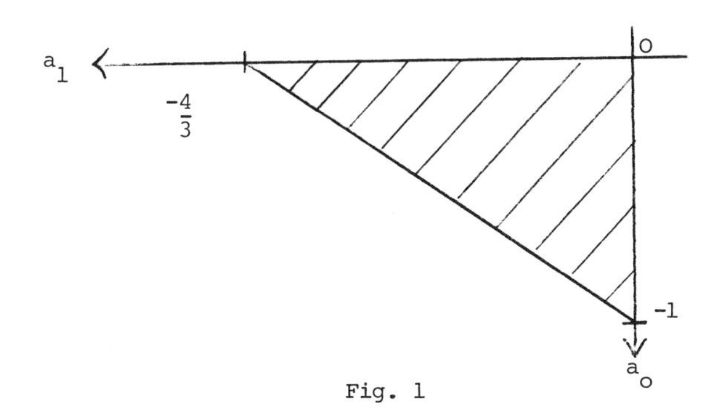

Fig. 1

Clearly this set is not the translate of a balanced set in $\mathbb{R}^2$ and consequently neither is Y_x in Y. We now prove two rather more general theorems.

II. THE CONTINUOUS CASE

We take X = C(T), T a closed interval of $\mathbb{R}$ with the usual supremum norm, and Y an n-dimensional Haar subspace of C(T). Then we have

Theorem 2.1

If X and Y are as defined above, and Y has dim Y = n > 1, then there does not exist a central proximity map from X to Y.

Proof

By the Haar property, we may choose points $t_1, t_2, \ldots t_{n-1} \in T$ and $y \in Y$ such that $y(t_i) = 0$ for $i = 1, 2, \ldots, n-1$. Then y is of constant sign in each interval $I_i = (t_i, t_{i+1})$ where $\bigcup I_i = T$, and $0 \leq i \leq n$.† Now take $\alpha_1 \in T$ such that $|y(\alpha_1)| = ||y||$. We shall suppose $y(\alpha_1) = 1$ (if not redefine y as -y and/or normalise). Then $\alpha_1 \in I_k$ say. Choose $\alpha_2 \in I_k$ such that $y(\alpha_2) = 1 - \delta$ where $\delta > 0$. Note that $\delta \leq 1$. We may

† Here we have taken $T = [t_o, t_n]$.

take $\alpha_1 < \alpha_2$. Then on the set $U = \{t_1, t_2, \ldots, t_k, \alpha_1, \alpha_2, t_{k+1}, \ldots, t_{n-1}\}$ define $x \in C(T)$ by $x(t_i) = (-1)^{k-i+1}$, $x(\alpha_1) = -x(\alpha_2) = 1$. For all $t \in T \setminus U$ define x so that $|x(t)| < 1$. Then by the characterisation theorem x has closest point 0 in Y. Furthermore $||x+y|| = ||x|| + ||y||$ since $||x|| + ||y|| \geq ||x+y|| \geq |(x+y)(\alpha_1)| \geq (x+y)(\alpha_1) = ||x|| + ||y||$. Now surround each point x_i, α_1, α_2 by disjoint open sets O_{x_i}, O_{α_1}, O_{α_2}. Set $S = (\bigcup_i O_{x_i}) \cup O_{\alpha_2}$. Now we consider the decomposition of T into $S \cup O_{\alpha_1} \cup C$ where $C = T \setminus O_{\alpha_1} \cup S$, and a standard compactness argument gives the result, $||x-y|| < ||x+y||$. Hence the proximity map is not central.

III. THE L_p CASE

The corresponding result in L_p is obtainable : here the setting is $X = L_p[0,1]$, $1 < p < \infty$, $Y = P_o[0,1]$, the functions constant almost everywhere on $[0,1]$. Let $x \in X$ be defined by

$$x_\epsilon^h(t) = \begin{cases} 1 & t \in [0,\tfrac{1}{2}) \\ h & t \in [\tfrac{1}{2}, \tfrac{1}{2} + \epsilon/2) \\ 0 & t \in [\tfrac{1}{2} + \epsilon/2, 1]. \end{cases}$$

Then $\int |x|^{p-1} \operatorname{sgn} x = \tfrac{1}{2} - h^{p-1} \frac{\epsilon}{2}$, and if we take $h^{p-1}\,\epsilon = 1$ we obtain the closest point to x as 0 from $P_o[0,1]$. Now for centrality of the proximity map we need

$$\int |x_\epsilon^h + b|^p = \int |x_\epsilon^h - b|^p \quad \forall\ b \in P_o[0,1].$$

Now

$$\int |x_\epsilon^h + b|^p = (1+b)^p.\ \tfrac{1}{2} + (h-b)^p\ \epsilon/2 + b^p.\ \tfrac{1}{2}(1-\epsilon),\ h > b$$

$$\int |x_\epsilon^h - b|^p = (1-b)^p . \tfrac{1}{2} + (h+b)^p . \epsilon/_2 + b^p . \tfrac{1}{2}(1-\epsilon),\ b < 1.$$

So for $0 < b < \min(1,h)$ we have

$$2\int\{|x_\epsilon^h + b|^p - |x_\epsilon^h - b|^p\} = bp\ (1-\epsilon h^{p-1}) + \frac{b^3 p(p-1)(p-2)}{3!}\ (1-\epsilon h^{p-3}) + \ldots$$

The expansion converges since $b < 1$ and $\frac{b}{h} < 1$.

Now $\epsilon h^{p-r} = \dfrac{\epsilon h^{p-1}}{h^{r-1}} = \dfrac{1}{h^{r-1}} < 1$ since $h > b > 1$, so that $\epsilon h^{p-r} < 1$ for all

$r \geqslant 1$.

Hence $2\int\{|x_\epsilon^h + b|^p - x_\epsilon^h - b|^p\} > 0$, unless $p = 2$. Thus we can conclude:

<u>Theorem 3.1</u>

The proximity map $P : L_p\ [0,1] \to P_o\ [0,1]$ is not central, unless $p = 2$.

Although easy to see, this theorem is important since it shows that Golomb's theory is not applicable to L_p spaces. However, work of Sullivan [4], shows that the simple generalisation of the algorithm of Diliberto and Straus converges (i.e. the conclusion of theorem 1.1 holds although the hypotheses do not).

IV. REMARKS

These results are with the exception of theorem 2.1, results about individual subspaces of given Banach spaces. It is possible in some cases to state theorems which hold for all subspaces of particular spaces. Work in this vein will appear in [3], where it is shown in particular that the property of central proximity in the space ℓ_1 is extremely rare.

The conclusion of Theorem 2.1 is connected with some work carried out recently by N. Dyn [5]. This paper shows that the natural generalisation of the Diliberto-Straus procedure to the subspace of $C(S\times T)$ given by $P_1(S) \otimes C(T) + C(S) \otimes P_0(T)$, where $S = T = [0,1]$, fails to converge even in norm. Theorem 2.1 can be easily extended to show that Golomb's argument fails in this setting due to the lack of central proximity.

REFERENCES

1. Diliberto, S.P. and Straus, E.G., "On the approximation of Functions of several variables by sums of functions of fewer variables", Pacific J. Math 1 (1951) pp. 195-210.
2. Golomb, M., "Approximation by Functions of Fewer variables", pp. 275-327 in "On Numerical Approximation" (R. Langer, ed.), University of Wisconsin Press, Madison, Wisconsin, 1959.
3. Light, W.A. and Sulley, L.J., "Central Proximity Maps", University of Lancaster Technical Report, to appear.
4. Sullivan, F., "A Generalisation of Best Approximation Operators", Ann. Mat. Pura. Appl. 107 (1975) pp. 245-261.
5. Dyn, N. "A Straightforward Generalization of Diliberto and Straus' Algorithm Does Not Work", University of Wisconsin MRC Technical Report No. 1916 (1979).

INTERPOLATION AND PROBABILITY

Rudolph A. Lorentz

Gesellschaft für Mathematik
und Datenverarbeitung

For an $m \times n$ incidence matrix $E = (e_{ik})_{i=1,k=1}^{m \quad n}$ with elements that are zeros and ones, and with exactly n ones, we consider the Birkhoff interpolation: $P^{(k-1)}(x_i) = c_{ik}$ if $e_{ik} = 1$, for polynomials P of degree not exceeding n-1. If the problem is solvable for all real knots X: $x_1 < \ldots < x_m$ and all real data c_{ik}, the matrix is regular. Otherwise it is singular. For an exposition of known results in Birkhoff interpolation see (3).

Necessary for the regularity of E is the Pólya condition, which means that for k = 1,...,n the first k columns of E contain at least k ones. Then E is called a Pólya matrix. Equivalent to this is the backward Pólya condition: any last n-k columns contain at most n-k ones.

The purpose of this paper is to prove that Pólya matrices with large m,n are singular with probability close to one. In (2), it has been shown that if

$$m \geq (1 + \delta)\frac{n}{\log n} \qquad (1.1)$$

Copyright © 1980 by Academic Press, Inc.
All rights of reproduction in any form reserved.
ISBN: 0-12-171050-5

for some $\delta > 0$, then for all large n, most Pólya matrices are singular. We wish to use another property of E (a less geometric one) to establish a similar theorem without restriction (1.1).

The property mentioned depends on a selection of any three different rows F_1, F_2, F_3 of E. By $(F_1)_2$, we denote the precoalescence of row F_1 with respect to row F_2, by $(F_1)_{23}$ -that of F_1 with respect to the coalescence of F_2 and F_3. If F and G are two disjoint sequences of integers, F,G is a sequence listing (in this order) both of them. Then

$$((F_1)_2, F_2)_3 \quad \text{and} \quad (F_1)_{23}, (F_2)_3 \tag{1.2}$$

are two sequences, consisting of the same integers, but perhaps in a different order. Let Δ(mod 2) be the number of permutations which transform one of them into the other. The following is a special case of a criterion of singularity in (1):

Proposition A. The matrix is singular if $\Delta \equiv 0$.

For example, for the two 3 x 3 matrices

$$\text{a)} \begin{bmatrix} 1 & 0 & 0 \\ 1 & 0 & 0 \\ 1 & 0 & 0 \end{bmatrix} \qquad \text{b)} \begin{bmatrix} 1 & 0 & 0 \\ 0 & 1 & 0 \\ 1 & 0 & 0 \end{bmatrix} \tag{1.3}$$

we have $\Delta \equiv 0$ and $\Delta \equiv 1$ respectively, hence the second matrix is singular.

An important problem is that of counting classes of matrices (see 2).. We often denote a class of matrices and their number by the same symbol. Thus, the class M(m,n;p) of all m x n incidence matrices with p ones has $\binom{mn}{p} = M(m,n;p)$ matrices. If p = n, we denote this number by M(m,n). For the number of all Pólya matrices with n ones, it has been proved in (2), that

$$P(m,n) = \frac{1}{n+1}\binom{m(n+1)}{n} \leq cn^{-1}M(m,n). \qquad (1.4)$$

We will show that for m, n large, the distribution of ones in Pólya matrices is uniform. Let p be an integer with $m \geq 3p$. We divide the integers $1,\ldots,m$ into p groups consisting of $[m/p]$ or $[m/p] + 1$ consecutive integers. For each $E \in M(m,n)$, let E_k be the submatrix of E consisting of the rows numbered by G_k. We will show that for almost all Pólya matrices, each E_k has approximately $n(1/p)$ ones.

Lemma 1. Let $\varepsilon > 0$ be given. Then there is an r, $0 < r < 1$ for which for all m, n sufficiently large, all $E \in M(m,n)$ except for at most $r^n M(m,n)$, have at most $n(p^{-1} + \varepsilon)$ ones in each E_k of the above decomposition. The number r is independent of m and n.

Proof. Let g_k be the number of rows of E_k. Then the exact probability that the slice E_k has exactly i ones is given by the hypergeometric distribution

$$h(n,ng_k,mn;i) = \frac{\binom{mg_k}{i}\binom{mn-mg_k}{n-i}}{\binom{mn}{n}}. \qquad (1.5)$$

This may be evaluated directly or by means of (5). The direct method consists of using Stirling's formula in 1.5 and then estimating the asymptotic value of a product of functions of the form

$$f(x) = x^x(1-x)^{1-x}$$

for diverse values of the argument.

The other alternative is to use (5) or similar estimates. Then one obtains directly

$$P[|X_k - n/p| \geq n\varepsilon] \leq 2\exp(-2nt^2) \qquad (1.6)$$

where X_k is the number of ones falling into E_k. Lemma 1 is a consequence of this bound.

Lemma 2. Let $\varepsilon > 0$. Then there is an r, $0 < r < 1$ not depending on m and n such that for all m, n sufficiently large, all $P \in P(m,n)$, except for at most $r^n P(m,n)$, have at most $n(p^{-1} + \varepsilon)$ ones in each P_k of the decomposition of P into p submatrices.

This lemma follows from lemma 1 and the inequality 1.4.

We introduce the following terminology. A column k is special for a horizontal submatrix E' of E if the matrix formed by the first k - 1 columns of E' satisfies the backward Pólya condition. A triplet of E' are three ones located one each in three rows i < i' < i" of E' and in adjacent columns 3l, 3l + 1, 3l + 2 provided the triplet itself satisfies the Pólya condition and that these columns contain no other ones in E'. The three rows and three columns define a cage C: thus there are 16 possible triplets in a cage C. A cage C or a triplet is special if the column 3l is special in E'. In this case the column 3l + 3 is also special.

Lemma 3. Let E' be a fixed m x n matrix with $N \leq n$ ones. Then there are at most N positions that are not special in E'.

In our decomposition, each E_k has, by lemma 2, few ones so that many of the above special positions are followed by three empty columns. If one places a triplet into this vacancy, a matrix in $P(m,n;n+3)$ is generated. By means of a counting argument for this mapping of $P(m,n)$ into $P(m,n;n+3)$, the following lemma is proved.

Lemma 4. Let $P'(m,n)$ denote the set of non-exceptional matrices of lemma 2. Then almost all matrices $E \in P'(m,n)$ satisfy the following condition: there exist p disjoint sets of three columns $3l_k$, $3l_k + 1$, $3l_k + 2$, $k = 1,\ldots,p$ with the property that the k-th set contains a special triplet of E_k. The number of exceptional matrices does not exceed $c(p)n^{-1}P(m,n)$ for $n = 7$.

We denote by $\overline{P}(m,n)$ the set of all non-exceptional matrices of lemma 4.

Theorem 5. For each $\varepsilon > 0$ and all sufficiently large m, n all but $\varepsilon P(m,n)$ Pólya matrices are singular.

Proof. $\overline{P}(m,n)$ will be divided into equivalence classes each of which consists of matrices all having a fixed set of cages according to lemma 4 and the same entries outside the cages. In each cage, at least one of the two possible triplets 1.3a,b will lead to a singular matrix. Thus at most $(15/16)^p P(m,n)$ of the matrices can be regular. Althogether, at most

$$\{cr^n + c(p)n^{-1} + (15/16)^p\}P(m,n)$$

of all Pólya matrices can be regular. Selecting first a $p \geq 7$, taking $m \geq 3p$ and then a large n, we see that at most $\varepsilon P(m,n)$ of all Pólya matrices are regular.

ACKNOWLEDGMENTS

The results obtained here are due to joint work with G. G. Lorentz to whom I am greatly indebted. Thanks are also due to H.-G. Galbas.

REFERENCES

1. Lorentz, G. G., J. Approx. Th. 20, 178 (1977).
2. Lorentz, G. G.,and Riemenschneider, S. D., Acta Math. Sci. Hungar. 33, 127 (1979).
3. Lorentz, G. G., and Riemenschneider, S. D., In "Approximation Theory and Functional Analysis" (J. B. Prolla,ed.) p. 187, North Holland Pub. Co., Amsterdam(1979).
4. Schoenberg, I. J., J. Math. Anal. Appl. 16,538(1966).
5. Serfling, R. J., Ann. of Stat. 2,39(1974).

ON COMPUTING THE POLYNOMIALS IN PADÉ APPROXIMANTS BY SOLVING LINEAR EQUATIONS

Yudell L. Luke[1]

Department of Mathematics
University of Missouri
Kansas City, Missouri

Let the Padé approximation to a given function $h(x)$ be notated as $A_m(x)/B_n(x)$ where $A_m(x)$ and $B_n(x)$ are suitably normalized polynomials in x of degree m and n respectively. If $h(x)$ has an at least formal infinite series in powers of x, then the coefficients in these polynomials are found by requiring that $B_n(x)h(x) - A_m = 0(x^{m+n+1})$. The purpose of this paper is to explore questions concerning the computation of the coefficients in $A_m(x)$ and $B_n(x)$. As is well known, there are several methods to get Padé approximations. For recent treatments of the subject, see Graves-Morris (1), Wuytack (2), and Bultheel and Wuytack (3). In particular, Graves-Morris discusses criteria for a good numerical method and proposes in order of importance (1) reliability, (2) discrimination (stability), (3) accuracy, (4) efficiency, (5) storage and (6) generalisability. Analyses of these questions for the various methods are rather scanty.

In this paper we examine the first three criteria when the method employed is that of solving systems of linear equations to determine the coefficients in $B_n(x)$. Once these coefficients are known the coefficients in $A_m(x)$ are readily found. Our approach in the main is heuristic. For the numerics five models are used for which the coefficients in $B_n(x)$ (and in some cases the coefficients in $A_m(x)$) are integers known a priori. The functions treated are e^{-x}, $x^{-1}\ln(1+x)$, $(1+x)^{\frac{1}{2}}$, $(1+x)^{-\frac{1}{2}}$ and the divergent but asymptotic expansion which emanates from the exponential integral $ze^z \int_z^\infty t^{-1}e^{-t}dt$. We take $m = n$.

---

[1]Supported by NSF grant MCS-78-0137

Copyright © 1980 by Academic Press, Inc.
All rights of reproduction in any form reserved.
ISBN: 0-12-171050-5

All computations were done on an Amdahl 470/V7 operating under OS MVS Release 3.7 and using the FORTRAN IV H-Extended Compiler. The precision used is quadruple where the numbers are in floating arithmetic with 32 digits in base 10. The machine code for solving the linear equation system is taken from IMSL Library V2. It is called subroutine LEQTIF, which is based on Gaussian elimination (Crout algorithm) with equilibration and partial pivoting. In our application, the matrix was of the Toeplitz type.

For each model the maximum number of incorrect figures in the coefficients which define the polynomials $A_n(x)$ and $B_n(x)$ is about n, $n \leq 20$. From the computational point of view, we are concerned about the errors $\delta_n(x)$ in $A_n(x)$ and $\omega_n(x)$ in $B_n(x)$, the corresponding relative errors and more importantly, the relative error in $A_n(x)/B_n(x)$. In each case for a given x, the errors and the relative errors in the polynomials increase as n increases. However, the relative error in $A_n(x)/B_n(x)$ is virtually nil to within the precision of the machine. The reason for this has a theoretical basis and can be explained in a general way. It can be shown that $\omega_n(x)h(x)-\delta_m(x) = O(x^{m+1})$ and so $\delta_m(x)/\omega_n(x)$ mimics the Padé form $A_m(x)/B_n(x)$ in that both are approximations to $h(x)$. Thus if $\delta_m(x)/\omega_n(x) - A_m(x)/B_n(x)$ and $\omega_n(x)/B_n(x)$, the relative error in $B_n(x)$, are sufficiently small, then the relative error in $A_m(x)/B_n(x)$ is inconsequential. Under these circumstances, I conjecture that computation of the polynomials $A_m(x)$ and $B_n(x)$ based on solving a system of linear equations rates high on any such scale assigned to the criteria noted above. A more detailed analysis replete with numerical examples will be given elsewhere.

REFERENCES

1. Graves-Morris, P.R., in "Padé Approximation and its Applications" (L. Wuytack, ed.), p. 231. Springer-Verlag, New York (1979).
2. Wuytack, L., in "Padé Approximation and its Applications" (L. Wuytack, ed.), p. 375. Springer-Verlag, New York (1979).
3. Bultheel, A., and Wuytack, L., "Stability of Numerical Methods for Computing Padé Approximants," these Proceedings.

ON A THEOREM OF MONTESSUS DE BALLORE FOR (μ, ν) - TYPE RATIONAL APPROXIMANTS IN $\mathbb{C}^n$

Clement H. Lutterodt

Department of Mathematics
University of South Florida
Tampa, Florida

I. INTRODUCTION

Recently Karlsson and Wallin (1977) have proved a generalization of the classical theorem of Montessus in two variables using rational approximants constructed from homogeneous polynomials. They pointed out by means of counter-examples in their formulation how sensitive the selection of the coefficients of $Q_n f$ (and consequently those of $Q_n Q f$) equated to zero is, in computing P_n/Q_n which converges to f in the Montessus sense.

The purpose of this paper is to extend Montessus'result of Karlsson and Wallin to the case of rational approximants constructed from nonhomogeneous polynomials in $\mathbb{C}^n$, avoiding the difficulties of Chisholm and Graves-Morris (1975).

The main results of this paper are embodied in theorems 1 and 2. Theorem 1 is the generalized Montessus' theorem in $\mathbb{C}^n$ whereas theorem 2 generalizes a special case of a result of Karlsson and Saff (1979). We have omitted the proofs of theorem 1 and Lemma 1 in order to cut down the

Copyright © 1980 by Academic Press, Inc.
All rights of reproduction in any form reserved.
ISBN: 0-12-171050-5

length of this paper. We shall however present them elsewhere, Lutterodt (1980).

A. Notation and Definitions

Let $z = (z_1,\ldots,z_n)$ be an n-tuple point in $\mathbb{C}^n$; let $\mu:=(\mu_1,\ldots,\mu_n)$ and $\nu:=(\nu_1, \ldots, \nu_n)$ be n-tuples of non-negative integers, i.e. in $\mathbb{N}^n$.

Let $\mathcal{R}_{\mu\nu}$ be the class of all rational functions of the form

$$R_{\mu\nu}(z) = P_\mu(z)/Q_\nu(z), \quad Q_\nu(0) \neq 0$$

where $P_\mu(z)$ and $Q_\nu(z)$ are polynomials of multiple degree at most μ and ν, respectively, and $(P_\mu(z), Q_\nu(z)) = 1$.

<u>Definition 1</u>. Suppose f(z) is analytic at the origin and $f(0) \neq 0$. An $R_{\mu\nu}(z)\epsilon\mathcal{R}_{\mu\nu}$ is said to be a (μ,ν)-type rational approximant to f(z) at z = 0 if

$$\frac{\partial^{|\lambda|}}{\partial z^\lambda}(Q_\nu(z)f(z) - P_\mu(z))\Big|_{z=0} = 0 \qquad (I.1)$$

for $\lambda \in E^{\mu\nu} \subset \mathbb{N}^n$, a lattice interpolation set with the following properties:

(i) $0 \in E^{\mu\nu}$

(ii) $\lambda \in E^{\mu\nu} \Rightarrow \gamma \in E^{\mu\nu} \quad \gamma_i \leq \lambda_i \quad i = 1,\ldots,n$

(iii) $E_\mu:=\{\lambda\epsilon\mathbb{N}^n: 0 \leq \lambda_i \leq \mu_i, \quad i = 1,\ldots,n\} \subset E^{\mu\nu}$

(iv) $|E^{\mu\nu}| \leq \prod_{i=1}^{n}(\mu_i + 1) + \prod_{i=1}^{n}(\nu_i + 1) - 1$

(v) Each projected variable has the Padé index set

(vi) Each $\nu_i \leq \mu_i \quad i=1,\ldots,n.$

Here $|E^{\mu\nu}|$ is the cardinality of $E^{\mu\nu}$. There are many $E^{\mu\nu}$'s fulfilling the above properties. We shall assume that a suitable $E^{\mu\nu}$ has been selected for the construction of the approximants. See for instance, Lutterodt (1976).

It was shown by Karlsson and Wallin (1977) that in order to get some kind of uniqueness for the rational approximants, a maximality condition must be invoked on the interpolation set, $E^{\mu\nu}$ in our case.

Definition 2. The interpolation set $E^{\mu\nu}$ is said to be maximal if

$$|E^{\mu\nu}| \geq \prod_{i=1}^{n} (\mu_i+1) + \prod_{i=1}^{n} (\nu_i+1) - 1 \ .$$

Now if we require that our chosen $E^{\mu\nu}$ be maximal and use the form $P_\mu(z) = \sum_{\alpha\in E_\mu} a_\alpha z^\alpha$, $Q_\nu(z) = \sum_{\beta\in E_\nu} b_\beta z^\beta$ and $f(z) = \sum_{\lambda\in\mathbf{N}^n} c_\lambda z^\lambda$ analytic at $z = 0$, then we extract from definition 1 the following sets of linear equations:

$$\sum_{\beta=0}^{\min(\lambda,\nu)} b_\beta \, c_{\lambda-\beta} = a_\lambda \qquad \lambda\in E_\mu \tag{I.2}$$

$$\sum_{\beta=0}^{\min(\lambda,\nu)} b_\beta \, c_{\lambda-\beta} = 0 \qquad \lambda\in E^{\mu\nu}\setminus E_\mu \tag{I.3}$$

The sums and indices are all multiple. Equation (I.3) provides a non-trivial solution for the b's. However, by taking $b_{0\ldots0} = Q_\nu(0) \neq 0$ to be unity, we can change (I.3) into an inhomogeneous matrix equation. If the rank of the augmented matrix is $\prod_{i=1}^{n} (\nu_i+1) - 1$ or

$$\det(c_{\lambda - r}) \neq 0, \quad r \neq 0, \quad \lambda \in E^{\mu\nu} \backslash E_{\mu}$$

then the solution set of the remaining b's is said to be <u>unisolvent</u>. These b's can then be used to determine the a's from (I.2). The (μ,ν)-type rational approximant thus determined shall be called <u>unisolvent</u> w.r.t the chosen maximal $E^{\mu\nu}$. We shall denote such a solution by

$$\pi_{\mu\nu}(z) = P_{\mu\nu}(z)/Q_{\mu\nu}(z).$$

B. Some Remarks

(i) The above definition of rational approximants may be used to generate rational approximants for function without convergent power series but with a cetain high degree of smoothness.

(ii) If $f(z)$ is analytic at $z = 0$ and $f(0) = 0$ and if $f(z)$ is <u>regular</u> order m in the j-th variable, then the Weierstrass Preparation theorem guarantees the existence of a unique pseudo-polynomial in z_j, say $W(\hat{z},z_j)$ where $\hat{z} = (z_1,\ldots,z_{j-1},z_{j+1},\ldots,z_n)$, such that $W(0,0) = 0$ and

$$f(z) = W(\hat{z},z_j)g(z)$$

Here $g(z)$ is a unit at $z = 0$, i.e. $g(0) \neq 0$. Our definition of a rational approximant then applies to the unit $g(z)$ of $f(z)$.

II. CONVERGENCE

<u>Theorem 1</u>. Let $\nu = (\nu_1,\ldots,\nu_n)$ be fixed. Suppose $f(z)$ is analytic at origin in $\mathbb{C}^n$ and meromorphic with finite pole set given by $G_\nu := \{z \in \mathbb{C}^n : q_\nu(z) = 0\}$ where $q_\nu(z)$

is a polynomial of multiple degree ν. Let $\rho > 0$ be fixed and let $\Delta_\rho := \{z \in \mathbb{C}^n : |z_i| < \rho, \quad i = 1,\ldots,n\}$ such that $\Delta_\rho \cap G_\nu \neq \phi$.

Suppose $\pi_{\mu\nu}(z)$ is a unisolvent (μ,ν)-rational approximant to $f(z)$, with a pole set $Q_{\mu\nu}^{-1}(0)$. Then as $\mu \to (\infty,\ldots,\infty)$

(i) $\Delta_\rho \cap Q_{\mu\nu}^{-1}(0) \to \Delta_\rho \cap G_\nu$

(ii) $\pi_{\mu\nu}(z) \to f(z)$ uniformly on compact subsets of $\Delta_\rho \setminus G_\nu$

<u>Theorem 2</u>. Suppose $f(z)$ is analytic at the origin and is at most meromorphic with a finite pole set in $\mathbb{C}^n$. Suppose for each fixed $\nu = (\nu_1,\ldots,\nu_n)$, the pole set of each unisolvent (μ,ν)-rational approximant $\pi_{\mu\nu}(z)$ to $f(z)$ tends to infinity as $\mu \to (\infty,\ldots,\infty)$. Then $f(z)$ must be entire in $\mathbb{C}^n$.

<u>Lemma 1</u>. Let $\nu = (\nu_1,\ldots,\nu_n)$ be fixed. The pole set $Q_{\mu\nu}^{-1}(0)$ of $\pi_{\mu\nu}(z)$ tends to infinity as $\mu \to (\infty,\ldots,\infty)$ if and only if given any $\rho > 0$ and polydisk Δ_ρ

$$Q_{\mu\nu}^{-1}(0) \cap \Delta_\rho = \phi$$

for sufficiently large μ.

<u>Proof of Theorem 2.</u> By hypothesis the pole set of $\pi_{\mu\nu}(z)$ tends to infinity as $\mu \to (\infty,\ldots,\infty)$ for each fixed ν. By Lemma 1, given any $\rho > 1$ and polydisk Δ_ρ, $\exists$ $\mu_0 = (\mu_{10},\ldots,\mu_{n0})$ such that for $\mu_i > \mu_{i0}$ $i = 1,\ldots,n$

$$\Delta_\rho \cap Q_{\mu\nu}^{-1}(0) = \phi$$

for each fixed ν.

Since f(z) is at most meromorphic with a finite pole set, we let G_σ be this finite pole set and let σ be the multiple degree of q_σ in

$$G_\sigma := \{z \in \mathbb{C}^n : q_\sigma(z) = 0\}$$

Then by theorem 1, if we choose $\nu = \sigma$, we must have on Δ_ρ as $\mu \to (\infty,\dots,\infty)$,

$$\Delta_\rho \cap Q^{-1}_{\mu\sigma}(0) \to \Delta_\rho \cap G_\sigma .$$

But for $\mu_i > \mu_{i0}$, $i = 1,\dots,n$, we have above

$$\Delta_\rho \cap Q^{-1}_{\mu\sigma}(0) = \phi .$$

Hence $\Delta_\rho \cap G_\sigma$ must be null. Since $\rho > 1$ is arbitrary and therefore so is Δ_ρ, G_σ must be at infinity. Hence f(z) must be entire.

ACKNOWLEDGMENTS

I would like to thank Professor Saff for fruitful discussions and for drawing my attention to his theorem with Karlsson. I would also like to express my sincere appreciation to the CIES for the Fulbright-Hays Fellowship and its subsequent extension to cover the period during which this work was completed. Finally I would like to express my gratitude to the Department of Mathematics, University of South Florida, for their hospitality.

REFERENCE

1. Chisholm, J.R.S. & Graves-Morris, P. (1975) 'Generalizations of the theorem of de Montessus to two-variable approximants.' Proc. Roy. Soc. Ser. A 342, 341-372.

2. Karlsson, J. & Saff, E.B. (1979). Preprint.

3. Karlsson, J. & Wallin, H. (1977) 'Padé and Rational Approximation' (Saff and Varga, eds.), p. 83, Academic Press, NY.

4. Lutterodt, C.H. (1976) 'On Boundaries of Rational Approximants in Several Variables.' Ghana Sci. Journal 16.

5. Lutterodt, C.H. (1980) 'Montessus de Ballore theorem for (μ, ν)-type rational approximants in $\mathbb{C}^n$ and some consequences.' Preprint.

A FORMULA FOR THE DEGREE OF APPROXIMATION BY TRIGONOMETRIC POLYNOMIALS

Tom Lyche

Institutt for informatikk
Universitetet i Oslo
Oslo, Norway

I. INTRODUCTION

Let for $m \in N$ S_m be trigonometric functions of order m on $I = [-\pi,\pi]$ given by

$$S_{2n+1} = \{a_o + \sum_{k=1}^{n} (a_k \cos kx + b_k \sin kx): \quad a_k, b_k \in \mathbb{R}\}$$

$$S_{2n} = \{ \sum_{k=1}^{n} (a_k \cos(k-\tfrac{1}{2})x + b_k \sin(k-\tfrac{1}{2})x): a_k, b_k \in \mathbb{R}\} .$$

Let L_m be the differential operators

$$L_{2n+1} = D(D^2 + 1^2) \dots (D^2 + n^2)$$

$$L_{2n} = (D^2 + \tfrac{1}{2}^2)(D^2 + (2-\tfrac{1}{2})^2) \dots (D^2 + (n-\tfrac{1}{2})^2)$$

where $D = d/dx$. We note that $L_m f = 0$ for every $f \in S_m$. Given $g \in C^m(I)$ we let $\text{dist}(g,S_m)$ be the degree of approximation of g by functions from S_m using the norm $||g|| = ||g||_I = \max\{|g(x)| : x \in I\}$. We show that

$$\text{dist}(g,S_m) = \begin{cases} \pi|L_{2n+1}g(\xi)|/(n!)^2 & m=2n+1 \\ |L_{2n}g(\xi)|/\prod_{k=1}^{n}(k-\tfrac{1}{2})^2 & m=2n \end{cases} \tag{1}$$

Copyright © 1980 by Academic Press, Inc.
All rights of reproduction in any form reserved.
ISBN: 0-12-171050-5

for some $\xi \in I$. The proof is based on an integral representation for the error in trigonometric interpolation ([1],[3],[4]), and the fact that although the best approximation on I is not unique g has an interpolating best approximation. No assumptions of periodicity are made in this paper.

A formula corresponding to (1) for approximation by algebraic polynomials is known, see [5] p. 78 .

II. TRIGONOMETRIC BEST APPROXIMATION

(1) will follow from proposition 4 below. We first give two lemmas.

LEMMA 1. _To each_ $g \in C^1(I)$ _there exists a best approximation_ $f \in S_m$ _and_ m _distinct points_ $x_1, x_2, \ldots, x_m$ _in_ $(-\pi,\pi)$ _such that_ $g(x_k) = f(x_k)$ $k=1,2,\ldots,m$.

Proof. Since S_m is finite dimensional there exists for each α_o with $0<\alpha_o<\pi$ a constant C_m depending only on m and α_o such that $||f||_{[-\pi,\pi]} \leqq C_m||f||_{[-\alpha,\alpha]}$ for any $f \in S_m$ and any α with $\alpha_0 \leqq \alpha \leqq \pi$ (see [6] p. 90 for an actual constant when $m=2n+1$). Let $k \in N$. The natural basis for S_m forms a Chebyshev system on $I_k = [-\pi+1/k, \pi-1/k]$. We let f_k be the unique best approximation to g on I_k. Since $||f_k||_I \leqq C_m||f_k||_{I_k} \leqq C_m(||f_k-g||_{I_k} + ||g||_{I_k}) \leqq 2C_m||g||_I$ the sequence $\{f_k\}$ is uniformly bounded on I. Hence ([2] p.14) $\{f_k\}$ has a subsequence which converges uniformly on I to an $f \in S_m$. This f is a best approximation to g on I . Let $r_k = g - f_k$ and let $||r_k||_{I_k} = |r_k(y_{j,k})|$ $j = 0,1,\ldots,m$. Then there are m sequences $\{x_{jk}\}$ such that $f_k(x_{jk}) = g(x_{jk})$ $j=1,\ldots,m$. Moreover

$y_{j-1,k} < x_{j,k} < y_{j,k}$ $j=1,2, \dots ,m$. Since $||r_{k+1}||_{I_{k+1}} \geq ||r_k||_{I_k}$ and $\{dr_k/dx\}$ is uniformly bounded on I there are constants μ and M such that $\mu \leq |r_k(y_{jk})|$ and $||dr_k/dx|| \leq M$ all j and k. Thus $\mu \leq |r_k(y_{jk})| = |r_k(y_{jk}) - r_k(x_{jk})| \leq M|y_{jk} - x_{jk}|$. It follows that $|x_{j+1,k} - x_{j,k}| \geq \mu/M$ for all j and k . Now if x_j is a limit point of $\{x_{jk}\}$ $j = 1,2, \dots ,m$ then $x_1,x_2, \dots ,x_m$ are distinct and $f(x_j) = g(x_j)$ $j=1,2, \dots ,m$. Moreover $x_j \varepsilon (-\pi,\pi)$.

LEMMA 2. *Let for* $m \varepsilon N$, $g \varepsilon C(I)$, *and distinct points* $x_1,x_2, \dots ,x_m$ *in* $[-\pi,\pi)$ f *be the unique element in* S_m *such that* $f(x_k) = g(x_k)$ $k=1,2, \dots ,m$. *Then for* $x \varepsilon I$

$$g(x) - f(x) = \prod_{k=1}^{m} \sin\tfrac{1}{2}(x-x_k) \int_{-\pi}^{\pi} T_m(x,y)L_m g(y)\,dy \qquad (2)$$

where T_m *is nonnegative for* $x,y \varepsilon I$.

Proof. See [1], [3] , and [4] .

PROPOSITION 3. *Let* $m \varepsilon N$ *and suppose* $g,h \varepsilon C^m(I)$. *If* $|L_m g(x)| \leq L_m h(x)$, $x \varepsilon I$ *then* $dist(g,S_m) \leq dist(h,S_m)$

Proof. Let f be the interpolating best approximation to h. If $z \varepsilon I$ is such that $||g-f|| = |g(z) - f(z)|$ then using (2) $dist(g,S_m) \leq |g(z) - f(z)| \leq \Pi|\sin\frac{1}{2}(z-x_k)| \cdot \int T_m(z,y)|L_m g(y)|dy \leq \Pi|\sin\frac{1}{2}(z-x_k)|\int T_m(z,y)L_m h(y)dy = |h(z) - f(z)| \leq dist(h,S_m)$.

PROPOSITION 4. *Let* $m \varepsilon N$ *and the function* e *be given by* $e(x) = 1$ *if* m *is even and* $e(x) = x$ *if* m *is odd. If* $g \in C^m(I)$ *then*

$$\mathrm{dist}(g,S_m) = \mathrm{dist}(e,S_m)\,|L_m g(\xi)|/c_m \quad , \ \xi \in I \tag{3}$$

where $c_m = L_m e(x)$, $x \in I$; *i.e.* $c_m = \Pi_{k=1}^{n}(k-\frac{1}{2})^2$ *if* $m=2n$ *and* $c_m = (n!)^2$ *if* $m=2n+1$.

Proof. We follow the proof of theorem 60 in [5] by using L_m instead of $(d/dx)^m$ and e instead of x^m .

Since $\mathrm{dist}(e,S_m)$ is equal to π if m is odd and equal to 1 if m is even (1) follows from (3) .

We remark that (3) is valid on any interval of length less than 2π . Bounds for $\mathrm{dist}(e,S_m)$ for such intervals can be obtained from [1] .

REFERENCES

1. Koch, P.E. and Lyche, T., Bounds for the error in trigonometric Hermite interpolation, in "Quantitative Approximation", to be published by Academic Press.

2. Lorentz, G.G., "Approximation of Functions", Holt, Rinhart and Winston Inc., New York, 1966.

3, Lyche, T., A Newton form for trigonometric Hermite interpolation, BIT 19(1979),229-235.

4. Lyche, T. and Winther, R., A stable recurrence relation for trigonometric B-splines, J. Appr. Th. 25(1979),266-279.

5. Meinardus, G., "Approximation of Functions: Theory and Numerical Methods", Springer Verlag, Berlin, 1967.

6. Timan, A.F., "Theory of Approximation of Functions of a Real Variable", Pergamon Press, Oxford, 1963.

DEGREE OF APPROXIMATION IN COMPUTERIZED TOMOGRAPHY

W. R. Madych[1]

Department of Mathematics
Iowa State University
Ames, Iowa

1. <u>Introduction</u>. Suppose f is a reasonable complex valued function defined on the plane, R^2, and let u be a point on the unit circle centered at the origin of R^2. We will usually call such a point, u, a direction. Consider the mapping $f \to P_u f$, where $P_u f$ is the function defined by the formula

$$(1) \qquad P_u f(t) = \int_{-\infty}^{\infty} f(tu + sv)\,ds$$

and where v is a unit vector orthogonal to u so that $\{u, v\}$ form a right-handed coordinate system. For fixed u, $P_u f$ is a function on the real line, R, and is called a <u>radiograph</u> of f. Since $P_u f(t) = P_{-u} f(-t)$, a fact which follows immediately from (1), once $P_u f$ is known, $P_{-u} f$ can be easily computed. Hence, we say that a set of n directions, $\{u_1, \ldots, u_n\}$ is <u>distinct</u> if $|\langle u_i, u_j \rangle| < 1$ whenever $i \neq j$, where $\langle , \rangle$ denotes the usual inner (scalar, dot) product in R^2.

The basic mathematical problem in computerized tomography

---

[1] This work was partially supported by the U.S. Department of Energy, contract No. W-7405-Eng-82 Division of Basic Energy Sciences, Applied Mathematical Sciences, AK-01-04.

Copyright © 1980 by Academic Press, Inc.
All rights of reproduction in any form reserved.
ISBN: 0-12-171050-5

is the following: <u>Given n distinct directions, $\{u_1, \ldots, u_n\}$, and the n radiographs, $\{P_{u_1} f, \ldots, P_{u_n} f\}$, find f.</u>

It is obvious that a general function, f, cannot be determined from a finite set of radiographs. This fact raises the following question: <u>How well can one approximate f using only n of its radiographs</u>?

The point of this report is to announce some of our results concerning the last question. However before doing so, let us set up some notation and technical definitions.

The symbols x and y denote points in the plane, R^2, and s, t denote points on the real line, R. The notation $\langle x,y \rangle$ denotes the usual inner product in R^2. $|\xi|$ denotes the Euclidean length of the quantity ξ, namely, if ξ is a point in R^2 then $|\xi| = \sqrt{\langle \xi,\xi \rangle}$, if ξ is a real or complex number then $|\xi|$ denotes the absolute value or modulus; the meaning of this notation should be clear from the context. The symbol dx or dt denotes Lebesque measure in R^2 or R respectively. D denotes the disk of radius one centered at the origin in R^2, namely, $D = \{x: |x| \leq 1\}$.

Given a measurable subset, K, of R^2 or R and a number p, $1 \leq p \leq \infty$, $L^p(K)$ denotes the usual Banach space of equivalence classes of complex-valued Lebesque measurable functions on K. The norm in $L^p(K)$ of an element, f, is denoted by $||f||_{L^p(K)}$. If $0 < \alpha \leq 1$ and p is as above, $\Lambda^p_\alpha(D)$, denotes the class of those Schwartz distributions, f, on R^2 which are supported in D with the property that $f_y - f$ is in $L^p(R^2)$ for all y and $||f||_{\Lambda^p_\alpha}$ is finite, where

$$||f||_{\Lambda^p_\alpha} = \sup_{|y|>0} \{|y|^{-\alpha}||f_y - f||_{L^p(R^2)}\}$$

and f_y denotes the y translate of f. Clearly $\Lambda^p_\alpha(D) \subset L^p(D)$, $||f||_{L^p(D)} \leq 2^{1-\alpha}||f||_{\Lambda^p_\alpha(D)}$ whenever $f \in \Lambda^p_\alpha(D)$ and, if $1 \leq p < \infty$, then $\Lambda^p_\alpha(D)$ is dense in $L^p(D)$.

Generic constants which appear in certain estimates and whose meaning should be clear from the context will always be denoted by the symbol C. C need not be the same at different occurrences.

2. Some negative results. We regard elements, f, in $L^p(D)$ as Schwartz distributions on R^2 with support in D. In particular, given f in $L^p(D)$ and a direction u, the mapping $f \to P_u f$ is well defined. In fact, $P_u f(t)$ exists for almost all t and is in $L^p(I)$, where I is the interval $[-1,1]$.

Let $U = \{u_1, \ldots, u_n\}$ be a set of n distinct directions and consider the mapping $f \to P_U f$ defined by $P_U f = (P_{u_1} f, \ldots, P_{u_n} f)$. Let I^p_U denote the image of $L^p(D)$ under this mapping. We say that the transformation $R\colon I^p_U \to L^p(D)$ is a reconstruction if is homogeneous. Namely, R satisfies

$$R(cg) = cR(g) \tag{2}$$

for all g in I^p_U and all scalars c. Observe that (2) implies that $R(0) = 0$. Also note that this property is rather mild, in the sense that any reasonable "reconstruction method" should enjoy it.

Now, given a set of n distinct directions, U, a reconstruction, R, and a subspace, M, of $L^p(D)$ with seminorm $||\cdot||_M$, we define $E_p(M, R, U)$ by the following formula:

(3) $E_p(M,\mathcal{R},U) = \sup\{||f - \mathcal{R}P_U f||_{L^p(D)} : f \in M \text{ and } ||f||_M \leq 1\}$.

The above quantity measures how well the mapping $\mathcal{R}: P_U f \to \mathcal{R}P_U f$ "reconstructs" f. Namely, using property (2) of $\mathcal{R}$, it is easy to see that

$$||f - \mathcal{R}P_U f||_{L^p(D)} \leq E_p(M,\mathcal{R},U) \cdot ||f||_M$$

for all f in M and, given a positive ε, there are f's in M for which

$$||f - \mathcal{R}P_U f|| > (E_p(M,\mathcal{R},U) - \varepsilon)||f||_M.$$

Similarly we define

(4) $$E_p(M,U) = \inf_{\mathcal{R}} E_p(M,\mathcal{R},U)$$

and

(5) $$E_p(M,n) = \sup_{U} E_p(M,U)$$

where the infimum in (4) is taken over all possible reconstruction methods and the supremum in (5) is taken over all possible collections of n distinct directions. The meanings of these quantities should be clear from the definitions.

The quantities defined above can be readily estimated in many cases. For example, if M is finite dimensional and $U = \{u_1\}$, then $E_p(M,U) = 0$ for almost all such U's. In fact in many such cases it is not difficult to find a reconstruction $\mathcal{R}$ and a direction u_1 such that $\mathcal{R}P_{u_1} f = f$ for all f in M. Unfortunately these facts are not very interesting from the point of view of practical applications since such reconstructions are highly unstable when the dimension of M gets large, in the sense that slight perturbations of f will generate large perturbations in $\mathcal{R}P_{u_1} f$. Here we are primarily interested in the case $M = L^p(D)$ or $\Lambda_\alpha^p(D)$, $0 < \alpha \leq 1$.

<u>Theorem 1</u>: <u>If U is a finite collection of distinct directions then $E_p(L^p(D),U) = 1$ for all p, $1 \leq p \leq \infty$.</u> This result, of course, is not at all surprising. However in the case of $\Lambda_\alpha^p(D)$ we get a somewhat more encouraging result.

<u>Theorem 2</u>: <u>Suppose U is any set of n equally spaced directions. Then $E_p(\Lambda_\alpha^p(D),U) \geq Cn^{-\alpha}$, where C is a constant independent of n. This is true for any p, $1 < p < \infty$, and any α, $0 < \alpha \leq 1$.</u>

A set of n distinct directions $U = \{u_1, \ldots, u_n\}$ is said to be equally spaced, of course, if

$u_j = (\cos\theta_j, \sin\theta_j)$ $j = 1, \ldots, n$ where $\theta_j = \theta_{j-1} + \frac{\pi}{n}, j = 2, \ldots, n.$

3. <u>Some positive results</u>. Our positive results are based on two elementary observations:

(i) If f is integrable and supported in D and g is a locally integrable function with the property that $g(x) = h(\langle x,u\rangle)$ for some direction, u, and function of one real variable, h, then $g*f(x) = \int_{-\infty}^{\infty} P_u f(t) h(\langle x,u\rangle - t)\,dt$.

(ii) If $\{u_1, \ldots, u_n\}$ are distinct directions then the polynomials $P_{jk}(x)$, where $P_{jk}(x) = \langle x, u_k\rangle^j$, $u_k = 1, \ldots, j+1$, $j = 0, \ldots, n-1$, span the space, P_{n-1}, of polynomials in two variables of total degree less than or equal to $n-1$. (See [2])

If one can now construct a polynomial, Q, of two variables which satisfies appropriate moment conditions, using (i) and (ii) above it is not difficult to define a reconstruction, R, which will give the "correct" degree of approximation for $\Lambda_\alpha^p(D)$. We do this as follows:

(a) Given a positive integer n, choose another integer, m, so that $n - 4 < 4m - 3 \leq n$ and let $q(t) = c(t-\xi_1)^{-2}(P_m(t))^2$

where P_m is the classical Legendre polynomial of degree m on [0,3], $0 < \xi_1 < \ldots < \xi_m < 3$ are its zeros, and c is chosen so that $\int_0^3 q(t)dt = 1$.

(b) Define the two variable polynomial Q_n by the formula

$$Q_n(x) = \pi^{-1}q(|x|^2) \tag{5}$$

Observe that Q_n is a radially symmetric, non-negative polynomial of degree no greater than n - 1. The Gaussian quadrature formula and estimates on the nodes $\xi_1, \ldots, \xi_n$ (see [1] and [5]) together with a polar change of variables yield

$$\int_{|x|\leq 3} Q_n(x)dx = 1 \tag{6}$$

and if $0 < \alpha \leq 2$

$$\int_{|x|\leq 3} |x|^\alpha Q_n(x)dx \leq Cn^{-\alpha} \tag{7}$$

where C is a constant independent of n.

Now given any collection $U = \{u_1, \ldots, u_n\}$ of n distinct directions, using (ii), it follows that

$$Q_n(x) = \sum_{j=0}^{n-1} \sum_{k=1}^{j+1} a_{jk} \langle x, u_k \rangle^j \tag{8}$$

where the a_{jk}'s are constants which are, in principle, easily calculated.

We are now ready to write down the formula for the reconstructions. Given the radiographs $P_U f = (P_{u_1} f, \ldots, P_{u_n} f)$ we define the mapping $\mathcal{R}_n : P_U f \to \mathcal{R}_n P_U f$ by the formula

$$\mathcal{R}_n P_U f(x) = \sum_{j=0}^{n-1} \sum_{k=1}^{j+1} a_{jk} \int_{-1}^{1} P_{u_k} f(t)(\langle x, u_k \rangle - t)^j dt \tag{9}$$

where the a_{jk}'s are determined by (8). Observe that if f is in $L^p(D)$ then

$$\mathcal{R}_n P_U f(x) = Q_n * f(x). \tag{10}$$

Using (6) and (7) it is not difficult to show the following:

Theorem 3: Suppose U is a collection of n distinct directions and f is in $\Lambda_\alpha^p(D)$, $0 < \alpha \leq 1$, $1 \leq p \leq \infty$. Given $P_U f$ and the reconstruction $\mathcal{R}_n$ defined by (9) it follows that

$$\|f - \mathcal{R}_n P_U f\|_{L^p(D_1)} \leq C n^{-\alpha} \|f\|_{\Lambda_\alpha^p}$$

where C is a constant independent of f and n.

Note that the above estimate is independent of the directions, U, as long as they are distinct. Also observe that the above theorem implies the following.

Corollary 1: If U is a collection of n distinct directions and $\mathcal{R}_n$ is the reconstruction defined by (9). Then, if $0 < \alpha \leq 1$ and $1 \leq p \leq \infty$, $E_p(\Lambda_\alpha^p, \mathcal{R}_n, U) \leq C n^{-\alpha}$ where C is a constant independent of U and n.

Theorems 2 and 3 also imply

Corollary 2: If $0 < \alpha \leq 1$ and $1 \leq p \leq \infty$ then $E_p(\Lambda_\alpha^p, n) = O(n^{-\alpha})$. Furthermore, O cannot be replaced by o in the above formula.

Similar results hold for spaces defined in terms of more general moduli of smoothness. Details will appear elsewhere.

REFERENCES

[1] R. P. Feinerman and D. J. Newman, Polynomial Approximation, The Williams and Wilkens Co., Baltimore, 1974.

[2] B. F. Logan and L. A. Shepp, Optimal reconstruction of a function from its projections, Duke Math. J., 42(1975), 645-659.

[3] L. A. Shepp and J. B. Kruskal, Computerized tomograph: the new medical X-ray technology, Amer. Math. Monthly, 85(1978) 420-439.

[4] K. T. Smith, D. C. Solomon, and S. L. Wagner, Practical and mathematical aspects of the problem of reconstructing objects from radiographs, Bull. Amer. Math. Soc., 83(1977) 1227-1270.

[5] G. Szego, Orthogonal Polynomials, A.M.S. Colloquium Publications, Vol. 23, Ed. 2, 1959.

INTERPOLATION TO SCATTERED DATA IN THE PLANE BY LOCALLY DEFINED C^1 FUNCTIONS

Lois Mansfield[1]

Department of Applied Mathematics and Computer Science
University of Virginia
Charlottesville, Virginia

Suppose that the data (x_i, y_i, z_i), $i = 1,\ldots,n$, is given and one wants to find a smooth surface satisfying the interpolation conditions

$$z_i = f(x_i, y_i), \quad i = 1,\ldots,n.$$

The data points (x_i, y_i) are assumed to be distinct but are not assumed to lie in any special pattern such as at the nodes of a rectangular grid.

Applications of this interpolation problem include the representation of single-valued surfaces such as contour or sub-surface structure maps. One is given a set of (x_i, y_i, z_i) data representing measured or computed values and wants to obtain the visual impression of a smooth surface $z = f(x,y)$ interpolating the data. This data comes from terrain information gathered in the field or from air-photographs, or, in the case of sub-surface structure maps, from bore-hole data. The surface interpolation algorithm must then be interfaced with algorithms for contour plotting or surface perspective plotting.

To determine the interpolating surface, we suggest that one first determine the convex hull of the points

---

[1]Research supported by the National Science Foundation

Copyright © 1980 by Academic Press, Inc.
All rights of reproduction in any form reserved.
ISBN: 0-12-171050-5

(x_i,y_i), $i = 1,\ldots,n$, and then construct a triangular grid covering the convex hull having the given set (x_i,y_i), $i = 1,\ldots,n$, as vertices. Good recent algorithms to do this include Green and Sibson [5], Lawson [6], and Brassel and Reif [3].

It is easy to determine a locally defined continuous function which interpolates the given z_i at the points (x_i,y_i) using a piecewise linear polynomial. However, piecewise linear interpolation gives a rough appearance to the final surface or a rather jagged appearance to the final contour map. In the applications mentioned above the surfaces one wants to represent are smooth so that it is desirable to use an interpolant with at least C^1 continuity.

A good way to determine a locally defined C^1 - interpolant is to use functions originally invented for use in the finite element method to solve fourth order elliptic partial differential equations. It is possible to define C^1 - piecewise quintic polynomial interpolants, but the number of parameters required is far in excess of what can reasonably be estimated from the given data in applications like those we have described above. It was shown in [7] that locally defined C^1 - piecewise polynomial interpolants on a triangular grid must be of degree at least five. More appropriate interpolants are constructed by augmenting the set of quadratic or cubic polynomials on each triangle by rational or piecewise cubic functions so that these functions can be pieced together to form a function in C^1. See [2], [4], and [9].

The functions we recommend and shall describe are obtained by piecing together the 9-parameter rational interpolants discussed in [2]. Let the triangle T have vertices $P_i = (x_i, y_i)$, i = 1,2,3. Let $\lambda_i(x,y)$, i = 1,2,3, denote the barycentric coordinates of a point $(x,y) \varepsilon R^2$ with respect to the vertices of T. Equivalently, suppose that the edge e_i of T opposite the vertex P_i has the equation $\lambda_i(x,y) = 0$, normalized so that $\lambda_i(P_i) = 1$. Let $T(T)$ be the set of all polynomials which are cubic along parallels to the edges of T. Elements of $T(T)$ have been called tricubic polynomials by Birkhoff [1] and include all cubic polynomials along with the quartic polynomials $\lambda_1^2\lambda_2\lambda_3$ and $\lambda_1\lambda_2^2\lambda_3$. Note that $\lambda_1^2\lambda_2\lambda_3 + \lambda_1\lambda_2^2\lambda_3 + \lambda_1\lambda_2\lambda_3^2 = \lambda_1\lambda_2\lambda_3$ since $\lambda_1 + \lambda_2 + \lambda_3 = 1$.

We define the rational functions $\hat{\phi}_i$ by

$$(1) \qquad \hat{\phi}_i = \frac{\lambda_i^2\lambda_{i+1}\lambda_{i+2}}{1-\lambda_i}, \qquad i = 1,2,3,$$

where the subscripts are taken cyclically so for i = 3, $\lambda_{i+1} = \lambda_1$. The $\hat{\phi}_i$ are zero on the edges of T and have cubic normal derivatives on two of the edges. Let $\Phi_{15}(T)$ be the 15-parameter family of functions obtained by adjoining the three rational functions $\hat{\phi}_i$, i = 1,2,3, defined by (1) to the set $T(T)$ of tricubic polynomials. Finally we let $\Phi_9(T)$ be the 9-parameter family obtained from $\Phi_{15}(T)$ be including only those elements of $\Phi_{15}(T)$ where normal derivatives reduce to linear polynomials on the edges of T.

Suppose $\lambda_i(x,y)$ is given by

$$(2) \qquad \lambda_i(x,y) = \tilde{b}_i x - \tilde{a}_i y + \tilde{c}_i,$$

where

$$(3) \qquad \tilde{a}_i = \frac{a_j}{2\Delta}, \quad \tilde{b}_i = \frac{b_i}{2\Delta}, \quad \tilde{c}_i = \frac{c_i}{2\Delta},$$

where

$$(4) \qquad a_i = x_{i+2} - x_{i+1}, \quad b_i = y_{i+2} - y_{i+1}, \quad c_i = x_{i+2}y_{i+1} - x_{i+1}y_{i+2},$$

with $\Delta = \text{area } (T) = \frac{1}{2}\{(x_{i+2}-x_i)(y_{i+1}-y_i) - (y_{i+2}-y_i)(x_{i+1}-x_i)\}$.

Let

$$d_{ij} = a_j f_x(P_i) + b_j f_y(P_i) \tag{5}$$

be the directional derivative at P_i along the edge e_j scaled by multiplying by the length of the edge e_j. The orientation of T is taken so that the boundary of T is traversed in the direction of increasing subscripts of the vertices. The function Q_T given by

$$\begin{aligned} Q_T = & \sum_{i=1}^{3} (3\lambda_i^2 - 2\lambda_i^3 - 6\, A_{i,i+1}\phi_{i+1} - 6\, A_{i,i+2}\phi_{i+2}) z_i \\ & + \sum_{i=1}^{3} (\lambda_i^2 \lambda_{i+1} + B_{i,i+2}\phi_{i+2} + \phi_{i+1}) d_{i,i+2} \\ & + \sum_{i=1}^{3} (\lambda_i^2 \lambda_{i+2} + B_{i,i+1}\phi_{i+1} + \phi_{i+2}) d_{i,i+1} \end{aligned} \tag{6}$$

where

$$A_{i,i+1} = \frac{b_{i+1} b_i + a_{i+1} a_i}{a_{i+1}^2 + b_{i+1}^2}, \tag{7}$$

$$B_{i,j} = \frac{b_{i+1} b_{i+2} + a_{i+1} a_{i+2} - 2(b_i b_j + a_i a_j)}{a_j^2 + b_j^2}, \tag{8}$$

and

$$\phi_i = \frac{\lambda_{i+1}^2 \lambda_{i+2}^2 \lambda_i}{1-\lambda_{i+1}} + \frac{\lambda_{i+1}^2 \lambda_{i+2}^2 \lambda_i}{1-\lambda_{i+2}}, \quad i = 1,2,3, \tag{9}$$

has given values z_i and directional derivatives d_{ij} at the vertices of T. The ϕ_i of (9) differ from the $\hat{\phi}_i$ defined by (1) by a tricubic polynomial. By piecing together the functions Q_T over the triangular grid, a C^1-interpolant is obtained which interpolates to given values and first partial derivatives at the vertices of the triangular grid.

In most applications only values of these grid points are known. We now describe a method of estimating the partial derivatives at each data point by means of a weighted average of directional divided differences. The basic idea of this procedure can be motivated by piecewise Bessel interpolation on a regular grid on a line where cubic Hermite interpolation is used with the derivatives estimated by centered divided differences. If an irregular grid is used instead of a regular grid, one would want to replace the centered divided differences by an average of one-sided divided differences.

For each data point P_0, let the set of neighboring points be given by $N(P_0) = \{P_1,\ldots,P_k\}$, where $P_i = (x_i,y_i)$ is a neighbor of $P_0 = (x_0,y_0)$ if the line segment $\overline{P_oP_i}$ is an edge in the triangulation. For each $P_i \varepsilon N(P_0)$,

$$(10) \quad f(x_i,y_i) \approx f(x_0,y_0) + \alpha_i f_x(x_0,y_0) + \beta_i f_y(x_0,y_0),$$

where $\alpha_i = x_i - x_0$ and $\beta_i = y_i - y_0$.
By forcing equality in (10) we have a system of k equations for the two unknowns $f_x(x_0,y_0)$ and $f_y(x_0,y_0)$. We force (10) to hold in least squares sense and take as our estimates of $f_x(x_0,y_0)$ and $f_y(x_0,y_0)$, the values of s and t which minimize the weighted sum of squares

$$(11) \quad G(s,t) = \sum_{i=1}^{k} w_i(\alpha_i s + \beta_i t - c_i)^2,$$

where $w_i = (\alpha_i^2 + \beta_i^2)^{-1}$ and $c_i = f(x_i,y_i) - f(x_0,y_0)$. This amounts to solving the linear system

$$(12) \quad \begin{aligned} as + dt &= e \\ ds + bt &= f \end{aligned}$$

where

$$a = \sum_{i=1}^{k} w_i\alpha_i^2 \quad , \quad b = \sum_{i=1}^{k} w_i\beta_i^2 \quad , \quad d = \sum_{i=1}^{k} w_i\alpha_i\beta_i,$$

$$e = \sum_{i=1}^{k} w_i\alpha_i c_i \quad , \quad f = \sum_{i=1}^{k} w_i\beta_i c_i.$$

It is pointed out in an M.S. report [8] by David McSweeney at the University of Kansas that the coefficients a,b,d,e,f, for each vertex can be accumulated very efficiently in one pass through the list of triangles so that partial derivative estimates at all n data points may be calculated in time proportional to n. More details are given in [8] along with a program and several computer examples.

References

1. Birkhoff, G., "Tricubic polynomial interpolation," Proc. Natl. Acad. Sci., 68 (1971), 1162-1164.

2. Birkhoff, G., and Mansfield, L., "Compatible triangular finite elements", J. Math. Anal. Appl., 47 (1974), 531-553.

3. Brassel, K. E. and Reif, "A procedure to generate Thiessen polygons", Geog. Analysis, 11 (1979).

4. Goël, J. J., "Construction of basis functions for numerical utilization in Ritz's method," Numer. Math., 12 (1968), 435-447.

5. Green, P. J. and Sibson, R., "Computing Dirichlet tessellations in the plane," Computer Journal, 21 (1978), 168-173.

6. Lawson, C. L., "Software for C^1 surface interpolation," in Mathematical Software III, ed. J. R. Rice, Academic Press, New York, 1977.

7. Mansfield, L., "Interpolation to boundary data in triangles with application to compatible finite elements," in Approximation Theory II, ed. G. G. Lorentz, et al., Academic Press, New York, 1976, 449-455.

8. McSweeney, D., University of Kansas Computer Science Dept. Technical Report, (to appear).

9. Zienkiewicz, O. C., The Finite Element Method in Engineering Science, 2nd ed., McGraw Hill, New York, 1971.

RECENT ADVANCES IN NEAR-BEST APPROXIMATION

J.C. Mason

Department of Mathematics and Ballistics,

Royal Military College of Science,

Shrivenham, Swindon, England

1. INTRODUCTION

The term "near-best" was introduced by Mason (1) in 1970 to provide a theoretical basis for certain practical approximation methods. If a function f in a normed linear space X has a best approximation f_n^B in a subspace Y of dimension n+1, then any other approximation f_n is near-best within a relative distance ρ_n if

$$\|f - f_n\| \leqslant (1 + \rho_n)\|f - f_n^B\|. \tag{1}$$

In the case of the L_∞ norm, the term "near-minimax" may be used in place of near-best. The definition has practical implications if ρ_n is suitably small; for example, if $\rho_n \leqslant 9$ for, say, $n \leqslant 20$, then f_n is not more than one decimal place less accurate than f_n^B.

Also in 1970 Cheney and Price (2) gave an important review of "minimal projections". Since any projection P_n of f on a subspace Y of dimension n+1 can be shown to satisfy (1) for any $\rho_n \geqslant \sigma_n$, where

$$\sigma_n = \|P_n\|,$$

projections can be thought of as realisations of near-best approximations. They have practical significance as long as $\|P_n\|$ is suitably small.

Copyright © 1980 by Academic Press, Inc.
All rights of reproduction in any form reserved.
ISBN: 0-12-171050-5

In the present paper we summarise results about algebraic polynomial approximations devised from projections based on series expansion and interpolation criteria, and we raise a number of open questions. The emphasis is on establishing that σ_n is "small" and that a sequence of approximations $\{f_n\}$ converges in norm.

Define the "canonical" polynomial $\phi_n(x)$ for X to be the monic polynomial of degree n of minimum norm. Polynomial approximations f_n^S and f_n^I are then defined as follows:

f_n^S is the partial sum of degree n of the expansion of f in $\{\phi_k\}$,

f_n^I is the polynomial which interpolates f in the n+1 zeros of ϕ_{n+1}.

2. REAL L_∞ APPROXIMATION

For the space $C[-1, 1]$ with the L_∞ norm, $\phi_n(x) = 2^{1-n} T_n(x)$ where T_n is the Chebyshev polynomial of the first kind. Results due to Hardy (3), Ehlich and Zeller (4), Luttmann and Rivlin (5), Jackson (6) and others may be summarised as follows:

(i) For f_n^S, $\sigma_n = \lambda_n = \frac{1}{\pi}\int_0^\pi \frac{|\sin(n+\frac{1}{2})x|}{\sin\frac{1}{2}x}\,dx = \frac{4}{\pi^2}\log n + 1.27033\ldots + o(1)$

(ii) For f_n^I, $\sigma_n = \gamma_n = \frac{1}{n+1}\sum_{k=0}^{n} \cot\frac{(2k+1)\pi}{4(n+1)} = \frac{2}{\pi}\log n + .9625\ldots + o(1)$

(iii) Both f_n^S and f_n^I converge in norm to f, if f is Dini-Lipschitz continuous.

The constants λ_n and γ_n do not exceed 5 for n < 500, and so both f_n^S and f_n^I are practical near-minimax approximations.

Result (ii) is based on a bound for interpolation at $\{x_k\}$:

$$\sigma_n \leqslant \hat{\sigma}_n = \|\Sigma|\ell_k|\|, \text{ where } \ell_k(x) = \prod_{i\neq k}\left(\frac{x-x_i}{x_k-x_i}\right).$$

This bound also holds for complex L_∞ interpolation with x replaced by z.

3. REAL L_1 APPROXIMATION

For L_1 $[-1, 1]$, $\phi_n(x) = 2^{-n} U_n(x)$, where U_n is the Chebyshev polynomial of the second kind. Freilich and Mason (7) proved the following results:

(i) For f_n^S, $\sigma_n \leqslant \lambda_{n+1} \sim \frac{4}{\pi^2} \log n$

(ii) f_n^S converges in norm if f is L_2 integrable.

A rather restrictive classical result (see (1)) holds for f_n^I:

(iii) $f_n^I = f_n^B$ if, for all n, $f - f_n^I$ vanishes only at the zeros of $U_{n+1}(x)$.

<u>Open Questions</u>

1. For f_n^I determine σ_n.
2. Prove that f_n^I converges in norm.

4. REAL L_2 AND L_p APPROXIMATION

For functions in L_2 $[a, b]$ with weight function $w(x)$, $\{\phi_k\}$ is the monic orthogonal polynomial system corresponding to $w(x)$. The classical theory for f_n^S and early results by Erdös and Turán (8) and Erdös and Feldheim (9) for f_n^I may be summarised as follows:

(i) $f_n^S = f_n^B$ and converges in norm to f.

(ii) f_n^I converges in norm to f.

(iii) For the weight function $(1 - x^2)^{-\frac{1}{2}}$ and for f in $C[-1, 1]$, f_n^I converges to f in every L_p norm $(0 < p < \infty)$.

<u>Open Questions</u>

3. Determine ρ_n for f_n^I (in L_2).
4. Is there a ρ_n for f_n^I valid in all L_p norms for either of the weight functions $(1 - x^2)^{\frac{1}{2}}$ and $(1 - x^2)^{-\frac{1}{2}}$?

5. REAL MULTIVARIATE APPROXIMATION

The results of §2 and §3 have been extended to multivariate approximation on a hypercube by Mason (10). If $f^S_{n_1,\ldots,n_N}$ and $f^I_{n_1,\ldots,n_N}$ denote relevant series expansion and interpolation polynomials of degrees $n_1, \ldots, n_N$ in N variables $x_1, \ldots, x_N$, then, in L_∞,

$$\sigma_{n_1,\ldots,n_N} = \prod_{k=1}^{N} \lambda_{n_k} \quad \text{for } f^S_{n_1,\ldots,n_N}$$

$$\sigma_{n_1,\ldots,n_N} = \prod_{k=1}^{N} \gamma_{n_k} \quad \text{for } f^I_{n_1,\ldots,n_N}$$

where λ_n and γ_n are the constants of §2 for the univariate case. Moreover $f^S_{n_1,\ldots,n_N}$ and $f^I_{n_1,\ldots,n_N}$ converge in norm if f is Dini-Lipschitz continuous in the sense that

$$\sum_{k=1}^{N} \omega_k(\delta_k) \prod_{k=1}^{N} \log \delta_k \to 0 \text{ as } \{\delta_k\} \to 0,$$

where ω_k is a "partial modulus of continuity" in x_k.

In the case of the L_1 norm, the results of §3 generalise without complications. It is also clear that result (i) of §4 for f^S_n in L_2 extends to multivariate approximation.

6. COMPLEX L_∞ APPROXIMATION

For a function f(z) analytic inside a contour C and continuous on C, canonical polynomials $\{\phi_k(z)\}$ may be defined and will lead to a series expansion polynomial f^S_n. However, the zeros of ϕ_{n+1} do not necessarily lie on C, and so in the definition of f^I_n the zeros of $\phi_{n+1}(z) - \phi_{n+1}(z_0)$, where z_0 is some chosen point of C, are often used instead. A number of contours have been considered. In each case the Lebesgue function $\hat{\sigma}_n$ defined in §2 gives a useful bound on σ_n for f^I_n.

(a) Circle $|z| = 1$

In this case $\phi_n(z) = z^n$ and f_n^S is the truncated Taylor series. Choose interpolation points for f_n^I to be zeros of $z^{n+1} - 1$. The following results are due to Landau (see (11)), Gronwall (12), and Geddes and Mason (13):

(i) For f_n^S, $\sigma_n = \sum_{k=0}^{n} \binom{-\frac{1}{2}}{k}^2 \sim \frac{1}{\pi} \log n$

and $\sigma_n \leq \tau_n = \frac{1}{\pi}\int_0^{\pi} \frac{|\sin(n+1)x|}{\sin x}\,dx = \frac{4}{\pi^2} \log n + .98941 \ldots + o(1)$

(ii) For f_n^I, $\sigma_n \leq \frac{1}{n+1} \sum_{k=0}^{n} \operatorname{cosec} \frac{(2k+1)\pi}{2n+2} = \frac{2}{\pi} \log n + .07998 + o(1)$

(b) Ellipse $|z + \sqrt{z^2 - 1}| = \rho \quad (\rho > 1)$

Here $\phi_n(z) = 2^{1-n}\, T_n(z)$, and f_n^S is the truncated Chebyshev series in z. Interpolation points are chosen to be the zeros of

$T_{n+1}(z) - i \sinh((n+1) \log \rho)$,

and are the images of the zeros of $w^{n+1} - i$ under the transformation $\rho w = z + \sqrt{z^2 - 1}$ of the ellipse on the unit circle. Geddes (14), (15) and Mason (16) have obtained the results:

(i) For f_n^S, $\sigma_n \leq \lambda_n$ (the Lebesgue constant of §2)

(ii) For f_n^I, $\hat{\sigma}_n$ behaves numerically like log n for $1 \leq \rho \leq 25$.

Open Questions

5. Determine σ_n as a function of log n for f_n^I.

(c) General Contour

For both circle and ellipse, the canonical polynomials $\{\phi_k\}$ are also the Faber polynomials (see (17)) for the contour. Since Faber polynomials relate (Faber) mappings of a given contour on the unit circle, it is justifiable to use them for series expansions on a general contour in place of canonical polynomials. If f_n^S is the truncated Faber series, the following results hold. (The first is due to Kovari and Pommerenke (18) and Elliott (19) and the second is classical (see (17)).

(i) For f_n^S, $\sigma_n \leq \left\| \frac{1}{2\pi} \int_{|t|=1} \left| \sum_{k=0}^{n} \frac{\phi_k(z)}{t^{k+1}} \right| |dt| \right\| \leq \frac{V}{\pi} \tau_n$ (2)

where V is the total rotation of C and τ_n is the constant of §6(a).

(ii) f_n^S converges in norm on the interior of C.

For f_n^I, interpolation points are chosen to be images of the zeros of $w^{n+1} - w_0^{n+1}$ (for some $|w_0| = 1$) under a Faber mapping of C on $|w| = 1$. The following results have been obtained by Féjer (20) and Elliott (19).

(iii) f_n^I converges in norm to f if C is a Jordan curve and $w_0 = 1$.

(iv) For f_n^I, $\hat{\sigma}_n$ behaves numerically like log n if C is a semi-disc and w_0 is an (n+1)st root of ± 1.

Open Questions

6. For f_n^I find σ_n for an arbitrary contour.

7. For f_n^I find σ_n as a function of log n for a semi-disc.

(d) Regular Polygon

On the polygon, define f_n^S from a Faber series and f_n^I by interpolation at images of roots of unity. The Faber polynomials may be determined for an arbitrary polygon (Nepritvorennaja (21)), and Elliott (19) has obtained specific numerical results for a square as well as computing the general bound (2) on σ_n. Afolabi and Geddes (22) have computed f_n^I and $\hat{\sigma}_n$ for a regular polygon and their result is as follows:

(i) For f_n^I, $\hat{\sigma}_n$ behaves numerically like log n (specifically $\hat{\sigma}_n \leq 5$ for $n \leq 10$ for polygons with between 3 and 25 sides).

Open Questions

8. For f_n^S, determine a sharp bound on σ_n as a function of log n.

9. For f_n^I, determine σ_n as a function of log n.

7. COMPLEX L_1 APPROXIMATION

Results for truncated series expansions f_n^S based on canonical polynomials, have been obtained by Mason (23), (16) for L_1 approximation on both a contour C and the region R enclosed in C, for the cases of both the circle and the ellipse.

For the circular contour and region, $\phi_n = z^n$, and results are:

(i) On both the contour and the region, $\sigma_n \leq \tau_n$ (defined above), and f_n^S converges in norm to f.

For the elliptical contour and region, $\phi_n = 2^{-n} U_n(z)$, and results are:

(ii) On the contour, $\sigma_n < \lambda_{n+1}$ and f_n^S converges in norm to f.

(iii) On the region, with the L_1 norm weighted by $|z^2-1|^{-\frac{1}{2}}$, $\sigma_n \leq \lambda_{n+1}$ and f_n^S converges in norm to f.

Open Questions

10. For f_n^S, find σ_n on an elliptical region in an unweighted norm.
11. For f_n^I, find σ_n on circular and elliptical contours and regions.

REFERENCES

1. Mason, J.C. In: "Approximation Theory", A. Talbot (Ed.), Academic Press, 1970, pp 7-33.
2. Cheney, E.W. and Price, K.H., Ibid pp 261-290.
3. Hardy, G.H., J. London Math. Soc. 17 (1942), pp 4-13.
4. Ehlich, H. and Zeller, K., Math. Ann. 164 (1966), pp 105-112.
5. Luttmann, F.W. and Rivlin, T.J., IBM J. Develop. 9 (1965), pp 187-191.
6. Jackson, D. "The Theory of Approximation", AMS Colloquium Publications XI, 1930.
7. Freilich, J.H. and Mason, J.C., J. of Approx. Th. 4 (1971), pp 183-193.
8. Erdős, P. and Turán, P., Annals of Math. 38 (1937), pp 142-155.

9. Erdös, P. and Feldheim, E., Comptes Rendus 203 (1936), pp 913-915.

10. Mason, J.C. "Near-best multivariate approximation by Fourier series, Chebyshev series, and Chebyshev interpolation", J. of Approx. Th. (1980), to appear.

11. Dienes, P. "The Taylor Series - An Introduction to the Theory of Functions of a Complex Variable", Oxford, 1931.

12. Gronwall, T.H., Bull. Amer. Math. Soc. 27 (1921), pp 275-279.

13. Geddes, K.O., and Mason, J.C., SIAM J. Numer. Anal. 12 (1975), 111-120.

14. Geddes, K.O., SIAM J. Numer. Anal. 15 (1978), pp 1225-1233.

15. Geddes, K.O. In: "Theory of Approximation with Applications", A.G.Law and B.N. Sahney (Eds.) Academic Press, 1976, pp 155-170.

16. Mason, J.C. In: "Multivariate Approximation", D.C. Handscomb (Ed.), Academic Press, 1978, pp 115-135.

17. Markushevich, A.I. "Theory of Functions of a Complex Variable. Vol 3". Prentice-Hall, 1967.

18. Kovari, T. and Pommerenke, C.H., Math. Z. 99 (1967), pp 193-206.

19. Elliott, G.H. "The Construction of Chebyshev Approximations in the Complex Plane", Ph.D. Thesis, University of London, 1979.

20. Féjer, L., Göttinger Nachrichten (1918), pp 319-331.

21. Nepritvorennaja, L.M., Prikl. Mat. I. Programmirovanie 2 (1970), pp 47-56. (In Russian).

22. Afolabi, M.O. and Geddes, K.O., Proc 7th Manitoba Conf on Num. Math. and Computing, U of Manitoba, Winnipeg, 1977, pp 163-176.

23. Mason, J.C., J. of Approx. Th. 24 (1978), pp 330-343.

CONTINUED FRACTIONS ASSOCIATED WITH TWO-POINT PADÉ APPROXIMATIONS

John McCabe[1]

The Mathematical Institute
University of St. Andrews
St. Andrews
Fife, Scotland, U.K.

1. INTRODUCTION

The last decade has been a period of rapidly increasing interest in the theory and application of Padé approximations. There have been several conferences devoted either entirely to the subject or else with a strong emphasis on Padé approximations. (See (1-4) for the proceedings of some of these meetings). New texts on Padé approximations and books that contain chapters on them have recently appeared or will soon be available. See, for example (5-10). A comprehensive list of relevant articles up to 1976 is given in (11). The growth of interest in Padé methods has resulted in the development of several generalizations of the Padé concept. One such generalization yields the two-variable Canterbury approximants of Chisholm and his colleagues at the University of Kent at Canterbury, and another, the approximants obtained from series of orthogonal series rather than power series. (See (11) for references). A third generalization has led to the so called two-point Padé approximations. It is these rational functions that are the subject of this short

---

[1]Present address: Dept. of Mathematics and Computer Science, College of William and Mary, Williamsburg, Virginia, USA.

Copyright © 1980 by Academic Press, Inc.
All rights of reproduction in any form reserved.
ISBN: 0-12-171050-5

paper.

2. Two-Point Padé Approximations

Padé approximations are based on a single power series expansion. Baker, in (12), was the first to extend the concept and form approximations from series expansions at two points, and then called them two-point Padé approximants. Particular interest has been given to the special case in which the points of expansion are zero and infinity.

Consider the two series

$$c_0 + c_1 z + c_2 z^2 + c_3 z^3 + \dots\dots\dots , \quad c_0 \neq 0 \qquad \text{(A)}$$

and

$$\frac{b_1}{z} + \frac{b_2}{z^2} + \frac{b_3}{z^3} + \frac{b_4}{z^4} + \dots\dots\dots\dots , \quad b_1 \neq 0. \qquad \text{(B)}$$

We can construct a doubly infinite rectangular array of rational functions $M_{r,s}(z)$, $r = 0, 1, 2, \dots$, $s = 0, \pm 1, \pm 2, \dots$, the properties of $M_{r,s}(z)$ being

a) For $s = 0, \pm 1, \pm 2, \dots$, and $r > |s|$, $M_{r,s}(z)$ is a ratio of polynomials of degree $r - 1$ and r respectively and 'fits' $r + s$ terms of the series (A) and $r - s$ terms of (B) when expanded accordingly.

b) For $s > 0$ and $r \leq s$, $M_{r,s}(z)$ is a ratio of polynomials of degree $s - 1$ and r and 'fits' $r + s$ terms of (A). Hence $M_{r,s}(z)$ is the usual Padé approximant $P_{s-1,r}(z)$ for the series (A).

c) For $s < 0$ and $r \leq |s|$, $M_{r,s}(z)$ is the element $E_r^{-(s+r)}(z)$ of Wynn's E array for the series (B), that is $M_{r,s}(z)$ is of the form

$z^{s+r} \sum_{k=0}^{-(s+1)} A_k z^k \Big/ \sum_{k=0}^{r} B_k z^k$ and fits $r - s$ terms of (B). The array of these functions $M_{r,s}(z)$ is known as the M table and is the interlocking of the E array of (B), the major part of the Padé table for (A) and the two-point Padé table for (A) and (B). A detailed description of the M table is given in (13). There are many continued fractions whose convergents are sequences of elements of the complete M table, and an account of these will appear in a subsequent paper. Restricting ourselves to the two-point Padé table, any sequence that is obtained by moving along the central row, and then up or down in a staircase fashion, are the convergents of a continued fraction known as an M fraction, see (13). Of special interest is

$$\frac{c_0}{1+d_1z} + \frac{n_2z}{1+d_2z} + \frac{n_3z}{1+d_3z} + \cdots\cdots$$

whose convergents are the sequence $M_{k,0}(z)$, $k = 1, 2, 3, \ldots$. Continued fractions of the above form are alternatively called general T fractions because of their similarity with T fractions introduced by Thron, see (14). Perron (15) observed that T fractions had a correspondence property with two series expansions.

Many authors have developed the theory and applications of the two-point Padé approximants about the points at zero and infinity. See (13) and (16-20) for some theory, and (21-24) for some applications. Two-point Padé approximations can be particularly useful when only one or a few terms of one series are available. The advantage in using continued fraction representations of sequences of Padé approximations, whether one or two point, is that successive members of the sequence can be obtained by calculating at most two more coefficients. A special case where this is particularly advantageous is when the series expansions are series solutions of certain ordinary differential equations.

3. Differential Equations

In a fundamental paper (25), Laguerre considered functions which satisfy differential equations of the form

$$WE' = VE + U, \quad E' = dE/dx$$

where U,V and W are polynomials in x.

He studied properties of the Padé approximants for the (formal) series solution of the equation obtained by assuming a series solution in powers of $1/x$. Among the many relationships and properties of the numerator and denominator polynomials of the Padé approximations that were obtained was the following.

Let $A_n(x)/B_n(x)$ be the main diagonal Padé approximation for this series solution. The polynomial $B_n(x)$ and the function

$$\left\{A_n(x) - E(x)B_n(x)\right\}\exp\left\{\int^x \frac{V}{W}\, dx\right\}$$

are the two solutions of a certain easily obtainable second order linear differential equation. Hence, the ratio of the two solutions is a constant multiple of the integrating factor of the equation multiplied by the error of the Padé approximant. If W and V are quadratic and linear respectively, the polynomial $B_n(x)$ and the second solution are hypergeometric functions and error estimates can be obtained. See (26). Laguerre and then Luke (26) and Elliot (27) derived the Padé approximants without using continued fractions explicitly, the two latter authors giving approximations and error estimates for many transcendental functions. The present author (28) and Drew and Murphy (29), obtained the continued fractions whose convergents are the Padé approximations. These can be very elegantly expressed when W is quadratic and V is linear. In particular cases the continued fractions that correspond to one or both series solutions of the equations can be obtained directly from the equation either by Lagrange's method or by the method of successive differentiation first used by Euler to solve Riccati equations. Laguerre's results easily generalize to the two-point Padé approximants for the two series solutions of the equation, as to certain other rational functions. Of course, the continued fractions that correspond to both series do not necessarily provide suitable approximations to the solution of the equation. However, if this continued fraction is known, then the Padé approximations for each individual series can quite easily be obtained. The details of the relationships between the various Padé approximants will appear in a subsequent paper.

REFERENCES

1. Graves Morris, P.R., Padé Approximations and their applications, Academic Press, London, 1973.
2. Rocky Mountain Journal of Mathematics, Vol. 4, No. 2, 1974.
3. Saff, E.B., Varga, R.S., Padé and Rational Approximations, Academic Press, New York, 1977.
4. Wuytack, L., Lecture Series in Mathematics No. 765, Springer Verlag, 1979.
5. Henrici, P., Applied and Computational Complex Analysis, Vol. 2, Wiley, New York, 1977.
6. Baker, G.A., Essentials of Padé Approximants, Academic Press, New York, 1975.
7. Gilewicz, J., Approximants de Padé, Lecture notes in Mathematics No. 667, Springer Verlag, 1978.
8. Bender, C.M., and Orzag, S.A., Advanced Mathematical Methods, McGraw-Hill, New York, 1978.
9. Graves-Morris, P.R., Baker, G.A., Addison, Wesley, to appear in 1980.
10. Jones, W.B., Thron, W.J., Continued Fractions: Analytic Theory and Applications, Addison-Wesley. To appear in 1980.
11. Chui, C.K., Recent results on Padé Approximants and related problems. Approximation Theory II, Academic Press, New York, 1976.
12. Baker, G.A., Physical Review 135, 1964.
13. McCabe, J.H., A formal extension of the Padé table to include two-point Padé approximants, J. Inst. Math. Applic., Vol. 15, 1975.
14. Thron, W.J., Some Properties of the continued fraction $1 + d_0 z + k(z/1+d_n z)$. Bull. Amer. Math. Soc., Vol. 54, 1948.
15. Perron, O., Die Lehre von den Kettenbruchen, Band II, B.G. Tenbuer, Stuttgart, 1957.
16. Drew, D.M., Murphy, J.A., Branch Points, M fractions and rational approximants generated by linear equations, J. Inst. Math. Applic., Vol. 19, 1977.
17. McCabe, J.H., Murphy, J.A., Continued fractions which correspond to

series expansions at two points. J. Inst. Math. Applic., Vol. 17, 1976.

18. Sidi, A., Some Aspects of two-point Padé Approximants. Technical report, Israel Inst. of Tech., 1978.
19. Jones, W.B. and Thron, W.J., Two-point Padé tables and T fractions. Bu. Amer. Math. Soc., Vol. 83, 1977.
20. Jones, W.B. and Magnus, A., Computation of poles of two-point Padé approximants and their limits. To appear.
21. McCabe, J.H., A continued fraction expansion, with a truncation error estimate, for Dawson's integral. Math. of Comp., Vol. 28, 1974.
22. Frost, P.A. and Harper, E.Y., An extended Padé procedure for constructing global approximations from asymptotic expansions: an explication with examples. S.I.A.M. Review, Vol. 18, 1976.
23. Guttman, A.J., Derivation of "mimic functions" from regular perturbation expansions in fluid mechanics. J. Inst. Math. Applic., Vol. 15, 1975.
24. Grundy, R.E., Laplace transform inversion using two-point rational approximants. J. Inst. Math. Applic., Vol. 20, 1977.
25. Laguerre, E., Sur le reduction en fractions continues d'une fraction qui satisfait a une equation differentielle lineare du premier ordre dont les coefficients sont rationnels. J. Math. Pures. Appl., Vol. 1, 1885.
26. Luke, Y.L., The Padé table and the τ method. J. Maths. and Phys., Vol. 37, 1958.
27. Elliot, D., Truncation errors in Padé approximations to certain functions. An alternative approach. Math. of Comp., Vol. 21, 1967.
28. McCabe, J.H., Ph.D. Thesis, Brunel University, 1971.
29. Drew, D.M. and Murphy, J.A., Continued fraction solutions of differential equations. Tech. Report No. 26, Brunel University, 1973.

BERNSTEIN THEOREMS FOR A PARABOLIC EQUATION IN ONE SPACE VARIABLE

Peter A. McCoy

Mathematics Department
United States Naval Academy
Annapolis, Maryland

One classical theorem of S. N. Bernstein (1) relates a domain of regularity of an analytic function of a single complex-variable with a sequence of best local polynomial approximates. Extensions arise for analytic solutions of the non-characteristic Cauchy problem

$$\begin{cases} L(u) = u_{xx} + b(x,t)u - c(x,t)u_t = 0; \ (x,t) \in \Omega \\ u(0,t) = f(t), \ u_x(0,t) = g(t); \ t \in \Delta \end{cases}$$

The coefficients $b,c: cl(\Omega) \to C$ are analytic on the closure of the domain $\Omega = E_\delta \times E_\rho$; $E_r = \{t \in C : |t-1| + |t+1| < r+\frac{1}{r}\}$, $0 < r < 1$ and the data $\{f,g\}$ is analytic on the set $\Delta = \bigcup_{k=1}^{m} [a_k, b_k]$ located in $\{t \in C : Re(t) = 0\} \bigcap E_\rho$.

Central to the development is D. L. Colton's (3-4) invertible integral operator $P_1: \{f,g\} \to u$ that is defined for unique associated analytic function pairs $\{f,g\}$ on a disk $D \supset \Delta = [a_1, b_1]$. The operator is modified by a majorant argument that here and throughout replaces the initial circle of integration ∂D by the Ellipse ∂E_ρ and the segment

Copyright © 1980 by Academic Press, Inc.
All rights of reproduction in any form reserved.
ISBN: 0-12-171050-5

$[a_1, b_1]$ by the set Δ. Reasoning as in Hadamard's multiplication of singularities theorem (the Envelope Method (5-6)) produces an essential regularity theorem that applies to the principal branch of $P_1\{f,g\}$.

<u>Theorem 1</u>. Let $b(x,t)$ and $c(x,t)$ be analytic on the domain $cl(E_\delta \times E_\rho)$. Then $u = P_1\{f,g\}$ is the analytic solution of the non-characteristic Cauchy problem

$$\left\{ \begin{array}{l} L(u) = 0; \ (x,t) \in E_\delta \times (E_\rho/\{\lambda_1, \ldots, \lambda_m\}) \\ u(0,t) = f(t), \ u_x(0,t) = g(t); \quad t \in \Delta \end{array} \right\}$$

if, and only if, the P_1 associates $\{f,g\}$ are analytic on $E_\rho/\{\lambda_1, \ldots, \lambda_m\}$.

It is this theorem that now allows the development to proceed along the lines of the analytic function theory. Let the sets P_n and Q_n be defined as

$$P_n = \text{Span}\ \{1, t, t^2, \ldots, t^n\}$$

$$Q_n = \text{Span}\ \{1/(t-\lambda_1), \ldots, 1/(t-\lambda_1)^{n+1}; \ldots; 1/(t-\lambda_m), \ldots, 1/(t-\lambda_m)^{n+1}\}$$

with $K_n = P_n$ when $\Delta = [a_1, b_1]$, otherwise $K_n = P_n \quad Q_n$, $n=0, 1, 2, \ldots$. The n-th order Chebyshev approximates to the Cauchy data are given by

$$e(u, K_n, \Delta) = \inf\{\sup\{|u(0,t) - r(t)| : t \in \Delta\}\ r \in K_n\}, \ n=0,$$

$1, 2, \ldots$ where the poles of the functions in the set Q_n are located at the zeros λ_j, $1 \le j \le m$, of the polynomial

$$\prod_{k=1}^{m} (a_k - z) = \prod_{k=1}^{m} (b_k - z).$$

The fundamental measures in the analysis are the approximates

$$\sigma(u, K_n, \Delta) = \max\ \{e(u, K_n, \Delta),\ e(u_x, K_n, \Delta)\}, \ n=0, 1, 2, \ldots .$$

Using combinations of Theorem 1 and results (1,2) from classical analytic function theory, it is easy to see that Bernstein theorems for the non-characteristic Cauchy problem readily unlock. The following two theorems are typical.

Theorem 2. Let $b(x,t)$ and $c(x,t)$ be analytic on the domain $cl(E_\delta \times E_\rho)$. Then $u = P_1\{f,g\}$ is the analytic solution of the non-characteristic Cauchy problem

$$\left\{\begin{array}{l} L(u) = 0; \quad (x,t) \in E_\delta \times E_\rho \\ u(0,t) = f(t), \; u_x(0,t) = g(t); \; t \in [-1,+1] \end{array}\right\}$$

if, and only if,

$$\lim_n \sup[\sigma(u,P_n,[-1,+1])]^{1/n} \leq \rho, \; 0 < \rho < 1 .$$

Theorem 3. Let $b(x,t)$ and $c(x,t)$ be analytic on the domain $C \times C$. Then $u = P_1\{f,g\}$ is the analytic solution of the non-characteristic Cauchy problem

$$\left\{\begin{array}{l} L(u) = 0; \quad (x,t) \in C \times (C/\{\lambda_1,\ldots,\lambda_m\}) \\ u(0,t) = f(t), \quad u_x(0,t) = g(t); \; t \in \Delta \end{array}\right\}$$

if, and only if,

$$\lim_n [\sigma(u,K_n,\Delta)]^{1/n} = 0$$

Remarks: (I) The operators

$$\underset{\sim}{L}(w) = w_{xx} + a(x,t)w_x + b(x,t)w - c(x,t)u_t$$

and

$$L(u) = u_{xx} + b(x,t)u - c(x,t)u_t$$

are equivalent under the change of dependent variable $w(x,t) = u(x,t) \exp\{-1/2 \int_0^x a(s,t)\, ds\}$.

(II) Examples of Bernstein theorems that are developed for second order elliptic equations via integral transforms can be found in (7-8).

REFERENCES

1. Bernstein, S. N., Lecon sur les proprietes extremales et la meilleure approximation des functions analytiques d'une variable réelle, Gauthier-Villars, Paris, (1926).

2. Chandler, J. D., A generalization of a theorem of S. N. Bernstein, *Proc. Amer. Math. Soc., vol. 6* , 95-100 (1977).

3. Colton, D. L., The noncharacteristic Cauchy problem for parabolic equations in one space variable, *SIAM J. Math. Anal. 5*, 263-72 (1974).

4. Colton, D. L., Solution of boundary value problems by the method of integral operators, Research notes in Math., vol. 6, Pitman Publ. Co., Belmont, CA (1976).

5. Colton, D. L. and Gilbert, R. P., Singularities of solutions to elliptic partial differential equations with analytic coefficients, *Quar. J. Math., vol.19*, 391-396 (1968).

6. Gilbert, R. P., Constructive methods for elliptic equations, Lecture notes in Math., vol. 365, Springer-Verlag, New York (1974).

7. McCoy, P. A., Polynomial approximation of generalized biaxisymmetric potentials, *J. Approx. Theory, vol. 25*, 153-167 (1979).

8. McCoy, P. A., Best L^p approximation of generalized biaxisymmetric potentials, *Proc. Amer. Math. Soc.* (in press).

SEGMENTED APPROXIMATION

H. W. McLaughlin
and
J. J. Zacharski

Department of Mathematical Sciences
Rensselaer Polytechnic Institute
Troy, New York 12181

DEDICATED TO PROFESSOR G. G. LORENTZ
ON THE OCCASION OF HIS SEVENTIETH BIRTHDAY

In this paper we present an algorithm for partitioning data into segments which allows, in turn, one to optimally approximate the data. The algorithm has been tested and computer results will appear elsewhere.

The segmented approximation problem about which the algorithm is concerned is described first. We assume given a finite set of points $X = \{x_1 < x_2 < \ldots < x_n\}$ with corresponding real data $y_1, \ldots, y_n$. A positive integer, $m (\leq n - 1)$, is fixed and $m + 1$ of the points of X, denoted $K = \{k_0, k_1, \ldots, k_m\}$ are chosen such that $x_1 = k_0 \leq k_1 \leq \ldots \leq k_m = x_n$. The symbol $[k_i]$ denotes $\{x \in X: k_{i-1} < x \leq k_i\}$ if $2 \leq i \leq m$ and denotes $\{x \in X: k_0 \leq x \leq k_1\}$ if $i = 1$. (The symbol $[k_0]$ is not defined.) We call $[k_i]$ the ith segment of X, $1 \leq i \leq m$, and the points k_i, $0 \leq i \leq m$, are called knots. The knots, K, partition X into m segments some of which may be empty.

For each K the ith segment, $[k_i]$, is assigned a nonnegative number, denoted by $E(k_i)$, $1 \leq i \leq m$, such that: (1) if $K = \{k_0, k_1, \ldots, k_m\}$ and $K^1 = \{k_0^1, k_1^1, \ldots, k_m^1\}$ are two choices of knots such that for some i $(1 \leq i \leq m)$, $[k_i] \subset [k_i^1]$, then $E(k_i) \leq E(k_i^1)$; and (2) $E[\phi] = 0$. It is

Copyright © 1980 by Academic Press, Inc.
All rights of reproduction in any form reserved.
ISBN: 0-12-171050-5

understood that $E(k_i)$ depends only on the points of $[k_i]$ so that in (1), if for some i, $1 \leq i \leq m$, $[k_i] = [k_i^1]$ then $E(k_i) = E(k_i^1)$.

For our application we have in mind that the number $E(k_i)$ is determined by approximating the data on the ith segment in an optimal manner, from a given class of approximating functions using a given norm.

Purpose of Algorithm. The algorithm presented here is designed to find a set of knots, $K^{opt} = \{k_0^{opt}, \ldots, k_m^{opt}\}$, such that for every choice of the knots $K = \{k_0, \ldots, k_m\}$, $\max_{1 \leq i \leq m} E(k_i^{opt}) \leq \max_{1 \leq i \leq m} E(k_i)$. We think of the set of knots K^{opt} as an optimal set of knots and of the corresponding number $E^{opt} = \max_{1 \leq i \leq m} E(k_i^{opt})$, as the optimal (global) error.

The algorithm uses the notion of E-maximal knots.

Definition. Let E be a nonnegative real number and $1 \leq i \leq m-1$. The knot k_i in $K = \{k_0, \ldots, k_i, \ldots, k_m\}$ is E-maximal if $E(k_i) \leq E$ and if when k_i is replaced in K by k_i^1 such that $k_i < k_i^1$ then $E(k_i^1) > E$. (In words, k_i is E-maximal in K if $E(k_i) \leq E$ and when k_i is moved to the right the resulting error exceeds E.) If every knot k_i, $1 \leq i \leq m-1$ in K is E-maximal then the set of knots, K, is said to be E-maximal.

A short induction argument can be used to prove the following remark.

Remark 1. For every nonnegative number, E, one can construct an E-maximal set of knots, K.

Segmented Approximation Algorithm

MAIN STEP	Input:	j ,	An integer, $1 \leq j \leq m-1$.
		E^t,	A nonnegative number.
		K ,	A set of knots, $k_0 \leq k_1 \leq \ldots \leq k_m$.
	Computation:		redefine $k_j, k_{j+1}, \ldots, k_{m-1}$ such that each k_i is E^t maximal in K, $j \leq i \leq m-1$.
	Output:		K, a set of knots.

STEP 0. Choose knots, K, arbitrarily. Set $E^t = \max_{1 \leqslant i \leqslant m} E(k_i)$, (or choose E^t arbitrarily such that $E^t \geqslant E^{opt}$).

STEP 1. Set $j = 1$. Do Main Step. This step results in a new set of knots, K.

The input for main step is: $j = 1$, K and E^t from Step 0.

STEP 2. Set $E^t = \max_{1 \leqslant i \leqslant m} E(k_i)$. Set $K^t = K$. If $E(k_m) = E^t$ go to Step 8, otherwise go to Step 3.

The knots, K, are being stored in K^t for future use. If $E(k_m) = E^t$ then $E^{opt} = E^t$ and algorithm terminates.

STEP 3. Set $\ell = \min\{i \mid j \leqslant i \leqslant m-1$ and $E(k_i) = E^t\}$. Compute a new value of the knot, k_ℓ, such that $k_{\ell-1} \leqslant k_\ell \leqslant k_{\ell+1}$ and such that $[k_\ell]$ is the largest interval such that $E(k_\ell) < E^t$. This step results in a new set of knots, K.

Find the left most interval on which the error E^t is achieved. On this interval move the right end point to the left just enough to reduce the error.

STEP 4. If $\ell = m-1$, go to Step 6, otherwise go to Step 5.

STEP 5. Set $j = \ell + 1$. Do Main Step. This step results in a new set of knots, K.

The input for Main Step is: $j = \ell + 1$, E^t from Step 2, K from Step 3.

STEP 6. If $E(k_m) \geqslant E^t$ go to Step 8, otherwise, set $E = \max_{\ell+1 \leqslant i \leqslant m} E(k_i)$.

If $E(k_m) \geqslant E^t$ then $E^{opt} = E^t$, and algorithm terminates.

STEP 7. If $E < E^t$, set $j = 1$ and go to Step 2, otherwise, go to Step 3.

If $E < E^t$ then $E^{opt} < E^t$. If $E = E^t$ it isn't clear if $E^{opt} < E^t$.

STEP 8. Set $K^{opt} = K^t$. Set $E^{opt} = E^t$.

E^t and K^t were stored at Step 2.

STEP 9. END.

Remark 2. The algorithm terminates.

Proof. Step 7 is the only step where routing to previous steps is encountered. If Step 2 follows Step 7 a new value of E^t is computed which is less than the previous value of E^t. Since there are only a finite number of knots K there are only a finite number of values of E^t which can be computed. Thus Step 2 can be encountered at most a finite number of times.

If Step 3 follows Step 7 a new value of ℓ is computed which is greater than the previous value. Since there are at most a finite number of values of ℓ which can be computed for each E^t, Step 3 can be encountered at most a finite number of times while E^t is fixed.

Remark 3. When the algorithm terminates one of the outputs is a set of knots K^t and the other output is a real number E^t. Each of the knots k_i^t in K^t is E^t maximal in K^t, $1 \leqslant i \leqslant m-1$.

Proof. The algorithm terminates at Step 8. Here K^{opt} is set equal to K^t which was stored in Step 2. Both in the case when Step 2 follows Step 1 and in the case when Step 2 follows Step 7, the set of knots, K^t, was computed so that each k_i^t is E^t maximal, $1 \leqslant i \leqslant m-1$.

Remark 4. If $E^t > E^{opt}$, the knots, K, which are the output of Step 3 have the property that $k_i^{opt} \leqslant k_i$, $1 \leqslant i \leqslant m-1$.

Remark 5. If $E^t > E^{opt}$, the knots, K, which are the output of Step 5 have the property that $k_i^{opt} \leqslant k_i$, $1 \leqslant i \leqslant m-1$.

Lemma 1. Let $E^\beta \leqslant E^\alpha$. Let $K^\alpha = \{k_0^\alpha, k_1^\alpha, \ldots, k_m^\alpha\}$ be E^α maximal and let $K^\beta = \{k_0^\beta, k_1^\beta, \ldots, k_m^\beta\}$ be E^β maximal. Then $k_i^\beta \leqslant k_i^\alpha$, $1 \leqslant i \leqslant m-1$.

Proof. If not there exists a smallest i, $1 \leqslant i \leqslant m-1$ such that $k_i^\alpha < k_i^\beta$; call such an i, j. Then both $k_{j-1}^\beta \leqslant k_{j-1}^\alpha$ and $k_j^\alpha < k_j^\beta$. Thus

$[k_j^{\alpha}] \subset [k_j^{\beta}]$ where the inclusion is proper. Since K^{α} is E^{α} maximal, $E^{\alpha} < E(k_j^{\beta}) \leqslant E^{\beta}$.

Theorem (Convergence) When the algorithm terminates, one of the outputs is a set of knots K^t and the other is a nonnegative number E^t. The value of E^t is the optimal error (E^{opt}) and the knots K^t are optimal knots (K^{opt}). Further, the knots K^t are E^t maximal.

Proof. To show that $E^t = E^{opt}$, i.e., that when the algorithm terminates the current value of E^t is indeed the optimal error, and that $K^t = K^{opt}$, it suffices to show that if $E^t > E^{opt}$ then $E(k_m) < E^t$ where K is the set of knots current at Step 2 or Step 6. That this is sufficient follows from the fact that the algorithm terminates at Step 8 which follows either Step 2 or Step 6. To show that $E(k_m) < E^t$, under the assumption that $E^t > E^{opt}$, it suffices to show that $k_{m-1}^{opt} \leqslant k_{m-1}$, since in this case $E(k_m^{opt}) \geqslant E(k_m)$ and hence $E^t > E^{opt} \geqslant E(k_m^{opt}) \geqslant E(k_m)$.

If K is the set of knots current at Step 2 then K is E^t maximal in both the case Step 2 follows Step 1 and the case Step 2 follows Step 7 (see Remark 3). And by Lemma 1, $k_{m-1}^{opt} \leqslant k_{m-1}$. If K is the set of knots current at Step 6, then by Remark 5, $k_{m-1}^{opt} \leqslant k_{m-1}$.

PREVIOUS WORK

Segmented approximation with variable knots has been examined from many different viewpoints. The variety of choices for norm, knot placement criteria, and smoothness conditions at the knots will result in different formulations of the segmented problem.

Bellman [1] suggests dynamic programming as a method of solution of the problem with the knots chosen to minimize the least squares error on the entire interval. Dynamic programming also may be used for problems using other norms regardless of whether the data is continuous or discrete. For the discrete case, it has been the only known method, other than exhaustive search, which is guaranteed to converge to a global minimum. Cox [2] argues that dynamic programming is superior to exhaustive enumeration. Guthery [3] uses dynamic programming to solve the partition regression problem (which is a least squares segmented approximation on discrete data). He states this method is better than other methods in that it can handle many knots with less problems and that a global optimum is guaranteed. However, Pavlidis [4] points out that dynamic programming has excessive computational requirements which grow very fast with the number of segments.

Lawson [5] uses the idea of dynamic programming to prove the existence of a balanced error solution for the continuous data problem in which the knots are chosen to minimize the maximum of the errors on the subintervals. In addition, he proves that balanced errors is a sufficient condition for optimality. Rather than use dynamic programming as a method of solution, he proposes an algorithm [6] which seeks to balance the errors. The fault of this algorithm is it will fail if an error of zero on a subinterval is encountered. Pavlidis and Maika [7] also give an algorithm based on the balanced error property. It is a functional iteration and will converge locally, possibly to a local minimum.

The lack of a balanced error property for discrete data leaves exhaustive enumeration and dynamic programming as the two methods guaranteed to converge globally for the discrete data case. Pavlidis [4] has a discrete optimization algorithm which is quicker but may converge to a local minimum, instead of global minimum.

Pavlidis and Horowitz [8] have a "split and merge" algorithm for a modified version of the problem. Here the number of knots can be changed either by removing one from a segment with small error (merging adjacent segments) or by inserting one in a segment with large error (splitting a segment). Another approach without a given number of knots is Rice's adaptive approximation [9]. Here a tolerance is given and knots are inserted according to some scheme (say in the middle of a segment whose error needs to be decreased) until the total error is less than the tolerance. Asymptotic estimates are given for the number of steps needed to reach a given tolerance. McGee and Carleton [10] and Ertel and Fowlkes [11] also consider a problem in which the number of knots is allowed to vary.

Other criteria for knot selection can be employed. Quandt [12-14] and others use hypothesis testing and maximum likelihood estimates to decide whether a regression is from two different regimes (segments) on a time axis and, if so, where the switch (knot) occurs.

The segmented problem with smoothness requirements at the knots is studied in the literature on splines with free knots, e.g. Esch and Eastman [15], [6], Schumaker [16], Handscomb [17], Braess [18], and Burchard [19]. In the statistical literature see Robison [20], Hudson [21], Hinkley [22], Gallant and Fuller [23], Bacon and Watts [24], and Ertel and Fowlkes [11].

Two dimensional analogs have been considered but are not referenced here.

Other interesting work is found in [25] - [31].

REFERENCES

1. Bellman, R. On the Approximation of Curves by Line Segments Using Dynamic Programming, Comm. ACM 4 (1961), 284.
2. Cox, M. G., Curve Fitting with Piecewise Polynomials, J. Inst. Maths. Applic. 8 (1971), 36-52.
3. Guthery, S. B., Partition Regression, J. Am. Stat. Assn. 69 (1974), 945-947.
4. Pavlidis, T., Waveform Segmentation Through Functional Approximation, IEEE Trans. Comput. C-22 (1973), 689-697.
5. Lawson, C. L., Characteristic Properties of the Segmented Rational Minmax Approximation Problem, Numer. Math. 6 (1964), 293-301.
6. Esch, R. E. and Eastman, W. L., Computational Methods for Best Approximation, Sperry Rand Res. Cen., Sudbury, Mass., Tech. Rep. SEG-TR-67-30, Dec. 1967.
7. Pavlidis, T. and Maika, A. P., Uniform Piecewise Polynomial Approximation with Variable Joints, J. Approx. Th. 12 (1974), 61-69.

8. Pavlidis, T. and Horowitz, S. L., Segmentation of Plane Curves, IEEE Trans. Comput. C-23 (1974), 860-870.
9. Rice, J. R., Adaptive Approximation, J. Approx. Th. 16 (1976), 329-337.
10. McGee, V. E. and Carleton, W. T., Piecewise Regression, J. Am. Stat. Assn. 65 (1970), 1109-1124.
11. Ertel, J. E. and Fowlkes, E. B., Some Algorithms for Linear Spline and Piecewise Multiple Regression, J. Am. Stat. Assn. 71 (1976), 640-648.
12. Quandt, R. E., The Estimation of the Parameters of a Linear Regression System Obeying Two Separate Regimes, J. Am. Stat. Assn. 53 (1958), 873-880.
13. Quandt, R. E., Tests of the Hypothesis that a Linear Regression Obeys Two Separate Regimes, J. Am. Stat. Assn. 55 (1960), 324-330.
14. Quandt, R. E., A New Approach to Estimating Switching Regressions, J. Am. Stat. Assn. 67 (1972), 306-310.
15. Esch, R. E. and Eastman, W. L., Computational Methods for Best Spline Function Approximation, J. Approx. Th. 2 (1969), 85-96.
16. Schumaker, L., Uniform Approximation by Chebyshev Spline Functions. II: Free Knots, SIAM J. Numer. Anal. 5 (1968), 647-656.
17. Handscomb, D. C., Characterization of Best Spline Approximation with Free Knots, in Approximation Theory, A. Talbot, Ed., London-New York: Academic Press, 1970.
18. Braess, D., Chebyshev Approximation by Spline Functions with Free Knots, Numer. Math. 17 (1971), 357-366.
19. Burchard, H. G., Splines (with optimal knots) are Better, Applic. Anal. 3 (1974), 309-319.
20. Robison, D. E., Estimates for the Points of Intersection of Two Polynomial Regressions, J. Am. Stat. Assn. 59 (1964), 214-224.
21. Hudson, D. J., Fitting Segmented Curves Whose Join Points Have to be Estimated, J. Am. Stat. Assn. 61 (1966), 1097-1129.
22. Hinkley, D. V., Inference in Two-Phase Regression, J. Am. Stat. Assn. 66 (1971), 736-743.
23. Gallant, A. R. and Fuller, W. A., Fitting Segmented Polynomial Regression Models Whose Join Points have to be Estimated, J. Am. Stat. Assn. 68 (1973), 144-147.
24. Bacon, D. W. and Watts, D. G., Estimating the Transition Between Two Intersecting Straight Lines, Biometrika 58 (1971), 525-534.
25. Bellman, R. and Roth, R., Curve Fitting by Segmented Straight Lines, J. Am. Stat. Assn. 64 (1969), 1079-1084.
26. Burchard, H. G. and Hale, D. F., Piecewise Polynomial Approximation on Optimal Meshes, J. Approx. Th. 14 (1975), 128-147.
27. Collatz, L., Einschliessungssatz Für Die Minimalabweichung Bei Der Segmentapproximation. Simposio Internationale Sulle Applicationi Dell'Analisi Alla Fisica Matematica, 1964.
28. Gluss, B., Further Remarks on Line Segment Curve-Fitting Using Dynamic Programming, Comm. ACM 5 (1962), 441-443.
29. Pavlidis, T., Linguistic Analysis of Waveforms, in Software Engineering, J. T. Tou, Ed., Vol. 2, New York: Academic Press, 1971, 203-225.
30. Rice, J. R., General Purpose Curve Fitting, in Approximation Theory, A. Talbot, Ed., New York-London: Academic Press, 1970, 191-204.
31. Stone, H., Approximation of Curves by Line Segments, Math. Comput. 15 (1961), 40-47.

This list is not exhaustive.

AN ALGORITHMIC APPROACH TO FINITE LINEAR INTERPOLATION

G. MÜHLBACH

University of Hannover
Hannover, Fed. Rep. of Germany

Let $\mathbb{K}$ be a commutative field of characteristic zero, let n be a natural number and let F_n be a linear space over $\mathbb{K}$ of dimension n. By F_n^* we denote its dual. The general problem of finite linear interpolation reads as follows [1]:

Given n-tuples $f^n := (f_1,\ldots,f_n) \in F_n^n$, $L^n := (L_1,\ldots,L_n) \in F_n^{*n}$, $w^n := (w_1,\ldots,w_n) \in \mathbb{K}^n$, find an element $h = \sum_{k=1}^{n} a_k f_k \in F_n$ such that

$$L_i h = w_i \qquad (i=1,\ldots,n) . \tag{1}$$

We will refer to this problem by $H[w^n] = H[w^n;f^n,L^n]$ and by $h = h[w^n] = h[w^n;f^n,L^n] = \sum_{k=1}^{n} a_k f_k$ to its solution whenever a unique solution exists. Then the coefficient $a_k = a_k[w^n] = a_k[w^n;f^n,L^n]$ will be called the k-th divided difference of w^n with respect to the systems f^n and L^n. When $I = \{i_1,\ldots,i_m\}$ is any subset of $\{1,\ldots,n\}$ and $u^n = (u_1,\ldots,u_n)$ is any n-tuple we set $u^I := (u_{i_1},\ldots,u_{i_m})$ and $L^I g := (L_{i_1}g,\ldots,L_{i_m}g)$ if $g \in F_n$. For simplification, when $I = \{1,2,\ldots,m\}$ we write u^m instead of u^I. Clearly, problem $H[w^n;f^n,L^n]$ has a unique solution iff

Copyright © 1980 by Academic Press, Inc.
All rights of reproduction in any form reserved.
ISBN: 0-12-171050-5

f^n and L^n are bases of F_n and F_n^*, respectively.

THEOREM 1: Let $1 \le m < n$ be integers. Suppose that f^n resp. L^n is a basis of F_n resp. F_n^*. Then for every $w^n \in \mathbb{K}^n$ the solution $h[w^n] = \sum_{k=1}^{n} a_k f_k$ of the problem $H[w^n;f^n,L^n]$ is

$$h[w^n] = h[w^m] + \sum_{k=m+1}^{n} a_k \cdot (f_k - h[L^m f_k]), \tag{2}$$

where $h[u^m]$ solves the problem $H[u^m;f^m,L^m]$. The coefficients a_k $(k=m+1,\ldots,n)$ are uniquely determined as solutions of

$$\sum_{k=m+1}^{n} a_k \cdot (L_i f_k - L_i h[L^m f_k]) = w_i - L_i h[w^m] \quad (i=m+1,\ldots,n). \tag{3}$$

Conversely, if f^m is a basis of $F_m :=$ span f^m and L^m is a basis of F_m^* and if (3) can be uniquely solved for the a_k $(k=m+1,\ldots,n)$ then also f^n, L^n are bases of F_n, F_n^*, respectively.

Proof. Consider the mappings $F_n \ni g \xrightarrow{A_\nu} h[L^\nu g] \in F_\nu$ $(\nu=m,n)$. By the assumptions, $A_n = \mathrm{id}_{F_n}$, $A_m|F_m = \mathrm{id}_{F_m}$. Hence, $(A_n - A_m)(f_j) = 0$ $(j=1,\ldots,m)$. Consequently, for all $w^n \in \mathbb{K}^n$

$$\begin{aligned} h[w^n] = (A_m + A_n - A_m)(h[w^n]) &= A_m(h[w^n]) + \sum_{k=m+1}^{n} a_k \cdot (f_k - A_m(f_k)) \\ &= h[w^m] + \sum_{k=m+1}^{n} a_k \cdot (f_k - h[L^m f_k]). \end{aligned}$$

Assume that f^m resp. L^m is a basis of F_m resp. F_m^*. The homogeneous system corresponding to (3) has a nontrivial solution iff there exist $c_k \in \mathbb{K}$ $(k=m+1,\ldots,n)$ not all equal to zero such that

$$\sum_{k=m+1}^{n} c_k \cdot (L_i f_k - L_i h[L^m f_k]) = 0 = L_i (g - h[L^m g]) \quad (i=m+1,\ldots,n)$$

where $g := \sum_{k=m+1}^{n} c_k f_k$. With $\sum_{k=1}^{m} c_k f_k := -h[L^m g]$ because of the interpolation properties of $h[L^m g]$ this is equivalent with

$$L_i(g - h[L^m g]) = 0 = \sum_{k=1}^{n} c_k L_i f_k \qquad (i=1,\ldots,n). \tag{4}$$

Evidently, (4) holds with $c_{m+1},\ldots,c_n$ not all equal to zero iff f^n or L^n fails to be a basis of F_n or F_n^*, respectively.

In terms of numerical linear algebra: computing the solution $h[w^n]$ of problem $H[w^n]$ according to theorem 1 from the solution $h[w^m]$ of the "smaller, m-dimensional problem" is equivalent to solving the system (1) for the a_k by block-elimination using $(L_i f_k)_{i,k=1,\ldots,m}$ as elimination block.

What reasons will make Newton's method superior to just solving the system (1) using Gauss-elimination? 1° The generalized Newton-formula (2) again has the welcome permanence property: when the set of data grows only some new terms have to be computed and added. 2° When the elements $f_k - h[L^m f_k]$ $(k=m+1,\ldots,n)$ are known or easily determined then computation of the interpolant according to (2) can reduce the number of algebraic operations. 3° When applied to Hermite-Birkhoff-interpolation Birkhoff's mean value representation of the interpolation error can easily be derived from (2) just in the same way as Cauchy has got his well known remainder formula [3].

THEOREM 2: _Let_ $0 \le r < m < n$ _be integers and let_ f^n _resp._ L^n _be a basis of_ F_n _resp._ F_n^*. _Suppose that there are_ s+1 _subsets_ $I_j = \{i_{j1},\ldots,i_{jm}\}$ _of_ $\{1,\ldots,n\}$ $(j=0,\ldots,s)$ _subject to_

(i) $\quad \text{card } I_j = m \qquad (j=0,\ldots,s)$

(ii) $\quad m > \text{card } I_j \cap I_{j-1} \ge r \qquad (j=1,\ldots,s)$

(iii) $\quad \bigcup_{j=0}^{s} I_j = \{1,\ldots,n\}$

(iv) $\quad \det(L_i f_k)_{\substack{i \in I_j \\ k=1,\ldots,m}} \ne 0 \qquad (j=0,\ldots,s)$

and, when $r > 0$,

(v) $\quad$ for $j=1,\ldots,s$ _there exists_ $K_j \subset I_j \cap I_{j-1}$ _with_

$$\text{card } K_j = r \text{ and } \det(L_i f_k)_{\substack{i \in K_j \\ k=1,..,r}} \neq 0.$$

For every $w^n \in \mathbb{K}^n$ *let* $a := (a_{m+1}[w^n],...,a_n[w^n])^T$. *Then* a *is uniquely determined as solution of the* $s\cdot(m-r) \geq n-m$ *equations*

$$C \cdot a = b , \tag{5}$$

$b := (b_{j,p})_{j=1,..,s;p=r+1,..,m}$, $A := (a_{j,p;k})_{\substack{j=1,..,s;p=r+1,..,m \\ k=1,..,n-m}}$

$$b_{j,p} = a_p[w^{I_j};f^m,L^{I_j}] - a_p[w^{I_{j-1}};f^m,L^{I_{j-1}}]$$

$$a_{j,p;k} = a_p[L^{I_j}f_{m+k};f^m,L^{I_j}] - a_p[L^{I_{j-1}}f_{m+k};f^m,L^{I_{j-1}}].$$

Conversely, when f^m *is a basis of* F_m *and when* (5) *has a unique solution for every choice of* $w^n \in \mathbb{K}^n$ *then* f^n *and* L^n *are bases of* F_n *and* F_n^*, *respectively.*

Proof. According to theorem 1 the divided differences $a_k = a_k[w^n]$ $(k=m+1,...,n)$ obey the s+1 linear equations

$$h[w^n] = h[w^{I_j}] + \sum_{k=m+1}^{n} a_k \cdot (f_k - h[L^{I_j}f_k]) \quad (j=0,...,s).$$

Eliminating $h[w^n]$ by subtracting from each equation its predecessor and comparing coefficients of f_p $(p=r+1,...,m)$ we arrive at (5). Assume that f^m is a basis of F_m and that (i)-(v) hold true. It remains to show that (5) has a unique solution for every choice of $w^n \in \mathbb{K}^n$ iff f^n and L^n are bases of F_n and F_n^*, respectively. Now, the corresponding homogeneous system has a nontrivial solution iff there exist $c_k \in \mathbb{K}$ $(k=m+1,...,n)$ not all equal to zero such that $\sum_{k=m+1}^{n} c_k \cdot \tau_{k-m} = 0$ where τ_q is the q-th column of C. Then $g := \sum_{k=m+1}^{n} c_k f_k$ is nontrivial, and the last equation is equivalent to $a_p[L^{I_j}g;f^m,L^{I_j}] = a_p[L^{I_{j-1}}g;f^m,L^{I_{j-1}}]$ $(j=1,..,s;p=r+1,..,m)$. Hence, when $r > 0$, for $j=1,...,s$

$g_j := h[L^{I_j}g; f^m, L^{I_j}] - h[L^{I_{j-1}}g; f^m, L^{I_{j-1}}] \in \text{span}\{f_1,\dots,f_r\}$. If for $j=1,\dots,s$ and $i \in K_j$ L_i is applied to g_j we get $L_i g_j = 0$. From (v) we infer that $g_j = 0$ $(j=1,\dots,s)$, hence $h[L^{I_o}g; f^m, L^{I_o}] = \dots = h[L^{I_s}g; f^m, L^{I_s}] =: h \in F_m$ which is trivial when $r = 0$ and which is equivalent with $L_i(g-h) = 0 = \sum_{k=1}^{n} c_k L_i f_k$ $(i=1,\dots,n)$ where $\sum_{k=1}^{m} c_k f_k := -h$. This holds with $c_{m+1},\dots,c_n$ not all equal to zero iff f^n or L^n fails to be a basis of F_n or F_n^*, respectively.

In terms of numerical linear algebra: computing the solution $h[w^n]$ according to theorem 2 is equivalent with applying blockwise the "method of division and subtraction" ([2],p.133).

Theorem 2 generalizes and sharpens theorem (3.1) of [4] giving also a partial converse. As a consequence, by an easy calculation, from theorem 2 for interpolation by trigonometric polynomials of degree n at most with simple knots $x_1,\dots,x_{2n+1} \in [0,2\pi)$ the following recurrence formulas for the divided differences

$$[x_1,\dots,x_{2n+1}|\,{}^{f}_{k}] := a_k[L^{2n+1}f; f^{2n+1}, L^{2n+1}] \qquad (k=2n,2n+1)$$

of a function $f:[0,2\pi)\longrightarrow \mathbb{R}$ with respect to $(f_1,\dots,f_{2n+1}) := (1, \cos(\cdot), \sin(\cdot), \cos 2(\cdot), \dots, \cos n(\cdot), \sin n(\cdot))$ and $L_i f := f(x_i)$ $(i=1,\dots,2n+1)$ are deduced which are simpler than that of [4], p.404:

$$[x_1,\dots,x_{2n+1}|\,{}^{f}_{2n}] = \frac{1}{N}[(C_1 + C_2)b_1 - (S_1 - S_2)b_2]$$

$$[x_1,\dots,x_{2n+1}|\,{}^{f}_{2n+1}] = \frac{1}{N}[(S_1 + S_2)b_1 - (-C_1 + C_2)b_2]$$

where

$$C_1 = \cos x_{2n+1} + \cos x_{2n} - \cos x_2 - \cos x_1 ,$$

$$S_1 = \sin x_{2n+1} + \sin x_{2n} - \sin x_2 - \sin x_1 ,$$

$$C_2 = \cos(\sum_{q=1}^{2n-1} x_q) - \cos(\sum_{q=3}^{2n+1} x_q),$$

$$S_2 = \sin(\sum_{q=1}^{2n-1} x_q) - \sin(\sum_{q=3}^{2n+1} x_q),$$

$$b_1 = [x_3,\ldots,x_{2n+1}|{}^{f}_{2n-2}] - [x_1,\ldots,x_{2n-1}|{}^{f}_{2n-2}] ,$$

$$b_2 = [x_3,\ldots,x_{2n+1}|{}^{f}_{2n-1}] - [x_1,\ldots,x_{2n-1}|{}^{f}_{2n-1}] ,$$

$$N = C_1^2-C_2^2-S_2^2+S_1^2 = 16\cdot\sin\frac{x_{2n+1}-x_1}{2}\sin\frac{x_{2n+1}-x_2}{2}\sin\frac{x_{2n}-x_1}{2}\sin\frac{x_{2n}-x_2}{2},$$

$$[x_1,x_2,x_3|{}^{f}_{2}] = \frac{\dfrac{f(x_2)-f(x_1)}{\sin\frac{x_2-x_1}{2}}\cos(\frac{x_2+x_3}{2}) - \dfrac{f(x_3)-f(x_2)}{\sin\frac{x_3-x_2}{2}}\cos(\frac{x_1+x_2}{2})}{2\sin\frac{x_3-x_1}{2}},$$

$$[x_1,x_2,x_3|{}^{f}_{3}] = \frac{\dfrac{f(x_2)-f(x_1)}{\sin\frac{x_2-x_1}{2}}\sin(\frac{x_2+x_3}{2}) - \dfrac{f(x_3)-f(x_2)}{\sin\frac{x_3-x_2}{2}}\sin(\frac{x_1+x_2}{2})}{2\sin\frac{x_3-x_1}{2}}.$$

REFERENCES

1 Davis, Ph.J., Interpolation and Approximation, Blaisdell Publishing Company, 1963

2 Faddeev, D.K., Faddeeva, V.N., Computational Methods of Linear Algebra, Freeman a. Company, 1963

3 Mühlbach, G., An Algorithmic Approach to Hermite-Birkhoff Interpolation, to appear

4 Mühlbach, G., Num. Math. 32, 293-408, 1979

AN EXTENSION OF THE FREUD-POPOV LEMMA

Manfred W. Müller

Lehrstuhl Mathematik VIII
Universität Dortmund
Dortmund, Germany

A lemma of Freud-Popov which has been very useful in deriving Jackson type theorems for restricted and also nonrestricted approximation will be extended here to an estimation measuring simultaneous approximation by higher order moduli of smoothness.

I. INTRODUCTION AND MAIN RESULT

Combining a theorem of Popov-Sendov (5) with a lemma of Freud-Popov (3) by means of intermediate space techniques De Vore (2) obtains the following statement on simultaneous approximation:

For each function $f \in C^k(I)$, $k \in \mathbb{N}$, $I=[0,1]$, there exists a spline S_r of order $r > k$ (degree $r-1$) to an equidistant partition of the interval I into n subintervals, such that

$$\| f^{(j)} - S_r^{(j)} \|_\infty \leq C_1 n^{-k+j} \omega(f^{(k)}; n^{-1}), \quad j = 0,1 \;, \tag{1.1}$$

where the constant C_1 is depending only on r and $\omega(g;\cdot)$ denotes the ordinary modulus of continuity of the function $g : I \to \mathbb{R}$.

For the proof of (1.1) he needs the fact that to each function $f \in C^1(I)$, an arbitrary and fixed $\varepsilon > 0$ and to each natural number $r \geq 2$ there exists a function $F_{\varepsilon,r} \in C^r(I)$, such that

Copyright © 1980 by Academic Press, Inc.
All rights of reproduction in any form reserved.
ISBN: 0-12-171050-5

$$\| f^{(j)} - F_{\varepsilon,r}^{(j)} \|_\infty \leq C_2\, \omega_{r-j}(f^{(j)}; \varepsilon),\ j = 0,1 \tag{1.2}$$

and

$$\| F_{\varepsilon,r}^{(r)} \|_\infty \leq C_2\, \varepsilon^{-r}\, \omega_r(f;\varepsilon), \tag{1.3}$$

where the constant C_2 is depending only on r and $\omega_s(g;\cdot)$, $s \geq 2$, denote the s-th order moduli of smoothness of g.

Inequality (1.2) for $j = 0$ and inequality (1.3) are exactly the subject of the above mentioned Freud-Popov lemma. However one of my students observed (see Mönch-Pattberg (4)), that the function $F_{\varepsilon,r}$ constructed in (3), p. 170 (there denoted by $f_{k,h}$) does not satisfy (1.2) for j=1 and (1.3). This will be shown in section IV.

The main purpose of this paper is to give a constructive proof of Freud-Popov's lemma in the following generalized form, containing (1.2) and (1.3) as a spezial case.

THEOREM. <u>For</u> $f \in C^k(I)$, $k \in \mathbb{N}_0$, <u>an arbitrary</u> $\varepsilon \in (0,1]$ <u>and each</u> natu-<u>ral number</u> $r > k$ <u>there exists a function</u> $F_{\varepsilon,r} \in C^r(I)$, <u>such that</u>

$$\| f^{(j)} - F_{\varepsilon,r}^{(j)} \|_\infty \leq C_3\, \omega_{r-j}(f^{(j)}; \varepsilon),\ j = 0(1)k \tag{1.4}$$

<u>and</u>

$$\| F_{\varepsilon,r}^{(r)} \|_\infty \leq C_3\, \varepsilon^{-r} \omega_r(f;\varepsilon), \tag{1.5}$$

<u>with</u> C_3 <u>a constant depending only on</u> r.

The proof is using B-spline methods and will follow ideas in De Vore's paper (1).

II. PRELIMINARIES AND NOTATION

$\Delta_h^r(f;x)$, $r \in \mathbb{N}$, $h > 0$, denotes the r-th forward difference of the function f with step size h at the point $x \in \mathbb{R}$. Let M_r denote the B-spline of order r with knots at the points $0,1,\dots, r$. M_r is non-negative and vanishes outside $(0,r)$. We take a normalization so that

$\int_0^r M_r(u)du = 1$. Since M_r is the Peano kernel for $\Delta_1^r(f;o)$, we easily obtain the representation

$$\Delta_h^r(f;x) = h^r \int_0^r f^{(r)}(x+u)M_r(h^{-1}u)h^{-1}du \ , \ x \in I \ , \tag{2.1}$$

if $f \in C^r[o,1 + rh]$.

A function $f \in C^k(I)$ can be extended from I to a larger interval $I = [o,b]$, $b > 1$, with no loss of smoothness (measured by the r-th order modulus of smoothness, $r > k$). More precisely there is a linear operator $T : C^k(I) \ni f \to Tf \in C^k(J)$ such that $Tf = f$ on I and

$$\omega_r((Tf)^{(j)};\ h)_J \leq C_4\ \omega_r(f^{(j)};\ h)_I \ ,\ j = O(1)k, \tag{2.2}$$

with C_4 a constant depending only on r and b. (The subscripts J resp. I are indicating that the corresponding moduli are calculated on J resp. I.) Tf is called the Whitney extension of f (6).

III PROOF OF THE THEOREM

Let $J: = [o,1 + r^2]$. For $f \in C^k(I)$ and $\varepsilon \in (o,1]$ we define

$$F_{\varepsilon,r}(x): = f(x) + (-1)^{r+1} \int_0^r \Delta_{\varepsilon u}(Tf;x)M_r(u)du,\ x \in I, \tag{3.1}$$

which can also be written in the form

$$F_{\varepsilon,r}(x) = \sum_{i=1}^{r} \binom{r}{i} (-1)^{i+1} \int_0^r Tf(x+i\varepsilon u)M_r(u)du,\ x \in I. \tag{3.2}$$

In order to obtain from (3.1) the derivatives $F_{\varepsilon,r}^{(j)}$, $j = O(1)k$, it is allowed to exchange differentiation with respect to x and integation, since the integrand has continuous partial derivatives with respect to x up to the k-th order. Thus

$$F_{\varepsilon,r}^{(j)}(x) = f^{(j)}(x) + (-1)^{r+1} \int_0^r \Delta_{\varepsilon u}^{r} ((Tf)^{(j)};x)M_r(u)du,\ x \in I\ ,$$

$j = O(1)k$. Fix $x \in I$. Then with (2.2)

$$|f^{(j)}(x) - F_{\varepsilon,r}^{(j)}(x)| \leq \int_0^r |\Delta_{\varepsilon u}^r((Tf)^{(j)};x)| M_r(u)du$$

$$\leq \omega_r((Tf)^{(j)};\ r\varepsilon)_J \leq C_4\ \omega_r(f^{(j)};\ r\varepsilon)_I$$

$$\leq C_4 r^r 2^j \omega_{r-j}(f^{(j)};\varepsilon) \leq : K_1\omega_{r-j}(f^{(j)};\varepsilon),$$

with K_1 a constant depending only on r (considering that in our case $b = 1 + r^2$ and 2^j can independantly of j be estimated by 2^k). Since $x \in I$ was arbitrary, we have proved (1.4).

To estimate the r-th derivative of $F_{\varepsilon,r}$, let T_1 be a primitive of Tf and in general T_s a primitive of T_{s-1} for $s \geq 2$. Then for the i-th term in (3.2), we have by substituting $v = i\varepsilon u$

$$\int_0^{2r\varepsilon} Tf(x+v)M_r((i\varepsilon)^{-1}v)(i\varepsilon)^{-1}dv = (i\varepsilon)^{-r}\ \Delta_{i\varepsilon}^r(T_r;x),$$

$x \in I$, because of (2.1). Thus (3.2) can be written in the form

$$F_{\varepsilon,r}(x) = \sum_{i=1}^r \binom{r}{i}(-1)^{i+1}(i\varepsilon)^{-r}\Delta_{i\varepsilon}^r(T_r;x),\ x \in I. \tag{3.3}$$

Fix $x \in I$. Then with (2.2)

$$|F_{\varepsilon,r}^{(r)}(x)| = |\sum_{i=1}^r \binom{r}{i}(-1)^{i+1}(i\varepsilon)^{-r}\Delta_{i\varepsilon}^r(Tf;x)|$$

$$\leq (2^r-1)\varepsilon^{-r}\max_i|\Delta_{i\varepsilon}^r(Tf;x)| \leq (2^r-1)\varepsilon^{-r}\omega_r(Tf;r\varepsilon)_J$$

$$\leq K_2\varepsilon^{-r}\omega_r(f;\varepsilon)_I,$$

with K_2 a constant depending only on r. Since $x \in I$ was arbitrary, we have proved the theorem completely if we are choosing $C_3 := \max(K_1,K_2)$.

IV FINAL REMARKS

From (3.3) it can be seen that $F_{\varepsilon,r}$ is even $r + k$ - times conti - nuously differentiable.

In the original proof of Freud-Popov's lemma the function $f \in C(I)$ is extended first of all to a function $f^* \in C[o,1+r\varepsilon]$ by means of the polynomial of degree at most r interpolating f at the points $1-i\varepsilon, i=0(1)r$ (here with an $\varepsilon \in (0,r^{-1})$). (3.1) is then replaced by

$$F^*_{\varepsilon,r}(x) = \frac{(-1)^{r+1}}{\varepsilon^r} \{ \int_o^{\varepsilon} \}^r \sum_{i=1}^{r} \binom{r}{i} (-1)^{r-i} f^*(x + \frac{i}{r}(t_1 + \ldots + t_r)) \cdot$$

$$\cdot dt_1 \ldots dt_r \, , \; x \in I.$$

Obviously $F^*_{\varepsilon,r} \in C(I)$, but in trying to calculate derivatives there are arising difficulties, since f^* will in general not be differentiable at the point $x = 1$, for what reason it is not allowed to exchange differentiation with respect to x and integration as it was possible in the above proof.

REFERENCES

1. De Vore, R., Degree of approximation, in "Approximation Theory II" (G.G. Lorentz, C.K. Chui, L.L. Schumaker, ed.), pp. 117-161. Academic Press, New York, (1976).
2. De Vore, R., Monotone approximation by splines, SIAM Math. Anal. 8, 891-905 (1977).
3. Freud, G. and Popov, V., On approximation by spline functions, in "Proc. Conf. on Constructive Theory of Functions" (G. Alexits, S.B. Stečkin, ed.), pp. 163-172. Akademia Kiadó, Budapest, (1969).
4. Mönch-Pattberg, H., Über monotone Approximation durch Splines, Diplomarbeit, Universität Dortmund, (1978).
5. Popov V.A. and Sendov, Bl.K., The classes that are characterized by the best approximation by splines functions, Math. Notes 8, 550-557 (1970).
6. Whitney, H., Analytic extensions of differentiable functions defined on chlosed sets, Trans. Amer. Math. Soc. 36, 63-89 (1934).

BEST APPROXIMATION PROBLEMS ARISING FROM GENERALIZED INVERSE OPERATOR THEORY

M. Z. Nashed[1]
Department of Mathematical Sciences
University of Delaware
Newark, Delaware 19711

ABSTRACT

It is well known that generalized inverses of linear operators in Hilbert spaces can be characterized by extremal properties. In Banach spaces the proximal approach to generalized inverses is not equivalent to other definitions of generalized inverses. Also, the perturbation theory for proximal generalized inverse is quite distinct from the perturbation theory for generalized inverses based on topological complementation. All these differences lead to several interesting problems in *best approximation theory* with respect to various *vector* and *operator norms*, as well as to problems involving geometry of Banach spaces, complementation and gaps between subspaces. We highlight some recent results with emphasis on open problems in this area.

1. Let V and W be (real or complex) vector spaces and let $L: V \to W$ be a linear transformation. The range and nullspace of L are denoted by $R(L)$ and $N(L)$ respectively. A linear map $M: W \to V$ such that $LML = L$ is called an *inner inverse* of L. If M is an inner inverse, then LM and ML are idempotent and

$$V = N(L) \dot{+} R(ML) \ , \quad W = R(L) \dot{+} N(LM) \ . \tag{1}$$

Analogously, M is an *outer inverse* to L if $MLM = M$. Each outer inverse induces the direct sum decompositions

$$V = R(M) \dot{+} N(ML) \quad \text{and} \quad W = N(M) \dot{+} R(LM) \ . \tag{2}$$

If M is both an inner inverse and an outer inverse of L, then M is called an *algebraic generalized inverse* (AGI) of

[1]Partially supported by NSF Grant MCS-79-04408.

Copyright © 1980 by Academic Press, Inc.
All rights of reproduction in any form reserved.
ISBN: 0-12-171050-5

L. In this case

$$V = N(L) \dot{+} R(M) \quad \text{and} \quad W = R(L) \dot{+} N(M) \quad . \tag{3}$$

Every linear map has an AGI (in particular if M is any inner inverse, then MLM is an AGI).

Conversely, suppose

$$V = N(L) \dot{+} M \quad \text{and} \quad W = R(L) \dot{+} S \tag{4}$$

and let P be the projector of V onto $N(L)$ along M, and Q be the projector of W onto $R(L)$ along S. Then there exist a unique AGI $L^{\#} = L^{\#}_{P,Q}$ which is the unique solution of the equations:

$$LXL = L, \; XLX = X, \; LX = Q, \; XL = I-P \; . \tag{5}$$

The algebraic theory of generalized inverse is virtually complete (see [9] for a comprehensive treatment). However, if we inquire about the existence of generalized inverses with certain properties in the framework of analysis, then there are some open problems. For simplicity, assume that L is a bounded or closed densely defined linear operator acting between Banach spaces V and W. L has a *bounded* inner inverse if and only if $N(L)$ and $R(L)$ have topological complements (see, e.g. [9]). No such simple criterion exists for bounded outer inverses. A necessary and sufficient condition for $M \neq 0$ to be a bounded outer inverse for L is that $R(M)$ and $N(M)$ be closed complemented subspaces of V and W respectively, i.e. the algebraic decompositions (2) must also be *topological* decompositions. If L has a bounded inner inverse, then L has a bounded outer inverse, but not conversely. *Problems*: (a) *Find a criterion in terms of* L *for the existence of a bounded outer inverse*; (b) *Does every* L *have a bounded outer inverse*? (c) *When does* L *have an inner (outer, generalized) inverse belonging to the Schatten class* C_p, $1 \leq p < \infty$? (See Section 5.)

Note that criteria for the generalized inverse $L^{\dagger}$ in Hilbert and Banach spaces to be compact are easy to derive using series (or sequence) representations of $L^{\dagger}$ (see e.g. [4, pp. 378-386],[6]). Two examples: $L^{\dagger}y = \lim_{n\to\infty} (L^*L + \frac{1}{n} I)^{-1}L^*y$

$L^{\dagger}y = \sum_{k=1}^{\infty} (I+L^*L)^{-k} L^*y$, where convergence is in the operator norm if and only if $R(L)$ is closed. Thus, e.g., a necessary and sufficient condition for $L^{\dagger}$ to be compact is that $R(L)$ be closed and $(I+L^*L)^{-1} L^*$ be compact.

2. The Moore-Penrose inverse of an mxn matrix A is the generalized inverse $A^{\dagger} = A^{\#}_{P,Q}$, where P and Q in (4) are chosen to be orthogonal (equivalently, Hermitian) projectors. Let $||C||_F = (\text{trace } C^*C)^{1/2}$ denote the Frobenius norm of a matrix C. There are three extremal characterizations of $A^{\dagger}$ due to Penrose [10]:

(E1) For every $m \times p$ matrix B, $A^{\dagger}B$ is the unique matrix which minimizes $||AX-B||_F$ and for which $||X||$ is minimal among such minimizers. In particular, $A^{\dagger}$ is the unique best-approximate solution (with respect to the Frobenius norm) of the matrix equation $AX = I$; similarly for $XA = I$.

(E2) For each $b \in R^m$, $A^{\dagger}b$ is the unique vector which minimizes $||Ax-b||$ (in the Euclidean norm) and is of minimum Euclidean norm among such minimizers. (E2) follows, of course, immediately from (E1).

(E3) Let $J(A) := \{X: AXA = A\}$. Then $A^{\dagger}$ is the unique matrix which minimizes $||X||_F$ over $J(A)$.

The *vector version* of the extremal characterization of $A^{\dagger}$, i.e. (E2), generalizes immediately to Hilbert spaces. In fact, it was established by Tseng for closed densely defined linear operators in Hilbert spaces before Penrose's paper appeared. Some *operator-extremal* characterizations of $A^{\dagger}$, namely analogs of (E1) and (E3) in Hilbert spaces have recently been obtained by Engl and Nashed [2]. Generalized inverses in Banach spaces do not have in general extremal properties analogous to (E1)-(E3) since proximinal mappings for Chebyshev subspaces are very rarely "orthogonal" projectors. Therefore, the study of the interplay between generalized inverse operator theory in normed spaces and best approximate solutions of linear operator equations (i.e., the minimization problems (E1)-(E3)) leads to two distinct approaches (see [8] and [9]) to generalized inverses, which happen to coincide in Hilbert spaces.

3. From the viewpoint of analysis and approximation theory, the algebraic theory of generalized inverses is inadequate. For those purposes, it is not enough that P and Q be linear idemponent; we want them to be continuous, even when neither L nor $L^{\#}$ is bounded. This has led to the theory of so-called left-topological and right-topological inner inverses ([8],[9]). In particular, let X and Y be normed linear spaces and let $T\colon D(T) \subset X \to Y$ be a linear operator. Let T be domain decomposable with respect to the projector $P \in L(X)$, i.e., $Px \in N(T)$ for all $x \in D(T)$, $N(T) \subset R(P)$, and $D(T) \cap N(P)$ is dense in $N(P)$. Suppose also that there exists a continuous projector Q onto $\overline{R(T)}$. Then (see [9;Theorem 2.10]) T has a unique generalized inverse $T^{\dagger} = T^{\dagger}_{P,Q}$ with the properties: $D(T^{\dagger}) = R(T) + N(Q)$, $R(T) = D(T) \cap N(P)$, $N(T^{\dagger}) = N(Q)$, $T^{\dagger}T = I - P$ on $D(T)$ and $TT^{\dagger} = Q$ on $D(T^{\dagger})$.

When does $T^{\dagger}_{P,Q}y$ provide a "best approximate" solution (in the sense of (E2)) to the equation $Tx = y$? This and related questions for extremal solutions and for minimal-norm solution when $y \in R(T)$ have been thoroughly studied in [9], and have motivated the introduction of the notions of left-orthogonal, right-orthogonal, and orthogonal inner inverses. In particular, it is shown that if $||I-Q|| = 1$, and $||I-P|| = 1$ and if in Y nearest points from $\overline{R(T)}$ are unique, then for all $y \in D(T^{\dagger})$, $T^{\dagger}_{P,Q}$ is a (not necessarily unique) best approximate solution of $Tx = y$. These conditions are rarely satisfied in normed spaces without inner products. In general, $||I-Q||$ is a measure of the quality of complementation $\overline{R(T)}$: the larger $||I-Q||$, the "worse" is the complementation. If $||I-Q|| = 1$ we say that $\overline{R(T)}$ has an orthogonal complement; this represents the best complementation. Note that if $||I-Q|| = 1$ it is not necessarily true that $||Q|| = 1$, i.e., "orthogonality" is not symmetric in general. To dramatize the severity of the situation, we recall that while every finite-dimensional subspace has a topological complement, even a two-dimensional space need not have an orthogonal complement in normed spaces. This suggests that we compromise on the notion of "best approximate solution". One approach would be to consider "optimal" complements instead of "orthogonal" complements. Define the projection constant $\gamma(R(T);Y) = \inf||I-Q||$, where Q runs through all projectors from Y onto $\overline{R(T)}$. We say that a complement to $\overline{R(T)}$ is ε-optimal if $\left| ||I-\tilde{Q}||-\gamma \right| \leq \varepsilon$, $\varepsilon > 0$. Similarly for $||I-\tilde{P}||$. The generalized inverse induced by $\tilde{P}$ and $\tilde{Q}$ leads to "approximate" extremal properties. The theory in [9 ;Section 3] can be adapted to this situation.

4. Another approach to generalized inverses in normed spaces is via metric (nearest point) projectors. Define $T^{\delta}y = \{u \in X: \ u \text{ minimizes } ||Tx-y||\}$. Here $D(T^{\delta})$ consists of all $y \in Y$ for which $\inf||Tx-y||$ is attained. The set-valued mapping $y \to T^{\delta}(y)$ is called the *metric generalized inverse*. A (in general nonlinear) function $T^{\sigma}: D(T^{\delta}) \to X$ such that $T^{\sigma}(y) \in T^{\delta}(y)$ is called a selection for the metric generalized inverse. Given a *linear* selection T^{σ}, when does there exist *continuous* and *linear* idempotents P and Q such that $T^{\sigma} = T^{\dagger}_{P,Q}$? Here P,Q and $T^{\dagger}_{P,Q}$ have the same meaning as in section 3 . That is, when can a linear selection of a metric generalized inverse be realized as a *topological* generalized inverse? Some sufficient conditions are given in [9]. Also, some aspects when T^{δ} is single-valued are considered in [3]. However, the interplay between (nonlinear) metric decompositions $M \oplus M^{\theta}$, where M^{θ} is an (not necessarily unique) orthogonal complement of M, and topological decom-

positions $M \oplus N$, where M is either $\overline{R(T)}$ or $N(T)$, has not been fully explored.

5. Engl and Nashed [2] have recently obtained new extremal characterizations of generalized inverse operators in Hilbert spaces, which generalize and extend (E1) and (E3). Let $A \in L(H_1,H_2)$ with closed range and $B \in L(H_3,H_2)$, where H_1, H_2, and H_3 are Hilbert spaces. Let $F: L(H_3,H_1) \to L(H_3)$ be defined by $F(x) = (AX-B)^*(AX-B)$. Then the set $\{F(x): X \in L(H_3,H_1)\}$ has a smallest element F_0 with respect to the Hermitian order on $L(H_3)$. Let $\Omega = \{X: F(X) = F_0\}$. The set $\{X^*X: X \in \Omega\}$ has a smallest element F_1 with respect to the Hermitian order. The unique $X \in \Omega$ with $X^*X = F_1$ is $X = A^\dagger B$. Similar results are established in [2] for closed operators and for operators with nonclosed range. It is also shown that for a certain class of operators one can phrase these characterizations in terms of a whole class of norms (including Schatten norms [1], but not including the uniform operator norm). This provides new extremal characterizations even in the matrix case: it is shown that (E1) holds for a large class of norms, and not just the Frobenius norm. It would be of interest to *characterize all norms for which the extremal characterization* (E1) *holds*. (E3) is also generalized in [2].

The general linear matrix equation $\sum_{i=1}^{m} A_i X C_i = B$, where A_i, C_i and B are given matrices, has a unique best approximate solution. This equation is only apparently more general than the equation $AX = B$; in fact it can be reduced to it using tensor products. In particular, the matrix equation $AXC = B$ can be treated in this way, and its best approximate solution is $A^\dagger B C^\dagger$. The methods and results of [2] are restricted to the operators $AX-B$ or $XA-B$ and do not seem to generalize to the *operator equation* $\sum_{i=1}^{m} A_i X C_i = B$, where A_i, B_i and C are given bounded linear operators on Hilbert spaces. *When does this equation have a best approximate solution? In particular, does* $AX-XC = B$ *have a best approximate solution (with respect to some ordering or norm) if the intersection of the spectra of* A *and* C *is nonempty* (otherwise, it is well known (see [11]) that the equation has a unique solution)?

REFERENCES

[1] Dunford, N., and Schwartz, J. T., Linear Operators, Vol. II, Interscience, New York, 1963 (see XI.9).

[2] Engl, H. W., and Nashed, M. Z., New extremal characterizations of generalized inverses of linear operators, submitted.

[3] Holmes, R. B., A Course on Optimization and Best Approximation, Springer-Verlag, New York, 1972.

[4] Nashed, M. Z., ed., Generalized Inverses and Applications, Academic Press, New York, 1976.

[5] Nashed, M. Z., On the perturbation theory for generalized inverse operators in Banach spaces, in "Functional Analysis Methods in Numerical Analysis" (M. Z. Nashed, ed.), pp. 180-195, Springer-Verlag, New York, 1979.

[6] Nashed, M. Z., and Engl, H. W., Random generalized inverses and approximate solutions of random operator equations, in "Approximate Solution of Random Equations" (A. T. Bharucha-Reid, ed.), pp. 149-210, North Holland, New York, 1979.

[7] Nashed, M. Z., and Votruba, G. F., A unified approach to generalized inverses of linear operators: I. Algebraic, topological and projectional properties, Bull. Amer. Math. Soc. 80 (1974), 825-830.

[8] ____________, ibid: II. Extremal and proximal properties, 831-835.

[9] ____________, A unified operator theory of generalized inverses, in [4], pp. 1-109.

[10] Penrose, R., On best approximate solutions of linear matrix equations, Proc. Cambridge Philos. Soc. 52 (1956), 17-19.

[11] Radjavi, H., and Rosenthal, P., Invariant Subspaces, Springer-Verlag, New York, 1973.

MOMENTS OF B-SPLINES

Edward Neuman
Institute of Computer Science
University of Wrocław

I. INTRODUCTION AND NOTATION

Let $M_{j,k}(x) = k[t_j,t_{j+1},\ldots,t_{j+k}](\cdot-x)_+^{k-1}$ $(j \in Z,\ k \in N,\ t_j < t_{j+1} < \ldots < t_{j+k})$ denote a classical B-spline of Curry and Schoenberg. By $\mu_l(k,j,\underline{t})$ we denote the l-th moment of the B-spline $M_{j,k}$, i.e.,

$$\mu_l(k,j,\underline{t}) = \int_{-\infty}^{\infty} x^l M_{j,k}(x)dx \quad (l = 0, 1,\ldots),$$

where $\underline{t} = (t_j,t_{j+1},\ldots,t_{j+k})$ is a vector of the knots of the B-spline $M_{j,k}$. Moments of B-splines are useful in some problems which arise from numerical analysis. The first of them is the following one. For the numerical calculations of the integrals with the weight function $M_{j,k}$

$$\int_a^b M_{j,k}(x)f(x)dx$$

we may use a quadrature formula of the Gauss type. The nodes of such a formula are the zeros of the orthogonal polynomials on the interval [a,b] with the function $M_{j,k}$ as a weight. If the moments of $M_{j,k}$ are available then one can construct such a system of polynomials at least in a numerical way (Phillips and Hanson, 1974). The second problem is connected with some method of construction of a convex interpolating spline with an arbitrary order, of convexity. In this method the moments of B-splines play an important role (Neuman, 1980).

Without loss of generality and for the sake of brevity we put j = 0.

Copyright © 1980 by Academic Press, Inc.
All rights of reproduction in any form reserved.
ISBN: 0-12-171050-5

For this reason we write $\mu_l(k,\underline{t})$ instead of $\mu_l(k,0,\underline{t})$.

Proofs and further details will appear elsewhere.

II. ARBITRARY KNOTS

We begin this section with the well-known formula for the l-th moment of the B-spline $M_{0,k}$. Namely

$$\mu_l(k,\underline{t}) = \binom{k+l}{k}^{-1} \sum_{0\le j_1 \le j_2 \le \ldots \le j_l \le k} t_{j_1} t_{j_2} \cdots t_{j_l} \equiv \binom{k+l}{k}^{-1} S_{\underline{t}}(k+l,k)$$

$$(l = 0,1,\ldots;\ k \in N).$$

In the special case $t_j = j$ $(j = 0,1,\ldots,k)$ the numbers $S_{\underline{t}}(\ ,\ )$ coincide with the classical Stirling numbers of the second kind. The proof of the above formula is simple and follows immediately from

$$[t_0,t_1,\ldots,t_k]f = \frac{1}{k!} \int_{-\infty}^{\infty} M_{0,k}(x) f^{(k)}(x)dx \qquad (f \in C^k(R)).$$

(Curry and Schoenberg, 1966).

Below we give a few recurrence formulae for the calculation of the quantities $\mu_l(\ ,\ )$ in the case of arbitrary knots of the B-spline $M_{0,k}$. We have

$$\mu_l(k,\underline{t}) = \binom{k+l}{k}^{-1} \sum_{j=0}^{l} t_k^{l-j} \binom{k+j-1}{j} \mu_j(k-1,\underline{t}) \qquad (k = 2,3,\ldots;\ l = 0,1,\ldots),$$

where the initial values are equal to

$$\mu_j(1,\underline{t}) = (t_1^{j+1} - t_0^{j+1})/[(j+1)(t_1 - t_0)] \qquad (j = 0,1,\ldots).$$

Other recurrence formulae

$$\sum_{j=1}^{p} \binom{n}{j} \prod_{m=0}^{j-1} (t_p - t_m) \mu_{n-j}(j,\underline{t}) = t_p^n - t_0^n \qquad (p = 0,1,\ldots;\ p \le n),$$

$$\sum_{j=1}^{n} \binom{n}{j} \prod_{\substack{m=0\\m\neq p}}^{j-1} (t_p - t_m)\mu_{n-j}(j,\underline{t}) = nt_p^{n-1} \qquad (n \in N;\ p = 0,1,\ldots,n-1).$$

Now let $t_j = -\cos \pi j/k$ $(j = 0,1,\ldots,k)$. The B-spline $M_{0,k}$ with these knots is a perfect B-spline (Schoenberg, 1971). In this case we have the following simple formula for the moments

$$\mu_l(k,\underline{t}) = \binom{k+1}{k}^{-1} \frac{2^{k-1}}{k} \sum_{k=0}^{k}{}'' (-1)^{j+k}\, t_j^{k+1} \qquad l \text{ - even},$$

$$= 0 \qquad l \text{ - odd}.$$

III. EQUIDISTANT KNOTS

Different recurrence formulae hold in the case of equidistant knots. Let $t_j = a + jh$ $(a \in R,\ h > 0,\ j = 0,1,\ldots,k)$. We have

$$\mu_l(k,\underline{t}) = \sum_{m=0}^{l} \binom{l}{m} a^m h^{l-m} \mu_{l-m}(k,\underset{\sim}{t}),$$

where $\underset{\sim}{t} = (0,1,\ldots,k)$. The B-spline $M_{0,k}$ with the knots $t_j = j$ is called a forward B-spline. For this special choice of knots we have also further recurrences

$$\mu_l(k,\underset{\sim}{t}) = c_l D_l^{(-k)} 2^l + \sum_{m=1}^{l} (-1)^{m+1} \left(\frac{k}{2}\right)^m \binom{l}{m} \mu_{l-m}(k,\underset{\sim}{t}) \qquad (l = 0,1,\ldots;\ k \in N),$$

where

$$c_l = 0 \qquad l \text{ - odd},$$

$$= 1 \qquad l \text{ - even},$$

and $D_{2l}^{(-k)}$ are given by the formula

$$\left(\frac{t}{2\sin t/2}\right)^k = \sum_{l=0}^{\infty} (-1)^l \frac{t^{2l}}{(2l)!\,2^{2l}} D_{2l}^{(-k)}$$

(Nörlund, 1924).

Further we have also

$$\mu_l(k,\underset{\sim}{t}) = \frac{k}{l}\sum_{j=1}^{l}(-1)^j\binom{l}{j}B_j\mu_{l-j}(k,\underset{\sim}{t}) \qquad (k,l \in N),$$

where B_j denotes the j-th Bernoulli number.

Let $B_l^{(z)}$ denote the Nörlund polynomial of degree l in variable z. Then one can prove

$$\mu_l(k,\underset{\sim}{t}) = B_l^{(-k)} \qquad (l=0,1,\ldots;\ k\in N)\ .$$

Let $t_j = -k/2 + j$ $(j=0,1,\ldots,k)$. In this case we have the following explicit formula for $\mu_l(\ ,\ )$, namely

$$\mu_l(k,\underline{t}) = 2^{-l}D_l^{(-k)} \qquad l\text{ - even,}$$

$$= 0 \qquad l\text{ - odd.}$$

The B-spline with the above knots is called a central B-spline.

Now we assume that $t_j = jh$ $(j=0,1,\ldots,k;\ h>0)$. In this case we have further recurrence formulae

$$\mu_l(m+n,\underline{t}) = \sum_{j=0}^{l}\binom{l}{j}\mu_j(m,\underline{t})\mu_{l-j}(n,\underline{t})\ ,$$

$$\mu_{l+1}(m+n,\underline{t}) = \frac{m+n}{n}\sum_{j=0}^{l}\binom{l}{j}\mu_{l-j}(m,\underline{t})\mu_{j+1}(n,\underline{t}) \qquad (l=0,1,\ldots;\ m,n\in N),$$

$$\mu_l(k,\underline{t}) = \frac{1}{2}\sum_{m=0}^{l-1}(-1)^m\binom{l}{m}(kh)^{l-m}\mu_m(k,\underline{t}) \qquad (l=1,3,5,\ldots).$$

REFERENCES

Curry, H., and Schoenberg, I. (1966). *J.Analyse Math. 17*, 71.

Neuman, E. (1980). *In* "Numerische Methoden der Approximationstheorie," Vol. 5 (L. Collatz, G. Meinardus, H. Werner, eds.). Birkhäuser, Basel (to appear).

Nörlund, N. (1924). "Vorlesungen über Differenzenrechnung." Springer, Berlin.

Phillips, J., and Hanson, R. (1974). *Math. Comput. 28*, 666.

Schoenberg, I. (1971). *Israel J. Math. 57*, 291.

CONVERGENCE AND SUMMABILITY OF ORTHOGONAL SERIES

Paul G. Nevai

Department of Mathematics
The Ohio State University
Columbus, Ohio

Let us denote by S the class of positive measures $d\alpha$ such that $\mathrm{supp}(d\alpha) = [-1,1]$ and

$$\int_0^{\pi} \log \alpha'(\cos t)dt < \infty .$$

For $d\alpha \in S$ the corresponding orthonormalized polynomials $p_n(d\alpha,x)$ $(n = 0,1,2,\ldots)$ satisfy a three-term linear homogeneous difference equation. It will be convenient for us to write this difference equation as

$$(1)\quad p_n(d\alpha,x) - 2xp_{n-1}(d\alpha,x) + p_{n-2}(d\alpha,x) = $$
$$= a_n p_n(d\alpha,x) + b_{n-1}p_{n-1}(d\alpha,x) + a_{n-1}p_{n-2}(d\alpha,x) ,$$

$n = 0,1,2,\ldots,\ p_{-1} = 0\ ,\ p_0 = [\alpha(1) - \alpha(-1)]^{-1/2}$ and

$$a_n = 1 - 2\int_{-1}^{1} x\, p_{n-1}(d\alpha,x)\overline{p_n}(d\alpha,x)d\alpha(x) ,$$

$$b_n = -2\int_{-1}^{1} x\, p_n^2(d\alpha,x)d\alpha(x) .$$

The purpose of this note is to find applications of the following results concerning the Szegő class S.

THEOREM A. [3, Theorem 7.9] *If* $d\alpha \in S$ *then*

$$\sum_{n=0}^{\infty} (a_n^2 + b_n^2) < \infty .$$

THEOREM B. [2, Theorem 2] *Let* $d\alpha \in S$. *Then*

This material is based upon work supported by the National Science Foundation under Grant No. MCS 78-01868.

Copyright © 1980 by Academic Press, Inc.
All rights of reproduction in any form reserved.
ISBN: 0-12-171050-5

$$\limsup_{N \to \infty} \frac{1}{N} \sum_{n=0}^{N} p_n^2(d\alpha, x) \leq \frac{e}{\pi\alpha'(x)\sqrt{1-x^2}}$$

for almost every $x \in [-1,1]$.

THEOREM C. [2, Corollary 3] *Suppose* $d\alpha \in S$ *and*

$$\sum_{k=0}^{\infty} c_k^2 < \infty.$$

Then for every $\epsilon > 0$ *the series*

$$\sum_{k=0}^{\infty} c_k p_k(d\alpha, x)$$

is (C, ϵ) *summable almost everywhere in* $[-1,1]$.

THEOREM D. [1, Theorem 8.3] *If* $d\alpha \in S$ *then there exists a constant* $K > 1$ *such that*

$$|p_n(d\alpha, x)| \leq K^{\sqrt{n}} \qquad n = 1, 2, \ldots,$$

for every $x \in [-1,1]$.

These four powerful theorems will enable us to easily prove some important results about orthogonal series. Our first theorem deals with the generating function

$$(2)\quad f(d\alpha, z, x) = \sum_{k=0}^{\infty} p_k(d\alpha, x) z^k$$

of the orthonormalized system $\{p_k(d\alpha, x)\}$.

THEOREM 1. *Let* $d\alpha \in S$. *Then the series defining* $f(d\alpha, z, x)$ *converges uniformly for* $-1 \leq x \leq 1$ *and* $|z| \leq \rho < 1$. *Moreover, for every real* t

$$(3)\quad \lim_{r \uparrow 1} f(d\alpha, re^{it}, x) = f(d\alpha, e^{it}, x)$$

exists for almost every $x \in [-1,1]$. *For fixed* t, *the function* $f(d\alpha, e^{it}, x)(1 - 2xe^{it} + e^{2it})$ *belongs to* $L^2(d\alpha)$.

Proof. It follows immediately from Theorem D that the series in (2) converges uniformly for $-1 \leq x \leq 1$ and $|z| \leq \rho < 1$. The next step is to apply the identity

$$(4)\quad \sum_{k=0}^{n} z^k [p_k(d\alpha, x) - 2x p_{k-1}(d\alpha, x) + p_{k-2}(d\alpha, x)] =$$
$$= (1 - 2xz + z^2) \sum_{k=0}^{n-1} z^k p_k(d\alpha, x) + z^n p_n(d\alpha, x) - z^{n+1} p_{n-1}(d\alpha, x)$$

and Theorem D to obtain

$$f(d\alpha, re^{it}, x) = \frac{\sum_{k=0}^{\infty} r^k e^{ikt}[p_k(d\alpha,x) - 2xp_{k-1}(d\alpha,x) + p_{k-2}(d\alpha,x)]}{(1 - 2xre^{it} + r^2e^{2it})} .$$

By (1) and Theorems A and C the series

$$(5) \quad \sum_{k=0}^{\infty} e^{ikt}[p_k(d\alpha,x) - 2xp_{k-1}(d\alpha,x) + p_{k-2}(d\alpha,x)]$$

has the property that for every real t it is $(C,1)$ summable almost everywhere in $[-1,1]$. Hence, for every real t it is also Abel summable almost everywhere in $[-1,1]$. Finally, Theorem A and (1) imply that the Abel sum of (5) is a function in $L^2(d\alpha)$.

Remark 1. In the previous proof Theorem C could have been substituted by the joint application of Beppo Levi's theorem and L. Carleson's theorem about almost everywhere convergence of trigonometric Fourier series of L^2 integrable functions. But this substitution would have yielded only a weaker result. Namely, we would have proved (3) only for almost every t.

Remark 2. It is easy to see that if $d\alpha \in S$ then for every real t the series

$$\sum_{k=0}^{\infty} p_k(d\alpha,x)e^{ikt}$$

diverges for almost every $x \in [-1,1]$. Indeed, by Theorem 6.2.33 of [3]

$$\lim_{k\to\infty} p_k(d\alpha,x) = 0$$

for $x \in E$ can only be true if $\text{meas}(E) = 0$.

Our next result is going to deal with convergence of trigonometric series

$$(6) \quad \sum_{k=0}^{\infty} c_k e^{ikt}$$

when the coefficients c_k are given as Fourier coefficients of a function $g \in L^2(d\alpha)$ with respect to the orthogonal polynomial system

$\{p_k(d\alpha,x)\}$, that is

$$(7)\quad c_k = \int_{-1}^{1} g(x)\, p_k(d\alpha,x)\,d\alpha(x)\quad , \ k = 0,1,\ldots$$

It is evident that from the point of view of numerical computations it is much more convenient to work with series of the form (6) than with the orthogonal expansion

$$(8)\quad \sum_{k=0}^{\infty} c_k p_k(d\alpha,\cos t)$$

since asymptotic formulas for $p_k(d\alpha,\cos t)$ are of no help when trying to approximate the sum of (8). For instance, if the sequence $\{c_k\}$ is given then it is easy to compute any partial sum of (6); whereas the partial sums of (8) can only be found if the orthogonal polynomials are explicitly known. In the following, the function G_t will be defined by

$$G_t(x) = \frac{g(x) - g(\cos t)}{2(x-\cos t)}$$

THEOREM 2. Let $d\alpha \in S$ and $g \in L^2(d\alpha)$. Suppose that for a given real t the function G_t belongs to $L^2(d\alpha)$. Then the series (6) defined by (7) converges.

Proof. Put $z = e^{it}$ in (4) and divide both sides of it by $1-2xe^{it} + e^{2it} = 2e^{it}(\cos t - x)$. Now applying (1) we get

$$\sum_{k=0}^{n-1} e^{ikt} p_k(d\alpha,x) = \frac{e^{i(n-1)t}\, p_n(d\alpha,x)}{2(x-\cos t)} - \frac{e^{int} p_{n-1}(d\alpha,x)}{2(x-\cos t)} -$$

$$- \sum_{k=0}^{n} e^{i(k-1)t} \frac{a_k p_k(d\alpha,x) + b_k\, p_{k-1}(d\alpha,x) + a_{k-1}p_{k-2}(d\alpha,x)}{2(x-\cos t)}$$

Multiplying both sides of this identity by $g(x) - g(\cos t)$ and integrating it against $d\alpha$ we obtain

$$(9)\quad \sum_{k=0}^{n-1} c_k e^{ikt} = g(\cos t)\,[\alpha(1)-\alpha(-1)]^{-1/2} + d_n e^{i(n-1)t} - d_{n-1} e^{int} - \sum_{k=0}^{n} [a_k d_k + b_{k-1} d_{k-1} + a_{k-1} d_{k-2}] e^{i(k-1)t}$$

Where

$$d_k = \int_{-1}^{1} G_t(x) p_k(d\alpha, x)\, d\alpha(x) .$$

By the assumptions $G_t \in L^2(d\alpha)$ so that

$$\sum_{k=0}^{\infty} d_k^2 < \infty .$$

Thus by Theorem A the series

$$\sum_{k=0}^{\infty} [\,|a_k d_k| + |b_{k-1} d_{k-1}| + |a_{k-1} d_{k-2}|\,]$$

converges. Consequently, the theorem directly follows from (9).

Corollary 1. If $d\alpha \in S$ then

$$(10)\quad \sum_{k=0}^{\infty} \int_{-\infty}^{y} p_k(d\alpha, x)\, d\alpha(x)$$

converges for every real y.

Proof. If $y \notin [-1,1)$ then each term but the first one in (10) is zero so that (10) converges. If $-1 \le y < 1$ then we define g to be the characteristic function of $[-1,y]$. It is obvious that $G_0 \in L^2(d\alpha)$ so that (10) converges because by Theorem 2 the series (6) converges when $t = 0$.

It might be worthwhile to remark that by Bessel's inequality the series

$$\sum_{k=0}^{\infty} [\int_{-\infty}^{y} p_k(d\alpha, x)\, d\alpha(x)]^2$$

always converges.

Corollary 2. If $d\alpha \in S$ and the $d\alpha$-measure of the point x is positive then the series

$$\sum_{k=0}^{\infty} p_k(d\alpha,x)e^{ikt}$$

converges for every real t such that $\cos t \neq x$.

Proof. Apply Theorem 2 to the characteristic function of the point x.

Our last goal in this note is to establish some convergence and summability properties of orthogonal series

$$(11)\quad \sum_{k=0}^{\infty} e_k p_k(d\alpha,x)$$

when the second differences of the coefficients e_k satisfy some conditions. Let us point out that conditions (12) and (14) are always satisfied whenever the coefficients e_k form a monotonic sequence.

THEOREM 3. Suppose $d\alpha \in S$ and $\{e_k\}$ $(k = -1,0,1,\ldots, e_{-1} = 0)$ is a null-sequence. Then the series (11) is (C,1) summable almost everywhere in $(-1,1)$ if

$$(12)\quad \sum_{k=0}^{\infty} (e_{k+1} - 2e_k + e_{k-1})^2 < \infty .$$

Moreover, the series (11) converges for almost every $x \in (-1,1)$ if

$$(13)\quad \sum_{k=0}^{\infty} e_k^2 < \infty$$

and

$$(14)\quad \sum_{k=0}^{\infty} |e_{k+1} - 2e_k + e_{k-1}| < \infty .$$

Proof. Using the recurrence relation (1) it is easy to see that the identity

$$(15)\quad 2(1-x)\sum_{k=0}^{n} e_k p_k(d\alpha,x) = -\sum_{k=0}^{n} (e_{k+1} - 2e_k + e_{k-1})p_k(d\alpha,x) +$$

$$+ \Sigma_n + e_{n+1}p_n(d\alpha,x) - e_n p_{n+1}(d\alpha,x)$$

holds were Σ_n is defined by

$$\Sigma_n = \sum_{k=1}^{n+1} e_{k-1} a_k p_k(d\alpha,x) + \sum_{k=0}^{n} e_k b_k p_k(d\alpha,x) + \sum_{k=0}^{n-1} e_{k+1} a_{k+1} p_k(d\alpha,x) .$$

Now suppose that (12) is satisfied. Then by Theorem C the first sum on the right hand side of (15) is (C,1) summable almost everywhere in $(-1,1)$. Applying Theorems A and C we can also conclude that Σ_n is almost everywhere (C,1) summable in $(-1,1)$. Hence by (15) the first part of the theorem will be proved if we show that

$$\lim_{n\to\infty} e_{n+1}p_n(d\alpha,x) = 0 \quad (C,1)$$

and

$$\lim_{n\to\infty} e_n p_{n+1}(d\alpha,x) = 0 \quad (C,1)$$

for almost every $x \in (-1,1)$. Both of these relations follow immediately from Theorem B since e_n tends to 0 when $n\to\infty$. If (13) and (14) are satisfied then we can argue as follows. By (14) and Beppo Levi's theorem the first sum on the right side of (15) converges almost everywhere in (-1,1) when $n\to\infty$. Furthermore, by Theorem A and (13) the sequence Σ_n also converges as $n\to\infty$ for almost every $x \in (-1,1)$. Finally, condition (13) implies that

$$\lim_{n\to\infty} e_{n+1}\, p_n(d\alpha,x) = \lim_{n\to\infty} e_n p_{n+1}(d\alpha,x) = 0$$

almost everywhere in (-1,1). Therefore, in view of (15), the series (11) converges for almost every $x \in (-1,1)$.

In conclusion we remark that Theorem 3 is not completely analogous to the corresponding statement about trigonometric series since in case of trigonometric series one can obtain uniform convergence of

$$\sum_{k=0}^{\infty} e_k e^{ikt}$$

on $[\epsilon, 2\pi - \epsilon]$ if [14] holds; whereas we are only able to state almost everywhere convergence. We would certainly welcome any new method that would enable us to investigate the convergence of (11) at given points x when either (12) or (14) is satisfied.

REFERENCES

[1] Ya L. Geronimus, Orthogonal Polynomials, Consultants Bureau, New York, 1961.

[2] A. Mate and P. Nevai, Bernstein's inequality in L^p for $0<p<1$ and (C,1) bounds for orthogonal polynomials, Annals of Mathematics, 110 (1980).

[3] P. Nevai, Orthogonal Polynomials, Memoirs of the American Mathematical Society, Number 213, 1979.

UNIQUENESS OF BEST APPROXIMATIONS IN SPACES OF CONTINUOUS FUNCTIONS

G. Nürnberger

Institut für Angewandte Mathematik
Universität Erlangen-Nürnberg
Erlangen, West-Germany

I. Singer

Institutul de Matematica
Bucuresti, Romania

A complete characterization of those functions in C[a,b] which have a unique best approximation from linear spline spaces is given. Moreover it is shown that the set of functions in C(T) having a strongly unique best approximation is dense in the set of functions in C(T) having a unique best approximation by approximation with arbitrary linear subspaces.

I. INTRODUCTION

We investigate uniqueness and strong uniqueness of best approximations in C(T). If G is a linear subspace of C(T) and $f \in C(T)$ then $g_o \in G$ is called a best approximation (resp. a strongly unique best approximation) of f if for each $g \in G$ $\|f-g\| \geq \|f-g_o\|$ (resp. if there exists a constant $K_f > 0$ such that for each $g \in G$ $\|f-g\| \geq \|f-g_o\| + K_f\|g-g_o\|$).

In section II we investigate uniqueness of best approximations from $G=S_{n,k}$, the class of spline functions of degree n with k fixed knots. As a consequence of a general result on weak Chebyshev systems Nürnberger [4] has obtained a characterization of strong uniqueness of best approximations

Copyright © 1980 by Academic Press, Inc.
All rights of reproduction in any form reserved.
ISBN: 0-12-171050-5

from $S_{n,k}$ in terms of alternation properties of the error function, from which partial results of Schumacker [7], Schaback [6] and Nürnberger [3] follow. Strauß [8], including results of Rice [5] and Schumaker [7], has given necessary conditions and sufficient conditions for uniqueness of best approximations from $G=S_{n,k}$ involving only alternation properties of the error function. As Strauß [8] pointed out this is not enough for a characterization - in contrary to strong uniqueness. In this paper we characterize uniqueness of best approximations from $G=S_{n,k}$ by using alternation properties <u>and</u> "flatness" of the error function at the knots.

In general unique and strongly unique best approximations are not the same (see McLaughlin, Somers [2]). However, in section III we show that $S_G = \{f \in C(T): f$ has a strongly unique best approximation$\}$ is dense in $U_G = \{f \in C(T): f$ has a unique best approximation$\}$ for the case that G is an arbitrary subspace of C(T). Combining this result with a result of Garkavi [1] we obtain a complete characterization of those finite-dimensional subspaces G in C(T) for which U_G (respectively S_G) is dense in all of C(T).

The results are given without proofs but the proofs along with additional material will appear elsewhere.

II. UNIQUENESS OF BEST APPROXIMATIONS FOR SPLINE FUNCTIONS

For a compact Hausdorffspace T let C(T) be the space of all real-valued continuous functions f on T endowed with the norm $\|f\| = \sup\{|f(t)|: t \in T\}$. By $S_{n,k}$ we denote the (n+k+1)--dimensional subspace of spline functions of degree n with k fixed knots $x_1,\dots,x_k$ in $C[a,b]$. We call points

$a \leq t_1 < \ldots < t_p \leq b$ <u>alternating extreme points</u> of a function $f \in C[a,b]$ if $\varepsilon(-1)^i f(t_i) = \|f\|$, $i=1,\ldots,p$, $\varepsilon=\pm 1$.

First of all we state the following characterization of strongly unique best approximations for splines.

THEOREM 1 (Nürnberger [4]). For $f \in C[a,b] \setminus S_{n,k}$ and $g_o \in S_{n,k}$ the following statements (1) and (2) are equivalent:

(1) g_o is a strongly unique best approximation of f.

(2) (a) $f-g_o$ has at least n+k+2 alternating extreme points in $[a,b]$. (b) $f-g_o$ has at least j+1 alternating extreme points in each interval

$$[a,x_j)\ ,\ (x_{k-j+1},b]\ ,\ j=1,\ldots,k$$

$$(x_i,x_{i+j+n})\ ,\ j \geq 1\quad (\text{if } k>n+1)$$

Concerning uniqueness of best approximations for splines we first consider the following

EXAMPLE. We consider $S_{1,1}$ as a subspace of $C[-1,1]$, where the knot $x_1=0$. Let $f_1 \in C[-1,1]$ be defined by $f_1(-1)=1$, $f_1(-1/2)=-1$, $f_1(0)=1$, $f_1(1)=-1$ and linear elsewhere. Given a small $\varepsilon>0$ let $f_2 \in C[-1,1]$ be defined by $f_2(t)=f_1(t)$ if $t \notin (0,\varepsilon)$ and $f_2(t)=1-(2/\varepsilon)t^2$ if $t \in (0,\varepsilon)$. It is easy to verify that zero is a unique best approximation of f_2 and zero is a best approximation of f_1, but not a unique one.

This example shows that in general uniqueness of best approximations for splines depends on alternation properties and "flatness" of the error function at the knots. Therefore we use the following notation.

DEFINITION 2. A function $f \in C[a,b]$ is called <u>flat from the right</u> (resp. left) at $t_o \in (a,b)$, if for each $\varepsilon>0$ there exists a sequence (t_m) in (a,b) converging to t_o such that $t_m > t_o$ (resp. $t_m < t_o$) and $|f(t_o)-f(t_m)| < \varepsilon |t_o - t_m|^n$ for each m. (n is the integer which corresponds to $S_{n,k}$.)

Now we have the following characterization of unique best approximations for splines.

THEOREM 3. For $f \in C[a,b] \setminus S_{n,k}$ and $g_o \in S_{n,k}$ the following statements (1) and (2) are equivalent:

(1) g_o is a unique best approximation of f.

(2) (a) $f-g_o$ has at least n+k+2 alternating extreme points in $[a,b]$. (b) $f-g_o$ has at least j+1 alternating extreme points in each interval

$$[a,x_j] \ , \ [x_{k-j+1},b] \ , \ j=1,\dots,k$$
$$[x_i,x_{i+j+n}] \ , \ j \geq 1 \quad (\text{if } k > n+1)$$

(c) If $f-g_o$ has only j alternating extreme points in some interval $[a,x_j)$ resp. $[x_i,x_{i+j+n})$ (resp. $(x_{k-j+1},b]$ resp. $(x_i,x_{i+j+n}]$) then $f-g_o$ is flat from the left at x_j resp. x_{i+j+n} (resp. from the right at x_{k-j+1} resp. x_i).

Strauß [8] has shown that (1) in Theorem 3 implies (2b) in Theorem 3 and moreover that (2b) in Theorem 1 together with the fact that g_o is a best approximation of f implies (1) in Theorem 3.

III. UNIQUENESS AND STRONG UNIQUENESS OF BEST APPROXIMATIONS

Mc Laughlin, Somers [2] have shown that for each finite--dimensional non-Chebyshev subspace G of $C[a,b]$ there exists a function in $C[a,b]$ having a unique best approximation which is not a strongly unique best approximation. (A subspace G of

C(T) is called <u>Chebyshev</u>, if each $f \in C(T)$ has a unique best approximation.) However, we have the following

THEOREM 1. If G is a subspace of C(T), where T is a metric space, then $\{f \in C(T)$: f has a strongly unique best approximation$\}$ is dense in $\{f \in C(T)$: f has a unique best approximation$\}$.

(This result is also true for $C_o(T)$, where T is a locally compact metric space.)

For the space c_o we can even prove more. (Here c_o denotes the space of all real sequences (α_m) converging to zero and endowed with the norm $\|(\alpha_m)\| = \sup\{|\alpha_m|: m=1,2,...\}$.)

THEOREM 2. If G is a subspace of c_o then each unique best approximation of an element in c_o is a strongly unique best approximation of this element.

Now using our Theorem 1 and Theorem I in Garkavi [1] we obtain the following

THEOREM 3. If G is a finite-dimensional subspace of C[a,b] then the following three statements are equivalent:

(1) $\{f \in C[a,b]$: f has a unique best approximation$\}$ is dense in C[a,b].

(2) $\{f \in C[a,b]$: f has a strongly unique best approximation$\}$ is dense in C[a,b].

(3) No $g \in G$, $g \neq 0$, vanishes on an interval.

The equivalence of (1) and (3) has been proved by Garkavi [1]. A more general result holds for C(T), where T is a metric space, but for shortness it must be omitted here.

REFERENCES

1. Garkavi A.L., Almost Chebyshev systems of continuous functions, Amer. Math. Soc. Transl. 96 (1970), 177-187

2. Mc Laughlin H.W., Somers K.B., Another characterization of Haar subspaces, J. Approximation Theory 14 (1975), 93-102

3. Nürnberger G., Unicity and strong unicity in approximation theory, J. Approximation Theory 26 (1979), 54-70

4. Nürnberger G., A local version of Haar's theorem in approximation theory, preprint

5. Rice J.C., Characterization of Chebyshev approximation by splines, J. SIAM Num. Anal. (1967), 557-565

6. Schaback R., On alternation numbers in nonlinear Chebyshev approximation, J. Approximation Theory 23 (1978), 379-399

7. Schumaker L.L., Uniform approximation by Tschebyscheffian spline functions, J. Math. Mech. 18 (1968), 369-378

8. Strauß H., Eindeutigkeit bei der gleichmäßigen Approximation mit Tschebyscheffschen Splinefunktionen, J. Approximation Theory 15 (1975), 78-82

PENALTY-FINITE ELEMENT APPROXIMATIONS OF UNILATERAL PROBLEMS IN ELASTICITY

J. T. Oden[1]

The University of Texas at Austin
Austin, Texas

I. INTRODUCTION

Let V and W be Hilbert spaces, $a: V \times V \to \mathbb{R}$ and $b: W \times V \to \mathbb{R}$ continuous bilinear forms, and $f \in V'$. The following unilateral contact of a linearly elastic body with a rigid frictionless foundation leads to a variational boundary-value problem of the type

$$\left.\begin{aligned} &(u,\sigma) \in V \times W : \quad a(u,v) - b(\sigma,v) = f(v) \quad \forall\, v \in V \\ &b(\tau-\sigma,u) \geq \langle \tau-\sigma, s\rangle \quad \forall\, \tau \in W', \quad \tau \leq 0 \end{aligned}\right\} \tag{1}$$

Here s is given and $\langle\cdot,\cdot\rangle$ denotes duality on $W' \times W$. In particular, for the elasticity problem on a Lipschitz region $\Omega \subset \mathbb{R}^3$,

$$\left.\begin{aligned} &V = \{v \in (H^1(\Omega))^3 \mid v = 0 \text{ on } \Gamma_D\}, \quad W = H^{1/2}(\Gamma_C) \\ &a(u,v) = \int_\Omega \sum_{i,j=1}^{3} \sigma_{ij}(u)\varepsilon_{ij}(v)\,dx, \quad b(\sigma,v) \\ &= \langle \sigma, \gamma_n(v)\rangle \\ &f(v) = \sum_{i=1}^{3}\left(\int_\Omega f_i v_i\,dx + \int_{\Gamma_F} s_i v_i\,ds\right) \end{aligned}\right\} \tag{2}$$

Here v is the displacement vector, the boundary Γ of Ω is of the form $\Gamma = \overline{\Gamma}_C \cup \overline{\Gamma}_D \cup \overline{\Gamma}_F$, Γ_C being the contact

[1]Supported by U.S. Air Force Office of Scientific Research, Contract No. F-49620-78-C-0083.

Copyright © 1980 by Academic Press, Inc.
All rights of reproduction in any form reserved.
ISBN: 0-12-171050-5

surface and Γ_F the portion of the boundary where tractions s_i are applied; $\gamma_n(v) = v_i n_i$ is the normal trace of v on Γ_C and σ is the contact pressure; $\sigma_{ij}(u)$ and $\varepsilon_{ij}(v)$ are the stress and strain tensors due to displacements u and v respectively, and f_i are the components of the given body force. The contact pressure σ plays the role of a Lagrange multiplier associated with the constraint

$$\gamma_n(u) - s \leq 0 \quad \text{on } \Gamma_C \tag{3}$$

We wish to regularize this problem by introducing a penalty term to handle the constraint (3).

II. PENALTY FORMULATION

Instead of (1), we consider the penalized problem

$$u_\varepsilon \in V : \quad a(u_\varepsilon, v) + \frac{1}{\varepsilon} b((\gamma_n(u_\varepsilon)-s)_+, v) = f(v) \quad \forall \; v \in V \tag{4}$$

where $\varepsilon > 0$ and $(\gamma_n(u_\varepsilon)-s)_+$ is the positive part of $\gamma_n(u_\varepsilon)-s$. We easily prove the following:

<u>Theorem 1</u>. Let mes $\Gamma_D > 0$, $\Gamma_C \cap \Gamma_D = \emptyset$. Then

(i) $\exists$ a unique solution u_ε to (4) $\forall \; \varepsilon > 0$

(ii) $u_\varepsilon \to u$, the solution of (1), strongly on V as $\varepsilon \to 0$

(iii) $\sigma_\varepsilon \to \sigma$, the solution of (1), strongly in W' as $\varepsilon \to 0$,

where

$$\sigma_\varepsilon = - \frac{1}{\varepsilon}(\gamma_n(u_\varepsilon)-s)_+ \tag{5}$$

(iv) Indeed, there exists $C > 0$, independent of ε , such that $||u-u_\varepsilon||_V + ||\sigma-\sigma_\varepsilon||_{W'} \leq C\varepsilon$. $\square$

III. APPROXIMATIONS

We now use conforming finite element approximations to construct a family of finite-dimensional subspaces $\{V_h\}$ of V, h being the mesh parameter, with the usual interpolation properties. Our finite element approximations of (4) assumes the form

$$u_\varepsilon^h \in V_h : \quad a(u_\varepsilon^h, v^h) + \varepsilon^{-1} J((u_\varepsilon^h - s)_+, \gamma_n(v^h)) = f(v^h)$$
$$\forall \; v^h \in V_h \tag{6}$$

Here J denotes an approximation of $b(\cdot,\cdot)$ produced by numerical integration of the penalty term; e.g.

$$J(f) = \sum_{e=1}^{E} J_e(f) \; ; \quad J_e(f) = \sum_{j=1}^{G} Q_j f(\xi_j) \tag{7}$$

where E is the number of elements modelling Γ_C, Q_j and ξ_j are numerical quadrature weights and points.

The major result of this work is the following theorem:

<u>Theorem 2</u>. In addition to the hypotheses and conventions stated above, let there exist a finite-dimensional subspace W_h of W' such that

$$\alpha \|\tau^h\|_{W'} \leq \sup_{v^h \in V_h - \{\underset{\sim}{0}\}} \frac{|J(\tau^h \gamma_n(v^h))|}{\|v^h\|_V} \quad \forall \; \tau^h \in W_h \tag{8}$$

where α is independent of h, and $\bigcup_h W_h$ is dense in W'. In addition, let $\sigma_\varepsilon^h \in W_h$ be uniquely defined by

$$\sigma_\varepsilon^h(\xi_j) = -\varepsilon^{-1}(\gamma_n(u_\varepsilon^h(\xi_j)) - s_h \lambda(\xi_j))_+ \tag{9}$$

with $s_h = \gamma_n(s_h)$, $s_h \in V_h$. Then, $\forall \; \varepsilon > 0$, there exists a unique solution $(u_\varepsilon^h, \sigma_\varepsilon^h)$ to (6) and the sequence of solutions obtained as $\varepsilon \to 0$ converges strongly in $V \times W'$ to the solution of (1). □

IV. REMARKS

1. Condition (8) is critical and represents a condition for numerical stability of the scheme (6). Remarkably, it is found that (8) is, in general, only satisfied if a reduced (inexact) numerical integration scheme (7) is employed: exact integration of the penalty term will, in general, produce an unstable scheme with α in (8) zero or dependent on h .

2. Condition (8) serves to define an appropriate approximation space W_h for the contact pressure σ . Let $\Omega \subset \mathbb{R}^2$ and suppose tensor products of quadratics (9-node Lagrange quadrilateral elements) are used in the construction of V_h . Then one can verify the following:

Integration Formula for J	The space W_h
3-point Gauss	Method fails
Simpson's Rule	Conforming piecewise quadratures
2-point Gauss	Method fails
Trapezoid Rule	Conforming piecewise linear

3. It is clear from remark 2 that, as $\varepsilon \to 0$ for fixed h , $(u_\varepsilon^h, \sigma_\varepsilon^h)$ converges to a solution $(u^h, \sigma^h) \in V_h \times W_h$ of a mixed finite element approximation of (1)

4. A priori error estimates can be obtained for the approximations $(u_\varepsilon^h, \sigma_\varepsilon^h)$ in terms of h and ε . These indicate a strong sensitivity of the results to the location of nodal points on the contact surface Γ_C . If a node is located precisely at the contact interface, a substantial improvement in accuracy is experienced.

5. Problem (6), of course, is nonlinear. However, the conditioning of penalty formulations using elements satisfying (8) is such that the nonlinear systems of equations generated by (6) can be solved very efficiently in practical problems by a variety of iterative schemes. Additional details on the study of various theoretical questions, possible solution algorithms, and a wide collection of numerical results can be found in [1].

ACKNOWLEDGMENTS

The work summarized here is the result of joint work with Dr. N. Kikuchi and Mr. Y. J. Song. Our efforts were supported by Contract F-49620-78-C-0083 from the U.S. Air Force Office of Scientific Research.

REFERENCE

1. Oden, J. T., N. Kikuchi, and Y. J. Song, "Reduced Integration and Exterior Penalty Methods for Finite Element Approximations of Contact Problems in Incompressible Elasticity," TICOM Report 80-2, Austin, Texas, 1980.

ON APPROXIMATIONS IN SUBNORMS

Gerhard Opfer

Institut für Angewandte Mathematik
Universität Hamburg
Hamburg, Federal Republic of Germany

There exists a well known literature on approximation problems with generalized weight functions (cf. Moursund [5], Dunham [3]). Here we are going to treat approximation problems in still another type of generalization: we generalize the ordinary sup-norm. Therefore the subnorms we are going to introduce could also be called generalized sup-norms.

One motivation for investigating such problems comes from the treatment of conformal mapping problems as complex approximation problems. For illustration we mention an example: The conformal mapping of any simply connected region R in $\mathbb{C}$ onto a square renders the functional

$$\mu(f) = \sup_{z \in R} \max(|\mathrm{Re}\, f(z)|, |\mathrm{Im}\, f(z)|) \tag{1}$$

minimum within all holomorphic functions f defined on R and normalized by $f(z_0) = 0$, $f'(z_0) = 1$, where $z_0 \in R$ is a given point (cf. Opfer [6]).

It is even true that the functional

$$\mu(f) = \sup_{z \in R} \nu(f(z)) \tag{2}$$

has a unique minimum within the mentioned class of functions provided ν is a non-negative, positively homogeneous, upper semi-continuous, and non identically vanishing functional on $\mathbb{C}$ (cf. Opfer [8]).

Copyright © 1980 by Academic Press, Inc.
All rights of reproduction in any form reserved.
ISBN: 0-12-171050-5

To mention another motivation let T be a compact set and $C = C(T,\mathbb{K}^n)$ be the set of continuous functions defined on T with values in $\mathbb{K}^n$ ($\mathbb{K} = \mathbb{C}$ or $\mathbb{K} = \mathbb{R}$), $n \in \mathbb{N}$. If $|\ |$ is the Euclidean norm in $\mathbb{K}^n$ and $\hat{|}\ \hat{|}$ is any other norm in $\mathbb{K}^n$ (over $\mathbb{R}$) then the ordinary sup-norm $\|f\| = \sup_{t\in T} |f(t)|$ is of course equivalent to the norm $\hat{\|}f\hat{\|} = \sup_{t\in T} \hat{|}f(t)\hat{|}$. As a consequence there are constants k, K with $0 < k \leq K$ such that

$$k\hat{\|}f\hat{\|} \leq \|f\| \leq K\hat{\|}f\hat{\|} \quad \text{for all} \quad f \in C \ . \tag{3}$$

Let $v \in V \subset C$ be a best approximation of a given function $f \in C$ in the norm $\|\ \|$. It may be that v is difficult to compute; therefore we compute a best approximation $\hat{v} \in V$ of f in the norm $\hat{\|}\ \hat{\|}$, instead, which is possibly easier to obtain. From (3) we deduce then

$$k\hat{\|}f-\hat{v}\hat{\|} \leq \|f-v\| \leq K\hat{\|}f-\hat{v}\hat{\|} \ , \tag{4}$$

$$\|v-\hat{v}\| \leq 2K\hat{\|}f-\hat{v}\hat{\|} \ . \tag{5}$$

In case $\hat{|}\ \hat{|}$ is a p-norm on $\mathbb{R}^n$ (the case $\mathbb{C}^n$ could be regarded as $\mathbb{R}^{2n}$) with $p \geq 1$ or $p = \infty$, the constants k, K are easily available:

$$k = \begin{cases} n^{\frac{1}{2}-\frac{1}{p}} & \text{for } 1 \leq p \leq 2 \ , \\ 1 & \text{for } p \geq 2 \ , \end{cases} \qquad k = \begin{cases} 1 & \text{for } 1 \leq p \leq 2 \ , \\ n^{\frac{1}{2}-\frac{1}{p}} & \text{for } p \geq 2 \ . \end{cases} \tag{6}$$

This was used by Barrodale [1] and Barrodale-Delves-Mason [2] in the complex case ($\mathbb{K}^n = \mathbb{C}$) for some special values of p.

We keep the already introduced notation $C = C(T,\mathbb{K}^n)$ where $\mathbb{K}^n$ has the usual topology. By $\mathbb{R}^+$ we understand the non-negative reals.

Definition. A functional $\nu: \mathbb{K}^n \to \mathbb{R}^+$ is called a subnorm on $\mathbb{K}^n$ if it has the following two properties:

(I) ν is upper semi-continuous, i.e.

$$\{x \in \mathbb{K}^n : \nu(x) < \alpha\} \text{ is open for all } \alpha \in \mathbb{R}^+ . \tag{7}$$

(II) ν is nontrivial, i.e.

$$\text{there is an } x_0 \in \mathbb{K}^n \text{ with } \nu(x_0) > 0 . \tag{8}$$

If in addition

(III) ν is positively homogeneous, i.e.

$$\nu(\alpha x) = \alpha\nu(x) \quad \text{for all} \quad (x,\alpha) \in \mathbb{K}^n \times \mathbb{R}^+ , \tag{9}$$

then ν will be called a positively homogeneous subnorm on $\mathbb{K}^n$. If ν is a (positively homogeneous) subnorm the functional

$$\mu(f) = \sup_{t \in T} \nu(f(t)) , \quad f \in C \tag{10}$$

will be called a (positively homogeneous) subnorm on C. ■

Standard analysis arguments show that μ from (10) is a finite functional on C. The principal difference between a norm and a subnorm is the lack of convexity and boundedness of the sets defined in (7). For positively homogeneous subnorms these sets are starlike, however. Examples are given in Opfer [2].

It is evident now that we are interested in approximation problems with respect to subnorms: In C there is a given subset V and a given function f. We are looking for best approximations $\hat{v} \in V$ of f which are defined by

$$\mu(f-\hat{v}) \leq \mu(f-v) \quad \text{for all} \quad v \in V . \tag{11}$$

where μ is a given subnorm on C. In this situation the number

$$\rho_V(f) = \inf_{v\in V} \mu(f-v) \tag{12}$$

will be called the minimal distance between f and V, too.

We shall give one result here which can be regarded as predecessor of a characterization theorem. It is a general form of a theorem given by Meinardus [4, Theorem 16].

<u>Theorem</u>. Assume an approximation problem with respect to a subnorm as described above is given. Let there be a function $v_0 \in V$ and a subset $T_0 \subset T$ with the properties:

$$\nu(f(t)-v_0(t)) \neq 0 \quad \text{for all} \quad t \in T_0 , \tag{13}$$

to each $v \in V$ there is a $t_v \in T_0$ with

$$\nu(f(t_v)-v_0(t_v))(\nu(f(t_v)-v_0(t_v)) - \nu(f(t_v)-v(t_v))) \leq 0 , \tag{14}$$

then

$$\inf_{t\in T_0} \nu(f(t)-v_0(t)) \leq \rho_V(f) . \tag{15}$$

<u>Proof</u>. The inequality (15) is true if the left hand side is zero. Therefore we may assume that

$$\inf_{t\in T_0} \nu(f(t)-v_0(t)) > 0 . \tag{16}$$

The remainder of the proof is by contradiction. If (15) is not true there is a $v_1 \in V$ with

$$\rho_V(f) \leq \mu(f-v_1) < \inf_{t\in T_0} \nu(f(t)-v_0(t))$$

where the second inequality now implies that

$$\nu(f(t)-v_1(t)) < \nu(f(t)-v_0(t)) \quad \text{for all} \quad t \in T_0 . \tag{17}$$

From (13) and (17) it follows that

$$\nu(f(t)-v_0(t))(\nu(f(t)-v_0(t))-\nu(f(t)-v_1(t))) > 0 \quad \text{for all} \quad t \in T_0 ,$$

a contradiction to (14).

It is clear that further assumptions on the subnorm ν lead to more specific results. E.g. for the (sub)norm defined in (1) a characterization theorem for best approximations can be deduced from the above Theorem when V is convex in C (cf. Opfer [7], Theorem 2]).

REFERENCES

[1] I. Barrodale: Best Approximation of Complex-Valued Data, G.A. Watson (ed.): Numerical Analysis (Proceedings of a Conference, Dundee, 1977), Springer, Berlin, Heidelberg, New York, 1978, pp. 14-22.

[2] I. Barrodale - L.M. Delves - J.C. Mason: Linear Chebyshev Approximation of Complex-Valued Functions, Math. Comp., 32(1978), pp. 853-863.

[3] C.B. Dunham: Chebyshev Approximation with Respect to a Weight Function, J. Approximation Theory, 2(1969), pp. 223-232.

[4] G. Meinardus: Approximation of Functions: Theory and Numerical Methods, Springer, Berlin, Heidelberg, New York, 1967.

[5] D.G. Moursund: Chebyshev Approximation Using a Generalized Weight Function, J. SIAM Numer. Anal., 3(1966), pp. 435-450.

[6] G. Opfer: New Extremal Properties for Constructing Conformal Mappings, Numer. Math., 32(1979), pp. 423-429.

[7] G. Opfer: On Certain Approximations of Vector-Valued Functions, W. Schempp/K. Zeller (eds.): Multivariate Approximation Theory (Proceedings of a Conference, Oberwolfach, 1979), ISNM51 (1979), Birk-Hauser, Basel, pp. 265-271.

[8] G. Opfer: Conformal Mappings onto Prescribed Regions Via Optimization Techniques, to appear in Numer. Math.

AN ALGORITHMIC APPROACH TO NONLINEAR APPROXIMATION PROBLEMS

M.R. Osborne

Department of Statistics
Institute of Advanced Studies
Australian National University
Canberra, ACT

I. INTRODUCTION

Consider the problem given $F : R^P \to B$ determine $a \in R^P$ to

$$\min_{a \in R^P} ||F(a)||_B \tag{1.1}$$

It is assumed that F satisfies any necessary smoothness properties but similar assumptions are not made about $||\cdot||_B$. In fact it is likely that the results presented here will be most useful when $||\cdot||_B$ is not smooth. These results are of two kinds. First it is noted that nonlinearity presents little additional formal complication to the discussion of the first algorithm of Remes presented in [2], and the corresponding development is sketched. This algorithm requires the solution of a discrete nonlinear problem at each step, and suitable generalisations of the Gauss-Newton method are described.

2. THE EXCHANGE ALGORITHM

An exchange algorithm is a procedure for successively refining estimates of a best approximation by solving adaptively defined, finite dimensional subproblems. It requires that a sequence of mappings

Copyright © 1980 by Academic Press, Inc.
All rights of reproduction in any form reserved.
ISBN: 0-12-171050-5

$$\mu_n : B \rightarrow B_n \ , \ n = 0,1,2\ldots \tag{2.1}$$

be generated having the properties :

(i) The problem

$$\min_{a \in R^P} ||\mu_n F(a)||_{B_n} \tag{2.2}$$

has a solution a_n uniformly bounded in norm $\forall n$.

(ii) For each $x \in B$, $\forall n$, $\exists \mu_n = \mu_n(x) \ni$

$$||\mu_n(x)\ x||_{B_n} = ||x||_B \ . \tag{2.3}$$

(iii) For $\forall x \in B$ and $i > j$

$$||x||_B \geq ||\mu_i x||_{B_i} \geq ||\mu_j x||_{B_j} \tag{2.4}$$

The algorithm now takes the form:

(i) Solve $\min_{a \in R^P} ||\mu_n F(a)||_{B_n}$ to obtain a_n .

(ii) If $||F(a_n)||_B \leq ||\mu_n F(a_n)||_{B_n}$ then stop.

(iii) Otherwise define $\mu_{n+1} \ni ||\mu_{n+1}F(a_n)||_{B_{n+1}} = ||F(a_n)||_B$ and repeat (i) .

The argument in [2] now gives a convergence result.

<u>Theorem</u> 2.1 Limit points of the sequence $\{a_n\}$ solve (1.1) .

<u>Example</u> 2.1 Consider

$$< e_i, \mu_0 F >_{B_0} = < v_i, F >_B \ , \ i = 1,2,\ldots,n_0$$

where e_i is the i'th coordinate functional, and the v_i, $i = 1,2,\ldots,n_0$ are chosen such that $v_i \in \partial||x_i||_B$ (so that $< v_i, x_i >_B = ||x_i||_B$) for selected x_i, $i = 1,2,\ldots,n_0 \in B$. Define

$$\mu_{n+1}F = [\mu_n F^T, < v_n, F >_B]^T$$

where $v_n \in \partial\ ||F(a_n)||_B$, and let $||.||_{B_n}$ be the corresponding maximum norm. It is readily verified that (2.3) and (2.4) are satisfied.

<u>Remark</u> (i) The hard assumption to justify is (i). It is satisfied if F is linear and $\nabla\mu_n F$ has full column rank [2]. For nonlinear problems the natural generalisation (at least locally) is that $\exists\ K > 0 \ni$

$$||x - y||_{R^P} \leq K||\mu_0(F(x) - F(y))||_{B_0}.$$

(ii) When F is linear in a and $||.||_{B_n}$ is the maximum norm, the minimization problem can be solved by linear programming. Step (iii) of the algorithm is now equivalent to selecting the norm to enter the basis in the simplex algorithm applied to the dual of the linear program. This suggests an exchange algorithm implemented as a sequence of simplex steps applied to an adaptively defined dual program. It is equivalent to the algorithm described here only if each simplex step leads to the optimum of the corresponding $||\mu_n F||_{B_n}$. Such an algorithm has been described for polyhedral norms in the case B finite dimensional in [1].

(iii) In [5] the term singular is used to describe linear best approximation problems in the maximum norm when the optimum is achieved with less than $p+1$ points of extremal deviation. In this case the exchange algorithm generates confluent rows in the corresponding sequence of linear programs so that the basis matrices become increasingly singular. This behaviour is characterised by very slow convergence so the algorithm cannot be considered a good computing strategy. Such behaviour would be anticipated when $||.||_B$ is smooth.

(iv) If <u>strong uniqueness</u> obtains for each of the subproblems, at least for n sufficiently large, so that, $\forall\, n \geq n_0$, $\exists\, \gamma > 0 \ni ||\mu_n F(a)||_{B_n} \geq ||\mu_n F(a_n)||_{B_n} + \gamma||a-a_n||_{R^P}$ then rank $\{\partial||\mu_n F(a_n)||_{B_n}\}$ is at least $p+1$ [4]. This implies $\exists$ $p+1$ residuals equal in magnitude to the extremal deviation when B_n is the maximum norm. Here confluence does not occur, and the exchange algorithm is convergent.

3. THE GENERALISED GAUSS-NEWTON ALGORITHM

At each step of the exchange algorithm a discrete, nonlinear subproblem must be solved. In this section a generalised Gauss-Newton algorithm is presented for a somewhat wider problem class where the objective function $G(.)$, $B_n \to R$ is convex, stable, and bounded below, and we show that limit points of the sequence of iterations are stationary points of (2.2) in the sense of [6].

Definition a_n^* is a stationary point for (2.2) iff $\exists\, v \in \partial G(\mu_n F(a_n^*))$ $\ni \langle v, \nabla\mu_n F(a_n^*) \rangle_{B_n} = 0$.

The main steps in the algorithm are:

(i) Solve the linear subproblem (LSP)

$$\min_h \; G(r);\; r = \mu_n F(a_n^j) + \nabla\mu_n F(a_n^j)h.$$

Note that h_j, r_j solve the LSP iff $\exists\, v_r \in \partial G(r_j) \ni \langle v_r, \nabla\mu_n F_j \rangle_{B_n} = 0$ where $F_j = F(a_n^j)$. A necessary condition for the LSP to have a bounded solution set H_j is that $\nabla\mu_n F_j$ have full column rank.

(ii) Select $h_j \in H_j$ as a direction for a descent step. That is

$$a_n^{j+1} = a_n^j + \gamma_j h_j$$

where the descent step γ_j is chosen such that

(a) $\gamma_j = 1$ if $0 < \sigma \le \psi(\mu_n F, a_n^j, 1)$ where $0 < \sigma < \frac{1}{2}$,

(b) γ_j satisfies $\sigma \le \psi(\mu_n F, a_n^j, \gamma) \le 1-\sigma$ otherwise

$$\text{where } \psi = \frac{G(\mu_n F_j) - G(\mu_n F(a_n^j + \gamma h_j))}{\gamma\, \phi(\mu_n F, a_n^j)}$$

and ϕ is chosen such that

(α) $\phi\,(\mu_n F, a) \ge 0$,

(β) $\phi\,(\mu_n F, \hat{a}) = 0 \Rightarrow \hat{a}$ stationary point, and

(γ) $\lim_{\gamma \to 0+} (\mu_n F, a, \gamma) = \dfrac{-G'(\mu_n F; \nabla\mu_n Fh)}{\phi(\mu_n F, a)} \ge 1$

where G' is the directional derivative [7]. These conditions require h_j to be a descent direction for minimizing $G(\mu_n F_j)$ and imply $\exists\, \gamma > 0$ satisfying (ii) if a_n^j is not a stationary point.

Theorem 3.1 Assume a (sub)sequence $\{a_n^j\} \to \bar{a}_n$. Then $\bar{a}_n$ is a stationary point of (2.2) provided inf $\{\gamma_j\} > 0$ or $\{\gamma_j \|h_j\|^2_{R^p}\} \to 0$ and $\phi(\mu_n F, a)$ is lower semicontinuous at $\bar{a}_n$.

Two possibilities for ϕ have been considered.

(i) $\phi = -G'(\mu_n F_j;\ \nabla\mu_n F_j h_j) = \min - < v,\ \nabla\mu_n F_j h_j >_{B_n},\ v \in \partial\ G(\mu_n F_j)$

which clearly gives $\lim\limits_{\gamma\to 0+} \psi(\mu_n F_j,\ a_n^j, \gamma) = 1$, and

(ii) $\phi = G(\mu_n F_j) - G(r_j)$

which has advantages in that it uses quantities which occur naturally in the computation and it suggests possibilities when derivatives are not available. In this case $\lim\limits_{\gamma\to 0+} \psi \geq 1$.

To justify these choices we have:

Theorem 3.2 $G(\mu_n F_j) - G(r_j) = 0 \Leftrightarrow a_n^j$ is a stationary point of (2.2)

Corollary $G'(\mu_n F_j\ ;\ \nabla\mu_n F_j h_j) = 0 \Leftrightarrow a_n^j$ is a stationary point of (2.2).

Remark (i) The presentation in terms of G(.) is useful. For example it gives a local convergence result for Huber's M-estimator

$$G(.) = \sum_{j=1}^{\dim(B_n)} \rho(< e_j\ ,\ .\ >_{B_n})$$

[3] where $\rho(u) = \tfrac{1}{2} u^2\ ,\ |u| \leq c$.

$= c|u| - \tfrac{1}{2} c^2\ ,\ \ |u| > c$.

(ii) There is no problem in extending the characterisation result to take account of constraints. This gives 'a_n^j a stationary point $\Leftrightarrow G(\mu_n F_j) = G(r_j)$ and a_n^j feasible.' The result follows easily from a multiplier relation for the constrained problem. Recently Watson has used this result and an exact (non-differentiable) penalty objective function to solve constrained problems (private communication).

(iii) Example 2.1 shows that the maximum norm is natural on B_n. In this case the LSP is a linear programming problem and the generalised Gauss-Newton algorithm is second order convergent provided the optimal basis matrices have uniformly bounded inverses iff

$$\partial||r_j||_{B_n} \subseteq \partial||\mu_n F(\bar{a}_n)||_{B_n}$$

$\forall_j$ large enough [4]. It is perhaps of equal consequence that if this condition does not hold then necessarily $||h_j|| \nrightarrow 0$ for at least one $h_j \in H_j$ and convergence is very slow.

REFERENCES

1. Anderson, D.H. and Osborne M.R.; "Discrete, linear approximation problems in polyhedral norms', *Num. Math.*, *26*, 179-189, (1976).
2. Cheney, E.W.; 'Introduction to Approximation Theory', McGraw Hill, New York, (1966).
3. Huber, P.J.; 'Robust Statistical Procedures', SIAM Regional Conference Series 27, (1977).
4. Jittorntrum, K. and Osborne, M.R.; 'Strong uniqueness and second order convergence in nonlinear discrete approximation', *Num. Math.*, to appear.
5. Osborne, M.R. and Watson G.A.; 'A note on singular minimax approximation problems', *J. Math. Anal. and Applic.*, *25*, 692-700, (1969).
6. Osborne, M.R. and Watson G.A.; 'Nonlinear approximation problems in vector norms', in Numerical Analysis, ed. G.A. Watson, Lecture Notes in Mathematics 630, Springer, (1978).
7. Rockafellar, R.T.; 'Convex Analysis', Princeton Univ. Press, (1970).

INCREASING THE DISTANCE FROM THE POINTS OF A SET

P. L. Papini[1]

Istituto Matematico
Università di Bologna
Bologna, Italy

We define maximal points for sets in a normed space, a concept akin to that of minimal points introduced by Beauzamy. We say that a point is maximal for a given set, when moving from it along any direction we decrease the distance from some point of the set itself. Some properties of these points are indicated.

1. Let X be a normed space over the real field R; C will always denote a non-empty subset of X: we denote by $\overline{C}$ the closure, by $\mathring{C}$ the interior and by δC the boundary of C. Given in X two points x, y, we denote by $\langle x,y\rangle$ the segment joining them; also, we set

$$N'(x,y)=\lim_{t\to 0^+}(\|x+ty\|-\|x\|)/t.$$

Note that $-N'(x,y)\leqslant N'(x,-y)$ for every pair x, y.

The following notion was considered in [1]: the point $x\in X$ is called <u>minimal</u> for the set C when no point $x'\in X-\{x\}$ exists such that

(1) $\|x'-y\|\leqslant\|x-y\|$ for every $y\in C$.

We set <u>min(C)</u>$=\{x\in X;\ x \text{ is minimal for } C\}$.

If x' satisfies (1), then all the points of $\langle x,x'\rangle$ are pointwise closer (abbreviated "p.c.") than x to C, that is

(1') $\|tx'+(1-t)x-y\|=\|x-y+t(x'-x)\|=\|t(x'-y)+(1-t)(x-y)\|\leqslant\|x-y\|$ for every $y\in C$, $0\leqslant t\leqslant 1$.

Thus the condition (1) implies the following:

(2) $N'(x-y,x'-x)\leqslant 0$ for every $y\in C$.

---

[1] Supported by the GNAFA of the CNR (National Council of Research of Italy).

Copyright © 1980 by Academic Press, Inc.
All rights of reproduction in any form reserved.
ISBN: 0-12-171050-5

By setting $\bar{x}=2x-x'$, we obtain from (2):

$$0\leq -N'(x-y,x-\bar{x})\leq N'(x-y,\bar{x}-x)$$

thus

(3) $N'(x-y,\bar{x}-x)\geq 0$ for every $y\in C$.

Condition (3) is equivalent to

(3') $|x-y|\leq|x-y+t(\bar{x}-x)|=|t\bar{x}+(1-t)x-y|$ for every $y\in C$ and $0\leq t\leq 1$.

We want to put into evidence the situation indicated (and characterized) by the negation of (3') (or (3)): so we are led to the following

DEFINITION. We call a point $x\in X$ maximal for a set C when no point $\bar{x}\in X$, $\bar{x}\neq x$ can be found such that all the points of $\langle x,\bar{x}\rangle$ are pointwise farther (abbreviated "p.f.) than x from the points of C. We set:

$\max(C)=\{x\in X;$ x is maximal for $C\}=\{x\in X;$ for no point $\bar{x}\neq x$ (3) holds$\}$.

REMARK 1. A point $\bar{x}\neq x$ satisfying $\|\bar{x}-y\|\geq\|x-y\|$ for every $y\in C$ can always be found when C is bounded: for this reason we define maximal points by considering segments of points (and not only points) p.f. from C.

REMARK 2. Given a set C and a point $p\in X$, the approximation region A(p,C) between p and C was defined in [2, p. 345] in this way:

$$A(p,C)=\{x\in X;\ N'(x-y,p-x)\geq 0 \text{ for every } y\in C\}.$$

So a point x is maximal for C in our definition if and only if $x\notin A(p,C)$ for any $p\in X$, while $\max(C)=X-\bigcup_{p\in X}A(p,C)$. Also, we note that a notion of the same kind was studied in [3].

2. The implications stated before defining maximal points show that

(4) $\max(C)\subset\min(C)$ for every set $C\subset X$.

The following properties follow easily from the definition:

$$\max(C)=\max(\bar{C});\quad \mathring{C}\subset\max(C);\quad C\subset C' \text{ implies } \max(C)\subset\max(C').$$

Also, we can state the following two simples lemmata (the proof of the first one being trivial).

LEMMA 1. If X is strictly convex and $\bar{x}\neq x$ satisfies (3), then for $x_t=t\bar{x}+(1-t)x$, $0<t\leq 1$, we have $\|x_t-y\|>\|x-y\|$ for every $y\in C$.

LEMMA 2. If $\bar{x}$ satisfies (3), then $\langle x,\bar{x}\rangle\cap\bar{C}\subset\{x\}$.

Proof. Let $x_t=t\bar{x}+(1-t)x\in\bar{C}$ for some $t\in[0,1]$; by using (3') with $y=x_t$, we obtain $x_t=x$.

An easy adaptation of the proof of [1, lemma III.4] implies that, for every $C\subset X$, we have:

(5) $\min(\delta C)\cup C=\min(C)$;

moreover, it is easy to see that the following is true:

(6) if C is bounded, then $\overset{\circ}{C} \subset \max(\delta C) \subset \min(\delta C)$.

Thus we have:

PROPOSITION 1. For every bounded set C, we have $\min(C)=\min(\delta C)$.

Proof. If C is bounded, we obtain from (6): $C=\overset{\circ}{C} \cup \delta C \subset \min(\delta C)$, thus (5) implies the conclusion.

By using lemma 2, we are going to prove that a similar result holds for maximal points.

PROPOSITION 2. For every bounded set C we have $\max(C)=\max(\delta C)$.

Proof. The inclusion $\max(\delta C) \subset \max(C)=\max(\overline{C})$ is known. To prove the reverse inclusion, assume $x \in \max(C)$. If $x \in \overset{\circ}{C}$, then $x \in \max(\delta C)$ by (6); let $x \notin \overset{\circ}{C}$, and suppose $x \notin \max(\delta C)$: then there exists $\overline{x}$ such that

(Δ) $\|x_t - y\| \geq \|x-y\|$ for $x_t = t\overline{x}+(1-t)x$, $0 \leq t \leq 1$, and every $y \in \delta C$.

It is easy to see (by using the boundedness of C) that $x_t \notin \overset{\circ}{C}$ (for $0 \leq t \leq 1$). Since $x \in \max(C)$, there exist $\overline{t} \in (0,1]$ and $\overline{y} \in C$ such that $\|x_{\overline{t}} - \overline{y}\| < \|x-\overline{y}\|$; but then $\overline{y} \in C - \delta C$, while $x_{\overline{t}} \in \overline{\delta C}$ for lemma 2: thus $y \in \overset{\circ}{C}$, $x_{\overline{t}}$ $\widehat{(X-\overset{\circ}{C})}$, therefore there exists $y' \in \langle x_{\overline{t}}, \overline{y} \rangle \cap \delta C$, and we obtain:

$$\|x_{\overline{t}} - y'\| = \|x_{\overline{t}} - \overline{y}\| - \|\overline{y} - y'\| < \|x-\overline{y}\| - \|\overline{y} - y'\| \leq \|x-y'\|$$

against (Δ). So a contradiction, proving the proposition.

Next result is concerned with finite dimensional spaces.

PROPOSITION 3. Suppose $\dim(X)<\infty$; then $\max(C)$ is always an open set.

Proof. Consider in X a sequence (x_n), and let be $x=\lim_{n\to\infty} x_n$; we have to prove that if $x_n \notin \max(C)$ for every $n \in N$, then $x \notin \max(C)$. Suppose that there exists $\overline{x}_n \neq x_n$ (for every n) satisfying (3): fixed $\varepsilon>0$, it is not restrictive to assume $\|\overline{x}_n - x_n\| = \varepsilon$. Then the sequence $(\overline{x}_n)$ is bounded, so there exists a subsequence $(\overline{x}_{n_k})$ convergent to some $\overline{x} \in X$; of course $\overline{x} \neq x$ since $\|\overline{x}-x\| = \lim_{n\to\infty} \|\overline{x}_{n_k} - x_{n_k}\| = \varepsilon \neq 0$; moreover (for every $y \in C$) we have:

$$N'(x-y,\overline{x}-x) \geq \overline{\lim_{n\to\infty}}\, N'(x_{n_k}-y, \overline{x}_{n_k} - x_{n_k}) \geq 0$$

since $N'(x,y)$ is upper semi-continuous as a function from $X \times X$ into R. Therefore $x \notin \max(C)$, and the theorem is proved.

By a compactness argument, it is also simple to prove the following:

PROPOSITION 4. If $N'(x-y,\overline{x}-x)>0$ for every $y \in C$, then $\overline{x}$ is p.f. than x to the points of C.

It has been shown that min(C) can be expressed in a very nice way when X is an inner product spaces; also, these spaces can be characterized by using min(C), where C is the unit ball of X. A result of the same kind is the following.

PROPOSITION 5. Denote by $\delta^S(C)$ the set of all boundary points of a convex set C, which are not support points for C; then, if X is an inner product space, we have:

(7) $\max(C)=\overset{\circ}{C}\cup\delta^S(\overline{C})$ for every convex set C.

Conversely, if (7) holds and $\dim(X)\geqslant 3$, then X is an inner product space.

The proof of this and other related results will be given in a paper, now in preparation.

A consequence of proposition 5 is that, if $\dim(X)=\infty$, then max(C) is not necessarily closed (cf. proposition 4).

The following question was raised in [2, pp. 119-120]
given a convex set C, is min(C) convex?
Similarly, we can ask the following:
is max(C) convex when C is convex? Also, while it is easy to construct a non convex set such that max(C) is not convex in R^2 with the sup norm, we know no example of such a type in strictly convex spaces.

References

1. Beauzamy, B. and B. Maurey, Points minimaux dans les espaces de Banach, J. Functional Analysis 24 (1977), 107-139.
2. Bruck, E. Jr., Nonexpansive projections on subsets of Banach spaces, Pacific J. Math. 47 (1973), 341-355.
3. Godini, G., On suns, Rev. Roumaine Math. Pures Appl. 21 (1976), 869-871.

UNIFORM LIPSCHITZ AND STRONG UNICITY CONSTANTS ON SUBINTERVALS

S. O. Paur

J. A. Roulier[1]

Department of Mathematics

North Carolina State University

Raleigh, North Carolina

I. INTRODUCTION

Let Π_n denote the algebraic polynomials of degree n or less. For a given closed interval I of the real line let C(I) denote the continuous real-valued functions on I endowed with the uniform norm $||\cdot||_I$. For a fixed positive integer n, we define the best uniform approximation $P_{f,I}$ to $f \in C(I)$ from Π_n by $||f - P_{f,I}||_I = \inf\{||f - P||_I \mid P \in \Pi_n\}$. The classical strong unicity and Lipschitz conditions are found in Cheney [2, pp. 80-82], and are presented here. We assume n is fixed (here and throughout the paper) and that approximation is from Π_n, except in Theorem 2.3 where we replace Π_n by a finite dimensional Haar subspace M. The definition of best approximation from M is defined analogously.

[1]Supported in part by NASA Grant NSG 1549-S1

Copyright © 1980 by Academic Press, Inc.
All rights of reproduction in any form reserved.

ISBN: 0-12-171050-5

<u>Theorem 1.1.</u> Let $f \in C(I)$. Then there are constants $\lambda_{f,I} > 0$ and $\gamma_{f,I} > 0$ such that

$$\|P_{f,I} - P_{g,I}\|_I \leq \lambda_{f,I}\|f - g\|_I \tag{1.1}$$

for all $g \in C(I)$, and

$$\|f - P_{f,I}\|_I \leq \|f - Q\|_I - \gamma_{f,I}\|Q - P_{f,I}\|_I \tag{1.2}$$

for all $Q \in \Pi_n$.

We note that $\lambda_{f,I}$ is called a uniform Lipschitz constant and $\gamma_{f,I}$ is called a strong unicity constant. Expression (1.2) is called the strong unicity inequality. We further note that if $\gamma_{f,I} > 0$ is known, then an acceptable value of $\lambda_{f,I}$ is $\dfrac{2}{\gamma_{f,I}}$. See Cheney [2, p. 82].

In [4], Henry and Roulier investigate the existence of uniform Lipschitz constants on all symmetric intervals of the form $[-\alpha,\alpha] \subset [-1,1]$ for a given $f \in C[-1,1]$. Sufficient conditions on f are obtained to guarantee the existence of a constant $\lambda_f > 0$ so that

$$\|P_{f,J} - P_{g,J}\|_J \leq \lambda_f\|f - g\|_J \tag{1.3}$$

for all $g \in C(J)$ and for all $J \subset [-1,1]$ of the form $J = [-\alpha,\alpha]$. Examples are also given of functions $f \in C(I)$ which fail to have such λ_f.

In this paper we present sufficient conditions on $f \in C(I)$ to insure the existence of a strong unicity constant $\gamma_f > 0$ valid for all closed subintervals of I. This, in turn, guarantees that (1.3) is valid for all closed subintervals of I.

II. THE MAIN THEOREMS

We state our results without proof since these will appear in a subsequent paper. The proofs of Theorems 2.2 and 2.5 employ techniques similar to those used in the proof of Theorem 3.1 [4, p. 228]. However, all three theorems use the following lemma due to Cline [3]. (See also [1]).

<u>Lemma 2.1.</u> Let $h \in C[-\theta,\theta]$ with $h \notin \Pi_n$. Let $P \in \Pi_n$ be the best approximation to h on $[-\theta,\theta]$ and for each Chebyshev alternation $E = \{t_j\}_{j=1}^{n+2}$ for $h - P$, define $q_i \in \Pi_n$ by $q_i(t_j) = \text{sgn}[h(t_j) - P(t_j)]$, $j = 1,2,\ldots,n+2$, $j \neq i$ and $i = 1,2,\ldots,n+2$. Let $\Omega(E) = \max_{1\leq i\leq n+2} \{\|q_i\|_\theta\}$. Then there exists a Chebyshev alternation E^* for $h - P$ so that

$$\lambda_h(\theta) \leq 2\Omega(E^*) \tag{2.1}$$

where $\lambda_h(\theta)$ is the Lipschitz constant for h on $[-\theta,\theta]$ and so that

$$[\gamma_h(\theta)]^{-1} \leq \Omega(E^*) \tag{2.2}$$

where $\gamma_h(\theta)$ is the strong unicity constant for h on $[-\theta,\theta]$.

<u>Theorem 2.2.</u> If $f \in C^{n+1}[-1,1]$ with $f^{(n+1)}(x) > 0$ on $[-1,1]$, then there are positive constants λ_f and γ_f so that for all closed subintervals $J \subset [-1,1]$,

$$\|P_{f,J} - P_{g,J}\|_J \leq \lambda_f\|f - g\|_J \tag{2.3}$$

for all $g \in C(J)$, and

$$\|f - P_{f,J}\|_J \leq \|f - Q\|_J - \gamma_f\|Q - P_{f,J}\|_J \tag{2.4}$$

for all $Q \in \Pi_n$.

Theorem 2.3 as well as Lemma 2.4 is used in the proof of Theorem 2.5. The first part (2.5) is due to Henry and Schmidt [5] and we have proved (2.6) which is a similar result for strong unicity constants. For the following theorem, we assume that approximation is from a finite dimensional Haar subspace M (see [4], p. 224). $P_{f,I}$ is defined as above with Π_n replaced by M.

<u>Theorem 2.3.</u> If Γ is a compact subset of C(I) and $\Gamma \cap M = \emptyset$ where $M \subset C(I)$ is a finite-dimensional Haar subspace, then there are constants $\lambda_\Gamma > 0$ and $\gamma_\Gamma > 0$ so that

$$\|P_{f,I} - P_{g,I}\|_I \leq \lambda_\Gamma \|f - g\|_I \tag{2.5}$$

for all $f \in \Gamma$ and $g \in C(I)$, and

$$\|f - P_{f,I}\|_I \leq \|f - Q\|_I - \gamma_\Gamma \|Q - P_{f,I}\|_I \tag{2.6}$$

for $f \in \Gamma$ and $Q \in M$.

<u>Lemma 2.4.</u> Let $f \in C[-1,1]$. Suppose $\varepsilon > 0$ and there does not exist a closed interval $I \subset [-1,1]$ so that $\ell(I) \geq \varepsilon$ and f restricted to I is in Π_n. Then there are constants $\lambda_f(\varepsilon) > 0$ and $\gamma_f(\varepsilon) > 0$ so that for every closed interval $J \subset [-1,1]$ satisfying $\ell(J) \geq \varepsilon$,

$$\|P_{f,J} - P_{g,J}\|_J \leq \lambda_f(\varepsilon) \|f - g\|_J \tag{2.7}$$

for all $g \in C(J)$ and

$$\|f - P_{f,J}\|_J \leq \|f - Q\|_J - \gamma_f(\varepsilon) \|Q - P_{f,J}\|_J \tag{2.8}$$

for all $Q \in \Pi_n$.

<u>Theorem 2.5.</u> Let $f \in C^{n+1}[-1,1]$ so that $f^{(n+1)}(x) \neq 0$ for $x \in [-1,0)$ or $x \in (0,1]$. Suppose there are $m,M \in \mathbb{R}$, $0 < m \leq M$ and $p \in \Pi_k$, so that

$$0 \leq m|p^{(n+1)}(x)| \leq |f^{(n+1)}(x)| \leq M|p^{(n+1)}(x)|$$

on $[-\delta,\delta]$ for some $\delta > 0$. Then there are constants $\lambda_f > 0$ and $\gamma_f > 0$ so that for all closed intervals $J \subset [-1,1]$

$$||P_{f,J} - P_{g,J}||_J \leq \lambda_f ||f - g||_J \tag{2.9}$$

for all $g \in C(J)$ and

$$||f - P_{f,J}||_J \leq ||f - Q||_J - \gamma_f ||Q - P_{f,J}||_J \tag{2.10}$$

for all $Q \in \Pi_n$.

III. CONCLUSIONS

The examples in [4] show that the hypotheses in the theorems of section II cannot be weakened although Theorem 2.5 could be stated for a function having n + 1 continuous derivatives whose (n + 1)st derivative has a finite number of zeroes in [-1,1].

A potential application of theorems such as these is in the study of convergence of some of the adaptive curve fitting methods (e.g. see [6], [7]). With these techniques best approximations are computed on various subintervals by Remez type algorithms. The availability of a global strong unicity constant for all subintervals could be used to show convergence properties of the Remez algorithm independent of the subinterval on which it is applied.

REFERENCES

1. Bartelt, W. M., J. Approximation Theory 14, 245-250 (1975).
2. Cheney, E. W., "Introduction to Approximation Theory," McGraw-Hill, New York (1966).
3. Cline, A. K., J. Approximation Theory 8, 160-172 (1973).
4. Henry, M. S., and Roulier, J. A., J. Approximation Theory 21, 224-235 (1977).
5. Henry, M. S., and Schmidt, D., in "Theory of Approximation with Applications" (A. G. Law and B. N. Sahney, Ed.) pp. 24-42, Academic Press, New York (1976).
6. Hull, J. A., and Taylor, G. D., submitted.
7. Taylor, G. D., submitted.

COMPUTING NONLINEAR SPLINE FUNCTIONS

Dennis D. Pence

Department of Mathematics
University of Vermont
Burlington, Vermont

A nonlinear interpolating spline is a curve in the (x,y)-plane passing through proscribed points in a certain order for which the energy functional, defined as the integral of the curvature squared with respect to arc length, takes on a local minimum. Assuming that the curve can be represented by y as some function of x, this problem takes the form of seeking $y(x)$, a local minimum for

$$E(y) = \int_{X_0}^{X_N} \frac{(y'')^2}{(1 + (y')^2)^{5/2}}\, dx, \quad y(X_j) = Y_j, \quad j=0,1,\ldots,M.$$

This models the bending of a thin beam or elastica required to pass through the proscribed points. Malcolm (1977) gives an excellent explanation of the physical assumptions of this model and a historical survey of efforts to compute these.

Malcolm also presents a finite difference approximation for $E(y)$ and a new procedure for finding a local minimum for the discretized problem. The purpose of this note is to suggest an alternative procedure which is essentially a finite element method for computing these nonlinear spline functions using linear splines.

Parabolic splines with simple knots are conforming elements for this variational problem. Using B-splines (see de Boor (1978,p.108)), let $\underline{t} = (t_i)$ be some distinct knot sequence with $t_0 = X_0$ and $t_N = X_M$ and suppose that

Copyright © 1980 by Academic Press, Inc.
All rights of reproduction in any form reserved.
ISBN: 0-12-171050-5

$$y(x) = \alpha_{-2} B_{-2,3,\underline{t}} + \alpha_{-1} B_{-1,3,\underline{t}} + \ldots + \alpha_{N-1} B_{N-1,3,\underline{t}}.$$

The energy functional for such a y is exactly

$$E(\underline{\alpha}) = \sum_{i=0}^{N-1} \frac{\beta_i - \beta_{i-1}}{3(t_{i+2} - t_i)} \left(\frac{3\beta_i + 2\beta_i^2}{(1+\beta_i^2)^{3/2}} - \frac{3\beta_{i-1} + 2\beta_{i-1}^2}{(1+\beta_{i-1}^2)^{3/2}} \right)$$

where $\beta_i = 2(\alpha_i - \alpha_{i-1})/(t_{i+2} - t_i)$, i.e. β_i is the corresponding coefficient for the B-spline representation of y' (see de Boor (1978,p.138)).

Briefly, a Picard iteration procedure, very similar to that of Malcolm (1977), can be used to find a local minimum for $E(\underline{\alpha})$ subject to the interpolation constraints. A good initial choice for the coefficient vector $\underline{\alpha}$ is that of the parabolic spline which minimizes

$$\tilde{E}(y) = \int_{x_0}^{x_N} (y'')^2 \, dx.$$

In other words, a good first estimate is simply the parabolic spline which best approximates the natural cubic interpolating spline.

Further work is anticipated considering linear cubic splines as elements together with some numerical method for approximating the energy functional. Note that no error analysis is presented, and that even the question of the existence of such nonlinear interpolating splines is largely open (see Golomb (1978)).

REFERENCES

de Boor, C. (1978). A Practical Guide to Splines, Springer-Verlag, New York.

Golomb, M. (1978). Stability of interpolating elastica, MRC Tech. Sum. Rep. #1852, University of Wisconsin-Madison.

Malcolm, M. A. (1977). On the computation of nonlinear spline functions, SIAM J. Numer. Anal. 14, 254-282.

EXTREMAL BASES FOR NORMED VECTOR SPACES

Manfred Reimer

Abteilung Mathematik
Universität Dortmund
Dortmund, Germany

I. INTRODUCTION

Recently, W. Gautschi (7), (8) was concerned with the concept of well conditioned representations for polynomials. His issue is that the basis should provide for a map onto the coefficients with low condition number.

Fortunately, the numerically most stable evaluation schemes for polynomials use such a basis (M. Reimer (11), R. Kusterer and M. Reimer (9)) so that the originally quite different concepts of proper representation and stable evaluation fit very well together.

However, we believe that the quality of a basis cannot depend on how it has been normed, as is true in Gautschi's concept. Hence, the concept of a good basis should be "projectively invariant", as in the following

Definition. Let X denote a real or complex normed space with basis $\{x_i\}_{i \in I}$. For $i \in I$ define X_i to be the hyperplane of X where the coefficient with x_i vanishes. Then the basis is called extremal basis, iff every x_i, $i \in I$, is best-approximated in X_i by zero.

In other words, in an extremal basis, no element can be approximated by means of other ones. Obviously, in inner product spaces, all orthogonal bases are extremal, but our concept is applicable even in spaces with an

Copyright © 1980 by Academic Press, Inc.
All rights of reproduction in any form reserved.
ISBN: 0-12-171050-5

uniform norm, which are our concern. In this case, any two elements of an extremal basis are orthogonal in the sense of G. Birkhoff (2). See also I. Singer (14, p.91).

II. EXISTENCE OF EXTREMAL BASES

In what follows, let X denote any n-dimensional vector space over $\mathbb{K} \in \{\mathbb{R}, \mathbb{C}\}$, consisting of all $\mathbb{K}$-valued continuous functions defined on a compact set B. X shall be provided with $\|x\| := \max\{|x(t)| : t \in B\}$, $x \in X$.

Theorem 1. In X there is one extremal basis at least.

Proof. By the arguments of E. W. Cheney and K. H. Price (3), it follows from Auerbach's Theorem in its dual form that there exist elements $x_1,\dots,x_n \in X$ and points $t_1,\dots,t_n \in B$ such that

$$x_j(t_k) = \delta_{jk} \tag{1}$$

and

$$\|x_j\| = 1 \tag{2}$$

for $j,k = 1,\dots,n$. Obviously,

$$\|x_j + \sum_{\substack{k=1 \\ k \neq j}}^{n} \xi_k x_k\| \geqq |x_j(t_j)| = 1 = \|x_j\|$$

is valid for arbitrary $\xi_k \in \mathbb{K}$, $k \in \{1,\dots,n\} \setminus \{j\}$, while $j \in \{1,\dots,n\}$ is fixed. Hence the x_j - s perform an extremal basis.

In general, uniqueness is not valid. For instance, in the space of all real homogeneous polynomials of a certain degree in a fixed number of variables and with B to be the unit-sphere, we obtain a whole family of extremal bases from a fixed one by rotation.

III. CONDITION OF EXTREMAL BASES

Extremal bases find their justification for instance in the fact that they are well conditioned in the sense of W. Gautschi. As can easily be proved, the following theorem is valid:

Theorem 2. Let X_n denote any normed vector space over $\mathbb{K} \in \{\mathbb{R}, \mathbb{C}\}$ with basis $\{x_1,\ldots,x_n\}$ where $\|x_j\| = 1$ for $j = 1,\ldots,n$. Let $\phi_n : X_n \to \mathbb{K}^n$ be the mapping onto the coefficients with respect to this basis. Then, if the basis is extremal and if $\mathbb{K}^n$ is provided with the maximum norm, the following estimate holds:

$$\text{cond } \phi_n = \|\phi_n\| \, \|\phi_n^{-1}\| \leq n.$$

Note that we estimate cond ϕ_n by the trace-norm in the sense of A. F. Ruston (13).

IV. EXAMPLES

Example 1. Let $X_n := \mathbb{P}_{n-1}$ denote the space of all polynomials in one variable and with degree not exceeding n-1 and let $B := [-1,1]$. Due to L. Fejér (5), the "extremal" points of the Legendre-polynomial P_{n-1}, which are the roots of $(x^2-1)P'_{n-1}(x)$, furnish Lagrangians with property (2), which, therefore, perform an extremal basis. By a foot-note (5, p. 6), this is the only extremal Lagrangian basis. These extremal points can also be identified to be the Fekete-points (see M. Fekete (6) and E. W. Cheney and K. H. Price (3)).

Now, let L_n denote the corresponding Lebesgue-constant, then we have

$$\text{cond } \phi_n = \|\phi_n^{-1}\| = L_n \quad \text{where} \quad L_n = O(\log n)$$ [1]

[1] Private communication of B. Sündermann, Universität Dortmund

as in case of the Chebyshev polynomial zeros. But, at least for $3 < n \leq 40$, the Lebesgue-constants of the Fekete-points are less than in the Chebyshev-case (though not optimal for fixed n). Hence we are led by our definition of extremal bases to a polynomial representation which is nearly optimal in the sense of W. Gautschi.

Example 2. Let $B := B_r := \{z \mid z \in \mathbb{C}^r, |z| \leq 1\}$ denote the complex r-dimensional unit-ball. For $m = (m_1,\ldots,m_r) \in \mathbb{N}_0^r$ and $z\ (z_1,\ldots,z_r) \in \mathbb{C}^r$ let

$$|m| := m_1 + \ldots + m_r, \quad z^m := z_1^{m_1} \ldots z_r^{m_r},$$

$$\tilde{\Pi}_m^r := \mathrm{span}\{z^n \mid n_j < m_j \text{ for one } j \in \{1,\ldots,r\}\}.$$

Moreover, we furnish the direct sum $\Pi_m^r := \tilde{\Pi}_m^r \oplus [z^m]$ with the maximum-norm with respect to B_r. Then, at first, we can prove

Theorem 3. Let $m \in \mathbb{N}_0^r$. Then z^m is best approximated in $\tilde{\Pi}_m^r$ by zero.

Proof. Let $z = (z_1,\ldots,z_r) \in \mathbb{C}^r$. Each z_j can be represented in the form

$$z_j = \tau_j e^{i\phi_j} \text{ where } \tau_j \geq 0, \ 0 \leq \phi_j < 2\pi$$

so that

$$z^m = \tau^m e^{i\pi \sum_{j=1}^{r} m_j \phi_j} .$$

From this we obtain

$$\|z^m\| = \sqrt{\frac{m^m}{\mu^\mu}} \quad \text{with } \mu := |m| .$$

The norm is attained if and only if

$$\tau_j = \sqrt{\frac{m_j}{\mu}} \quad \text{for } j = 1,\ldots,r. \tag{3}$$

In what follows, the τ_j - s are chosen according to (3). Then the points

$$z_k := (\tau_1 e^{k_1\pi i/m_1}, \ldots, \tau_r e^{k_r\pi i/m_r}), \tag{4}$$

$$k = (k_1, \ldots, k_r) \in K_m^r := \{\ell \mid \ell \in \mathbb{N}_o^r,\ 1 \leqq \ell_j \leqq 2m_j \text{ for } 1 \leqq j \leqq r\} \tag{5}$$

are extreme points of z^m, which are different in pairs, where

$$z_k^m = (-1)^{|k|} \|z^m\| \quad \text{for} \quad k \in K_m^r . \tag{6}$$

Finally we define the linear functional

$$\lambda P := \sum_{k \in K_m^r} \bar{z}_k^m P(z_k) \quad \text{for} \quad P \in \Pi_m^r . \tag{7}$$

Then, if z^n is an element of Π_m^r, there exists an $\ell \in \{1,\ldots,r\}$ such that $n_\ell < m_\ell$ so that $n_\ell \not\equiv m_\ell \bmod (2m_\ell)$. Hence

$$\lambda z^n = \sum_{k \in K_m^r} \bar{z}_k^m z_k^n = \tau^{m+n} \sum_{k \in K_m^r} e^{i\pi \sum_{j=1}^{r} (n_j - m_j) k_j / m_j}$$

vanishes since we have

$$\sum_{k_\ell = 1}^{2m_\ell} e^{i\pi (n_\ell - m_\ell) k_\ell / m_\ell} = 0$$

(Berenzin-Zhidkov (1, p. 448)). Hence λP vanishes in $\tilde{\Pi}_m^r$ and from this, together with (6), the statement of Theorem 3 follows (see Rivlin (12, p. 64, Theorem 2.6)).

Theorem 3 is an extension of the well known fact that, in the one-dimensional case, the norm of the polynomial z^n+ terms of lower degree is minimum for z^n itself, see Davis (4), e. g. It is equivalent with the statement that among all polynomials $P \in \Pi_m^r$ with $\|P\| \leqq \|z^m\|$, the monomial z^m is one for which the coefficient $c_m(P)$ occuring with z^m is maximum in absolute value, the maximum being one. Hence the norm of this coefficient functional is

$$\|c_m\| = 1/\|z^m\| = \sqrt{\frac{\mu^\mu}{m^m}} \quad \text{where} \quad \mu = |m| . \tag{8}$$

We now consider the complex space $X := \mathbb{P}_\mu^{r*}$ consisting of all complex homogeneous polynomials of degree $\mu \in \mathbb{N}_o$. For any $m \in \mathbb{N}_o^r$ with $|m| = \mu$ let $\tilde{\mathbb{P}}_m^{r*}$ denote the subspace of $\mathbb{P}_\mu^{r*}$, where $c_m(P) = 0$. Then $\tilde{\mathbb{P}}_m^{r*}$ is a subspace of $\tilde{\Pi}_m^r$. Hence, by Theorem 3, zero is a best approximation to z^m in $\tilde{\mathbb{P}}_m^{r*}$.

From this it follows that the monomials $\{z^m : |m| = \mu\}$ perform an extremal basis for the space of all complex homogeneous polynomials of degree μ. Note that this is an example of an extremal basis, which is not Lagrangian.

Note also that $\mathbb{P}_\mu^{r*}$ is a subspace of Π_m^r for all m with $|m| = \mu$. Hence (8) is valid even if c_m is restricted to $\mathbb{P}_\mu^{r*}$. This means that among all homogeneous polynomials of degree μ which do not exceed one in absolute value on the complex unit-ball, the coefficient occuring with z^m is maximum in case of the polynomial $z^m/\| z^m\|$. The maximum is given by (8).

Remark. Since the values of (6) are real, our results can be interpreted in real terms, and this with respect to the space of all real homogeneous harmonic polynomials which are linear combinations of the monomials

$$x_1^{p_1} \cdots x_r^{p_r} y_1^{q_1} \cdots y_r^{q_r} \text{ where } p_j + q_j = m_j \quad \text{for } j = 1,\dots r.$$

We do not persue this fact in details here.

REFERENCES

1 Berezin, I. S., and Zhidkov, N. P., "Computing Methods" v. II. Pergamon Press, Oxford (1965).

2 Birkhoff, G., Duke Math. J. 1, 169 (1935).

3 Cheney, E. W., and Price, K. H., in "Approximation Theory" (A. Talbot, ed.), p. 261. Academic Press, New York (197o).

4 Davis, Ph. J., "Interpolation and Approximation". Blaisdell, New York (1963).

5 Fejér, L., Math. Ann. 1o6, 1 (1932).

6 Fekete, M., Math. Zeitschr. 17, 228 (1923).

7 Gautschi, W., in "Recent Advances in Numerical Analysis" (C. de Boor and G. H. Golub, ed.), p. 45. Academic Press, New York (1978).

8 Gautschi, W., Math. Comp. 33, 343 (1979).

9 Kusterer, R., and Reimer, M., Math. Comput. 33, 1o19 (1979).

1o Luttmann, F. W., and Rivlin, Th. J., IBM J. 187 (1965).

11 Reimer, M., Numer. Math. 23, 321 (1975).

12 Rivlin, Th. J., "The Chebyshev Polynomials". John Wiley & Sons, New York (1974).

13 Ruston, A. F., Proc. Cambridge Phil. Soc. 58, 476 (1964).

14 Singer, I., "Best Approximations in Normed Linear Spaces by Elements of Linear Subspaces". Springer, Berlin (197o).

APPROXIMATION PROBLEMS IN TENSOR PRODUCT SPACES

John Respess
E. W. Cheney

Department of Mathematics
The University of Texas
Austin, Texas

I. INTRODUCTION

A typical problem in bivariate approximation is as follows: Let S and T be compact Hausdorff spaces and let G and H be subsets of S and T respectively. These subsets can be identified with subsets $\overline{G}$ and $\overline{H}$ of $C(S \times T)$ in a natural way. E.g., $g \in G$ is identified with the function $\overline{g} \in C(S \times T)$ such that

$$\overline{g}(s,t) = g(s) , \quad \forall \ s, \ t .$$

Given $f \in C(S \times T)$, one problem is to approximate f by a sum

$$f \approx \overline{g} + \overline{h} ,$$

where $\overline{g} \in \overline{G}$ and $\overline{h} \in \overline{H}$. A natural question is whether a best approximation exists; i.e., whether there exists a function $z \in \overline{G} + \overline{H}$ such that

$$\|f-z\| = \text{dist}(f, \overline{G} + \overline{H}) .$$

If the answer is affirmative for all f in $C(S \times T)$, then $\overline{G} + \overline{H}$ is said to be <u>proximinal</u>. A mapping $A: C(S \times T) \to \overline{G} + \overline{H}$ taking a function f to one of its best approximations is a <u>proximity map</u> (or metric selection).

Copyright © 1980 by Academic Press, Inc.
All rights of reproduction in any form reserved.
ISBN: 0-12-171050-5

In case $G = C(S)$ and $H = C(T)$, $\overline{G} + \overline{H}$ is known to be proximinal [1]. However, there exist cases in which the sum is not proximinal [2]. It may be asked then, what conditions on G and H will ensure the proximinality of $\overline{G} + \overline{H}$.

II. TENSOR PRODUCT APPROXIMATION

A slightly different problem involves the construction from G and H of tensor product subspaces of $C(S \times T)$. For example, instead of embedding G in $C(S \times T)$ as above, we consider functions of the form

$$\sum_{i=1}^{n} y_i(t) g_i(s)$$

where the g_i's are from G and the y_i's are continuous coefficient functions. If λ denotes the supremum norm on $C(S \times T)$, we denote the completion (with respect to λ) of the set of all such functions by $G \otimes_\lambda C(T)$. Our notation follows [3]. We ask, under what conditions will it be true that the subspace

$$M = G \otimes_\lambda C(T) + C(S) \otimes_\lambda H$$

is proximinal?

We cannot expect M to be proximinal unless it is closed. It can be established that if G and H are complemented subspaces of $C(S)$ and $C(T)$ respectively, then M is complemented and therefore closed. In fact, in general,

THEOREM 1. *If G and H are complemented subspaces of Banach spaces X and Y respectively and α is any uniform cross-norm on $X \otimes Y$, then $G \otimes_\alpha Y + X \otimes_\alpha H$ is complemented in $X \otimes_\alpha Y$.*

A proof of the preceding makes use of the fact that $G \otimes_\alpha Y$ and $X \otimes_\alpha H$ are separately complemented. Indeed, given bounded linear projections

$$P: X \to G, \quad Q: Y \to H,$$

the equations

$$\bar{P}\left(\sum_{i=1}^{n} x_i \otimes y_i\right) = \sum_{i=1}^{n} Px_i \otimes y_i$$

$$\bar{Q}\left(\sum_{i=1}^{n} x_i \otimes y_i\right) = \sum_{i=1}^{n} x_i \otimes Qy_i$$

define bounded linear projections $\bar{P}$ and $\bar{Q}$ on dense subsets of $G \otimes_\alpha Y$ and $X \otimes_\alpha H$ respectively. These projections can be extended to projections on the respective completions. The Boolean sum

$$\bar{P} \oplus \bar{Q} = \bar{P} + \bar{Q} - \bar{P}\bar{Q}$$

is a bounded linear projection from $X \otimes_\alpha Y$ onto $G \otimes_\alpha Y + X \otimes_\alpha H$. The latter is therefore complemented in $X \otimes_\alpha Y$.

Returning to the more specific setting of $C(S \times T) = C(S) \otimes_\lambda C(T)$, we have the following result.

LEMMA 1: <u>If G is a subspace of $C(S)$ which has a continuous proximity map, then $G \otimes_\lambda C(T)$ is proximinal in $C(S \times T)$.</u>

Proof: By a theorem of Grothendieck,

$$C(T) \otimes_\lambda G \cong C(T,G) \subset C(T,C(S)) \cong C(T,S).$$

Thus, it suffices to prove $C(T,G)$ proximinal in $C(T,C(S))$. If A is a continuous proximity map from $C(S)$ onto G, and if $f \in C(T,C(S))$, then $A \circ f$ is a best approximation of f in $C(T,G)$. Indeed, if $g \in C(T,G)$ then

$$\|f - A \circ f\| = \sup_t \|f(t) - A(f(t))\| = \sup_t \operatorname{dist}(f(t), G)$$
$$\leq \sup_t \|f(t) - g(t)\| = \|f - g\|.$$

(Note: a generalization of this occurs in [2].)

Our principal result is a consequence of the following lemma:

LEMMA 2: *If G and H are subspaces of a Banach space E such that*

(1) *there exist bounded linear projections P_G and P_H onto G and H respectively, and $P_G P_H = P_H P_G$,*

(2) *G and H have proximity maps A_G and A_H respectively, and*

(3) *for a given $f \in E$ the map $\Gamma : G \to H$ defined by $\Gamma(g) = A_H(f-g)$ is compact,*

then f has a best approximation in $G+H$.

Proof: Let $\{z_n\}_{n=1}^{\infty}$ be a sequence from $G+H$ such that $\|f-z_n\| \downarrow \operatorname{dist}(f, G+H)$. Then

$$z_n = (P_G \oplus P_H) z_n = P_G(z_n - P_H z_n) + P_H z_n = g_n + h_n$$

where $g_n = P_G(z_n - P_H z_n)$ and $h_n = P_H z_n$. It follows from the fact that $\{z_n\}$ is bounded that $\{g_n\}$ is bounded. Thus the sequence of functions $h_n' = A_H(f-g_n)$ lies in a compact set and therefore has a cluster point h^*. Let g^* be a best approximation to $f-h^*$ in G. Given $\varepsilon > 0$ we pick n so that $\|f-z_n\| < \operatorname{dist}(f, G+H) + \varepsilon$ and $\|h_n - h^*\| < \varepsilon$. Then

$$\begin{aligned} \|f-h^*-g^*\| &\leq \|f-h^*-g_n\| \leq \|f-h_n-g_n\| + \|h_n-h^*\| \\ &= \|f-z_n\| + \|h_n-h^*\| \leq \operatorname{dist}(f, G+H) + 2\varepsilon . \end{aligned}$$

Thus $g^* + h^*$ is a best approximation in $G+H$ to f.

By verifying that the conditions of Lemma 2 are met, the following theorem can be established.

THEOREM 2: <u>Let</u> S <u>and</u> T <u>be compact Hausdorff spaces</u>. <u>Let</u> G <u>be a finite-dimensional subspace of</u> $C(S)$ <u>with a continuous proximity map and let</u> H <u>be a finite-dimensional subspace of</u> $C(T)$ <u>with a proximity map satisfying a Lipschitz condition</u>. <u>Then</u> $G \otimes C(T) + C(S) \otimes H$ <u>is proximinal in</u> $C(S \times T)$.

Proximity maps from $C(S \times T)$ onto $G \otimes C(T)$ and $C(S) \otimes H$ can be constructed from the given proximity maps. For example, if $A_G: C(S) \to G$ is a proximity map, then the map $\bar{A}_G$ defined for $f \in C(S \times T)$ by

$$(\bar{A}_G f)(s,t) = (A_G f^t)(s)$$

is a proximity map from $C(S \times T)$ onto $G \otimes C(T)$. The verification that the proximity map $\bar{A}_H: C(S \times T) \to C(S) \otimes H$ satisfies condition (3) of Lemma 2 is the major difficulty in the proof of Theorem 2. Although straightforward, the proof is involved and will not be reproduced here.

EXAMPLE: If G is as described in Theorem 2 and H is spanned by a zero-free function, it can be shown by the use of the strong unicity theorem and Freud's theorem that the proximity map onto H satisfies a Lipschitz condition. Thus $G \otimes C(T) + C(S) \otimes H$ is proximinal in $C(S \times T)$.

REFERENCES

1. Diliberto, S.P. and Straus, E.G., On the approximation of a function of several variables by the sum of functions of fewer variables, Pacific J. Math., 1(1951), 195-210.

2. Franchetti, C. and Cheney, E.W., Best approximation problems for multivariate functions, Center for Numerical Analysis Report 155, (1980), The University of Texas at Austin.

3. Schatten R., A Theory of Cross Spaces, Annals of Mathematics Studies No. 26, Princeton University Press, Princeton, N.J., 1950.

ADDITIONAL REFERENCES

1. Attlestam, B. and Sullivan, F.E., Iteration with best approximation operators, Rev. Roumaine Math. Pures Appl., 21(1976), #2, 125-131.

2. Aumann, Georg, Über approximative nomographie. II, Bayer. Akad. Wiss. Math. - Nat. Kl. S. - B. (1959), 103-109.

3. Babaev, M. - B.A., The approximation of polynomials in two variables by functions of the form $\phi(x)+\psi(y)$, Soviet Math. II (1970), 1034-1036.

4. Buck, R.C., Alternation theorems for functions of several variables, J. Approximation Theory, 1(1968), 325-334.

5. Deutsch, F. and Lambert, J.M., On the continuity of metric projections, preprint (1979).

6. Dyn, N., A straightforward generalization of Diliberto and Straus' algorithm does not work, Math Research Center, Madison, Wis., (1978).

7. Franchetti, C., On the alternating approximation method, Bollettino U.M.I., 7(1973), 169-175.

8. von Golitschek, M., An algorithm for scaling matrices and computing the minimum cycle mean in a diagraph, preprint, (1979).

9. Golomb, M., Approximation by functions of fewer variables, in On Numerical Approximation, University of Wisconsin Press, Madison, (1959).

10. Light, W.A. and Cheney, E.W., On the approximation of a bivariate function by the sum of univariate functions, Center for Numerical Analysis Report 140, (1978), The University of Texas at Austin.

11. Rivlin, T.J. and Sibner, R.J., The degree of approximation of certain functions of two variables by a sum of functions of one variable. Amer. Math. Monthly, 72(1965), 1101-1103.

SOME APPLICATIONS OF STRONG REGULARITY TO MARKOV CHAINS AND FIXED POINT THEOREMS

B. E. Rhoades

Department of Mathematics
Indiana University
Bloomington, Indiana

A sequence $\{x_n\}$ is said to be almost convergent, ac, if

(1) $$\lim_n |(x_i + x_{i+1} + \cdots + x_{i+n-1})/n - s| = 0 ,$$

uniformly in i. An infinite matrix A is called regular if it is limit preserving over c, the space of convergent sequences. Necessary and sufficient conditions for regularity are: (i) $\|A\| = \sup_n \sum_{k=0}^{\infty} |a_{nk}| < \infty$, (ii) $\lim_n a_{nk} = 0$ for each k, and (iii) $\lim_n \sum_k a_{nk} = 1$. A is called strongly regular if c_A, the convergence domain of A, contains ac. Let $A_n(x) = \sum_k a_{nk} x_k$.

Theorem 0.[7]. A is strongly regular if and only if $\lim_n \sum_k |a_{nk} - a_{n,k+1}| = 0$.

We shall call A a hump matrix if each row of A contains a hump; i.e., for each n there exists an integer $p(n)$ such that $a_{nk} \le a_{n,k+1}$ for $0 \le k \le p$ and $a_{nk} \ge a_{n,k+1}$ for $k \ge p$.

Let $H = \{A | A$ is a regular hump matrix and $\lim_n \sup_k |a_{nk}| = 0\}$.

Copyright © 1980 by Academic Press, Inc.
All rights of reproduction in any form reserved.
ISBN: 0-12-171050-5

Lemma. $ac = \bigcap_{A \in H} c_A$

It is easy to show that each A in H is strongly regular, hence $ac \subseteq \bigcap_{A \in H} c_A$.

To show the opposite inclusion let $x \notin ac$. We shall show there is a member of H which does not sum x . Without loss of generality we may assume $s = 0$ in (1). If (1) is not satisfied, then there exists an $\epsilon > 0$ and a strictly increasing sequence $\{p_k\}$ satisfying

$$\sup_n \frac{1}{p_k} |x_n + \ldots + x_{n+p_k-1}| \geq \epsilon/2 \quad \text{for all } k .$$

Thus, for each k one can find an integer $n = n(k)$ such that

$$(2) \quad |x_n + \ldots + x_{n+p_k-1}| \,/\, p_k \geq \epsilon/4.$$

Construct A as follows. For $k = 0$, choose the smallest integer $n = n(0)$ such that (2) is satisfied.

Define $a_{nj} = 0$ for $0 \leq n < n(0) + p_0 - 1$. Set

$$a_{n+p_0-1,j} = \begin{cases} 1/p_0 \ , & n \leq j \leq n + p_0 - 1 \\ \\ 0 \ , & \text{all other } j . \end{cases}$$

For $k = 1$, pick $n(1)$ the smallest integer such that (2) is satisfied.

Define $a_{nk} = a_{n(0)+p_0-1,k}$ for $n(0) + p_0 - 1 < n < n(1) + p_1 - 1$. Define

$$a_{n(1)+p_1-1,k} = \begin{cases} 1/p_1 \ , & n(1) \leq j \leq n(1) + p_1 - 1 \\ \\ 0 \ , & \text{all other } j . \end{cases}$$

Continuing in this manner we obtain a matrix A in H such that $A_{n(k)+p_k-1}(x) \geq \epsilon/4$, so that A does not sum x.

The Lemma remains true in any sequentially complete Banach space.

Theorem 1. Let A be any strongly regular matrix, $\{s_k^{(\ell)}\}_{k=0}^{\infty}$ a convergent sequence for $\ell = 0, 1, \ldots, d-1$. Let $s^{(\ell)} = \lim_k s_k^{(\ell)}$, $\ell = 0, 1, \ldots, d-1$. Define $\{t_n\}$ by $t_{md+\ell} = s_m^{(\ell)}$, $m = 0, 1, 2, \ldots$, $\ell = 0, 1, \ldots, d-1$. Let $y_n = \sum_k a_{nk} t_k$. Then

$$\lim y_n = \frac{1}{d} \sum_{\ell=0}^{d-1} s^{(\ell)} .$$

Theorem 2. Let A be a strongly regular matrix, P a Markov matrix for a discrete-time stationary Markov chain with countable state space. Then

$$\lim_n \sum a_{nk} P^k = \lim_n \frac{1}{n+1} \sum_{k=0}^{n} P^k .$$

In [6] Theorems 1 and 2 were proved for regular matrices such that each row is a monotone decreasing sequence. In [9] these theorems were extended to matrices in H. By the Lemma they are true for every strongly regular matrix.

Let $\{X_t\}_{t=1}^{\infty}$ be an infinite Markov chain with transition probabilities $\{P_t\}$ defined on a countable state space $S = \{1,2,\ldots\}$. Let $P^{m,m+n} = P_{m+1} P_{m+2} \cdots P_{m+n}$, or $P^{m,m+n} = P^n$ if P is stationary. Let Q be a matrix, each row of which is the same. The sequence $\{P_t\}_{t=1}^{\infty}$ is said to be strongly ergodic, with constant stochastic matrix Q, if, for every $m \geq 0$, $\lim_n \|P^{m,m+n} - Q\| = 0$.

An irreducible stochastic matrix P, of period $d(d \geq 1)$ partitions the state space S into d disjoint subspaces $C_0, C_1, \ldots, C_{d-1}$, and P^d yields d stochastic matrices $\{T_\ell\}_{\ell=0}^{d-1}$, where T_ℓ is defined on C_ℓ. If P is finite, then each T_ℓ is automatically strongly ergodic, but if P is infinite, the strong ergodicity of T_ℓ is not guaranteed. A stochastic matrix will be called periodic strongly ergodic if it is an irreducible stochastic matrix, of period d, in which T_ℓ is strongly ergodic for $\ell = 0, 1, \ldots, d-1$.

Theorem 3. Let P be a periodic strongly ergodic matrix of period d, Q the matrix each row of which is the eigenvector ψ of P which solves uniquely the system of equations $\psi P = \psi$, $\sum_{i \in S} \psi_i = 1$. Then, for any strongly regular matrix A,

$$\lim_n \left\| \sum_{k=1}^{\infty} a_{nk} P^k - Q \right\| = 0 .$$

Theorem 4. Let $\{P_t\}_{t=1}^{\infty}$ be a nonstationary Markov chain, P a periodic strongly ergodic matrix with left eigenvector ψ, and Q a matrix each row of which is ψ. Let A be any strongly regular matrix. If $\lim_t \|P_t - P\| = 0$, then

$$\lim_n \sup_{m \geq 0} \left\| \sum_{k=1}^{\infty} a_{nk} P^{m,m+k} - Q \right\| = 0 .$$

Theorems 3 and 4 were established, for the Cesàro matrix of order one by Bowerman, David and Isaacson [2]. In [9] these results were extended to matrices A in $\mathcal{H}$. By the Lemma, the theorems are true for strongly regular matrices.

Let H be a real Hilbert space, C a closed convex subset of H, T a selfmapping of C. Let $A_n x$ denote the nth term of the Cesàro transform of the sequence of iterates $\{T^k x\}$. Baillon [1] proved that, if T is a nonexpansive selfmapping of C which has a fixed point, then $\{A_n x\}$ converges weakly to a fixed point of T. This result has been extended to strongly regular matrices by Brézis and Browder [2], Bruck [4], and Reich [8]. In a recent paper [5] Hirano and Takahashi extended the Baillon result to asymptotically nonexpansive mappings.

A mapping T is said to be asymptotically nonexpansive over C if, for each $x, y \in C$, $||T^i x - T^i y|| \leq (1 + \alpha_i)||x - y||$, $i = 1, 2, \ldots,$ where $\lim_i \alpha_i = 0$.

Theorem 5. Let C be a closed convex subset of a real Hilbert space H, T an asymptotically nonexpansive selfmap of C such that $\{T^n x\}$ is bounded for each $x \in C$. Let A be a strongly regular matrix. Then, for each $X \in C$, $\{A_n x\}$ converges weakly to a fixed point p, which is the asymptotic center of C.

By examining the proof in Hirano and Takahashi and by the Lemma of this paper, it is sufficient to show that Lemma 1 of [5] is true for each matrix $A \in \mathcal{H}$. This is indeed the case, and the detail of the proof will appear elsewhere.

REFERENCES

1. J.B. Baillon, Un Théorème de type ergodique pour les contractions non lineare dans un space de Hilbert, C. R. *Acad. Sci. Paris Ser. A-B*, *280* (1975), A1511-A1514.

2. H.T. Bowerman, H.T. David, and D. Isaacson, The convergence of Cesàro averages for certain nonstationary Markov chains, *Stochastic Processes Appl.*, *5* (1977), 221-230.

3. H. Brézis and F. Browder, Remarks on Nonlinear Ergodic Theory, *Advances in Math.*, *25* (1977), 165-177.

4. R.E. Bruck, On the almost-convergence of iterates of a nonexpansive mapping in Hilbert space and the structure of the weak ω-limit set, to appear.

5. N. Hirano and W. Takahashi, Nonlinear Ergodic theorems for nonexpansive mappings in Hilbert spaces, *Kodai Math. J.*, *2* (1979), 11-25.

6. Y. Kadota, A limit theorem for Markov chains and Hordijk-Tijms iterative method, *Mem. Fac. Sci. Kyushu Univ. Ser. A30* (1976), 169-182.

7. G.G. Lorentz, A contribution to the theory of divergent sequences, *Acta. Math.*, *80* (1948), 167-190.

8. S. Reich, Almost Convergence and Nonlinear Ergodic Theorems, *J. Approx. Theory*, *24* (1978), 269-272.

9. B.E. Rhoades, The convergence of matrix transforms for certain Markov chains, *Stochastic Processes Appl.*, *9* (1979), 85-93.

LACUNARY TRIGONOMETRIC INTERPOLATION: CONVERGENCE

S.D. Riemenschneider[1]
A. Sharma[1]
P.W. Smith[2]

Department of Mathematics
University of Alberta
Edmonton, Alberta
Canada T6G 2G1

The problem of existence and uniqueness of $(0,m_1,\ldots,m_q)$ interpolation $(m_0 = 0 < m_1 < \ldots < m_q)$, by trigonometric polynomials at the equidistant nodes $2k\pi/n$ $(k = 0,1,\ldots,n-1)$ has been solved by Cavaretta, Sharma and Varga [1] (see also [4]). For a precise statement of the results we introduce the class $\mathcal{T}_M$ of trigonometric polynomials

$$T(\theta) = a_0 + \sum_{\nu=1}^{M} (a_\nu \cos\nu\theta + b_\nu \sin\nu\theta) \tag{1}$$

and the class $\mathcal{T}_{M,\varepsilon}$ of trigonometric polynomials of the form

$$T(\theta) = a_0 + \sum_{\nu=1}^{M-1} (a_\nu \cos\nu\theta + b_\nu \sin\nu\theta) + a_M \cos(M + \pi\varepsilon/2) \tag{2}$$

$\varepsilon = 0$ or 1. The interpolation problem will be solvable only for an appropriate choice of approximating class $\mathcal{T}_M, \mathcal{T}_{M,\varepsilon}$ which will depend on the composition of $(m_1,\ldots,m_q)$. Let E_q

---

[1] Supported in part by NSERC Grants A7687 and A3094.
[2] Present Address: Old Dominion University, Norfolk, VA. U.S.A.

Copyright © 1980 by Academic Press, Inc.
All rights of reproduction in any form reserved.
ISBN: 0-12-171050-5

(respectively O_q) denote the number of even (respectively odd) integers in the set $(m_1, \ldots, m_q)$.

The results of [1], [4] can be summarized as follows: The $(0, m_1, \ldots, m_q)$ interpolation problem is uniquely solvable

I. for $T(\theta) \in \mathcal{T}_M$ only if $n = 2m + 1$, $q = 2r$ and $M = nr + m$;

II. for $T(\theta) \in \mathcal{T}_{M,0}$ only if (a) $n = 2m + 1$, $q = 2r + 1$ $E_q - O_q = 1$ and $M = nr + n$, or (b) $n = 2m$, $q = 2r$, $E_q - O_q = 0$ and $M = nr + m$;

III. for $T(\theta) \in \mathcal{T}_{M,1}$ only if $n = 2m + 1$ or $2m$, $q = 2r + 1$, $E_q - O_q = -1$ and $M = nr + n$.

Here we shall consider the corresponding convergence problem. Let $\rho_{k,m_j}(\theta)$ denote the fundamental polynomials for the $(0, m_1, \ldots, m_q)$ interpolation problem. It turns out that $\rho_{k,m_j}(\theta)$ is in the appropriate class $\mathcal{T}_M$ or $\mathcal{T}_{M,\varepsilon}$ and by periodicity

$$\rho_{k,m_j}(\theta) = \rho_{0,m_j}(\theta - (2k\pi/n)) \tag{3}$$

where ρ_{0,m_j} is uniquely determined by

$$\rho_{0,m_j}^{(m_s)}(2k\pi/n) = \delta_{s,j} \cdot \delta_{0,k} \tag{4}$$

(See [1] for a complete description of $\rho_{k,m_j}(\theta)$).

The modified interpolation operator $R_n(f;\theta)$ is defined by

$$\begin{aligned} R_n(f;\theta) &= \textstyle\sum_{k=0}^{n-1} f(2k\pi/n)\,\rho_{0,0}(\theta - 2k\pi/n) \\ &\quad + \textstyle\sum_{j=1}^{q} \sum_{k=0}^{n-1} \beta_{k,j}^{(n)} \rho_{0,m_j}(\theta - 2k\pi/n) \end{aligned} \tag{5}$$

where $\{\beta_{k,j}^{(n)}\}$ are given numbers. A natural question is: when does $R_n(f;\theta)$ converge uniformly to $f(\theta)$? We announce the following theorems partially answering this question.

Theorem 1. Let f be a 2π-periodic function which satisfies the Dini-Lipschitz condition $\omega(f,1/\delta)\log 1/\delta \to 0$ as $\delta \to 0$. If the interpolation is of the type I, II(b) or III and if

$$\beta_{kj}^{(n)} = o(n^{m_j}/\log n), \quad 0 \le k \le n-1; 1 \le j \le q \tag{6}$$

as $n \to \infty$, then the modified interpolation operator $R_n(f;\theta)$ converges uniformly to $f(\theta)$ as $n \to \infty$.

Theorem 2. Let the interpolation be of the type II(a), and suppose that the numbers $0 = m_0 < m_1 < \ldots < m_q$ are alternately even and odd; m_{1+2j}, $j = 0,\ldots,r$, are even and m_{2+2j}, $j = 0,\ldots,r-1$ are odd. If $f(\theta)$ is 2π-periodic and satisfies the Zygmund condition

$$f(\theta+h) - 2f(\theta) + f(\theta-h) = o(h), \quad h \to 0 \tag{7}$$

and if

$$\beta_{kj}^{(n)} = o(n^{m_j-1}), \quad 1 \le j \le q; 0 \le k \le n-1 \tag{8}$$

as $n \to \infty$, then $R_n(f;\theta)$ converges uniformly to $f(\theta)$ as $n \to \infty$.

For the proofs one first estimates the fundamental polynomials, then the standard technique introduced by Kiš [2] is applied to complete the result. We shall sketch here a proof of the simplest case, case I, by the methods introduced in [3].

In case I, $M = nr + m$, so that we may write the fundamental polynomials in the form

$$\rho_{0,m_j}(\theta) = \sum_{\lambda=-r}^{r} T_{\lambda,j}(\theta)e^{i\lambda n\theta} \tag{9}$$

where the $T_{\lambda,j}(\theta)$ are trigonometric polynomials of degree $n = 2m + 1$. Using the notation $D = \frac{d}{d\theta}$, the interpolatory conditions (4) leads by way of Leibnitz's formula to the system of equations

$$\rho_{0,m_j}^{(m_s)}(2k\pi/n) = \sum_{\lambda=-r}^{r}(D+i\lambda n)^{m_s}T_{\lambda,j}(\theta)\Big|_{\theta=2k\pi/n}$$

$$= \begin{cases} L(2k\pi/n) & s = j \\ 0 & \text{otherwise} \end{cases} \tag{10}$$

$k = 0,1,\ldots,n-1$ and $1 \le s,j \le q$ and where $L(\theta) = \sum_{\nu=-m}^{m} e^{i\nu\theta}/n$ is the Lagrange polynomial with $L(0) = 1$ and $L(2k\pi/n) = 0$ $k \ne 0$. The n points $\{2k\pi/n\}$ determine the trigonometric polynomial uniquely so that the right side of (10) is a polynomial identity. This means that the polynomials $T_{\lambda,j}$ satisfy an inhomogeneous system of differential equations with constant coefficients. Therefore, using standard techniques, we obtain

$$(H(D)T_{\lambda,j})(\theta) = (H_{\lambda,j}(D)L)(\theta) \tag{11}$$

$-r \le \lambda \le r, j = 0,\ldots,q$, where $H(D)$ is the differential operator defined by the determinant

$$\det\{(D-irn)^{m_s},\ldots,D^{m_s},\ldots,(D+irn)^{m_s};\ s= 0,\ldots,2r\} \tag{12}$$

and $H_{\lambda,j}(D)$ is the cofactor of the entry $(D+i\lambda n)^{m_j}$ in $H(D)$.

If we define the determinant $V(\alpha)$ by

$$V(\alpha) = \det\{(\alpha-r)^{m_s},\ldots,\alpha^{m_s},\ldots,(\alpha+r)^{m_s};\ s=0,\ldots,2r\} \tag{13}$$

then for any trigonometric polynomial $T(\theta) = \sum_{\nu=-m}^{m} a_\nu e^{i\nu\theta}$ we have

$$(H(D)T)(\theta) = (in)^{\Sigma m_s}\sum_{\nu=-m}^{m} a_\nu V(\nu/n)e^{i\nu\theta}. \tag{14}$$

The conditions in I guarantee that $V(\alpha) \ne 0$ for $-\frac{1}{2} \le \alpha \le \frac{1}{2}$ (see [4,lemma 4]). Moreover, $V(\alpha)$ is an even polynomial in α. Since $V(\nu/n) \ne 0$, the operator $H(D)$ is invertible on the trigonometric polynomials and this inverse is given by

$$[H(D)^{-1}T](\theta) = (in)^{-\Sigma m_s} \sum_{\nu=-m}^{m} (a_\nu / V(\nu/n)) e^{i\nu\theta} \tag{15}$$

The operators $H_{\lambda,j}(D)$ and $H(D)^{-1}$ commute when acting on trigonometric polynomials. Therefore from (11) and (15), we can deduce that

$$T_{\lambda,j}(\theta) = H_{\lambda,j}(D)[H(D)^{-1}L](\theta) . \tag{16}$$

The estimates for the fundamental polynomials require the following lemma.

Lemma 1. There is a constant C depending only on $(0,m_1,\ldots,m_q)$ such that

$$\sum_{k=0}^{n-1} |[H(D)^{-1}L](\theta-2k\pi/n)| \leq Cn^{-\Sigma m_s} \log n \tag{17}$$

when the conditions of case I are fulfilled.

Proof. From summation by parts and the evenness of $V(\alpha)$, we obtain from (15).

$$\begin{aligned} [H(D)^{-1}L](\theta)| &= |\textstyle\sum_{\nu=-m}^{m} e^{i\nu\theta}/V(\nu/n)|n^{-1-\Sigma m_s} \\ &\leq 2n^{-1-\Sigma m_s} \sup_r |\textstyle\sum_{\nu=-r}^{r} e^{i\nu\theta}| \{\textstyle\sum_{\nu=0}^{m-1} |V(\nu/n)^{-1} - V\left(\frac{\nu+1}{n}\right)^{-1}| \\ &\quad + |V(\tfrac{m}{n})|\} . \end{aligned} \tag{18}$$

Since $V(\alpha)$ is a polynomial which does not vanish on $[-\tfrac{1}{2},\tfrac{1}{2}]$, $1/V(\alpha)$ is of bounded variation on $[-\tfrac{1}{2},\tfrac{1}{2}]$. Therefore, from inequality (18) we obtain

$$|H(D)^{-1}(L)(\theta)| \leq Cn^{-\Sigma m_s} \min\{1,(n\sin\theta/2)^{-1}\} .$$

Inequality (17) is now obtained by means of the known relation

$$\sum_{k=0}^{n-1} \min\{1,[n\sin(\theta-2k\pi/n)]^{-1}\} = \mathcal{O}(\log n) . \qquad \square$$

It is now possible from (16),(17) and Bernstein's inequality to obtain the estimate

$$\sum_{k=0}^{n-1} |T_{\lambda,j}(\theta - 2k\pi/n)| \le Cn^{-m_j} \log n \ . \tag{19}$$

From relations (3),(9) and (19), we have

Theorem 3. In the case I, the fundamental polynomials for $(0,m_1,\ldots,m_q)$ interpolation satisfy the inequality

$$\sum_{k=0}^{n-1} |\rho_{k,m_j}(\theta)| \le Cn^{-m_j} \log n \ , \quad j = 0,1,..,q \ .$$

As was mentioned earlier, case I of Theorem 1 now follows by a standard technique. Complete details and remarks to the related literature will appear elsewhere.

REFERENCES

1. A.S. Cavaretta, A. Sharma, R.S. Varga, (1979). Lacunary trigonometric interpolation on equidistant nodes. Preprint.
2. O. Kiš, (1960). On trigonometric (0,2) interpolation (Russian), *Acta Math. Sci. Hungar. 11, 255-276.*
3. S.D. Riemenschneider and A. Sharma, (1979). Birkhoff interpolation at the nth roots of unity: convergence. *Canad. J. Math. to appear.*
4. A. Sharma, P.W. Smith and J. Tzimbalario, (1979) Polynomial interpolation in roots of unity with applications. Preprint.

The research of P. W. Smith was partially supported by the U. S. Army Research Office under Grant #DAHC04-75-G-0816.

GENERALIZED PADÉ APPROXIMANTS

H. van Rossum

Department of Mathematics
Universiteit van Amsterdam
Amsterdam, The Netherlands

SUMMARY

In recent years a generalized Padé approximant (GPA) to a formal expansion in polynomials (FEP), was introduced in a number of papers and in several different ways (see section 1 and References).

In this paper we propose still another approach (section 2). In section 1 we review (very briefly) the definitions and some results for the different GPA's.

1. INTRODUCTION

In the following $f(x) = \sum_{n=0}^{\infty} c_n p_n(x)$ denotes an arbitrary FEP, where $(c_n)_{n=0}^{\infty}$ is a normal sequence of complex numbers i.e., all determinants

$$\begin{vmatrix} c_i & c_{i-1} & \cdots & c_{i-j} \\ c_{i+1} & c_i & \cdots & c_{i-j+1} \\ \cdot & \cdot & \cdots & \cdot \\ c_{i+j} & c_{i+j-1} & \cdots & c_i \end{vmatrix}, (i,j = 0,1,\ldots;\ c_k = 0 \text{ if } k < 0),$$

differ from zero.

Copyright © 1980 by Academic Press, Inc.
All rights of reproduction in any form reserved.
ISBN: 0-12-171050-5

$p_n(x)$ is a polynomial of degree n $(n = 0,1,\ldots)$ in the indeterminate x, with complex coefficients.

a. Linear (m,k)-GPA for f.

The linear (m,k)-GPA for f is the generalized rational function $U_{m,k}(x)/V_{m,k}(x)$, where $U_{m,k}(x) = \sum_{r=0}^{k} a_r p_r(x)$, $V_{m,k}(x) = \sum_{s=0}^{m} b_s p_s(x)$ and

$$V_{m,k}(x)f(x) - U_{m,k}(x) = O(p_{m+k+1}(x)) \tag{1.1}$$

where $O(p_{m+k+1}(x))$ is an FEP such that $p_0, p_1, \ldots, p_{m+k}$ are absent. The equations determining the coefficients a_r $(r = 0,1,\ldots,k)$ and b_s $(s = 0,1,\ldots,m)$ are linear.

Usually x is taken to be a complex variable. These GPA's are often called Newton-Padé approximants. They were first considered by Saff [20] in 1972, then by Karlsson [17] in 1972, by Warner [23], [24] in 1974, by Fleisher [12] in 1972 for an FEP in Legendre polynomials. A detailed account is given in Galluci and Jones [13], see also the references given there. This approximation is often referred to as Hermite interpolation [23]. See also Claessens [5], [6] 1976. The case where $(c_n)_{n=0}^{\infty}$ is a Stieltjessequence is treated by Bransley [1], [2]. Constructions of the algorithms for the Newton-Padé table are given in Claessens [5], [6] and in [23]. A collection of papers on linear GPA's is mentioned by Chui [4] and Brezinski [7], 1978. Compare also a recent paper by Claessens [7], 1978.

For an FEP in orthogonal polynomials Holdeman [16] has defined a GPA of linear type in 1969, as follows:
The (m,k)-GPA to $f(x) = \sum_{n=0}^{\infty} c_n g_n(x)$, where $(g_n(x))_{n=0}^{\infty}$ is a sequence of polynomials orthogonal on [a,b], is to be found from

$$f(x)\sum_{r=0}^{m} b_r p_r(x) - \sum_{s=0}^{k} a_s g_s(x) = \sum_{l=k+m+1}^{\infty} \rho_l g_l(x)$$

where $(p_n)_{n=0}^{\infty}$ is another system of polynomials orthogonal on [a,b].

b. Nonlinear (m,k)-GPA for f.

This is the generalized rational function $U_{m,k}(x)/V_{m,k}(x)$, where

$$U_{m,k}(x) = \sum_{r=0}^{k} a_r p_r(x), \quad V_{m,k}(x) = \sum_{s=0}^{m} b_s p_s(x)$$

and

$$f(x) - \frac{U_{m,k}(x)}{V_{m,k}(x)} = O(p_{m+k+1}(x)) \tag{1.2}$$

The nonlinear GPA need not to exist and if it does it is not unique in general. However, for a special case a uniqueness result was established by Avram Sidi [22] in 1977. It reads: "If the (m,k)-GPA to $f(x) = \sum_{n=0}^{\infty} c_n p_n(x)$ exists (where the p_n's are orthogonal on a real interval [a,b]) and is put in its irreducible form, the approximant $U_{m,k}(x)/V_{m,k}(x)$ is unique if $V_{m,k}(x)$ has no zeros in [a,b]". The nonlinear case hase been treated by Clenshaw and Lord [8] in 1974 and by Avram Sidi [21] in 1975, for the case of an FEP in Chebyshew polynomials of the first kind. Fleisher [13] in 1974 considered nonlinear GPA's for an FEP in Legendre polynomials.

Remark 1. If in what precedes we take $p_n(x) = x^n$ $(n = 0,1,...)$ the conditions (1,1) en (1.2) both lead to the same (m,k)-ordinary Padé approximant (OPA). In general however (1.1) and (1.2) obviously lead to different GPA's.

Remark 2. Closely related to the preceding material is the work of Coates [9], in 1966.

c. Other types of GPA's to an FEP.

Up till now the different types of GPA joined the following two properties:

I The GPA is a rational function.

II The GPA satisfies a relation either of type (1.1) or of type (1.2).

We will now consider GPA's which only satisfy II in a form similar to (1.2).

1. Zakian [25] in 1972 defined an approximant $I_{MN}f$ for a function $f(t)$, $t \in [0,\infty)$ by

$$I_{MN}f = \int_0^\infty f(\lambda) \sum_{i=1}^{N} \frac{K_i}{t} e^{-\frac{\lambda a_i}{t}} d\lambda, \quad (M < N),$$

where a_i, K_i are constants. The approximants of a certain subclass of the class of approximants I_{MN} have been shown to be of Padé type. We refer to [4] for related material on this subject.

2. Common and Stacey [9] in 1977 introduced GPA's for FEP's in Legendre polynomials (in a complex variable), called LPA's [1]). See also [10]. Let the series $f(w) = \sum_{n=0}^{\infty} c_n w^n$ converge on the disc $|w| < r$, with $r > 1$, in the complex w-plane. Then the FEP in Legendre polynomials, $g(z) = \sum_{n=0}^{\infty} c_n P_n(z)$ converges on an elliptical disc with foci at $z = \pm 1$ and semi major axis $r+1/r$ in the z-plane, as is well known. It is shown that

$$g(z) = \frac{1}{2\pi i} \oint_{\Gamma} \frac{f(w)dw}{\sqrt{w^2 - 2zw + 1}} \tag{1.3}$$

where Γ is a scroc enclosing the branch points $w = z \pm \sqrt{z^2 - 1}$ The (m,k)-LPA, $T_{m,k}^* g$ is defined by

$$T_{m,k}^* g(z) = \frac{1}{2\pi i} \oint_{\Gamma} \frac{T_{m,k} f(w)dw}{\sqrt{w^2 - 2zw + 1}}, \tag{1.4}$$

where $T_{m,k}f(w)$ is the (m,k)-OPA for f, and Γ suitably chosen. The LPA's are shown to satisfy

$$g(z) - T_{m,k}^* g(z) = O(P_{m+k+1}(z)).$$

The convergence of sequences of LPA's is compared to the convergence of similar sequences of Fleisher's nonlinear LPA's. In the example chosen, Common and Stacey's LPA's perform better. The idea to transform the approximant in the same way as the function it approximates, as exemplified in (1.3) and (1.4) is also present in our definition of GPA's (Definition 2.1)

[1]) We are indebted to Prof. Common for drawing our attention to this work at an early stage.

2. A NEW DEFINITION OF A GPA TO AN FEP

Definition 2.1. Let $g(x) = \sum_{n=0}^{\infty} c_n p_n(x)$ be an FEP where $\deg p_n = n$ $(n = 0,1,\ldots)$, $(c_n)_{n=0}^{\infty}$ is a normal sequence of complex numbers and x is an indeterminate.
$T_{m,k}{}^{*}g$ is said to be the (m,k)-GPA to g if

$$T_{m,k}{}^{*}g(x) = \sum_{n=0}^{\infty} c_n^{(m,k)} p_n(x), \text{ where}$$

$$\sum_{n=0}^{\infty} c_n^{(m,k)} x^n = T_{m,k}f(x), \text{ and}$$

$T_{m,k}f(x)$ is the (m,k)-OPA to $f(x) = \sum_{n=0}^{\infty} c_n x^n$.

Remark 1. Obviously $c_n^{(m,k)} = c_n$, $n = 0,1,\ldots,m+k$.

Remark 2. $g(x) - T_{m,k}{}^{*}g(x) = O(p_{m+k+1}(x))$.

We will now describe the underlying mapping in Definition 2.1., in the case where $z \in \mathbb{C}$ and $(p_n(z))_{n=0}^{\infty}$ is a complete orthonormal sequence (CON sequence) of polynomials, in a certain Hilbert space of holomorphic functions.

Let R denote a region in the complex plane and consider the set $L_h^2(R)$ of the functions holomorphic in R and such that for every $f \in L_h^2(R)$ we have

$$\iint_R |f(z)|^2 dxdy < \infty \quad (z = x + iy; \quad x,y \in \mathbb{R}; \quad i = \sqrt{-1}).$$

$L_h^2(R)$ is a Hilbert space with repect to the inner product $\langle .,. \rangle$ defined as follows

$$\forall f,g \in L_h^2(R), \ \langle f,g \rangle = \iint_R f(z)\overline{g(z)}dxdy.$$

Hence

$$||f||^2 = \iint_R |f(z)|^2 dxdy.$$

It is a well-known result (for instance [15]), that the sequence

$$\left(\sqrt{\frac{n+1}{\pi}}\, z^n \right)_{n=0}^{\infty}$$

is a CON-sequence in $L_h^2(D)$ where D is the unit disc (centered at 0) in the complex z-plane. We remark, that if $f \in L_h^2(D)$ and $f(z) = \sum_{n=0}^{\infty} c_n z^n$, its Fourier coefficients b_n with respect to the CON sequence above are related to the c_n as follows:

$$b_n = c_n \sqrt{\frac{\pi}{n+1}}\ , \quad (n = 0,1,\dots). \tag{2.1}$$

Another well-known result we need is:
Let R' be a region in the complex ζ-plane ($\zeta = \xi + i\eta$; $\xi,\eta \in \mathbb{R}$; $i = \sqrt{-1}$) mapped conformally by $z = w(\zeta)$ onto the region R in the z-plane ($z = x + iy$; $x,y \in \mathbb{R}$; $i = \sqrt{-1}$). This conformal mapping induces an isometric isomorphism between the Hilbert spaces $L_h^2(R)$ and $L_h^2(R')$, as follows $g \in L_h^2(R)$, $(g \circ w)w' \in L_h^2(R')$ and

$$\iint_R |g(z)|^2 dxdy = \iint_{R'} |g(w(\zeta))|^2\ w'(\zeta)|^2 d\zeta d\eta$$
$$\Rightarrow \quad \|g\|_R = \|(g \circ w)w'\|_{R'}$$

Then the above mentioned mapping is defined by $g \leftrightarrow (g \circ w)w'$. See Higgings [15], p.67.

Now we prove

<u>Theorem 2.1.</u> Let $(p_n(z))_{n=0}^{\infty}$ be a CON sequence for $L_h^2(R)$, where R is a region in the complex plane and $p_n(z)$ a polynomial of degree n with complex coefficients, $(n = 0,1,\dots)$. The mapping χ defined by $\chi(z^n) = p_n(z)$, $(n = 0,1,\dots)$ satisfies $\chi = \psi\tau$, where ψ is an isometric mapping from $L_h^2(D)$ onto $L_h^2(R)$ and τ is a linear mapping from $L_h^2(D)$ on itself, satisfying

$$\tau(z^n) = \sqrt{\frac{n+1}{\pi}}\, z^n \quad (n = 0,1,\dots).$$

<u>Proof.</u> First map the region R in the z-plane by $\zeta = w(z)$ onto the unit disc D in the ζ-plane. This induces an isometric mapping ϕ_1 from $L_h^2(D)$ onto $L_h^2(R)$ such that with the CON sequence

$$\left(\sqrt{\frac{n+1}{\pi}}\,\zeta^n\right)_{n=0}^{\infty} \quad \text{in } L_h^2(D)$$

is associated (by $\zeta = w^n(z)$), the CON sequence,

$$\left(\sqrt{\frac{n+1}{\pi}}\,w^n(z)w'(z)\right)_{n=0}^{\infty} \quad \text{in } L_h^2(R).$$

Next $\phi_2 : L_h^2(R) \to L_h^2(R)$ maps this sequence on the CON sequence $(p_n(z))_{n=0}^{\infty}$ isometrically, as follows

$$\phi_2\left(\sqrt{\frac{n+1}{\pi}}\,w^n(z)w'(z)\right) = p_n(z), \quad (n = 0,1,\ldots).$$

Hence $\chi = \psi\tau$ with $\psi = \phi_1\phi_2$ isometric and τ given by

$\tau(z^n) = \sqrt{\frac{n+1}{\pi}}\,z^n$, $(n = 0,1,\ldots)$ □

In some cases it is possible to relate the convergence of the GPA's to the convergence of the OPA's. This is the content of the following

Theorem 2.2. Let $(p_n(z))_{n=0}^{\infty}$ be a CON sequence of polynomials ($\deg p_n = n$, $(n = 0,1,\ldots)$) for the Hilbert space $L_h^2(R)$, where R is a region in the complex plane. Let $g(z) = \sum_{n=0}^{\infty} c_n p_n(z)$, $z \in R$ and $g \in L_h^2(R)$. Furthermore let χ be the mapping used in Theorem 2.1., i.e., $\chi(z^n) = p_n(z)$, $(n = 0,1,\ldots)$, and $\chi f = g$, $f \in L_h^2(D)$ (D is the unit disc).

Then $\lim_{m+k\to\infty} T_{m,k}f = f$ uniformly on $D \Rightarrow \lim_{m+k\to\infty} T^*_{m,k} = g$ uniformly on any compact subset of R.

Proof. By assumption $T_{m,k}f \in L_h^2(D)$ if m and k are large enough. χ maps $T_{m,k}f$ into $T^*_{m,k}g$, hence for m and k large enough, $T^*_{m,k}g \in L_h^2(R)$. Uniform convergence of $T_{m,k}f$ to f on D, if $m+k \to \infty$ is equivalent to:

$$\max_{z\in D}\left|\sum_{n=0}^{\infty}(c_n - c_n^{(m,k)})z^n\right| \to 0, \text{ if } m+k\to\infty.$$

From this it follows: $\sum_{n=0}^{\infty}\left|c_n - c_n^{(m,k)}\right| \to 0$, if $m+k\to\infty$.

Let $\|\cdot\|_R$ denote the norm on $L_h^2(R)$. Now

$$\left\|\sum_{n=0}^{\infty}(c_n - c_n^{(m,k)})p_n(z)\right\|_R^2 = \sum_{n=0}^{\infty}\left|c_n - c_n(m,k)\right|^2 \to 0 \text{ if}$$

$m+k \to \infty$.

Hence $T^*_{m,k}g$ converges in norm to g.
Now the kernel function $K(t,z) = \sum_{n=0}^{\infty} (\pi/n+1)t^n z^{-n}$ is continuous on $(D \times D)$.
Hence $\sum_{n=0}^{\infty} p_n(t)\overline{p_n(z)}$ is continuous on $R \times R$ so $\sum_{n=0}^{\infty} \left| p_n(z) \right|^2$ is bounded on any compact subset of R. Since $L_h^2(R)$ is a reproducing kernel space it follows that the convergence of the sequence $T^*_{m,k}g \to g$, $m+k \to \infty$ is uniform on any compact subset of R. □

References

1. Barnsley, M., The bounding properties of the multipoint Padé approximant to a series of Stieltjes on the real line, *J. Math. Phys. 14* (1973), 299-313.
2. Barnsley, M., The bounding properties of the multipoint Padé approximant to a series of Stieltjes, *Rocky Mountains J. Math., 4* (1974), 331-334.
3. Brezinski, C., Accélération de la Convergence en Analyse numérique, Lecture Notes in Mathematics, no. 584 (1977)
4. Chui, C.K., Recent results on Padé approximants and related problems in Approximation Theory II, G.G. Lorentz, C.K. Chui and L.L. Schumaker, Eds., Academic Press, N.Y. (1976), 79-115.
5. Claessens, G., The rational Hermite interpolation problem and some related recurrence formulas, *J. Comp. and Appl. Math., 2* (1976), 117-123.
6. Cleassens, G., A new algorithm for osculatory rational approximation, *Num. Mathematik, 27* (1976), 77-83.
7. Claessens, G., On the Newton-Padé approximation problem, *J. Appr. Theory, 22* (1978), 150-160.
8. Coates, J., On the algebraic approximation of functions, *Indag. Math.* (1966) I, II, III, 421-461.
9. Common, K., and Stacey, T., The convergence of Legendre Padé approximants to the Coulomb and other scattering amplitudes, (1977) *UKC preprint.*

10. Common, K., and Stacey, T., Legendre Padé approximants and their Application in potential Scattering, (1977) *UKC preprint.*

11. Fleisher, J., Analytic continuation of scattering amplitudes and Padé approximants, *Nucl. Phys. B 37* (1972), 59-76.

12. Fleisher, J., Nonlinear Padé approximants for Legendre series, *J. Math. Phys. 14* (1973), 246-248.

13. Galluci, M.A., and Jones, W.B., Rational approximation corresponding to Newton-Padé series, *J. Appr. Theory, 17* (1976), 366-392.

14. Higgins, J.R., Completeness and basis properties of sets of special functions, (1977) Cambridge University Press.

15. Holdeman, J.T., A method for the approximation of functions defined by formal series expansions in orthogonal polynomials, *Math. Comput., 23* (1969), 275-288.

16. Karlson, J., Rational interpolation with free poles in the complex plane, *Univ. of UMEA preprint* (1972).

17. Karlson, J., Rational interpolation and best rational interpolation, *Univ. of UMEA preprint* (1974).

18. Kober, H., Dictionary of Conformal representations, (1952), Dover Publications.

19. Saff, E.B., An extention of Montessus de Ballore's theorem on the convergence of the interpolation rational functions, *J. Appr. Theory, 6* no 1 (1972), 63-67.

20. Avram Sidi, Computation of the Chebyshev-Padé table, *J. Comp. and Appl. Math. 1* no 2 (1975), 69-71.

21. Avram Sidi, Uniqueness of Padé approximants from series of orthogonal polynomials, *Msth. Comput. 31,* no 139 (1977), 738-739.

22. Warner, D.D., Hermite interpolation with rational functions, Doctoral dissertation, Univ. of Calif. San Diego (1974).

23. Warner, D.D., An extention of Saff's theorem on the convergence of interpolating rational functions, to appear.

24. Zakian, V., Properties of I_{MN} approximants in Padé approximants and their Applications. Graves-Morris, ed., Acad. Press London (1973).

H. van Rossum,
Instituut voor Propedeutische Wiskunde,
Unversiteit van Amsterdam,
Roetersstraat 15,
1018 WB Amsterdam.
The Netherlands.

APPROXIMATION BY CONVEX QUADRATIC SPLINES

J. A. Roulier[1]

Department of Mathematics
North Carolina State University
Raleigh, North Carolina

D. F. McAllister[1]

Department of Computer Science
North Carolina State Univeristy
Raleigh, North Carolina

I. INTRODUCTION

In [1], McAllister and Roulier present an algorithm for the calculation of an osculatory quadratic spline. The parameters of the spline depend on the points to be interpolated $\{(x_i,y_i)\}_{i=0}^{N}$ and the associated first derivatives $\{m_i\}_{i=0}^{N}$ at each point. The spline S is then determined and evaluated using subroutine MEVAL. This subroutine computes a first degree spline L which interpolates the data and such that the slope of the linear segment of L passing through (x_i,y_i) is m_i for $i = 0,1,\ldots,N$. L is then used to determine a piecewise quadratic Bernstein polynomial S such that

[1] Supported in part by NASA grant NSG 1549-S1

Copyright © 1980 by Academic Press, Inc.
All rights of reproduction in any form reserved.
ISBN: 0-12-171050-5

$S(x_i) = y_i$, $S'(x_i) = m_i$ for $i = 0,1,\ldots,N$, and $S \in C^1[x_o,x_N]$. That is, S is an osculatory quadratic spline. In addition, the knots of S will consist of the abscissae of the original data as well as at most two additional knots in each interval (x_i,x_{i+1}) for $i = 0,1,\ldots,N-1$. S will preserve the monotonocity and convexity of the data where this is consistent with the slopes $\{m_i\}_{i=0}^N$.

The subroutine SLOPES calculates values $\{m_i\}_{i=0}^N$ based on data $\{(x_i,y_i)\}_{i=0}^N$ so that the spline S produced by MEVAL using these values will preserve the monotonicity and convexity of the data. S will, of course, also preserve local extrema of the data. The spline S produced using these slopes will have knots at the abscissae of the original data and at exactly one additional point in each interval (x_i,x_{i+1}) for $i = 0,1,\ldots,N-1$.

In this report, we present the approximation properties of the spline S produced by SLOPES and MEVAL for the case of equally spaced abscissae $x_i = x_o + ih$ for $i = 0,1,\ldots,N$ and ordinates corresponding to function values of a differentiable function f on $[x_o,x_N]$. Thus $y_i = f(x_i)$ for $i = 0,1,\ldots,N$. In particular we will present estimates for

$$\max_{x_o \le x \le x_N} |f^{(j)}(x) - S^{(j)}(x)|, \quad j = 0,1,$$

and for

$$\left| \int_{x_o}^{x_N} f(x)\,dx - \int_{x_o}^{x_N} S(x)\,dx \right|.$$

II. THE MAIN THEOREMS

The following theorems are the principal results of this paper. The proofs will appear in a subsequent publication.

In what follows, we assume that f is a function defined on a closed interval [a,b], and that $x_i = a + ih$ for $i = 0,1,\ldots,N$, with $h = \frac{b-a}{N}$. Thus, $x_N = b$.

Theorem 2.1. Let $f \in C^3[a,b]$ with $f'(x) \neq 0$ on [a,b], and let $x_i = a + ih$ for $i = 0,1,\ldots,N$ where h is given by $h = \frac{b-a}{N}$. Let S_h be the shape preserving quadratic spline interpolating the data

$$\{(x_i, f(x_i))\}_{i=0}^{N}$$

and produced by subroutine MEVAL using slopes provided by subroutine SLOPES. Let X_h be the union of all subintervals $[x_i, x_{i+1}]$ for which neither endpoint is a or b. Then

$$\max_{x \in X_h} |f^{(j)}(x) - S_h^{(j)}(x)| = O(h^{3-j}) \text{ for } j = 0,1 \qquad (2.1)$$

and

$$\max_{x \in [a,b] \setminus X_h} |f^{(j)}(x) - S_h^{(j)}(x)| = O(h^{2-j}) \text{ for } j = 0,1 \qquad (2.2)$$

Theorem 2.2. Let f be as in Theorem 2.1. Let $I_h = \int_a^b S_h(x)dx$ and $I = \int_a^b f(x)dx$. Then

$$I_h = I + O(h^3). \qquad (2.3)$$

using a theorem in Meinardus [2, p. 78] applied to an interval of length h/2 or more for the case f is a cubic polynomial with a non-zero coefficient on the x^3 term.

The only reason that (2.1) is not valid on all of [a,b] or for functions with relative extrema is that subroutine SLOPES chooses a slope of zero at each relative extremum of the data and chooses slopes at the endpoints a and b in a different manner than at other data points. This is necessary if the algorithm is to preserve the local monotonicity and convexity of the data, which is the purpose for which the algorithm was originally designed. For example, such a property would be desirable in computer aided geometric design.

As is shown in Theorem 2.3 a modification of the choice of slopes gives an estimate like (2.1) on all of [a,b] and a reduction in the number of knots of nearly 50%. On the other hand the preservation of local monotonicity and convexity is not guaranteed except in cases such as $f'(x) > 0$, $f''(x) > 0$ on [a,b] and for h sufficiently small.

Numerical calculations have been carried out which verify these results and will be presented elsewhere. We also point out that Richardson type extrapolation is possible for improving some of the estimates, but this also will be presented elsewhere.

REFERENCES

1. McAllister, D. F., and Roulier, J. A., An Algorithm for Computing a Shape Preserving Osculatory Quadratic Spline, Trans. Mathematical Software (submitted).

2. Meinardus, G., "Approximation of Functions: Theory and Numerical Methods," Springer-Verlag, New York, (1967), p. 78.

ON CONSTANTS OF APPROXIMATION IN THE THEOREMS OF JACKSON AND JACKSON-TIMAN

P.O. Runck
H.F. Sinwel [1]

Institut für Mathematik
Johannes Kepler Universität
Linz, Austria

I. INTRODUCTION

Let $E_n(f)$ be the distance from $f \in C[-1,1]$ to the best approximation polynomial of degree $\leq n$ in the uniform norm and $E_n^*(g)$ the distance from $g \in C_{2\pi}$ to the best trigonometric approximation polynomial. Then we have the following two Jackson theorems:

[1] Supported by the Austrian "Fonds zur Förderung der wissenschaftlichen Forschung".

Copyright © 1980 by Academic Press, Inc.
All rights of reproduction in any form reserved.
ISBN: 0-12-171050-5

Theorem A: For all $r \in \mathbb{N}$, $f \in C^r[-1,1]$ and $n \geq r-1$ we have the inequality

$$E_n(f) \leq \left(\frac{\pi}{2}\right)^r \frac{\|f^{(r)}\|}{(n+1)n\ldots(n-r+2)}$$

and (due to Favard, Achieser, Krein 1937)

Theorem B: For all $r \in \mathbb{N}$, $g \in C_{2\pi}^r$ and $n \in \mathbb{N}_o$ we have

$$E_n^*(g) \leq K_r \frac{\|g^{(r)}\|}{(n+1)^r}$$

with $\frac{\pi^2}{8} = K_2 < K_4 < \ldots < \frac{4}{\pi} < \ldots < K_3 < K_1 = \frac{\pi}{2}$

With $W_r := \{f \in C[-1,1] \mid f^{(r-1)}$ abs.const., $|f^{(r)}| \leq 1$ a.e.$\}$ and $E_n(W_r) := \sup_{f \in W_r} E_n(f)$ we can find a Corollary of theorem A:

Theorem A': For all $r \in \mathbb{N}$ and $n \geq r-1$ we have

$$E_n(W_r) \leq \left(\frac{\pi}{2}\right)^r \frac{1}{(n+1)n\ldots(n-r+2)}$$

Bernstein has proved in 1947 the asymptotic formula

Theorem C: $\lim_{n\to\infty} n^r E_n(W_r) = K_r$.

II. THE THEOREM OF JACKSON

In his "Diplomarbeit" H.F. Sinwel succeeded in refining Theorem A:

Theorem 1: For all $r \in \mathbb{N}$, $f \in C^r[-1,1]$ and $n \geq r-1$ we have

$$E_n(f) \leq K_r \frac{\|f^{(r)}\|}{(n+1)n\ldots(n-r+2)}$$

or

Theorem 1': For all $r \in \mathbb{N}$ and $n \geq r-1$ we have

$$E_n(W_r) \leq K_r \frac{1}{(n+1)n\ldots(n-r+2)}$$

In the periodic case the given upper bound $K_r.(n+1)^{-r}$ is the best possible for all r and n, but in theorem 1 K_r is only the best possible constant in the limit for $n \to \infty$.

Now we shall give some steps of the proof of theorem 1'. The results are obtained by going back to the following problem of periodic approximation:

If $f \in W_r$ then for $\tilde{f}(x) := f(x) - \sum_{k=0}^{r-1} \frac{1}{k!} f^{(k)}(0)\, x^k$ we have $\tilde{f}(x) = \frac{1}{(r-1)!} \int_0^x f^{(r)}(u)(x-u)^{r-1}du.$

For the 2π-periodic function $g := \tilde{f} \circ \sin$ we have $g(t) = \frac{1}{(r-1)!} \int_0^{\sin t} f^{(r)}(u)(\sin t - u)^{r-1}du$ and $E_n(f) = E_n(\tilde{f}) = E_n^*(\tilde{f} \circ \sin) = E_n^*(g).$

Now we have to solve the best approximation problem for g:

Case r = 1: $g(t) = \int_0^{\sin t} f'(u)du \Longrightarrow g'(t) = f'(\sin t)\cos t$

$\Longrightarrow \|g'\| \leq 1 \Longrightarrow E_n(f) = E_n^*(g) \leq \frac{K_1}{n+1} \quad (n \in \mathbb{N}_0)$

Case r = 2: $g(t) = \int_0^{\sin t} f''(u)(\sin t - u)du \Longrightarrow$

$\Longrightarrow g''(t) = f''(\sin t).\cos^2 t - \int_0^{\sin t} f''(u)du.\sin t \Longrightarrow$

$$\Longrightarrow \|g''\| \leq \|\cos^2 t + \sin^2 t\| = 1 \Longrightarrow$$

$$\Longrightarrow E_n(f) = E_n^*(g) \leq \frac{K_2}{(n+1)^2} < \frac{K_2}{(n+1)n} \quad (n \in \mathbb{N})$$

<u>Case r ≧ 3</u>: For all $j \in \mathbb{N}_0$ we define

$$D_{r+j}(t) := \frac{1}{\pi} \sum_{k=1}^{\infty} k^{-r-j} \cos(kt - \frac{r+j}{2}\pi)$$

$$g_j(t) := \frac{1}{(r-1)!} \int_0^{2\pi} D_{r+j}(t-u) f^{(r)}(\sin u)\cos u . \{(\frac{d}{du})^{r+j-1}(\sin u - v)^{r-1}|_{v=\sin u}\}du$$

$$\tilde{g}_j(t) := \frac{1}{(r-1)!} \int_0^{2\pi} D_{r+j}(t-u) . (\int_0^{\sin u} f^{(r)}(\sin v)(\frac{d}{du})^{r+j}(\sin u - v)^{r-1} dv)du$$

$$K_r := \frac{4}{\pi} \sum_{k=0}^{\infty} (-1)^{k(r+1)} (2k+1)^{-r-1}$$

$$B_{r,j} := \frac{1}{(r-1)!} \|(\frac{d}{dt})^{r+j-1}(\sin t - u)^{r-1}|_{u=\sin t}\|$$

Then we get by induction for all $s \in \mathbb{N}_0$

$$g = \sum_{j=0}^{s} g_j + \tilde{g}_s + \frac{1}{2\pi}\int_0^{2\pi} g \quad \text{and therefore}$$

$$E_n(f) = E_n^*(g) \leq \sum_{j=0}^{s} E_n^*(g_j) + E_n^*(\tilde{g}_s)$$

Using theorem B we get $E_n^*(g_j) \leq \frac{K_{r+j}}{(n+1)^{r+j}} B_{r,j}$

and $E_n^*(\tilde{g}_s) \leq \frac{K_{r+s}}{(n+1)^{r+s}} \frac{(r-1)^{r+s}}{r!} \xrightarrow[s\to\infty]{} 0$

Therefore we obtain $E_n(f) \leq \sum_{j=0}^{\infty} \frac{K_{r+j}}{(n+1)^{r+j}} B_{r,j}$ and using the inequality $B_{r,j} \leq \sum_{k=0}^{j} (r-1)^k B_{r-1,j-k}$ we get

$$E_n(f) \leq \frac{K_r}{(n+1)n\ldots(n-r+2)}$$

For details of the proof see (4),(6).

<u>Remark</u>: If $n = r-1$, we have $E_{r-1}(W_r) = 2^{1-r} \cdot \frac{1}{r!}$ (see (2)), while theorem 1' gives only $E_{r-1}(W_r) < K_r \cdot \frac{1}{r!}$.

III. THE THEOREM OF JACKSON AND TIMAN

For the Jackson-Timan-Theorem we can obtain very good approximation constants. In (5) we give the proof of the following

Theorem 2: For all $r \in \mathbb{N}$, $f \in W_r$ and $n > 2r$ there exists a polynomial p of degree less than n, so that for all $x \in [-1,1]$

$$|f(x) - p(x)| < \frac{K_r}{(n-2)(n-4)\dots(n-2r)} \left(\sqrt{1-x^2} + \frac{2r}{n}|x|\right)^r$$

For the class W_1 we can obtain

Theorem 3: For all $f \in W_1$ and $n > 1$ there exists a polynomial p of degree less than n, so that for all $x \in [-1,1]$

$$|f(x) - p(x)| \leq \left(\tan \frac{\pi}{2n}\right).\left(\sqrt{1-x^2} + \frac{3}{n}|x|\right)$$

The term $\tan \frac{\pi}{2n}$ is the best possible because for all $n > 1$ there exists a function $f \in W_1$, so that for all polynomials p of degree less than n there exists an $x \in]-1,1[$, so that

$$|f(x) - p(x)| \geq \left(\tan \frac{\pi}{2n}\right)\sqrt{1-x^2}$$

The idea of the proof is as follows:
We set $g := f \circ \cos - \frac{1}{2\pi}\int_0^{2\pi} f \circ \cos$ and get

$$g(t) = -\sin t \int_0^{2\pi} D_1(t-u)\cos(t-u).f'(\cos u)du +$$
$$+ \cos t \int_0^{2\pi} D_1(t-u)\sin(t-u).f'(\cos u)du$$

We construct the approximation polynomial by replacing the kernels $D_1\cos$ and $D_1\sin$ by suitable trigonometric polynomials.

REFERENCES

1. Cheney, E.W., Introduction to Approximation Theory, Mc Graw-Hill, New York (1966).
2. Fisher, S.D., J. Approx. Th. 21, p. 43-59 (1977).
3. Lorentz, G.G., Approximation of Functions, Holt, Rinehart and Winston, New York (1966).
4. Sinwel, H.F., Konstanten im Satz von Jackson, Institutsbericht Nr. 156, Universität Linz (1980).
5. Sinwel, H.F., Konstanten im Satz von Jackson-Timan, Institutsbericht Nr. 157, Universität Linz (1980).
6. Sinwel, H.F., Uniform approximation of differentiable functions by algebraic polynomials, submitted to J. Approx. Th.

INCOMPLETE POLYNOMIALS: AN ELECTROSTATICS APPROACH

E. B. Saff[1]

Department of Mathematics
University of South Florida
Tampa, Florida

J. L. Ullman[2]

Department of Mathematics
University of Michigan
Ann Arbor, Michigan

R. S. Varga[3]

Department of Mathematics
Kent State University
Kent, Ohio

Respectfully dedicated to Professor G. G. Lorentz, whose pathfinding results and whose open problems and conjectures in mathematics have greatly inspired us all.

I. INTRODUCTION

In 1976, G. G. Lorentz [4] introduced the study of certain constrained polynomials which are referred to as <u>incomplete polynomials</u>. By an <u>incomplete polynomial of type</u> θ, $0 < \theta < 1$, we mean any real or complex polynomial (of any degree) which can be written in the form

$$P(x) = \sum_{k=s}^{n} a_k x^k \text{ , where } s \geq \theta n \text{ , } s > 0. \tag{1.1}$$

---

[1]Research supported in part by AFOSR.

[2]Research supported in part by NSF MCS-77-27117.

[3]Research supported in part by AFOSR, and by the Dept. of Energy.

Copyright © 1980 by Academic Press, Inc.
All rights of reproduction in any form reserved.
ISBN: 0-12-171050-5

For the most part, research interest has focused upon the behavior, relative to the interval [0, 1], of such polynomials $P(x)$ for θ fixed. For example, on denoting the supremum norm over a set $B \subset \mathbb{C}$ by

$$\|g\|_B := \sup\{|g(z)| : z \in B\},$$

and letting

$$I_\theta := \{P : P \text{ is an incomplete polynomial of type } \theta\}, \quad 0 < \theta < 1, \tag{1.2}$$

we have the following fundamental property of incomplete polynomials:

<u>Theorem 1.1</u>. (Kemperman and Lorentz [2], Saff and Varga [6], [7]). <u>If</u> $P \in I_\theta$, $P \not\equiv 0$, <u>and if</u> $\xi \in [0, 1]$ <u>is any point for which</u> $|P(\xi)| = \|P\|_{[0,1]}$, <u>then</u> $\xi \geq \theta^2$.

It is moreover shown in [6] that $\xi > \theta^2$, and in [7] that this lower bound θ^2 is <u>sharp</u> in the sense that if $\xi(P)$ is the smallest such ξ in $[0, 1]$ with $|P(\xi)| = \|P\|_{[0,1]}$ for each $P \not\equiv 0$ in I_θ, then

$$\inf\{\xi(P) : P \not\equiv 0 \text{ in } I_\theta\} = \theta^2.$$

In this regard, letting π_r denote the set of all real polynomials of degree at most r (with $\pi_{-1} := \{0\}$), consider the following extremal problem. Given any pair (s, m) of nonnegative integers, set

$$\begin{aligned} \mathcal{E}_{s,m} &:= \min\{\|x^s(x^m - g_{m-1}(x))\|_{[0,1]} : g_{m-1} \in \pi_{m-1}\} \\ &= \|x^s(x^m - \hat{g}_{m-1}(x))\|_{[0,1]}. \end{aligned} \tag{1.3}$$

Then, the incomplete polynomial

$$\mathcal{T}_{s,m}(x) := x^s(x^m - \hat{g}_{m-1}(x))/\mathcal{E}_{s,m}, \tag{1.4}$$

which is of type s/n with $n := s + m$, is called the <u>constrained Chebyshev polynomial of degree</u> n, having a constrained zero of order s at $x = 0$.

Several properties of these constrained Chebyshev polynomials were obtained in [7]. In particular, we remarked there that $\mathcal{T}_{s,m}(x)$ attains its maximum absolute value on [0, 1] in precisely m + 1 points, with necessarily alternating signs. Hence, if $s/(s+m) \geq \theta$, then, by Theorem 1.1, these alternation points must lie in $(\theta^2, 1]$, and consequently, $(\theta^2, 1)$ contains all nontrivial (i.e., nonzero) zeros of $\mathcal{T}_{s,m}(x)$.

One purpose of this note is to obtain (cf. Theorem 3.6) the precise asymptotic distribution of these zeros for any sequence $\{\mathcal{T}_{s_i,m_i}(x)\}_{i=1}^{\infty}$ for which $s_i/n_i \to \theta$ and $n_i \to \infty$ (where $n_i := s_i + m_i$). Our approach, which is to study the electrostatics analogue of the problem, also provides a streamlined method for proving several of the fundamental properties of incomplete polynomials.

The outline of this paper is as follows. In Section II, we discuss a generalization of the incomplete polynomials of (1.1), and give some known results. In Section III, we state and prove our main results on the asymptotic distribution of zeros, and in Section IV, we mention two related problems.

II. POLYNOMIALS VANISHING AT BOTH ENDPOINTS

Note that an incomplete polynomial P(x) in (1.1) has a zero of order at least θn at the left endpoint of the interval [0, 1]. In [3], Lachance, Saff, and Varga studied the more general possibility of polynomials vanishing at <u>both</u> endpoints of an interval. For reasons that will be subsequently clear, we take this interval to be [-1, 1]. Then, by an <u>incomplete polynomial of type</u> (θ_1, θ_2), where $0 \leq \theta_1$, $0 \leq \theta_2$, and $0 < \theta_1 + \theta_2 < 1$, we mean any real or complex polynomial which can be written in the form

$$p(t) = (t-1)^{s_1}(t+1)^{s_2} \sum_{k=0}^{n-s_1-s_2} \alpha_k t^k,$$

where (2.1)

$$s_1 \geq \theta_1 n, \quad s_2 \geq \theta_2 n, \quad s_1+s_2 > 0.$$

Furthermore, we set

$$I_{\theta_1,\theta_2} := \{p : p \text{ is an incomplete polynomial of type } (\theta_1,\theta_2)\}. \tag{2.2}$$

We remark that the collection I_{θ_1,θ_2} contains polynomials of arbitrarily large degree, and is closed under ordinary multiplication, but not under addition.

With the above notation, the generalization of Theorem 1.1, relative to the interval $[-1, 1]$, is as follows.

Theorem 2.1. (Lachance, Saff, and Varga [3]). _If_ $p \in I_{\theta_1,\theta_2}$, $p \not\equiv 0$, _and if_ $\xi \in [-1, 1]$ _is any point for which_ $|p(\xi)| = \|p\|_{[-1,1]}$, _then_

$$a(\theta_1, \theta_2) \leq \xi \leq b(\theta_1, \theta_2), \tag{2.3}$$

where, with $\sigma := \theta_2 + \theta_1$, $\delta := \theta_2 - \theta_1$, a _and_ b _are given by_

$$\left.\begin{aligned} a = a(\theta_1, \theta_2) &:= \sigma\delta - \sqrt{(1-\sigma^2)(1-\delta^2)} \\ b = b(\theta_1, \theta_2) &:= \sigma\delta + \sqrt{(1-\sigma^2)(1-\delta^2)}\,. \end{aligned}\right\} \tag{2.4}$$

Note that when $\theta_1 = 0$ and $\theta_2 = \theta$, we find $a = 2\theta^2 - 1$, and $b = 1$, so that the interval $2\theta^2 - 1 \leq t \leq 1$ of (2.3), after the transformation $x = (t+1)/2 : [-1, 1] \rightarrow [0, 1]$, becomes $\theta^2 \leq x \leq 1$, which agrees with Theorem 1.1. Unlike Theorem 1.1, however, the sharpness of both endpoints a and b for the general case of Theorem 2.1 has not been previously established. In the next section, we prove that the interval in (2.3) _is_, in general, best possible. For this purpose, we study the properties of the <u>two-endpoint constrained Chebyshev polynomials</u> $T_{s_1,s_2,m}(t)$ which are defined as follows (cf. [3]).

Let (s_1, s_2, m) be any triple of nonnegative integers, and set

$$E_{s_1,s_2,m} := \min\{\|(t-1)^{s_1}(t+1)^{s_2}(t^m - h_{m-1}(t))\|_{[-1,+1]} : h_{m-1} \in \pi_{m-1}\}$$
$$= \|(t-1)^{s_1}(t+1)^{s_2}(t^m - \hat{h}_{m-1}(t))\|_{[-1,+1]}, \tag{2.5}$$

and

$$T_{s_1,s_2,m}(t) := (t-1)^{s_1}(t+1)^{s_2}\{t^m - \hat{h}_{m-1}(t)\}/E_{s_1,s_2,m}. \tag{2.6}$$

Note that $T_{s_1,s_2,m}$ is an incomplete polynomial of type $(s_1/n, s_2/n)$, where $n := s_1 + s_2 + m$ is its total degree.

III. MAIN RESULTS

Our approach to describing the behavior of incomplete polynomials of type (θ_1, θ_2), relative to $[-1, 1]$, is to consider the following

<u>Electrostatics Problem</u>. Let $0 \le \theta_1$, $0 \le \theta_2$, satisfy $0 < \theta_1 + \theta_2 < 1$. Suppose that on the interval $[-1, 1]$, a fixed charge of amount θ_1 is placed at $t = 1$, a fixed charge of amount θ_2 is placed at $t = -1$, and a continuous charge of amount $1 - \theta_1 - \theta_2$ is placed on $[-1, 1]$, allowing it to reach equilibrium, the only constraint being that the charges remain confined to the interval $[-1, 1]$. Here and throughout, the logarithmic potential and its corresponding force is assumed. Then, the problem is to describe the distribution of the continuous charge.

This is answered in

<u>Theorem 3.1</u>. <u>For the electrostatics problem described above, the continuous charge of amount</u> $1 - \theta_1 - \theta_2$ <u>lies entirely in the interval</u>

$a(\theta_1, \theta_2) \le t \le b(\theta_1, \theta_2)$, _where_ a _and_ b _are defined in_ (2.4). _Moreover, the point density of the charge_ $1 - \theta_1 - \theta_2$ _on this interval is given by_

$$\frac{\sqrt{(t-a)(b-t)}}{\pi(1-t^2)}, \quad a \le t \le b. \tag{3.1}$$

We shall prove Theorem 3.1 by considering the limit of the corresponding problem for discrete point charges.

Lemma 3.2. For each sufficiently large integer n, let $s_1(n)$, $s_2(n)$ be positive integers such that $s_1(n) + s_2(n) < n$ and such that

$$\lim_{n\to\infty} \frac{s_1(n)}{n} = \theta_1, \quad \lim_{n\to\infty} \frac{s_2(n)}{n} = \theta_2, \qquad \theta_1 + \theta_2 < 1. \tag{3.2}$$

Suppose that a charge of amount $s_1(n)/n$ is placed at $t = 1$, and a charge of $s_2(n)/n$ is placed at $t = -1$, and let $m(n) := n - s_1(n) - s_2(n)$ point charges, each of amount $1/n$, be placed on $(-1, 1)$ so that equilibrium is reached. Let $-1 < t_{n,1} < t_{n,2} < \cdots < t_{n,m(n)} < 1$ denote the location of these point charges, and set

$$f_{m(n)}(z) := \prod_{k=1}^{m(n)} (z - t_{n,k}). \tag{3.3}$$

Then, uniformly on each closed set in $\mathbb{C}\setminus[a(\theta_1, \theta_2), b(\theta_1, \theta_2)]$, we have

$$\lim_{n\to\infty} \frac{1}{m(n)} \frac{f'_{m(n)}(z)}{f_{m(n)}(z)} = \psi(z), \tag{3.4}$$

where

$$\psi(z) := \frac{\delta - \sigma z + \sqrt{(z-a)(z-b)}}{(1-\sigma)(z^2-1)}, \tag{3.5}$$

the quantities σ, δ, a, and b being defined in (2.4).

We remark that, in (3.5), $\sqrt{(z-a)(z-b)}$ denotes the branch with cut $[a, b]$ that behaves like $+z$ near ∞.

Proof. From the equilibrium equations

$$\frac{1}{n}\left[\sum_{\substack{j=1\\ j\neq k}}^{m(n)} \frac{1}{t_{n,k}-t_{n,j}} + \frac{s_1(n)}{t_{n,k}-1} + \frac{s_2(n)}{t_{n,k}+1}\right] = 0, \quad k = 1, \cdots, m(n),$$

it follows, as in Szegö [10, p. 141], that $f_{m(n)}(z)$ satisfies the differential equation

$$(z^2-1)f''_{m(n)} + 2\{[s_1(n) + s_2(n)]z + s_1(n) - s_2(n)\}f'_{m(n)}$$
$$= \{m(n)[m(n)-1]+2m(n)[s_1(n) + s_2(n)]\}f_{m(n)}. \tag{3.6}$$

In fact, $f_{m(n)}(z)$ is just a constant multiple of the Jacobi polynomial $P^{(\alpha,\beta)}_{m(n)}(z)$, where $\alpha := 2s_1(n) - 1$ and $\beta := 2s_2(n) - 1$ (cf. [10, p. 140]). Now, with the assumption of (3.2), it is proved in [5] that the sequence of these Jacobi polynomials, as $n \to \infty$, has no limit point of zeros exterior to $[a, b]$. Hence, the functions

$$\psi_n(z) := \frac{1}{m(n)} \frac{f'_{m(n)}(z)}{f_{m(n)}(z)} \tag{3.7}$$

form a normal family of analytic functions in the complement of $[a, b]$, relative to the extended complex plane $\mathbb{C}^*$.

On dividing (3.6) by $f_{m(n)}$ and using the easily verified relations

$$\frac{f'_{m(n)}(z)}{f_{m(n)}(z)} = m(n)\psi_n(z), \quad \frac{f''_{m(n)}(z)}{f_{m(n)}(z)} = m(n)\psi'_n(z) + m^2(n)\psi_n^2(z),$$

we find that $\psi_n(z)$ of (3.7) satisfies

$$(z^2 - 1)[m(n)\psi'_n + m^2(n)\psi_n^2] + 2\{[s_1 + s_2]z + s_1 - s_2\}m(n)\psi_n$$
$$= m(n)[m(n) - 1] + 2m(n)(s_1 + s_2), \tag{3.8}$$

where, for convenience, we have written s_1 for $s_1(n)$, s_2 for $s_2(n)$. Now, let $\psi(z)$ be any limit function (as $n \to \infty$) in $\mathbb{C}^*\backslash[a, b]$ of any subsequence of the normal family $\{\psi_n(z)\}$. Since $\psi'(z)$ is the limit of the corresponding sequence of derivatives, it follows from (3.8), on dividing by n^2 and using the assumptions (3.2), that $\psi(z)$ satisfies the quadratic equation

$$\psi^2(z) - \frac{2(\delta-\sigma z)}{(1-\sigma)(z^2-1)}\psi(z) - \frac{1+\sigma}{(1-\sigma)(z^2-1)} = 0, \tag{3.9}$$

where $\sigma = \theta_2 + \theta_1$, $\delta = \theta_2 - \theta_1$. Hence,

$$\psi(z) = \frac{\delta - \sigma z \pm \sqrt{(z-a)(z-b)}}{(1-\sigma)(z^2-1)}. \tag{3.10}$$

Since $z\psi_n(z) \to 1$ as $z \to \infty$ for each n, then necessarily also $z\psi(z) \to 1$ as $z \to \infty$, which implies that the plus sign must be taken for the radical in (3.10). Hence, $\psi(z)$ is given by (3.5).

Finally, as $\psi(z)$ represents an arbitrary limit function of any subsequence of $\{\psi_n(z)\}$ in $\mathbb{C}^*\setminus[a, b]$, the conclusion (3.4) of the lemma follows. ∎

Concerning the limiting distribution of the point charges in Lemma 3.2, we prove

Lemma 3.3. Let $q_m(z) = \prod_{k=1}^{m}(z - \tau_{m,k})$ be a sequence of polynomials, each having all its zeros on $[-1, 1]$, and suppose that

$$\lim_{m\to\infty} \frac{1}{m}\frac{q_m'(z)}{q_m(z)} = \psi(z), \text{ for } z \in \mathbb{C}\setminus[-1, 1], \tag{3.11}$$

where $\psi(z)$ is defined in (3.5) with $0 \leq \theta_1 + \theta_2 < 1$. For each m, let ν_m denote the atomic measure (on the Borel sets of $[-1, 1]$) having mass $1/m$ each point $\tau_{m,k}$, $k = 1, 2, \cdots, m$. Then, there exists a measure ν^* such that $\nu_m \to \nu^*$ weakly as $m \to \infty$, where the support of the measure ν^* is exactly $[a(\theta_1, \theta_2), b(\theta_1, \theta_2)]$ (cf. (2.4)), and where

$$\nu^*\{(\alpha, \beta)\} = \frac{1}{\pi(1-\theta_1-\theta_2)}\int_\alpha^\beta \frac{\sqrt{(t-a)(b-t)}}{1-t^2}\,dt, \ \forall(\alpha,\beta)\subset[a,b]. \tag{3.12}$$

Proof. We first note that

$$\frac{1}{m}\frac{q_m'(z)}{q_m(z)} = \int_{-1}^{+1}\frac{d\nu_m(t)}{z-t}, \ \forall\, z \notin [-1, 1]. \tag{3.13}$$

Since $\nu_m\{[-1, 1]\} = 1$ for all m, it follows from Helly's theorem (cf. [9], [12]) that there exists a subsequence ν_{m_i} and a measure ν^* such that $\nu_{m_i} \to \nu^*$ weakly. Hence,

$$\lim_{i\to\infty} \int_{-1}^{+1} \frac{d\nu_{m_i}(t)}{z - t} = \int_{-1}^{+1} \frac{d\nu^*(t)}{z - t}, \ \forall\ z \notin [-1, 1].$$

But, from (3.13) and (3.11), this means that

$$\int_{-1}^{+1} \frac{d\nu^*(t)}{z - t} = \psi(z), \ \forall\ z \notin [-1, 1]. \tag{3.14}$$

Now, by applying the Stieltjes inversion formula (cf. [11, p. 250]) to (3.14), we have for each interval $(\alpha, \beta) \subset [-1, 1]$ that

$$\nu^*\{(\alpha,\beta)\} + \frac{\nu^*\{\alpha\}}{2} + \frac{\nu^*\{\beta\}}{2} = \lim_{y\to 0^+} \frac{-1}{\pi} \int_{\alpha}^{\beta} \operatorname{Im} \psi(t+iy)dt. \tag{3.15}$$

From (3.5), we see on the other hand that

$$\hat{\psi}(t) := \lim_{y\to 0^+} \operatorname{Im} \psi(t+iy) = \begin{cases} 0, & \text{for } t \notin [a, b] \\ \dfrac{\sqrt{(t-a)(b-t)}}{(1-\sigma)(t^2-1)}, & \text{for } t \in (a, b). \end{cases}$$

Thus, (3.15) implies that the support of ν^* is exactly the interval $[a, b]$, and since $\hat{\psi}(t)$ is continuous on $(-1, 1)$, ν^* is absolutely continuous with respect to Lebesgue measure. Consequently, (3.15) yields

$$\nu^*\{(\alpha, \beta)\} = \frac{1}{\pi(1-\sigma)} \int_{\alpha}^{\beta} \frac{\sqrt{(t-a)(b-t)}}{1 - t^2} dt, \ \forall\ (\alpha, \beta) \subset [a, b].$$

Since this is true for every weak limit ν^*, Lemma 3.3 is proved. ∎

Proof of Theorem 3.1. Combining Lemmas 3.2 and 3.3, we see that the limit, as $n \to \infty$, of the discrete electrostatics problems described in Lemma 3.2, has a charge of θ_1 at $t = 1$, a charge of θ_2 at $t = -1$, and a charge of $1 - \theta_1 - \theta_2$ distributed continuously over the interval

$[a(\theta_1, \theta_2), b(\theta_1, \theta_2)]$, with density (cf. (3.12)) given by

$$(1 - \theta_1 - \theta_2)d\nu^*(t) = \frac{\sqrt{(t-a)(b-t)}}{\pi(1 - t^2)}\,dt\ ,\quad t \in [a, b]. \quad \blacksquare$$

We remark that the function

$$V(z;\ \theta_1,\ \theta_2) := \theta_1 \ln|z-1| + \theta_2 \ln|z+1| + (1 - \theta_1 - \theta_2)\int_a^b \ln|z-t|\,d\nu^*(t) \tag{3.16}$$

gives the potential for the electrostatics problem of Theorem 3.1. In fact, it can be shown that

$$V(z;\ \theta_1,\ \theta_2) \equiv \ln \Delta + \ln G(z;\ \theta_1,\ \theta_2), \tag{3.17}$$

where

$$\Delta := \frac{1}{2}\sqrt{(1+\sigma)^{1+\sigma}(1-\sigma)^{1-\sigma}(1+\delta)^{1+\delta}(1-\delta)^{1-\delta}}, \tag{3.18}$$

and (cf. [3, p. 425])

$$G(z;\ \theta_1,\ \theta_2) := |\varphi(z)|\left|\frac{\varphi(z)-\varphi(1)}{\varphi(1)\varphi(z)-1}\right|^{\theta_1}\left|\frac{\varphi(z)-\varphi(-1)}{\varphi(-1)\varphi(z)-1}\right|^{\theta_2}, \tag{3.19}$$

where

$$w = \varphi(z) := (\sqrt{z-a} + \sqrt{z-b})/(\sqrt{z-a} - \sqrt{z-b})$$

maps $\mathbb{C}^*\setminus[a(\theta_1, \theta_2), b(\theta_1, \theta_2)]$ onto the exterior of the unit circle $|w| = 1$. To prove (3.17), one need only verify that the difference $V - (\ln \Delta + \ln G)$ is harmonic in $\mathbb{C}^*\setminus[a, b]$, equal to zero at ∞, and approaches a constant as z approaches the boundary segment $[a, b]$. Thus, this harmonic function is identically zero in $\mathbb{C}^*\setminus[a, b]$.

Now in [3, p. 434], it is proved that any sequence of two-endpoint constrained Chebyshev polynomials (cf. (2.6)) $T_{s_1(n),s_2(n),m(n)}(t)$ of respective degrees n for which

$$s_1(n)/n \to \theta_1,\ s_2(n)/n \to \theta_2,\ n := s_1(n)+s_2(n)+m(n) \to \infty, \tag{3.20}$$

satisfies

$$\lim_{n\to\infty} |T_{s_1(n),s_2(n),m(n)}(z)|^{1/n} = G(z;\ \theta_1,\ \theta_2),\ z \in \mathbb{C}\backslash[a,\ b]. \tag{3.21}$$

Hence, we claim that the sequence of polynomials

$$q_{m(n)}(z) := T_{s_1(n),s_2(n),m(n)}(z)/(z-1)^{s_1(n)}(z+1)^{s_2(n)},\quad n = 1, 2, \cdots, \tag{3.22}$$

satisfies the hypotheses of Lemma 3.3. Indeed, on writing $T_n(z)$ for $T_{s_1(n),s_2(n),m(n)}(z)$ and $V(z)$ for $V(z;\ \theta_1,\ \theta_2)$ for simplicity, (3.17) and (3.21) imply that

$$\frac{1}{n}\frac{T_n'(z)}{T_n(z)} \to V_x(z) - i\,V_y(z),\quad z = x+iy \in \mathbb{C}\backslash[a,\ b].$$

Hence, from (3.16) and (3.14), it follows that

$$\frac{1}{m(n)}\frac{q_{m(n)}'(z)}{q_{m(n)}(z)} \to \frac{1}{(1-\theta_1-\theta_2)}\left\{V_x(z) - i\,V_y(z) - \frac{\theta_1}{z-1} - \frac{\theta_2}{z+1}\right\} = \int_a^b \frac{d\nu^*(t)}{z-t} = \psi(z)$$

for all $z = x+iy \in \mathbb{C}\backslash[a,\ b]$. Consequently, we have

Theorem 3.4. <u>Let</u> $\hat{p}_n(t) := T_{s_1(n),s_2(n),m(n)}(t)$ <u>be any sequence of two-endpoint constrained Chebyshev polynomials of respective degrees</u> n <u>for which</u> (3.20) <u>holds</u>. <u>Then, the zeros of the sequence</u> $\hat{p}_n(t)$, <u>other than those at</u> $t = \pm 1$, <u>have no limit points exterior to</u> $[a,\ b]$, <u>and are dense in</u> $[a,\ b]$, <u>where</u> a <u>and</u> b <u>are defined in</u> (2.4). <u>Furthermore, if</u> $N_n(\alpha,\ \beta)$ <u>denotes the total number of zeros of</u> $\hat{p}_n(t)$ <u>in any interval</u> $(\alpha,\beta) \subset [a,b]$, <u>then</u>

$$\lim_{n\to\infty} \frac{N_n(\alpha,\ \beta)}{n} = \frac{1}{\pi}\int_\alpha^\beta \frac{\sqrt{(t-a)(b-t)}}{1-t^2}\,dt. \tag{3.23}$$

Theorem 3.4 immediately implies the sharpness of Theorem 2.1. Indeed, if we consider any sequence of constrained Chebyshev polynomials from I_{θ_1,θ_2} for which (3.20) holds, then their respective alternation points all lie on $[a,\ b]$ and are interlaced by their respective zeros. As these zeros are, from Theorem 3.4, dense in $[a,\ b]$, we have

Corollary 3.5. *The interval* $[a, b]$ *of Theorem 2.1 is sharp in the sense that, for any pair* (θ_1, θ_2) *with* $0 \le \theta_1$, $0 \le \theta_2$, *and* $0 < \theta_1 + \theta_2 < 1$, *there exists a sequence* $\{p_n(t)\} \subset I_{\theta_1,\theta_2}$ *of nonconstant polynomials and sequences of points* $\xi_{n,1}$, $\xi_{n,2}$ *in* $[-1, 1]$ *such that* $|p_n(\xi_{n,1})| = |p_n(\xi_{n,2})| = \|p_n\|_{[-1,1]}$ *with* $\lim_{n\to\infty} \xi_{n,1} = a$; $\lim_{n\to\infty} \xi_{n,2} = b$.

We conclude this section by describing the distribution of zeros for the (one-endpoint) constrained Chebyshev polynomials of (1.4). This is just a special case of Theorem 3.4 when $\theta_1 = 0$, $\theta_2 = \theta$, and the interval $[-1, 1]$ is mapped to $[0, 1]$.

Theorem 3.6. *Let* $\hat{P}_n(x) := \mathcal{T}_{s(n),m(n)}(x)$ *be any sequence of constrained Chebyshev polynomials* (cf. (1.4)) *of respective degrees* $n = s(n)+m(n)$ *for which* $s(n)/n \to \theta$ as $n \to \infty$, *where* $0 < \theta < 1$. *Then, the zeros of* $\hat{P}_n(x)$, *other than those at* $x = 0$, *have no limit point exterior to* $[\theta^2, 1]$, *and are dense in* $[\theta^2, 1]$. *Furthermore, if* $\mathfrak{N}_n(\alpha, \beta)$ *denotes the total number of zeros of* $\hat{P}_n(x)$ *in any interval* $(\alpha, \beta) \subset [\theta^2, 1]$, *then*

$$\lim_{n\to\infty} \frac{\mathfrak{N}_n(\alpha,\beta)}{n} = \frac{1}{\pi}\int_\alpha^\beta \frac{1}{x}\sqrt{\frac{x-\theta^2}{1-x}}\,dx. \tag{3.24}$$

IV. RELATED QUESTIONS

Concerning the possibility of uniform approximation of continuous functions on $[0, 1]$ by incomplete polynomials of type θ (cf. (1.1)), the following result is known.

Theorem 4.1. (Saff and Varga [8], v. Golitschek [1]). *Let* $0 < \theta < 1$ *be fixed, and let* $F \in C[0, 1]$ *with* $F \notin I_\theta$ (cf. (1.2)). *Then, a necessary and sufficient condition that* F *be the uniform limit on* $[0, 1]$ *of a sequence of incomplete polynomials of type* θ *is that* $F(x) = 0$ *for all* $0 \le x \le \theta^2$.

For the more general two-endpoint case, the characterization of the uniform limits on $[-1, 1]$ of incomplete polynomials of type (θ_1, θ_2) remains an open question. However, the results of our investigations suggest the following

Conjecture 4.2. Let $0 < \theta_1$, $0 < \theta_2$, with $0 < \theta_1 + \theta_2 < 1$, and let $F \in C[-1, 1]$ with $F \notin I_{\theta_1,\theta_2}$ (cf. (2.2)). Then, a necessary and sufficient condition that F be the uniform limit on $[-1, 1]$ of a sequence of incomplete polynomials of type (θ_1, θ_2) is that $F(t) = 0$ for all $t \in [-1, a(\theta_1, \theta_2)] \cup [b(\theta_1, \theta_2), 1]$, where a and b are given in (2.4).

Another related question concerns the behavior of incomplete polynomials that have a high-order zero at an interior point λ of $[0, 1]$, namely, the collection

$$I_\theta^{(\lambda)} := \{P\colon P(x) = \sum_{k=s}^{n} a_k (x-\lambda)^k,\ s \geq \theta n,\ s > 0\},\ 0 < \lambda < 1. \tag{4.1}$$

Here, the related electrostatics problem involves placing a fixed charge of amount θ at $x = \lambda$, and the remaining charge of amount $1 - \theta$ on $[0, 1]$, so that equilibrium is reached. It seems likely that the methods employed in this paper can be adapted to find the distribution of such charges, part of which will lie on $[0, x_1(\lambda)]$ and part on $[x_2(\lambda), 1]$, where $x_1(\lambda) < \lambda < x_2(\lambda)$. However, the analysis is more involved in that elliptic integrals arise. This generalization will be reserved for a later occasion.

REFERENCES

1. Golitschek, M. v., Approximation by incomplete polynomials, (to appear).

2. Kemperman, J. H. B. and Lorentz, G. G., Bounds for polynomials with applications, Nederl. Akad. Wetensch. Proc. Ser. A. 82(1979), 13-26.

3. Lachance, M, E. B. Saff, and R. S. Varga, Bounds for Incomplete Polynomials Vanishing at Both Endpoints of an Interval, in Constructive Approaches to Mathematical Models (C. V. Coffman, and G. J. Fix, eds.), Academic Press, New York, 1979, pp. 421-437.

4. Lorentz, G. G., Approximation by incomplete polynomials (problems and results), in Padé and Rational Approximation: Theory and Applications (E. B. Saff and R. S. Varga, eds.), Academic Press, New York, 1977, pp. 289-302.

5. Moak, D. S., E. B. Saff, and R. S. Varga, On the zeros of Jacobi polynomials $P_n^{(\alpha_n, \beta_n)}(x)$, Trans. Amer. Math. Soc. 249(1979), 159-162.

6. Saff, E. B. and Varga, R. S., The sharpness of Lorentz's theorem on incomplete polynomials, Trans. Amer. Math. Soc. 249(1979) 163-186.

7. Saff, E. B. and Varga, R. S., On incomplete polynomials, Numerische Methoden der Approximationstheorie, Band 4 (L. Collatz, G. Meinardus, and H. Werner, eds.), ISNM 42, Birkhäuser Verlag, Basel and Stuttgart, 1978, pp. 281-298.

8. Saff, E. B. and Varga, R. S., Uniform approximation by incomplete polynomials, Internat. J. Math. Math. Sci. 1(1978), 407-420.

9. Shohat, J. and Tamarkin , J., The Problem of Moments, Mathematical Surveys, No. 1, American Mathematical Society, New York, 1943.

10. Szegö, G., Orthogonal Polynomials, Colloquium Publication, Vol. XXIII, 4th Ed., Amer. Math. Soc., Providence, Rhode Island, 1975.

11. Wall, H. S., Analytic Theory of Continued Fractions, Van Nostrand, New York, 1948.

12. Widder, D. V., The Laplace Transform, Princeton University Press, Princeton, 1946.

ON BEST SIMULTANEOUS APPROXIMATION

B. N. Sahney

Department of Mathematics
University of Calgary
Calgary, Alberta

S. P. Singh

Department of Mathematics and Statistics
Memorial University of Newfoundland
St. John's, Newfoundland

In this paper, a few results on best simultaneous approximation have been proved under relaxed conditions. These results extend the work of Holland, Sahney and Tzimbalario. In the end, using fixed point theory, a result related to the distance between two sets has been given.

INTRODUCTION

Several mathematicians have studied the problem of simultaneous approximation. Dunham [8] and Diaz and McLaughlin [6, 7] considered simultaneous Chebyshev approximation of two real valued functions defined on [0, 1]. The problem of a best simultaneous approximation of two functions in normed linear spaces has been discussed by Phillips and Sahney [16] and others. Goel et. al. [10] and Brondsted [3] studied this problem in normed linear spaces.

Copyright © 1980 by Academic Press, Inc.
All rights of reproduction in any form reserved.
ISBN: 0-12-171050-5

Recently, Holland, Sahney and Tzimbalario [11] proved a few results in this area in a more general setting. Milman [14], Bosznay [2], and Mach [13] have also studied the problem of best simultaneous approximation. Dunham [9] recently proved further results in this field. Blatt [1] studied the problem of non-linear best simultaneous approximation.

The aim of this paper is to extend results given in [11].

We need the following terminologies:

Let C be a subset of a normed linear space X. Given any bounded subset F of X, define

$$d(F, C) = \inf_{x \in C} \sup_{f \in F} \|f - x\|.$$

An element $\underline{x}$ in C is said to be a best simultaneous approximation to F if

$$d(F, C) = \sup_{f \in F} \|f - \underline{x}\|.$$

A bounded subset C of a normed linear space X is said to be remotal if for each $x \in X$ there exists a point c in C farthest from x, i.e., for each $x \in X$, the set

$$F_C(x) = \{y \in C / \|x - y\| = \sup_{c \in C} \|x - c\|\} \neq \phi .$$

A compact subset of a normed linear space is remotal.

C is said to be remotal with respect to $B \subset X$ if $F_C(x) \neq \phi$ for each $x \in B$. C is uniquely remotal if each point of the space has a unique farthest point in C.

In [11], it is given that if C is a finite dimensional subspace of a strictly convex normed linear space X, then there exists one and only one best simultaneous approximation from the elements of C to any given compact subset F of X.

The following question naturally arises. Is the hypothesis of finite dimension necessary? We prove the following theorem where finite dimension condition is relaxed, but we take C as a compact subset of a normed linear space X. Convexity on C is not needed.

THEOREM 1. *Let X be a normed linear space, C a compact subset of X. Let $F \subset X$ be a uniquely remotal set with respect to C. Then there exists a unique best simultaneous approximation from the elements of C to F.*

PROOF. In [11], it has been shown that the function $g : C \to R$ defined by

$$g(x) = \sup_{y \in F} \|x - y\|$$

is continuous.

Since C is compact, g attains its infimum at $x_0 \in C$ say, i.e.,

$$g(x_0) = \sup_{y \in F} \|y - x_0\|.$$

Since F is a uniquely remotal set with respect to C, there exists a unique $y_0 \in F$ such that

$$d(F, C) = \|y_0 - x_0\|.$$

Thus the proof.

Now, we prove the following:

THEOREM 2. *Let* X *be a strictly convex normed linear space and* C *a reflexive subspace of* X. *Then there exists one and only one best simultaneous approximation from the elements of* C *to any set* F *that is remotal with respect to* C.

PROOF. Since F is remotal with respect to C, F is bounded, i.e.

$$\|f\| \leq M \quad \text{for all} \quad f \in F.$$

Take a ball $B(O, M) \subset C$. Then

$$\inf_{c \in C} \sup_{f \in F} \|c - f\| = \inf_{c \in B} \sup_{f \in F} \|c - f\|.$$

Since B is weakly compact and $g : C \to R$ given by $g(c) = \sup_{f \in F} \|c - f\|$, is a weakly lower semicontinuous function, attains its infinimum say at c_0. Thus

$$d(C, F) = \sup_{f \in F} \|c_0 - f\|.$$

Uniqueness is proved by using the strict convexity.

Suppose that c_1 and c_2, $c_1 \neq c_2$, are two best simultaneous approximations by elements of F, i.e.,

$$\sup_{f \in F} \|f - c_1\| = \sup_{f \in F} \|f - c_2\| = r, \quad \text{say}.$$

Then $\sup_{f \in F} \|f - \frac{c_1 + c_2}{2}\| = r$. [$g$ is convex, see [11]].

Since F is remotal with respect to C, there is a $f_0 \in F$ such that

$$\sup_{f \in F} \|f - \frac{c_1 + c_2}{2}\| = \|f_0 - \frac{c_1 + c_2}{2}\| = r,$$

it then follows that

$$\|f_0 - c_1\| = r \quad \text{and} \quad \|f_0 - c_2\| = r.$$

Since X is strictly convex, we get that $c_1 = c_2$.

Now, we study the distance between two sets. Let A and B be two subsets of X. The distance between A and B is

$$d(A, B) = \inf_{a \in A} \inf_{b \in B} \|a - b\|.$$

Cheney and Goldstein [5] proved the following:

Let A_1 and A_2 be two closed, convex sets in Hilbert space. Let P_i denote the proximity map for A_i. Any fixed point of $P_1 P_2$ is a point of A_1 closest to A_2, and conversely.

Pai [15] gave the following theorem for uniformly convex Banach spaces.

Let X be a uniformly convex Banach space and A, B be two closed, convex sets such that one of them is compact, then $\underline{a} \in A$ and $\underline{b} \in B$ exist such that

$$d(A, B) = \|\underline{a} - \underline{b}\|.$$

We prove the following for strictly convex Banach space.

THEOREM 3. *Let* X *be a strictly convex Banach space and* A *a closed, convex, locally compact subset of* X, *and let* B *be a compact, convex subset of* X. *Then there exist* $\underline{a} \in A$ *and* $\underline{b} \in B$ *such that*

$$d(A, B) = \|\underline{a} - \underline{b}\|.$$

(*The points* $\underline{a} \in A$ *and* $\underline{b} \in B$ *are also called proximal if* $d(A, B) = \|\underline{a} - \underline{b}\| = \inf_{a \in A, b \in B} \|a - b\|$).

PROOF. Let $P_A : B \to A$, and $P_B : A \to B$ be proximity maps. Then P_A is continuous [12] and so is P_B [4]; thus, the composition map $T = P_B P_A : B \to B$ is also continuous. By Schauder fixed point theorem T has a fixed point in B.

Let $T\underline{b} = \underline{b}$ for $\underline{b} \in B$. Then

$$\underline{b} = P_B P_A \underline{b} = P_B \underline{a} \quad \text{where} \quad \underline{a} = P_A \underline{b} \quad \text{say, } \underline{a} \in A.$$

Thus, $\|\underline{b} - \underline{a}\| = d(A, B)$.

We give the following simple example to illustrate the theorem.

Let $X = R$ be a strictly convex Banach space and let $A = [0, 1] \subset X$ and $B = [3, 4] \subset X$. Then $P_A(x) = 1$ for all $x \in B$; and $P_B(y) = 3$ for all $y \in A$. The mapping $T : P_B P_A : B \to B$ has a fixed point. In this case,

$$T(3) = P_B P_A(3) = P_B(1) = 3 \quad \text{and} \quad \|3 - 1\| = d(A, B).$$

REFERENCES

1. Blätt, H. P. (1970), Nicht-linear Gleichmässige simultan-approximation, Dissertation, Univ. des Saarlandes, Saarbrücken.
2. Bosznay, A. P. (1978), A remark on simultaneous approximation, J. Approx. Theory, 23, 296-298.
3. Brondsted, A. (1976), A note on best simultaneous approximation in normed linear spaces, Canad. Math. Bulletin, 19, 359-360.
4. Cheney, E. W. (1966), Introduction to Approximation Theory, McGraw-Hill, New York.
5. Cheney, E. W. and Goldstein, A. A. (1959), Proximity maps for convex sets, Proc. Amer. Math. Soc., 10, 448-450.
6. Diaz, J. B. and McLaughlin, H. W. (1969), Simultaneous approximation of a set of bounded real functions, Math. Comp., 23, 583-593.
7. Diaz, J. B. and McLaughlin, H. W. (1972), On simultaneous Chebyshev approximation and Chebyshev approximation with an additive function, J. Approx. Theory, 6, 68-71.
8. Dunham, C. B. (1967), Simultaneous Chebyshev approximation of functions on an interval, Proc. Amer. Math. Soc., 18, 472-477.
9. Dunham, C. B. (1977), Simultaneous approximation of a set, J. Approx. Theory, 21, 205-208.
10. Goel, D. S., Holland, A. S. B., Nasim, C. and Sahney, B. N. (1974), On best simultaneous approximation in normed linear spaces, Canad. Math. Bull., 17, 523-527.
11. Holland, A. S. B., Sahney, B. N. and Tzimbalario, J. (1976), On best simultaneous approximation, J. Indian Math. Soc., 40, 69-73.
12. Köthe, G. (1969), Topological vector spaces, I, Springer-Verlag, New York.
13. Mach, J. (1979), On the existence of best simultaneous approximation, J. Approx. Theory, 25, 258-265.
14. Milman, P. D. (1977), On best simultaneous approximation in normed linear spaces, J. Approx. Theory, 20, 223-238.
15. Pai, D. V. (1974), Proximal points of convex sets in normed linear spaces, Yokohama Math. Jour., 22, 53-78.
16. Phillips, G. M. and Sahney, B. N. (1973), Best simultaneous approximation in the L_1 and L_2 norms. Theory of Approximations with Applications, Academic Press, 213-219.

NUMERICAL ERROR EVALUATION IN LINEAR APPROXIMATION
- THE LEAST-SQUARES CASE -

Robert Schaback

Lehrstühle für Numerische und
Angewandte Mathematik der
Universität Göttingen
D-3400 Göttingen
West Germany

Using the linear functionals given by duality, a general strategy for computing explicit a-posteriori error bounds for nearly optimal linear approximations is proposed. The amount of additional calculations needed is kept within reasonable limits and the practical applicability is shown by a series of numerical examples. Due to space limitations only the least-squares case is worked out here in detail.

1. INTRODUCTION

Let V be an n-dimensional linear subspace of C(T) with T compact in $\mathbb{R}^d$, $1 \le d < \infty$. We assume to know some candidate $\hat{v}$ for the best approximation v^* to some fixed function $f \in C(T)$ with respect to a norm $\| \cdot \|$ on C(T). Let

$$E(f,V,\| \cdot \|) = \| f-v^* \| = \inf_{v \in V} \| f-v \| \tag{1}$$

denote the minimum deviation and let $F : \mathbb{R}^n \to V$ with

Copyright © 1980 by Academic Press, Inc.
All rights of reproduction in any form reserved.

ISBN: 0-12-171050-5

$$F(a_1,\ldots,a_n) := a_1v_1 + a_2v_2 + \ldots + a_nv_n \quad (a_i \in \mathbb{R}) \quad (2)$$

be a canonical parametrization of V using a basis $v_1,\ldots,v_n$.

The main objective of this contribution is to give bounds for $\hat{v} = F(\hat{a})$, $v^* = F(a^*)$ of the following types:

$$\rho_1(f,\hat{v},V) \le \| f-v^* \| \le \| f-\hat{v} \| \text{ (deviation error)}, \quad (3)$$

$$\| \hat{v}-v^* \| \le \rho_2 \ (f,\hat{v},V) \quad \text{(function space error)}, \ (4)$$

$$\| \hat{a}-a^* \| \le \rho_3 \ (f,\hat{v},V) \quad \text{(parameter space error)}, (5)$$

with <u>numerically accessible</u> functions ρ_i.

2. ERROR FUNCTIONALS

Let $\hat{\varphi}$ be a linear continous functional on C(T) with norm not exceeding 1 and let η be a real constant with

$$\hat{\varphi}(f-\hat{v}) \ge \eta \ . \quad (6)$$

Then, for any other approximation v to f from V, we have

$$\| f-v \| \ge \eta + \| \hat{v}-v \| \cdot \frac{\hat{\varphi}(\hat{v}-v)}{\| \hat{v}-v \|}$$

This inequality is fundamental for the rest of this contribution. A simple application is the de la Valleé-Poussin type

<u>Theorem 1:</u> If the restriction of $\hat{\varphi}$ to V has norm $\varepsilon \ge 0$ (theoretically $\varepsilon = 0$ is possible in case $\hat{v} = v^*$), then

$$\frac{1}{1+\varepsilon}(\eta - \varepsilon\| f-\hat{v} \|) \le E(f,V,\| \cdot \|) \le \| f-\hat{v} \| \ . \quad (7)$$

This inequality for $E(f,V,\| \cdot \|)$ is practically applicable in almost every case, since all the items appearing are numerically accessible whenever algorithms based on duality in the sense of optimization theory are used.

The straightforward way of estimating ε starts with the evaluation of constants

$$\varepsilon_j = \frac{|\varphi(v_j)|}{\| v_j \|} \qquad (1 \le j \le n) , \tag{8}$$

where $v_1,\dots,v_n$ generate the linear space V.If, in addition, a constant c is known satisfying

$$\| (a_1 \| v_1 \|,\dots,a_n \|v_n\|) \|_p \le c \| \sum_{j=1}^{n} a_j v_j \| \tag{9}$$

for any set of parameters $a_1,\dots,a_n$, one can choose ε according to

$$\varepsilon = c \; \| (\varepsilon_1,\dots,\varepsilon_n) \|_q \; . \tag{10}$$

The norms $\| \;\; \|_p$ and $\| \;\; \|_q$ should be complementary L_p and L_q norms on $\mathbb{R}^n$, i.e. with $p^{-1} + q^{-1} = 1$. The constant c, however, measures the quality of the basis and is the main obstacle to be overcome in the sequel.

3. LINEAR L_2 APPROXIMATION (see also Sautter [1])

If $\hat{v}$ is a candidate for an optimal approximation, the functional

$$\hat{\varphi}(g) := \frac{(g, f-\hat{v})}{\| f-\hat{v} \|_2} \tag{11}$$

can be used in conjunction with the value

$$\eta = \| f-\hat{v} \|_2 = \hat{\varphi}(f-\hat{v}) \; . \tag{12}$$

The estimate (7) then takes the form

$$\frac{1-\varepsilon}{1+\varepsilon} \| f-\hat{v} \|_2 \le E(f,V) \le \| f-\hat{v} \|_2 , \tag{13}$$

where ε is the norm of $\hat{\varphi}$ on V. If V is generated by an orthonormal basis consisting of functions $v_1,\ldots,v_n$, one has $c = 1$ in (9) and (10) for $p = q = 2$. If $v_1\ldots,v_n$ form a non-orthonormal basis, c varies with the "condition" of the basis and should be estimated numerically, if a reasonable error evaluation is required.

If the domain T consists only of a finite number $m \geq n$ of points, one can take v_j as columns of a $m\times n$ matrix U, which usually is transformed by an orthogonal matrix Q into

$$QU = \begin{pmatrix} R \\ 0 \end{pmatrix}, \tag{14}$$

where R usually is a nonsingular $n\times n$ upper triangular matrix. Then c can be estimated via

$$c \leq \| R^{-1} \|_{2,2} \max_{1\leq j\leq n} \| Re_j \|_2 \leq \sqrt{n} \| R^{-1} \|_{\infty,\infty} \max_{1\leq j\leq n} \| Re_j \|_2,$$

where e_j stands for the j-th unit vector and the norm

$$\| A \|_{p,q} = \sup_{x\neq 0} \frac{\| Ax \|_q}{\| x \|_p} . \tag{16}$$

is used. The constants ε_j are zero in theory and should therefore be calculated without using Q and R by direct evaluation of inner products. The roundoff in this computation may roughby be accounted for by simply adding m times the machine precision eps to ε_j. Likewise it is easy to introduce additional terms for errors committed by imprecise evaluation of $\| f-\hat{v} \|$, for instance.

The knowledge of ε is basic also for the error estimation in function space:

<u>Theorem 2</u>: Let ε and $\hat{v}$ be defined as above. Then for any $v \in V$ with $\| f-v \|_2 \leq \| f-\hat{v} \|_2$ we have

$$\| v-\hat{v} \|_2 \leq 2\, \varepsilon\, \| f-\hat{v} \|_2 . \tag{17}$$

The proof simply follows from

$$\begin{aligned} \| v-\hat{v} \|_2^2 &\le \| v-\hat{v} \|_2^2 - \| f-v \|_2^2 + \| f-\hat{v} \|_2^2 \\ &= 2(f-\hat{v}, v-\hat{v}) \\ &= 2\ \hat{\varphi}(v-\hat{v}) \cdot \| f-\hat{v} \|_2 \qquad (18) \\ &\le \varepsilon \cdot 2 \cdot \frac{\| v-\hat{v} \|_2}{\sqrt{2\varepsilon}} \cdot \| f-\hat{v} \|_2 \sqrt{2\varepsilon} \\ &\le \frac{1}{2} \| v-\hat{v} \|_2^2 + 2\varepsilon^2 \| f-\hat{v} \|_2^2 . \end{aligned}$$

<u>Corollary:</u> For a discrete domain T one can combine (17) and (9) to give the estimate

$$\begin{aligned} \| a-\hat{a} \|_2 &\le 2\varepsilon \| R^{-1} \|_{2,2} \| f-\hat{v} \|_2 \\ &\le 2\varepsilon \sqrt{n} \| R^{-1} \|_{\infty,\infty} \| f-\hat{v} \|_2 \end{aligned} \qquad (19)$$

in parameter space.

For nondiscrete domains T one should avoid a posteriori calculations of c by choosing orthonormalization right from the start as the basic numerical strategy. That is, first evaluate a triangular n×n matrix L converting a non-orthonormal basis $v_1 \ldots, v_n$ into a (near-)orthonormal basis $u_1, \ldots, u_n$ via

$$\begin{pmatrix} u_1 \\ \vdots \\ u_n \end{pmatrix} = L \begin{pmatrix} v_1 \\ \vdots \\ v_n \end{pmatrix} . \qquad (20)$$

This can be done by any of the well-known (stabilized) orthonormalization procedures. Then the Gram matrix $((u_j, u_k))$ closely approximates the unit matrix and it is easy to use Gerschgorin's theorem to give a positive lower bound B^2 for its eigenvalues. Then

$$c = B^{-1} \max_{1 \le i \le n} \| u_i \|_2 \qquad (1 \le i \le n)$$

can be taken as a reliable value in (10), even when orthonormalization is not perfect. This admits the estimates (13) and (17), whereas for (19) one should use

$$\| a-\hat{a} \|_2 \leq \| L \|_{2,2} \| (a-\hat{a})^T L^{-1} \|_2 \leq B^{-1} \| L \|_{2,2} \| v-\hat{v} \|_2$$

$$\leq B^{-1} \sqrt{n} \| L \|_{\infty,\infty} \| v-\hat{v} \|_2 \qquad (21)$$

in conjunction with (17). In any case, the extra cost of a reasonable a posteriori error evaluation still is comparable to the cost of the basic calculation itself.

4. NUMERICAL EXAMPLES: L_2 CASE

On a 21×21-grid on $[-1,+1]^2$ spaced like the extrema of the Chebyshev polynomial T_{20} in each direction we approximate $f(x,y) = \sqrt{4+x+2y}$ by the functions generated by

$$1, x, y, xy, x^2, y^2, x^2y, xy^2, x^2y^2$$

(see Watson [4] in the Chebyshev case).
Single precision calculation of the least-squares approximation by the IBM subroutine LLSQ [3] yields a function $\hat{v}$ with

$$\| f-\hat{v} \|_2^2 = 0.010082 .$$

Application of Theorem 1 leads to $\varepsilon = 1.75 \cdot 10^{-4}$ and

$$\| f-v^* \|_2^2 \geq 0.010075 ,$$

while Theorem 2 implies

$$\| v^*-\hat{v} \|_2^2 \leq 0.1235 \cdot 10^{-8} ,$$

$$\| a^*-\hat{a} \|_2^2 \leq 0.2296 \cdot 10^{-8}$$

Double precision recalculation (with DLLSQ) gives

$$\| a^*-\hat{a} \|_2^2 = 0.66 \cdot 10^{-12}$$

for comparison. In addition, the double precision result $\hat{v}$ satisfies

$$\| v^* - \hat{v} \|_2^2 \le 0.58 \cdot 10^{-27}$$

$$\| a^* - \hat{a} \|_2^2 \le 0.11 \cdot 10^{-26} .$$

With the machine precision eps we used (on a UNIVAC 1100/82 computer) one can see in both cases that approximately the overestimation is governed by the rule of thumb

$$\| a^* - \hat{a} \|_2 \sim \text{eps} \cdot \varepsilon^{-1} .$$

From a series of examples within the area of exponential approximation we select the approximation of $f(x) = (1+x)^{-1}$ from the space generated by $\exp(\lambda_i x)$, $i = 1,2,3$ with

$$\lambda_1 = -\ 4.336\ 157\ 075\ 161\ 31$$

$$\lambda_2 = -\ 1.546\ 985\ 019\ 769\ 86$$

$$\lambda_3 = -\ 0.227\ 576\ 311\ 531\ 917$$

on 51 points of [0,1] spaced like the extrema of T_{50} on [-1,+1]. The above frequencies were chosen by an application of Prony's method (see e.g. [2]). The approximation $\hat{v}$ given in double precision by the IBM program DLLSQ on a UNIVAC 1100/82 satisfies the error bounds

$$\| f - \hat{v} \|_2 = 0.17831151 \cdot 10^{-4}$$

$$\| f - v^* \|_2 \ge 0.17831145 \cdot 10^{-4}$$

$$\| v^* - \hat{v} \|_2 \le 0.678 \cdot 10^{-11}$$

$$\| a^* - \hat{a} \|_2 \le 0.270 \cdot 10^{-10}$$

5. CONCLUSION

The error estimation method proposed here needs an extra save storage for the initial data of the problem ($O(n \cdot m)$ locations for m points and n functions) and additional $O(n^3+n \cdot m)$ operations. This compares favorably with the usual storage of $O(n \cdot m)$ and the usual $O(n^2 \cdot m)$ operations needed for calculating discrete least squares approximations. The overestimation factor appears to stay within reasonable bounds for all examples we tested.

Similar error estimates can be derived for the Chebyshev case as well. The major drawback, however, is due to nonuniqueness and related degeneracies occurring when the Haar condition is not satisfied. We postpone this case to a later publication.

ACKNOWLEDGMENT

Calculations were performed on the UNIVAC 1100/82 of the Gesellschaft für wissenschaftliche Datenverarbeitung m b H, Göttingen.

REFERENCES

1. Sautter, W., A-posteriori-Fehlerabschätzungen für die Pseudoinverse und die Lösung minimaler Länge, Computing 14 (1975), 37-44.
2. Schaback, R., Suboptimal Exponential Approximations, SIAM J. Numer. Analysis 16 (1979), 1007-1018.
3. Scientific Subroutine Package, Version III, System/360/ IBM Publ. No. GH 20-0205-4, 1970.
4. Watson, G.A., A Multiple Exchange Algorithm for Multivariate Chebyshev Approximation, SIAM J. Numer. Analysis 12 (1975), 46-52.

WIENER-HOPF TECHNIQUES IN APPROXIMATING HOLOMORPHIC FUNCTIONS

Walter Schempp

Lehrstühl für Mathematik I
University of Siegen
Siegen, Germany

INTRODUCTION

Let $D = \{z \in \mathbb{C} \mid |z|<1\}$ denote the open unit disc in the complex plane and $(\mathcal{H}^p(D))_{p \in [1,\infty[}$ the usual scale of Hardy spaces modelled on D. By Bochner's trick the radial limit $\tilde{f}$ exists for each function $f \in \mathcal{H}^p(D)$. Moreover, f is uniquely determined by the restriction of $\tilde{f}$ to a subset Ω of Lebesgue measure $\lambda(\Omega)>0$ of the unit circle $T = \{z \in \mathbb{C} \mid |z| = 1\}$ bounding D. In the cases $p \in]1,\infty[$ the knowledge of $\tilde{f}|\Omega$ is sufficient to construct explicitly approximations of the functions $f \in \mathcal{H}^p(D)$ with respect to the Hardy norm $||?||_{\mathcal{H}^p(D)}$. Since there is a significant difference between the cases $p>1$ and $p=1$ the following problem arises:

Problem (p=1)

Construct Patil type approximations (in the sense of [3])

Copyright © 1980 by Academic Press, Inc.
All rights of reproduction in any form reserved.
ISBN: 0-12-171050-5

in the "biggest" normed Hardy space $\mathcal{H}^1(D)$ with respect to its norm topology.

The problem is solved by using Wiener-Hopf (Toeplitz) operators with $\mathcal{C}^2$-symbols smooth enough to multiply the (strong) dual BMO(T) (= vector space of all functions with bounded mean oscillation on T) of the Hardy subspace $\mathcal{H}^1(T) = \{\tilde{f} \mid f \in \mathcal{H}^1(D)\}$ of the Lebesgue space $L^1(T;\lambda)$.

TOEPLITZ OPERATORS

Since $\mathcal{H}^1(T)$ admits no closed complementary vector subspace of $L^1(T;\lambda)$ by Newman's theorem [2], the Toeplitz operators $P_\varphi: \mathcal{H}^p(T) \to \mathcal{H}^p(T)$ $(p \in]1,\infty[)$ with symbols $\varphi \in L^\infty(T;\lambda)$ defined by $P_\varphi: f \rightsquigarrow Q(\varphi f)$ (Q = natural projector of $L^p(T;\lambda)$ onto $\mathcal{H}^p(T)$) do not necessarily have extensions $\mathcal{H}^1(T) \to \mathcal{H}^1(T)$ which are continuous with respect to the norm topology. Therefore, the main idea is to assure the existence and the norm continuity of the Toeplitz operator P_φ on $\mathcal{H}^1(T)$ by some smoothness condition on its symbol φ.

Let Ω denote a non-empty open subset of T and let the function $\rho \in \mathcal{C}^2(T)$ be real-valued with support contained in Ω. In addition, suppose that ρ satisfies the following conditions:

(i) $\rho(w) \in [0,1]$ for each $w \in \Omega$;
(ii) $\lambda(\{w \in \Omega \mid \rho(w) = 1\}) > 0$.

For each value of the parameter $t \in]0,1[$ define the function $\rho_t : T \ni w \rightsquigarrow t\rho(w)/(1-t\rho(w)) \in \mathbb{R}_+$. Furthermore, introduce the functions

$$\varphi_t = 1+\rho_t = 1/(1-t\rho), \quad \theta_t = \log \varphi_t \quad (t \in]0,1[).$$

If $C_z: T \ni w \rightsquigarrow 1/(1-\bar{z}w) \in \mathbb{C}$ denotes the Cauchy-Szegö kernel of the disc D located at the point $z \in D$ then the Cauchy transform

$$h_t: D \ni z \rightsquigarrow \langle \Theta_t, C_z \rangle \in \mathbb{C}$$

of Θ_t belongs to the Hardy space $\mathcal{H}^\infty(D)$. Our first result reads as follows.

THEOREM 1. For $t \in]0,1[$ the boundary-value function $\tilde{h}_t$ of $h_t \in \mathcal{H}^\infty(D)$ satisfies

$$\operatorname{Re} \tilde{h}_t \in \mathcal{C}^2(T), \qquad \operatorname{Im} \tilde{h}_t \in \mathcal{C}^1(T).$$

Given $f \in L^1(T;\lambda)$ and any subarc $I \neq \emptyset$ of T, denote by

$$f_I = \frac{1}{\lambda(I)} \int_I f(w)\,d\lambda(w)$$

the average of f on I and let the quantity

$$M(f) = \sup_{\lambda(I) \leq 1} \left(\frac{1}{\lambda(I)} \int_I |f(w)-f_I|\,d\lambda(w)\right) \in \mathbb{R}_+ \cup \{+\infty\}$$

be the mean oscillation of f on T. By Theorem 1 the functions

$$\psi_t = e^{-\frac{1}{2} h_t} \qquad (t \in]0,1[)$$

admit boundary-value functions $(\tilde{\psi}_t)_{t \in]0,1[}$ that belong to the space $\operatorname{Lip}_1(T)$. This fact enables us to prove that the following multiplier result holds:

THEOREM 2. For $t \in]0,1[$ the functions $\operatorname{Re} \tilde{\psi}_t$ and $\operatorname{Im} \tilde{\psi}_t$ are multipliers of the complex vector space $BMO(T) = \{f \in L^1(T;\lambda) \mid M(f) < +\infty\}$ of all complex-valued functions with bounded mean oscillation on T, i.e., we have the inclusions

$$(\mathrm{Re}\ \tilde{\psi}_t).\ \mathrm{BMO}(T) \subseteqq \mathrm{BMO}(T), \qquad (\mathrm{Im}\ \tilde{\psi}_t).\ \mathrm{BMO}(T) \subseteqq \mathrm{BMO}(T)$$

for all $t \in]0,1[$.

This result combined with the closed graph theorem and a duality argument that is based on the identity $\mathrm{BMO}(T) = \mathcal{H}^1(T)'$ (Fefferman-Stein duality [1]) then yields

THEOREM 3. The Toeplitz operators $P_{\tilde{\psi}_t}$, $P_{\overline{\tilde{\psi}}_t}$ $(t \in]0,1[)$ are continuous endomorphisms of the complex Banach space $\mathcal{H}^1(T)$ with respect to the norm topology such that

$$\sup_{t \in]0,1[} ||P_{\tilde{\psi}_t}|| < +\infty, \quad \sup_{t \in]0,1[} ||P_{\overline{\tilde{\psi}}_t}|| < +\infty$$

holds.

Since $\varphi_t = |\tilde{\psi}_t|^{-2}$ holds for $t \in]0,1[$ we may deduce from the preceeding result the following

THEOREM 4. The Toeplitz operators $(P_{\varphi_t})_{t \in]0,1[}$ are topological automorphisms of the Hardy space $\mathcal{H}^1(T)$ with the inverses

$$P_{\varphi_t}^{-1} = P_{\tilde{\psi}_t} \circ P_{\overline{\tilde{\psi}}_t} \qquad (t \in]0,1[).$$

Moreover, the family $(P_{\varphi_t}^{-1})_{t \in]0,1[}$ of operators on $\mathcal{H}^1(T)$ is uniformly bounded.

PATIL TYPE APPROXIMATIONS

An application of Harnack's inequality shows that the family $(\psi_t)_{t \in]0,1[}$ converges pointwise in D to 0 as $t \to 1-$.

Thus, the identities

$$P_{\varphi_t}^{-1}(C_z) = P_{\tilde{\psi}_t} \circ P_{\overline{\tilde{\psi}}_t}(C_z) = \overline{\psi_t(z)}\tilde{\psi}_t C_z \qquad (t \in]0,1[)$$

for $z \in D$ combined with $\sup_{t \in]0,1[} \|\tilde{\psi}_t\|_\infty \leq 1$ and the fact that $\{C_z | z \in D\}$ forms a total family in the Banach space $\mathcal{H}^1(T)$ imply by Theorem 4 the relations

$$\lim_{t \to 1-} \|P_{\varphi_t}^{-1} f\|_1 = 0 \qquad (f \in \mathcal{H}^1(T)).$$

In view of the equalities

$$P_{\varphi_t}^{-1} = P_1 - P_{\varphi_t}^{-1} \circ P_{\rho_t} \qquad (t \in]0,1[)$$

the following approximation theorem obtains (cf. Patil [3]):

THEOREM 5. Let $f \in \mathcal{H}^1(D)$ be given. Starting with a function $\rho \in \mathcal{C}^2(T)$ that satisfies (i), (ii) construct the families $(\rho_t)_{t \in]0,1[}$, $(\psi_t)_{t \in]0,1[}$. Define the family $(f_t)_{t \in]0,1[}$ by means of $\tilde{f}|\Omega$ according to

$$f_t : D \ni z \rightsquigarrow \psi_t(z) \int_\Omega \frac{\overline{\tilde{\psi}}_t(w)\rho_t(w)\tilde{f}(w)}{1-z\overline{w}}\,dw \in \mathbb{C} \qquad (t \in]0,1[).$$

Then the approximation property

$$\lim_{t \to 1-} \|f-f_t\|_{\mathcal{H}^1(D)} = 0$$

holds with respect to the norm topology of $\mathcal{H}^1(D)$.

As a consequence we conclude that $\tilde{f}|\Omega = 0$ implies $f=0$ as mentioned above.

For a different approach based on a theorem of Hardy and Littlewood, see the recent paper by Zarantonello [6]. Also see the papers [4] and [5].

ACKNOWLEDGMENTS

I would like to express my thanks to the Deutsche Forschungsgemeinschaft (Bonn) for financial support. Special thanks are due to Professor M.Z. Nashed (Newark, DE/USA) for having invited me to give a lecture at the University of Delaware on some topics of approximation theory.

REFERENCES

1. Fefferman, C. and E.M. Stein, H^p spaces of several variables, Acta Math., 129 (1972), 137-193.
2. Newman, D.J., The non-existence of projections from L^1 to H^1. Proc. Amer. Math. Soc., 12 (1961), 98-99.
3. Patil, D.J., Representation of H^p-functions, Bull. Amer. Math. Soc. 78 (1972), 617-620.
4. Schempp, W., Approximations in the Hardy space $\mathcal{H}^1(D)$ with respect to the norm topology (to appear).
5. Schempp, W., Approximations of holomorphic functions from boundary-values (to appear).
6. Zarantonello, S.E., A representation of H^p-functions with $0<p<\infty$. Pacific J. Math. 79 (1978), 271-282.

A CHARACTERIZATION OF STRONG UNICITY CONSTANTS

Darrell Schmidt

Department of Mathematical Sciences
Oakland University
Rochester, Michigan

I. INTRODUCTION

Let X be a closed subset of $I = [a,b]$, and let $C(X)$ denote the space of all continuous, real-valued functions on X endowed with the uniform norm $\|\cdot\|$ over X. Let P be an n-dimensional Haar subspace of $C(I)$. It is assumed that $\text{card}(X) \geq n + 1$. For $f \in C(X)$, let p_f denote the unique best uniform approximation to f from P. The strong unicity constant $\gamma(f)$ for f is defined to be the largest positive number r such that

$$\|f - p\| \geq \|f - p_f\| + r\|p - p_f\| \qquad (1)$$

for all $p \in P$. Several recent papers [1,2,4,6-15] have studied the dependences of $\gamma(f)$ on f, X, and P. Characterizations of $\gamma(f)$ have been important in these investigations. In [1,3] it was noted that

$$\gamma(f) = \inf_{\substack{p \in P \\ \|p\| = 1}} \max_{x \in E} [\text{sgn}(f - p_f)(x)]p(x) \qquad (2)$$

Copyright © 1980 by Academic Press, Inc.
All rights of reproduction in any form reserved.
ISBN: 0-12-171050-5

where $E = \{x \in X : |(f - p_f)(x)| = \|f - p_f\|\}$. The characterization (2) has proven to be rather unwieldy in analyzing $\gamma(f)$. A more convenient (although restricted) characterization was obtained in [6,11]. If $x_0 < \dots < x_n$ is an alternant for $f - p_f$, Cline [6] showed that

$$r = \{\max_{0 \leq i \leq n} \|q_i\|\}^{-1} \tag{3}$$

satisfies (1) for all $p \in P$ and thus is a lower estimate for $\gamma(f)$ where q_i is the element of P such that $q_i(x_j) = \operatorname{sgn}(f - p_f)(x_j)$, $j = 0, \dots, n$, $j \neq i$. Henry and Roulier [11] observed that if $E = \{x_0, \dots, x_n\}$, then $\gamma(f)$ is in fact equal to the expression in (3). If $\operatorname{card}(E) > n + 1$, however, then $f - p_f$ has more than one alternant, and there may not be an alternant for $f - p_f$ for which (3) yields $\gamma(f)$. An example to this effect appears in [2]. In this note, we develop an interpolative characterization of $\gamma(f)$ similar to (3) which yields $\gamma(f)$ in all cases.

II. COMPUTATION OF $\gamma(f)$

Let $f \in C(X)$. It can easily be shown using (2) that

$$\gamma(f)^{-1} = \max\{\|p\| : p \in P,\ \sigma(x)p(x) \leq 1 \text{ for all } x \in E\} \tag{4}$$

where $\sigma(x) = \operatorname{sgn}(f - p_f)(x)$ (see [4]). We show that the maximization in (4) can be taken over a subset which interpolates $\sigma(x)$ at an appropriate set of n points in E.

LEMMA 1. *Let* $q \in P$ *satisfy* $\sigma(x)q(x) \leq 1$ *for all* $x \in E$ *and* $\|q\| = \gamma(f)^{-1}$. *Let* $A = \{x \in E : \sigma(x)q(x) = 1\}$ *and select* $x^* \in X$ *such that* $|q(x^*)| = \|q\|$. *Then there does not exist an* $h \in P$ *such that* $\sigma(x)h(x) < 0$ *for all* $x \in A$ *and* $[\text{sgn } q(x^*)]h(x^*) > 0$.

Proof. Assume such an $h \in P$ exists. For $\varepsilon > 0$, let $p_\varepsilon = q + \varepsilon h$. Then

$$\|p_\varepsilon\| \geq |p_\varepsilon(x^*)| = |q(x^*)| + \varepsilon|h(x^*)| > \|q\|.$$

Since $\sigma(x)h(x) < 0$ for $x \in A$, $\sigma(x)h(x) < 0$ in an open neighborhood N of A. For $x \in N \cap E$,

$$\sigma(x)p_\varepsilon(x) = 1 + \varepsilon\sigma(x)h(x) < 1.$$

Let $\mu = \max\limits_{x \in E \setminus N} \sigma(x)p(x) < 1$. For $x \in E \setminus N$,

$$\sigma(x)p_\varepsilon(x) \leq \mu + \varepsilon\|h\| < 1$$

for ε sufficiently small. For ε sufficiently small, (4) implies that $\gamma(f)^{-1} \geq \|p_\varepsilon\| > \|q\|$ which contradicts the hypothesis $\gamma(f)^{-1} = \|q\|$ and (4).

LEMMA 2. *Let* q, A, *and* x^* *be as in* Lemma 1. *Then there exists* n *points* $x_1 < \cdots < x_n$ *in* A *such that* (1) $x^* < x_1$ *and* $-\text{sgn } q(x^*), \sigma(x_1), \ldots, \sigma(x_n)$ *alternate in sign,* (2) $x^* > x_n$ *and* $\sigma(x_1), \ldots, \sigma(x_n), -\text{sgn } q(x^*)$ *alternate in sign, or* (3) $x_{i-1} < x^* < x_i$ *for some* $i = 2, \ldots, n$, *and* $\sigma(x_1), \ldots, \sigma(x_{i-1}), -\text{sgn } q(x^*), \sigma(x_i), \ldots, \sigma(x_n)$ *alternate in sign.*

Proof. Let $\{p_1, \ldots, p_n\}$ be a basis for P, and for $x \in X$, let $\hat{x} = (p_1(x), \ldots, p_n(x)) \in R^n$. By Lemma 1, the system of inequalities

$$-[\text{sgn } q(x^*)]\, v \cdot \hat{x}^* < 0$$

$$\sigma(x)v \cdot \hat{x} < 0, \; x \in A,$$

is inconsistent over $v \in R^n$ ($u \cdot w$ denotes the dot product of u and w). Since $\{x^*\} \cup A$ is compact, the theorem on linear inequalities [5, p. 19] implies that the n-dimensional zero vector 0_n is in the convex hull of $\{-[\operatorname{sgn} q(x^*)]\hat{x}^*\} \cup \{\sigma(x)\hat{x} : x \in A\}$. By the Theorem of Caratheodory [5, p. 17], 0_n is in the convex hull of some set of $n + 1$ or fewer points in $\{-[\operatorname{sgn} q(x^*)]\hat{x}^*\} \cup \{\sigma(x)\hat{x} : x \in A\}$. One of these points must be $-[\operatorname{sgn} q(x^*)]\hat{x}^*$. Otherwise, we could find $n + 1$ points $x_0 < \cdots < x_n$ in I such that 0_n is in the convex hull of $\{q(x_i)\hat{x}_i : i = 0, \ldots, n\}$. (If fewer than $n + 1$ points of A are involved, we merely add points of I at which q does not vanish.) By the lemma on p. 74 in [5], the $q(x_i)$ must alternate in sign. Thus q has at least n zeros and therefore vanishes identically which contradicts $\|q\| = \gamma(f)^{-1} > 0$. Thus there are n points $x_1 < \cdots < x_n$ in A such that 0_n is in the convex hull of $\{-[\operatorname{sgn} q(x^*)]\hat{x}^*\} \cup \{\sigma(x_i)\hat{x}_i : i = 1, \ldots, n\}$. That A contains at least n points follows from interpolation and Lemma 1. Lemma 2 now follows by inserting x^* into $\{x_1, \ldots, x_n\}$ so as to preserve order and from the lemma on p. 74 in [5].

The following theorem is an immediate consequence of Lemma 2.

THEOREM. *Let* $f \in C(X)$, p_f *be the best uniform approximation to* f *from* P, *and* $E = \{x \in X : |(f - p_f)(x)| = \|f - p_f\|\}$. *For* $x \in X$, *let* $\sigma(x) = \operatorname{sgn}(f - p_f)(x)$. *Let*

S <u>be the set of all ordered sequences</u> $Y : x_1 < \ldots < x_n$ <u>of</u> n <u>points in</u> E <u>such that</u> (1) $\sigma(x_1), \ldots, \sigma(x_n)$ <u>alternate in sign or</u> (2) <u>for some</u> $i = 2, \ldots, n$, $\sigma(x_1), \ldots, \sigma(x_{i-1})$ <u>alternate in sign</u>, $\sigma(x_i), \ldots, \sigma(x_n)$ <u>alternate in sign, and</u> $\sigma(x_{i-1}) = \sigma(x_i)$. <u>For</u> $Y : x_1 < \ldots < x_n$ <u>in</u> S <u>let</u> p_Y <u>be the element of</u> P <u>such that</u> $p_Y(x_i) = \sigma(x_i)$, $i = 1, \ldots, n$. <u>Then</u>

$$\gamma(f)^{-1} = \max\{\|p_Y\| : Y \in S, \sigma(x)p_Y(x) \leq 1 \text{ for all } x \in E\}. \tag{5}$$

We remark that if $\mathrm{card}(E) = n + 1$, then (5) reduces precisely to the characterization of Cline [6] and Henry and Roulier [10]. If E is finite (which is the case if X is finite) (5) provides a convenient computation of $\gamma(f)^{-1}$. In particular, (5) reduces computing $\gamma(f)^{-1}$ to a finite number of interpolation problems.

References

1. M.W. Bartelt, On Lipschitz conditions, strong unicity and a theorem of A.K. Cline, J. Approximation Theory 14(1975), 245-250.

2. M.W. Bartelt and M.S. Henry, Continuity of the strong unicity constant on C(X) for changing X, J. Approximation Theory, to appear.

3. M.W. Bartelt and H.W. McLaughlin, Characterizations of strong unicity in approximation theory, J. Approximation Theory 9(1973), 255-266.

4. M.W. Bartelt and D. Schmidt, On Poreda's problem for strong unicity constants, preprint.

5. E.W. Cheney, "Introduction to Approximation Theory," McGraw-Hill, New York, 1966.

6. A.K. Cline, Lipschitz conditions on uniform approximtion operators, J. Approximation Theory 8(1973), 160-172.

7. C.B. Dunham, A uniform constant of strong uniqueness on an interval, J. Approximation Theory, to appear.

8. O. Hájék, Uniform approximation: continuity properties, preprint.

9. M.S. Henry and L.R. Huff, On the behavior of the strong unicity constant for changing dimension, J. Approximation Theory, to appear.

10. M.S. Henry and J.A. Roulier, Uniform Lipschitz constants on small intervals, J. Approximation Theory 21(1977), 244-235.

11. M.S. Henry and J.A. Roulier, Lipschitz and strong unicity constants for changing dimension, J. Approximation Theory 22(1978), 85-94.

12. M.S. Henry and D. Schmidt, Continuity theorems for the product approximation operator, in "Theory of Approximation with Applications," A.G. Law and B.N. Sahney, editors, Academic Press, New York, 1976.

13. A. Kroó, The continuity of best approximations, Acta Math. Acad. Sci. Hung. 30(1977), 175-188.

14. S.J. Poreda, Counterexamples in best approximation, Proc. Amer. Math. Soc. 56(1976), 167-171.

15. D. Schmidt, On an unboundedness conjecture for strong unicity constants, J. Approximation Theory 24(1978), 216-223.

THE LANDAU PROBLEM FOR MOTIONS ON CURVES AND TIME-OPTIMAL CONTROL PROBLEMS

I. J. Schoenberg[1]

Mathematics Research Center
University of Wisconsin-Madison
Madison, Wisconsin

1. Statement of the Problem. Let $f(t)$ be real-valued, $-\infty < t < \infty$; as we think of t as time, we denote its derivatives by $\dot{f}(t)$, $\ddot{f}(t)$. We assume that $f \in C^1(R)$, hence $\dot{f} \in C(R)$, and that $\ddot{f}(t)$ is piecewise continuous with discontinuities only of the first kind. Using the supremum norm on R, we have the well known inequality of Landau

$$||\dot{f}|| \leqq \sqrt{2||f||\,||\ddot{f}||} \quad \text{(See [1], [2])} . \tag{1.1}$$

If $a > 0$ and $f(t)$ satisfies

$$-a \leqq f(t) \leqq a \quad \text{for all } t , \tag{1.2}$$

then (1.1) shows that

$$||\dot{f}|| < \sqrt{2a}\sqrt{||\ddot{f}||} . \tag{1.3}$$

Here $\sqrt{2a}$ is the best constant, as shown by the special motion defined by

$$\tilde{f}(t) = 4at(1-t) \quad \text{if } 0 \leqq t \leqq 1,$$

$$\tilde{f}(t+1) = -\tilde{f}(t) \quad \text{for all real } t . \tag{1.4}$$

We find that $||\tilde{f}|| = a$, $||\dot{\tilde{f}}|| = 4a$, $||\ddot{\tilde{f}}|| = 8a$, and these values produce the equality sign in (1.3).

[1] Sponsored by the United States Army under Contract Nos. DAAG29-75-C-0024 and DAAG29-80-C-0041.

Copyright © 1980 by Academic Press, Inc.
All rights of reproduction in any form reserved.
ISBN: 0-12-171050-5

We may interpret (1.2) by saying that $f(t)$ describes a motion in the interval

$$I_a = \{-a \leqq x \leqq a\} . \tag{1.5}$$

In terms of the functional

$$F(f) = \frac{\|\dot{f}\|}{\sqrt{\|\ddot{f}\|}} \tag{1.6}$$

we may restate the result (1.3) by writing

$$\sup_f F(f) = \sqrt{2a} \tag{1.7}$$

for all motions f within the interval I_a.

It seems natural to replace the interval (1.5) by a prescribed closed and connected set S of the complex plane $\mathbb{C}$, or belonging to a finite-dimensional euclidean space. Accordingly, we consider the class of motions

$$(S) = \{f(t);\ f(t) \in S \text{ for all } t, \quad f(t) \not\equiv \text{constant}\} . \tag{1.8}$$

We now formulate

Landau's problem for the set S. To determine the quantity

$$L(S) = \sup_{f \in (S)} F(f) . \tag{1.9}$$

We call $L(S)$ the Landau constant of the set S.

For a preliminary version of this paper see [3]; a first part has already appeared in [4], while the remaining part will appear in [5].

2. A Few Results from [4]. The result [1.7] may now be stated as

$$L(I_a) = \sqrt{2a}. \tag{2.1}$$

The Landau constants of a few more simple sets are known. Thus far the disc

$$D_a = \{|z| \leqq a\} \tag{2.2}$$

we have

$$L(D_a) = \sqrt{2a} \; . \tag{2.3}$$

For the circular ring

$$R_{a,b} = \{b \leqq |z| \leqq a\}, \qquad (0 \leqq b \leqq a) \tag{2.4}$$

we find that

$$L(R_{a,b}) = \sqrt{a + \sqrt{a^2 - b^2}} \; , \tag{2.5}$$

which reduces to (2.3) for $b = 0$. The same value (2.5) is answered by the Landau constant of the solid spherical shell

$$\{b \leqq \sqrt{x^2 + y^2 + z^2} \leqq a\} \; .$$

A more general result from [4] is

Theorem 1. *If the set S is a closed, bounded and convex set in a finite dimensional euclidean space, then*

$$L(S) = \sqrt{\text{diameter of } S} \; . \tag{2.6}$$

Special cases of this are (2.1) and (2.3). Theorem 1 is proved by finding motion $\tilde{f} \in (S)$ such that $F(\tilde{f}) = L(S)$.

3. The Main Results of [5]. As indicated by the title of [5], we specialize the investigation to the case when the set S reduces to a curve Γ. As expected, we get in this case simpler and more definite results. To start with, we assume that Γ is in the complex plane, that it is rectifiable with a continuously turning tangent, and that it has a piecewise continuous radius of curvature $R(s)$ as a function of the arc-length s. On some portions of Γ we may well have $R(s) = \infty$, meaning that Γ is there straight.

We classify arcs Γ into four cases.

Case 1. Γ is a bi-infinite arc, a parabola being an example. However, we exclude the case when Γ is an infinite straight line. Measuring s from a point O on Γ we have $-\infty < s < \infty$.

Case 2. Γ is a half-infinite arc. If O is its end-point, then we think of Γ as a bi-infinite arc (Case 1) doubled-up on itself at O. Thus again $-\infty < s < \infty$, with $\pm s$ corresponding to the same point of Γ.

Case 3. Γ is a closed curve of total length 2ℓ. A circle, or an ellipse are good examples. It is convenient to think of the s-axis as wrapped around Γ so that s and $s + 2\ell k$ represent the same point. Thus again $-\infty < s < \infty$.

Case 4. Γ is a finite arc of length ℓ. It is convenient to think of Γ as a closed curve (Case 3), with the two halfs $0 \leqq s \leqq \ell$ and $0 \geqq s \geqq -\ell$ identified. Now the three values $s, s + 2\ell$, and $-s$ correspond to the same point of Γ. Again $-\infty < s < \infty$.

As in the case of sets, we are concerned with the class of motions

$$(\Gamma) = \{f(t);\ f(t) \in \Gamma \quad \text{for all} \quad t,\ f(t) \not\equiv \text{const.}\}. \tag{3.1}$$

We select an arbitrary, but fixed constant

$$A > 0\ , \tag{3.2}$$

and consider the subclass of motions

$$(\Gamma)_A = \{f(t);\ f(t) \in (\Gamma),\ |\ddot{f}(t)| \leqq A \quad \text{for all} \quad t\}\ . \tag{3.3}$$

The main result of [5] is

Theorem 2. There is in $(\Gamma)_A$ a unique motion

$$\tilde{f}(t) = \tilde{f}_A(t)\ , \tag{3.4}$$

called the Landau motion on Γ, corresponding to A, *and having the following four characteristic properties*.

(i)

$$|\ddot{f}(t)| = A \quad \textit{for all real} \quad t\ . \tag{3.5}$$

(ii) *In all four cases the arc-length* $\tilde{s} = \tilde{s}(t)$, *corresponding to the point* $\tilde{f}(t)$, *increases steadily with* t *from* $-\infty$ *to* $+\infty$. *We assume that*

$$\tilde{f}(0) = 0\ . \tag{3.6}$$

(iii) *In* Case 2 *we have*

$$\tilde{f}(-t) = \tilde{f}(t) \tag{3.7}$$

and therefore $\tilde{f}(t)$ *is an even function. Observe that the continuity of* $\dot{\tilde{f}}(t)$ *implies that*

$$\dot{\tilde{f}}(0) = 0\ . \tag{3.8}$$

In Case 3 *the motion* $\tilde{f}(t)$ *along the closed curve* Γ *is periodic*. *If* $2T$ *is its smallest positive period, we have*

$$\tilde{f}(t + 2T) = \tilde{f}(t)\ . \tag{3.9}$$

In Case 4 *we have again the periodicity* (3.9), *and* $\tilde{f}(t)$ *is even so that*

$$\tilde{f}(-t) = \tilde{f}(t)\ . \tag{3.10}$$

Again the continuity of $\dot{\tilde{f}}(t)$ *implies that in this case*

$$\dot{\tilde{f}}(0) = \dot{\tilde{f}}(T) = 0\ . \tag{3.11}$$

It follows from (3.7), (3.9), *and* (3.10), *that if for* $t = t_0$ *we have* $\tilde{f}(t_0) = P \in \Gamma$. *Then* $|\dot{\tilde{f}}(t_0)|$ *assumes the same value independent of* t_0, *and depending only on* P. *We denote this value by*

$$|\dot{\tilde{f}}(t_0)| = |\dot{\tilde{f}}_P|\ , \tag{3.12}$$

to indicate that it depends on P *only*.

(iv) <u>This is the decisive characteristic property of</u> $\tilde{f}$: <u>If</u>

$$f(t) \in (\Gamma)_A \tag{3.13}$$

<u>and</u> t_1 <u>is such that</u>

$$f(t_1) = P , \tag{3.14}$$

<u>then</u>

$$|\dot{f}(t_1)| \leqq |\dot{\tilde{f}}_P| . \tag{3.15}$$

In words: <u>The speed</u> $|\dot{\tilde{f}}_P|$ <u>of the Landau motion at the point</u> $P \in \Gamma$ <u>is never exceeded by the speed at</u> P, <u>of any motion</u> $f(t)$, <u>on</u> Γ, <u>such that</u> $|\ddot{f}(t)| \leqq A$ <u>for all</u> t. <u>We may say that</u> $\tilde{f}$ <u>maximizes the speed at every</u> $P \in \Gamma$, <u>within the class of motions in</u> $(\Gamma)_A$.

As a consequence we have the

<u>Corollary 1</u>. <u>We have</u>

$$L(\Gamma) = F(\tilde{f}) = \frac{\|\dot{\tilde{f}}\|}{\sqrt{A}} . \tag{3.16}$$

<u>Example</u>. $\Gamma = I_a$ corresponds to Case 4 and the equations (1.4) define its Landau motion $\tilde{f}(t)$ corresponding to $A = 8a$. From $\|\dot{\tilde{f}}\| = 4a$ and (3.16) we obtain $L(I_a) = 4a/\sqrt{8a} = \sqrt{2a}$ which agrees with (2.1).

A proof of Theorem 2, hence a proof of the existence of the Landau motion $\tilde{f}(t)$ and its extremum property, depends on the following circumstances: For a motion $f(t) \in (\Gamma)$ we denote by $s = s(t)$ the arc-length at $f(t)$, and by $v = \frac{ds}{dt}$ the speed, so that $|v| = |\dot{f}(t)|$. If we pass from the variable v to the new variable

$$u = v^4 \tag{3.17}$$

we find the following: <u>If</u> $v \neq 0$ <u>then the inequality</u>

$$|\ddot{f}(t)| \leqq A$$

is equivalent to the differential inequality

$$\left(\frac{du}{ds}\right)^2 \leqq 16u\left(A^2 - \frac{u}{R^2(s)}\right) . \tag{3.18}$$

This is the reason why a proof of Theorem 2 is dominated by a discussion of the properties of the solutions of the diff. equation

$$\left(\frac{du}{ds}\right)^2 = 16u\left(A^2 - \frac{u}{R^2(s)}\right) . \tag{3.19}$$

4. The Landau Motions for the Simplest Curves. Besides the case of $\Gamma = I_a$, whose Landau motion for $A = 8a$ is described by (1.4), we have

Theorem 3. If $\Gamma = C_R$ is a circle of radius R, then the Landau motion for A is the uniform circular motion

$$\tilde{f}(t) = Re^{it\sqrt{A/R}} . \tag{4.1}$$

Since $\dot{\tilde{f}}(t) = \sqrt{AR}\ i\ \exp(i\sqrt{A/R})$, (3.16) shows that

$$L(C_R) = \|\dot{\tilde{f}}\|\ \sqrt{A} = \sqrt{AR}/\sqrt{A} = \sqrt{R} \tag{4.2}$$

in agreement with (2.5) for $b = a$.

More general is

Theorem 4. The Landau constant of the circular arc $C_{R,\ell}$, of radius R and length ℓ, is given by

$$L(C_{R,\ell}) = \begin{cases} \sqrt{R\sin\frac{\ell}{R}} & \text{if } \ell \leqq R\pi/2 , \\ \sqrt{R} & \text{if } \ell \geqq R\pi/2 . \end{cases} \tag{4.3}$$

If we let $R \to \infty$ then the arc $C_{R,\ell}$ becomes the segment $I_{\ell/2}$ and (4.3) shows that $L(I_{\ell/2}) = \sqrt{\ell}$, which is Landau's result (2.1).

As an example of Case 1 we state

Theorem 5. *The Landau motion on a parabola*

$$\Pi : y = \frac{1}{2p} x^2 \tag{4.4}$$

is identical with the Galilean motion

$$\tilde{f}(t) = t\sqrt{Ap} + i \frac{1}{2} At^2 \tag{4.5}$$

having the constant acceleration $\ddot{\tilde{f}}(t) = A_i$.

As a last example we consider the ellipse

$$E = E_{a,b} : \frac{x^2}{a^2} + \frac{y^2}{b^2} = 1 . \tag{4.6}$$

Observe that $f(t) \in (E_{a,b})$ iff $f_1(t) = f(t)/a \in (E_{1,b/a})$, and therefore $F(f) = \sqrt{a}\, F(f_1)$. This implies that

$$L(E_{a,b}) = \sqrt{a}\, L(E_{1,r}), \quad \text{where} \quad r = \frac{b}{a} . \tag{4.7}$$

A table of values of $L(E_{1,r})$ was computed by C. Vargas, of the MRC-Madison Computing Staff, by numerical integrations of the D.E. (3.19).

TABLE 1

r	$L(E_{1,r})$
.0	1.41421 = $\sqrt{2}$
.1	1.40978
.2	1.39643
.3	1.37394
.4	1.34199
.5	1.30011
.6	1.24811
.7	1.18609
.8	1.11782
.9	1.05409
1.0	1.00000

5. Time-Optimal Control Problems on Finite Arcs (Case 4). The simplest such problem is the classical case when $\Gamma = \{0 \leqq s \leqq \ell\}$ is a straight segment of length ℓ. Let $f(t)$ $(0 \leqq t \leqq T)$ be a motion on Γ restricted by

$$|\ddot{f}(t)| \leqq A . \tag{5.1}$$

We require that

$$f(0) = 0, \quad \dot{f}(0) = 0, \quad \text{and} \quad f(T) = \ell, \quad \dot{f}(T) = 0 , \tag{5.2}$$

and that the time T should be as small as possible. The well known result is that the shortest time is given by

$$T = 2\sqrt{\ell/A}, \qquad \text{(See [7, pp. 233-236])}. \tag{5.3}$$

For a solution of the similar problem when the constraint (5.1) is replaced by $|f^{(n)}(t)| \leqq A$, while the boundary conditions are

$$f(0) = \dot{f}(0) = \cdots = f^{(n-1)}(0) = 0, \ f(T) = L,$$
$$\dot{f}(T) = \cdots = f^{(n-1)}(T) = 0 , \tag{5.4}$$

see [6].

Theorem 6. Let Γ be a finite arc of length ℓ. Let $f(t)$ be a motion along Γ from $s = 0$ to $s = \ell$, such that we have the constraint (5.1) and the boundary conditions (5.2). The least value of T is achieved when $f(t) = \tilde{f}(t)$ is the Landau motion on Γ for the constant A.

Intuitively this result is clear, since the Landau motion is the fastest at every point of Γ among all motions $f(t)$ subject to (5.1). Noteworthy seems the special case when $\Gamma = C_{R,\ell}$ is a circular arc of radius R and length ℓ.

Theorem 7. The shortest time $T_{R,\ell}$ of describing $\Gamma = C_{R,\ell}$, for $\ell \leqq R\pi/2$, under the conditions (5.1) and (5.2) is given by

$$T_{R,\ell} = \frac{2}{\sqrt{A}} \int_0^{\ell/2} \frac{ds}{\sqrt{R \sin \frac{2s}{R}}} \,. \tag{5.5}$$

Letting here $R \to \infty$, when the arc becomes a segment, we obtain from (5.5) that $T_{\infty,\ell} = 2\sqrt{\ell/A}$, in agreement with (5.3). It would seem difficult to extend the results of [6] to the case of a finite arc Γ.

6. Time-Optimal Control Problems for Closed Curves (Case 2). We finally turn to the case when Γ is a closed curve of length 2ℓ. Let $\tilde{f}(t)$ denote its Landau motion of Theorem 2 corresponding to the control constant A. By Theorem 2 we know that $\tilde{f}(t)$ is a periodic function of period $2T$.

Theorem 8. *The period* $2T$ *of the Landau motion* $\tilde{f}(t)$ *is the shortest time in which a motion* $f(t)$ *on* Γ, *subject to*

$$|\ddot{f}(t)| \leqq A \quad \text{for all} \quad t \,, \tag{6.1}$$

can perform one complete revolution along the closed curve Γ.

Appropriate integrations of the D.E. (3.19) would furnish a table of values of $\tilde{f}(t)$, and a numerical value for $2T$.

REFERENCES

1. Landau, E., Einige ungleichungen fur zweimal differentiierbase Funktionen, Proc. London Math. Soc. (2), 13 (1913), 43-49.

2. Schoenberg, I. J., The elementary cases of Landau's problem on inequalities between derivatives, Amer. Math. Monthly, 80 (1973), 121-158.

3. Schoenberg, I. J., The Landau problem for motions on curves, MRC Technical Summary Report #1809, November 1977, University of Wisconsin-Madison, Madison, Wisconsin.

4. Schoenberg, I. J., The Landau problem I. The case of motions on sets, Indagationes Math., (1978), 276-286.

5. Schoenberg, I. J., The Landau problem II. The case of motions on curves, to appear in Indagationes Math.

6. Schoenberg, I. J., The perfect B-splines and a time-optimal control problem, Israel J. of Math., 10 (1971), 261-274.

7. Young, L. C., Lectures on the calculus of variations and optimal control theory, W. B. Saunders Co., Philadelphia, 1969.

EXACT DEGREES OF APPROXIMATION FOR TWO-DIMENSIONAL BERNSTEIN OPERATORS ON C^1

F. Schurer
F.W. Steutel

Department of Mathematics
Eindhoven University of Technology
Eindhoven, The Netherlands

Dedicated to Professor G.G. Lorentz on the occasion of his seventieth birthday

Exact degrees of approximation for two-dimensional Bernstein operators on C^1 are obtained in terms of moduli of continuity of first-order derivatives.

1. Introduction and Summary

1.1. Let f be a real function defined on [0,1]. The well-known Bernstein polynomial of order n corresponding to f is defined by

$$B_n^{(1)}(f;x) = \sum_{k=0}^{n} f(\tfrac{k}{n})p_{n,k}(x) \ ,$$
$$p_{n,k}(x) = \binom{n}{k}x^k(1-x)^{n-k} \qquad (n \in \mathbb{N};\ k = 0,1,\dots,n;\ x \in [0,1]) \ . \tag{1.1}$$

The $B_n^{(1)}$ are called the one-dimensional Bernstein operators.

Let $C^1([0,1])$ be the set of functions defined on [0,1] and having a continuous first derivative. In (1) the following result is proved: For $f \in C^1([0,1])$ and $\| \ \|$ denoting the supremum norm

$$\| B_n^{(1)}(f) - f \|_{[0,1]} \le \tfrac{1}{4}n^{-\frac{1}{2}}\ \omega_1(f;n^{-\frac{1}{2}}) \qquad (n \in \mathbb{N}), \tag{1.2}$$

where ω_1 denotes the modulus of continuity of f'. This improves on an inequality due to Lorentz (2, p.21), in the right-hand side of which the constant $\frac{3}{4}$ occurs instead of the best possible $\frac{1}{4}$. This constant is also best if in (1.2) we replace $n^{-\frac{1}{2}}\omega_1(f;n^{-\frac{1}{2}})$ by $\omega_1(f;n^{-1})$ (cf. (3)).

In this paper we give analogues of (1.2) for two-dimensional Bernstein operators.

Copyright © 1980 by Academic Press, Inc.
All rights of reproduction in any form reserved.
ISBN: 0-12-171050-5

Let $\Delta := \{(x,y);\ x \ge 0,\ y \ge 0,\ x+y \le 1\}$ and let $\square := [0,1]^2$. By $C^1(\Delta)$ and $C^1(\square)$ we denote the set of functions that are continuous together with their first-order derivatives. The Bernstein operators $B_n^{(2)}$ on $C^1(\Delta)$ and $B_{n,m}^{(2)}$ on $C^1(\square)$ are defined by their associated polynomials as follows:

$$B_n^{(2)}(f;x,y) = \sum_{k=0}^{n} \sum_{\ell=0}^{n-k} f(\frac{k}{n}, \frac{\ell}{n}) p_{n,k,\ell}(x,y) ,$$

$$p_{n,k,\ell}(x,y) = \binom{n}{k}\binom{n-k}{\ell} x^k y^\ell (1-x-y)^{n-k-\ell} \qquad (n \in \mathbb{N};\ k+\ell \le n;\ (x,y) \in \Delta), \tag{1.3}$$

and (cf. (1.1))

$$B_{n,m}^{(2)}(f;x,y) = \sum_{k=0}^{n} \sum_{\ell=0}^{m} f(\frac{k}{n}, \frac{\ell}{m}) p_{n,k}(x) p_{m,\ell}(y) . \tag{1.4}$$

Both $B_n^{(2)}(f;x,y)$ and $B_{n,m}^{(2)}(f;x,y)$ tend to $f(x,y)$ uniformly on Δ and $\square$ as their orders tend to infinity. In (4) a detailed account is given of the rate of convergence for continuous functions, in terms of their modulus of continuity ω defined by

$$\omega(f;\delta_1,\delta_2) = \sup\{|f(x',y') - f(x",y")|;\ |x'-x"| \le \delta_1,\ |y'-y"| \le \delta_2\} .$$

A first inequality for $f \in C^1(\square)$ was given by Stancu (5), who proved

$$\| B_{n,m}^{(2)}(f) - f \|_\square \le n^{-\frac{1}{2}} \omega_{10}(f;n^{-\frac{1}{2}},m^{-\frac{1}{2}}) + m^{-\frac{1}{2}} \omega_{01}(f;n^{-\frac{1}{2}},m^{-\frac{1}{2}}) , \tag{1.5}$$

where ω_{10} and ω_{01} are the moduli of continuity of $f_{10} := \frac{\partial f}{\partial x}$ and $f_{01} := \frac{\partial f}{\partial y}$.

A somewhat different inequality was recently obtained by Marlewski (6). Using inequalities similar to the ones used in (3) he proved the following result. Let w_{10} (and similarly w_{01}) be defined by

$$w_{10}(f;\delta_1) = \sup_{y \in [0,1]} \sup_{|x'-x"| \le \delta_1} |f_{10}(x',y) - f_{10}(x",y)| ,$$

then

$$\| B_{n,m}^{(2)}(f) - f \|_\square \le (\frac{1}{8} + \frac{1}{2} n^{-\frac{1}{2}}) w_{10}(f;n^{-1}) + (\frac{1}{8} + \frac{1}{2} m^{-\frac{1}{2}}) w_{01}(f;m^{-1}) . \tag{1.6}$$

1.2. In section 3 best-possible inequalities of type (1.5) will be given for $C^1(\Delta)$ and $C^1(\square)$. In order to do so, in section 2 we determine extremal functions (i.e. functions maximizing $|B^2(f;x,y) - f(x,y)|$ under suitable restrictions). The inequalities are then obtained by use of the results of section 2 and the one-dimensional results in (1) and (3).

2. The Extremal Functions

2.1. Let $(x,y) \in \Delta$ be fixed and let $\delta_1 = \delta_2$ be positive. Concentrating on the (more delicate) case of the operators $B_n^{(2)}$ on $C^1(\Delta)$, we are interested in

$$\sup\{|B_n^{(2)}(f;x,y) - f(x,y)|;\ f \in C^1(\Delta),\ \omega_{10}(f) = \bar{\omega}_{10}, \omega_{01}(f) = \bar{\omega}_{01}\}, \quad (2.1)$$

where $\omega_{ij}(f) := \omega(f_{ij};\delta_1,\delta_2)$, and $\bar{\omega}_{10}, \bar{\omega}_{01}$ are given. The object is to obtain an extremal function for this problem.

We recall (cf. (1)) that for the Bernstein operator $B_n^{(1)}$ on $C^1([0,1])$ an extremal function $\tilde{g}$ is given by $\tilde{g}(x) = 0$ and

$$\tilde{g}'(s) = j + \frac{1}{2} \qquad (j\delta \le s - x < (j+1)\delta;\ j = 0, \pm 1, \pm 2, \ldots) .$$

That is, $\tilde{g}(s) \equiv \tilde{g}(s;x;\delta)$ is of the form (we put $a_+ = \max(a,0)$)

$$\tilde{g}(s) = \frac{1}{2}|s-x| + \sum_{j=1}^{\infty} (|s-x| - j\delta)_+ \qquad (s \in \mathbb{R}) . \quad (2.2)$$

We proceed to construct an extremal function for (2.1), where we may take $f(x,y) = 0$, $f_{10}(x,y) = \frac{1}{2}\bar{\omega}_{10}$, $f_{01}(x,y) = \frac{1}{2}\bar{\omega}_{01}$. Without restriction (changing the sign of f if necessary) we have (cf. (1.3))

$$|B_n^{(2)}(f;x,y)-f(x,y)| = \sum_{k=0}^{n}\sum_{\ell=0}^{n-k} p_{n,k,\ell}(x,y)\{f(\tfrac{k}{n},\tfrac{\ell}{n})-f(x,y)\} =$$

$$= \sum_{k=0}^{n}\sum_{\ell=0}^{n-k} p_{n,k,\ell}(x,y) \int_{(x,y)}^{(k/n,\ell/n)} \{f_{10}(s,t)ds+f_{01}(s,t)dt\}.$$

Abbreviating we write

$$|B_n^{(2)}(f;x,y) - f(x,y)| = E(f_{10},f_{01}) = E_1(f_{10},f_{01}) = E_2(f_{10},f_{01}) ,$$

where E_1 and E_2 are obtained by taking the line integral along paths parallel to the axes, i.e.

$$E_1(f_{10},f_{01}) := \sum_{k=0}^{n}\sum_{\ell=0}^{n-k} p_{n,k,\ell}(x,y) \{\int_x^{k/n} f_{10}(s,y)ds + \int_y^{\ell/n} f_{01}(\tfrac{k}{n},t)dt\} ,$$

$$E_2(f_{10},f_{01}) := \sum_{k=0}^{n}\sum_{\ell=0}^{n-k} p_{n,k,\ell}(x,y) \{\int_y^{\ell/n} f_{10}(x,t)dt + \int_x^{k/n} f_{01}(s,\tfrac{\ell}{n})ds\} . \quad (2.3)$$

Forgetting for a moment that f_{10} and f_{01} are derivatives of one function f, we maximize $E(f_{10},f_{01})$ under the (weaker) conditions that $f_{10} \in D_{10}$

and $f_{01} \in D_{01}$, where (x and y are fixed)

$$D_{10} := \{h \in C(\mathbb{R}^2) \; ; \; h(x,y) = \frac{1}{2}\bar{\omega}_{10},\ \omega(h;\delta_1,\delta_2) = \bar{\omega}_{10}\} ,$$

with a similar definition for D_{01}.

By the one-dimensional result, for each fixed f_{01} the quantity E_1, and hence E, is maximal for $f_{10}(s,t) = \bar{\omega}_{10}\, \tilde{g}'(s) \in \bar{D}_{10}$ (closure under pointwise convergence) (cf. (2.2)). Similarly, E_2 and hence E is maximized by $f_{01}(s,t) = \bar{\omega}_{01}\, \tilde{g}'(t) \in \bar{D}_{01}$ for each fixed f_{10}. As $\bar{\omega}_{10}\, \tilde{g}'(s)$ and $\bar{\omega}_{01}\, \tilde{g}'(t)$ (as functions of (s,t)) are elements of $\bar{D}_{10}$ and $\bar{D}_{01}$ respectively, it follows that E is maximal for $f_{10}(s,t) = \bar{\omega}_{10}\, \tilde{g}'(s)$ and $f_{01}(s,t) = \bar{\omega}_{01}\, \tilde{g}'(t)$. As finally $\bar{\omega}_{10}\, \tilde{g}'(s) = \tilde{f}_{10}(s,t)$ and $\bar{\omega}_{01}\, \tilde{g}'(t) = \tilde{f}_{01}(s,t)$, with $\tilde{f}$ given by (2.4) below, and as $\tilde{f}$ is the pointwise limit of functions in $C^1(\Delta)$, we now have proved the following theorem:

THEOREM 2.1. Let $n \in \mathbb{N}$, $(x,y) \in \Delta$ and $\delta_1 = \delta_2 > 0$ be fixed. Then

$$\sup\{|B_n^{(2)}(f;x,y) - f(x,y)| \; ; \; f \in C^1(\Delta),\ \omega_{10}(f) = \bar{\omega}_{10},\ \omega_{01}(f) = \bar{\omega}_{01}\} = B_n^{(2)}(\tilde{f};x,y),$$

where $\tilde{f}$ is given by (cf. (2.2))

$$\tilde{f}(s,t) \equiv \tilde{f}(s,t;x,y;\delta_1,\delta_2) = \frac{1}{2}\bar{\omega}_{10}|s-x| + \frac{1}{2}\bar{\omega}_{01}|t-y| +$$

$$+ \sum_{j=1}^{\infty} \{\bar{\omega}_{10}(|s-x| - j\delta_1)_+ + \bar{\omega}_{01}(|t-y| - j\delta_2)_+\} = \tag{2.4}$$

$$= \bar{\omega}_{10}\, \tilde{g}(s;x;\delta_1) + \bar{\omega}_{01}\, \tilde{g}(t;y;\delta_2) .$$

From the proof it follows that theorem 2.1 also holds if we replace $B_n^{(2)}$ by $B_{n,m}^{(2)}$, and make the corresponding changes; in particular, δ_1 and δ_2 need not be equal.

REMARK

For $f \in C^1(\Delta)$ some of the integrands in (2.3) may be undefined for values of (s,t) with $s+t > 1$. It is not hard to see that the function

$$f(s,1-s) + \int_{1-s}^{t} f_{01}(1-v,v)\,dv$$

extends f, with all of its required properties, to $\square$.

3. Degrees of Approximation for $B_n^{(2)}$ and $B_{n,m}^{(2)}$

3.1. The object of this section is to obtain results analogous to (1.2) for $B_n^{(2)}$ with $f \in C^1(\Delta)$ and $B_{n,m}^{(2)}$ with $f \in C^1(\Box)$. As these results follow in a straightforward manner from theorem 2.1 together with results from (1) and (3), we state them without proof.

THEOREM 3.1. For $n \in \mathbb{N}$

$$|B_n^{(2)}(f;x,y) - f(x,y)| \leq c_n(x)\omega_{10}(f;n^{-1},n^{-1}) + c_n(y)\omega_{01}(f;n^{-1},n^{-1}), \quad (3.1)$$

where

$$c_n(u) = \frac{1}{2}u(1-u) + \frac{1}{2}n^{-1}\alpha(1-\alpha) \qquad (0 \leq u \leq 1;\ \alpha = nu - [nu]), \quad (3.2)$$

[nu] denoting the largest integer not exceeding nu. Moreover, the values of $c_n(x)$ and $c_n(y)$ as given by (3.2) are best possible in (3.1).

COROLLARY. For $n \in \mathbb{N}$

$$\| B_n^{(2)}(f) - f \|_\Delta \leq \frac{1}{4}\{\omega_{10}(f;n^{-1},n^{-1}) + \omega_{01}(f;n^{-1},n^{-1})\} .$$

THEOREM 3.2. For $n \in \mathbb{N}$

$$\| B_n^{(2)}(f) - f \|_\Delta \leq \frac{1}{4}n^{-\frac{1}{2}}\{\omega_{10}(f;n^{-\frac{1}{2}},n^{-\frac{1}{2}}) + \omega_{01}(f;n^{-\frac{1}{2}},n^{-\frac{1}{2}})\} , \quad (3.3)$$

where the constant $\frac{1}{4}$ is best possible. If we replace $n \in \mathbb{N}$ by $n \geq 2$ in (3.3), then the best constant drops to $(2\sqrt{5}-1)/16 = 0.217$.

We remark that (3.3) improves on an inequality due to Stancu (5).

3.2. Similar results hold for the operators $B_{n,m}^{(2)}$. In fact, one has

THEOREM 3.3. For $(n,m) \in \mathbb{N}^2$

$$|B_{n,m}^{(2)}(f;x,y) - f(x,y)| \leq c_n(x)\omega_{10}(f;n^{-1},m^{-1}) + c_m(y)\omega_{01}(f;n^{-1},m^{-1}) ,$$

with $c_n(x)$ and $c_m(y)$ given by (3.2). The values of $c_n(x)$ and $c_m(y)$ are best possible.

COROLLARY. For $(n,m) \in \mathbb{N}^2$

$$\| B_{n,m}^{(2)}(f) - f \|_\Box \leq \frac{1}{8}\{(1+n^{-1})\omega_{10}(f;n^{-1},m^{-1}) + (1+m^{-1})\omega_{01}(f;n^{-1},m^{-1})\} .$$

As this inequality remains true if the ω's are replaced by w's, we also obtain an improvement on Marlewski's estimate (1.6).

Finally we state an improvement of Stancu's inequality (1.5), viz.

THEOREM 3.4. For $(n,m) \in \mathbb{N}^2$

$$\| B_{n,m}^{(2)}(f) - f \|_{\square} \le \frac{1}{4}\{n^{-\frac{1}{2}}\omega_{10}(f;n^{-\frac{1}{2}},m^{-\frac{1}{2}}) + m^{-\frac{1}{2}}\omega_{01}(f;n^{-\frac{1}{2}},m^{-\frac{1}{2}})\} ,$$

where the constant $\frac{1}{4}$ is best possible.

ACKNOWLEDGMENT

We thank H.G. ter Morsche for helpful comments, especially on section 2.

REFERENCES

1. Schurer, F., and Steutel, F.W., *J. Approximation Theory*. *19*, 69 (1977).
2. Lorentz, G.G., "Bernstein Polynomials". Univ. of Toronto Press, Toronto, (1953).
3. Schurer, F., Sikkema, P.C., and Steutel, F.W., *Nederl. Akad. Wetensch. Proc. Ser. A*. *79*, 231 (1976).
4. Schurer, F., and Steutel, F.W., *in* "Multivariate Approximation Theory" (W. Schempp and K. Zeller, ed.), p. 413. Birkhäuser Verlag, Basel, (1979).
5. Stancu, D.D., *An. Şti. Univ. "Al. I. Cuza" Iaşi Sect. I a Mat.* *9*, 49 (1963).
6. Marlewski, A., Personal Communication.

SOME REMARKS ON APPROXIMATION IN $C(\Omega)$

Boris Shekhtman

Department of Mathematics
Kent State University
Kent, Ohio

INTRODUCTION

Let $F(\Omega)$ be a space of continuous functions defined on a compact Hausdorff space Ω. For example, $F(\Omega)$ could be $C(\Omega)$, $W_{m,p}(\Omega)$, $C^k(\Omega)$, $Lip_\alpha(\Omega)$, etc. Next, let $\Delta_n = \{t_j^{(n)}\}_{j=1}^n$ be some finite set of points in Ω. We are interested in studying approximation operators $p_n = p_n(\Delta_n)$ which can be factored as follows:

For every $x(t) \in F(\Omega)$,

$$p_n: x(t) \xrightarrow{i_n} \{x(t_j^{(n)})\} \xrightarrow{v_n} (p_n x)(t). \tag{1}$$

We wish to show that approximation in $C(\Omega)$ is significantly different from approximation in the other spaces mentioned above.

I. CONTRACTIVE APPROXIMATION

In this section, we shall show that the assumption $\|p_n\| \le 1$ (which is desirable in most of the spaces $F(\Omega)$) unfortunately implies a very slow convergence of $p_n \to 1$ in $C(\Omega)$.

More precisely, let Ω be a measurable convex compact domain in $\mathbb{R}^s$. We then establish

Copyright © 1980 by Academic Press, Inc.
All rights of reproduction in any form reserved.
ISBN: 0-12-171050-5

Theorem 1. For any choice of $\Delta_n = \{t_j^{(n)}\}_{j=1}^{n^s}$ and of $\beta > 0$, there exist functions $x_0(t) \in C(\Omega)$; $x_1(t) \in C^1(\Omega)$; $x_2(t) \in C^\infty(\Omega)$ such that, for any choice of v_n with

$$\|v_n\| \leq 1, \tag{2}$$

there are positive constants c_i, $i = 1, 2, 3$, for which

$$\begin{cases} n^{\beta}\|x_0 - P_n x_0\| \geq c_1, \\ n^{\beta+1}\|x_1 - P_n x_1\| \geq c_2, \\ n^2\|x_2 - P_n x_2\| \geq c_3, \text{ for infinitely many } n. \end{cases} \tag{3}$$

The proof of this theorem is based on the following

Lemma 2. For any choice of $\Delta_n = (t_j^{(n)})_{j=1}^{n^s} \subset \Omega$ and of $\varepsilon \in (0, \frac{1}{2}\text{mes}(\Omega))$, there exists a point $t^* \in \Omega$ such that

$$\text{dist}(t^*, \Delta_n) \geq \frac{\varepsilon^{\frac{1}{s}}}{A^{\frac{1}{s}} n} \text{ for infinitely many } n, \tag{4}$$

where A is the volume of the s-dimensional unit ball.

Proof. Suppose that the conclusion (4) is not true. Then, for every $t \in \Omega$, there exists an integer $N = N(t)$ such that

$$\text{dist}(t^*, \Delta_n) < \frac{\varepsilon^{\frac{1}{s}}}{A^{\frac{1}{s}} n} \text{ for all } n \geq N.$$

Then, the map $t \to N(t)$ defines a measurable function on Ω. Indeed, for every integer K

$$\{t: N(t) = k\} = \{t: \operatorname{dist}(t, \Delta_k) < \frac{\varepsilon^{\frac{1}{s}}}{A^{\frac{1}{s}} k}, \forall k \geq K\}$$

$$= \bigcap_{k>K} \bigcup_{j=1}^{k^s} B(t_j^{(k)}, \frac{\varepsilon^{\frac{1}{s}}}{A^{\frac{1}{s}} k}),$$

where $B(t, r)$ is the ball with the center t and radius r. By Lusin's theorem, we can conclude that there exists a set $L \subset \Omega$ such that $\operatorname{meas}(\Omega \setminus L) < \varepsilon$ and that $N(t)$ is continuous on L. Thus, there exists a number $\tilde{N}$ such that

$$N(t) \leq \tilde{N} \text{ for all } t \in L.$$

This shows that, for all $t \in L$,

$$\operatorname{dist}(t, \Delta_n) \leq \frac{\varepsilon^{\frac{1}{s}}}{A^{\frac{1}{s}} n} \text{ for all } n \geq \tilde{N}.$$

On the other hand,

$$L \subset \{t \in \Omega: \operatorname{dist}(t, \Delta_n) \leq \frac{\varepsilon^{\frac{1}{s}}}{A^{\frac{1}{s}} n} \; \forall n \geq \tilde{N}\}$$

$$\subset \{t \in \Omega: \operatorname{dist}(t, \Delta_{\bar{N}}) \leq \frac{\varepsilon^{\frac{1}{s}}}{A^{\frac{1}{s}} \tilde{N}}\}.$$

Thus, $\operatorname{meas}(L) \leq \operatorname{meas}\{t \in \Omega: \operatorname{dist}(t, \Delta_{\tilde{N}}) \leq \frac{\varepsilon^{\frac{1}{s}}}{A^{\frac{1}{s}} \tilde{N}}\}$

$$\leq \sum_{j=1}^{\bar{N}^s} \operatorname{meas}(B(t_j^{\bar{N}}, \frac{\varepsilon^{\frac{1}{s}}}{A^{\frac{1}{s}} \tilde{N}})) = \tilde{N}^s A \left(\frac{\varepsilon^{\frac{1}{s}}}{A^{\frac{1}{s}} \tilde{N}}\right)^s = \varepsilon,$$

which contradicts $\operatorname{meas}(\Omega \setminus L) < \varepsilon$. ∎

Proof of Theorem 1. To prove this theorem, it is sufficient to consider the function $x_2(t) = 1 - [\text{dist}(t, t^*)]^2$. We have

$$\|x_2(t) - v_n i_n x_2(t)\| \geq \|x_2(t)\| - \|v_n i_n x_2(t)\|$$

$$\geq 1 - \|i_n x_2(t)\| \geq 1 - 1 + (\varepsilon/A)^{2/s}/n^2,$$

for infinitely many n. Analogously, for every integer m, the functions

$$x_1(t) = 1 - [\text{dist}(t, t^*)]^{\frac{2m-1}{2m}}$$

and

$$x_0(t) = 1 - [\text{dist}(t, t^*)]^{\frac{2}{m}}$$

give us the rest of Theorem 1. ∎

II. DIVERGENCE OF INTERPOLATION PROJECTIONS

In this section, we will assume that p_n are interpolation projections with respect to Δ_n, i.e.,

$$(p_n x)(t_j^{(n)}) = x(t_j^{(n)}) \text{ for all } x \in F(\Omega).$$

This can happen iff (cf. [1]) the operators v_n are generalized inverses with respect to the operators i_n. We will also assume that the mesh points Δ_n satisfy the condition:

$$\text{for every } t \in \Omega, \text{ there exist } t_n \in 0_n\text{: } t_n \to t. \tag{5}$$

Condition (5) by itself cannot guarantee that

$$p_n x \to x \text{ pointwise for all } x \in F(\Omega). \tag{6}$$

Thus, we further assume that

$$\|p_n\| \leq \text{const.} \tag{7}$$

Do (5) and (7) imply (6)?

Proposition 3. If the imbedding $F(\Omega) \hookrightarrow C(\Omega)$ is compact, and if $\|p_n\|_F$ are uniformly bounded, then

$$p_n x \to x \text{ pointwise for each } x \in F(\Omega).$$

Proof. For each t, let $t_n \in \Delta_n$ be such that

$$t_n \to t.$$

Given any function $x(t) \in F(\Omega)$, we obtain

$$x(t) - (p_n x)(t) = x(t) - x(t_n) + (p_n x)(t_n) - (p_n x)(t).$$

Clearly, $x(t_n) \to x(t)$ since x is continuous. The set of functions $\{p_n x\}$ is uniformly bounded in $F(\Omega)$, and thus, it is compact in $C(\Omega)$. By the Arzela-Ascoli Theorem, the set $\{p_n x\}$ is equicontinuous, and thus,

$$(p_n x)(t) - (p_n x)(t_n) \to 0 \quad \blacksquare$$

The following example shows that the conclusion of Proposition 1 does not hold if we assume that $F(\Omega) = C(\Omega)$.

Example 4. First, we introduce the numbers

$$\bar{\varepsilon}_k^{(n)} := \frac{t_{k+1}^{(n)} - t_k^{(n)}}{2^n} \qquad \underline{\varepsilon}_k^{(n)} := \frac{t_k^{(n)} - t_{k-1}^{(n)}}{2^n};$$

$$\bar{t}_k^{(n)} := t_k^{(n)} + \bar{\varepsilon}_k^{(n)} \qquad \underline{t}_k^{(n)} := t_k^{(n)} - \underline{\varepsilon}_k^{(n)}.$$

For every function $x(t)$, let

$$(p_n x)(t) := \begin{cases} \dfrac{x(t_k^{(n)})}{\bar{\varepsilon}_k^{(n)}} (\bar{t}_k^{(n)} - t) & \text{if } t \in [t_k^{(n)}, \bar{t}_k^{(n)}]; \\ \dfrac{x(t_{k+1}^{(n)})}{\underline{\varepsilon}_{k+1}^{(n)}} (t - \underline{t}_{k+1}^{(n)}) & \text{if } t \in [\underline{t}_{k+1}^{(n)}, t_{k+1}^{(n)}]; \\ 0 & \text{if } t \in [\bar{t}_k^{(n)}, \underline{t}_{k+1}^{(n)}], \end{cases} \tag{8}$$

which is schematically shown in the figure below.

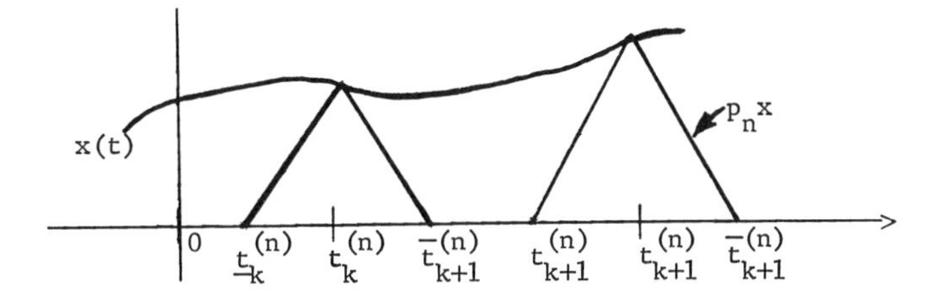

Now, the projections of (8) are uniformly bounded, and they interpolate with respect to Δ_n. However, they do not converge for any non-zero function, for any choice of Δ_n. Moreover, divergence takes place in all points of the interval $[0, 1]$, with the exception of those points which belong to the intersections $\bigcap_{n>N(t)} \Delta_n$, for some $N(t)$.

III. WHY IS THE SPACE OF PIECEWISE-LINEAR FUNCTIONS DENSE IN C[0, 1]?

Our next item is to analyze the reason for the divergence of projections of (8), and to determine which additional assumptions on the range p_n, together with the assumptions of (5), (6) and (7), can guarantee convergence.

Since the projections p_n are interpolation projections, the restriction ran $p_n | [t_k^{(n)}, t_{k+1}^{(n)}]$ depends on two parameters, and thus forms a two-dimensional subspace:

$$\text{ran } p_n | [t_k^{(n)}, t_{k+1}^{(n)}] = \text{span } \{s_k^{(n)}, \sigma_k^{(n)}\},$$

where $s_k^{(n)}, \sigma_k^{(n)} \in C[t_k^{(n)}, t_{k+1}^{(n)}]$. Set $s_k^{(n)}(t) \equiv 1$ for all k, n. Then, for each $t \in [t_k^{(n)}, t_{k+1}^{(n)}]$, the interpolation projection p_n can be written as

$$(p_n x)(t) = x(t_k^{(n)}) + [x(t_{k+1}^{(n)}) - x(t_k^{(n)})] z_k^{(n)}(t), \tag{9}$$

where

$$z_k^{(n)}(t) := \frac{\sigma_k^{(n)}(t) - \sigma_k^{(n)}(t_k^{(n)})}{\sigma_k^{(n)}(t_{k+1}^{(n)}) - \sigma_k^{(n)}(t_k^{(n)})}$$

(in case the denominator does not vanish). The expression $z_k^{(n)}(t)$ does not depend on $x(t)$. It is either bounded uniformly in n, in which case $(p_n x)(t) \to x(t)$ (since $x(t_{k+1}^{(n)}) - x(t_k^{(n)}) \to 0$), or it is unbounded. In the latter case, there exists a function $x(t)$ such that

$$[(t_{k+1}^{(n)}) - x(t_k^{(n)})]\; z_n^{(k)}(t) \to \infty,$$

which implies that the norms $\|p_n\|$ are also unbounded.

Unlike the projections of (8), the ranges of the projections of (9) do separate closed disjoint subsets of $[0, 1]$.

The convergence of the projections of (9) can be partially explained by the following result.

Theorem 4. Let K be a compact Hausdorff space, let $C(K)$ be the space of continuous functions on K, and let $X \subset C(K)$ be a subspace such that, for every two closed disjoint sets $F_1, F_2 \subset K$ and every $\epsilon > 0$, there exists a function $x = x(F_1, F_2, \epsilon)$ with the properties

$$\|x\| \leq 1 \;;\; 1 - \epsilon < x|_{F_1} < 1 + \epsilon \;;\; -\epsilon < x|_{F_2} < \epsilon. \tag{10}$$

Then, X is dense in $C(K)$.

Proof. We shall show that if μ is a regular Borel measure on K which annihilates X (i.e., $\mu \in X^{\perp}$), then $\mu = 0$. By regularity, it suffices to show that $\mu(F) = 0$ for every compact $F \subset K$. Let $F^C := K \setminus F$. Then, we can choose compact sets $F_n \subset F^C$ such that

$$\mu(F^C \setminus F_n) < \frac{1}{n}.$$

Now, we can also choose $x_n \in X$ such that

$$\|x_n\| \leq 1 \;;\; \|x_n|_F\| < \frac{1}{n} \;;\; \|x_n|_{F_n} - 1\| < \frac{1}{n}.$$

Clearly, the functions x_n converge to the characteristic function $\varkappa_F$ of F, almost everywhere with respect to $|\mu|$. Thus, we obtain

$$0 = \int_K x_n \, d\mu \to \int_K \varkappa_F \, d\mu = \mu(F). \quad \blacksquare$$

Corollary 5. Let $X \subset C(K)$ be such that, for every two closed disjoint sets $F_1, F_2 \subset K$, there exists a function $x \in X$ such that

$$\|x\| \leq 1 \; ; \; x|_{F_1} \equiv 0 \; ; \; x|_{F_2} \equiv 1.$$

Then, X is dense in C(K).

The author wishes to express his appreciation to Professor Varga for his assistance and interest in this work.

REFERENCES

1. Shekhtman, B., Abstract interpolation theory, Uch. Zap. Tartus. Univ. 498(1978), 82-93. (Russian).

VORONOVSKAYA TYPE FORMULAE FOR CONVOLUTION OPERATORS APPROXIMATING WITH GREAT SPEED

P.C. Sikkema

Department of Mathematics
University of Technology
Delft, Netherlands

I. INTRODUCTION

The convolution operators U_ρ to be considered are defined by

$$(U_\rho f)(x) = I_\rho^{-1} \int_{-\infty}^{\infty} f(x-t)\, \beta^\rho(t)dt \qquad (\rho \geq 1),$$

where f is defined, continuous and bounded on R and $\beta(t)$ satisfies on R the following four conditions:

(i) $0 \leq \beta(t) \leq 1$, $\beta(0) = 1$;

(ii) β is continuous at $t = 0$;

(iii) for all $\delta > 0$, $\sup_{|t|\geq\delta} \beta(t) < 1$;

(iv) $\beta \in L_1(R)$.

I_ρ is a normalizing factor: $I_\rho = \int_{-\infty}^{\infty} \beta^\rho(t)dt$.

The aim of this paper is to show that by choosing $\beta(t)$ in a neighbourhood of $t = 0$ in a special way the order of approximation of f(x) by $(U_\rho f)(x)$ if $\rho \to \infty$ at a point x where $f''(x)$ exists, is very high: it involves $\exp(\rho^{\frac{1}{2}})$. Full proofs and more relevant literature will be published elsewhere.

II. THE APPROXIMATION

THEOREM I. *Under the above conditions concerning* $\beta(t)$ *and* $f(t)$

$$(U_\rho f)(x) - f(x) \tag{1}$$

tends to zero if $\rho \to \infty$ *at all points* x *of* R.

The proof of theorem I is given in [1].

The *speed* with which (1) tends to zero if $\rho \to \infty$ depends upon the behaviour of $\beta(t)$ in a neighbourhood of $t = 0$. In this paper it is assumed

Copyright © 1980 by Academic Press, Inc.
All rights of reproduction in any form reserved.
ISBN: 0-12-171050-5

that for a certain $\delta > 0$ $\beta(t)$, satisfying conditions (i) - (iv) possesses the additional property

$$\text{(v)} \quad \beta(t) = \begin{cases} 1 + (\log ct^{\alpha})^{-1} + r\,|\log ct^{\alpha}|^{-q} & (0 \le t \le \delta) \\ 1 + (\log c'|t|^{\alpha'})^{-1} + r'\,|\log c'|t|^{\alpha'}|^{-q'} & (-\delta \le t < 0). \end{cases}$$

with $\alpha > \alpha' > 0$, $c > 0$, $c' > 0$, $q > 1$, $q' > 1$ and δ is chosen so small that $\beta > 0$ if $|t| \le \delta$.

If at a point $x \in R$ $f''(x)$ exists then

$$\text{(2)} \quad f(x-t) - f(x) = -t\,f'(x) + \tfrac{1}{2}t^2\,f''(x) + t^2 h_x(t),$$

where $h_x(t)$ is bounded on R and by defining $h_x(0) = 0$, $h_x(t)$ is continuous at $t = 0$. Then, using (2)

$$\text{(3)} \quad (U_\rho f)(x) - f(x) = I_\rho^{-1}\ \{-I_{1\rho}(\delta)\,f'(x) + \tfrac{1}{2}I_{2\rho}(\delta)\,f''(x) + J_\rho(\delta) + K_\rho(\delta)\},$$

with

$$I_{\nu\rho}(\delta) = \int_{-\delta}^{\delta} t^{\nu}\,\beta^{\rho}(t)dt \quad (\nu=0,1,2), \qquad J_\rho(\delta) = \int_{-\delta}^{\delta} t^2 h_x(t)\,\beta^{\rho}\,(t)dt,$$

$$\text{(4)} \qquad K_\rho(\delta) = \int_{|t|\ge\delta} \{f(x-t)-f(x)\}\,\beta^{\rho}\,(t)dt.$$

In order to determine the asymptotic behaviour of $I_{\nu\rho}(\delta)$ if $\rho \to \infty$ we set

$$I_{\nu\rho}(\delta) = \int_0^{\delta} t^{\nu} e^{\rho\log\beta(t)}\,dt + (-1)^{\nu} \int_0^{\delta} t^{\nu} e^{\rho\log\,\beta(-t)}\,dt$$

$$\text{(5)} \qquad = A_{\nu\rho}(\delta) + (-1)^{\nu}\,B_{\nu\rho}(\delta).$$

If δ is small enough we can write

$$A_{\nu\rho}(\delta) = \int_0^{\delta} t^{\nu}\,\exp\,\rho\,\{(\log ct^{\alpha})^{-1} + r|\log ct^{\alpha}|^{-q}$$

$$- \tfrac{1}{2}((\log ct^{\alpha})^{-1} + r|\log ct^{\alpha}|^{-q})^2 + s(\log ct^{\alpha})^{-3}\}\,dt,$$

with $s = s(t)$, $|s| \le 1$ on $[0,\delta]$. Substitution of $-\log ct^{\alpha} = u$ then gives

$$\text{(6)} \quad A_{\nu\rho}(\delta) = (\alpha c^{a})^{-1} \int_{-\log c\delta^{\alpha}}^{\infty} \exp(-\lambda(\rho,u))du$$

where

$$\lambda(\rho,u) = au + \rho\,\{\frac{1}{u} - \frac{r}{u^q} + \frac{1}{2}\,(\frac{1}{u} - \frac{r}{u^q})^2 + \frac{s^*}{u^3}\}\,,$$

s^* being bounded on $[-\log c\delta^\alpha, \infty)$ and $a = (\nu + 1)/\alpha$. On the interval of integration lies one absolute maximum, say at $u_0 = u_0(\rho, a)$ of the integrand. In order to determine u_0 the equation $\frac{\partial \lambda}{\partial u} = 0$ will be solved. By considering it, it is clear that the development of u_0 as function of ρ begins with $(a^{-1}\rho)^{\frac{1}{2}}$. Hence we need only consider the equation

$$(7) \quad a - \rho\{u^{-2} - qru^{-q-1} + u^{-3}\} + O(\rho^{-\frac{1}{2}}) = 0.$$

If $q > 2$ then $u_0 = (a^{-1}\rho)^{\frac{1}{2}} + \frac{1}{2} + o(1)$ and

$$(8) \quad \lambda(\rho, u_0) = 2(a\rho)^{\frac{1}{2}} + \tfrac{1}{2}a + o(1).$$

If $1 < q \le 2$, let n be that positive integer with

$$1 + (n+1)^{-1} < q \le 1 + n^{-1}.$$

Substitution in (7) of

$$u_0 = (a^{-1}\rho)^{\frac{1}{2}} + b_1\rho^{\mu_1} + \ldots + b_m\rho^{\mu_m} + o(1)$$

with $\frac{1}{2} > \mu_1 > \ldots > \mu_m \ge 0$ gives $m = n$, $\mu_i = 2^{-1}(1 - i(q-1))$ $(i = 1, \ldots, n)$ and the coefficients $b_1, \ldots, b_n$ are uniquely determined. Hence

$$(9) \quad \lambda(\rho, u_0) = 2(a\rho)^{\frac{1}{2}} + c_1\rho^{\mu_1} + \ldots + c_n\rho^{\mu_n} + \tfrac{1}{2}a + o(1).$$

For the determination of the asymptotic behaviour of $A_{\nu\rho}(\delta)$ the integral in the right-hand side of (6) is written as

$$(10) \quad \int_{-\log c\delta^\alpha}^{\infty} = \int_{-\log c\delta^\alpha}^{u_0-\rho^\sigma} + \int_{u_0-\rho^\sigma}^{u_0+\rho^\sigma} + \int_{u_0+\rho^\sigma}^{\infty} \quad ,$$

σ satisfying the inequalities

$$\max(\tfrac{1}{4}, \mu_1) < \sigma < \tfrac{1}{2}.$$

Of the three integrals in the right-hand side of (10) the middle one is written as

$$\exp(-\lambda(\rho, u_0)) \int_{u_0-\rho^\sigma}^{u_0+\rho^\sigma} \exp\{\lambda(\rho, u_0) - \lambda(\rho, u)\}\, du$$

$$= \exp(-\lambda(\rho, u_0)) \int_{u_0-\rho^\sigma}^{u_0+\rho^\sigma} \exp\{(u - u_0)\left(\frac{\partial \lambda}{\partial u}\right)_{u_0} + \tfrac{1}{2}(u - u_0)^2 \left(\frac{\partial^2 \lambda}{\partial u^2}\right)_{u^*}\}\, du,$$

where u^* lies in between u and u_0. It can be proved that the latter expression is equivalent to $a^{-1}\pi^{\frac{1}{2}}(a\rho)^{\frac{1}{4}}\exp(-\lambda(\rho, u_0))$ and that the first and third integral in the right-hand side of (10) are of a lower order. Consequently

(11) $A_{\nu\rho}(\delta) \sim \pi^{\frac{1}{2}} ((\nu + 1)c^{a})^{-1} (a\rho)^{\frac{1}{4}} \exp(-\lambda(\rho, u_0)) \qquad (\rho \to \infty).$

If, as is assumed, $\alpha > \alpha'$ then in (5) $B_{\nu\rho}(\delta) = o(A_{\nu\rho}(\delta))$ and hence

(12) $\qquad I_{\nu\rho}(\delta) \sim A_{\nu\rho}(\delta) \qquad (\rho \to \infty).$

Combining (11) and (12) setting $\nu = 0, 1, 2$ and substituting the results in (3) yields

(13) $(U_\rho f)(x) - f(x) = -(2^{\frac{3}{4}} c^{1/\alpha})^{-1} \exp(\psi(\rho,\alpha)).f'(x)(1+o(1))$
$\qquad + I_\rho^{-1} \{ J_\rho(\delta) + K_\rho(\delta)\}$ with

$\psi(\rho,\alpha) = -\lambda(\rho, 2/\alpha, u_0(\rho, 2/\alpha)) + \lambda(\rho, 1/\alpha, u_0(\rho, 1/\alpha))$

(14) $\quad = d_0(\alpha^{-1}\rho)^{\frac{1}{2}} + d_1\rho^{\mu_1} + \ldots + d_n\rho^{\mu_n}$ with $d_0 = 2(2^{\frac{1}{2}} - 1).$

as is shown by substitution of (9). Because the latter term in (13) is small o as compared with the coefficient of $f'(x)$ there holds

THEOREM 2. *If $\beta(t)$ satisfies conditions (i) - (v), if $f(t)$ is continuous and bounded on R and if $f''(x)$ exists then*

$$\lim_{\rho\to\infty} \exp(\psi(\rho,\alpha)). \{ (U_\rho f)(x) - f(x) \} = \gamma_1 f'(x),$$

with $\psi(\rho,\alpha)$ defined by (14) and $\gamma_1 = -(2^{\frac{3}{4}} c^{1/\alpha})^{-1}$.

III. EXAMPLES

1. If $q = 1\frac{1}{2}$ theorem 2 gives
   $$\lim_{\rho\to\infty} \exp \{d_0(\alpha^{-1}\rho)^{\frac{1}{2}} + d_1\rho^{\frac{1}{4}} + d_2 \} . \{ (U_\rho f)(x) - f(x) \} = \gamma_1 f'(x)$$
   with d_0 given by (14), $d_1 = -(2^{\frac{3}{4}}-1)\alpha^{-\frac{3}{4}}r$, $d_2 = (16\alpha)^{-1}(8-3r^2)$.

2. If $q \geq 2$ theorem 2 shows that
   $$\lim_{\rho\to\infty} \exp \{d_0(\alpha^{-1}\rho)^{\frac{1}{2}} + d_1\} .\{ (U_\rho f)(x) - f(x) \} = \gamma_1 f'(x)$$
   with d_0 given by (14), $d_1 = (2\alpha)^{-1}$ if $q > 2$, $d_1 = (2\alpha)^{-1}(1-2r)$ if $q = 2$.

REFERENCES

1. Sikkema, P.C., Approximation Formulae of Voronovskaya-Type for Certain Convolution Operators. J. Approximation Theory 26(1979) 26-45.

A CONSTRUCTIVE METHOD OF APPROXIMATING THE RIEMANN MAPPING FUNCTION

Annette Sinclair

Department of Mathematics
Purdue University
West Lafayette, Indiana

Suppose Ω is a convex, or almost convex, region bounded by a very smooth Jordan curve Γ and that a point of Ω, say zero, is assigned. Then the Riemann mapping function f^* mapping Ω conformally onto the unit disk is continuous on $\bar{\Omega}$ and is uniquely determined if it is required that $f(0) = 0$ and $f'(0) > 0$. Methods are described here for determination of a polynomial approximate, also an approximate of the form $z \exp(P_N(z))$.

In the 1950's much work was done on approximation of the mapping function by other methods. (See Gaier's "Konstructive Methoden der konformen Abbildung", Springer, 1964; also Applied Math. Series Nat. Bureau of Standards, vol. 18, 1952, edit. by E. F. Beckenbach, and vol. 42, 1955, edit. by John Todd.) Usually an integral equation was used or an expansion was found of the mapping in terms of an integral kernel. At the present conference, S. W. Ellacott has reported on "Applications of Approximation Theory in Numerical Conformal Mapping".

The method the author is suggesting for determination of an approximate for f^* involves mapping trial points z_k on Γ into preassigned points w_k on $|w| = 1$. Equally spaced

Copyright © 1980 by Academic Press, Inc.
All rights of reproduction in any form reserved.
ISBN: 0-12-171050-5

points, say the Nth roots of unity, are assigned on $|w| = 1$ and are kept fixed. Points z_k on Γ, at first spaced by dividing the parameter interval for Γ into N equal parts, are tentatively assigned.

To determine a polynomial approximate, first the polynomial P_N is found such that $P_N(0) = 0$ and $P_N(z_k) = w_k^*$. The z_k are then varied, one at a time, over a small subarc of Γ, with a new z_k being substituted for the old if so doing decreases the maximum deviation of $|P_N(z)|$ from $|w| = 1$. There are other appropriate criteria for a better position of the z_k, for example increase of $[P_N'(0) / \max|P_N(z)|, z \in \Gamma]$. For conformality everywhere interior to Γ, it is necessary that P_N' have no zero on the region Ω.

There is no reason to suppose that a good polynomial approximate of low degree does exist. The plan was to begin with a low degree polynomial, then insert new z_k on Γ between those z_k which, by the variation procedure, have been moved far apart, also to insert additional w_k between the corresponding w_k on $|w| = 1$. Such a program for change of degree was not written.

A program for determining a polynomial approximate P_N of degree N was written and executed by a Purdue undergraduate Paul W. Slatin, who has since graduated. Starting again for each N, the program was used to determine polynomial approximates of degrees 3, 4, 5, and 6, for mapping a given Jordan curve onto the unit circle.

If the derivative of an approximate has zeros interior to Γ, then conformality is lost at those points. If such zeros of P_N' are near Γ and if the image of Γ is close to the circle, the map can be blown up a bit, so that $C\ P_N(z)$ may still map Γ onto a curve near the circle, with its intersections exterior to the unit circle and so that Ω is mapped one-to-one onto the component of $C(P_N(\Gamma))$ containing the unit disk.

In the class of functions univalent on Ω such that $|f(z)| < 1$, with $f(0) = 0$ and $f'(0) > 0$, the mapping function f^* is that function which maximizes $f'(0)$. If this class is enlarged to include those functions not one-to-one, the sup of $f'(0)$ is not increased. (See Rudin's "Real and Complex Analysis", p. 275.) Thus, under appropriate criteria for selecting the best location of the z_k on Γ, it seems reasonable to expect that for large N, an approximate would yield a conformal map.

Our computer results were intended only to test this method. Points on the boundary of an ellipse with major and minor axes in the ratio of $\sqrt{7}: 1$ were mapped into points on the circle. On Purdue's Control Data Corporation 6500 system, for a rather complete print-out and a Versatec graph, the time was 6 seconds, at an estimated cost of 95 cents. At each step of the program, closeness of fit of the image of Γ to the circle was checked at 10N points. As the objective was to test the method, odd degree polynomials were usually used in order to offset advantages of symmetry.

After this program for a polynomial approximate was working, a program for an exponential type approximate was considered. Since the function, say g^*, for mapping Ω onto a circular disk of radius R, can be written in the form $\exp(\log z + h(z))$, an approximate g_N of the form

$\exp(\log z + P_N(z))$ might be used. If one requires that $P_N(0) = 0$, then $g_N'(0) = 1$. Although the radius R of the disk $|w| < R$ is then unknown, it is known to be between Min $|z|$ and Max $|z|$ for z on Γ. (Apply Bieberbach, "Conformal Mapping", pp. 134-5.) For the first trial R, Max $|z|$ would be an appropriate approximate. Then P_N should be determined so that $P_N(z_k) = \log R + \log(w_k/z_k)$ and $P_N(0) = 0$. After one lap around Γ for varying the z_k, the R used should be replaced by $\exp(\max E(z))$, (where $E(z) = \log|z| + \operatorname{Re} P_N(z)$) only if doing so decreases R.

A computer program for determination of such an exponential type approximate has not been written. The author would be glad to have any useful ideas presented here put into effect.

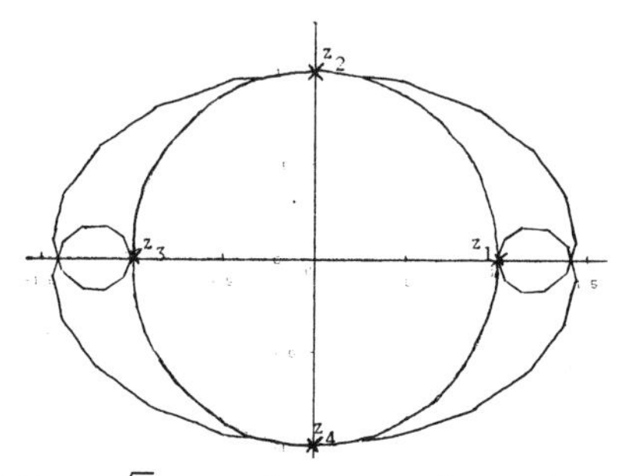

Fig. 1. Ellipse $x = \sqrt{7}\cos\theta$, $y = \sin\theta$ into Circle by $P_4(z)$

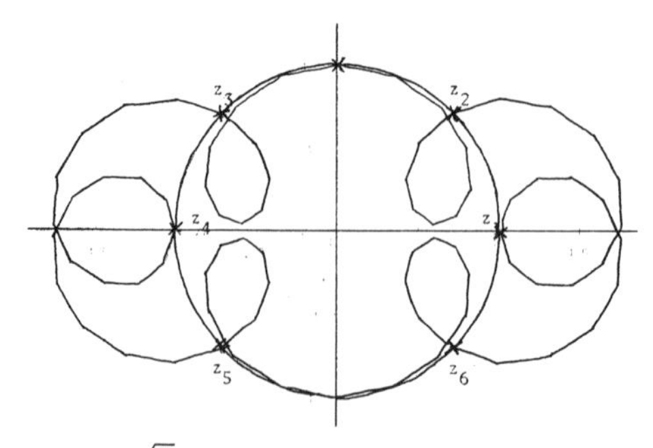

Fig. 2. Ellipse $x = \sqrt{7}\cos\theta$, $y = \sin\theta$ into Circle by $P_6(z)$

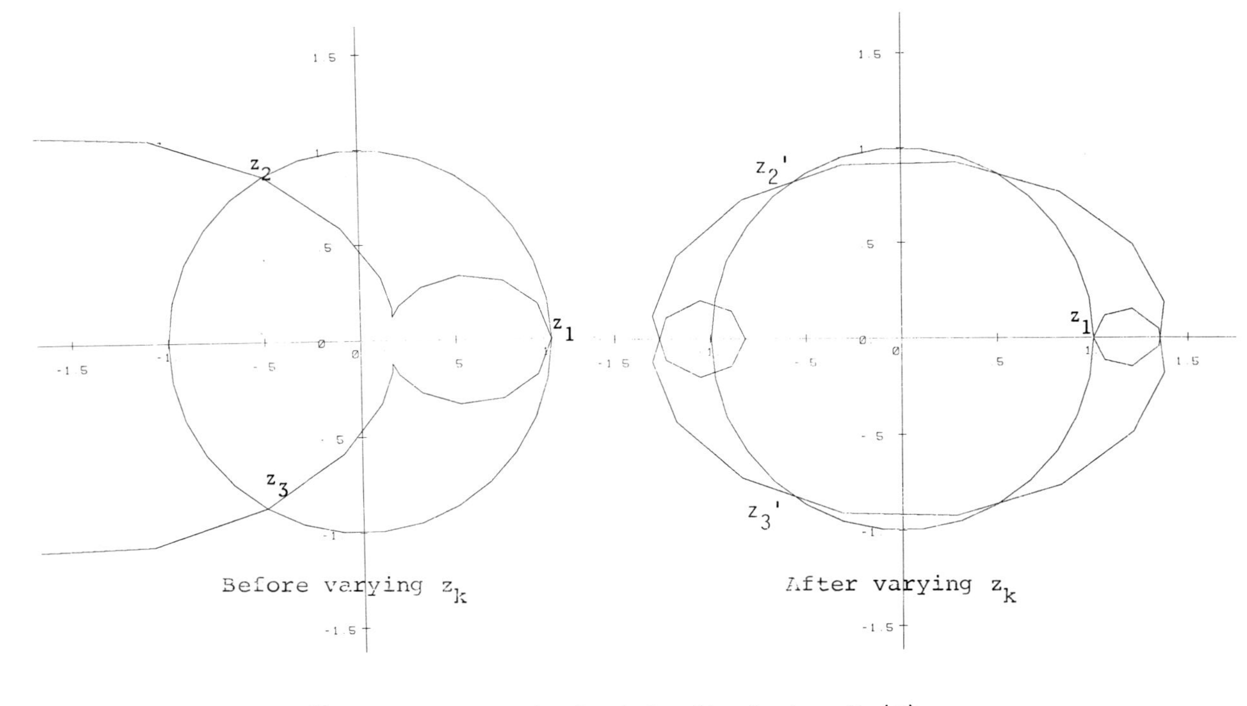

Fig. 3. Ellipse $x = \sqrt{7} \cos \theta$, $y = \sin \theta$ into Circle by $P_3(z)$.

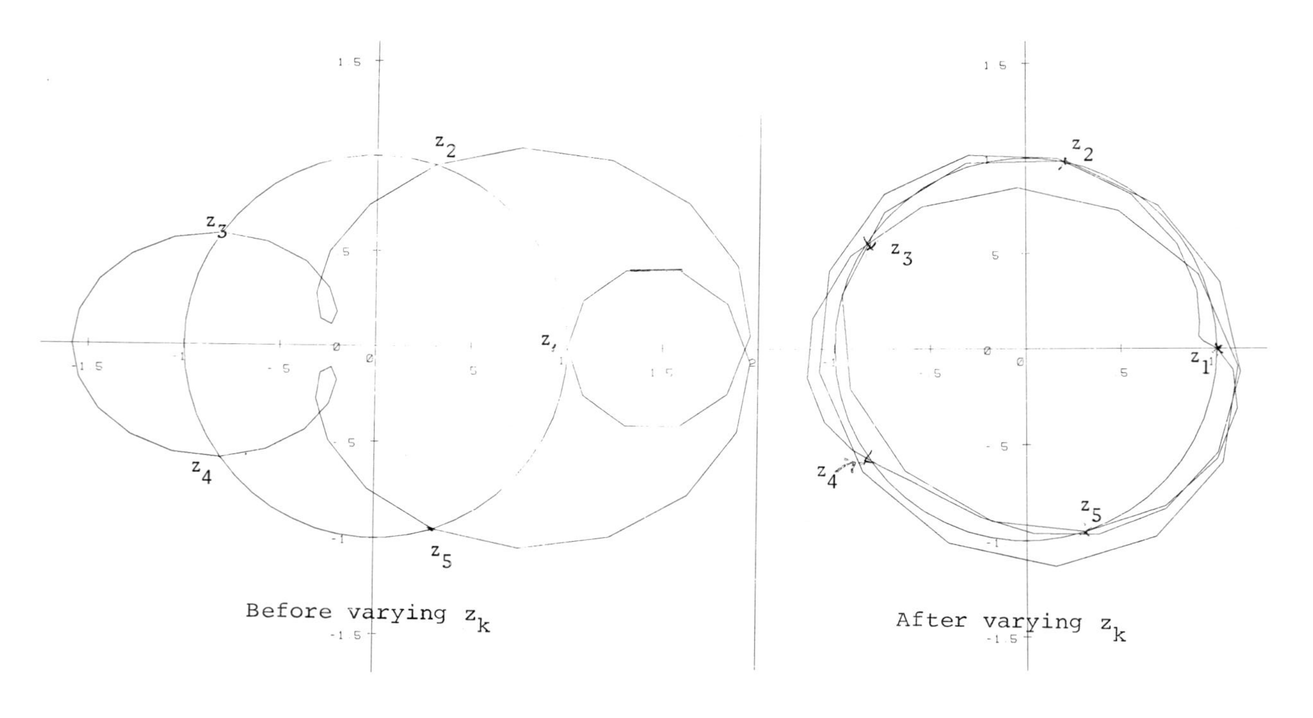

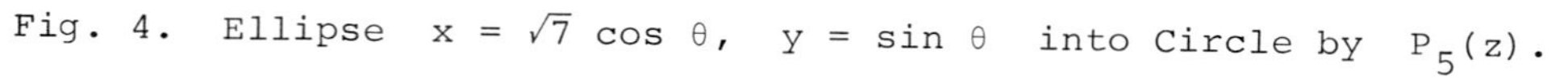

Fig. 4. Ellipse $x = \sqrt{7} \cos \theta$, $y = \sin \theta$ into Circle by $P_5(z)$.

APPROXIMATION BY DISCRETE TOTALLY POSITIVE SYSTEMS

P.W. Smith[1]
S.D. Riemenschneider[2]

Department of Mathematics
University of Alberta
Edmonton, Alberta
Canada T6G 2G1

Professor G.G. Lorentz recently conjectured that in order to approximate the power x^N by polynomials utilizing only k powers less than N it would be best to use the k nearest powers $x^{N-k},\ldots,x^{N-1}$. The first author [5] gave a proof of this conjecture by showing that an actual pointwise improvement could be accomplished by moving to a system with powers closer to x^N. In section 1, we give an analog to this result in the discrete setting, while in section 2 we look at best approximation in ℓ_1 for discrete systems to obtain some results analogous to those of C. Micchelli in [2].

1. An Improvement Theorem

A discrete analog of Lorentz's problem is the following: Given an $m\times n$ totally positive matrix A of rank m, we wish to approximate the last row of A by a linear combination of k other rows. Is this best done with the k

---

[1]Present Address: Old Dominion University, Norfolk, Va U.S.A.
[2]Supported in part by NSERC Grant A7687.

Copyright © 1980 by Academic Press, Inc.
All rights of reproduction in any form reserved.
ISBN: 0-12-171050-5

"nearest" rows, rows $m-k,\ldots,m-1$? To give the affirmative answer to this question we first need to set notation.

Let $\underline{\gamma} = (\gamma_i,\ldots,\gamma_k)$ and $\underline{\lambda} = (\lambda_1,\ldots,\lambda_k)$ be index sets with $\lambda_i < \lambda_{i+1}$, $\gamma_i < \gamma_{i+1}$, $i = 1,\ldots,k$. If $\underline{a}$ and $\underline{b}$ are n-dimensional vectors, then $\underline{a} \leq \underline{b}$ if and only if $a_i \leq b_i$ for $i = 1,\ldots,n$. Also, $|\underline{a}|$ is the vector $(|a_1|,\ldots,|a_n|)$. Furthermore, $S^-(\underline{a})$ will denote the number of strong sign changes in the vector $\underline{a}$, and $S^+(\underline{a})$ will denote the number of weak sign changes in $\underline{a}$ (found by replacing 0's by +1 or -1 so as to maximize the number of sign changes in $\underline{a}$). Define $S(\underline{\lambda}) = \text{span}\{\underline{r}^{\lambda_1},\ldots,\underline{r}^{\lambda_k}\}$ and $\text{dist}_p(\underline{r}^m, S(\underline{\lambda})) = \min\{\|\underline{r}^m - \underline{s}\|_p : \underline{s} \in S(\underline{\lambda})\}$, where $\|\cdot\|_p$ is the ℓ^p-norm. We shall employ the abbreviations STP and TP for strictly totally positive and totally positive respectively.

Lemma 1. Let A be an $m\times n$ STP matrix with rows $\underline{r}^1,\ldots,\underline{r}^m$, and let k be a positive integer less than $\min(m,n) - 1$. Let $\underline{\gamma} \leq \underline{\lambda}$, $\lambda_k < m$, $\underline{\gamma} \neq \underline{\lambda}$, be index sets. If $\underline{p} = \underline{r}^m + \underline{s}$, $\underline{s} \in S(\underline{\lambda})$ and $\underline{q} = \underline{r}^m + \underline{s}^*$, $\underline{s}^* \in S(\underline{\lambda})$, are such that $p_{i_\ell} = q_{i_\ell} = 0$, $\ell = 1,\ldots,k$, then $|\underline{p}| \leq |\underline{q}|$ with strict inequality for all indices i for which $q_i \neq 0$.

Proof: The proof of Theorem 1 in [5] carries over without difficulty.

Lemma 2. Let A, $\underline{\gamma}$, and k be as in Lemma 1 . Suppose that $\underline{q} = \underline{r}^m + \underline{s}$, $\underline{s} \in S(\underline{\gamma})$, satisties $S^+(\underline{q}) = k$. Let $K(\underline{q},\underline{\gamma})$ be the closure in $\mathbb{R}^n$ of the set $\{\underline{r} = \underline{r}^m + \underline{s}' : \underline{s}' \in S(\underline{\gamma})$ and $\text{sgn}\, r_i = \text{sgn}\, q_i\}$. Then $\underline{r}^*$ is an extreme point of the convex set $K(\underline{q},\underline{\gamma})$ if and only if $\underline{r}^* = \underline{r}^m + \underline{s}^*$, $\underline{s}^* \in S(\underline{\gamma})$

, $\underline{r}^*$ has k-zeros, and sgn $r_i^* =$ sgn q_i if either $q_i = 0$ or $r_i^* \neq 0$. Furthermore, $K(\underline{q},\underline{\gamma})$ is compact.

Corollary. $K(\underline{q},\underline{\gamma})$ has a finite number of extreme points.

Proof. First suppose that the extreme point $\underline{r}^*$ has fewer than k zeros. Then there is a nontrivial element $\underline{s}' \in S(\underline{\gamma})$ which vanishes at the zeros of $\underline{r}^*$. Thus for sufficiently small $\varepsilon > 0$, $\underline{r}^* \pm \varepsilon \underline{s}'$ belongs to $K(\underline{q},\underline{\gamma})$; contradicting the extremality of $\underline{r}^*$. Since $\underline{r}^*$ has k-zeros then its sign structure is completely determined and sgn $r_i^* =$ sgn q_i when $q_i = 0$ or $r_i^* \neq 0$. On the other hand, if $\underline{r}^*$ has k zeros and is in $K(\underline{q},\underline{\gamma})$, then it must be extreme since the k zeros determine $\underline{r}^*$ uniquely.

For the compactness we observe that if $\{\underline{r}^j\}$ is unbounded in $K(\underline{q},\underline{\gamma})$, then upon normalizing $\{\underline{r}^j-\underline{r}^m\} = \{\underline{s}^j\}$ and passing to the limit, we obtain an element of $S(\underline{\gamma})$ with too many weak sign changes. □

We are now in a position to give the "improvement" theorem.

Theorem 1. Let A, k, $\underline{\gamma}$, $\underline{\lambda}$ be as in Lemma 1. Suppose that $\underline{q} = \underline{r}^m + \underline{s}$, $\underline{s} \in S(\underline{\gamma})$, has k weak sign changes $S^+(\underline{q}) = k$. Then there exists $\underline{p} = \underline{r}^m + \underline{s}^*$, $\underline{s}^* \in S(\underline{\lambda})$, so that $|\underline{p}| \leq |\underline{q}|$ with strict inequality except when $q_i = 0$.

Proof. Clearly $\underline{q} \in K(\underline{q},\underline{\gamma})$. Let $\underline{q}^\alpha$, $\alpha = 1,\ldots,M$, be the extreme points of $K(\underline{q},\underline{\gamma})$. Then $\underline{q} = \sum_{\alpha=1}^M c_\alpha \underline{q}^\alpha$ with $c_\alpha \geq 0$, $\sum c_\alpha = 1$. By Lemma 2 each $\underline{q}^\alpha$ is determined by k zeros, say $\alpha_1,\ldots,\alpha_k$. If $\underline{p}^\alpha - \underline{r}^m$ is chosen in $S(\underline{\lambda})$ so that $p^\alpha_{\alpha_i} = 0$, $i = 1,\ldots,k$, then by Lemma 1 we have $|\underline{p}^\alpha| \leq |\underline{s}^\alpha|$

with strict inequality except where $\underline{s}^\alpha$ vanishes. The desired result is obtained by taking $\underline{p} = \sum_{\alpha=1}^{M} c_\alpha \underline{p}^\alpha$. □

Before applying this theorem to the approximation problem above we need a lemma that transforms the TP case to the STP case.

Lemma 3. If A is a $m \times n$ TP matrix of rank m, $m \le n$ and $B_\varepsilon = e^{-\frac{1}{\varepsilon}(i-j)^2}$ $1 \le i,j \le n$, then $A_\varepsilon \equiv AB_\varepsilon$ is STP and $A_\varepsilon \to A$ as $\varepsilon \to 0$.

Proof. See Gantmacher and Krein [1,p. 86].

Theorem 2. If A is a $m \times n$ TP matrix of rank m, $k + 1 < m \le n$, then $\text{dist}_p(\underline{r}^m, S(\underline{\lambda}))$, $1 \le p \le \infty$, is minimized by choosing $\underline{\lambda} = (m-k,\ldots,m-1)$. The minimum is strict if A is an STP matrix.

Proof. First assume that A is STP . If $\underline{\lambda} \ge \underline{\mu}$, then let $\underline{q} = \underline{r}^m - \underline{s}$, $\underline{s} \in S(\underline{\mu})$, be the error of best approximation to $\underline{r}^m$ from $S(\underline{\mu})$. It is known that $S^+(\underline{q}) = k$, and by Theorem 1 there exists a $\underline{p} \in S(\underline{\lambda})$ such that $|\underline{p}| \le |\underline{q}|$. The monotonicity of the ℓ^p norm now implies the result.

If A is TP , we apply the above argument to A_ε of Lemma 3 and take the limit. □

§2. Best ℓ^1 approximation

In this section we shall consider the problem of the best ℓ^1 approximation of an element $\underline{f}$ of the convexity cone of $S \equiv S((1,2,\ldots,m))$. Let $n > m$. The convexity cone of S, denoted by $K_C(S)$, is the set of all vectors $\underline{f}$ in $\mathbb{R}^n$

such that for the matrix with rows $\underline{r}^1,\ldots,\underline{r}^m$, $\underline{f}$, all $(m+1)\times(m+1)$ minors are non-negative. Here the matrix A with rows $\underline{r}^1,\ldots,\underline{r}^m$ will be either STP or TP of rank m .

If $\underline{f} \in K_C(S)\backslash S$, then there is a certain sign structure for the error of best approximation. If $\underline{e} = \underline{f} - \underline{s}$ is the error of b.a. and A is STP , then $S^+(\underline{e}) = m$. Indeed, if $S^+(\underline{e}) < m$, then we could choose $\underline{s}^* \in S$ which vanishes at the weak sign changes of $\underline{e}$ and $\underline{f} - \underline{s} \pm \varepsilon\underline{s}^*$ would give a better approximation error. Similarly, a smoothing argument (Lemma 3) shows that $S^-(\underline{e}) \leq m$ if A is TP and has rank m . By duality, there is a $\underline{\phi} \in \mathbb{R}^n$ with $\|\underline{\phi}\|_\infty = 1$ such that $\underline{\phi} \perp S$ and $\underline{\phi} \cdot \underline{e} = \|\underline{e}\|_1$. This implies $\phi_i = \operatorname{sgn} e_i$ except when $e_i = 0$. If A is STP , then $\underline{\phi} \perp S$ implies that $S^+(\underline{\phi}) \geq m$. Therefore, when A is STP the dual functional $\underline{\phi}$ satisfies $S^+(\underline{\phi}) = m$ ($\underline{\phi}$ can have no more weak sign changes than $\underline{e}$). Thus we are lead to consider a concept of alternation introduced by Micchelli and Pinkus [3]: Given $0 = j_0 < j_1 <\ldots< j_m < j_{m+1} = n + 1$ and a vector $\underline{\lambda} \in \mathbb{R}^n$, $\underline{\lambda}$ is said to alternate between $j_1,\ldots,j_m$ provided there exists a single sign σ such that $\lambda_\ell = (-1)^i\sigma$ for $j_{i-1} < \ell < j_i$, $i = 1,\ldots,m+1$.

Our discussion above shows a special case of the following theorem due to Pinkus.

Theorem 3. [4] Let A be a real $m\times n$ matrix, $m < n$. Then there exists $\underline{\lambda} \in \mathbb{R}^n$ which alternates between some $\{j_i\}_{i=1}^m$, $1 \leq j_i <\ldots< j_m < n$ such that $\|\underline{\lambda}\|_\infty = \max|\lambda_i| = 1$ and $A\underline{\lambda} = \underline{0}$.

We note that the alternation indices $j_1,\ldots,j_m$ for $\underline{\lambda}$ need not be uniquely defined. Some questions immediately come to mind: When is the linear functional $\underline{\lambda}$ given in Theorem 2 unique? Does it always correspond to the error of best ℓ^1 approximation of $\underline{f} \in K_C(S)\backslash S$ by elements in S, and if so, can a best approximation be constructed from $\underline{\lambda}$?

To obtain suitable conditions that will guarantee uniqueness, we need the following lemma.

Lemma 4. If the span of $K_C(S)$ contains a m-dimensional STP subspace and if $\underline{\lambda}_1 \neq \underline{\lambda}_2$ both alternate m-times and annihilate S, then there exists $\underline{f} \in K_C(S)$ so that $\underline{\lambda}_1\underline{f} \neq \underline{\lambda}_2\underline{f}$.

Proof. Let $\underline{\phi} = \underline{\lambda}_1 - \underline{\lambda}_2$. We claim that $\underline{\phi}$ has at most $m - 1$ strong sign changes. Now $(\underline{\phi})_i = +2, -2$ or 0 if i is not an alternation index for either $\underline{\lambda}_1$ or $\underline{\lambda}_2$, and $\underline{\phi}$ can only change sign when it crosses an alternation index. Further note that between adjacent alternation indices of $\underline{\lambda}_1$ (or of $\underline{\lambda}_2$), $\underline{\phi}$ cannot change sign no matter how often alternation indices of $\underline{\lambda}_2$ (resp. $\underline{\lambda}_1$) are crossed. Thus the number of sign changes are maximized when the alternation indices of $\underline{\lambda}_1$, $\underline{\lambda}_2$ coincide and in this case there are at most $m - 1$ sign changes.

If the alternation indices of $\underline{\lambda}_1$, $\underline{\lambda}_2$ coincide, then $(\underline{\lambda}_1)_{j_\ell} \neq (\underline{\lambda}_2)_{j_\ell}$ for some ℓ. We select $\underline{f} \in K_C(S)$ such that $(\underline{f})_{j_i} = 0$ $i \neq \ell$, and is 1 at $i = \ell$. If the indices do not coincide, then select $\underline{f} \in K_C(S)$ that vanishes at the sign changes of $\underline{\phi}$. In both cases, $\underline{\lambda}_1\underline{f} \neq \underline{\lambda}_2\underline{f}$. □

Theorem 4. Let A be a $m \times n$ TP matrix of rank m, $m < n$ with rows $r^1, \ldots, r^m$, and let $S = \text{span}\{\underline{r}^1, \ldots, \underline{r}^m\}$. If the span of the convexity cone $K_C(S)$ contains a m-dimensional STP system, then the $\underline{\lambda}$ given in Theorem 2 is unique.

Proof. Suppose that $\underline{\lambda}_1 \neq \underline{\lambda}_2$ are two functionals given by Theorem 2. By Lemma 4, there exists $f \in K_C(S)\backslash S$ such that $\underline{\lambda}_i\underline{f} \neq \underline{\lambda}_i\underline{f}$. Then there is a c, $0 < c < 1$, so that (say) $(\underline{\lambda}_1 - c\underline{\lambda}_2)\underline{f} = \underline{0}$. The vector $\underline{\phi} = \underline{\lambda}_1 - c\underline{\lambda}_2$ has the same sign structure as $\underline{\lambda}_1$. If $\underline{f} \cup S$ is STP, then we find $\underline{g} = \sum_{j=1}^m a_j\underline{r}^j + a_{m+1}\underline{f}$ which interpolates zero at the alternation indices of $\underline{\lambda}_1$. Then

$$\underline{\phi} \cdot \underline{g} = 0 = \pm \sum_{i \neq j_\ell} |(\lambda_1 - c\lambda_2)_i \cdot g_i|$$

This is impossible.

If $\underline{f} \cup S$ is not STP, then we smooth the matrix A' (formed by adding the $m + 1^{st}$ row $\underline{f}$ to A) by lemma 3. Then we find $\underline{g}_\varepsilon = \sum_{j=1}^m a_j^\varepsilon \underline{r}_\varepsilon^j + a_{m+1}^\varepsilon \underline{f}_\varepsilon$ which interpolates 0 at the j_ℓ's and is normalized so that $\|\underline{g}_\varepsilon\|_\infty = 1$. Letting $\varepsilon \downarrow 0$, we obtain a nontrivial $\underline{g} = \sum a_i \underline{r}^i + a_{m+1}\underline{f}$ with the same strong sign structure as $\underline{\lambda}_1$. The argument proceeds as before. □

A best approximation can be constructed from $\underline{\lambda}$ by interpolation at the alternation indices. First, we need a lemma.

Lemma 5. Let A be a $m \times n$ TP matrix of rank m, $m < n$, with rows $\underline{r}^1, \ldots, \underline{r}^m$. If the span of the convexity cone $K_C(S)$ contains a m-dimensional STP system, then $\underline{s} \in S$

and $(\underline{s})_{j_\ell} = 0$ for the alternation indices $j_1 < \cdots < j_m$ of $\underline{\lambda}$ in Theorem 2 implies $\underline{s} \equiv \underline{0}$.

Proof. If there is a non-trivial $\underline{s} \in S$ for which $(\underline{s})_{j_\ell} = 0$, $\ell = 1,\ldots,m$, then the $m \times m$ minor of A, $A\begin{pmatrix}1,\ldots,m\\ j_1,\ldots,j_m\end{pmatrix}$ is zero. Therefore, there are constants c_{j_ℓ}, not all zero and with $\max|c_{j_\ell}| = 1$, so that $\sum_{\ell=1}^m c_{j_\ell}(\underline{s})_{j_\ell} = 0$ for all $\underline{s} \in S$. Let $(\underline{\phi})_i = c_{j_\ell}$ if $i = j_\ell$, $\ell = 1,\ldots,m$, and be zero elsewhere. There is an $\underline{f} \in K_C(S)$ such that $\underline{\phi} \cdot \underline{f} \neq 0$ and $\underline{\phi} \cdot \underline{f} \neq \underline{\lambda} \cdot \underline{f}$. We may select a number c so that $(\underline{\lambda} - c\underline{\phi}) \cdot \underline{f} = 0$. Then $\underline{\lambda} - c\underline{\phi} \perp \underline{f} \cup S$. As in the proof of the last theorem, it is possible to find a nontrivial $\underline{g}$ in span $\{\underline{f} \cup S\}$ which interpolates 0 at $i = j_\ell$, $\ell = 1,\ldots,m$. But then

$$(\underline{\lambda} - c\underline{\phi}) \cdot \underline{g} = 0 = \pm \sum_{i \neq j_\ell} |(\underline{\lambda})_i \cdot (\underline{g})_i| .$$

This contradicts the nontriviality of $\underline{g}$. □

Theorem 5. Let A and $K_C(S)$ be as in Lemma 5. For any $\underline{f} \in K_C(S) \backslash S$, a best ℓ^1 approximation to $\underline{f}$ can be found by interpolating $\underline{f}$ at the alternation indices of the $\underline{\lambda}$ given in Theorem 2.

Proof. By Lemma 5 there is a unique $\underline{s}_o \in S$ which interpolates $\underline{f}$ at the alternation indices j_ℓ, $\ell = 1,\ldots,m$. From Cramer's rule it is easy to see that $\underline{f} - \underline{s}_o$ alternates in sign i.e. $(-1)^{\sigma+\ell}(\underline{f}-\underline{s})_i \geq 0$ for $j_\ell < i < j_{\ell+1}$, $\ell = 0,\ldots,m$. Therefore $|\underline{f}-\underline{s}_o| = \underline{\lambda} \cdot (\underline{f}-\underline{s}_o) = \underline{\lambda} \cdot (\underline{f}-\underline{s}) \leq |\underline{f}-\underline{s}|$, $\underline{s} \in S$. □

Remark. The best approximation to $\underline{f}$ constructed in Theorem 5 need not be unique since the alternation indices are not unique: e.g. for $1,1,-1,-1$ one has the choice of $j = 2$ or $j = 3$ for the alternation index. More concretely, $\underline{r}' = (1,1,1,1)$ $\underline{f}' = (1,4,9,16)$ then $\|f'-\alpha r'\|$ is minimized for $4 \le \alpha \le 9$. Hence $\underline{\lambda} = (1,1,-1,-1)$ and there are many best approximations including two that can be found by interpolation.

REFERENCES

(1) F.R. Gantmacher and M.G. Krein, (1960). "Oszillationsmatrizen, Oszillationskerne und kleine Schwingungen Mechanischer Systeme:, *Akademie Verlag, Berlin.*

(2) Charles A. Micchelli, (1977). Best L^1 approximation by weak Chebychev systems and the uniqueness of interpolating perfect splines. *J. Approximation Theory 19, 1-14.*

(3) Charles A. Micchelli and Allan Pinkus, (1977). On n-widths in L^∞ . *Trans. Amer. Math. Soc. 234, 139-174.*

(4) Allan Pinkus, (1976). A simple proof of the Hobby-Rice Theorem. *Proc. Amer. Math. Soc. 60, 82-84.*

(5) Philip W. Smith, (1978). An improvement theorem for Descartes systems. *Proc. Amer. Math. Soc. 70, 26-30.*

The research of P. W. Smith was partially supported by the U. S. Army Research Office under Grant #DAHC04-75-0816.

NOTE ON A SUBGRADIENT PROJECTION ALGORITHM

V.P. Sreedharan

Michigan State University
East Lansing, Michigan

1. INTRODUCTION

This note summarizes the main points of a paper to appear in full elsewhere. An algorithm for minimizing a certain type of non-differentiable convex function constrained to a convex polytope in $\mathbb{R}^n$ is presented. The particular form of this problem was motivated by some questions in the equilibrium theory of mathematical economics. The economic motivation and some other mathematical considerations appear in Rubin and Sreedharan [8]. The problem is a generalization of the minimum norm problem of approximation theory [10]. Our algorithm is a generalization of Rosen's well known gradient projection method [6]. Rosen required that the objective function be twice continuously differentiable. The algorithms of Wolfe [12] and Lemarechal [2] handle the optimization of unconstrained non-differentiable convex functions. However, they do not apply to the constrained case. Rosen's gradient projection method can run into the problem of "jamming" or "zigzagging" and converge to a non-optimal point. The algorithm we propose is prompted by Rosen [6], Polak [3], [4] and Sreedharan [9], [10]. The algorithm of Bertsekas and

Copyright © 1980 by Academic Press, Inc.
All rights of reproduction in any form reserved.
ISBN: 0-12-171050-5

Mitter [1] is theoretically applicable to constraints even more general than linear ones. In spite of the ease of proof of convergence by these authors, their algorithm requires the computation of the "ϵ - subdifferential" of the objective function, a prohibitive task even in the linearly constrained case. Our algorithm, in contrast to the Bertsekas - Mitter algorithm, requires only the computation of a subset of the ϵ - subdifferential and is computationally feasible for a large class of problems. The actual computational details and experience with a Fortran program for our algorithm on a CDC 6500 computer will be given in the paper by Rubin [7].

2. PROBLEM

In this paper we denote the standard Euclidean inner product of two vectors in $\mathbb{R}^n$ by simply juxtaposing them. The corresponding Euclidean length of a vector is denoted by $|\cdot|$. Here is the problem: Let $X \subset \mathbb{R}^n$ be a non-empty convex polytope defined by

$$X = \{x \in \mathbb{R}^n \mid a_i x \leq b_i \ , \ i = 1, \cdots, m\}$$

where a_i , $x \in \mathbb{R}^n$, $b_i \in \mathbb{R}$. Note that by definition a polytope is bounded. Let f be a smooth strictly convex function on X; i.e. a strictly convex function which is of class C^1 in a neighborhood of X. Further, let v be a piecewise affine convex function on X; i.e.

$$v(x) = \max\{v_j(x) \mid 1 \leq j \leq r\} ,$$

where

$$v_j(x) = g_j x + c_j \ , \ g_j \ , \ x \in \mathbb{R}^n \ , \ c_j \in \mathbb{R}.$$

Such a v is not differentiable everywhere except in trivial cases. The problem is to minimize $f(x)+v(x)$ subject to the constraint $x \in X$. Symbolically we have

$$(P) \quad \begin{cases} a_i x \leq b_i & , \ i = 1,\cdots,m \\ f(x)+v(x) \ (\min). \end{cases}$$

We shall refer to this as problem (P). This form of the objective function arises naturally, when v is obtained by solving a linear programming problem whose right hand side, the constraint constants, depends linearly on a parameter x. In fact, this particular form arose from the economic model referred to earlier.

3. NOTATION

Before we state the algorithm we need some notation. For any non-empty closed convex subset $S \subset \mathbb{R}^n$ there is a unique point $a \in S$ nearest to the origin. We denote this point by $N[S]$. The point $a = N[S]$ is characterized by the inequality

$$a(x-a) \geq 0 \ , \ \forall \ x \in S. \tag{3.1}$$

Let x be a point in the constraint set X and $\epsilon \geq 0$. We first define the following sets of indices:

$$I_\epsilon(x) = \{1 \leq i \leq m \mid a_i x \geq b_i - \epsilon\} ; \tag{3.2}$$

$$J_\epsilon(x) = \{1 \leq j \leq r \mid v_j(x) \geq v(x) - \epsilon\} . \tag{3.3}$$

Note that

$$I_0(x) = \{1 \leq i \leq m \mid a_i x = b_i\} , \tag{3.4}$$

and

$$J_0(x) = \{1 \leq j \leq r \mid v_j(x) = v(x)\} . \tag{3.5}$$

Using the above index sets we define the following convex subsets of $\mathbb{R}^n$:

$$C_\epsilon(x) = \text{cone}\{a_i \mid i \in I_\epsilon(x)\} \; ; \tag{3.6}$$

and

$$K_\epsilon(x) = \text{conv}\{g_j \mid j \in J_\epsilon(x)\} \; . \tag{3.7}$$

Here and elsewhere we denote by cone S the convex cone generated by S with apex at the origin and by conv S the convex hull of the set S.

4. ALGORITHM

Here then is the algorithm.

<u>Step</u> 0. Start with arbitrary $x_0 \in X$, $\epsilon_0 > 0$ and $k = 0$.

<u>Step</u> 1. Compute $y_0 = N[K_0(x_k) + \nabla f(x_k) + C_0(x_k)]$.

If $y_0 = 0$, STOP; x_k is the solution of problem (P).

If $y_0 \neq 0$, set $\epsilon = \epsilon_0$.

<u>Step</u> 2. Compute $y_\epsilon = N[K_\epsilon(x_k) + \nabla f(x_k) + C_\epsilon(x_k)]$.

<u>Step</u> 3. If $|y_\epsilon|^2 > \epsilon$ set $\epsilon_k = \epsilon$, $s_k = y_\epsilon$ and GO TO Step 5.

<u>Step</u> 4. Replace ϵ by $\epsilon/2$ and GO TO Step 2.

<u>Step</u> 5. Compute $\bar{\alpha}_k = \max\{\alpha \in \mathbb{R} \mid x_k - \alpha s_k \in X\}$. (It can be shown that $\bar{\alpha}_k$ is positive.) Find $\alpha_k \in [0, \bar{\alpha}_k]$ such that there exists

$$z_k \in K_0(x_k - \alpha_k s_k) + \nabla f(x_k - \alpha_k s_k) \; ,$$

with

$$z_k s_k = 0.$$

If no such α_k exists, set $\alpha_k = \bar{\alpha}_k$.

<u>Step</u> 6. Define $x_{k+1} = x_k - \alpha_k s_k$. Increment k by 1 and GO TO Step 1.

Steps 1 and 2 can be implemented as special quadratic programs. Step 5 requires a properly constructed line search and some comparisons. In practice, Step 1 would be replaced by the statement: STOP if $|y_0|$ is sufficiently small. We refer to the paper of Rubin [7] for the computational details and experience. With small problems even hand calculations using the above algorithm yielded good answers.

5. CONVERGENCE OF THE ALGORITHM

The proof that the algorithm converges is somewhat involved and depends on a series of lemmas some of which are of independent interest. Here we shall summarize the statements of some of the key lemmas. We then state the main theorem. The complete proofs of the lemmas and the main theorem can be found in a paper [11] under preparation by the present author.

We need some terminology and notation. When $F:\mathbb{R}^n \to [-\infty,\infty]$ is a convex function its ϵ-subdifferential $\partial_\epsilon F(x)$, where $\epsilon \geq 0$, is defined by saying $u \in \partial_\epsilon F(x)$ iff $F(y) \geq F(x) + u(y-x) - \epsilon$, $\forall\ y \in \mathbb{R}^n$. $\partial_0 F(x)$ is the subdifferential of F at x and is denoted by $\partial F(x)$. See Rockafeller [5] for all these and related notions. $\partial F(x)$ can be empty. The constrained problem (P) is converted into an unconstrained one via the indicator function δ of the constraint set X. $\delta(x) = 0$ if $x \in X$ and $\delta(x) = \infty$ if $x \notin X$. Then $F = f + v + \delta$ is convex on $\mathbb{R}^n$ and the problem minimize $F(x)$, $x \in \mathbb{R}^n$ is equivalent to (P). We can show that the following lemma holds.

5.1 Lemma: $\partial F(x) = \nabla f(x) + K_0(x) + C_0(x)$.

This shows that our stopping criterion in Step 1 is well-chosen; for $0 \in \partial F(\bar{x})$ iff $\bar{x}$ is a minimizer of F. We state some more lemmas.

5.2 Lemma: Step 4 is not executed infinitely often in any one iteration.

5.3 Lemma: Step 5 is feasible, i.e. if $s_k \neq 0, -s_k$ is a feasible direction of strict descent at the point x_k.

5.4 Corollary: $\bar{\alpha}_k$ is positive.

5.5 Lemma: Let $\varphi(\alpha) = F(x_k - \alpha s_k)$, $s_k \neq 0$. If $\bar{\alpha}_k$ is not a minimizer of φ on $[0, \bar{\alpha}_k]$, then z_k satisfying Step 5 exists.

5.6 Corollary: α_k is positive.

5.7 Lemma: If 0 is a cluster point of the sequence (s_k), then (x_k) converges to $\bar{x}$, the minimizer of F.

5.8 Lemma: If 0 is a cluster point of (ϵ_k), then (x_k) converges to $\bar{x}$.

5.9 Lemma: The sequence (s_k) is bounded.

5.10 Lemma: If the sequence (s_k) is bounded away from 0, then the sequence (α_k) converges to zero.

5.11 Lemma: If there exists $\epsilon > 0$ such that $\epsilon_k \geq \epsilon$ for every k, and if (α_k) converges to zero, then there is a convergent subsequence $(x_{k'})$ of the sequence (x_k), such that $I_0(x_{k'}) = I_0(x)$, for every k', where $x = \lim_{k'} x_{k'}$.

We finally state the theorem.

5.12 Theorem: The algorithm either generates a terminating sequence with its last term as the solution of problem (P) or else the algorithm generates a convergent infinite sequence with its limit as the solution of problem (P).

REFERENCES

1. D.P. Bertsekas and S. Mitter, A descent method for optimization problems with nondifferentiable cost functionals, SIAM J. on Control 11 (1973) 637-652.

2. C. Lemarechal, An extension of Davidon methods to non differentiable problems, Math. Programming Study 3 (1975) 95-109.

3. E. Polak, On the convergence of optimization algorithms, Rev. Fr. Inform. Rech. Operation (16-R1) (1969) 17-34.

4. E. Polak, Computational Methods in Optimization, Academic Press, New York, New York (1971).

5. R.T. Rockafeller, Convex Analysis, Princeton Univ. Press, Princeton, NJ, 1970.

6. J.B. Rosen, The gradient projection method for nonlinear programming, Part I: Linear constraints, J. SIAM 8 (1) (1960) 181-217.

7. P. Rubin, Implementation of a subgradient projection algorithm (under preparation).

8. P. Rubin and V.P. Sreedharan, Existence of equilibria in a PIES-type economic model (under preparation).

9. V.P. Sreedharan, Least-squares algorithms for finding solutions of overdetermined linear equations which minimize error in an abstract norm, Num. Math. 17 (1971) 387-401.

10. V.P. Sreedharan, Least-squares algorithms for finding solutions of overdetermined systems of linear equations which minimize error in a smooth strictly convex norm, J. of Approx. Theory 8 (1) (1973) 46-61.

11. V.P. Sreedharan, A subgradient projection algorithm (under preparation).

12. P. Wolfe, A method of conjugate subgradients for minimizing nondifferentiable functions, Math. Programming Study 3 (1975) 145-173.

UNIQUENESS IN L_1-APPROXIMATION FOR CONTINUOUS FUNCTIONS

Hans Strauss

Institut für Angewandte Mathematik
Universität Erlangen-Nürnberg
Erlangen, West-Germany

I. INTRODUCTION

This paper deals with the problem of uniqueness in L_1-approximation for continuous functions.

Let I be an interval [a,b] and let C(I) be the space of all real-valued continuous functions f on I under the L_1-norm $||f||: = \int_I |f| = \int_I |f(x)|dx$. If G is an n-dimensional subspace of C(I) then we denote by $P_G(f): = \{g_o \in G: ||f-g_o|| = \inf_{g \in G} ||f-g||\}$ the set of best approximations for a function f in C(I). If every function f in C(I) has a unique best approximation then G is called a <u>Unicity space</u>.

Krein [5] proved the following Theorem: Let the space $L_1(I)$ be given. Then there exists no finite dimensional subspace which admits for every f in $L_1(I)$ exactly one best approximation.

On the other hand Jackson showed that the subspace G spanned by the powers $1,\ldots,x^n$ is a Unicity space in C(I). Krein extended this result to Chebyshev spaces. Galkin [4] and Strauss [7] proved that every subspace G of polynomial splines with fixed knots is a Unicity space. For a special

Copyright © 1980 by Academic Press, Inc.
All rights of reproduction in any form reserved.
ISBN: 0-12-171050-5

class of spline functions this result was established by Carroll-Braess [1]. Thus it is natural to ask for conditions insuring L_1-uniqueness. DeVore established a nice condition which is very useful for applying it to special function spaces. But the condition is not necessary for L_1-uniqueness. Cheney-Wulbert [2] gave a Characterization Theorem for Unicity spaces. In this result first of all it is necessary to characterize the functions having 0 as a best approximation.

We want to give another Characterization Theorem for Unicity spaces. The condition of DeVore follows from our Theorem. Using these results it is easy to prove uniqueness for some special function spaces.

II. CHARACTERIZATION THEOREM

We define the following class of functions corresponding to a subspace G in C(I):

Let H_G be the set of all functions in C(I) such that for every h in H_G there exists a g_h in G satisfying $|h(x)| = |g_h(x)|$ for $x \in I$.

The subspace G is a subset of H_G and H_G is a subset of C(I). Every function $h \in H_G$ corresponds to a function g_h in G. The functions h and g_h have the same zero intervals. Let l be the maximal number of separated zeros of g_h then h has at most l sign changes.

THEOREM 1. The subspace G is a Unicity space, if and only if, every function h in H_G has a unique best approximation in G.

This Theorem reduces the problem of characterizing Unicity spaces to the subclass H_G of $C(I)$. H_G is derived from G in a simple way.

THEOREM 2. Every function h in H_G has a unique best approximation, if and only if, there is no nontrivial h in H_G such that $0 \in P_G(h)$.

It is shown in the proof that $g_h \in P_G(h)$ if $0 \in P_G(h)$ where h corresponds to g_h.

Using the Characterization Theorem for best approximations we can summarize the preceding results:

THEOREM 3. The following conditions are equivalent:

(a) The subspace G is a Unicity space

(b) There exists no nontrivial function h in H_G satisfying

$$\left|\int_I g \operatorname{sgn} h\right| \leq \int_{Z(h)} |g|$$

for every g in G where $Z(h)$ is the set of all zeros of h.

Therefore it is very important to characterize the sign functions of the elements of H_G. We denote by $\Sigma_P = \bigcup_{p \in P} \{\operatorname{sgn} p\}$ the set of sign functions corresponding to the elements of a subset P of $C(I)$. Two sign functions s_1, s_2 are called equal, $s_1 = s_2$, if $s_1(x) = s_2(x)$ almost everywhere in I.

COROLLARY 4. Let $\Sigma_G = \Sigma_{H_G}$ then G is a Unicity space.

Proof. Let a nontrivial $s \in \Sigma_{H_G}$ be given then there exists a g in G satisfying $s = \operatorname{sgn} g$. We obtain

$$\left|\int_I g\, s\right| > \int_{Z(s)} |g| = 0$$

and conclude from Theorem 3 that G is a Unicity space.

The condition $\Sigma_G = \Sigma_{H_G}$ is the condition of DeVore (see Strauss [8]).

III. EXAMPLES

(1) Let G be an n-dimensional Chebyshev subspace then $\Sigma_G = \Sigma_{H_G}$.

Every g in G has at most n-1 separated zeros. It follows that every s in Σ_{H_G} has at most n-1 sign changes. We conclude from a Theorem of Krein that there exists a function g in G such that s = sgn g and we obtain $\Sigma_{H_G} \subset \Sigma_G$. The converse is trivial. This completes the proof.

(2) Let G be an n-dimensional weak Chebyshev space that is, every function g in G has at most n-1 sign changes. We call a zero f(x) = 0 an <u>essential</u> zero with respect to G if there is a g in G with $g(x) \neq 0$. If each $x \in I$ is an essential zero it can be shown that every s in Σ_{H_G} has at most n-1 sign changes.

(3) L_1-uniqueness for spline functions with fixed knots can be proved in the following way:

Let G be a subspace spanned by the functions $1, \ldots, x^{n-1}$, $(x-x_1)_+^{n-1}, \ldots, (x-x_k)_+^{n-1}$, $a<x_1<\ldots x_k<b$. Every g in G has at most n+k-1 separated zeros. It follows that every s in Σ_{H_G} has at most n+k-1 sign changes. We conclude from properties of weak Chebyshev spaces that there exists a function g in G, $g \neq 0$, satisfying $s(x)g(x) \geq 0$, $x \in I$, and

$$|\int_I g\, s| > \int_{Z(s)} |g| = 0$$

for every nontrivial s in Σ_{H_G}. We obtain from Theorem 3 that G is a Unicity space.

(4) Sommer [6] considered <u>generalized</u> <u>spline</u> <u>spaces</u>. We denote by $V_n(G)$ the following class of n-dimensional subspaces: There exist knots $a=x_o<x_1<\ldots<x_s=b$ such that the spaces $G_i = G|[x_{i-1},x_i]$, $i=1,\ldots,s$, are Chebyshev spaces of dimension n_i. Sommer was able to prove L_1-uniqueness for an important subclass of $V_n(G)$ by using the methods of (3). It can be shown by simple examples that L_1-uniqueness is not true in general.

REMARK. In many important cases we obtain L_1-uniqueness by proving $\Sigma_G = \Sigma_{H_G}$ or similiar conditions. To prove that G is not a Unicity space it is necessary to find a nontrivial function h in H_G satisfying the relation

$$|\int_I g \text{ sgn } h| \leq \int_{Z(h)} |g|$$

for every g in G.

REFERENCES

1. Carroll, M.P., and Braess, D., On Uniqueness of L_1-Approximation for certain Families of Spline Functions, J. Approx. Theory 12 (1974), 362-364

2. Cheney, E.W., and Wulbert D.E., The Existence and Unicity of Best Approximations, Math. Scand. 24 (1969), 113-140

3. DeVore, R., Private communication

4. Galkin, R.V., The Uniqueness of the Element of Best Approximation to a Continuous Function Using Splines with Fixed Knots, Math. Notes 15 (1974), 3-8

5. Krein, M., The L-problem in abstract linear normed space, in " Some Questions in the Theory of Moments" (N.I. Akiezer, M. Krein ed.), Translations of Mathematical Monographs, Vol. 2, Amer. Math. Soc., Providence (1962)

6. Sommer, M., Weak Chebyshev Spaces and Best L_1-Approximation, J. Approx. Theory, to appear

7. Strauss, H., L_1-Approximation mit Splinefunktionen, in "Numerische Methoden der Approximationstheorie" (L. Collatz, G. Meinardus ed.), 151-162, Birkhäuserverlag, Stuttgart, 1975

8. Strauss, H., Best L_1-Approximation, preprint

WEIGHTED APPROXIMATION
AND LOCALIZATION

W.H. Summers

Department of Mathematics
University of Arkansas
Fayetteville, Arkansas

I. THE WEIGHTED APPROXIMATION PROBLEM

Setting the notation, X will denote a completely regular Hausdorff space, and V will be a set of nonnegative upper semicontinuous (u.s.c.) functions on X with the elements of V being referred to as weights. Associated with the pair (X, V), we have the weighted locally convex space

$CV_0(X) = \{f \in C(X):$ fv vanishes at infinity on X for every $v \in V\}$ equipped with the weighted topology ω_V generated by the seminorms

$$p_v(f) = ||fv|| = \sup\{|f(x)|v(x): \ x \in X\}$$

determined by all $v \in V$, where C(X) denotes the algebra of all complex valued continuous functions on X. Finally, A will denote a subalgebra of C(X) which contains the constant functions, while W will be a vector subspace of $CV_0(X)$ such that $AW \subseteq W$; i.e., W is a module over A.

The basic problem is to describe the closure of W in $CV_0(X)$ under these circumstances; the type of description normally sought is one given in terms of a localization. To make this more precise, let K be a closed partition of X

Copyright © 1980 by Academic Press, Inc.
All rights of reproduction in any form reserved.
ISBN: 0-12-171050-5

(i.e., $\mathcal{K}$ is a collection of pairwise disjoint closed sets which cover X), and assume that $f \in CV_0(X)$ satisfies $f|K \in \overline{W|K}^{\omega_{V|K}}$ for every $K \in \mathcal{K}$. Then does it follow that $f \in \overline{W}^{\omega_V}$? If so, we will say that W is $\mathcal{K}$-localizable.

The so called "strict weighted approximation problem" as posed by L. Nachbin (cf. [3]) asks for necessary and sufficient conditions for W to be $\mathcal{K}$-localizable when $\mathcal{K}$ is the closed partition $\mathcal{N}_A$ of equivalence classes obtained by putting $x \sim y$ whenever $f(x) = f(y)$ for all $f \in A$. Among a number of sufficient conditions given by Nachbin, the one he termed the bounded case of the weighted approximation problem has been particularly useful; the bounded case occurs when every $f \in A$ is bounded on the support of each $v \in V$.

1.1. THEOREM (Nachbin [3]). In the bounded case of the weighted approximation problem, if A is selfadjoint, then W is $\mathcal{N}_A$-localizable.

A set $K \subseteq X$ is said to be A-antisymmetric if $f \in A$ with $f|K$ real valued implies $f|K$ is a constant function. Every A-antisymmetric set is contained in a maximal one; moreover, the collection $\mathcal{B}_A$ of all maximal A-antisymmetric sets is a closed partition of X which agrees with $\mathcal{N}_A$ whenever A is selfadjoint. The next result is an extension of E. Bishop's well known generalization of the Stone-Weierstrass theorem [2].

1.2. THEOREM ([4]). In the bounded case of the weighted approximation problem, W is always $\mathcal{B}_A$-localizable.

In general, $\mathcal{B}_A$ can be a much coarser partition than $\mathcal{N}_A$. Indeed, if X is the closed unit disc (in $\mathbb{C}$) and if A is the disc algebra, then $\mathcal{B}_A = \{X\}$ while $\mathcal{N}_A = \{ \{x\}: x \in X\}$.

II. THE WORK OF ARENSON

Assume, for the time being, that X is compact. A set $F \subseteq X$ is called an A-peak set if there exists $f \in A$ such that $f(F) = 1$ and $|f(x)| < 1$ for $x \in X \setminus F$; a closed set $F \subseteq X$ will be called weakly A-analytic if every $\overline{A|F}$ - peak set either coincides with F or is nowhere dense in F.

2.1. THEOREM (Arenson [1]). Every weakly A-analytic set is contained in a maximal weakly A-analytic set, the collection $\mathcal{A}_A$ of all maximal weakly A-analytic sets covers X, and W = A is $\mathcal{A}_A$-localizable (where V consists of only the function 1_X which is identically 1 on X).

The collection $\mathcal{A}_A$ need not be a partition of X. However, among all closed partitions $\mathcal{K}$ of X for which $F \in \mathcal{A}_A$ implies there exists $K \in \mathcal{K}$ such that $F \subseteq K$, there is a finest one; denote it by $\mathcal{K}_A$.

2.2. THEOREM (Arenson [1]). The closed partition $\mathcal{K}_A$ is finer than $\mathcal{B}_A$ and W = A is $\mathcal{K}_A$-localizable (where $V = \{1_X\}$).

2.3. EXAMPLE (Arenson [1]). Let Y be the prism $\{(x, y, z) \in \mathbb{R}^3: 0 \leq x \leq 1, 0 \leq z \leq 1, |y| \leq z\}$, and take B to be all $f \in C(Y)$ which are holomorphic in each z-section

$$D_z = \{(x, y, z) \in Y: z \text{ is fixed}\}, z \in (0, 1].$$

Now, taking X to be the compact quotient of Y obtained by identifying $(x, 0, 0)$ with $(0, 0, x)$ for each $x \in (0, 1]$, let

ϕ be the natural projection of Y onto X, and put

$$A = \{f \in C(X): \; f \circ \phi \in B\}.$$

Then $\mathcal{B}_A = \{X\}$, $\mathcal{N}_A = \{ \{x\}: \; x \in X\}$, but

$$\mathcal{K}_A = \{D_z: \; z \in (0, 1]\} \bigcup \{(0, 0, 0)\}.$$

III. A WEIGHTED ANALOGUE OF ARENSON'S RESULT

Henceforth, we will assume that X is locally compact. To obtain a viable replacement for the notion of weakly A-analytic sets in this context, we proceed by first putting

$$B(A, F) = \overline{A|F \bigcap C_b(F)},$$

where $F \subseteq X$, $C_b(F) = \{f \in C(F): \; f \text{ is bounded}\}$, and the closure is taken with respect to the topology of uniform convergence on F. A closed set $F \subseteq X$ will be called A-<u>analytic</u> if the following two conditions are satisfied:

i) F is A-antisymmetric;

ii) every B(A, F) - peak set either coincides with F or is nowhere dense in F.

Our first result, which follows as a consequence of the Weierstrass polynomial approximation theorem, sheds some light on the relationship between these two conditions.

3.1. THEOREM. Let F be a closed subset of X. If $A|F \subseteq C_b(F)$ and if every B(A, F) - peak set either coincides with F or is nowhere dense in F, then F is A-analytic. In particular, if X is compact, then every weakly A-analytic set is A-analytic.

3.2. LEMMA. Every A-analytic set is contained in a maximal A-analytic set, and the collection $\mathcal{A}_A$ of all maximal A-analytic sets covers X.

3.3. LEMMA. Among all closed partitions K of X for which $F \in A_A$ implies there exists $K \in K$ such that $F \subseteq K$, there is a finest (which will be denoted by K_A). Moreover, K_A is finer than B_A.

An application of Zorn's lemma suffices to establish Lemma 3.2. In the case of Lemma 3.3, existence is clear, while the fact that K_A is finer than B_A follows because each $F \in A_A$ is A-antisymmetric.

We are now in a postion to state our main result.

3.4. THEOREM. In the bounded case of the weighted approximation problem, assuming that X is locally compact, W is alway A_A- and K_A-localizable.

It is well known that every $T \in CV_0(X)'$ can be identified in a natural way with some Radon measure on X, and the proof hinges upon showing that support sets of certain of these Radon measures are A-analytic. Detailed arguments for this and the other assertions in the present section will appear in a subsequent article.

REFERENCES

1. E.L. Arenson, Certain properties of algebras of continuous functions, Dokl. Akad. Nauk SSSR 171 (1966), 767-769; Soviet Math. Dokl. 7 (1966), 1522-1524.

2. E. Bishop, A generalization of the Stone-Weierstrass theorem, Pacific J. Math. 11 (1961), 777-783.

3. L. Nachbin, Weighted approximation for algebras and modules of continuous functions: real and self-adjoint complex cases, Ann. of Math. 81 (1965), 289-302.

4. W.H. Summers, Weighted approximation for modules of continuous functions, Bull. Amer. Math. Soc. 79 (1973), 386-388.

L_p SATURATION OF POSITIVE CONVOLUTION OPERATORS

J. J. Swetits

Department of Mathematical Sciences
Old Dominion University
Norfolk, Virginia

B. Wood

Department of Mathematics
University of Arizona
Tucson, Arizona

I. INTRODUCTION

Let ϕ be a positive, even continuous function on $[-R,R]$, $\phi(0) = 1$ and ϕ decreasing on $[0,R]$. For $f \in L_p[0,R]$, $1 \le p \le \infty$, and $n = 1,2,\ldots$, let

$$K_n(f,x) = \rho_n \int_0^R f(t)\phi^n(t-x)dt, \tag{1.1}$$

where

$$\rho_n^{-1} = 2\int_0^R \phi^n(t)dt.$$

In addition, let

$$\mu_n^2 = \rho_n \int_0^R t^2\phi^n(t)dt.$$

The sequence, $\{K_n\}$, defined by (1.1) was studied by Korovkin (2), who showed that if f is continuous on $[0,R]$, then

$$\lim_{n\to\infty} K_n(f,x) = f(x)$$

Copyright © 1980 by Academic Press, Inc.
All rights of reproduction in any form reserved.
ISBN: 0-12-171050-5

uniformly on intervals of the form $[\delta, R-\delta]$, $0 < \delta < R/2$, provided $\mu_n^2 \to 0 (n \to \infty)$.

Bojanic and Shisha (1) obtained a quantitative version of Korovkin's result. Under the assumption

$$\lim_{x \to 0^+} (1-\phi(x)x^{-\alpha} = c \tag{1.2}$$

for some $\alpha > 0$, $c > 0$, it's shown in (1) that, for $f \in C[0,R]$

$$||K_n(f)-f||_{I_\delta} \le C(||f||n^{-2/\alpha} + w(f,n^{-1/\alpha})),$$

where $||\cdot||_{I_\delta}$ denotes the uniform norm on the interval $[\delta, R-\delta]$ and $w(f,\cdot)$ is the modulus of continuity of f on $[0,R]$. It's also shown in (1) that (1.2) implies that μ_n^2 is asymptotically equivalent to $n^{-2/\alpha}$.

Wood (3) showed, for $f \in L_p[0,R]$, $1 \le p \le \infty$, that

$$||K_n(f) - f||p \le C_p(||f||_p \mu_n^{1/p} + w_{2,p}(f, \mu_n^{1/2p})). \tag{1.3}$$

In (1.3) $||\cdot||_p$ denotes the L_p norm on $[0,R]$, and $w_{2,p}(f,\cdot)$ is the second order integral modulus of smoothness of f measured in $L_p[0,R]$. Furthermore, it can be shown that, for $f(x) \equiv 1$, $||K_n(f) - f||p$ is asymptotically equivalent to $\mu_n^{1/p}$. Thus (1.3) is best possible for the space, $L_p[0,R]$.

The purpose of this note is to point out that (1.3) can be substantially improved if the behavior of f at the endpoints of $[0,R]$ is suitably restricted.

II. MAIN RESULT

For each $\gamma > 0$, let

$$\mu_{n,\gamma} = \rho_n \int_0^R t^\gamma \phi^n(t)dt.$$

The assumption (1.2) implies that $\mu_{n,\gamma}$ is asymptotically equivalent to μ_n^γ, which is asymptotically equivalent to $n^{-\gamma/\alpha}$.

$L_p^{(2)}[0,R]$ denotes the space of functions, f, with an absolutely continuous first derivative and $f'' \in L_p[0,R]$.

Our main result is the

THEOREM. <u>Let</u> $f \in L_p[0,R]$, $1 \le p < \infty$. <u>Let</u> $\{K_n\}$ <u>be defined by</u> (1.1) <u>and assume</u> (1.2) <u>is satisfied</u>. <u>Then</u>

(i) <u>If</u> $1 < p < \infty$, <u>then</u> $||K_n(f) - f||_p = O(\mu_n^2)\ (n \to \infty)$ <u>if and only if</u> $f \in L_p^2[0,R]$ <u>and</u> $f(0) = f(R) = f'(0) = f'(R) = 0$.

(ii) <u>If</u> $p = 1$, <u>then</u> $||K_n(f) - f||_1 = O(\mu_n^2)\ (n \to \infty)$ <u>if and only if</u> f' <u>is of bounded variation on</u> $[0,R]$ and $f(0) = f(R) = 0$.

(iii) <u>If</u> $1 \le p < \infty$, <u>then</u> $||K_n(f) - f||_p = o(\mu_n^2)\ (n \to \infty)$ <u>if and only if</u> $f(x) = 0$ <u>almost everywhere</u>.

<u>Remark</u>: In the statement of the theorem, all norms are measured in $L_p[0,R]$.

The proof of the theorem is too lengthy for the space provided here. However, we can indicate the role played by (1.2) in establishing the necessity of the end point conditions.

Assume $f \in L_p^{(2)}[0,R]$, $1 < p < \infty$, $f(0) = f(R) = 0$ and $f'(0) \neq 0$. Then it can be shown that a necessary condition for

$$||K_n(f) - f||_p = O(\mu_n^2)\ (n \to \infty)$$

is

$$\rho_n \int_0^\delta |g(x)| \int_x^R (t-x)\phi^n(t)\,dt\,dx = O(\mu_n^2)\ (n \to \infty) \tag{2.1}$$

for every $g \in L_q[0,R]$, $p + q = pq$, where $\delta > 0$ is chosen so that f' is bounded away from 0 on $[0,\delta]$.

Choose $g(x) = x^{-1/2q}$. Then the left side of (2.1) becomes

$$C_q \rho_n \int_0^\delta t^{2-\frac{1}{q}} \phi^{(n)}(t)dt,$$

where $C_q > 0$ depends only on q.

It can then be shown that (2.2) is asymptotically equivalent to $\mu_n^{2-\frac{1}{q}} \neq 0(\mu_n^2)$, $(n \to \infty)$.

REFERENCES

1. Bojanic, R. and Shisha, O., On the precision of uniform approximation of continuous functions by certain linear positive operators of convolution types, J. Approximation Theory 8(1973), 101-113.
2. Korovkin, P. P., "Linear Operators and Approximation Theory," Delhi, 1960.
3. Wood, B., Degree of L_p approximation by certain positive convolution operators, J. Approximation Theory 23(1978), 354-363.

INEQUALITIES FOR DERIVATIVES OF POLYNOMIALS HAVING REAL ZEROS

Josef Szabados
Hungarian Academy of Sciences
Budapest, Hungary

A. K. Varma
Department of Mathematics
University of Florida
Gainesville, Florida

Dedicated to Prof. G. G. Lorentz

Let us denote by $S_n[a,b]$ the class of algebraic polynomials of degree $\le n$ which have only real roots and no roots in (a,b) and $\Pi_n[-1,1]$ will denote the set of polynomials of the form $P(x) = \sum_{k=0}^{n} a_k(1+x)^k(1-x)^{n-k}$ with $a_k \ge 0$. Elements of Π_n are called polynomials with positive coefficients in $(1-x)$ and $(1+x)$ by Prof. G. G. Lorentz. Here in this work we are concerned with the following theorems.

Theorem A (P. Erdös). Let $P_n \in S_n[a,b]$ then

$$\max_{a\le x\le b} |P_n'(x)| \le \frac{e\,n}{b-a} \max_{a\le x\le b} |P_n(x)|. \tag{1.1}$$

Theorem B (G. G. Lorentz). Let $P_n \in \Pi_n$ then for $r = 1,2,\ldots$ there exist a constant c_r for which

Copyright © 1980 by Academic Press, Inc.
All rights of reproduction in any form reserved.
ISBN: 0-12-171050-5

$$\max_{-1\le x\le 1} |P_n^{(r)}(x)| \le c_r n^r \max_{-1\le x\le 1} |P_n(x)|. \qquad (1.2)$$

The following theorem is in a certain sense, a generalization of Theorem A.

Theorem 1. If $P_n(x)$ is a polynomial of degree at most n which has only real roots all but 1 outside of the interval (-1,1) then

$$\max_{-1\le x\le 1} |P_n'(x)| \le c\, n \max_{-1\le x\le 1} |P_n(x)|, \qquad (1.3)$$

where c is a positive constant independent of n.

An application of Theorem 1 is the following

Theorem 2. If $P_n(x) \in S_n[-1,1]$ then

$$\max_{-1\le x\le 1} |P_n''(x)| \le c\, n^2 \max_{-1\le x\le 1} |P_n(x)|, \qquad (1.4)$$

where c is a positive constant independent of n.

It is known that if $P_n \in S_n[-1,1]$ then $P_n \in \Pi_n$ and so from theorem B (1.4) follows as well.

Let us also denote by $H_n[-1,+1]$ the class of all algebraic polynomials of degree n having all real zeros lying inside [-1,1] subject to the condition $P_n(1) = 1$. Next, we now state

Theorem 3. Let $P_n \in H_n[-1,+1]$ then

$$\| P_n' \|^2_{L_2[-1,1]} = \int_{-1}^{1} (P_n'(x))^2 dx > \frac{n}{4}.$$

Moreover there exists a polynomial $P_0(x) \in H_n[-1,1]$ such that

$$\| P_0' \|^2_{L_2[-1,1]} = \frac{n}{4} + \frac{1}{8} + \frac{1}{8(2n-1)}.$$

2. Proof of Theorem 1. We may assume that $\max_{-1\le x\le 1} |P_n(x)| \le 1$.
Let $x_1, x_2, \ldots, x_n$ be the roots of $P_n(x)$ such that $-1 \le x_1 \le 0$, $|x_i| \ge 1$, $(i = 2,3,\ldots,n)$ (the case $0 < x_1 < 1$ can be treated similarly). Further let $P_n'(x_1) < 0$ (the case $P_n'(x_1) > 0$ can be treated similarly), and $\alpha \in [-1,x_1)$ the smallest argument in this subinterval for which $P_n'(\alpha) \le 0$ (i.e., either α is a local maximum place for $P_n(x)$ and then $P_n'(\alpha) = 0$ or $\alpha = -1$). Similarly, let $\beta \in (x_1,1)$ be the largest argument for which $P_n'(\beta) \le 0$. Then by Laguerre's theorem [[4], page 106]

$$x_1 + \frac{1}{n} \le \beta \le 1 \tag{2.1}$$

(because the maximum place cannot be closer to a root than the nth part of the length of the interval considered and this length is, by assumption $1-x_1 \ge 1$).

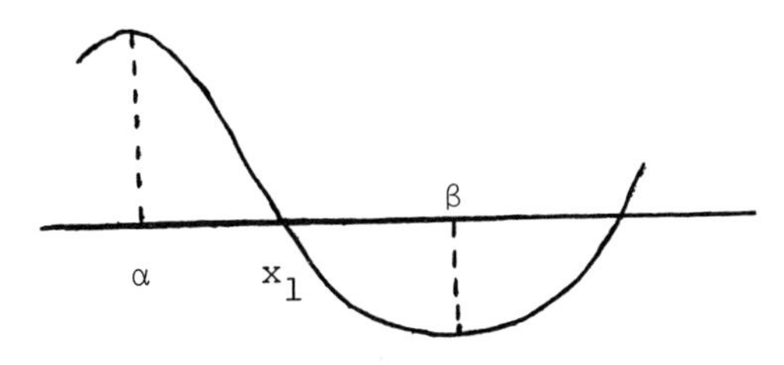

Being $P_n(\beta) < 0$, $P_n'(\beta) \le 0$, we have

$$0 \le \frac{P_n'(\beta)}{P_n(\beta)} = \frac{1}{\beta - x_1} + \sum_{i=2}^{n} \frac{1}{\beta - x_i} . \tag{2.2}$$

Thus by the inequality $1 + x \le e^x$ (2.1) and (2.2) we obtain

$$\frac{P_n(\alpha)}{|P_n(\beta)|} = \frac{x_1-\alpha}{\beta-x_1} \prod_{i=2}^{n} \frac{\alpha-x_i}{\beta-x_i} \leq n(x_1-\alpha) \prod_{i=2}^{n} (1 - \frac{\beta-\alpha}{\beta-x_i})$$

$$\leq n(x_1-\alpha) \exp -(\beta-\alpha) \sum_{i=2}^{n} \frac{1}{\beta-x_i} \leq n(x_1-\alpha) \exp \frac{\beta-\alpha}{\beta-x_1}$$

$$\leq n(x_1-\alpha) \exp \left(1 + \frac{x_1-\alpha}{\beta-x_1}\right) \leq n(x_1-\alpha) \exp \left(1 + n(x_1-\alpha)\right).$$

i.e.,

$$P_n(\alpha) \leq \min(1, e^2 n(x_1-\alpha)). \tag{2.3}$$

Now, in estimating $|P_n'(x)|$ we may restrict ourselves to the interval $[-1,x_1]$ because applying Theorem A to the interval we get

$$\max_{x \in [x_1 1]} |P_n'(x)| \leq \frac{e\,n}{1-x_1} \cdot 1 \leq e\,n\,.$$

We distinguish two cases:

Case 1. $-1 \leq x < \alpha$. Then by the definition of α, $P_n'(\alpha) = 0$, i.e.,

$$0 = \frac{P_n'(\alpha)}{P_n(\alpha)} = \frac{1}{\alpha-x_n} + \sum_{i=1}^{n-1} \frac{1}{\alpha-x_i}\,,$$

where x_n is a greatest root such that $x_n \leq -1$. Proceeding as before,

$$\frac{P_n(x)}{P_n(\alpha)} = \frac{x-x_n}{\alpha-x_n} \prod_{i=1}^{n-1} \frac{x-x_i}{\alpha-x_i} = \frac{x-x_n}{\alpha-x_n} \prod_{i=1}^{n-1} (1 - \frac{\alpha-x}{\alpha-x_i}).$$

Therefore, $\displaystyle\frac{P_n(x)}{P_n(\alpha)} \leq \frac{x-x_n}{\alpha-x_n} \exp \left(-(\alpha-x) \sum_{i=1}^{n-1} \frac{1}{\alpha-x_i}\right)$

$$= \frac{x-x_n}{\alpha-x_n} \exp \frac{\alpha-x}{\alpha-x_n}\,.$$

Since $\alpha-x \le \alpha-x_n$ it follows that

$$\frac{P_n(x)}{P_n(\alpha)} \le e \frac{x-x_n}{\alpha-x_n} . \tag{2.4}$$

Therefore, by $P_n'(\alpha) = 0$, (2.3) and (2.4)

$$0 < P_n'(x) = P_n(x) \sum_{i=1}^{n} \frac{1}{x-x_i} \le P_n(x) \sum_{x_i \le -1} \frac{1}{x-x_i}$$

$$\le P_n(x) \frac{\alpha-x_n}{x-x_n} \sum_{x_i \le -1} \frac{1}{\alpha-x_i} = P_n(x) \frac{\alpha-x_n}{x-x_n} \left(\frac{1}{x_1-\alpha} + \sum_{x_i \ge 1} \frac{1}{x_i-\alpha}\right)$$

$$\le e\, P_n(\alpha) \left(\frac{1}{x_1-\alpha} + n\right) \le c\, n .$$

Case 2. $\alpha \le x \le x_1$. Then by $P_n'(\alpha) \le 0$ we have

$$\frac{P_n'(\alpha)}{P_n(\alpha)} = \frac{1}{\alpha-x_1} + \sum_{i=2}^{n} \frac{1}{\alpha-x_i} \le 0,$$

and thus

$$\frac{P_n(x)}{P_n(\alpha)} = \frac{(x_1-x)}{(x_1-\alpha)} \prod_{i=2}^{n} \frac{x-x_i}{\alpha-x_i} = \frac{x_1-x}{x_1-\alpha} \prod_{i=2}^{n} \left(1 - \frac{\alpha-x}{\alpha-x_i}\right)$$

$$\le \frac{x_1-x}{x_1-\alpha} \exp\left((x-\alpha) \sum_{i=2}^{n} \frac{1}{\alpha-x_i}\right) \le \frac{x_1-x}{x_1-\alpha} \exp \frac{x-\alpha}{x_1-\alpha} \le e \frac{x_1-x}{x_1-\alpha} .$$

Thus by (2.3) we obtain

$$0 \le -P_n'(x) = P_n(x) \sum_{i=1}^{n} \frac{1}{x_i-x} \le P_n(x) \left(\frac{1}{x_1-x} + \sum_{x_i \ge 1} \frac{1}{x_i-x}\right)$$

$$\le e \frac{P_n(\alpha)}{x_1-\alpha} + n \le C\, n .$$

This completes the proof of Theorem 1.

Proof of Theorem 2. Let $P_n(x) \in S_n[-1,+1]$ then in general $P_n'(x)$ does not belong to $S_n[-1,+1]$, but can have at most one zero inside $(-1,+1)$. Thus by using Theorem 1 and Theorem A we obtain

$$\max_{-1\le x\le 1} |P_n''(x)| \le C(n-1) \max_{-1\le x\le 1} |P_n'(x)| \le C\, n\,(n-1) \max_{-1\le x\le 1} |P_n(x)|.$$

This proves theorem 2 as well.

3. Proof of Theorem 3. On integrating by parts we have

$$\int_{-1}^{1} (1+x)(P_n'(x))^2 dx = \int_{-1}^{1} (1+x)\; P_n'(x)\; P_n'(x)\,dx$$

$$= (1+x)P_n(x)P_n'(x) \Big|_{-1}^{1} - \int_{-1}^{1} P_n(x)\{P_n'(x) + (1+x)\; P_n''(x)\}dx$$

$$= 2P_n(1)P_n'(1) - \int_{-1}^{1} ((1+x)\; P_n(x)\; P_n''(x) + P_n(x)\; P_n'(x))dx.$$

Therefore

$$2\int_{-1}^{1} (1+x)(P_n'(x))^2 dx = 2P_n(1)P_n'(1) - \int_{-1}^{1} P_n(x)P_n'(x)\,dx + \int_{-1}^{1} (1+x)((P_n'(x))^2 - P_n(x)P_n''(x))\,dx. \tag{3.1}$$

Let $P_n \in H_n[-1,+1]$, and we denote by $x_1, x_2, \ldots, x_n$ the real zeros of $P_n(x)$. It is well known that

$$(P_n'(x))^2 - P_n(x)P_n''(x) = P_n^2(x) \sum_{k=1}^{n} \frac{1}{(x-x_k)^2}\,, \text{ and} \tag{3.2}$$

$$P_n(x)P_n'(x) = P_n^2(x) \sum_{k=1}^{n} \frac{1}{x-x_k}\,. \tag{3.3}$$

From (3.1) - (3.3) we have

$$2\int_{-1}^{1} (1+x)(P_n'(x))^2 dx = 2P_n(1)P_n'(1)$$

$$+ \int_{-1}^{1} \sum_{k=1}^{n} \left(\frac{1+x}{(x-x_k)^2} - \frac{x-x_k}{(x-x_k)^2}\right) P_n^2(x)\,dx = 2P_n(1)P_n'(1)$$

$$+ \int_{-1}^{1} \sum_{x=1}^{n} \frac{1-x_k}{(x-x_k)^2}\, P_n^2(x)\,dx$$

since $|x_k| \leq 1$ we have

$$2\int_{-1}^{1}(1+x)(P_n'(x))^2dx \geq 2P_n^2(1)\sum_{k=1}^{n}\frac{1}{1-x_k}.$$

Since $1 + x \leq 2$ and $\frac{1}{1-x_k} \geq \frac{1}{2}$, $k = 1,2,\ldots,n$ we obtain

$$4\int_{-1}^{1}(P_n'(x))^2dx \geq n\ P_n^2(1) \geq n.$$

This proves theorem 3 as well. Let us note that if $P_0(x) = (\frac{1+x}{2})^n$ then we have

$$\int_{-1}^{1} P_0'(x)^2dx = \frac{n^2}{2(2n-1)} = \frac{n}{4} + \frac{1}{8} + \frac{1}{8(2n-1)}.$$

REFERENCES

(1) P. Erdos, Extremal properties of polynomials, Ann. of Math., Vol. 41, #2(1940), 310-313.

(2) G. G. Lorentz, Degree of approximation by polynomials with positive coefficients, Math. Ann. 151(1963), 239-251.

(3) J. T. Scheick, Inequalities for derivatives of polynomials of special type, Jour. of Approx. theory, Vol. 6, No. 4, Dec. 1972, pp. 354-358.

(4) P. Turan, On rational polynomials, Acta. Univ. Szeged Sect. Sci. Math. 11, 106-113 (1946).

ON ORTHOGONAL POLYNOMIALS ASSOCIATED WITH THE INFINITE INTERVAL

Joseph L. Ullman[1]

Department of Mathematics
University of Michigan
Ann Arbor, Michigan

1. INTRODUCTION

In [2], Freud investigated the behavior of the parameters of the three term recursion relationship

(1) $$xp_n(x) = c_{n+1}p_{n+1}(x) + c_n p_{n-1}(x)$$

for the orthonormal polynomials $\{p_n(x)\}$ associated with the weight function

$$W_{\rho,m} = |x|^{\rho} \exp(-|x|^m), \rho > -1 \; m > 0 .$$

Consider the expression

(2) $$\lim_{n\to\infty} \frac{c_n}{n^{1/m}} = \left(\frac{\Gamma(m+1)}{\Gamma\left(\frac{m}{2}\right)\Gamma\left(\frac{m}{2}+1\right)} \right)^{-1/m} = e_m .$$

He proved that (2) was true for $m = 2, 4, 6$. He conjectured that (2) was true for all positive even values of m. We refer to this as conjecture 1. He also proved that if $\lim_{n\to\infty} c_n/n^{1/m}$ exists for positive even values of m, then the value of the limit would be that the contained in (2). He expressed the thought that (2) might possibly be valid for

[1]Supported by N.S.F. MCS 77-27117

Copyright © 1980 by Academic Press, Inc.
All rights of reproduction in any form reserved.
ISBN: 0-12-171050-5

$m > 0$. We refer to the simpler statement that $\lim_{n\to\infty} c_n/n^{1/m}$ exists for $m > 0$ as conjecture 2.

For a given $W_{\rho,m}$, let $x_{1,n} < x_{2,n} < \ldots < x_{n,n}$ be the zeros of $p_n(x)$, and let $y_{i,n} = A_n x_{i,n} + B_n$, where A_n, B_n are chosen so that $-1 = A_n x_{1,n} + B_n$, $1 = A_n x_{n,n} + B_n$. Let $N_n^{\rho,m}(\alpha,\beta)$ be the number of $y_{i,n}$ between α and β, [1]. It is shown in [4] that if m is any positive even number for which conjecture 1 is true, then for any α, β, $-1 \leqq \alpha < \beta \leqq 1$,

$$\lim_{n\to\infty} \frac{N_n^{\rho,m}(\alpha,\beta)}{n} = \int_\alpha^\beta d\nu_{\rho,m}, \tag{3}$$

where

$$d\nu_{\rho_m} = \frac{q_m(x)}{\pi\sqrt{1-x^2}}\, dx, \tag{4}$$

$$q_m(x) = 1 - e_m \sum_0^{\frac{m}{2}-1} (m-2-k)\binom{m}{k} T_{m-2k}(x),$$

e_m is defined in (2) and $T_j(x)$ is the Tchebycheff polynomial normalized so that $T_j(1) = 1$. We call $\nu_{\rho,m}$ the contracted zero distribution measure. In particular (4) is true for $m = 2, 4, 6$. On the other hand, (4) does not have a readily apparent meaning for m positive. but not an even integer.

The main contribution of this paper is Theorem 1 in Section 4 which provides a generalization of (4) which is meaningful for all positive values of m, and will give the correct contracted zero distribution measure for any positive value of m for which conjecture 2 is true. Section 2 contains needed results stated and proved in [3].

Section 3 contains new extensions of the material in Section 2. In particular we find a general formula for contracted zero distribution measures. Theorem 1 is then derived as a special case of this formula. The author is grateful to Paul Nevai for several helpful conversations related to the contents of this paper. We also wish to thank Ward Cheney for his kind invitation to present these results at the Conference on Approximation Theory in honor of George Lorentz, held in Austin, Texas, January 8-12, 1980.

2. SOME ASYMPTOTIC RESULTS ON ZEROS

Let μ be a finite measure defined on the Borel subsets of $R = (-\infty,\infty)$ and for which $\int |x|^{2n} d\mu < \infty$, $n = 0,1,\ldots,$ and the support of μ, $S(\mu)$, is an unbounded set. Let $\{p_n(x)\}$ be the unique polynomials, $p_n(x) = \gamma_n x^n + \ldots$, which satisfy $\int p_m(x)\, p_n(x)\, d\mu = \delta_{m,n}$, and the recurrence relationship

$$xp_n(x) = \frac{\gamma_n}{\gamma_{n+1}} p_{n+1}(x) + \alpha_n p_n(x) + \frac{\gamma_{n-1}}{\gamma_n} p_{n-1}(x) . \tag{5}$$

Let $\phi(x)\colon R^+ \to R^+$, $(R^+ = [0,\infty))$, satisfy the condition that $\phi(x)$ is non-decreasing and $\lim_{x\to\infty} \frac{\phi(x+t)}{\phi(x)} = 1$ for each real and positive value of t. Then if

$$\lim_{n\to\infty} \frac{2\,\gamma_{n-1}}{\phi(n)\gamma_n} = b, \quad \lim_{n\to\infty} \frac{\alpha_n}{\phi(n)} = a , \tag{6}$$

we have the result

$$(7)\qquad \lim_{n\to\infty} \frac{\sum_{i=1}^{n} x_{i,n}^{k}}{\int_0^n \phi^k(t)dt} = \begin{cases} a^k , & \text{if } b = 0 , \\ \frac{1}{\pi} \int_{a-b}^{a+b} \frac{u^k}{\sqrt{b^2-(u-a)^2}} du , & \text{if } b \neq 0. \end{cases}$$

3. FROM ZERO ASYMPTOTICS TO CONTRACTED ZERO DISTRIBUTION MEASURES

We now consider a further extension of (7). Since $\phi(x)$ is non-decreasing on R^+, the same is true of $\phi_n(x) = \frac{\phi(nx)}{\phi(n)}$ for $0 \leqq x \leqq 1$. Hence by Helly's theorem, there is an increasing sequence of integers, say $\{k_n\}$, such that

$$(8)\qquad \lim_{n\to\infty} \phi_{k_n}(x) = \alpha(x)$$

holds for all x such that $0 \leqq x \leqq 1$. Let $\mu_\alpha(x)$ be the measure function of $\alpha(x)$, namely

$$(9)\qquad \mu_\alpha(x) = \text{meas}\{t | \alpha(t) \leqq x\} .$$

We now assume that

$$(10)\qquad a = 0, \quad b \neq 0$$

in (6) and make the following change of variables in (7). Let $b\phi(n)t_{i,n} = x_{i,n}$, so that

$$(11)\qquad \sum_{i=1}^{n} x_{i,n}^{k} = b^k \phi^k(n) \sum_{i=1}^{n} t_{i,n}^{k} .$$

Let $t = ns$, so that

$$(12)\qquad \int_0^n \phi^k(t)dt = n \int_0^1 \phi(ns)ds .$$

Also let $u = bv$ so that, with (10),

$$(13) \qquad \frac{1}{\pi}\int_{a-b}^{a+b} \frac{u^k}{\sqrt{b^2-(u-a)^2}}\, d\mu = \frac{b^k}{\pi}\int_{-1}^{1} \frac{v^k dv}{\sqrt{1-v^2}} \, .$$

Substituting (11), (12) and (13) into (7) yields

$$(14) \qquad \lim_{n\to\infty} \frac{\sum t_{i,n}^k}{n\int_0^1 \left(\frac{\phi(ns)}{\phi(n)}\right)^k ds} = \frac{1}{\pi}\int_{-1}^{1} \frac{v^k}{\sqrt{1-v^2}}\, dv \, .$$

Thus

$$(15) \qquad \lim_{n\to\infty} \frac{\sum_{i=1}^{k_n} t_{i,k_n}^k}{k_n\int_0^1 \phi_{k_n}^k(s)\,ds} = \frac{\lim_{n\to\infty} \frac{\sum_{i=1}^{k_n} t_{i,k_n}^k}{k_n}}{\int_0^1 \alpha^k(s)\,ds} = \frac{1}{\pi}\int_{-1}^{1} \frac{v^k}{\sqrt{1-v^2}}\, dv,$$

or

$$(16) \qquad \lim_{n\to\infty} \frac{\sum_{i=1}^{k_n} t_{i,k_n}^k}{k_n} = \frac{1}{\pi}\int_0^1 t^k d\mu_\alpha \int_0^1 \frac{v^k}{\sqrt{1-v^2}}\, dv \, .$$

We use the Lebesgue dominated convergence theorem in (15) and we write a Lebesgue integral as a Stieltjes integral in (16). We next assume that μ_α is absolutely continuous, so that $d\mu_\alpha = g(t)dt$, and we assume that $g(t)$ is continuous. The right side of (16) can then be written as

$$(17) \qquad \frac{1}{\pi}\int_0^1 t^k g(t)\,dt \int_{-1}^{1} \frac{v^k}{\sqrt{1-v^2}}\, dv = \frac{1}{\pi}\iint_R \frac{(tv)^k\, g(t)}{\sqrt{1-v^2}}\, dt\, dv \, ,$$

where R is the rectangle $0 \leq t \leq 1$, $-1 \leq v \leq 1$. If we make the change of variable $w = tv$, $y = t$, then since $|\partial(t,v)/\partial(w,y)| = |1/y|$, we find that the double integral in (17) is equal to

$$(18) \qquad \frac{1}{\pi}\int_{-1}^{1} w^k \left(\int_{|w|}^{1} \frac{g(y)}{\sqrt{y^2-w^2}}\, dy\right) dw \, .$$

Next, let ν_n be the unit measure which has an atom of mass $1/n$ at $t_{i,n}$, $i = 1,\ldots,n$. We can then write (16) as

$$\lim_{n\to\infty} \int t^{k_n} d\nu_{k_n} = \frac{1}{\pi}\int_{-1}^{1} w^k h(w)\,dw \tag{19}$$

where

$$h(w) = \frac{1}{\pi}\int_{|w|}^{1} \frac{g(y)}{\sqrt{y^2 - w^2}}\,dy\ . \tag{20}$$

It then follows from (19) that the measures ν_{k_n} converge weakly to a measure ν, which is absolutely continuous and can be written as $d\nu = h(w)dw$.

4. THE CONTRACTED ZERO DISTRIBUTION MEASURES ASSOCIATED WITH CONJECTURE 2

Theorem 1. In equation (1), say that $c_n/n^{1/m}$ converges for a value of $m > 0$. Then for all $\alpha, \beta, -1 \leqq \alpha < \beta \leqq 1$

$$\lim_{n\to\infty} \frac{N_n^{\rho,m}(\alpha,\beta)}{n} = \int_\alpha^\beta d\nu_{\rho,m}\ , \tag{21}$$

where

$$d\nu_{\rho,m} = \frac{m}{\pi}\int_{|w|}^{1} \frac{y^{m-1}dy}{\sqrt{y^2 - w^2}}\,dw\ . \tag{22}$$

Proof of Theorem 1. We first compare (1) with (5) and since $\alpha_n = 0$, (6) holds for $\phi(n) = n^{1/m}$. We then find for the sequence $\{n\}$, that $\alpha(x) = x^{1/m}$ in (8). Hence $\mu_\alpha = x^m$, and since $d\mu_\alpha = mt^{m-1}dt$, in (16), we find that the zero measures ν_n converge weakly to the measure $\nu = \nu_{\rho,m}$, given by substituting $g(y) = my^{m-1}$ in (20), as

$$d\nu_{\rho,m} = \frac{m}{\pi} \int_{|w|}^{1} \frac{y^{m-1}}{\sqrt{y^2 - w^2}}\, dw \, .$$

Since $\nu_{\rho,m}$ is absolutely continuous, it follows that (21) holds, and the proof of Theorem 1 is completed.

REFERENCES

1. Erdos, P. and Freud, G., On polynomials with regularly distributed zeros, Proc. London Math. Soc. (3) 29(1974), 521-537.

2. Freud, G., On the coefficients in the recursion formulae of orthogonal polynomials, Proc. Royal Irish Academy, Vol. 76, A, 1, (1976), 1-6.

3. Nevai, Paul G. and Dehesa, Jesus S., On Asymptotic Average Properties of zeros of orthogonal Polynomials, S.I.A.M. Journ. on Math. Anals., Vol. 10, 6, (1979), 1184-1192.

4. Ullman, J.L., Orthogonal Polynomials Associated with an Infinite Interval, accepted for publication by the Michigan Mathematical Journal.

A NUMERICALLY EXPEDIENT RATIONAL FINITE ELEMENT

E. L. Wachspress

Knolls Atomic Power Laboratory
Schenectady, New York

I. FINITE ELEMENT PARTITIONING

A finite element practitioner must partition the region of interest into elements. A prime consideration in making this early and crucial partition is the degree of difficulty in constructing appropriate basis functions and evaluating integrals associated with them in the discretization stage of the computation. Triangles and rectangles in global or isoparametric coordinates are particularly well-suited for finite element computation. A two-level partitioning of the region of interest may be envisioned. Each element of the coarse partition is further subdivided into elements belonging to a particular useful class. For example, one coarse element may be subdivided into triangles and another into rectangles. A coarse element with a curved boundary as one of its edges could be subdivided into isoparametric triangles. A new element well-suited for finite element computation is developed in this note. Basis functions are easily constructed and associated integrals in the discretization stage of the computation are evaluated readily.

Copyright © 1980 by Academic Press, Inc.
All rights of reproduction in any form reserved.
ISBN: 0-12-171050-5

II. INTEGRATION OF A RATIONAL FUNCTION OF ONE VARIABLE

Basis functions for the elements developed in this note are in general rational functions. The denominators are functions of only one variable. The practical utility of these elements derives from existence of simple closed form values for integrals of rational functions in one variable.

Let $Q_t(x)$ be a polynomial of degree t with simple roots r_i $(i = 1, 2, \ldots, t)$. Then

$$\frac{1}{Q_t(x)} = \sum_{i=1}^{t} \frac{A_i}{x-r_i} \quad ; \quad A_i = \frac{1}{q_0} \prod_{j \neq i} \frac{1}{r_i - r_j} \, ,$$
$$(Q_t(x) = q_0 x^t + q_1 x^{t-1} + \ldots) \quad . \tag{1}$$

More general expressions account for multiple roots of Q. Suppose r_k has multiplicity s_k. Then the coefficients B_{mk} in the expansion

$$\frac{1}{Q} \equiv \prod_k \frac{1}{(x-r_k)^{s_k}} = \sum_k \sum_{m=1}^{s_k} \frac{B_{mk}}{(x-r_k)^m} \tag{2}$$

are defined ([3], p. 14) by

$$B_{mk} = \frac{1}{(s_k - m)!} \left\{ \frac{d^{s_k - m}}{dx^{s_k - m}} \left[\frac{(x-r_k)^{s_k}}{Q(x)} \right]_{x=r_k} \right\} \quad . \tag{3}$$

For any polynomial P_0, synthetic division yields terms in the Euclidean algorithm

$$P_0(x) = (x-r_i)\, P_{1i}(x) + P_0(r_i)$$
$$P_1(x) = (x-r_i)\, P_{2i}(x) + P_1(r_i) \tag{4}$$
$$\vdots$$

One then obtains

$$\int^{x} \frac{P(x)}{Q(x)}\,dx = \sum_{i} \Bigg\{ B_{1i}\left[\int^{x} P_{1i}(x)dx + P_0(r_i)\ln(x-r_i)\right] + B_{2i}\left[\int^{x} P_{2i}(x)dx + P_{1i}(r_i)\ln(x-r_i) - \frac{P_0(r_i)}{x-r_i}\right] + \dots + B_{s_i i}\left[\int^{x} P_{s_i i}(x)dx + \dots - \frac{P_0(r_i)}{(s_i-1)(x-r_i)^{s_i-1}}\right]\Bigg\}. \tag{5}$$

It is significant that integration of a rational function in one variable is numerically expedient.

III. A NEW RATIONAL ELEMENT

Consider an element bounded below by $y=0$, on the left by $x=a$, on the right by $x=b$, and above by the curve on which $y-Q(x)=0$, where Q is a polynomial having no root in the interval $[a,b]$. This element may be subdivided into elements bounded by vertical lines and by curves on which $ny-sQ=0$ and $ny-(s+1)Q=0$ with s running from 0 to $n-1$. The exterior intersection points in the (finite) affine plane of the element sides are all on $y=0$. The x-values of these points are the roots of $Q(x)$. The basis function construction developed in [4] yields the denominator $Q(x)$ common to all basis functions of all the elements of this partition. This greatly simplifies finite element computation algebra. Integrals required for discretization in commonly used finite element formulations may be found in closed form.

Basis functions for degree-one approximation over the element are of the form

$$W_m(x,y) = \frac{(ny-gQ)P^{(m)}(x)}{Q(x)}\,; \qquad g=s,\quad s+1 \tag{6}$$

where $P^{(m)}$ and Q are polynomials of the same degree. Integrals in the finite element discretization are of the form

$$I = \int_{x_1}^{x_2} dx \int_{(s/n)Q(x)}^{(s+1/n)Q(x)} dy \frac{P(x,y)}{[Q(x)]^r} \quad , \quad r \text{ an integer} . \tag{7}$$

The y-integration is with polynomial integrands and the limits are such that the integrands for the x-integration are rational functions with powers of Q as denominators. Such integrals are evaluated by methods discussed in Section II. Only one partial fraction expansion must be determined since Q is common to all elements.

The lower boundary of the element need not be $y = 0$. The more general element with lower boundary $y - Q^*(x) = 0$ is handled by transforming to $y' = y - Q^*(x)$. The denominator for the basis functions becomes $Q'(x) = Q(x) - Q^*(x)$. Equations 6 and 7 are valid with y' replacing y and Q' replacing Q.

This new element is particularly useful in fluid problems with curved boundaries confining a flow and vertical element boundaries normal to the direction of net flow. Regions may in general be partitioned into coarse elements of this type together with a few more complicated regions for which other techniques may be appropriate. Once the partial fraction expansions are obtained for each coarse element, the numerical complexity of discretization with these elements should not be significantly greater than with the commonly used rectangular elements.

There are several advantages over use of isoparametric elements. Actual boundaries may be approximated more closely by algebraic curves. Higher degree approximation may be achieved within elements. The denominator Q in the basis functions remains fixed for all degree of approximation. There is no quadrature error since integrals are evaluated in closed form.

This new element thus broadens the class of problems that can be modeled conveniently for finite element computation in global coordinates.

IV. AN EXAMPLE

Collatz [1] discussed an element with $Q = 1 + \frac{x}{4} - \frac{x^3}{48}$. The lack of an x^2 term in Q leads to further simplification in that the sum of the roots of Q vanishes. The roots of Q are $r_1 = r = 4.7106$ $r_2 = -\frac{1}{2}r + r'i$, and $r_3 = -\frac{1}{2}r - r'i$ with $r' = 2.1547$. An element is shown in Figure 1 with midside nodes 5 and 6 introduced to achieve degree-one approximation.

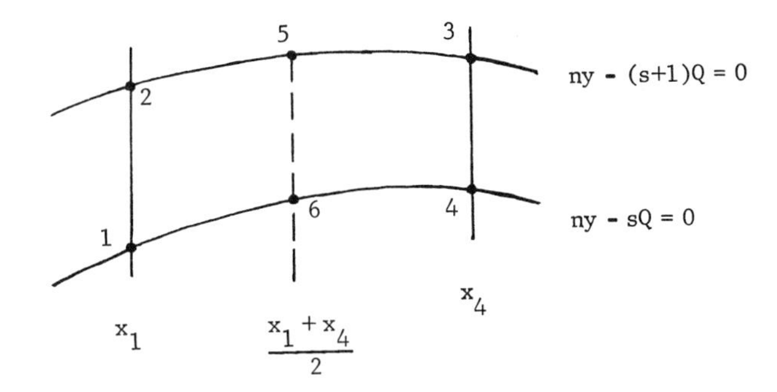

FIGURE 1. An element for the Collatz problem.

Consider, for illustrative purposes, basis function W_5. The adjacent factor (see [4], Ch. 5) must vanish at the nonvertex intersection of line (2;3) with $ny - (s+1)Q = 0$. The value of x at this point, say x_A, must satisfy $x_1 + x_4 + x_A = 0$. Thus, $x_A = -(x_1 + x_4)$ and the adjacent factor is $(x + x_1 + x_4)$. It follows that

$$W_5(x,y) = \frac{8(x-x_1)(x_4-x)(x+x_1+x_4)(ny-sQ(x))}{3(x_1+x_4)(x_4-x_1)^2 Q(x)},$$

and this defines $P(x)$ in

$$W_5(x,y) = \frac{P(x)(ny-sQ(x))}{Q(x)} . \tag{8}$$

Collatz considered Laplace's equation, for which finite element discretization requires evaluation of integrals of the form

$$I_{mn} = \iint_{\text{element}} dxdy\, \nabla W_m \circ \nabla W_n \,. \tag{9}$$

Carrying through the algebra, we discover that for this example the integrands in the x-integration (after the y-integration has been performed) are either polynomials or polynomials over $Q(x)$. Higher powers of Q do not appear in the denominator in this problem. This is illustrated by evaluation of one of the simpler integrands:

$$\begin{aligned} \left(\frac{\partial W_5}{\partial y}\right)^2 &= \frac{n^2 P^2}{Q^2} \\ \int_{\frac{sQ}{n}}^{\frac{(s+1)Q}{n}} dy \left(\frac{\partial W_5}{\partial y}\right)^2 &= \frac{Q}{n}\,\frac{n^2 P^2}{Q^2} = \frac{nP^2}{Q} \,. \end{aligned} \tag{10}$$

V. BLENDING

The "transfinite" blending technique of Gordon and Hall [2] may be used to transform a curved element into a unit square. An important consequence of this approach is, however, that the approximation space does not in general include polynomials of degree greater than zero. Finite element approximation errors with blending are not easily estimated. The Collatz example may be used to illustrate failure of blending to achieve degree one approximation. For this purpose, we choose the element in Figure 2.

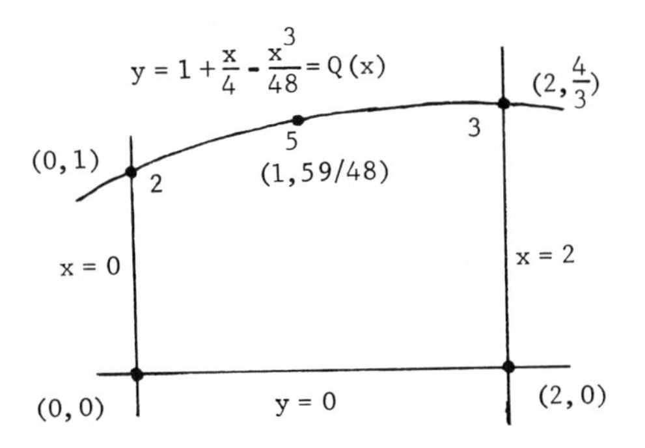

FIGURE 2. A specific element for the Collatz problem.

A transfinite bilinear Lagrangian mapping that transforms the unit square in (s,t) coordinates into this element is

$$x = 2s \qquad y = tQ(2s) = t(1 + s/2 - s^3/6) \ . \tag{11}$$

The basis functions in the s,t plane are

$$\begin{aligned} B_1(s,t) &= (1-s)(1-t) \ , & B_2(s,t) &= t(1-2s)(1-s) \ , \\ B_3(s,t) &= st(2s-1) \ , & B_4(s,t) &= s(1-t) \ , \\ B_5(s,t) &= 4st(1-s) \ . \end{aligned} \tag{12}$$

The expansion of the function y is

$$\begin{aligned} y \sim \sum_{i=1}^{s} y_i B_i(s,t) &= B_2 + \frac{4}{3} B_3 + \frac{59}{48} B_5 \\ &= t(1-2s)(1-s) + \frac{4}{3} st(2s-1) + \frac{59}{12} st(1-s) \\ &= t(1 + \frac{7}{12} \cdot s - \frac{s^2}{4}) = y \frac{(1 + (7/12)s - (s^2/4))}{(1 + (s/2) - (s^3/6))} \\ &= y \frac{1 + (7/24)x - (x^2/16)}{1 + (x/4) - (x^3/48)} \ . \end{aligned} \tag{13}$$

The approximation has no error at the nodes, but degree one is not attained with this approximation.

The rational basis functions for this element are

$$B_2(x,y) = y(2-x)(1-x)(3+x)/6Q(x)$$

$$B_3(x,y) = yx(x^2-1)/6Q(x) \tag{14}$$

$$B_5(x,y) = yx(4-x^2)/3Q(x) ,$$

and now

$$y \sim B_2 + \frac{4}{3}B_3 + \frac{59}{48}B_5$$

$$= \frac{y}{Q}\left[\frac{(2-x)(1-x)(3+x)}{6} + \frac{2}{9}x(x^2-1) + \frac{59}{3\cdot 48}x(4-x^2)\right] \tag{15}$$

$$= \frac{y}{Q}\left[1 + (x/4) - (x^3/48)\right] = y .$$

Similarly, $x = 2B_3 + 2B_4 + B_5$. Degree one approximation is attained.

It is interesting to note that in x,y coordinates the blending bases are

$$B_2[s(x,y),t(x,y)] = y(1-x)(2-x)/2Q(x) \tag{16}$$

$$B_3(x,y) = yx(x-1)/2Q(x) \qquad B_5(x,y) = yx(2-x)/Q(x) .$$

The blending basis functions lack the "adjacent" factors introduced in the rational bases to achieve degree one approximation and continuity along element interfaces. The blending transformation achieves continuity but not degree one approximation within the element. The new element may thus be considered as a corrected blending-function element.

REFERENCES

1. Collatz, L., "Discretization and chained approximation", Numerical Solution of Differential Equations (1973), Dundee Conference Proceedings, Springer-Verlag Lecture Notes in Mathematics #363, 32-43.
2. Gordon, W.J. and Hall, C.A., "Transfinite element methods", Numerische Mathematik, 21(1973), 109-129.
3. Pierce, B.O., A Short Table of Integrals, Ginn and Company (1929).
4. Wachspress, E.L., A Rational Finite Element Basis, Academic Press (1975).

SPLINE BASES, REGULARIZATION, AND GENERALIZED CROSS VALIDATION FOR SOLVING APPROXIMATION PROBLEMS WITH LARGE QUANTITIES OF NOISY DATA

Grace Wahba[1]

Department of Statistics
University of Wisconsin-Madison
Madison, Wisconsin

We consider the problem of estimating a smooth function $f(t)$, $t \varepsilon \Omega \subset R^d$ given noisy observations of linear functionals of f. In particular, one observes

$$z_i = L_i f + \varepsilon_i, \quad i = 1,2,\ldots,n,$$

where the ε_i are independent, zero mean random variables with common unknown variance σ^2. The L_i are assumed to be continuous linear functionals on a family H_m, $m \geq m_o$ of reproducing kernel Hilbert spaces (rkhs) where m indexes the highest square integrable derivative (not necessarily an integer) possessed by functions in H_m. In this setup, one seeks a regularized estimate $f_{n,\lambda}$ of f by solving the minimization problem: Find $f \varepsilon H_m$ to minimize

$$\frac{1}{n}\sum_{i=1}^{n}(L_i f - z_i)^2 + \lambda J_m(f) \qquad (*)$$

where $J_m(\cdot)$ is a norm or seminorm in H_m. The parameter λ controls the tradeoff between the infidelity

$$\frac{1}{n}\sum_{i=1}^{n}(L_i f - z_i)^2$$

of the approximate solution to the data, and the roughness $J_m(f_{n,\lambda})$ of

[1]This work supported by USARO Grant DAAG29-77-G-0207.

Copyright © 1980 by Academic Press, Inc.
All rights of reproduction in any form reserved.
ISBN: 0-12-171050-5

the solution. When a reproducing kernel for H_m is known, explicit solutions to this problem are well known. The method of generalized cross validation (GCV) has been shown to be an effective tool for choosing λ (as well as m) from the data. For applications to the solution of ill posed problems ($L_i f = \int K(t_i,s)f(s)ds$) see Wahba (1980) and references cited there, for applications to estimation of the derivative in one dimension, see Craven and Wahba (1979).

In this note we primarily consider the case $L_i f = f(t_i)$, $t_i \varepsilon R^2$, ($t_i = (x_i,y_i)$) and

$$J_2(f) = \int_{-\infty}^{\infty}\int_{-\infty}^{\infty} (f_{xx}^2 + 2f_{xy}^2 + f_{yy}^2)dxdy$$

or, more generally,

$$J_m(f) = \int_{-\infty}^{\infty}\int_{-\infty}^{\infty} \sum_{\nu=0}^{m} \binom{m}{\nu}\left(\frac{\partial^m}{\partial x^\nu \partial y^{m-\nu}} f(x,y)\right)^2 dxdy.$$

Then the minimizer, $f_{n,\lambda}$ of (*) is one of the "thin plate splines", so called because $J_2(f)$ is the bending energy of a thin plate.

Duchon (1976a) has given explicit formulae for $f_{n,\lambda}$ for the general d dimensional situation for m (integer or not) with $2m-d > 0$. (This condition is necessary for H_m to be an rkhs.) See also Meinguet (1978). Wahba (1979a) and Wahba and Wendelberger (1979) have shown how to use GCV to choose λ and m from noisy data, and have given a computational algorithm. Utreras Dias (1979) and Paihua Montes (1979), who along with Duchon are former students of P.J. Laurent, have also given computational procedures, and our work has benefited from the availability of their work. Wahba (1979b) has conjectured convergence results for the integrated mean square error (IMSE)

$$E \int_\Omega (f_{n,\lambda^*}(t) - f(t))^2 dt$$

in the noisy data case and has provided supporting numerical evidence for the conjectured rates. Here λ^* must tend to zero at the correct rate, and an estimate of the λ which minimizes the IMSE is provided by the GCV method.

Thin plate spline interpolating and smoothing methods (which involve minimization of a seminorm in an rkhs) are equivalent to a certain form of minimum variance unbiassed estimation on the homogeneous random fields of Matheron (1973). This was demonstrated in 1971 in Kimeldorf and Wahba (1971) in the one dimensional case for more general seminorms, and has been generalized by Duchon (1976b). Two dimensional interpolation and smoothing methods known as "kriging" in the mining industry are also based on minimum variance unbiassed estimation on Matheron's homogeneous random fields (see Delfiner (1978)). By comparing the form of the estimates in Delfiner (1978) and Duchon (1976a) one can see that, loosely speaking, the kriging estimates solve minimization problems in H_m for $m = 3/2, 5/2, \ldots$. The main difference between kriging as described by Delfiner and the thin plate splines we use below is in the choice of the free parameters (here λ and m) and how they are estimated. Delfiner uses (ordinary) cross validation as part of his estimation procedure.

Returning to interpolation case, where $f(t_i)$ is known exactly, Micchelli and Wahba (1979) have given lower bounds for best attainable convergence rates of $||f - f_{o,n}||_m$ for f "very smooth" in general rkhs, in terms of the rate of decay of the eigenvalues of the relevant reproducing kernel, and it appears to be establishable from Duchon (1978) that the best possible rates are achieved with thin plate spline interpolation for uniform rectangular or triangular arrays of points $t_1, t_2, \ldots, t_n$. This is in sharp contrast to interpolation in rkhs in two dimensions which are tensor products of two one dimensional rkhs.

In this latter case, uniform rectangular arrays of points do not lead to the best achieveable rates and better designs (ratewise) can be found. See Micchelli and Wahba (1979) and the examples in Delvos and Posdorf (1978) and Wahba (1979c).

In the remainder of this note we will discuss our computational experience and numerical results in smoothing 500 millibar height data from an irregular array of 120 North American weather stations, using thin plate splines and GCV. The numerical procedures we propose are very satisfactory on the University of Wisconsin-Madison UNIVAC 1110 for $n \le 130$ or so. After describing our present results, we will propose an approximate method which will allow the calculations to be made easily for much larger data sets. The numerical details for the $n \le 130$ case, and further numerical results may be found in Wahba (1979a) and Wahba and Wendelberger (1979). In summary, for $t = (x,y)$,

$$f_{n,\lambda}(t) = \sum_{i=1}^{n} c_i E_m(t - t_i) + \sum_{\nu=1}^{M} d_\nu \phi_\nu(t)$$

where $E_m(t) = |t|^{2m-2}\log|t|$, $|t| = \sqrt{x^2 + y^2}$; $\phi_1(t),\ldots,\phi_M(t) = 1, x, y, x^2, xy, \ldots, y^{m-1}$ where $M = \binom{m+1}{2}$ is the number of polynomials of total degree $\le m - 1$. The vectors $c = (c_1,\ldots,c_n)'$ and $d = (d_1,\ldots,d_m)'$ are solutions to $(K + n\lambda I)c + Td = z$, $z = (z_1,\ldots,z_n)'$, $T'c = 0$, where K is the $n\times n$ matrix with ij^{th} entry $E_m(t_i - t_j)$, and T is the $n\times M$ matrix with $j\nu^{th}$ entry $\phi_\nu(t_j)$. To have a unique solution we need that the rank of T be M. Let U be any $n\times n - M$ matrix whose $n - M$ columns are orthogonal unit vectors orthogonal to the M columns of T, and let $I - A(\lambda) = n\lambda U(U'KU + n\lambda I)^{-1}U'$. The GCV estimate of λ is the minimizer of $V(\lambda) = \frac{1}{n}||(I - A(\lambda))z||^2/(\frac{1}{n}\mathrm{Tr}(I - A(\lambda)))$. See Wahba (1979a), Craven and Wahba (1979). It is known that $U'KU$ is strictly positive definite.

$V(\lambda)$ is easily computed for many values of λ and c and d are easily computed once the eigenvalue decomposition $\Gamma D\Gamma'$ of the $n - M \times n - M$ matrix $U'KU$ is computed. We get this decomposition using double precision EISPACK. The condition number of the interpolation problem is the condition number of D.

Figure 1 shows the location of 120 North American weather stations. Using a mathematical model of 500 mb. heights provided by Koehler (1979) a model f was generated whose contours are given by the dashed lines in Figure 2. Data z was obtained by evaluating f at the weather stations and adding a pseudo random normal error with realistic standard deviation 10 meters. The value $\hat{\lambda}$ minimizing $V(\lambda)$ for $m = 2,3,\ldots,7$ was obtained by global search and $m = 5$ was selected because $V(\hat{\lambda})$ was smallest for $m = 5$. The resulting $f_{n,\hat{\lambda}}$ with $m = 5$ satisfied

$$\sum_{i=1}^{n} (f_{n,\hat{\lambda}}(t_i) - f(t_i))^2 \simeq 1.01 \min_{\lambda} \sum_{i=1}^{n} (f_{n,\lambda}(t_i) - f(t_i))^2$$

and so $\hat{\lambda}$ was an excellent choice of λ in this (typical!) example. The contours of $f_{n,\lambda}$ are given in Figure 2 by solid lines.

The condition number of D was between 10^6 and 10^{10} depending on m, for these 120 data points. The computation of an interpolant ($\lambda = 0$) will become unstable with increasing n as the condition number approaches the range of double precision accuracy. The presence of a sufficiently positive λ eliminates this problem. However, when this condition number is large it is likely that smoothing problems with very large n can be satisfactorily solved in some restricted subspaces of H_m.

Let $X_n = \text{span}\{\sum_{i=1}^{n} u_{ji} E_m(t - t_i),\ j = 1,2,\ldots,n - M\}$ where $u_j = (u_{ji},\ldots,u_{jn})'$ is the jth column of U. X_n is a proper $n - M$ dimensional subspace of M. If the Gram matrix $U'KU$ of this basis for X_n is machine rank deficient, it appears reasonable to minimize (*) in the union of an $N << n - M$ dimensional subspace of X_n and span $\{\phi_\nu\}_{\nu=1}^{M}$,

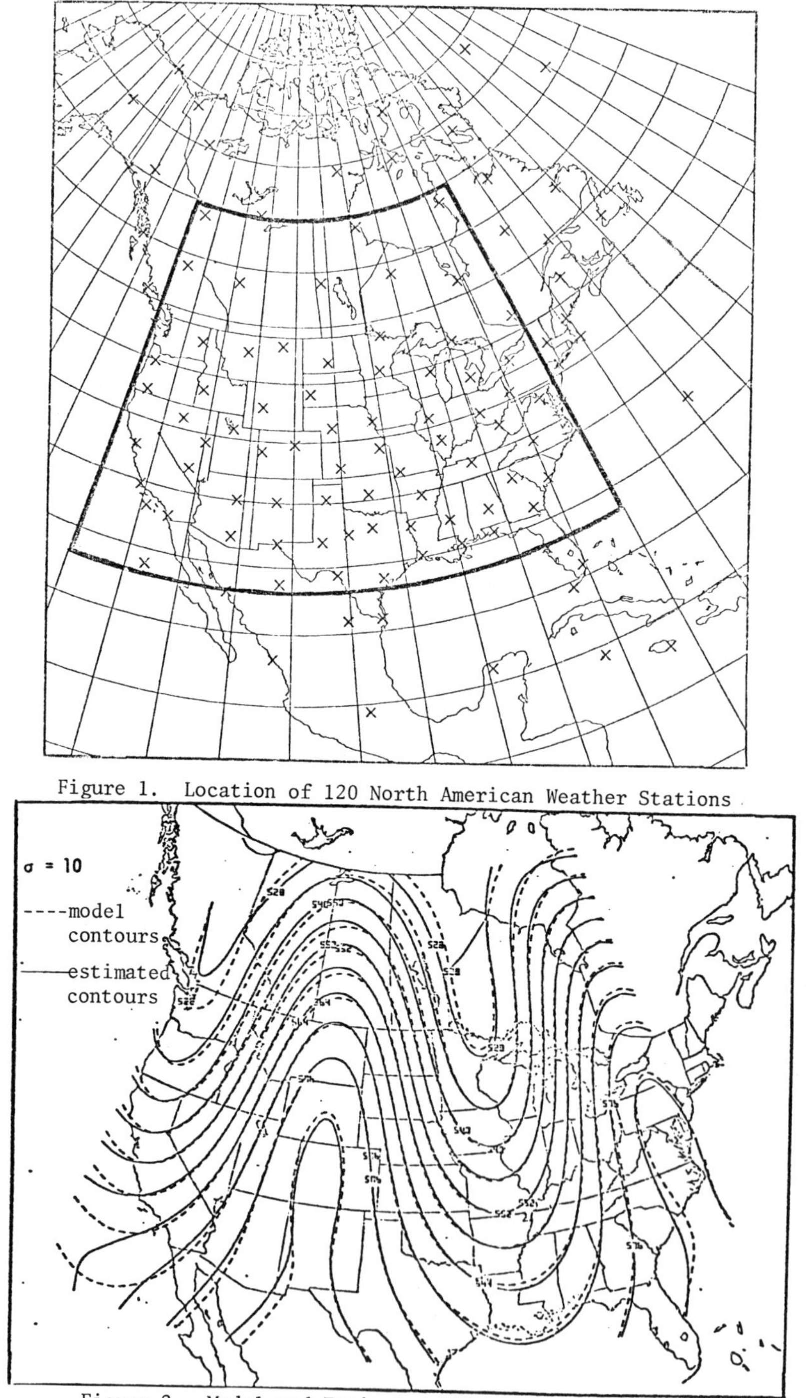

Figure 1. Location of 120 North American Weather Stations

Figure 2. Model and Estimated 500 Millibar Contours

or more generally, one may choose a set of points $s_1,\ldots,s_N \in \Omega$, possibly making the points denser, where the "right answer" is known to vary most, and minimize (*) in the subspace spanned by $\{\sum_{i=1}^{N} \tilde{u}_{ji} E_m(t - s_i),\ j = 1,2,\ldots,N - M\} \cup \{\phi_\nu\}_{\nu=1}^{M}$, where it is required that $\sum_{i=1}^{n} \tilde{u}_{ji}\phi_\nu(s_i) = 0,\ \nu = 1,\ldots,M,\ j = 1,\ldots, N - M$. This approach is likely to be reasonable, even when the L_i are not evaluation functionals, and in one dimension it is equivalent to approximately solving optimization problems in H_m in spaces of B-splines of degree $2m - 1$. Details of the equations may be found in Wahba (1980). Meteorological applications include $L_i f = f_y(t_i)$ and $L_i f = f_x(t_i)$ which are related to measured eastward and northward wind components through the geostrophic approximation to the wind. See Wahba and Wendelberger (1979).

REFERENCES

Craven, P. and Wahba, G. (1979). Smoothing noisy data with spline functions: estimating the correct degree of smoothing by the method of generalized cross validation. Numer. Math. 31, 377-403.

Delfiner, P. (1978). The intrinsic model of order k. Ecole des mines de Paris, Summer School notes.

Delvos, F.J. and Posdorf, H. (1978). N-th order blending, in Constructive Theory of Functions of Several Variables, W. Schempp and K. Zeller, eds., Lecture Notes in Mathematics, 571, 53-64.

Duchon, J. (1976a). Splines minimizing rotation-invariant semi-norms in Sobolev spaces. In Constructive Theory of Functions of Several Variables, W. Schempp and K. Zeller, eds., Lecture Notes in Mathematics, 571, 85-100, Springer.

Duchon, J. (1976b). Fonctions-spline et esperances conditionnelles de champs gaussiens. Annales Scientifiques de l'Universite de Clermont, No. 61.

Duchon, J. (1978). Sur l'erreur d'interpolation des fonctions de plusiers variables par les D^m-splines. R.A.I.R.O. analyse numerique, Vol. 12 No. 4, 325-334.

Kimeldorf, G. and Wahba, G. (1971). Some results on Tchebycheffian spline functions. J. Math. Anal. Applic. 33, 1, 82-95.

Koehler, T. (1979). A case study of height and temperature analysis derived from Nimbus-6 satellite soundings on a fine mesh model grid. University of Wisconsin-Madison, Department of Meteorology, Thesis.

Matheron, G. (1973). The intrinsic random functions and their applications. Adv. Appl. Prob. 5, 439-468.

Meinguet, J. (1978). An intrinsic approach to multivariate spline interpolation at arbitrary points. Proceedings of the NATO Advanced Study Institute on Polynomial and Spline Approximation, Calgary, B. Sahney, ed. 163-190. D. Reidell Publishing Co.

Micchelli, G. and Wahba, G. (1979). Design problems for optimal surface interpolation. University of Wisconsin-Madison, Department of Statistics Technical Report No. 565.

Montes, L. Paihua (1979). Quelques methodes numeriques pour le calcul de fonctions splines a une et plusiers variables. Thesis, University Scientifique et Medicale de Grenoble.

Utreras Dias, F. (1979). Cross validation techniques for smoothing spline functions in one or two dimensions. In Smoothing Techniques for Curve Estimation, T. Gasser and M. Rosenblatt, eds. Lecture Notes in Mathematics, No. 757, Springer-Verlag, Berlin.

Wahba, G. (1977). Practical approximate solutions to linear operator equations when the data are noisy, SIAM J. Numerical Analysis, 14, 4, 651-667.

Wahba, G. (1979a). How to smooth curves and surfaces with splines and cross-validation. In "Proceedings of the 24th Conference on the Design of Experiments" U.S. Army Research Office, Report 79-2, also University of Wisconsin-Madison Statistics Department Technical Report No. 555.

Wahba, G. (1979b). Convergence rates of "Thin Plate" smoothing techniques for curve estimation, T. Gasser and M. Rosenblatt, eds. Lecture notes in Mathematics, No. 757, Springer-Verlag, Berlin.

Wahba, G. (1979c). Interpolating surfaces: high order convergence rates and their associated designs, with application to X-ray image reconstruction. University of Wisconsin-Madison, Department of Statistics Technical Report No. 523.

Wahba, G. and Wendelberger, J. (1979). Some new mathematical methods for variational objective analysis using splines and cross validation. University of Wisconsin-Madison Statistics Department Technical Report No. 578.

Wahba, G. (1980). Ill-posed problems: Numerical and statistical methods for mildly, moderately, and severely ill posed problems with noisy data. University of Wisconsin-Madison Statistics Department Technical Report No. 595. To appear in Proceedings of the International Symposium on Ill-Posed Problems, M.Z. Nashed, ed.

SMOOTHNESS CONCEPTS ON GENERAL SETS AND EXTENSION OF FUNCTIONS

Hans Wallin

Department of Mathematics
University of Umeå
Umeå, Sweden

0. Introduction. Smoothness properties of functions f defined on a closed set $F\subset R^n$ are usually measured by means of moduli of continuity of different orders (moduli of smoothness) of f on F. However, it is often convenient to use other, closely related descriptions of smoothness properties, introduced for instance by means of local polynomial approximation or Peetre´s K-functional from interpolation theory. We refer to [4] for an excellent survey in the case $n=1$, $F=[0,1]$.

Closely related to the problem to measure the smoothness of f on F is the problem to extend f to a "nice" set $K\supset F$, for instance a rectangle K if F is bounded, without destroying the smoothness properties of f. If F is relatively regular in the sense that F is a domain with boundary ∂F locally in Lip 1, such an extension of f may be made by defining f outside F essentially by reflection in ∂F (Lichenstein, Hestenes, Seeley, Calderon, Stein; see for instance [1, Section 4.25] and [9, Ch. VI, 3]). This works for f in some L^p-space with $1\leq p\leq\infty$. If, on the other hand, F does not have this regularity a natural approach is to base an extension of f on Whitney´s extension theorem (see for instance [9, Ch. VI, 2]; see also [10]) which works for a general closed set F and corresponds to the case $p=\infty$. In this paper we are rather interested in working on a general set F than in a general space of functions. We let F be any closed set in R^n and stick to the

Copyright © 1980 by Academic Press, Inc.
All rights of reproduction in any form reserved.
ISBN: 0-12-171050-5

case $p=\infty$ and consider functions f in the class $C(F)$ of bounded, uniformly continuous functions on F. The general L^p-case has to be based on a Whitney extension theorem in L^p, $1\leq p\leq\infty$; such a theorem was proved in [6], [7], and [8].

We want to discuss the connection between different smoothness concepts and the extension problem. Proofs of the results in Section 1 and more details are found in [12]. The connection to earlier results is given in Section 2.

1. Definitions and results. 1.1. If $F\subset R^n$ we put $F(h) = \{x\in F:\ x+th \in F,\ \text{for } 0\leq t\leq 1\}$, $h\in R^n$, and

$$\Delta_h^r f(x) = \sum_{\ell=0}^{r} (-1)^{r-\ell}\binom{r}{\ell} f(x+\ell h),\ \text{for}\ x \in F(rh),$$

where r is a positive integer. The r-th modulus of smoothness ω_r of f on F is defined by

$$\omega_r(t) = \omega_r(t,f;F) = \sup_{0<|h|\leq t}\ \sup_{x\in F(rh)} |\Delta_h^r f(x)|,\ \text{for}\ t>0.$$

If $F(rh)$ is empty, the supremum over $F(rh)$ is interpreted as zero. We use the following notation: $d(x,F)$ is the distance from x to F; if $j = (j_1, \ldots, j_n)$ is a multi-integer, i.e. $j_1, \ldots, j_n$ are non-negative integers, then $|j| = j_1 + \ldots + j_n$, $x^j = x_1^{j_1} \ldots x_n^{j_n}$ if $x = (x_1, \ldots, x_n) \in R^n$, $j! = j_1! \ldots j_n!$, and D^j denotes the corresponding derivative of order $|j|$; if $E\subset R^n$ is closed, $C(E)$ denotes the class of bounded, uniformly continuous functions on E with supremum norm $\|\ \|_E$. We assume that k is a non-negative integer, $r = k+1$, $F\subset R^n$ is a closed set, and that $f \in C(F)$. In [11] we used Whitney's extension technique to prove the following theorem where j and ℓ denote multi-integers and $\omega_r(2^{-\nu}) = \omega_r(2^{-\nu},f;F)$; the "Taylor formula remainders" $R_{j\nu}$ in the theorem are typical ingredients in Whitney's theorem.

THEOREM 1. There exists, for every $a_1 > 0$ and $K = K(a_1) = \{x \in R^n: d(x,F) \le a_1\}$, an extension of f to a function $Ef \in C(K)$ satisfying, for some constant M_1,

$$\omega_r(t,Ef;K) \le M_1\, \omega_r(t,f;F), \quad \text{for} \quad 0 < t \le 1, \tag{1}$$

if and only if the following condition holds: For every $a_2 > 0$ there exist families $\{f_\nu^{(j)}\}_{|j| \le k}$, $\nu = 1, 2, \ldots$, of functions defined and bounded on F and satisfying, for some constant M_2, $\nu = 1, 2, \ldots$, and $x, y \in F$ (for $j=0$ we write $f_\nu^{(0)} = f_\nu$),

$$|f(x) - f_\nu(x)| \le M_2\, \omega_r(2^{-\nu}), \tag{2}$$

$$|R_{j\nu}(x,y)| \le M_2 |x-y|^{r-|j|}\, 2^{r\nu}\, \omega_r(2^{-\nu}),$$

$$\text{for} \quad |x-y| \le a_2 2^{-\nu}, \quad |j| \le k, \quad \text{if} \tag{3}$$

$$R_{j\nu}(x,y) = f_\nu^{(j)}(x) - \sum_{|j+\ell| \le k} \frac{f_\nu^{(j+\ell)}(y)}{\ell!}(x-y)^\ell, \quad \text{and}$$

$$|f_\nu^{(j)}(x) - f_\mu^{(j)}(x)| \le M_2\, 2^{|j|\nu}\, \omega_r(2^{-\mu}),$$

$$\text{for} \quad |j| \le k, \quad \nu \ge \mu \ge 1. \tag{4}$$

In the if-part, the constant M_1 depends only on a_1, M_2, k, and n. In the only if-part M_2 depends only on a_2, M_1, k, and n.

Theorem 1 is related to [10] where an analogous result was proved for Lipschitz spaces.

1.2. Theorem 1 gives one characterization of those $f \in C(F)$ for which there exist extensions $Ef \in C(K)$ satisfying (1). We now give an alternative characterization by means of local polynomial approximation. We consider open, half-open or closed n-dimensional cubes Q and Q' with sides parallel to the axes and put $\omega_r(t,f;F) = \omega_r(t)$.

THEOREM 2. There exists, for every $a_1 > 0$ and $K = K(a_1) = \{x: d(x,F) \le a_1\}$, an extension $Ef \in C(K)$ of f satisfying (1) if and only if the following condition holds for every $a_3 > 0$: For every cube Q, $Q \cap F \ne \phi$, with side of length $t \le a_3$ there exists a polynomial P_Q of degree $\leq k$ such that the polynomials P_Q are bounded on Q, uniformly in Q, and, for some constant M_3,

$$|f(x) - P_Q(x)| \leq M_3\, \omega_r(t), \quad \text{for} \quad x \in Q \cap F, \tag{5}$$

and, if Q', $Q' \cap Q \neq \phi$, $Q' \cap F \neq \phi$, is a cube with side $t' \leq a_3$ and $P_{Q'}$ the associated polynomial of degree $\leq k$, then

$$|P_Q(x) - P_{Q'}(x)| \leq M_3 \max(\omega_r(t), \omega_r(t'))\ , \quad \text{for} \quad x \in Q \cap Q'. \tag{6}$$

In the if-part M_1 depends only on a_1, M_3, k, and n. In the only-if part M_3 depends only on a_3, M_1, k, and n. It may be noted that it is trivial that the inequality (5) may be reversed in a certain sense (see for instance [4, p. 133]).

1.3. We now introduce the space $C^r(F)$ of Whitney functions of class C^r on F. We say that $\{f^{(j)}\}_{|j|\leq r} \in C^r(F)$ if $f^{(j)}$, $|j|\leq r$, are bounded, uniformly continuous functions on F such that if

$$R_j(x,y) = f^{(j)}(x) - \sum_{|j+\ell|\leq r} \frac{f^{(j+\ell)}(y)}{\ell!}(x-y)^\ell, \quad |j|\leq r,$$

then $R_j(x,y) = o(|x-y|^{r-|j|})$, for $x,y \in F$ and $|j|\leq r$, as $|x-y| \to 0$ (uniformly in $x,y \in F$). Hence, $C^0(F) = C(F)$. If $F = R^n$ it follows from Taylor´s formula that $\{f^{(j)}\}_{|j|\leq r} \in C^r(F)$ is uniquely determined by $f = f^{(0)}$ by the relation $f^{(j)} = D^j f$ and that the space of Whitney functions of class C^r on R^n may be identified with the space of functions f such that $D^j f$ are bounded and uniformly continuous on R^n for $|j|\leq r$. One version of Whitney´s extension theorem now says that there exists a linear mapping $E\colon C^r(F) \to C^r(R^n)$ such that for every $f = \{f^{(j)}\}_{|j|\leq r} \in C^r(F)$, $D^j(Ef) = f^{(j)}$ on F for $|j|\leq r$ (see [12, Theorem 3]).

We now define the K-functional on F for $f \in C(F)$ and $t>0$ by

$$K_r(t,f;F) = \inf_{C^r(F)} \{ \|f-g^{(0)}\|_F + t \max_{|j|=r} \|g^{(j)}\|_F\},$$

where the infimum is taken over all $\{g^{(j)}\}_{|j|\leq r} \in C^r(F)$.

As a complement of Theorem 1 we formulate

THEOREM 3. Assume that there exists, for every $a_1>0$ and $K = \{x: d(x,F) \leq a_1\}$, an extension $Ef \in C(K)$ of f satisfying (1). Then, for some positive constants M_4 and M_5,

$$M_4\ \omega_r(t,f;F) \leq K_r(t^r,f;F) \leq M_5\ \omega_r(t,f;F), \quad \text{for} \quad 0 < t \leq 1. \tag{7}$$

Here M_4 and M_5 depend only on M_1, k, and n. The converse of Theorem 3 does not hold (see [12]).

2. Connection to earlier results. We now assume that $F = \overline{\Omega}$ where Ω is a connected, bounded Lipschitz graph domain in the sense that the boundary of Ω is locally of class Lip 1 (see for instance [11, Section 4] for a precise statement). Then the results in this paper are essentially reduced to alternative proofs of known results. In [11, Theorem 3] Theorem 1 was used to prove that for these sets $F = \overline{\Omega}$ there exists, for every $a_1>0$ and $K = \{x: d(x,F) \leq a_1\}$, a bounded, linear extension operator $E: C(F) \to C(K)$ satisfying (1) for all $f \in C(F)$ with a constant M_1 independent of t and f. This is a result by Brudnyi [2, Theorem 2] and Johnen-Scherer [5, Cor. 1]. As a consequence of this and Theorem 2, (5) is true, with a constant M_3 independent of t and f, for all $f \in C(F)$, $F = \overline{\Omega}$, and this result can, of course, be localized in the sense that $\omega_r(t,f;F)$ in (5) may be replaced by $\omega_r(t,f;F\cap Q)$. We then obtain a result by Whitney and Brudnyi connecting ω_r to local polynomial approximation (see [13] and, in the general case, [3, Cor. 1]; see also [5, Cor. 3]). By combining Theorem 3 with the extension result mentioned above [11, Th. 3] we also obtain that the K-functional $K_r(t^r,f;F)$ is comparable in size, independently of f and t, $0 < t \leq 1$, to $\omega_r(t,f;F)$ for all $f \in C(F)$, $F = \overline{\Omega}$. This is, in the general case, a result by Brudnyi [2, Th. 3] and Johnen-Scherer [5, Th. 1] but goes back, for simpler F, at least to Peetre (see [4, Th. 2.1] and [12]).

REFERENCES

1. Adams, R.A., Sobolev spaces, Academic Press, New York, 1975.

2. Brudnyi, Ju.A., On a theorem of extension for one family of functional spaces (Russian), Zap. naucn. Sem. LOMI 56, 1976, 170-173.

3. Brudnyi, Ju.A., Piecewise polynomial approximation, imbedding theorems and rational approximation, Lect. Notes Math. 556, Springer-Verlag, Berlin, 1976, 73-98.

4. De Vore, R., Degree of approximation, Proc. Symp. Approx. theory II, Austin 1976, Academic Press, New York, 1976, 117-161.

5. Johnen, H. and K. Scherer, On the equivalence of the K-functional and moduli of continuity and some applications, Lect. Notes Math. 571, Springer-Verlag, Berlin, 1977, 119-140.

6. Jonsson, A. and H. Wallin, A Whitney extension theorem in L^p and Besov spaces, Ann. Inst. Fourier 28, 1978, 139-192.

7. Jonsson, A. and H. Wallin, A trace theorem for generalized Besov spaces with three indexes, Colloq. Math. Soc. J. Bolyai 19, North-Holland, 1979, 429-449.

8. Jonsson, A. and H. Wallin, The trace to subsets of R^n of Besov spaces in the general case, Dept. Math. Univ. Umeå, 1, 1979.

9. Stein, E.M., Singular integrals and differentiability properties of functions, Princeton Univ. Press, Princeton, 1970.

10. Wallin, H., The trace to closed sets of the Zygmund class of quasi-smooth functions in R^n, Proc. Durham Symp. 1977 on Multivariate approx., Academic Press, London, 1978, 83-93.

11. Wallin, H., Extensions of functions preserving moduli of smoothness, to appear in Proc. Conf. Approx. and function spaces, Gdansk, 1979.

12. Wallin, H., Local polynomial approximation, the K-functional, and extension of functions preserving moduli of smoothness, Dept. Math. Univ. Umeå, 4, 1980.

13. Whitney, H., On functions with bounded n-th differences, J. Math. Pures Appl. (9) 36, 1957, 67-95.

OBSERVATIONS ON THE COMBINED RECURSIVE COLLOCATION AND KERNEL APPROXIMATION TECHNIQUE FOR SOLVING VOLTERRA INTEGRAL EQUATIONS

B. Wood

Department of Mathematics
University of Arizona
Tucson, Arizona

I. INTRODUCTION

Consider the Volterra integral equation

$$u(x) = f(x) + \int_a^x K(x - t)g(u(t))dt, \qquad a \leq x \leq b. \tag{1.1}$$

It is assumed that $g \in C(-\infty,\infty)$, the convolution kernel $K \in L_1[0,b - a]$, $f \in C[a,b]$, and (1.1) has a unique continuous solution, u, on $[a,b]$. Let $\{L_n\}$ denote a sequence of linear operators from $L_1[0,b - a]$ into $C[0,b - a]$. For $K \in L_1[0,b - a]$, approximate $L_n(K)$ by its Fourier-Tschebycheff expansion

$$L_n(K,z) \doteq L_{m,n}(K,z) \equiv \sum_{j=0}^{m}{}' b_{j,n} T_j(-1 + \frac{2z}{b-a}), \qquad 0 \leq z \leq b - a, \tag{1.2}$$

where T_j denotes the jth Tschebycheff polynomial of the first kind and the coefficients $b_{j,n}$ can be approximated by

$$b_{j,n} \doteq \frac{2}{L}\sum_{\alpha=0}^{L}{}'' L_n(K,\frac{1}{2}(b - a)(\cos\frac{\pi\alpha}{L} + 1))\cos(\frac{j\pi\alpha}{L}), \tag{1.3}$$

$j = 0, 1, \ldots, m$ and $L \geq m$ (8). Here $\sum'$ denotes first summand to be halved. Equation (1.1) is replaced by

$$V(x) = f(x) + \sum_{j=0}^{m}{}' b_{j,n}\int_a^x T_j(-1 + \frac{2(x-t)}{b-a})g(V(t))dt, \qquad a \leq x \leq b. \tag{1.4}$$

Copyright © 1980 by Academic Press, Inc.
All rights of reproduction in any form reserved.
ISBN: 0-12-171050-5

Let N be a positive integer, $\delta = \frac{b-a}{N} > 0$, and partition $[a,b + \delta]$ with $a = x_0 < x_1 = x_0 + \delta < \ldots < x_N = b < x_{N+1} = b + \delta$. Let $\{\phi_{k,v}\}_{v=0}^{M}$ denote a basis in the space of M-th degree polynomials on $\sigma_k = [x_k, x_{k+2}]$, $k = 1, 2, \ldots, N - 1$. Define the polynomial ϕ_k on the interval σ_k by

$$\phi_k(x) = \sum_{v=0}^{M} \alpha_{k,v}\phi_{k,v}(x), \quad k = 1, 2, \ldots, N - 1, \tag{1.5}$$

where the $\alpha_{k,v}$'s will be determined below. For $a \leq x \leq b$ and v continuous on $[z,x]$, write

$$D_{m,n}(x;v) = \sum_{j=0}^{m}{}' b_{j,n}Y_j(x;v), \tag{1.6}$$

where

$$Y_j(x;v) = \int_a^x T_j\left(-1 + \frac{2(x-t)}{b-a}\right)g(v(t))dt. \tag{1.7}$$

Let u_0 denote an approximate continuous solution to (1.1) on $[a,a + 2\delta] = [x_0,x_2]$. This solution can be constructed by using an analytical method, such as a series expansion, which produces a good approximate solution near $x = a$. Solve the system

$$\begin{aligned} &\sum_{v=0}^{M} \alpha_{1,v}\phi_{1,v}(x_1) = u_0(x_1), \\ &\sum_{v=0}^{M} \alpha_{1,v}\phi_{1,v}(\xi_{1,j}) = f(\xi_{1,j}) + D_{m,n}(\xi_{1,j};u_0) \end{aligned} \tag{1.8}$$

at points $\xi_{1,j} = (1 - \eta_j)x_1 + \eta_j x_2$, $j = 1, \ldots, M$, with $0 < \eta_1 < \eta_2 < \ldots < \eta_M < 1$, for $\{\alpha_{1,v}\}$, $v = 0, 1, \ldots, M$. This provides the polynomial ϕ_1 on the interval $\sigma_1 = [x_1,x_3]$ and we define the continuous function $\hat{\phi}_1$ on $[x_0,x_3]$ by

$$\hat{\phi}_1(x) = \begin{cases} u_0(x), & x_0 \leq x \leq x_1, \\ \phi_1(x), & x_1 \leq x \leq x_3. \end{cases}$$

Continue this process. At the last step solve the system

$$\sum_{v=0}^{M} \alpha_{N-1,v}\phi_{N-1,v}(x_{N-1}) = \hat{\phi}_{N-2}(x_{N-1}),$$
$$\sum_{v=0}^{M} \alpha_{N-1,v}\phi_{N-1,v}(\xi_{N-1,j}) = f(\xi_{N-1,j}) + D_{m,n}(\xi_{N-1,j};\hat{\phi}_{N-2}), \tag{1.9}$$

at points $\xi_{N-1,j} = (1 - \eta_j)x_{N-1} + \eta_j x_N$, $j = 1, \ldots, M$, for $\{\alpha_{N-1,v}\}$, $v = 0, 1, \ldots, M$. The continuous function

$$\hat{\phi}_{N-1}(x) = \begin{cases} u_0(x), & a \leq x \leq a + \delta, \\ \phi_1(x), & a + \delta \leq x \leq a + 2\delta, \\ \vdots & \\ \phi_{N-1}(x), & a + (N - 1)\delta \leq x \leq b, \end{cases}$$

provides an approximate solution to (1.4).

The above is a modification of J. Bownds' (2) adaptation of recursive collocation (5). For a polynomial basis chosen on each subinterval σ_k, each of the systems (1.8), (1.9) is uniquely solvable for the $\alpha_{k,v}$, since, in each case, the coefficient matrix has a nonzero determinant. The situation is more complicated in (2) and (5) because the right-hand side of the associated linear system to be solved contains terms involving the $\alpha_{k,v}$. In (5) and (6) it is stated that a unique solution to the associated linear system exists in case g is linear and $|K|$ is bounded away from zero for $a \leq t \leq x \leq b$. In addition, Bownds requires that K be continuous on $(0,b - a]$ and both (2) and (5) require the evaluation, via quadrature, of terms of the form

$$\int_{x-\delta}^{x} K(x - t)g(u(t))dt.$$

The main advantage of (2) when compared with (5) is a reduction in the number of calculations required to carry out recursive collocation.

This is accomplished by utilizing the Tschebycheff approximation of a continuous kernel to evaluate certain terms as solutions of a simple differential system. This feature is retained in the present modification. Recall (1.6), (1.7) and let $Y(x;v) \equiv \mathrm{col}(y_0(x;v), \ldots, y_m(x;v))$. Using properties of Tschebycheff polynomials, we can show that for $a \leq x \leq b$ and v continuous on $[a,x]$, $\underline{Y}(x;v)$ solves the initial-value problem

$$y_0'(x;v) = g(v(x)), \qquad y_0(a;v) = 0,$$

$$y_1'(x,v) = \left(\frac{2}{b-a}\right)y_0(x;v) - g(v(x)), \qquad y_1(a;v) = 0,$$

$$y_j'(x;v) = \frac{1}{2}(z_j'(x;v) - z_{j-2}(x;v)), \qquad j = 2, 3, \ldots, m,$$

where

$$z_0(x;v) \equiv y_0(x;v),$$

$$z_1(x;v) \equiv 2y_1(x;v),$$

$$z_j'(x;v) = \left(\frac{4j}{b-a}\right)z_{j-1}(x;v) + z_{j-2}'(x;v) + 2^{-j}g(v(x)),$$

and

$$z_j(a;v) = 0, \qquad j = 2, 3, \ldots, m.$$

Two disadvantages of the present technique are the potential error involved in replacing K by $L_{m,n}(K)$ and possible complications in evaluating $L_n(K)$ for (1.3). In Section 2 we provide an example of operators which may minimize these disadvantages. We also note that, for certain g and L_n, there exists a unique continuous solution, $u_{m,n}$, of (1.4) which is arbitrarily close, in the uniform norm, to the solution of (1.1) for sufficiently large n and m.

II. CONVERGENCE

In (1.2) and (1.3) assume $\{L_n\}$ is a uniformly bounded sequence of positive linear operators mapping $L_1[0,b-a]$ into $C[0,b-a]$. Also assume $\|L_n(e_i) - e_i\|_1$ converges to zero, where $e_i(t) = t^i$ for $i = 0, 1, 2$.

We now demonstrate the stated convergence.

Theorem. Assuming that (1.1) has a unique continuous solution, u, on $[a,b]$, let $f \in C[a,b]$, $K \in L_1[0,b-a]$ and $g \in C(-\infty,\infty)$. Suppose g satisfies the Lipschitz condition

$$|g(z_2) - g(z_1)| \leq L|z_2 - z_1|$$

for $\min_{a \leq x \leq b} u(x) - 1 \leq z_1, \; z_2 \leq \max_{a \leq x \leq b} u(x) + 1$. Then, given $\epsilon > 0$, there exist sufficiently large n and m such that the approximating integral equation (1.4) has a unique continuous solution, $u_{m,n}$, on $[a,b]$, and

$$\max_{a \leq x \leq b} |u(x) - u_{m,n}(x)| \leq \epsilon.$$

Outline of Proof. The proof follows the method of (4). Namely, use a recent result on L_p-approximation [10], the Hölder inequality and standard results from the theory of Fourier-Tschebycheff series to establish, for $\epsilon > 0$,

$$\sup_{a \leq x \leq b} \int_a^x |K(x-t) - L_{m,n}(K, x-t)|dt < \epsilon,$$

for $n(\epsilon)$ and $m(n,\epsilon)$ sufficiently large. Now follow (4, pp. 136-139).

In the following example $L_n(K)$ is a polynomial. Therefore, the error involved in replacing K by $L_{m,n}(K)$ should be reduced.

Example. Let $\{P_n\}$ be a sequence of orthogonal polynomials on $[-1,1]$ whose weight function, w, is nonnegative, even, Lebesgue integrable on $[-1,1]$, and has the following properties:

$$0 < m \leq w(x), \qquad -r \leq x \leq r, \qquad 0 < r \leq 1;$$

$$w(x) \leq M < \infty, \qquad -\eta \leq x \leq \eta, \qquad 0 < \eta \leq 1.$$

Let the polynomial, R_n, of degree $4n - 8$ be defined by

$$R_n(x) = c_n \left[\frac{P_{2n}(x)}{(x^2 - \alpha_{2n}^2)(x^2 - \alpha_{2n-1}^2)} \right]^2,$$

where α_{2n-1}, α_{2n} are the two smallest positive zeros of P_{2n} and $c_n > 0$ is chosen so that

$$\int_{-r}^{r} R_n(x)\,dx = 1, \qquad n = 1, 2, \ldots .$$

Let

$$\hat{L}_n(f,y) = \int_0^r f(t) R_n(t - y)\,dt$$

for $f \in L_1[0,r]$, $0 \leq y \leq r$. This operator has been studied in (1), (7), (9), (10) and (11). For $z = h^{-1}(y) = (\frac{b-a}{r})y$, $0 \leq y \leq r$, and $K \in L_1[0,b - a]$, define

$$L_n(K,z) = \hat{L}_n(K \circ h^{-1}, \frac{rz}{b-a}).$$

REFERENCES

1. Bojanic, R., L'Enseignement Math. 15, 43-51 (1969).

2. Bownds, J. M., preprint.

3. Bownds, J. M., Applied Mathematics and Computation 4, 67-79 (1978).

4. Bownds, J. M. and Wood, B., J. of Approximation Theory 25, #2, 120-141 (1979).

5. Brunner, H., Mathematics of Computation 31, #139, 708-716 (1977).

6. Brunner, H., Computing 13, 67-79 (1974).

7. DeVore, R. A., The Approximation of Continuous Functions by Positive Linear Operators, Springer-Verlag, N. Y., 1972.

8. Snyder, M., Chebyshev Methods in Numerical Approximation, Prentice-Hall, N. J., 1966.

9. Swetits, J. J. and Wood, B., J. of Approximation Theory, 24, #4, 310-323 (1978).

10. Swetits, J. J. and Wood, B., preprint.

11. Wood, B., J. of Approximation Theory, 23, #4, 354-363 (1978).

THE FUNDAMENTALITY OF SEQUENCES OF TRANSLATES

R. A. Zalik

Department of Mathematics
Auburn University
Auburn, Alabama

Let t, x and y denote real numbers, z a complex number, let R stand for the set of real numbers, let f be a function in $L_1(R)$ $(L_2(R))$, let F denote its Fourier transform, and let $T(f)$ denote the linear span of the set of functions of the form $f(t+c)$, c real. Two classical theorems of N. Wiener affirm that $T(f)$ is dense in $L_1(R)$ $(L_2(R))$ if and only if $F(x) \neq 0$ for all real x $(F(x) \neq 0$, a. e.) (cf. [1, pp. 98, 100]). The question thus naturally arises as to under what conditions the linear span of a sequence of the form

$$\{f(t+c_k)\} \tag{1}$$

will be dense in various function spaces (such a sequence is called "fundamental"). Since the Fourier transform is an isometry in $L_2(R)$, the fundamentality of (1) thereon is readily seen to be equivalent to that of

$$\{F(t) \exp(c_k ti)\}. \tag{2}$$

We first study the fundamentality of (2). Given a complex sequence $\{c_k\}$, let $P(z)$ denote its canonical product (for a definition see [2]); given the numbers $p \geq 1$ and $\alpha \geq 1$, let q and β denote their conjugate exponents. With this notation we have:

Copyright © 1980 by Academic Press, Inc.
All rights of reproduction in any form reserved.
ISBN: 0-12-171050-5

Theorem 1. Let $F(t) \neq 0$ for all real t, let $1 < \alpha < \infty$ and $p \geq 1$. Let $E_p = L_p(R)$ if $p < \infty$, and $E_p = C_0(R)$ if $p = \infty$. Consider the following statements:

$$\text{For some } a > 0,\ \exp(a|t|^{\beta})F(t) \text{ is in } E_p \tag{3}$$

and

$$\text{For some } a > 0,\ \exp(a|t|^{\beta})/F(t) \text{ is in } L_q(R). \tag{4}$$

If (3) holds, for (2) to be fundamental in E_p it suffices that $P(z)$ be of order larger than α. Conversely, if (4) holds and α is not an odd integer, for (2) to be fundamental in E_p it is necessary that $P(z)$ be of order larger than α, or of order α and type larger than zero.

Sketch of Proof: Sufficiency. Let $p = \infty$, assume that (3) holds and (2) is not fundamental in $C_0(R)$. By an application of the Hahn-Banach theorem we see there is a nonzero linear functional that annihilates the elements of (2). Applying the Riesz representation theorem, we infer that there is a (bounded) complex-valued non-zero measure μ, such that

$$\int_R \exp(c_k ti)F(t)d\mu(t) = 0;\ k = 0,1,2,\ldots \tag{5}$$

By the theorems of Morera and Fubini, we see that

$$h(z) = \int_R \exp(zti)F(t)d\mu(t) \tag{6}$$

is an entire function. Since $\mu \neq 0$, we know from Bochner's theorem on the uniqueness of the Fourier-Stieltjes transform (cf. [3]), that $h(z)$ is not identically zero. The hypothesis implies that if $z = x+yi$,

$$|h(z)| \leq c_1 \int_R \exp(|yt|-a|t|^{\beta})\ d|\mu|(t).$$

Let $d > 0$ be a constant whose value will be determined "a posteriori". Applying Young's inequality, we have:

$$|yt| = (d|y|)(d^{-1}|t|) \leq d^{\alpha}\ |y|^{\alpha}\ \alpha^{-1} + d^{-\beta}|t|^{\beta}\beta^{-1}.$$

Hence,

$$|h(z)| \leq \exp(\alpha^{-1}d^{\alpha}|y|^{\alpha})\int_R \exp[d^{-\beta}\beta^{-1}-a)|t|^{\beta}]d|\mu|(t). \quad (7)$$

Setting now d so that $d^{-\beta}\beta^{-1}-a \leq 0$, the integral in the preceding inequality is finite. Thus $h(z)$ is an entire function of order α and finite type, and the conclusion follows from (5) and the basic properties of entire functions. The proof for $1 \leq p < \infty$ is similar, only that instead of Bochner's theorem, we use the uniqueness of the Fourier transform of an integrable function.

Necessity. Assume $P(z)$ is of order less than α or of order α and zero type. The hypotheses imply that there is a number δ such that for any $\eta > 0$ there is a linear combination $h(\theta)$ of $\cos \alpha\theta$ and $\sin \alpha\theta$ such that $h(0) = -2\eta$, $h(\pi) = -2\eta$, and $|h(\theta)| \leq \delta\eta$. By a theorem of V. Bernstein [4], there is an entire function $q_0(z,\eta)$ that has $h(\theta)$ as its indicator of growth. Thus $q(z,\eta) = q_0(z,\eta)P(z)$ is an entire function that vanishes at the points c_k, $|q(z,\eta)| \leq c_2 \exp(2\delta\eta\ |z|^{\alpha})$, and $|q(t,\eta)| \leq c_3\exp(-\delta\eta|t|^{\alpha})$. The conclusion now follows by a method similar to the one employed in the proof of the necessity in [5; Theorem 3].

From Theorem 1 we obtain:

Theorem 2. Let $f(t)$ be in $L_1(R)$, and let $F(t)$ denote its Fourier transform. Assume $F(t) \neq 0$ for all real t. If (3) holds for $p = 2$ (or $p = 1$), then for (1) to be fundamental in $L_2(R)$ (or $C_0(R)$), it suffices that $P(z)$ be of order larger than α. If (4) holds with $q = 2$, and α is not an odd integer, then for (1) to be fundamental in $L_2(R)$ or $C_0(R)$, it is necessary that $P(z)$ be of order larger than α or of order α and type larger than zero.

Sketch of Proof: The proof for $L_2(R)$, as well as that of the necessity for $C_0(R)$ follow from Theorem 1 and the isometric character of the Fourier transform. The proof of the sufficiency for $C_0(R)$ follows from Theorem 1 by an application of [6; p. 6, Thm. 3B].

Since the Fourier transform of $f(t) = (2\pi)^{-1/2}\exp(-t^2/2)$ is $f(t)$ itself, this function satisfies the conditions of Theorem 2 for $\alpha = 2$. Examples for other values of α can be obtained using the results of [7; Chapter 1].

The results stated above were announced in [8], and can be considerably refined. This will be the theme of a forthcoming paper.

If the $\{c_k\}$ are real, we can strengthen the sufficient condition in Theorem 1:

Theorem 3. Assume that (3) is satisfied, and let $\{c_k\}$ be an increasing sequence such that for some $\delta > 0$, $c_{k+1}^{\alpha} - c_k^{\alpha} \geq \delta$, let $c = \alpha^{-1}(a\beta)^{-\alpha/\beta}$, and assume that

$$\limsup_{r\to\infty} r^{-2c/\pi} \exp[2 \sum_{c_k<r^{1/\alpha}} c_k^{-\alpha}] = \infty .$$

Then (2) is fundamental in E_p.

Theorem 3 is sharper than a similar result of B. Faxén [9].

Proof of Theorem 3: Setting d in (7) so that $d^{-\beta}\beta^{-1} = a$, we readily infer that the function $g(z) = h(z^{1/\alpha})$ is holomorphic on $\operatorname{Re} z > 0$, $g(z) = O[\exp(c|z|)]$, $z \to \infty$, and vanishes at the points $\lambda_k = c_k^{\alpha}$; moreover the λ_k are positive and $\lambda_{k+1} - \lambda_k \geq \delta$ for all k, and the conclusion follows by applying Fuch's theorem (cf. [2]).

The reader will easily derive from Theorem 3 a counterpart of Theorem 2.

So far, we have studied the fundamentality of (1) and (2) for $\alpha < \infty$. We now turn to the case $\alpha = \infty$. If $c > 0$ is an arbitrary real number, by $T(c)$ we shall denote the series $\sum(1 - |\tanh(\pi c_k/4c)|)$. With this notation we have:

Theorem 4. Let $F(t) \neq 0$ for all real t, and assume that $|\mathrm{Im}(c_k)| < a$ for all k. If (3) holds (with $\beta = 1$), the divergence of $T(a)$ suffices for (2) to be fundamental in E_p; if (4) holds, the divergence of $T(a)$ is necessary for (2) to be fundamental in E_p. If $\exp(-a|x|)/F(x)$ is bounded, the divergence of $T(a)$ is necessary for (2) to be fundamental in C_0 or L_p $(1 \leq p \leq 2)$.

Corollary. If $F(x) = \exp(-a|x|)$, then for (2) to be fundamental in C_0 or L_p $(1 \leq p \leq \infty)$ it is sufficient that $T(a)$ be divergent. Conversely, the divergence of $T(a)$ is necessary for (2) to be fundamental in C_0 or L_p $(1 \leq p \leq 2)$. This corollary generalizes a theorem of Paley and Wiener (cf. [10; p. 766, Theorem 1]. For sequences of translates we have:

Theorem 5. Let $f(t)$ be a function in L_1 and L_2 whose Fourier transform $F(x)$ has no real zeros, and let $a > 0$ be given.

i) Assume that $|\mathrm{Im}(c_k)| < a$ for all k. Let $f(t)$ have a holomorphic extension to $|\mathrm{Im}(z)| < a$, and assume that for any real number c, $-a < c < a$, $f(t+ci)$ is in L_1. If $\exp(a|x|)F(x)$ is bounded, the divergence of $T(a)$ suffices for (1) to be fundamental in C_0 and every L_p, $1 \leq p < \infty$.

ii) If $\exp(-a|x|)/F(x)$ is bounded, $|\mathrm{Im}(c_k)| < a$ for all k, and $f(t)$ has a holomorphic extension to $|\mathrm{Im}(z)| < a$ such that for every real c, $-a < c < a$, $f(t+ci)$ is in L_1, the divergence of $T(a)$ is necessary for (1) to be fundamental in C_0 or L_2.

Proofs of Theorems 4 and 5 will appear elsewhere. For other results on the closure of translates, see [11] and [12].

REFERENCES

1. N. Wiener, The Fourier integral and certain of its applications. Cambridge U. Press, 1933; reprint, Dover, N. Y. 1958.
2. R. P. Boas, Jr., Entire Functions, Academic Press, N. Y., 1954.
3. M. Cotlar and R. Cignoli, An Introduction to Functional Analysis, American Elsevier, N. Y. 1974.
4. V. Bernstein, Sulle proprietà caratteristiche delle indicatrici di crescenza delle trascendenti intere d'ordine finito. Memorie R. Accad. d'Italia, Classe di Scienzefisiche, Math., e Nat. (Rome) 7 (1936), 131-189.
5. R. A. Zalik, On approximation by shifts and a theorem of Wiener, Transactions of the Amer. Math. Soc. 243 (1978), 299-308.
6. R. R. Goldberg, Fourier transforms, Cambridge Tracts, No. 52, Cambridge U. Press, London 1961.
7. I. M. Gel'fand and G. E. Šilov, Fourier transforms of rapidly increasing functions and questions of the uniqueness of the solution of Cauchy's problem, Amer. Math. Soc. Transl. (2) 5 (1957), 221-274. (Transl. of Uspehi Mat. Nauk 8 (1953), 3-54.)
8. R. A. Zalik, On some gap theorems and the closure of translates. Notices of the Amer. Math. Soc., No. 2 (1978)* 754-B35.
9. B. Faxén, On approximation by translates and related problems in function theory. Doctoral thesis, University of Göteborg, 1978.
10. R. E. A. C. Paley and N. Wiener, Notes on the theory and application of Fourier transforms; Note IV, a theorem on closure, Trans. Amer. Math. Soc. 35 (1933), 766-768.
11. R. A. Zalik, Approximation by nonfundamental sequences of translates, Proc. Amer. Math. Soc. 78 (1980), 261-266.
12. R. A. Zalik, On fundamental sequences of translates, Proc. Amer. Math. Soc. (in press).

GENERALIZATIONS OF KREIN'S THEOREM ON FUNCTIONS WITH PRESCRIBED ZEROS IN A HAAR SPACE

Roland Zielke

Universitaet Osnabrueck
Osnabrueck, West Germany

Let M be a totally ordered set, a = inf M, b = sup M, and U an n-dimensional Haar space in the set F of real-valued functions defined on M, $n \geq 1$, i.e., a linear space in which no nontrivial element has n zeros or a strong alternation of length n+1, or, equivalently, a weak alternation of length n+1 (a basis of U is called a Čebyšev system; see [6], lemma 3.1).

By [5] we may without loss of generality assume M to be a subset of the real line.

For intervals M and $U \subset C(M)$, KREIN [3] gave sufficient conditions for the existence of an $f \in U$ with prescribed zeros and sign properties (see also [2]). In [6], chapter 6, we extended these results to the situation described above. However, our results did not include sign changes without zeros; to remove this restriction is the aim of this note.

For $f \in F$, denote by Z(f), DZ(f), SZ(f) the sets of zeros, double zeros and simple zeros of f, i.e., let

$$DZ(f) = \{ x \in Z(f) \mid \bigvee_{u,v \in M} u < x < v \text{ and sign } f \Big|_{u,v \cap M \setminus \{x\}} = \text{const.} \neq 0\},$$

$SZ(f) = Z(f) \setminus DZ(f)$; also, let $\delta(f) = \text{card } DZ(f)$, $\sigma(f) = \text{card } SZ(f)$.

Copyright © 1980 by Academic Press, Inc.
All rights of reproduction in any form reserved.
ISBN: 0-12-171050-5

By an appropriate order-preserving mapping, all closed intervals in $G := (a,b) \setminus M$ may be shrunk to single points and all half-open intervals in G may be omitted completely. Let Q be the union of all (open) intervals in G, and $I = G \setminus Q$.

We distinguish four types of sign changes without zeros of an $f \in U$: Let

$$CH_1(f) = \{x \in I \mid \bigvee_{u,v \in M} u < x < v \text{ and } \operatorname{sign} f\big|_{[u,x)\cap M} = - \operatorname{sign} f\big|_{(x,v]\cap M} \neq 0\},$$

$$CH_2(f) = \{x \in M \mid \bigvee_{y \in (x,\infty)\cap M} (x,y)\cap M = \emptyset \text{ and } f(x)\cdot f(y) < 0\},$$

$$CH_3(f) = \{x \in M \mid x = \inf\{M\cap(x,\infty)\} \text{ and } \bigvee_{v \in (x,\infty)\cap M} f(x)\cdot \operatorname{sign} f\big|_{(x,v)\cap M} < 0\},$$

$$Ch_4(f) = \{x \in M \mid x = \sup\{M\cap(-\infty,x)\} \text{ and } \bigvee_{u \in (-\infty,x)\cap M} f(x)\cdot \operatorname{sign} f\big|_{(u,x)\cap M} < 0\},$$

and $\tau_i(f) = \operatorname{card} CH_i(f)$, $i = 1,2,3,4$.

Then $\tau(f) := \tau_1(f) + \ldots + \tau_4(f)$ is the number of sign changes without zeros of f.

It is known ([4] or [6], lemma 6.3) that for every nontrivial $f \in U$, one has $\tau(f) + \sigma(f) + 2\,\delta(f) \leq n - 1$. Our result is the following

<u>Theorem</u> : Let $s_1, \ldots, s_k, d_1, \ldots, d_e \in M$ and $c_1, \ldots, c_r \in I$ be pairwise distinct points with $s_1 < \ldots < s_k$, $a < d_1 < \ldots < d_e < b$ and $0 \leq k + r + 2e \leq n-1$. Then there is an $f \in U$ with $SZ(f) = \{s_1, \ldots, s_k\}$, $DZ(f) = \{d_1, \ldots, d_e\}$, $CH(f) = \{c_1, \ldots, c_r\}$ if one of the following conditions holds:

a) $n - k - r - 2e$ is odd;

b) $a,b \notin M$;

c) $a,b \in M$, $a < s_1$, $s_k < b$;

d) $a,b \in M$, $a = s_1$, $s_k = b$, $M \setminus \{a,b\}$ has no smallest and no largest element;

e) $a \in M$, $b \notin M$, $a = s_1$, $M \setminus \{a\}$ has no smallest element.

<u>Proof</u> : Let $f_1, \ldots, f_n$ be a basis of U and $p := \sum_{i=1}^{n} |f_i|$. As p is positive we may form the Čebyšev system $q_1, \ldots, q_n$ with $q_i = f_i/p$, $i=1,\ldots,n$. Let $x \in I$ and $j \in \{1,\ldots,n\}$ be fixed. As q_j is bounded, there exist

$\alpha := \liminf_{y \to x-} q_j(y)$ and $\beta := \limsup_{y \to x-} q_j(y)$. We claim $\alpha = \beta$. Suppose the contrary. Then there are strictly increasing sequences $y_1, y_2, \ldots$ and $z_1, z_2, \ldots$ in M converging to x and such that each f_i, $i = 1, \ldots, n$, has constant sign θ_i on $M \cap [\min\{y_1, z_1\}, x)$, and

$$q_j(y_m) < \frac{\alpha + \beta}{2} < q_j(z_m) \text{ for } m = 1,2, \ldots .$$

This implies $f_j(z_m) > \frac{\alpha + \beta}{2} \sum_{i=1}^{n} \theta_i f_i(z_m)$,

$$f_j(y_m) < \frac{\alpha + \beta}{2} \sum_{i=1}^{n} \theta_i f_i(y_m) \text{ for } m = 1,2, \ldots \text{, and so}$$

$f_j - \frac{\alpha+\beta}{2} \sum_{i=1}^{n} \theta_i f_i \in U$ changes sign infinitely often, a contradiction.

As an analogous statement holds for right-hand limits, for every $x \in I$ there exist $q_i(x-) := \lim_{y \to x-} q_i(y)$ and $q_i(x+) := \lim_{y \to x+} q_i(y)$.

Let $\bar{q}_i : M \cup I \to R$ be defined by $\bar{q}_i\big|_M = q_i$ and $\bar{q}_i(x) = (q_i(x-) + q_i(x+))/2$ for $x \in I$, $i = 1, \ldots, n$. Denote $V = \operatorname{span}\{q_1, \ldots, q_n\}$, $\bar{V} = \operatorname{span}\{\bar{q}_1, \ldots, \bar{q}_n\}$. Let $\bar{q} \in \bar{V} \setminus \{0\}$ and q the corresponding element of V. $\bar{q}$ has no strong alternation of length n+1, because otherwise the same would hold for q. Now suppose $\bar{q}$ has n zeros $x_1, \ldots, x_n$ with $x_1 < \ldots < x_n$. It is easy to see that we have to consider only the case where exactly one of these, say x_t, lies in I; so $SZ(q) = \{x_1, \ldots, x_{t-1}, x_{t+1}, \ldots, x_n\}$, and q has constant nonzero sign between each pair of neighbouring zeros. Without loss of generality assume $\operatorname{sign} q\big|_{(x_{t-1}, x_{t+1}) \cap M} = 1$. Let $u, v \in M$ with $x_{t-1} < u < v < x_t$, and $h \in V$ with $h(u) = h(v) = 0$ and

$$h(x_j) = (-1)^{j-t} \left\{ \begin{array}{ll} \text{for } j = 1,\ldots,t-1,t+1,\ldots,n-1 & \text{if } t \in \{2,\ldots,n-1\} \\ \text{for } j = 2,\ldots,n-1 & \text{if } t \in \{1,n\}. \end{array} \right\} (*)$$

The weak alternation condition implies that h is negative near x_t. From $\sum_{i=1}^{n} |q_i| \equiv 1$ follows that there is an $\tilde{h} \in V$ with $\tilde{h}(x_t-) < 0$. So for a sufficiently small $\gamma > 0$, $g := h + \gamma\tilde{h}$ has the same sign as h in the points x_j from (*), and $g(x_t-) < 0$. For sufficiently small $\lambda > 0$, $q + \lambda g$ then has a strong alternation of length n+1 in the points x_t from (*) and three suitably chosen points of M near x_t , a contradiction. So $\overline{V}$ is a Haar space.

By theorem 6.5 in [6] , under any of the conditions a) to e) of our theorem there is a $\overline{q} = \sum_{i=1}^{n} \alpha_i \overline{q}_i$ with $DZ(\overline{q}) = \{d_1, \ldots, d_e\}$ and $SZ(\overline{q}) = \{s_1, \ldots, s_k, c_1, \ldots, c_r\}$. Clearly $f = \sum_{i=1}^{n} \alpha_i f_i$ then has the desired properties.

Remarks : 1) We considered only sign changes without zeros in points of I because $CH_i(g) \subset M$ for all $g \in U$ and $i = 2,3,4$ together with theorem 6.5 in [6] implies that in all points outside I, not only can one require sign changes but also zeros.

2) If all the conditions a) to e) fail, the theorem no longer holds in general. If M is a half-open interval, counterexamples are known for all $n > 1$; if M is a closed interval, for all odd $n > 1$. In the case $a < \min\{M \setminus \{a\}\}$ counterexamples exist for $n = 3$ if $b \notin M$, and for $n = 4$ if $b \in M$ (see [6], chapter 10, and [1] for details).

REFERENCES

[1]: Haverkamp,R. : J.Approx.Th. 23 (1978), 104-107.

[2]: Karlin,S. and W.J.Studden :"Tchebycheff Systems". J.Wiley and Sons, 1966.

[3]: Krein,M.G. : Transl.Amer.Math.Soc. 2 (1951), 1-122.

[4]: Kurshan,R,P. and B.Gopinath : J.Approx.Th. 21 (1977),126-142.

[5]: Stockenberg,G. : Dissertation, Duisburg 1976.

[6]: Zielke,R. :"Discontinuous Čebyšev Systems". Lect.Notes in Math., vol. 707, Springer 1979.

GENERALIZED NONLINEAR FUNCTIONAL AND OPERATOR SPLINES IN FOCK SPACES

Rui J. P. de Figueiredo[1]

Dept. of Electrical Engineering and Dept. of Mathematical Sciences

Rice University

Houston, Texas

I. INTRODUCTION

In this note, the concepts of a generalized nonlinear functional spline and of a generalized nonlinear operator spline are presented based on the recent work of the author in collaboration with L. V. Zyla and T. A. W. Dwyer, III.

Generalized splines are usually defined in terms of their minimum norm property [1]-[3]. This and their other extremal properties make them particularly suitable not only for the solution of conventional curve fitting problems but also for application to problems of signal reconstruction [4] [5], signal estimation [6] [7] and system modeling [8] and identification [9]. One of the most general forms of a spline is the generalized operator spline proposed among others by Sard [10] for the interpolating case and extended to the smoothing case by other authors [11] [9]. These interpolating and smoothing splines will be called "Sard-type spline".

All the above generalized splines have been perceived and developed

[1]Supported in part by the ONR Contract NOOO 14-79-C-0442.

Copyright © 1980 by Academic Press, Inc.
All rights of reproduction in any form reserved.
ISBN: 0-12-171050-5

in the context of "linear" problems. In what follows, the notion of a Sard-type spline is extended to a form which makes it readily applicable to "nonlinear" problems. This is achieved via the Fock spaces $F_P(E)$ and $F_P(E,Y)$, where: E and Y are separable Hilbert spaces over the reals (the extension to the complex field case being clear and hence omitted for simplicity in presentation); $F_P(E)$ is a Hilbert space of appropriate nonlinear functionals on E ; and $F_P(E, Y)$ is a Hilbert space of appropriate nonlinear operators from E to Y . What we call _generalized nonlinear functional_ and _operator splines_ S_F and S_0 are simply Sard-type splines respectively in the Fock spaces $F_P(E)$ and $F_P(E, Y)$.

In the following section we introduce the reader to the above Fock spaces, and then give precise definitions of S_F and S_0 in section III. These splines are of considerable value in the solution of the problems of nonlinear system identification, modeling, and simulation as described in references [12] through [14], where mathematical as well as implementational details have been worked out.

II. THE SPACES $F_P(E)$ and $F_P(E,Y)$

Let $\{B_i: i=0, 1, \ldots\}$ be an orthonormal basis in E . Then a homogeneous Hilbert-Schmidt polynomial f_n of degree n in elements of E can be expressed as

$$f_n \stackrel{\Delta}{=} f_n(.) \stackrel{\Delta}{=} \sum_{i_1=0}^{\infty} \cdots \sum_{i_n=0}^{\infty} c_{i_1 \ldots i_n} (B_{i_1}, .)_E \ldots (B_{i_n}, .)_E \qquad (1)$$

where $c_{i_1 \ldots i_n}$ are appropriate real constants, symmetric in the indices, and $(., .)_E$ denotes the inner product in E. The space E^n of such polynomials is a Hilbert space [15] under the inner product

$$(f_n, g_n)_n = \sum_{i_1} \cdots \sum_{i_n} c_{i_1 \ldots i_n} d_{i_1 \ldots i_n}, \qquad (2)$$

where $c_{i_1 \dots i_n}$ and $d_{i_1 \dots i_n}$ are the constants, defined in the same way as $c_{i_1 \dots i_n}$ in (1), associated with any two elements f_n and g_n of E^n.

We now have:

Definition 1. Let p be a positive constant. The Fock space $F_p(E)$ is the linear space of nonlinear functionals f on E defined by

$$f = \sum_{n=0}^{\infty} \frac{1}{n!} f_n \tag{3}$$

where f_n, defined as above, satisfy

$$||f||_{F_p(E)} = [\sum_{n=0}^{\infty} \frac{p^n}{n!} ||f_n||_n^2]^{\frac{1}{2}} < \infty . \tag{4}$$

We will call (3) an "abstract Volterra functional series".

Proposition 1 [15]. $F_p(E)$ is a Hilbert space under the inner product:

$$(f,g)_{F_p(E)} = \sum_{n=0}^{\infty} \frac{p_n}{n!} (f_n, g_n)_n \quad \forall f, g \in F_p(E) . \tag{5}$$

Proposition 2 [12]. $F_p(E)$ is a reproducing kernel Hilbert space with the reproducing kernel K defined by

$$K(u, v) = \exp [(u,v)_E] . \tag{6}$$

Now, with the understanding that Y is a space of real-valued functions on an interval I of the real line (such as $L^2(I)$ or a Sobolev space of real-valued functions on I) we may make the constants $c_{i_1 \dots i_n}$ in (1) vary depending on a variable $t \in I$, and rewrite (1) as

$$f_n(.)(t) = \sum_{i_1} \dots \sum_{i_n} c_{i_1 \dots i_n}(t) \quad (B_{i_1}, .)_E \dots (B_{i_n}, .)_E . \tag{7}$$

By means of an orthonormal basis $\{C_j : j= 0, 1, \dots \}$ in Y, (7) may be expressed in the form

$$f_n(.)(t) = \sum_{i_1} \cdots \sum_{i_n} \sum_j c_{i_1 \ldots i_n j}\, C_j(t)\, (B_{i_1}, .)_E \cdots (B_{i_n}, .)_E . \quad (8)$$

We thus arrive at

Definition 2 [14]. Let E, Y, p, $\tilde{f}_n \triangleq f_n(.)(.)$ be as agreed upon above, then the Fock space $F_p(E,Y)$ is the Hilbert space of operators $\tilde{f} = f(.)(.):E \to Y$ defined by

$$\tilde{f} = f(.)(.) = \sum_{n=0}^{\infty} \frac{1}{n!} \tilde{f}_n , \quad (9)$$

with the inner product

$$(\tilde{f}, \tilde{g})_{F_p(E,Y)} = \sum_{n=0}^{\infty} \frac{p^n}{n!} (\tilde{f}_n, \tilde{g}_n)_n , \quad (10)$$

where

$$(\tilde{f}_n, \tilde{g}_n)_n = \sum_{i_1} \cdots \sum_{i_n} \sum_j c_{i_1 \ldots i_n j}\, d_{i_1 \ldots i_n j} : \quad (11)$$

and the meaning of the constants $c_{i_1 \ldots i_n j}$ and $d_{i_1 \ldots i_n j}$ is clear. We will call (9) an "abstract Volterra operator series".

Fock spaces $F_p(E)$ with p^n in (4) and (5) replaced by $(n+1)^p$ have been investigated by Dwyer [16]. The extension to the replacement of $\{p^n\}$ by an arbitrary weight sequence $\{p_n\}$ was made in [14].

It may be elucidating at this point to give a simple example of one of the above spaces. If E is $L^2(I)$, then $F_p(E)$ becomes the space of the well-known nonlinear functional series introduced by Vito Volterra [17]

$$f(u)(t) = h_0(t) + \int_I h_1(t; t_1)u(t_1)\, dt_1 + \frac{1}{2!} \int_I \int_I h_2(t;t_1 t_2)u(t_1) u(t_2)\, dt_1\, dt_2 + \cdots . \quad (12)$$

The Volterra functionals and their variants the Wiener functionals [18] have played a key role [19] as input-output maps of nonlinear systems. However, the multiple integrations associated with them have constituted a serious obstacle toward their utilization except in weakly nonlinear problems where terms higher than a very low order can be neglected.

The generalized splines defined in the following section permits one to overcome these difficulties by permitting an easily implementable global approximation to them as demonstrated in [14].

III. GENERALIZED NONLINEAR FUNCTIONAL AND OPERATOR SPLINES

To begin with, we introduce generalized splines in $F_p(E)$ and $F_p(E)$ whose associated structural operator L is the identity operator I_o.

Let $w = \{w_i \varepsilon F_p(E): i=1, \ldots, m\}$ be a set of m linearly independent continuous linear functionals on $F_p(E)$. Then, given the set $\tilde{y}(t)$ of m real numbers $\{\tilde{y}_1(t), \ldots, \tilde{y}_m(t)\}$, we define the generalized nonlinear functional spline $S_F(I_o; w; \tilde{y}(t))$ as that element of $F_p(E)$ which is the unique solution of the problem

$$\min_{\substack{V_t \varepsilon F_p(E) \\ w_i(V_t) = \tilde{y}_i(t),\ i=1, \ldots, m}} \|V_t\|_{F_p(E)} \quad . \tag{13}$$

Similarly, given a set W of continuous linear operators $W_i: F_p(E,Y) \to Y$, $i=1, \ldots, m$, all possessing closed ranges, and a set $\tilde{y}=\{\tilde{y}_i \varepsilon Y: i=1, \ldots, m\}$, we define the generalized nonlinear operator spline $S_O(I_o; W; \tilde{y})$ as that element of $F_p(E,Y)$ which solves the problem

$$\min_{\substack{V \varepsilon Fp(E,Y) \\ W_i(V) = \tilde{y}_i,\ i=1, \ldots, m}} \|V\|_{F_p(E,Y)} \quad . \tag{14}$$

As shown in [12], if w_i are evaluation functionals, i.e. if there are $\tilde{u}_i \varepsilon E$, $i=1, \ldots, m$, such that

$$w_i(V_t) = V_t(\tilde{u}_i) \ , \ i=1, \ldots, m, \tag{15}$$

then, according to (6), we may write the expression for $S_F(I; w; \tilde{y}(t))$

$$S_F(I; w; \tilde{y}(t))\ (u)$$

$$= \sum_{i=1}^{m} a_i(t) \exp[\frac{1}{p} (\tilde{u}_i, u)_E], \forall u \epsilon \in, \qquad (16)$$

where the constants $a_i(t)$ are chosen to satisfy the interpolating constraints. If in (16), we replace $a_i(t)$ by $a_i(.)$, (16) gives the expression for $S_0(I; W; \tilde{y})$, where the interpretation of W and $\tilde{y}$ is clear.

A smoothing spline is defined in the usual way. For example, in the case of the functional spline we would replace (13) by

$$\min_{V_t \epsilon F_p(E)} \|V_t\|^2_{F_p(E)} + \sum_{i=1}^{m} \lambda_i (w_i(V_t) - \tilde{y}_i(t))^2, \ \lambda_i = \text{positive constants} \qquad (17)$$

We can go on and generalize the definitions of the splines mentioned above by associating a structural operator L in the minimum norm definition. Thus in the case of a nonlinear functional spline, $S_F(L; w; \tilde{y}(t))$ would be obtained by replacing in (13) the norm functional $\|V_t\|_{F_p(E)}$ by

$$\|LV_t\|_{H_1} , \qquad (18)$$

where L is a continuous linear operator from $F_p(E)$ to a Hilbert space H_1 such that [9] [10]:

$$N(L) \cap N(w) = \emptyset, \ (N(.) = \text{Null space of } (.)), \qquad (19)$$

and there is a positive constant β such that

$$\|V_t\|_{Fp(E)} \leqq \beta \|LV_t\|^2_{H_1} + \sum_{i=1}^{m} \lambda_i (w_i(V_t))^2 . \qquad (20)$$

With a suitably defined linear operator $\tilde{L}: F(E,Y) \to H_2$, where H_2 is a Hilbert space, the nonlinear operator spline $S(\tilde{L}; W; \tilde{y})$ is similarly defined. In typical applications, L or $\tilde{L}$ would be a Frechet differential operator having a particular problem-dependent structure.

IV. CONCLUSION

It has been shown, above, how the "idea" of "spline", first proposed by I. J. Schoenberg [20] more than three decades ago, can be fruitfully applied to the approximation of important class of nonlinear functionals and nonlinear operators. Of the many potential applications, the one [14] to the approximation of Volterra and Wiener functionals appears to be most promising.

REFERENCES

1. De Boor, C., and Lynch, R. E., "On splines and their minimum properties", J. Math. & Mech., 15, 953-969 (1966).

2. Anselone, P. M., and Laurent, P. J., "A general method for the construction of interpolating and smoothing spline functions", Num. Math., 12, 66-82 (1968).

3. Jerome, J., and Schumaker, L., "On Lg splines", J. Approx. Theory, 2, 29-49 (1969).

4. de Figueiredo, R. J. P. and Netravali, A. N., "Optimal spline digital simulators of analog filters", IEEE Trans. on Circuit Theory, CT-18, 711-717 (1971).

5. Netravali, A. N., and de Figueiredo, R. J. P., "Optimal reconstruction of signals", Proc. of 1972 Princeton Conf. on Information Sciences and Computers, Princeton, N. J. (1972).

6. Weinert, H. L., and Kailath, T., "Stochastic interpretations and recursive algorithms for spline functions", Ann. Math. Stat., 2, 789-794 (1974).

7. de Figueiredo, R. J. P., "LM-g splines", J. Approx. Theory, 19, 332-360 (1977).

8. de Figueiredo, R. J. P., Caprihan, A., and Netravali, A. N., "On optimal modeling of systems", J. Optimization Theory, 11, 68-83 (1973).

9. de Figueiredo, R. J. P., and Caprihan, A., "An algorithm for the construction of the generalized smoothing spline with application to system identification", Proc. 11th Annual Conference on Information Sciences and Systems, the Johns Hopkins University, March 1977.

10. Sard, A., "Optimal approximation", J. of Functional Analysis, 1, 222-244 (1967); 3, 367-369 (1968).

11. Munteanu, M. J., "Generalized smoothing spline functions for operators", SIAM J. Num. Analysis, 10, 28-34 (1973).

12. Zyla, L. V., and de Figueiredo, R. J. P., "Nonlinear system identification based on a Fock space framework", Rice University Tech. Report EE-7901 (May 1979), submitted to SIAM J. of Control.

13. de Figueiredo, R. J. P., and Zyla, L. V., "An interpolation and smoothing approach for modeling large scale nonlinear multivariable systems", Proc. of the 12th Asilomar Conf. on Circuits, Systems, and Computers, Pacific Grove, Ca., IEEE Publ. no. 78CH1369-8C/CAS/CS (1978).

14. de Figueiredo, R. J. P., and Dwyer, T. A. W., III, "A best approximation framework and implementation for simulation of large-scale nonlinear systems", IEEE Trans. on Circuits and Systems, CAS-27, Nov. (1980).

15. Dwyer, T. A. W., III, "Partial differential equations in Fischer-Fock spaces for the Hilbert-Schmidt holomorphy type", Bull. Am. Math. Soc., 77, 725-730 (1971).

16. Dwyer, T. A. W., III, "Holomorphic representation of tempered distributions and weighted Fock spaces", in Analyse Fonctionelle et Applications (L. Nachbin, ed.), Hermann, Paris (1975).

17. Volterra, V., Theory of Functionals and of Integral and Integro-Differential Equations, Dover, New York (1959).

18. Wiener, N., Nonlinear Problems in Random Theory, The Technology Press, M. I. T., and J. Wiley and Sons, Inc., New York (1958).

19. Schetzen, M., The Volterra and Wiener Theories of Nonlinear Systems, Wiley-Interscience, New York (1980).

20. Schoenberg, I. J., "Contributions to the problem of approximation of equidistant data by analytic functions", Quart. Appl. Math., 4, 45-99, 112-141 (1946).